THE BOWHEAD WHALE

THE BOWHEAD WHALE

Balaena mysticetus: Biology and Human Interactions

Edited by

J.C. George
Department of Wildlife Management, North Slope Borough, Utqiaġvik, AK, United States

J.G.M. Thewissen
Department of Anatomy and Neurobiology, Northeast Ohio Medical University, Rootstown, OH, United States

ELSEVIER

ACADEMIC PRESS
An imprint of Elsevier

Academic Press is an imprint of Elsevier
125 London Wall, London EC2Y 5AS, United Kingdom
525 B Street, Suite 1650, San Diego, CA 92101, United States
50 Hampshire Street, 5th Floor, Cambridge, MA 02139, United States
The Boulevard, Langford Lane, Kidlington, Oxford OX5 1GB, United Kingdom

British Library Cataloguing-in-Publication Data
A catalogue record for this book is available from the British Library

Library of Congress Cataloging-in-Publication Data
A catalog record for this book is available from the Library of Congress

ISBN: 978-0-12-818969-6

For Information on all Academic Press publications
visit our website at https://www.elsevier.com/books-and-journals

Publisher: Charlotte Cockle
Acquisitions Editor: Anna Valutkevich
Editorial Project Manager: Sara Valentino
Production Project Manager: Sreejith Viswanathan
Cover Designer: Christian J. Bilbow

Typeset by MPS Limited, Chennai, India

Dedication

"To Harry Kupaaq Brower, Sr. and Dr. Thomas Frank Albert, Sr. Their collaboration and vision set the stage for making bowhead whales one of the best-studied cetaceans."

Nuimaruaġigivut:
Harry Kupaaq Brower, Sr.-munlu, Thomas Frank Albert, Sr.-munlu.
Isagunŋagaak aġviġum qimilġuuqtaulhaaġniŋa taġium niġrutipayaaŋiññi.

Contents

III

Interactions with humans

List of contributors

C.J. Ashjian Department of Biology, Woods Hole Oceanographic Institution, Woods Hole, MA, United States

A.B. Baird Department of Natural Sciences, University of Houston-Downtown, Houston, TX, United States

J.W. Bickham Department of Ecology and Conservation Biology, Texas A&M University, College Station, TX, United States

Susanna B. Blackwell Greeneridge Sciences, Inc., Santa Barbara, CA, United States; University of California, Santa Cruz, CA, United States

J.L. Bolton Environmental and Fisheries Sciences Division, Northwest Fisheries Science Center, National Marine Fisheries Service, National Oceanic and Atmospheric Administration, Seattle, WA, United States

Greg A. Breed Institute of Arctic Biology, University of Alaska Fairbanks, Fairbanks, AK, United States

A.A. Brower Alaska Fisheries Science Center, National Marine Fisheries Service, National Oceanic and Atmospheric Administration, Seattle, WA, United States; Joint Institute for the Study of the Atmosphere and Ocean, University of Washington, Seattle, WA, United States

H.K. Brower, Jr Alaska Eskimo Whaling Commission, Utqiaġvik, AK, United States; North Slope Borough, Mayors Office, Utqiaġvik, AK, United States

R.G. Campbell Graduate School of Oceanography, University of Rhode Island, Kingston, RI, United States

M.A. Castellini Graduate School, University of Alaska Fairbanks, Fairbanks, AK, United States

J.J. Citta Alaska Department of Fish and Game, Fairbanks, AK, United States

Christopher W. Clark Cornell Lab of Ornithology, Center for Conservation Bioacoustics, Cornell University, Ithaca, NY, United States

J.T. Clarke Alaska Fisheries Science Center, National Marine Fisheries Service, National Oceanic and Atmospheric Administration, Seattle, WA, United States; Joint Institute for the Study of the Atmosphere and Ocean, University of Washington, Seattle, WA, United States

Lisa Noelle Cooper Department of Anatomy and Neurobiology, Musculoskeletal Research Group, Northeast Ohio Medical University, Rootstown, OH, United States

S.L. Danielson College of Fisheries and Ocean Sciences, University of Alaska Fairbanks, AK, United States

Robert Elsner (deceased) College of Fisheries and Ocean Sciences, University of Alaska Fairbanks, Fairbanks, AK, United States

M.C. Ferguson Alaska Fisheries Science Center, National Marine Fisheries Service, National Oceanic and Atmospheric Administration, Seattle, WA, United States; School of Aquatic and Fishery Sciences, University of Washington, Seattle, WA, United States

S.H. Ferguson Fisheries and Oceans Canada, Central and Arctic Region, Winnipeg, MB, Canada

Erich Follmann (deceased) Institute of Arctic Biology, University of Alaska Fairbanks, Fairbanks, AK, United States

S. Fortune Institute for Oceans and Fisheries, Marine Mammal Research Unit, University of British Columbia, Vancouver, BC, Canada

John Gatesy Division of Vertebrate Zoology and Sackler Institute for Comparative Genomics, American Museum of Natural History, New York, NY, United States

J.C. George Department of Wildlife Management, North Slope Borough, Utqiaġvik, AK, United States

Geof H. Givens Givens Statistical Solutions LLC, Fort Collins, CO, United States

Vera Gorbunova Department of Biology, University of Rochester, Rochester, NY, United States

R.G. Hansen Greenland Institute of Natural Resources, Copenhagen, Denmark

Mads Peter Heide-Jørgensen Greenland Institute of Natural Resources, Copenhagen, Denmark

D.J. Hillmann Department of Comparative Biomedical Sciences, School of Veterinary Medicine, Louisiana State University, Baton Rouge, LA, United States

Patrick R. Hof Nash Family Department of Neuroscience and Friedman Brain Institute, Icahn School of Medicine at Mount Sinai, New York, NY, United States

Lara Horstmann College of Fisheries and Ocean Sciences, University of Alaska Fairbanks, Fairbanks, AK, United States

H.P. Huntington Ocean Conservancy, Eagle River, AK, United States

S. Inutiq Iqaluit, NU, Canada

N. Kanayurak Department of Wildlife Management, North Slope Borough, Utqiaġvik, AK, United States

William R. Koski LGL Limited, King City, ON, Canada

Olivier Lambert Directorate Earth and History of Life, Royal Belgian Institute of Natural Sciences, Brussels, Belgium

Jessica Lefevre Alaska Eskimo Whaling Commission, Utqiaġvik, AK, United States

Dennis Litovka ChukotTINRO, Anadyr, Chukotka, Russia

S.C. Lubetkin Seattle, WA, United States

B. Lyberth Kalallit Nunaanni Aalisartut Piniartullu Kattuffiat (KNAPK; Association of Fishers and Hunters of Greenland), Nuuk, Greenland

Felix G. Marx Museum of New Zealand Te Papa Tongarewa, Wellington, New Zealand

Michael R. McGowen Department of Vertebrate Zoology, Smithsonian National Museum of Natural History, Washington, DC, United States

M. Moore Biology Department, Woods Hole Oceanographic Institution, Woods Hole, MA, United States

Sue E. Moore Department of Biology, Center for Ecosystem Sentinels, University of Washington, Seattle, WA, United States

P.B. Nader Department of Anatomy, College of Veterinary Medicine, Lincoln Memorial University, Harrogate, TN, United States

G. Noongwook Savoonga Whaling Captains Association, Savoonga, AK, United States

S.R. Okkonen College of Fisheries and Ocean Sciences, University of Alaska Fairbanks, AK, United States

Brian T. Person Department of Wildlife Management, North Slope Borough, Utqiaġvik, AK, United States

P.J. Ponganis Center for Marine Biotechnology & Biomedicine, Scripps Institution of Oceanography, University of California San Diego, La Jolla, CA, United States

L. Quakenbush Alaska Department of Fish and Game, Fairbanks, AK, United States

Mary Ann Raghanti Department of Anthropology, School of Biomedical Sciences, and Brain Health Research Institute, Kent State University, Kent, OH, United States

Randall R. Reeves Okapi Wildlife Associates, Hudson, QC, Canada

Sam Ridgway Department of Pathology, University of California, San Diego and National Marine Mammal Foundation, San Diego, CA, United States

Rosalind M. Rolland Anderson Cabot Center for Ocean Life, New England Aquarium, Boston, MA, United States

D. Rotstein Marine Mammal Pathology Services, Olney, MD, United States

C. Sakakibara Oberlin College, Oberlin, OH, United States

I.R. Schultz Environmental and Fisheries Sciences Division, Northwest Fisheries Science Center, National Marine Fisheries Service, National Oceanic and Atmospheric Administration, Seattle, WA, United States

Todd L. Sformo Department of Wildlife Management, North Slope Borough, Utqiaġvik, AK, United States; Institute of Arctic Biology, University of Alaska Fairbanks, Fairbanks, AK, United States

Gay Sheffield Alaska Sea Grant, College of Fisheries and Ocean Sciences, University of Alaska Fairbanks, Nome, AK, United States; Marine Advisory Program, University of Alaska Fairbanks, Nome, AK, United States

O.V. Shpak A.N. Severtsov Institute of Ecology and Evolution of Russian Academy of Sciences, Moscow, Russia

V. Skhauge Sireniki, Chukotka, Russia

Kathleen M. Stafford Applied Physics Laboratory, University of Washington, Seattle, WA, United States

R. Stimmelmayr Department of Wildlife Management, North Slope Borough, Utqiaġvik, AK, United States; Institute of Arctic Biology, University of Alaska Fairbanks, Fairbanks, AK, United States

Robert Suydam Department of Wildlife Management, North Slope Borough, Utqiaġvik, AK, United States

Raymond J. Tarpley Department of Veterinary Integrative Biosciences, College of Veterinary Medicine and Biomedical Sciences, Texas A&M University, College Station, TX, United States

J.G.M. Thewissen Department of Anatomy and Neurobiology, Northeast Ohio Medical University, Rootstown, OH, United States

Aaron M. Thode Marine Physical Laboratory, Scripps Institution of Oceanography, University of California, San Diego, CA, United States

Barbara J. Tudor Department of Wildlife Management, North Slope Borough, Utqiaġvik, AK, United States

A. Von Duyke Department of Wildlife Management, North Slope Borough, Utqiaġvik, AK, United States

T.J. Weingartner College of Fisheries and Ocean Sciences, University of Alaska Fairbanks, AK, United States

A.J. Werth Department of Biology, Hampden-Sydney College, Hampden-Sydney, VA, United States

D. Wetzel Mote Marine Laboratory, Sarasota, FL, United States

A.L. Willoughby Alaska Fisheries Science Center, National Marine Fisheries Service, National Oceanic and Atmospheric Administration, Seattle, WA, United States; Joint Institute for the Study of the Atmosphere and Ocean, University of Washington, Seattle, WA, United States

Bernd Würsig Texas A&M University at Galveston, Galveston, TX, United States

G.M. Ylitalo Environmental and Fisheries Sciences Division, Northwest Fisheries Science Center, National Marine Fisheries Service, National Oceanic and Atmospheric Administration, Seattle, WA, United States

E. Zdor University of Alaska Fairbanks, Fairbanks, AK, United States

Judith E. Zeh Department of Statistics, University of Washington, Seattle, WA, United States

Preface

"The Greenland Whale is one of the most wonderful animals in the world, and the baleen, or whalebone, one of its greatest peculiarities." —*Charles Darwin*

The bowhead whale is known by many names: Greenland whale, *aġviq* (Iñupiat), *aghveq* (Siberian Yupik), *Balaena mysticetus*, among others, and the animal itself means different things to different people. Regardless of the name, all agree the bowhead is a remarkable and superbly adapted animal. Darwin never saw a living bowhead but he recognized the baleen rack as one of the most unusual features of any whale, and it was the only whale mentioned by name in *On the Origin of Species*. They have, by far, the longest baleen of any whale species: up to 320 plates suspended from each side of the upper jaw (about 640 total), and the longest plates reaching 480 cm (15 ft.) in large whales. This enables bowheads to thrive on the unpredictable and sometimes sparse prey in their sub-Arctic and Arctic feeding grounds. To support the huge baleen racks, their head is enormous, making up a third or more of the body length in adults. They also have the thickest blubber layer of any whale, which provides insulation and is also a food reserve and buffer against episodic scarcity of food, a common occurrence.

"The whales, they give themselves" —*Umialiq Captain Harry Brower, Sr.*

In Iñupiat society, the *umialiq* or whaling boat captain is highly revered, and Mr. Brower was among the most respected. Iñupiat hunters believe they have a connection with the whale's *inua* (spirit) and that a bowhead "gives itself" to a worthy hunter (Brewster, 2004). The significance of this animal to many northern indigenous societies along the sub-Arctic and Arctic coasts is hard to describe in words and hard to overstate. They have a deep reverence and respect for this animal that has sustained the survival of their people for thousands of years. Where accessible, bowheads have been and remain central to the whaling societies by providing food, light, heat, and construction materials, and building community ties as people work together to prepare, hunt, process, share, and celebrate the harvest of the whale.

In a sense, the bowhead played a pivotal role in the development of the North Slope Borough's Department of Wildlife Management's (NSB DWM) bowhead research program in Utqiaġvik (Barrow), Alaska, upon which many of the contributions in this book are rooted. Thus, we dedicate this book to *Umialiq* Captain Brower, and his friend NSB Senior Scientist Dr. Thomas F. Albert, Sr. The friendship, reliance, and respect that Brower and Albert had for each other has been the subject of several books, and is another example of how shared interests in bowheads and the challenges to harvest them helped build a community of subsistence hunters and scientists. As the Iñupiat say, there is no other single animal that you can share with an entire village.

"You're going to find out the bowhead is a thermos bottle" —*Umialiq Captain Edward Hopson, Sr.*

Many scientists who study bowheads also developed a deep fascination and reverence for this animal. Much knowledge about bowheads resides within indigenous communities, and local and "western" scientific knowledge are converging. A synergy has been developing since the days of Brower and Albert and continues to flourish, promoting research in productive and rewarding ways that consciously focuses on both the health and understanding of the stocks to determine sustainable harvest levels for bowheads. When a whale is landed, it is a joyous occasion but the whalers know it must be processed quickly to avoid spoilage, as bowheads retain heat and cool slowly. When we first proposed studies on bowhead body temperature, Captain Hopson (quoted above) offered us advice on the insulation properties based on a lifetime of whale hunting, observation, with generosity in sharing and humor. He was right, as the hunters often are. Our measurements indicated that harvested whales cool slowly. Confirming other indigenous knowledge, the discovery of a stone weapon in a harvested bowhead indicated that some bowheads live more than 200 years, longer than any other mammal. New research shows that bowheads seem unaffected by most of the progressive diseases of old age; cancer is rare to absent, making bowheads an animal of great interest to the medical community.

The four stocks or populations of bowheads live their entire life in the sub-Arctic and Arctic seas. They begin life in freezing seawater in spring within a complex of lead systems—the series of fractures and open water amidst shifting sea ice. They are the only baleen whale (mysticete) that gives birth in Arctic waters and feeds through the winter, foregoing a migration to temperate waters like other baleen whales. Young bowheads grow slowly, taking some 25 years to reach sexual maturity and then may reach 60 ft. (19 m) in length and weigh more than 100 tons.

While the indigenous peoples of North America, Greenland, and Chukotka hunted bowheads for at least the last 2000 years, all stocks of bowheads were hunted to near-extinction by European and American commercial hunters. Two of the four populations still face a tenuous recovery. However, all stocks face new threats from industry, shipping, commercial fishing, and ecological changes from climate warming. Commercial whaling was particularly devastating to indigenous peoples that depended on bowheads across their range.

It has been nearly 30 years since the landmark volume *The Bowhead Whale* was published (Burns et al., 1993). It summarized what was known about bowheads up to that time. Since then, research has elucidated many additional aspects of bowhead biology and

Drawing of the harvest of a bowhead whale on the sea ice at Utqiaġvik, Alaska. The drawing spans several hours as the hunters and community haul the whale onto the ice, butcher it into sections (shares), and transport it back to the village. Drawing by Jean C. George.

ecology: genetics, the relatedness of the four stocks, gene function, biooceanography, energetics, growth rates, sensory systems, gut fermentation to name a few. New methods for estimating chronological age, current population size, and rate of increase, as well as extensive satellite tracking to determine migration paths, feeding areas, wintering areas and concerns about climate warming all have arisen in the last 30 years as well. In fact, many researchers consider the bowhead a sentinel for the health of the arctic marine ecosystem.

Native communities continue to be intensely focused on issues of animal health and food safety, urging the DWM to continue research on contaminant levels, effects of industrial activities, climate warming, body condition, and disease. The principal reason the DWM was founded by the North Slope Borough government was to address local criticisms of the estimated bowhead population size and the major impact of the 1977 International Whaling Commission moratorium on bowhead whaling. The hunters strenuously argued, based on their own observations, that the population size was greatly underestimated. Starting in the 1980s, the abundance of surveys confirmed this as well as many of their other observations over the past 40 years. Such studies continue to this day.

Considering the above, it is clearly time to update and summarize the current state of knowledge about bowhead whales. Our goal is to reach a broad audience so that researchers and the interested public have an up-to-date summary of current bowhead science, administrators can find data needed to make solid management decisions, and students and laypeople can find, in language relatively free of jargon, answers to basic questions about this species. We also hope that the many color photographs of bowheads in their natural setting will further the readers' admiration and understanding of this unique animal and its environment.

The book is organized by research topics ranging from fossil origins, basic biology, to the harvest by indigenous peoples and their traditional knowledge, and the effects of climate warming. While the Bering-Chukchi-Beaufort stock is the largest and, in some regards, the most intensively studied, we intended to present research on all stocks, that together have a circumpolar distribution, including Okhotsk Sea stock, East Greenland-Svalbard-Barents Sea stock, and the East Canada-West Greenland stock.

Finally, we hope that this book is a way of giving something back to the indigenous communities who have shared their advice, knowledge, observations, and whales with us—while asking little in return. This book would not have been possible without them.

John Craighead "Craig" George and J. G. M. 'Hans' Thewissen

AK, United States; OH, United States

Acknowledgments

Hundreds of people and a massive amount of research form the foundation of this book, and this makes writing the acknowledgments a nearly impossible task. We did our best.

We thank the authors who contributed chapters to this book, not only for their long hours, but also for advice, and ideas on the book, and their collaborative attitude and friendship. We also thank the following individuals, institutions, and organizations who significantly contributed to this book in a variety of ways: Cycil Fish, Jacqueline Fernandez-Hamberg, David Waugh, Bailey McKenna, Katheryn Mars, Erica Scarpitti, Sharon Usip, Bill Hess, Vicki Beaver, Corey Accardo, Amelia Brower, Brenda Rone, Peter Duley, the Sheldon Jackson Museum, Bureau of Ocean Energy Management, the Hennecke Family Foundation, Sitka Sound Science Center, and the Inupiat History and Language Commission. The publication team at Elsevier have been excellent to work with and we especially thank Anna Valutkevich, Sara Valentino, and Sreejith Viswanathan.

Contributions from the indigenous whale hunters and community members have been enormous. This book would not have been possible without their collaboration and support and is described in more detail below.

Regarding the eastern arctic bowhead populations (East Canada–West Greenland and East Greenland–Svalbard–Barents Sea), we express our gratitude to the hunters of Qeqertarsuaq that for four decades patiently have assisted with studies of whales in Disko Bay. They have been instrumental in developing methods for approaching and tagging bowhead whales in West Greenland in icy winter months. Mikkel Villum Jensen has been an essential contributor to the initial design of both pole-systems and the Air Rocket Transmitter System now widely used for tagging large whales. Without the hunters and Mikkel's persistence and knowledge, the tagging of large whales would not have developed this far.

The Russian contributions span three bowhead populations, Okhotsk Sea, Bering–Chukchi–Beaufort Seas, and East Greenland, Svalbard, Barents Sea (EGSB) Stock. We thank all the scientists, organizations, and volunteers involved in the Russian bowhead whale research. The National Park "Russian Arctic" has been collecting bowhead whale sightings for years in the Northern Barents Sea. Alexey Paramonov's passion for the Okhotsk Sea bowheads is remarkable, and without him the Western Okhotsk Sea fieldwork would not be possible. For Bering–Chukchi–Beaufort Seas bowheads, significant scientific contributions started 30 years ago by several Chukotkan organizations including the Naukan Native Organization, Yupik Eskimo Society of Chukotka, and the Association of Traditional Marine Mammal Hunters of Chukotka.

Many chapters of this book rely heavily on 50 years of research on Bering–Chukchi–Beaufort Seas population. That research was initiated in the 1970s by the National Marine Fisheries Service. Since 1981, much of the research was conducted under the auspices of the North Slope Borough (NSB), in close associations with Inupiat and Siberian Yupik whaling captains, their wives, the other hunters and associated scientists in Alaska. The hunters gave their time, provided access to harvested whales for data collection, and shared their knowledge in return for little more than promoting good science and a desire to provide the best possible information for making informed management decisions.

We specifically thank some individuals that provided critical assistance at the start of the NSB bowhead science program. Harry K. Brower, Sr. played a crucial role in "opening the door" to bowhead science to gain the confidence of the senior whale hunters of Utqiaġvik. Jake Adams, the first Chairman of the Alaska Eskimo Whaling Commission (AEWC), hired Ray Dronenburg who brought in Thomas "Tom" Albert, Sr. to initiate and lead a locally based bowhead science program. At the same time, the responsibility for the ice-based bowhead survey effort and field examination of harvested whales was transferred from the US government to the NSB. During this time, collaborations with the AEWC were critical in the cooperative management program. Eugene Brower, President of the Barrow Whaling Captains Association (BWCA, 1917–2017) and NSB Mayor (1981–83) was a strong supporter and crucial to the success of the program. The support of the BWCA allowed for detailed postmortem exams of harvested whales and the ice-based bowhead whale survey at Point Barrow and was key to the success of the program. John Burns, Sr. (Alaska Department of Fish and Game, retired) edited the classic 1993 book *The Bowhead Whale* and historian John R. Bockstoce was particularly helpful in offering support and ideas through the development, writing, and editing of this book. Geoff and Marie Adams Carroll had critical roles in the bowhead program and the conception of ideas for this book. Geoff initiated the ice-based bowhead abundance survey at Utqiaġvik in the 1970s, and Marie played a major role in the development of the AEWC.

We thank the NSB Mayors for their support of the bowhead program; these were (in chronological order): Eben Hopson, Jake Adams, Sr., Eugene Brower, George Ahmaogak, Sr., Jeslie Kaleak, Sr., Benny Nageak, Edward Itta, Charlotte Brower, and Harry Brower, Jr. We also recognize the Directors for the NSB Department of Wildlife Management since 1981 (in order): Lester Suvlu, Ron Nalikak, Benny Nageak, Warren Matumeak, Charles D. N. Brower, and Taqulik Hepa. The support of the NSB Assembly was also instrumental in the success of the program.

In addition to those already mentioned, we hold the following people in high regard for their assistance and commitment to the conservation and management of the bowhead whale (alphabetically; chapter authors, coauthors, and those mentioned in this narrative, are not listed).

Mike Aamodt, Billy Adams, Ben Ahmaogak, Sr., Perry Anashugak, Maggie Ahmaogak, Jonathan Aiken, Sr., Johnny Aiken, Robert Aiken, Ludmilla Ainana, Herman Aishanna, Benny Akootchook, Isaac Akootchook, Frankie Akpik, Sr., Eric Archer, Walt Audi, John Banister, Barry Bodfish, Joe Burgener, Ross Burgener, Peter Best, Howard Braham, Stephan Braund, Gordon Broadhead, April Brower Brooks, Arnold Brower, Sr., Arnold Brower, Jr., Carl Brower, Fredrick Brower, Price Brower, Lewis Brower, Doug Butterworth, Mary Core, Randy Crosby, Bill Cummings, Les Dalton, Liza Dela Rosa, Doug DeMaster, Greg Donovan, Qunniq Donovan, Dennis Duffield, Sarah Ellis, Bill Ellison, Vladimir Etylin, Gerald Haldiman, Cyd Hanns, Bob Harcharek, Bill Henk, Bob Henry, Charles Hopson,

Eddie Hopson, John Hopson, Jr., Allen Ingling, Matt Irinaga, Clancy Itta, Edward Itta, Matthew Iya, Igor Krupnik, Gordon Jarrell, Bill Kaleak, Joe Kaleak, John Kelley, Carl Kippi, Merlin Koonooka, Bill Kopplin, Bruce Krogman, Oliver Leavitt, Luther Leavitt, Tom Lohman, Ned Manning, Rosemary McGuire, James Matumeak, Vladimir Melnikov, Paul Nader, Thomas Napageak, Sr., Mary Nerini, David Norton, Percy Nusunginya, Egil Oen, Todd O'Hara, John Oktollik, Forest Dean Olemaun, Nate Olemaun, Sr., Tommy Olemaun, Margaret Opie, Crawford Patkotak, Michael Pederson, Mike Philo, Leslie Pierce, Rossman Peetook, Andre Punt, Adrian Raftery, Dave Ramey, Burton Rexford, John Reynolds, III, John Richardson, Cheryl Rosa, Dave Rugh, Bobby Sarren, Glenn Sheehan, Roger Silook, John Smithhistler, Nolan Solomon, Ron Sonntag, Lynn Sutcliffe, Barb Taylor, John Tichotsky, Mike Tillman, Kenneth Toovak, Dolores Vinas, Michael Wald, Terrie Williams, Gennady Zelensky, Eduard Zdor, Michael Zelensky, and many others.

Funding for the US bowhead science program has largely come from the NSB, as well as the State of Alaska, the National Oceanic and Atmospheric Administration, and BP Alaska for specific bowhead surveys. BOEM provided partial funds through Cooperative Agreement M20AC00004 to assist in assembling current bowhead information and editing this book. The late Senator Ted Stevens provided critical support over many years for Alaskan bowhead comanagement and science.

Finally, we thank our immediate and extended families. Craig is particularly thankful to his wife Cyd, and sons Luke and Sam who were supportive during his countless hours away in the field on the bowhead surveys and sampling whales. Hans thanks his family for their patience and support during this project.

Our sincere thanks to all,

Craig and Hans

SECTION I

Basic biology

CHAPTER

1

Higher level phylogeny of baleen whales

John Gatesy[1] *and Michael R. McGowen*[2]

[1]Division of Vertebrate Zoology and Sackler Institute for Comparative Genomics, American Museum of Natural History, New York, NY, United States [2]Department of Vertebrate Zoology, Smithsonian National Museum of Natural History, Washington, DC, United States

The phylogenetic branching history of the bowhead whale

With technological breakthroughs, it is possible to survey the genomes of different species both rapidly and relatively cheaply. With this bounty of data, the hereditary molecule, DNA, can be used to reconstruct the phylogenetic history of life. Furthermore, by calibrating molecular trees with evidence from the fossil record (Chapter 2), divergences among species can be dated using various molecular clock approaches (Thorne et al., 1998; Springer et al., 2019). For Cetacea, the clade that includes whales, dolphins, and porpoises, such "timetrees" specify the closest evolutionary relatives of the bowhead whale (Figs. 1.1 and 1.2) and more distantly related clades of species (Gatesy et al., 2013).

Molecular trees, such as the recent genome-scale hypothesis of McGowen et al. (2019), broadly corroborate evolutionary trees derived from the analysis of morphological characters (e.g., Deméré et al., 2005; Marx, 2010; Boessenecker and Fordyce, 2017), but with some key differences that significantly impact interpretations of whale evolution. Modern cetaceans are divided into Odontoceti (toothed whales) and Mysticeti (baleen whales), clades that diverged from each other ~36.7 Ma (Fig. 1.2A). Mysticeti shows a basal split between Balaenidae, which includes the bowhead (*Balaena*) plus right whales (*Eubalaena*), and Plicogulae, a diverse group composed of Balaenopteroidea (rorquals [Balaenopteridae], gray whale [Eschrichtiidae]) and Neobalaenidae (pygmy right whale [*Caperea*]). The basal split in Mysticeti dates to ~25.7 Ma (Oligocene) according to the timetree of McGowen et al. (2019) (Fig. 1.2A). Morphological analyses commonly have grouped all skim-feeding mysticetes into a monophyletic group, Balaenoidea, that includes Balaenidae and Neobalaenidae (e.g., Deméré et al., 2005; Churchill et al., 2012; Boessenecker and Fordyce, 2017),

The Bowhead Whale
DOI: https://doi.org/10.1016/B978-0-12-818969-6.00001-7

FIGURE 1.1 Bowhead whale (*Balaena mysticetus* at bottom of photo) congregating with two North Atlantic right whales (*Eubalaena glacialis*) near Cape Cod, Massachusetts; well outside the typical Arctic geographic range of bowheads. *Source: From Accardo, C.M., Ganley, L.C., Brown, M.W., Duley, P.A., George, J.C., Reeves, R.R., et al., 2018. Sightings of a bowhead whale (*Balaena mysticetus*) in the Gulf of Maine and its interactions with other baleen whales. J. Cetacean Res. Manag. 19, 23–30. Image taken by Peter Duley under NOAA/NMFS MMPA permit number 17355.*

but the molecular clade Plicogulae (mysticetes with grooved throats) has been corroborated by some morphology-based trees (e.g., Marx, 2010). The latter hypothesis suggests that many of the anatomical features related to skim feeding, such as a strongly arched rostrum with proportionally long baleen plates, evolved convergently in Balaenidae and Neobalaenidae.

Within Plicogulae, Neobalaenidae split from Balaenopteroidea in the Early Miocene, ~22.1 Ma (Fig. 1.2A). Balaenopteroidea traditionally has been divided into the families Eschrichtiidae (gray whale) and Balaenopteridae (rorquals) in cladistic analyses of morphological characters (Deméré et al., 2005; Marx, 2010; Boessenecker and Fordyce, 2017), as well as by some molecular (Fig. 1.2B; Steeman et al., 2009) and total evidence analyses (Deméré et al., 2008; Gatesy et al., 2013). An emerging consensus from molecular work, however, supports the derivation of *Eschrichtius* (gray whale) from within a paraphyletic Balaenopteridae (Rychel et al., 2004; McGowen et al., 2009, 2019; Hassanin et al., 2012; Árnason et al., 2018; Lammers et al., 2019; Fig. 1.2A, C–E). Balaenopterids (*Megaptera* and *Balaenoptera*) are highly specialized engulfment feeders that can gulp large volumes of prey-laden water using a suite of integrated evolutionary novelties. These include a loose jaw joint that enables outward rotation of the curved mandibles, a well-pleated throat pouch that expands during feeding bouts, and a flaccid tongue that permits extreme posterior extension of the oral cavity with the influx of seawater (Goldbogen et al., 2017). Following the capture of entire schools of small fishes or invertebrates, engulfment feeders expel water through the baleen filter that effectively retains tiny prey items in the mouth. By contrast, the eschrichtiid gray whale lacks the highly derived rorqual feeding apparatus and instead suction feeds on benthic invertebrates, infrequently using both skimming and

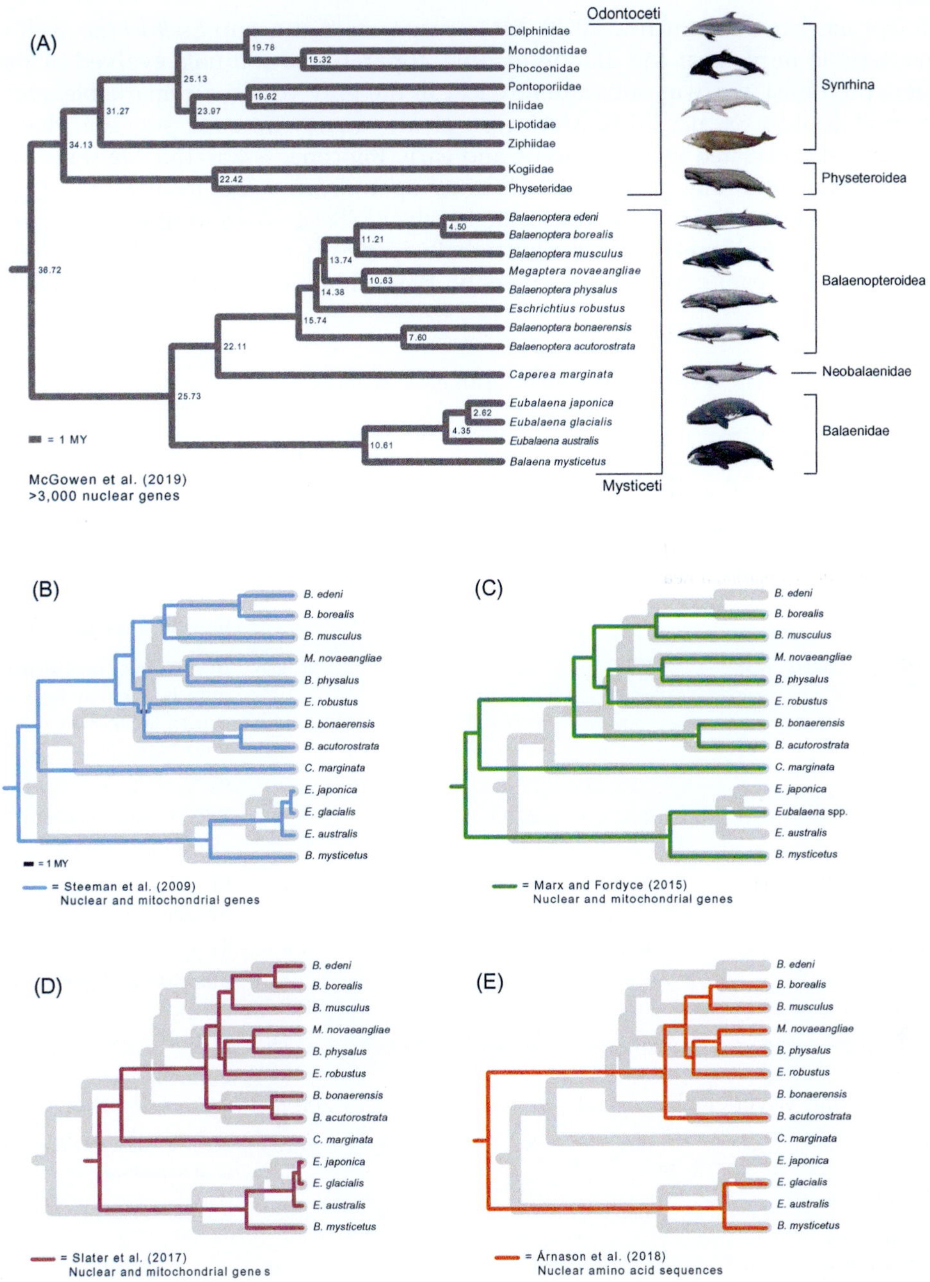

FIGURE 1.2 Phylogenetic relationships and divergence times of the bowhead whale, *Balaena mysticetus*, relative to other extant cetacean lineages. (A) The phylogenomic timetree of McGowen et al. (2019) shows estimated divergence times at nodes based on an autocorrelated rates model. Alternative timetrees for Mysticeti are shown in (B)–(E) (*thin colored branches*), with the timetree of McGowen et al. (2019) in the background of each panel for comparison (*thicker gray lineages*). Scale bars in millions of years (My) are indicated; timetrees in (B)–(E) are all to the same scale. For the timetree of Steeman et al. (2009), divergence times are taken from figure 3. DNA sequences from multiple species of *Eubalaena* were merged into a single operational taxonomic unit (*Eubalaena* spp.) in Marx and Fordyce (2015). Paintings of cetaceans are by Carl Buell (copyright John Gatesy).

engulfment mechanisms (Swartz, 2018). Molecular trees that group *Eschrichtius* well within Balaenopteridae imply that the unique feeding apparatus of rorquals evolved at the base of Balaenopteroidea and was subsequently lost in the gray whale, a remarkable evolutionary reversal (Gatesy et al., 2013). McGowen et al.'s (2019) timetree suggests that extant Balaenopteroids speciated in the Miocene and early Pliocene, ~4.5–15.7 Ma (Fig. 1.2A).

Evolutionary splits among extant balaenids span from ~2.6 to 10.6 Ma (Table 1.1; Fig. 1.2A). The closest living relatives of the bowhead (*Balaena*) are in the genus *Eubalaena*, which includes three closely related right whale species (southern, North Atlantic, North Pacific) (Fig. 1.3). Despite low extant diversity, Balaenidae has a rich evolutionary history

TABLE 1.1 List of recent molecular clock studies with mean estimated ages of five clades in millions of years (Ma).

	Mean (Ma)	Mean (Ma)	Mean (Ma)	Mean (Ma)	Mean (Ma)	
Study	**Balaenidae**	***Eubalaena***	**Plicogulae**	**Mysticeti**	**Cetacea**	**Method**
Sasaki et al. (2005)	17.1	4.4	23.3	27.3	35.4	Bayesian estimation using Thorne et al. (1998): mitochondrial genome
McGowen et al. (2009)	5.38	0.77	22.59	28.79	36.36	BEAST: tree constructed with mitochondrial and nuclear genes; dating with mitochondrial cytochrome *b* gene
Slater et al. (2010)	5.04	1.32	22.63	28.81	36.88	BEAST: mitochondrial cytochrome *b* gene
Hassanin et al. (2012): soft mean	11.5	NA	22.3	25.2	34.4	BEAST: complete mitochondrial genomes
Marx and Fordyce (2015): mean Z	9.82	NA	28.96	30.35	38.8	MrBayes: tip-dating with mitochondrial genes, nuclear genes, and fossils
Slater et al. (2017)	5.62	0.94	19.29	20.58	35.45	BEAST: tip-dating with mitochondrial genes, nuclear genes, and fossils
Árnason et al. (2018)	4.38	NA	NA	28.29	31.75	MCMCTREE: nuclear amino acid sequences
McGowen et al. (2019): 6-partitions, independent rates	4.79	1.88	21.65	25.29	36.63	MCMCTREE: nuclear genes
McGowen et al. (2019): 6-partitions, autocorrelated rates	10.61	4.35	22.11	25.73	36.72	MCMCTREE: nuclear genes

Methods used to make these trees include the Bayesian approach of Thorne et al. (1998), BEAST (Drummond et al., 2006), MrBayes (Ronquist and Huelsenbeck, 2003), and MCMCTREE (Yang, 2007).

FIGURE 1.3 Right whales are the closest relatives of the bowhead whale. Bowheads lack the callosities (*light-colored* in photo) seen on the head and lower jaw of this southern right whale, *Eubalaena australis*, off the coast of Argentina. *Source: Photo by Bernd Würsig.*

with various extinct species described from Miocene to recent deposits (Chapter 2). Relationships within *Eubalaena* conflict in various molecular phylogenetic estimates (Rosenbaum et al., 2000; Gaines et al., 2005; McGowen et al., 2009, 2019; Steeman et al., 2009; Slater et al., 2017) with the latest genome-scale analysis supporting the monophyly of the two Northern hemisphere species, *Eubalaena glacialis* and *Eubalaena japonica* (Fig. 1.2A).

Challenges for estimation of divergence times in Mysticeti

Given that Mysticeti is a well-studied taxonomic group with a rich fossil record, it is perhaps surprising that recent molecular clock studies have yielded an array of phylogenetic divergence times that differ sometimes markedly from each other (Fig. 1.2). For example, the basal split in Balaenidae between *Balaena* and *Eubalaena* ranges from 4.38 to 17.10 Ma in different estimates (Table 1.1), and dates from some studies (Slater et al., 2017; Árnason et al., 2018; Fig. 1.2D–E) are consistently younger than dates from others across Mysticeti (e.g., Marx and Fordyce, 2015; McGowen et al., 2019; Fig. 1.2A and C) (Table 1.1). Such discrepancies can directly impact key evolutionary inferences. For example, Slater et al. (2017) hypothesized that enormous body size (>10 m) evolved only very recently in Mysticeti. A clade-wide shift toward gigantism was directly linked to a Late Pliocene change in ocean dynamics that amplified the density and patchiness of prey at low trophic levels (~ 3 Ma to the present). This inference is predicated on the extremely shallow divergence times in their tip-dated timetree for Mysticeti (Fig. 1.2D).

Factors that influence inferred molecular clock dates can relate to the specific molecular and fossil data that are analyzed in different studies, usage of contrasting statistical approaches for inferring a timetree, and/or biological phenomena that challenge accurate estimation of divergence times. In terms of DNA sequence data, mitochondrial sequences that evolve very rapidly have been used in some studies (Sasaki et al., 2005; Slater et al., 2010; Hassanin et al., 2012), while others have focused on nuclear sequences that change at a significantly slower rate (Árnason et al., 2018; McGowen et al., 2019). Such differences in rate might impact divergence time estimates if models of molecular evolution do not correct adequately for overlapping mutations at the same sites. Which fossils are chosen for calibration can also drive estimated dates further into the past or pull dates toward the present. For example, most recent molecular clock estimates for the age of Cetacea (here defined as the last common ancestor of all extant forms) fall between 34.4 and 38.8 Ma (Table 1.1). However, extremely shallow dates for this clade (9.1–11.4 Ma) have been published (Phillips, 2016; Liu et al., 2017). Inadequate fossil calibrations with soft-bounded constraints were implemented in molecular clock analyses that included small-bodied mammals characterized by rapid molecular rates (e.g., rodents) and large-bodied mammals with extremely slow rates (e.g., whales). Although much less dramatic, discrepancies between molecular clock dates that are driven solely by choice of algorithm can occur when the exact same molecular and fossil data are analyzed (e.g., independent vs autocorrelated modeling of rates in McGowen et al., 2019; Table 1.1) or when the same set of extinct taxa are employed in tip-dated clock analyses (e.g., Marx and Fordyce, 2015 vs Slater et al., 2017; Fig. 1.2C and D; Table 1.1).

From a biological perspective, hybridization with introgression of genetic material between distant evolutionary relatives also can distort divergence times. If there is gene flow between extant species, the merging of genomes will reduce molecular divergence dates for these species (Springer et al., 2019). Interspecific aggregations of mysticetes are well documented (e.g., Accardo et al., 2018; Fig. 1.1), and genetic studies provide compelling evidence for both ancient and recent gene flow between mysticete species (Bérubé and Aguilar, 1998; Árnason et al., 2018). The partial mixing of gene pools in Mysticeti may be a significant impediment to estimation of divergence times because even the most advanced molecular clock methods assume that phylogeny has been a strictly bifurcating process.

Conclusions

Recent phylogenetic hypotheses for Mysticeti are generally congruent and robustly supported (Fig. 1.2) despite recent genomic evidence for the partial mixing of evolutionary lineages in this clade. Molecular clock studies show a range of divergence dates that are dependent on alternative analytical approaches (Table 1.1). Future work on mysticete phylogeny and evolution must grapple with the fact that gene flow among divergent lineages (e.g., blue whale and fin whale; Bérubé and Aguilar, 1998) can impact evolutionary inferences in this clade.

Acknowledgments

Funding was provided by NSF DEB-1457735, and paintings of cetaceans in Fig. 1.2A are by Carl Buell (copyright John Gatesy).

References

Accardo, C.M., Ganley, L.C., Brown, M.W., Duley, P.A., George, J.C., Reeves, R.R., et al., 2018. Sightings of a bowhead whale (*Balaena mysticetus*) in the Gulf of Maine and its interactions with other baleen whales. J. Cetacean Res. Manag. 19, 23–30.

Árnason, Ú., Lammers, F., Kumar, V., Nilsson, M.A., Janke, A., 2018. Whole-genome sequencing of the blue whale and other rorquals finds signatures for introgressive gene flow. Sci. Adv. 4, eaap9873.

Bérubé, M., Aguilar, A., 1998. A new hybrid between a blue whale, *Balaenoptera musculus*, and a fin whale, *B. physalus*: frequency and implications of hybridization. Mar. Mammal. Sci. 14, 82–98.

Boessenecker, R.W., Fordyce, R.E., 2017. A new eomysticetid from the Oligocene Kokoamu Greensand of New Zealand and a review of the Eomysticetidae (Mammalia, Cetacea). J. Syst. Palaeontol. 15, 429–469.

Churchill, M., Berta, A., Deméré, T.A., 2012. The systematics of right whales. Mar. Mammal. Sci. 28, 497–521.

Deméré, T.A., Berta, A., McGowen, M.R., 2005. The taxonomic and evolutionary history of fossil and modern balaenopteroid mysticetes. J. Mamm. Evol. 12, 99–143.

Deméré, T.A., McGowen, M.R., Berta, A., Gatesy, J., 2008. Morphological and molecular evidence for a stepwise evolutionary transition from teeth to baleen in mysticete whales. Syst. Biol. 57, 15–37.

Drummond, A.J., Ho, S.Y.W., Phillips, M.J., Rambaut, A., 2006. Relaxed phylogenetics and dating with confidence. PLoS Biol. 4, e5.

Gaines, C., Hare, M., Beck, S., Rosenbaum, H., 2005. Nuclear markers confirm taxonomic status and relationships among highly endangered and closely related right whale species. Proc. R. Soc. B: Biol. Sci. 272, 533–542.

Gatesy, J., Geisler, J.H., Chang, J., Buell, C., Berta, A., Meredith, R.W., et al., 2013. A phylogenetic blueprint for a modern whale. Mol. Phylogenet. Evol. 66, 479–506.

Goldbogen, J.A., Cade, D.E., Calambokidis, J., Friedlaender, A.S., Potvin, J., Segre, P.S., et al., 2017. How baleen whales feed: the biomechanics of engulfment and filtration. Annu. Rev. Mar. Sci. 9, 367–386.

Hassanin, A., Delsuc, F., Ropiquet, A., Hammer, C., Vuuren, B.J.V., Matthee, C., et al., 2012. Pattern and timing of diversification of Cetartiodactyla (Mammalia, Laurasiatheria), as revealed by a comprehensive analysis of mitochondrial genomes. C. R. Biol. 335, 32–50.

Lammers, F., Blumer, M., Rücklé, C., Nilsson, M.A., 2019. Retrophylogenomics in rorquals indicate large ancestral population sizes and a rapid radiation. Mob. DNA 10, 5.

Liu, L., Zhang, J., Rheindt, F.E., Lei, F., Qu, Y., Wang, Y., et al., 2017. Genomic evidence reveals a radiation of placental mammals uninterrupted by the KPg boundary. Proc. Natl. Acad. Sci. U. S. A. 114, E7282–E7290.

Marx, F.G., 2010. The more the merrier? A large cladistic analysis of mysticetes, and comments on the transition from teeth to baleen. J. Mamm. Evol. 18, 77–100.

Marx, F.G., Fordyce, R.E., 2015. Baleen boom and bust: a synthesis of mysticete phylogeny, diversity and disparity. R. Soc. Open. Sci. 2, 140434.

McGowen, M.R., Spaulding, M., Gatesy, J., 2009. Divergence date estimation and a comprehensive molecular tree of extant cetaceans. Mol. Phylogen. Evol. 53, 891–906.

McGowen, M.R., Tsagkogeorga, G., Álvarez-Carretero, S., Reis, M.D., Struebig, M., Deaville, R., et al., 2019. Phylogenomic resolution of the cetacean tree of life using target sequence capture. Syst. Biol. Adv. Access. 69, 479–501. Available from: https://doi.org/10.1093/sysbio/syz068.

Phillips, M.J., 2016. Geomolecular dating and the origin of placental mammals. Syst. Biol. 65, 546–557.

Ronquist, F., Huelsenbeck, J.P., 2003. MrBayes 3: Bayesian phylogenetic inference under mixed models. Bioinformatics 19, 1572–1574.

Rosenbaum, H.C., Brownell, R.L., Brown, M.W., Schaeff, C., Portway, V., White, B.N., et al., 2000. World-wide genetic differentiation of *Eubalaena*: questioning the number of right whale species. Mol. Ecol. 9, 1793–1802.

Rychel, A.L., Reeder, T.W., Berta, A., 2004. Phylogeny of mysticete whales based on mitochondrial and nuclear data. Mol. Phylogen. Evol. 32, 892–901.

Sasaki, T., Nikaido, M., Hamilton, H., Goto, M., Kato, H., Kanda, N., et al., 2005. Mitochondrial phylogenetics and evolution of mysticete whales. Syst. Biol. 54, 77–90.

Slater, G.J., Price, S.A., Santini, F., Alfaro, M.E., 2010. Diversity versus disparity and the radiation of modern cetaceans. Proc. R. Soc. B: Biol. Sci. 277, 3097–3104.

Slater, G.J., Goldbogen, J.A., Pyenson, N.D., 2017. Independent evolution of baleen whale gigantism linked to Plio-Pleistocene ocean dynamics. Proc. R. Soc. B: Biol. Sci. 284, 20170546.

Springer, M.S., Foley, N.M., Brady, P.L., Gatesy, J., Murphy, W.J., 2019. Evolutionary models for the diversification of placental mammals across the KPg boundary. Front. Genet. 10, 1241.
Steeman, M.E., Hebsgaard, M.B., Fordyce, R.E., Ho, S.Y.W., Rabosky, D.L., Nielsen, R., et al., 2009. Radiation of extant cetaceans driven by restructuring of the oceans. Syst. Biol. 58, 573–585.
Swartz, S.L., 2018. Gray whale. In: Würsig, B., Thewissen, J.G.M., Kovacs, K.M. (Eds.), Encyclopedia of Marine Mammals, third ed. Academic Press, London, pp. 422–428.
Thorne, J.L., Kishino, H., Painter, I.S., 1998. Estimating the rate of evolution of the rate of molecular evolution. Mol. Biol. Evol. 15, 1647–1657.
Yang, Z., 2007. PAML 4: phylogenetic analysis by maximum likelihood. Mol. Biol. Evol. 24, 1586–1591.

CHAPTER

2

Fossil record

Felix G. Marx[1] *and Olivier Lambert*[2]

[1]Museum of New Zealand Te Papa Tongarewa, Wellington, New Zealand [2]Directorate Earth and History of Life, Royal Belgian Institute of Natural Sciences, Brussels, Belgium

Introduction

Balaenids are an ancient lineage of baleen whales (Mysticeti), which today only survives in the form of two genera: *Balaena*, or bowhead whales; and three species of *Eubalaena*, which—like the family itself—are commonly referred to as right whales. Bowhead whales today occur exclusively in northern polar waters, and have gained fame as the Methuselah among mysticetes: some individuals are thought to be over 200 years old (George et al., 1999; Chapter 7; *Balaena* also holds the distinction of being the only mysticete genus named by Linnaeus (1758). Because of this long history, numerous species were referred to it over the past 250 years, but nearly all have since been reidentified or relegated to the status of nomen dubium (see McLeod et al., 1993 for a detailed review).

Mysticetes descend from toothed ancestors, as exemplified by the small but ferocious *Janjucetus hunderi*, which inhabited Australian waters 26.5–23 Ma (million years ago; Fig. 2.1) (Fitzgerald, 2006). By contrast, living mysticetes are toothless, and instead rely on a set of comb-like keratinous plates (baleen) to filter tiny prey directly from seawater (Pivorunas, 1979; Chapter 14). Compared to other mysticetes, balaenids are bulky and adapted for slow cruising rather than speed (Woodward et al., 2006). When foraging, they swim forwards with the mouth partially open, taking in prey-laden water at the front while simultaneously expelling excess water at the back of the mouth. This form of continuous "skim feeding" benefits from a large filtration area, which in turn has caused the baleen plates to become extremely elongate: in the case of bowheads, up to 4 m (Werth and Potvin, 2016; Chapter 14).

To accommodate the enormous baleen racks, the rostrum of right whales is narrow and notably arched (Fig. 2.2). This arrangement makes the skull appear remarkably tall, and is matched by both an equally tall lower lip and a large supraoccipital bone for the attachment of strong neck muscles. Other typical balaenid features include broad, paddle-like flippers with five fingers; the lack of a dorsal fin; hypertrophy of the periotic, which

The Bowhead Whale
DOI: https://doi.org/10.1016/B978-0-12-818969-6.00002-9

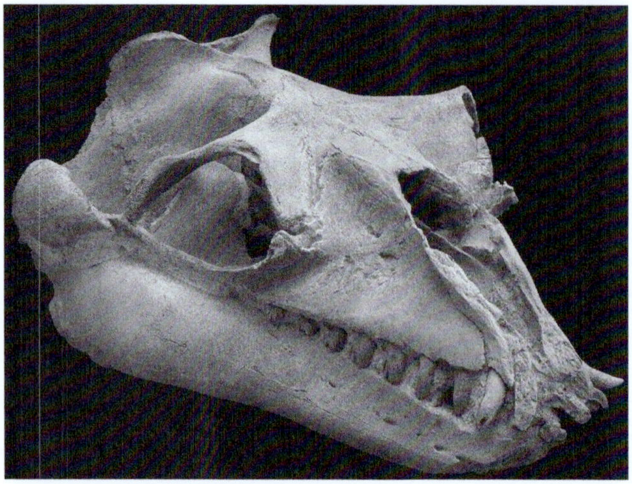

FIGURE 2.1 Skull of the archaic toothed mysticete *Janjucetus hunderi* (Museums Victoria, Melbourne, Australia, specimen P216929) from Australia. *Janjucetus* occurred around 26.5–23 Ma and, just like many other early mysticetes, had teeth but no baleen. *Source: Photograph by Erich M. G. Fitzgerald.*

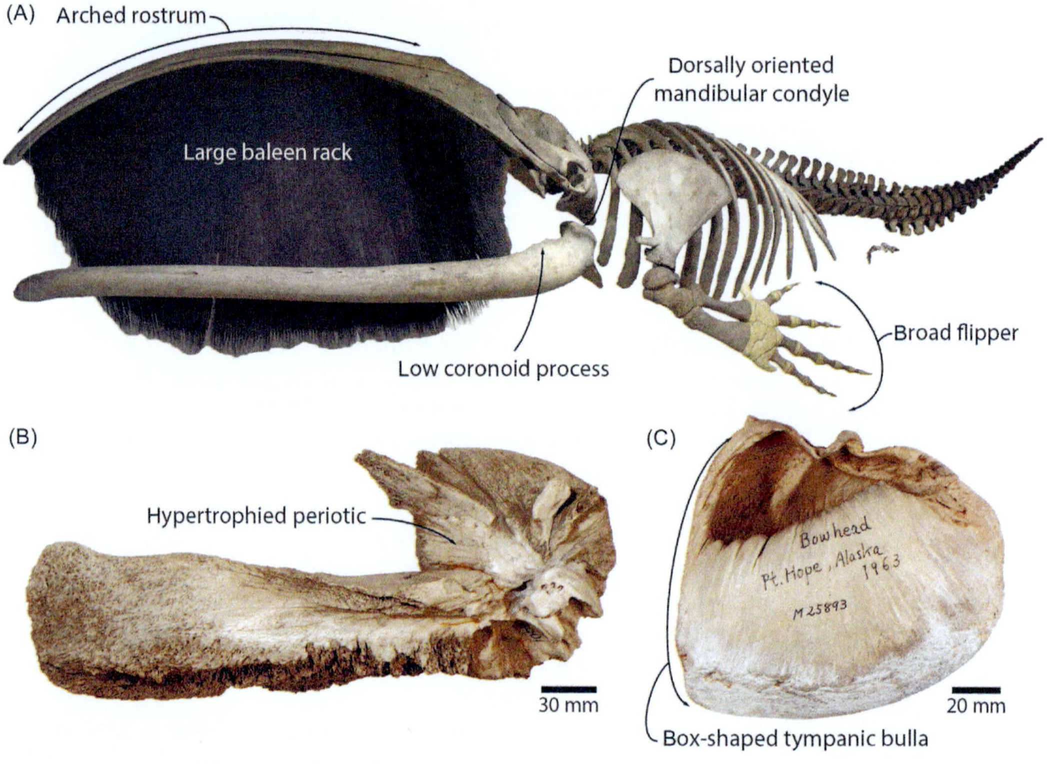

FIGURE 2.2 Skeleton of the bowhead whale (*Balaena mysticetus*). (A) The skeleton (Zoological Museum of the University of Copenhagen, Denmark, specimen CN1). (B) The right periotic (National Museum of Natural History, Smithsonian Institution, Washington, DC, USA, specimen 63301). (C) The right tympanic bulla (National Museum of Nature and Science, Tokyo, Japan, specimen M25893).

houses the organs of hearing and balance; box-shaped, posteriorly diverging tympanic bullae; a well-defined glenoid fossa housing the synovial craniomandibular joint; a robust mandible with a twisted symphyseal portion, a low coronoid process, a dorsally oriented articular condyle, and a well-developed mylohyoid sulcus; fused neck vertebrae; and, primitively, the retention of a comparatively well-developed hind limb comprising the pelvis, femur, and cartilaginous tibia.

In bowhead whales, the rostrum and neurocranium form a continuous arc, with the nasal bones being elevated above the level of the supraoccipital (Chapter 9). By contrast, *Eubalaena* has a more irregular skull outline, and a dome-like vertex that rises above the level of the rostrum. Bowhead whales further differ in having a straighter frontoparietal suture (in lateral view), and more slender forelimb bones that retain both the olecranon process on the ulna and the coracoid process on the scapula (Bisconti, 2000; Churchill et al., 2012; Westgate and Whitmore, 2002). Externally, the two genera appear relatively similar, but bowhead whales lack the callosities (patches of roughened skin infested by whale lice) characterizing *Eubalaena* and are somewhat larger, with a maximum body length of about 19 m.

Balaenid origins and the Miocene gap

The distinctive bauplan of balaenids has ancient origins, and fundamentally has remained almost unchanged since their first appearance in the fossil record. Within its scope, however, right whales once attained a far greater diversity of shapes and sizes than is apparent in the living species.

Right whales are generally considered to be basal to all other extant mysticetes (including the pygmy right whale, *Caperea marginata*), and may have evolved as early as 28 Ma (Fordyce, 2002; Marx and Fordyce, 2015; McGowen et al., 2009; Chapter 1); however, fossils from that time have not yet been unambiguously identified. The earliest definitive balaenid is *Morenocetus parvus* from the lower Miocene of Argentina (Fig. 2.3), which at an age of 20–18 Ma is by far the oldest member of any of the extant baleen whale families (Cabrera, 1926). *Morenocetus* was smaller (about 5–6 m) than its living relatives, but already had the tall skull and hypertrophied periotic typical of modern balaenids (Buono et al., 2017).

The second-oldest balaenid, dating to about 16–15 Ma, is *Peripolocetus vexillifer* from the middle Miocene of California, USA. The holotype of this species is fragmentary, and was only recognized as a right whale following the discovery of a more complete specimen from the same locality (Deméré and Pyenson, 2015). Together with *Morenocetus*, it is often considered basal to all other balaenids (Buono et al., 2017; Duboys de Lavigerie et al., 2020; Gol'din and Steeman, 2015), but the phylogenetic position of these early species remains in flux (e.g., Bisconti, 2005; Bisconti et al., 2017).

After *Peripolocetus*, there is a large gap in the balaenid fossil record lasting until about 7–6 Ma. Why this is so remains an enduring and largely underappreciated mystery (Buono et al., 2017; Deméré and Pyenson, 2015). Curiously, there are several Miocene rock formations across the globe that are rich in cetacean remains, yet have never yielded a right whale fossil. Pertinent examples include the Chesapeake Group of the eastern United States (Gottfried et al., 1994); the Pisco Formation of southern Peru (Di Celma et al., 2017); and the Bihoku Group of southern Japan (Otsuka and Ota, 2008).

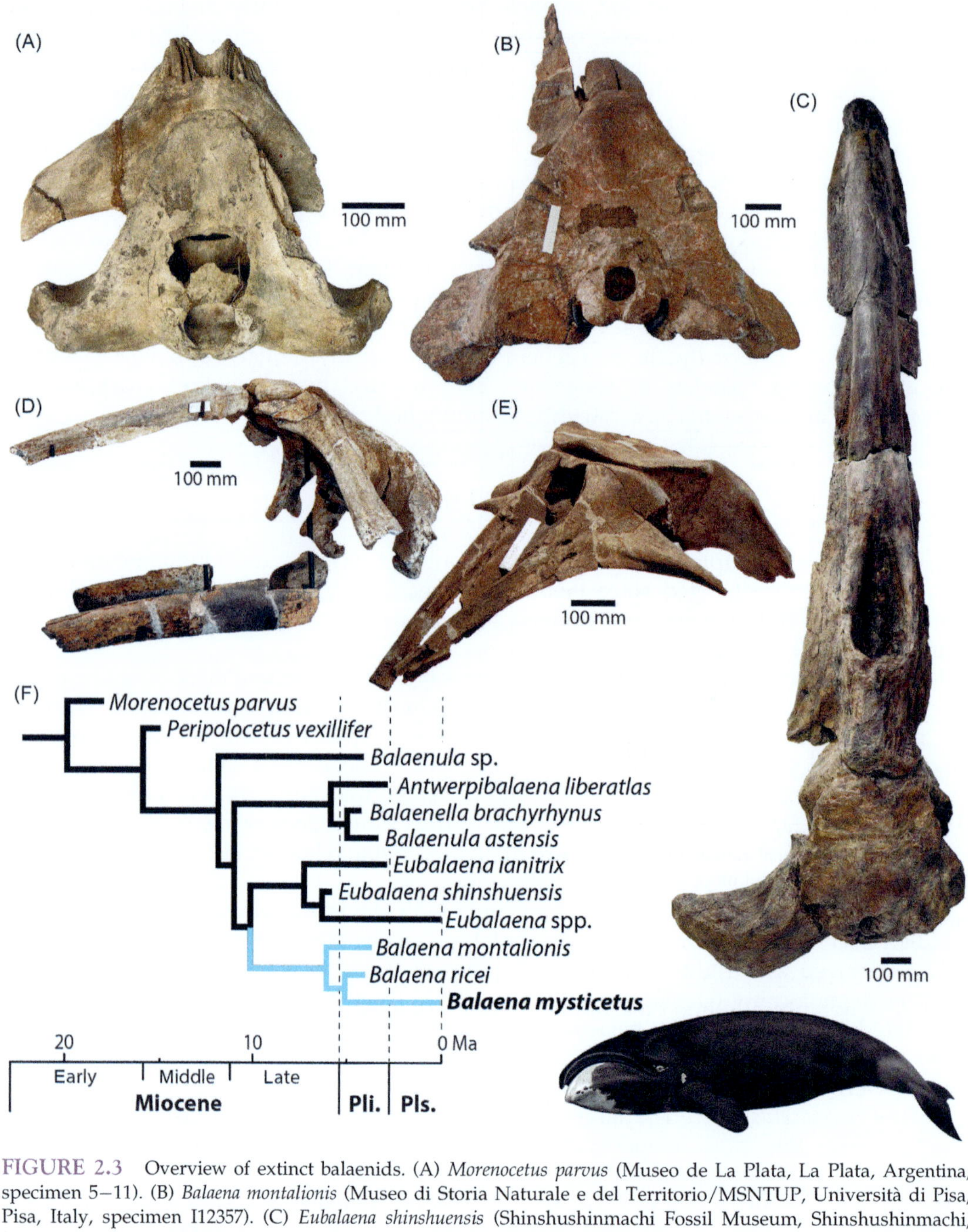

FIGURE 2.3 Overview of extinct balaenids. (A) *Morenocetus parvus* (Museo de La Plata, La Plata, Argentina, specimen 5–11). (B) *Balaena montalionis* (Museo di Storia Naturale e del Territorio/MSNTUP, Università di Pisa, Pisa, Italy, specimen I12357). (C) *Eubalaena shinshuensis* (Shinshushinmachi Fossil Museum, Shinshushinmachi, Japan, specimen CV0024). (D) *Balaenula astensis* (MSNTUP I12555). (E) *Balaenella brachyrhynus* (Natuurmuseum Brabant, Tilburg, The Netherlands, specimen 42001). (F) Time-calibrated phylogeny of living and extinct right whales, following Duboys de Lavigerie et al. (2020); bowhead whales are shown in blue. Drawing of *Balaena mysticetus* by Carl Buell. *Pli.*, Pliocene; *Pls.*, Pleistocene.

Given the worldwide distribution of these localities, there seems to be no obvious biogeographical explanation for the "balaenid gap." Perhaps early right whales were relatively rare and/or restricted to habitats not captured by the Miocene fossil localities explored to date. Alternatively, most of them may have been limited to the Southern Hemisphere, large swathes of which remain underexplored. Support for this idea comes from as yet undescribed specimens from Argentina dating to 12–9 Ma (Buono et al., 2009).

Late Neogene diversification and the emergence of bowheads

About 7–6 Ma, a new wave of right whale fossils abruptly appears across the globe. Oldest amongst them is *Eubalaena shinshuensis* from Japan (Kimura, 2009), which at an estimated length of 12–13 m was twice as large as all of its predecessors (Buono et al., 2009, 2017). Its appearance heralded a short-lived phase, lasting until about 3 Ma, during which balaenids diversified into a variety of species and body sizes: from the diminutive (6–8 m long) *Balaenella brachyrhynus*, *Balaenotus insignis*, *Balaenula balaenopsis*, and *Balaenula astensis*, to the medium-sized (9–10 m) *Antwerpibalaena liberatlas*, and the relatively large (>10 m) *Balaena ricei* and *Eubalaena ianitrix* (Bisconti, 2000, 2005; Bisconti et al., 2017; Duboys de Lavigerie et al., 2020; Trevisan, 1941; Van Beneden, 1880; Westgate and Whitmore, 2002).

The phylogenetic relationships of most of these species remain poorly resolved. Some analyses variously intersperse them with living right whales (Bisconti, 2005; Bisconti et al., 2017; Churchill et al., 2012), whereas others support a closely knit crown group comprising only *Balaena* and *Eubalaena* (Buono et al., 2017; Duboys de Lavigerie et al., 2020). All studies agree, however, that bowhead whales emerged as part of this late Neogene "explosion" in balaenid diversity, giving rise to a lineage with at least three species: *Balaena montalionis*, *B. ricei*, and the extant *Balaena mysticetus*. Of these, *B. ricei* is the oldest (c. 4.9–4.4 Ma), which is notably younger than some recent molecular estimates (10.6 Ma) for the split between *Balaena* and *Eubalaena* (Chapter 1). Initially, *Balaena* occurred as far south as 35°N–37°N and broadly overlapped with *Eubalaena* in its geographic range (Field et al., 2017).

About 3 Ma, small balaenids, along with most other small mysticetes, disappear from the fossil record in tandem with the onset of Northern Hemisphere glaciation and more patchy prey distributions (Marx and Fordyce, 2015; Slater et al., 2017). *Balaena* and *Eubalaena* survived, perhaps because of their relatively large size, and eventually started to separate geographically into their modern polar vs temperate habitats (Foote et al., 2013).

Acknowledgements

We thank Erich M.G. Fitzgerald for providing the photograph of *Janjucetus hunderi*, and Carl Buell for his drawing of *Balaena mysticetus*.

References

Bisconti, M., 2000. New description, character analysis and preliminary phyletic assessment of two Balaenidae skulls from the Italian Pliocene. Palaeontogr. Ital. 87, 37–66.

Bisconti, M., 2005. Skull morphology and phylogenetic relationships of a new diminutive balaenid from the Lower Pliocene of Belgium. Palaeontology 48, 793–816. Available from: https://doi.org/10.1111/j.1475-4983.2005.00488.x.

Bisconti, M., Lambert, O., Bosselaers, M., 2017. Revision of "*Balaena*" *belgica* reveals a new right whale species, the possible ancestry of the northern right whale, *Eubalaena glacialis*, and the ages of divergence for the living right whale species. PeerJ 5, e3464. Available from: https://doi.org/10.7717/peerj.3464.

Buono, M.R., Dozo, M.T., Cozzuol, M.A., 2009. Balaenidae (Mammalia, Cetacea, Mysticeti) del Mioceno de Patagonia: antecedentes, nuevos registros y proyecciones. Ameghiniana 46 (Suppl. 4), 66R.

Buono, M.R., Fernández, M.S., Cozzuol, M.A., Cuitiño, J.I., Fitzgerald, E.M.G., 2017. The early Miocene balaenid *Morenocetus parvus* from Patagonia (Argentina) and the evolution of right whales. PeerJ 5, e4148. Available from: https://doi.org/10.7717/peerj.4148.

Cabrera, A., 1926. Cetáceos fósiles del Museo de La Plata. Rev. Mus. La Plata 29, 363–411.

Churchill, M., Berta, A., Deméré, T.A., 2012. The systematics of right whales (Mysticeti: Balaenidae). Mar. Mamm. Sci. 28, 497–521. Available from: https://doi.org/10.1111/j.1748-7692.2011.00504.x.

Deméré, T.A., and N.D. Pyenson, 2015. Filling the Miocene 'balaenid gap'—the previously enigmatic *Peripolocetus vexillifer* Kellogg, 1931 is a stem balaenid (Cetacea: Mysticeti) from the middle Miocene (Langhian) of California, USA. J. Vertebr. Paleontol., Program and Abstracts, 115.

Di Celma, C., Malinverno, E., Bosio, G., Collareta, A., Gariboldi, K., Gioncada, A., et al., 2017. Sequence stratigraphy and paleontology of the Upper Miocene Pisco Formation along the western side of the lower Ica Valley (Ica Desert, Peru). Riv. Ital. Paleontol. Stratigr. 123, 255–274. Available from: https://doi.org/10.13130/2039-4942/8373.

Duboys de Lavigerie, G., Bosselaers, M., Goolaerts, S., Park, T., Lambert, O., Marx, F.G., 2020. New Pliocene right whale from Belgium informs balaenid phylogeny and function. J. Syst. Palaeontol. Available from: https://doi.org/10.1080/14772019.2020.1746422.

Field, D.J., Boessenecker, R., Racicot, R.A., Ásbjörnsdóttir, L., Jónasson, K., Hsiang, A.Y., et al., 2017. The oldest marine vertebrate fossil from the volcanic island of Iceland: a partial right whale skull from the high latitude Pliocene Tjörnes Formation. Palaeontology 60, 141–148. Available from: https://doi.org/10.1111/pala.12275.

Fitzgerald, E.M.G., 2006. A bizarre new toothed mysticete (Cetacea) from Australia and the early evolution of baleen whales. Proc. R. Soc. B 273, 2955–2963. Available from: https://doi.org/10.1098/rspb.2006.3664.

Foote, A.D., Kaschner, K., Schultze, S.E., Garilao, C., Ho, S.Y.W., Post, K., et al., 2013. Ancient DNA reveals that bowhead whale lineages survived Late Pleistocene climate change and habitat shifts. Nat. Commun. 4, 1677. Available from: https://doi.org/10.1038/ncomms2714.

Fordyce, R.E., 2002. Oligocene origins of skim-feeding right whales: a small archaic balaenid from New Zealand. J. Vertebr. Paleontol. 22, 54A.

George, J.C., Bada, J., Zeh, J., Scott, L., Brown, S.E., O'Hara, T., et al., 1999. Age and growth estimates of bowhead whales (*Balaena mysticetus*) via aspartic acid racemization. Can. J. Zool. 77, 571–580. Available from: https://doi.org/10.1139/z99-015.

Gol'din, P., Steeman, M.E., 2015. From problem taxa to problem solver: a new Miocene family, Tranatocetidae, brings perspective on baleen whale evolution. PLoS ONE 10, e0135500. Available from: https://doi.org/10.1371/journal.pone.0135500.

Gottfried, M.D., Bohaska, D.J., Whitmore, F.C., 1994. Miocene cetaceans of the Chesapeake Group. Proc. San. Diego Soc. Nat. Hist. 29, 229–238.

Kimura, T., 2009. Review of the fossil balaenids from Japan with a re-description of *Eubalaena shinshuensis* (Mammalia, Cetacea, Mysticeti). Quad. Mus. Stor. Nat. Livorno 22, 3–21.

Linnaeus, C., 1758. Systema Naturae. Laurentii Salvii, Holmiae.

Marx, F.G., Fordyce, R.E., 2015. Baleen boom and bust: a synthesis of mysticete phylogeny, diversity and disparity. R. Soc. Open. Sci. 2, 140434. Available from: https://doi.org/10.1098/rsos.140434.

McGowen, M.R., Spaulding, M., Gatesy, J., 2009. Divergence date estimation and a comprehensive molecular tree of extant cetaceans. Mol. Phylogenet. Evol. 53, 891–906. Available from: https://doi.org/10.1016/j.ympev.2009.08.018.

McLeod, S.A., Whitmore, F.C., Barnes, L.G., 1993. Evolutionary relationships and classification. In: Burns, J.J., Montague, J.J., Cowles, C.J. (Eds.), The Bowhead Whale. Society for Marine Mammalogy, Lawrence, KS, pp. 45–70.

Otsuka, H., Ota, Y., 2008. Cetotheres from the early Middle Miocene Bihoku Group in Shobara District, Hiroshima Prefecture, West Japan. Misc. Rep. Hiwa Mus. Nat. Hist. 49, 1–66.

Pivorunas, A., 1979. The feeding mechanisms of baleen whales. Am. Sci. 67, 432–440.

Slater, G.J., Goldbogen, J.A., Pyenson, N.D., 2017. Independent evolution of baleen whale gigantism linked to Plio-Pleistocene ocean dynamics. Proc. R. Soc. B 284, 20170546. Available from: https://doi.org/10.1098/rspb.2017.0546.

Trevisan, L., 1941. Una nuova specie di *Balaenula* Pliocenica. Palaeontogr. Ital. 40, 1–13.

Van Beneden, P.-J., 1880. Description des ossements fossiles des environs d'Anvers. Deuxième partie. Cétacés. Genres *Balaenula*, *Balaena* et *Balaenotus*. Ann. Mus. R. Hist. Nat. Belg. 4, 1–82.

Werth, A.J., Potvin, J., 2016. Baleen hydrodynamics and morphology of cross-flow filtration in balaenid whale suspension feeding. PLoS ONE 11, e0150106. Available from: https://doi.org/10.1371/journal.pone.0150106.

Westgate, J.W., Whitmore, F.C., 2002. *Balaena ricei*, a new species of bowhead whale from the Yorktown Formation (Pliocene) of Hampton, Virginia. Smithson. Contrib. Paleobiol. 93, 295–312.

Woodward, B.L., Winn, J.P., Fish, F.E., 2006. Morphological specializations of baleen whales associated with hydrodynamic performance and ecological niche. J. Morphol. 267, 1284–1294. Available from: https://doi.org/10.1002/jmor.10474.

CHAPTER

3

The stocks of bowheads

A.B. Baird[1] and J.W. Bickham[2]

[1]Department of Natural Sciences, University of Houston-Downtown, Houston, TX, United States [2]Department of Ecology and Conservation Biology, Texas A&M University, College Station, TX, United States

Introduction

Bowhead whales (*Balaena mysticetus*) are taxonomically placed in the family Balaenidae along with their closest relatives, the right whales (*Eubalaena australis*, *Eubalaena glacialis*, and *Eubalaena japonica*) (Fig. 3.1). *B. mysticetus* is subdivided into four stocks or populations currently recognized by the International Whaling Commission (IWC). These include the Okhotsk (OKS) Sea stock, the Bering-Chukchi-Beaufort (BCB) Seas stock, the East Canada-West Greenland (ECWG) stock, and the East Greenland-Svalbard-Barents (EGSB) Sea stock (for map see Fig. 3.2). For the most part, these populations have been recognized as distinct since the time of commercial whaling. Based on experience and observations, whalers well recognized the patterns of distribution and timing of migrations, key indicators of population structure that also guided their cruises. In recent years additional methods have been applied such as satellite tags and stable isotope analysis. But today, the gold standard for studies of population structure is genetics and genomics which provide insight into population differentiation, migration, adaptation, relatedness, levels of inbreeding and genetic diversity, historical demography, and evolutionary history.

Population differentiation in bowhead whales has likely been influenced by a variety of factors including Pleistocene climate oscillations, longevity, sea ice conditions, feeding strategy which involves an affinity for living near the ice, high fidelity to breeding and feeding grounds, high vagility, and the history of commercial whaling (Rugh et al., 2003). The historical environmental conditions during the Pleistocene likely provided recurrent range contractions and expansions, leading to the observed patterns of genetic subdivision. Nonetheless, we cannot discount whaling as a powerful force in shaping current patterns of population subdivision. All four of the recognized bowhead whale populations experienced extensive mortality because of commercial whaling, beginning in Labrador c.1540 and lasting into the early part of the 20th century (Ross, 1993) (Chapter 33). After nearly a

The Bowhead Whale
DOI: https://doi.org/10.1016/B978-0-12-818969-6.00003-0

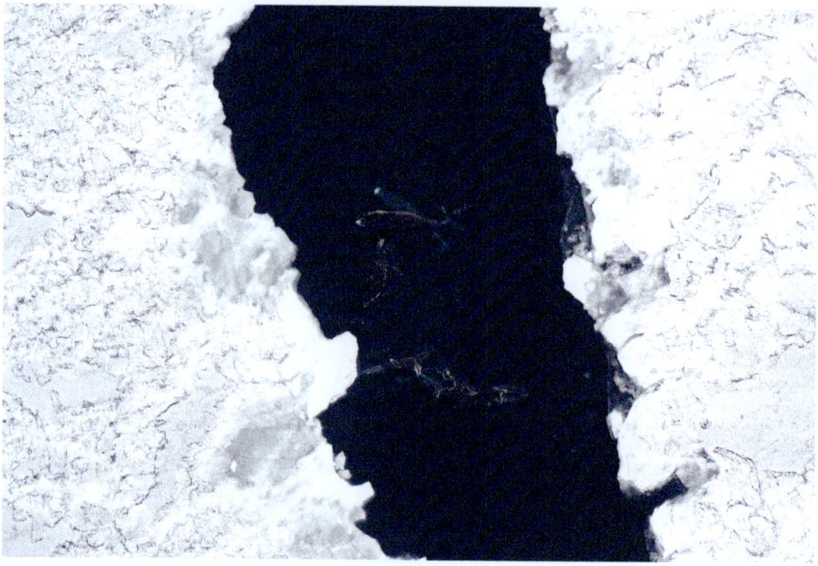

FIGURE 3.1 Five bowhead whales of the Bering-Chukchi-Beaufort Seas stock, interacting in a small lead, north of Point Barrow, Alaska. *Source: Photo by Vicki Beaver (NOAA/North Slope Borough, NMFS Permit No. 14245).*

century of recovery, two populations have rebounded nicely (BCB and ECWG) but both the OKS stock and the EGSB stock likely are comprised of only a few hundred individuals, each (Chapter 6).

The effects of whaling on the genetics of bowhead whales could include a reduction in genetic variability in populations termed a genetic bottleneck, as well as increased genetic divergence between populations. The latter would be due to genetic drift in a small population or populations as well as reduction in distributional ranges that could isolate previously connected populations. In this chapter, we first introduce various genetic techniques that have been used to study bowhead stock structure. Next, we review genetic studies of each stock of bowhead whales. Two important themes are population structure and what effects, if any, did whaling have on the genetics of this species. Finally, we explore the historical demography and evolutionary history of bowheads and explain the implications of population genetic studies of bowheads to the IWC.

Genetics of bowhead whales

A variety of genetic methods to examine bowhead stock structure have been employed over the past couple of decades, including microsatellites, mitochondrial DNA (mtDNA) sequencing, single nucleotide polymorphisms (SNPs), and whole-genome and transcriptome sequences. As technology has improved, these data have provided clearer insights into bowhead stock structure.

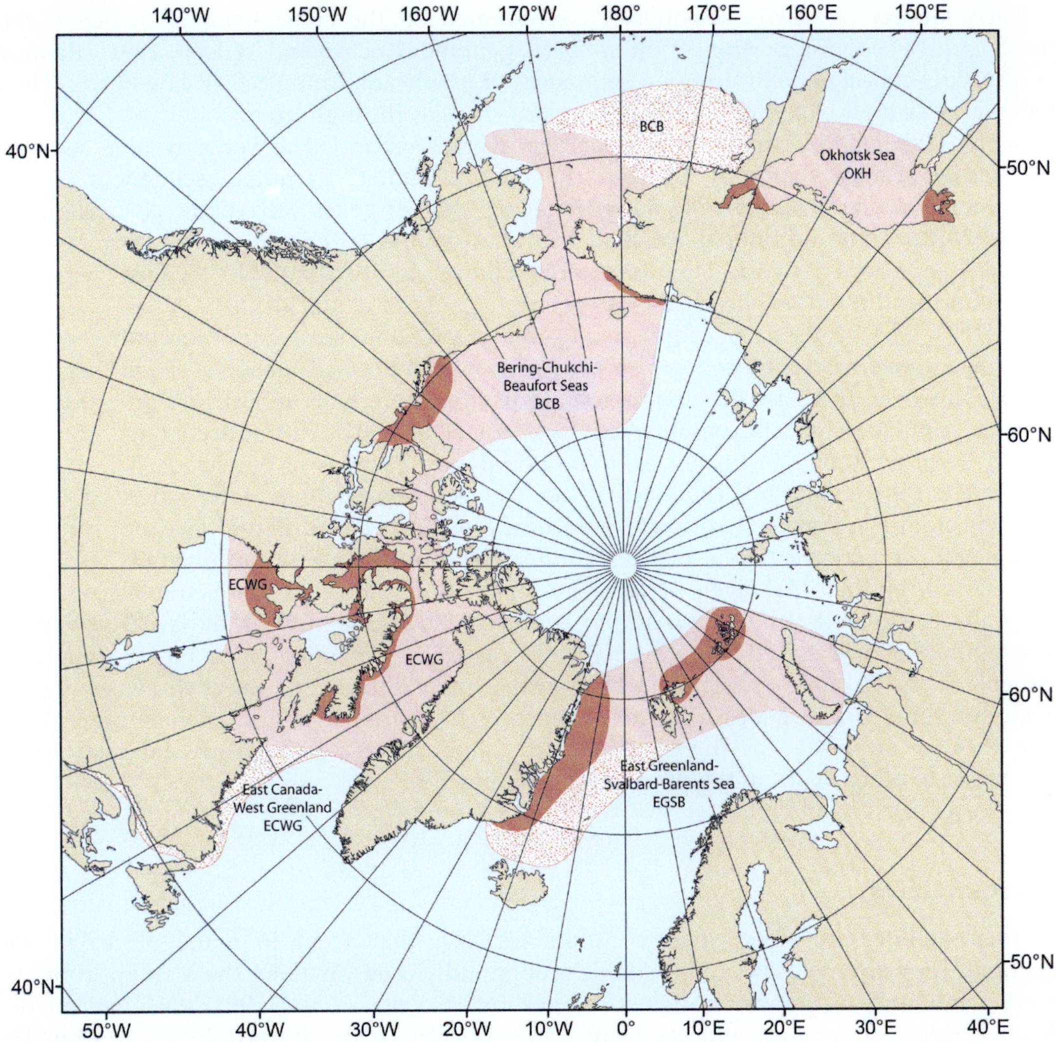

FIGURE 3.2 Current and historical ranges of bowhead whale stocks. *Pink*—current range; *Dark pink*—areas of high summer density; *Dotted*—historical distribution. *Source: Map by John Citta.*

Mitochondrial DNA

Mitochondrial DNA is a maternally inherited genetic marker and is present in the haploid (unpaired chromosome) state. It is often more useful in studying recent evolutionary history because it evolves quickly relative to the nuclear (biparentally inherited) genome. MtDNA sequences are easily obtained and techniques to utilize them have existed for several decades. As such, a great deal of mtDNA data exists for bowheads, and these have been used to study various aspects of their evolutionary and population genetic history.

Early studies on bowhead mtDNA concentrated on the control region, which is the most rapidly evolving portion of the mtDNA genome (Kocher and Wilson, 1991). Rooney et al. (2001) collected control region sequences of bowheads from 98 BCB bowheads. They used these sequences to model population size changes through time.

More recently, additional mtDNA markers have been added to better resolve some of the earlier studies that only utilized the control region. Complete sequences of the cytochrome-b (Cytb) and NADH dehydrogenase I (NDI) genes were obtained by Phillips et al. (2012). These additional mtDNA loci allowed the authors to obtain a more well resolved haplotype network. The practice of including multiple mtDNA loci has continued into recent studies on bowhead stock structure (e.g., Baird et al., 2018).

MtDNA can be used to estimate effective population size, and the control region described above has been used in this capacity for many cetacean species, including bowheads. However, there are drawbacks to only using a maternally inherited marker for estimating population genetic parameters. Bickham et al. (2013) introduced data from the X and Y chromosomes to further examine population genetics of bowheads. The X chromosome is biparentally inherited, while the Y chromosome is paternally inherited. Using markers from loci that have different inheritance patterns and different levels of recombination lead to more accurate estimates of effective population size and diversity. Bickham et al. (2013) reported that levels of diversity for the X chromosome were equivalent to theoretically predicted levels, whereas there was very low variability on the Y chromosome. The authors hypothesized that these results could be explained by a recent "selective sweep" of the Y chromosome. The biological explanation of the selective sweep in the Y chromosome is that males likely experience highly variable reproductive success. That hypothesis is supported by the occurrence of "super males" with greatly enlarged testes compared to other males in the population (Chapter 13).

Microsatellites

Microsatellites are short, repeated units of DNA that occur in many places in the genome. They are typically quite variable among individuals, making them an appropriate marker for studying population genetics questions. Rooney et al. (1999) developed microsatellite loci for bowheads and used previously developed loci from dolphins to examine demographic changes over time in BCB bowheads. They uncovered no genetic signatures of a recent genetic bottleneck in BCB bowheads using these data.

In addition to overall population size and among-population differentiation, substructuring of populations can be examined with microsatellites as well. Jorde et al. (2007) used microsatellites to examine temporal structuring of the BCB stock based on migrating whales collected at different times during the indigenous harvest. They observed groups of whales migrating at particular time intervals that were significantly less related than whales harvested during different time intervals. They suggested that there could be multiple stocks with different migration times migrating past Barrow. Another study using microsatellites suggested that animals harvested from Barrow and St. Lawrence Island (SLI) were genetically distinct (Givens et al., 2004). Both results were later refuted, as explained below.

Despite their usefulness in population genetic studies, there are potential pitfalls to microsatellite data. Microsatellites are analyzed by estimating the size of the repeated fragments of DNA. These estimates are dependent on the methods used to obtain and analyze the data and can vary between labs. They can also be unreliable if the species used to initially derive the microsatellite locus (and primers used to amplify it) is different from the species to which the microsatellite is applied in a study. These discrepancies make microsatellite work difficult to replicate across laboratories, and can result in generating flawed data (Givens et al., 2010).

The above microsatellite studies showing population substructuring used loci that were derived from species other than bowheads. When methods taking into account the previously mentioned pitfalls of microsatellite studies were applied to BCB bowheads, concerns about the existence of substructuring within the BCB stock disappeared. For example, when a panel of 22 microsatellites using bowheads as the focal species for development of the loci, along with a larger sample size of whales, the above patterns were no longer observed (Givens et al., 2010), and the hypothesis of substructuring within the BCB stock was not supported.

Whole genomes and single nucleotide polymorphisms

Genetic technology has advanced rapidly, allowing for improved methods and lowered costs associated with gathering larger amounts of data. Such advances have improved conservation studies on mammals, including bowheads and many other species of whales (Baird et al., 2019). Whole-genome sequences are now relatively easy and cost-effective to obtain. The acquisition of bowhead genome and transcriptome sequences has also improved population genetic studies on bowheads because it has allowed researchers to search the genome for SNPs.

SNPs are substitutions of single base-pairs in the genome. Because SNPs are analyzed by their sequence, they have a distinct advantage over microsatellites, are analyzed by estimation of their fragment size. Using certain methods for SNP genotyping, the data can be easily reproduced among different labs and thus a public database can be established and built upon, study by study, similar to what can be done with sequence data such as mtDNA control region sequences. Morin et al. (2012) compared the relative statistical power of SNPs and microsatellites. They concluded that a panel of 29 phased and unlinked bowhead SNP loci provided similar power as compared to a panel of 22 microsatellites in their ability to detect low levels of differentiation among bowhead populations when sample sizes were at least 20 individuals per population. The microsatellite panel performed better when used for estimates of effective population size (Ne) and for assigning samples to populations.

Baird et al. (2018) reported the results of stock structure analyses using a combination of 69 SNPs and 3 mtDNA loci. Their results showed that the SNP panel and mtDNA results were consistent with each other and with recent studies on bowhead stock structure using focal microsatellite data (Givens et al., 2010) and a smaller panel of SNPs (Morin et al., 2012).

Rapid advances in next-generation DNA sequencing methods (NGS) in recent years have driven down the cost of sequence analysis and made whole-genome sequencing readily available. This radical shift in methodology has been accompanied by advances in both computing capability and bioinformatics which provides the hardware and software necessary to analyze the huge datasets obtained with these methods. The bowhead whale was one of the first baleen whales to have both a whole-genome sequence (the entire DNA complement from a single individual) as well as transcriptome (RNA sequenced from a variety of tissues from multiple individuals) sequences published (Seim et al., 2014; Keane et al., 2015).

Interest in the bowhead genome and transcriptome for biomedical research stems from several unique attributes of this species, including longevity, lack of evidence of major age-related diseases like cancer, large body size, and absence of reproductive senility (Chapter 20). It is hypothesized that bowheads likely have adaptations protective of cancer and other age-related diseases. In fact, both of these studies found evidence of mutations in genes that could be involved in key adaptations for longevity, cancer and other disease resistance, thermoregulation, and adaptation for a lipid-rich diet. These studies have provided a rich resource for exploring many aspects of bowhead whale biology.

Bowhead stocks

Bering-Chukchi-Beaufort Seas stock

The BCB stock is currently the largest and most well-studied of the four bowhead stocks. Like the other stocks, it was subjected to intense commercial whaling estimated to have reduced its size to approximately 1000 individuals in the early 20th century (Woodby and Botkin, 1993; Rooney et al., 2001; Punt, 2006). The stock has since recovered and the most recent estimate of its size is >16,000 individuals (Givens et al., 2016; Chapter 6). It is still harvested annually for subsistence purposes by indigenous hunters in Alaska, United States, and Chukotka, Russia, and is the focus of an extensive research and monitoring program.

Data from a wide variety of genetic markers have been collected from BCB whales. Not surprisingly given its size, the BCB is the most genetically diverse of the bowhead stocks. Baird et al. (2018) studied SNP and mtDNA data from BCB, OKS, and ECWG whales and found that BCB shares some diversity with each of the other two, but was overall more similar to ECWG than to OKS.

East Greenland-Svalbard-Barents Sea stock

The largest prewhaling population of bowhead whales is thought to have been the EGSB stock that occupied the North Atlantic Ocean in the areas north and east of Greenland, around Spitsbergen (Svalbard) and east to Franz Josef Land and the Barents Sea. An estimate of abundance for the prewhaling population was 52,500 (Allen and Keay, 2006). This population was subjected to commercial hunting east of Greenland and in the area around Svalbard from 1610 to 1910, which reduced the population to such a low level as to be considered extinct or at least commercially extinct (Reeves, 1980). The stock was considered to number only in the tens of individuals (Wiig et al., 2007, 2010) but in the past few years, new sightings and detection by acoustic monitoring have provided new information about the abundance and distribution of this population (Moore et al., 2012; Stafford et al., 2012). A recent survey in an area of northeast Greenland known as the Northeast Water Polynya resulted in an abundance estimate of 318 individuals (Boertmann et al., 2015; Hansen et al., 2018). Given the limited area surveyed and the vast area in which at least sporadic sightings have occurred (e.g., Wiig, 1991), the total population must be considerably larger. Notwithstanding the slow pace of recovery of this population, there is presently evidence that it is increasing.

Borge et al. (2007) studied mtDNA control region sequences from 99 whales sampled from bones collected at Svalbard and presumed to represent the nearly extinct EGSB stock. The samples were ^{14}C dated and ranged from recent to more than 50,000 BP. They found high haplotype diversity, which will serve as a benchmark to search for potential diversity loss when data from the current postwhaling population become available. They compared their data to previously published data from the BCB population (Rooney et al., 2001) and noted a very low, but statistically significant, level of differentiation ($FST = 0.013$, $P < 0.0001$) between the populations. The biological significance of this was discounted because of the high number of low-frequency haplotypes in each population, the fact that the bone samples represent a "compilation of mitochondrial haplotypes over a period of approximately 50,000 years" (Borge et al., 2007, p. 2230) rather than a population survey as in the Rooney et al. (2001) dataset, and no phylogeographic structure is apparent in the haplotype network. Moreover, the statistical significance disappears when only the 25 most recent samples in the EGSB dataset are used. Rather, Borge et al. (2007) conclude that their evidence is indicative of circumpolar contact between Pacific and Atlantic bowhead stocks over time. This was also the conclusion of Alter et al. (2012) who compared mtDNA control region sequences from ECWG, BCB, and OKS whales with Borge et al.'s (2007) dataset and concluded that the data indicate contemporary and high gene flow between the Atlantic and Pacific Ocean basins.

Okhotsk Sea stock

The OKS stock of bowhead whales is a small population that is confined to the sea of Okhotsk in Russia. The most recent population estimate (Cooke et al., 2017) based on a genotypic mark-recapture method suggests a likely declining population of about 218. Although uncertain, this population is generally assumed to have been distinct from the larger BCB stock since before whaling (Moore and Reeves, 1993; Rugh et al., 2003). What is certain is that the high depletion of this population and the BCB caused by whaling has resulted in a greatly increased distributional hiatus making the OKS stock by far the most geographically isolated of the four currently recognized bowhead stocks. Consistent with that, it is also the genetically most distinct population. Alter et al. (2012), which is the only study to examine genetic differentiation among all four populations, confirmed previous studies (LeDuc et al., 2005) showing this population to be the most distinct as well as the least diverse, based on mtDNA control region sequences. Baird et al. (2018) compared the OKS population to the BCB and ECWG population and showed that Okhotsk was easily distinguished from both using nuclear SNP markers and as an extended mtDNA sequence of control region, cytochrome-b, and ND1 totaling 2494 bp. Morin et al. (2012) compared 29 phased SNPs and 22 microsatellite loci and likewise showed this population to be the most distinct compared to the BCB and ECWG populations. The ECWG and BCB populations differed only slightly which is indicative of relatively high levels of recent gene flow, which is also consistent with previous studies (Alter et al., 2012). And significantly, Foote et al. (2013), who studied ancient bowhead DNA samples from the Atlantic Ocean and modeled future climate and range shifts, suggest that "if populations respond to ongoing directional climate change by shifting their distribution northward to track the retreating sea ice, then the endangered Okhotsk Sea population may become increasingly isolated and vulnerable."

East Canada-West Greenland stock

The ECWG stock population of bowhead whales is currently considered to include the combined Davis Strait and Hudson Bay stocks (Rugh et al., 2003). It is a relatively robust population that has recovered well from the effects of commercial whaling. A recent estimate of abundance using genetic mark-recapture methods was obtained based on 1177 samples taken from 9 geographic areas covering well the distribution of the population. These samples were genotyped at 21 microsatellite loci which revealed 992 unique genotypes and 49 recaptures. The resulting abundance estimate was 11,747 (95% CI 8,169–20,043) (Frasier et al., 2020). However, the abundance estimate accepted by the IWC based on aerial surveys is 6,446 (95%, CI 3,876–10,721) (Doniol-Valcroze et al., 2015; see Chapter 3).

Genetic studies have consistently shown a slight but statistically significant level of population differentiation between the BCB and the ECWG population. For example, Givens et al. (2010) calculated an FST of only 0.006 based on an analysis of 22 microsatellite loci, that was significantly different from 0. Alter et al. (2012) calculated a statistically significant ΦST of 0.03 for mtDNA control region sequences between these two populations. The results of both Givens et al. (2010) and Alter et al. (2012) are indicative of a moderate level of gene flow between these two populations. Moreover, Alter et al. (2012) concluded that Arctic Sea ice is not as strong of a barrier to gene flow between these two populations as was previously assumed. Morin et al. (2012) similarly found low but significant differentiation between BCB and ECWG samples based on an analysis of 29 phased SNP loci and the same 22 microsatellite loci used in Givens et al. (2010). They concluded that these populations are demographically distinct and estimated a "trivial" level of gene flow (0.7 individuals per year). Morin et al. (2012) conclude that this estimated level of gene flow is consistent with historically sporadic gene flow between populations in different ocean basins in a species with a very long generation time.

Alter et al. (2012) also examined a set of ancient samples from Prince Regent Inlet (PRI) that predated the era of commercial whaling. The PRI samples were taken from an area that is within the present range of the ECWG population, but these samples were found to be genetically distinct from the current population. Surprisingly, the PRI samples were found to be more similar to the BCB population than the ECWG population, but still statistically significantly different. Thus, the PRI was likely a small isolated population located in a geographic cul-de-sac. According to the authors, it is likely that climate change, whaling, or a combination of the two led to the extinction of this surprisingly genetically distinct population.

Historical demography and evolutionary history

Genetic bottlenecks can have drastic consequences on populations. Of concern to conservationists, a bottleneck cannot only reduce the number of individuals in the population but also reduce genetic diversity. Whereas numbers of individuals can recover after a bottleneck, recovering genetic diversity can only occur through gene flow and new mutations, which can take many generations. This leaves populations vulnerable and unable to adapt to changing environments.

It is well documented that commercial whaling caused a dramatic drop in the size of the BCB stock (Woodby and Botkin, 1993; Rooney et al., 2001; Punt, 2006). Population geneticists have been interested in whether this bottleneck also affected genetic diversity of BCB bowheads.

Rooney et al. (1999), using microsatellites, and Rooney et al. (2001), based on mtDNA control region sequences, found that the genetic diversity of the BCB population was not affected by the bottleneck created due to commercial whaling. They also determined that the population expanded during the Middle to Late Pleistocene. When microsatellite data and three mtDNA loci were examined in combination, Phillips et al. (2012) also observed no evidence of a bottleneck caused by commercial whaling. They too found evidence of population expansion, and more recent analytical techniques were able to conclude that the expansion occurred around 70,000 years ago and subsequently declined around 15,000 years ago. These changes could be traced to climatic oscillations during those time periods. The population expansion occurred during the second previous glaciation and the decline coincided with the warming after the last glacial period. Given the timing of the decline, it is unlikely that there was an anthropogenic factor involved. Phillips et al. (2012) hypothesized that the short duration of commercial whaling for bowheads relative to their generation time prevented the loss of genetic diversity normally expected with such an event.

Acknowledgments

We thank the Alaska Eskimo Whaling Commission (AEWC), the Barrow Whaling Captains' Association, and other village Whaling Captains' Associations for their confidence, guidance, and support of our research, especially for providing access to samples. We gratefully acknowledge funding provided by the North Slope Borough Department of Wildlife Management.

References

Allen, R.C., Keay, I., 2006. Bowhead whales in the Eastern Arctic, 1611–1911: population reconstruction with historical whaling records. Environ. Hist. 12, 89–113.

Alter, E.S., Rosenbaum, H.C., Postma, L.D., Whitridge, P., Gaines, C., Weber, D., et al., 2012. Gene flow on ice: the role of sea ice and whaling in shaping Holarctic genetic diversity and population differentiation in bowhead whales (*Balaena mysticetus*). Ecol. Evol. 2 (11), 2895–2911.

Baird, A.B., Givens, G.H., George, J.C., Suydam, R.S., Bickham, J.W., 2018. Stock structure of bowhead whales inferred from mtDNA and SNP DNA. Paper SC/67b/SDDNA/01 presented to the IWC Scientific Committee (unpublished), 21pp. Available from: http://www.iwcoffice.org.

Baird, A.B., Robinson, M.J., Bickham, J.W., 2019. The role of the American Society of Mammalogists in mammalian conservation: from politics to conservation genetics. J. Mammal. 100, 774–785.

Bickham, J.W., Stuart, G.W., Downing, H.K., Patton, J.C., George, J.C., Suydam, R.S., 2013. Comparison of methods for molecular assessment of sex chromosome polymorphisms and levels of genetic diversity in the bowhead whale. Paper SC/65a/BRG22 presented to the International Whaling Commission Scientific Committee.

Boertmann, D., Kyhn, L.A., Witting, L., Heide-Jørgensen, M.P., 2015. A hidden getaway for bowhead whales in the Greenland Sea. Polar Biol. 38, 1315–1319.

Borge, T., Bachmann, L., Bjornstad, G., Wiig, O., 2007. Genetic variation in Holocene bowhead whales from Svalbard. Mol. Ecol. 16, 2223–2235.

Cooke, J.G., Shpak, O.V., Meschersky, I.G., Burdin, A.M., Maclean, S.A., Chichkina, A.N., 2017. Updated estimates of population and trend for Okhotsk Sea bowhead whales, 8pp. SC/67a/NH/10.

Doniol-Valcroze, T., Gosselin, J.-F., Pike, D., Lawson, J., Asselin, N., et al. 2015. Abundance estimate of the Eastern Canada – West Greenland bowhead whale population based on the 2013 High Arctic Cetacean Survey. DFO Can. Sci. Advis. Sec. Res. Doc. 2015/058.

Foote, A.D., Kaschner, K., Schultze, S.E., Garilao, C., Ho, S.Y.W., Post, K., et al., 2013. Ancient DNA reveals that bowhead whale lineages survived Late Pleistocene climate change and habitat shifts. Nat. Commun. 4, 1677.

Frasier, T.R., Petersen, S.D., Postma, L., Johnson, L., Heide-Jørgensen, M.P. and Ferguson, S.H., 2020. Abundance estimation from genetic mark-recapture data when not all sites are sampled: an example with the bowhead whale. *Global Ecology and Conservation*, p.e00903.

Givens, G.H., Bickham, J.W., Matson, C.W., Ozaksoy, I., Suydam, R.S., George, J.C., 2004. Examination of Bering-Chukchi-Beaufort Seas bowhead whale stock structure hypotheses using microsatellite data. Paper SC/56/BRG17 presented to the IWC Scientific Committee, June 2004.

Givens, G.H., Edmondson, S.L., George, J.C., Suydam, R., Charif, R.A., Rahaman, A., et al., 2016. Horvitz-Thompson whale abundance estimation adjusting for uncertain recapture, temporal availability variation and intermittent effort. Environmetrics 27, 134–146.

Givens, G.H., Huebinger, R.M., Patton, J.C., Postma, L.D., Lindsay, M., Suydam, R.S., et al., 2010. Population genetics of bowhead whales (*Balaena mysticetus*) in the Western Arctic. Arctic 63, 1–12.

Jorde, P.E., Schweder, T., Bickham, J.W., Givens, G.H., Suydam, R., Hunter, D., et al., 2007. Detecting genetic structure in migrating bowhead whales off the coast of Barrow, Alaska. Mol. Ecol. 16, 1993–2004.

Hansen, R.G., Borchers, D., Heide-Jørgensen, M.P., 2018. Summer surveys of marine mammals in the Greenland Sea and the Northeast Water and winter survey of marine mammals in the Northeast Water – preliminary report from field work in 2017 and 2018. Greenland Institute of Natural Resources.

Keane, M., Semeiks, J., Webb, A.E., Li, Y.I., Quesada, V., Craig, T., et al., 2015. Insights into the evolution of longevity from the bowhead whale genome. Cell Rep. 10, 112–122.

Kocher, T.D., Wilson, A.C., 1991. Sequence evolution of mitochondrial DNA in humans and chimpanzees: control region and a protein coding region. In: Osawa, S., Honjo, T. (Eds.), Evolution of Life: Fossils Molecules and Culture. Springer, Tokyo, pp. 391–413.

LeDuc, R.G., Dizon, A.E., Burdin, A.M., Blockin, S.A., George, J.C., Brownell Jr., R.L., 2005. Genetic analyses (mtDNA and microsatellites) of Okhotsk and Bering/Chukchi/Beaufort Seas populations of bowhead whales. J. Cetacean Res. Manage. 7, 107–111.

Moore, S.E., Reeves, R.R., 1993. Distribution and movement. In: Burns, J.J., Montague, J.J., Cowles, C.J. (Eds.), The Bowhead Whale. Allen Press, Lawrence, KS, pp. 313–386.

Moore, S.E., Stafford, K.M., Melling, H., Berchok, C., Wiig, Ø., Kovacs, K.M., et al., 2012. Comparing marine mammals acoustic habitats in Atlantic and Pacific sectors of the High Arctic: year-long records from Fram Strait and the Chukchi Plateau. Polar Biol. 35, 475–480.

Morin, P.A., Archer, F.I., Pease, V.L., Hancock-Hanser, B.L., Robertson, K.M., Huebinger, R.M., et al., 2012. Empirical comparison of single nucleotide polymorphisms and microsatellites for population demographic analyses of bowhead whales. Endanger. Species Res. 19, 129–147.

Phillips, C.D., Hoffman, J.I., George, J.C., Suydam, R.S., Huebinger, R.M., Patton, J.C., et al., 2012. Molecular insights into the historic demography of bowhead whales: understanding the evolutionary basis of contemporary management practices. Ecol. Evol. 3, 18–37.

Punt, A.E., 2006. Assessing the Bering-Chukchi-Beaufort Seas stock of bowhead whales using abundance data together with data on length or age. J. Cetacean Res. Manage. 8, 127–137.

Reeves, R.R., 1980. Spitsbergen bowhead stock: a short review. Mar. Fish. Rev. 42, 65–69.

Rooney, A.P., Honeycutt, R.L., Davis, S.K., Derr, J.N., 1999. Evaluating a putative bottleneck in a population of bowhead whales from patterns of microsatellite diversity and genetic disequilibria. J. Mol. Evol. 49, 682–690.

Rooney, A.P., Honeycutt, R.L., Derr, J.N., 2001. Historical population size change of bowhead whales inferred from DNA sequence polymorphism data. Evolution 55, 1678–1685.

Ross, W.G., 1993. Commercial whaling in the North Atlantic sector. In: Burns, J.J., Montague, J.J., Cowles, C.J. (Eds.), The Bowhead Whale. Society for Marine Mammalogy, Lawrence, KS, pp. 511–577.

Rugh, D., DeMaster, D., Rooney, A., Breiwick, J., Shelden, K., Moore, S., 2003. A review of bowhead whale (*Balaena mysticetus*) stock identity. J. Cetacean Res. Manage. 5, 267–279.

Seim, I., Ma, S., Zhou, X., Gerashchenko, M.V., Lee, S.-G., Suydam, R., et al., 2014. The transcriptome of the bowhead whale *Balaena mysticetus* reveals adaptations of the longest-lived mammal. Aging 6, 879–899.

Stafford, K.M., Moore, S.E., Berchok, C.L., Wiig, Ø., Lydersen, C., Hansen, E., et al., 2012. Spitsbergen's endangered bowhead whales sing through the polar night. Endanger. Species Res. 18, 95–103.

Wiig, Ø., 1991. Seven bowhead whales (*Balaena mysticetus* L.) observed at Franz Josef Land in 1990. Mar. Mamm. Sci. 7, 316–319.

Wiig, Ø., Bachmann, L., Janik, V.M., Kovacs, K.M., Lydersen, C., 2007. Spitsbergen bowhead whales revisited. Mar. Mamm. Sci. 23, 688–693.

Wiig, Ø., Bachmann, L., Øien, N., Kovacs, K.M., Lydersen, C., 2010. Observations of bowhead whales (*Balaena mysticetus*) in the Svalbard area 1940–2009. Polar Biol. 33, 979–984.

Woodby, D.A., Botkin, D.B., 1993. Stock sizes prior to commercial whaling. In: Burns, J.J., Montague, J.J., Cowles, C. (Eds.), The Bowhead Whale. Society for Marine Mammalogy, Lawrence, KS, pp. 387–407, 787pp.

CHAPTER

4

Distribution and behavior of Bering-Chukchi-Beaufort bowhead whales as inferred by telemetry

J.J. Citta[1], *L. Quakenbush*[1] *and J.C. George*[2]

[1]Alaska Department of Fish and Game, Fairbanks, AK, United States [2]Department of Wildlife Management, North Slope Borough, Utqiaġvik, AK, United States

Introduction

During the last three decades, many advances have been made in understanding the distribution, behavior, and ecology of the Bering-Chukchi-Beaufort (BCB) stock of bowhead whales. Many of these advances are due to telemetry studies.

The distribution of BCB bowhead whales was largely understood prior to the use of telemetry. The earliest information on the distribution of BCB bowhead whales comes from traditional knowledge held by indigenous subsistence whalers and the archeological evidence they left behind. Indeed, evidence suggests that aboriginal whaling for bowhead whales began in the Bering Sea approximately 2000 years ago and, by CE 900, bowhead whales were a significant part of people's diet in the Bering Strait and Chukchi Sea regions (see review in Stoker and Krupnik, 1993) and permanent villages established at locations where landform intercepted the bowhead's migratory path. Bowhead whale continues to be an important component in the diet of indigenous people over a significant portion of the whale's range (Suydam and George, 2018), resulting in a detailed local understanding of whale movements and behavior that has been handed down over generations (e.g., Noongwook et al., 2007; Quakenbush and Huntington, 2010; Huntington et al., 2016; Chapter 34). However, this understanding is largely limited to the vicinity of villages and hunting camps. Our understanding of the offshore distribution of bowhead whales was first provided by the Yankee whalers, who kept detailed records on the location of harvested whales. The commercial harvest lasted from 1849 to 1914; at first, whalers encountered bowheads in the southwestern Bering Sea (Fig. 4.1), but this area was quickly

The Bowhead Whale
DOI: https://doi.org/10.1016/B978-0-12-818969-6.00004-2

FIGURE 4.1 Bowhead tagging near Utqiaġvik, Alaska. The pole in the man's hand is used to attach the telemetry device to the whale. *Source: Photo by John Citta (Alaska Department of Fish and Game, NMFS Permit No. 18890).*

depopulated and the industry largely shifted to the Chukchi Sea by 1854. Although whales continued to be harvested in the Bering and Chukchi seas, by 1889, the industry shifted to the summer feeding grounds in the Canadian Beaufort Sea (Bockstoce and Botkin, 1983; Bockstoce et al., 2005; Chapter 33). Although the BCB stock now numbers more than 16,000 whales (Givens et al., 2016; Chapter 6), perhaps more than existed during the Yankee whaling period, they have not repopulated their historical summer range south of the shelf break in the Bering Sea. As the BCB population recovered from commercial whaling, information on distribution and behavior continued to come from the observations of subsistence whalers; however, scientists also began to observe whales from ships and aircraft (e.g., Moore and Reeves, 1993; Smultea et al., 2012), and use passive acoustics to detect whales (e.g., Clark and Johnson, 1984; Ljungblad et al., 1982). Such surveys continue to provide important information on the distribution and ecology of bowhead whales (e.g., Clarke et al., 2018; Okkonen et al., 2018; Clark et al., 2015).

Collectively, data indicated that bowhead whales wintered in the northern Bering Sea, most likely in Russian waters. BCB bowhead whales were thought to have an affinity for the ice edge and polynyas (e.g., Bogoslovskaya et al., 1982; Brueggeman et al., 1987; Zelensky et al., 1997). In April and May, most whales migrated north along the Alaska coast and then east through heavy Beaufort Sea ice to somewhere in the western Canadian Archipelago. Aerial surveys in summer and autumn (Moore and Reeves, 1993; Harwood et al., 2010) showed that bowhead whales were common near Cape Bathurst and on the continental shelf near Tuktoyaktuk, the Mackenzie Delta, and Hershel Island (Fig. 4.1). In late summer, whales migrated west, back toward Utqiaġvik (formerly Barrow), and then into the Chukchi Sea (Moore and Reeves, 1993). The path whales took between Utqiaġvik and the northern coast of Chukotka was largely unknown; however, large aggregations of whales were regularly observed in the autumn along the northern coast of Chukotka, Russia (e.g., Burns, 1993; Moore et al., 1995).

Hence, while the seasonal distribution of the BCB stock was broadly understood (see Figure 9.7 in Moore and Reeves, 1993), many of the specifics were unknown. The path whales took in the spring when crossing the Beaufort Sea was unknown, as was the path they took crossing the Chukchi Sea in autumn. Likewise, where the majority of whales summered prior to when aerial surveys occurred in autumn and where they wintered under the sea ice in the Bering Sea were also largely unknown. The prevailing thought was that whales wintered in areas of lighter ice, using the ice edge and large polynyas. The International Whaling Commission was concerned that the BCB stock may not be a single stock but instead be composed of two or more stocks (e.g., George et al., 2017; IWC 2007), which would have important implications for how subsistence harvest was managed (e.g., strike limits would have to be defined for each stock). Indeed, evidence suggested that multiple stocks may exist. Subsistence whalers at St. Lawrence Island suggested that bowhead whales following different paths around their island might belong to separate stocks (Noongwook et al., 2007) and, in 2000 and 2001, Russian observers documented ~500 bowhead whales migrating up the Russian coast within Bering Strait (Melnikov and Zeh, 2007). Based upon the timing of observations, they thought it unlikely that these whales were part of the spring migration past Utqiaġvik and were summer residents of the Chukchi Sea and, therefore, potentially members of a separate stock.

Every method of inference has limitations. The Yankee whalers removed the whales they documented, almost hunting the BCB stock to extinction, and observations of subsistence whalers are limited to areas near hunting camps and villages. Surveys by ship or airplane are expensive and limited in scope and by weather, typically only observing whales in particular seasons or regions, and provide no information on the behavior of individual whales over time or of behavior not visible under the surface. Acoustic surveys are typically limited to a few recording buoys, detections are restricted to vocal whales, and determining the number of whales detected can be problematic, even with complex acoustic arrays. Telemetry, especially satellite telemetry, has the capacity to overcome some of these issues (we discuss the limitations of telemetry later). Telemetry can collect data in foul weather, under sea ice, in the dark, and in foreign waters. By tracking individual whales, residence times for areas or regions can be calculated. Furthermore, modern satellite tags collect not only location data, but also information on dive behavior and the surrounding physical environment, allowing features that may aggregate zooplankton prey, such as oceanographic fronts and stratified layers to be identified. Note that what we refer to as a *tag* is composed of a transmitter (VHF or satellite-linked), hardware used to anchor the tag to the animal, and may also include sensors that measure the environment or animal behavior.

Telemetry studies on BCB bowhead whales began with VHF technology. As part of a feasibility study, two VHF radio tags were deployed on bowhead whales near

Tuktoyaktuk, Canada, in 1982 (Hobbs and Goebel, 1982). Although the tags appeared to be working, they had limited range and were difficult to relocate by ship and aircraft. Five bowhead whales were tagged with VHF transmitters in the western Canadian Beaufort Sea by Jeff Goodyear in 1985 (see Appendix 4 in Richardson, 1987); three tags were relocated, once each. Nine VHF transmitters were deployed on bowhead whales near Tuktoyaktuk in 1988 (Wartzok et al., 1989) and five more in 1990, also near Tuktoyaktuk (Wartzok et al., 1990). Using ships and aircraft, three were tracked to the vicinity of Point Barrow, one in 1989 and two in 1990. Again, tracking efforts were largely hampered by the difficulty in relocating VHF transmitters.

Hobbs and Goebel (1982) recommended investigating if using satellite-linked telemetry, where data are transmitted directly to satellites, would be more successful than VHF, where animals must be relocated to retrieve location data. Satellite telemetry was largely unproven until the mid-1980s (see review in Fancy et al., 1988); the first satellite transmitters were deployed on BCB bowhead whales during a study led by Bruce Mate in the autumn of 1992. Mate et al. (2000) tagged 12 juvenile bowheads with satellite tags. Eight tags yielded transmissions for 3–33 days ($\bar{x} = 14$ days). Most tags quit transmitting before leaving Canadian waters. However, four tags lasted long enough to document long-distance whale movements; three into the Alaskan Beaufort and one into the western Chukchi Sea, to within ~200 km of the northern coast of Chukotka, Russia. Dive summaries from these eight whales are reported in Krutzikowsky and Mate (2000).

These telemetry studies were funded to assess potential effects of oil and gas exploration in Beaufort Sea waters of both Canada and Alaska. The importance of the eastern Alaskan Beaufort Sea as a feeding ground for bowhead whales was of particular interest due to oil and gas development. Richardson and Thomson (2002) concluded that although bowhead whales feed in the eastern Alaskan Beaufort Sea, this area likely contributed little to the annual diet of bowhead whales. This finding was contentious with Alaska Natives subsisting on bowhead whales in Beaufort Sea communities and was questioned by Lowry et al. (2004) and others (see review in Lowry et al., 2004), who found food in the stomachs of harvested whales. Furthermore, interest in petroleum development within the Chukchi Sea was growing, leading to renewed concern over bowhead whale movements and distribution. In 2004, a new tagging study began as a collaboration between the Alaska Department of Fish and Game, the North Slope Borough, and the Alaska Eskimo Whaling Commission. In 2007, this collaboration was expanded to include Fisheries and Oceans Canada. The project was designed in close collaboration with Alaska Native subsistence whalers who helped define research priorities and deployed the majority of tags (see Table 1 in Citta et al., 2015).

Description of the tagged sample of Bering-Chukchi-Beaufort bowhead whales

During 2006–18, we deployed 77 satellite tags on bowhead whales that yielded location data; of these, 62 also provided dive data (see Citta et al., 2015, and Olnes et al., in press, for a description of tag and data types). Most tags were deployed using a pole as a harpoon, either thrown or jabbed, a technique bowhead and beluga hunters use for hunting.

On average, tags that transmitted did so over a span of 167 days. Thirty-one tags lasted more than 180 days and 10 lasted more than 365 days, thereby providing information on year-round movements. Most whales were tagged near Point Barrow, Alaska ($n = 51$), or Tuktoyaktuk, Canada ($n = 21$); three were tagged near St. Lawrence Island, Alaska, one at Herschel Island, Canada, and one at Shingle Point, Canada (Fig. 4.2).

The tagged sample is skewed toward immature males; however, this skew is unlikely to affect our results. Based upon the harvest at Utqiaġvik, the sex ratio should be approximately 50:50 (Suydam and George, 2018). Of 40 tagged whales of known sex, 26 (i.e., 65%) were male, possibly because females with calves were avoided during tagging. The sample is also skewed toward immature whales, defined as those with estimated lengths <13 m. Koski et al. (2006) used photogrammetry to estimate the length distribution of bowhead whales near Point Barrow; we recalculated this length distribution, removing calves, which we did not sample. Our recalculation of Koski et al. (2006) results in 58.9% immature whales and the tagged sample consisted of 65% immature whales. To achieve an equal sex ratio and a representative age ratio, we would have to tag approximately

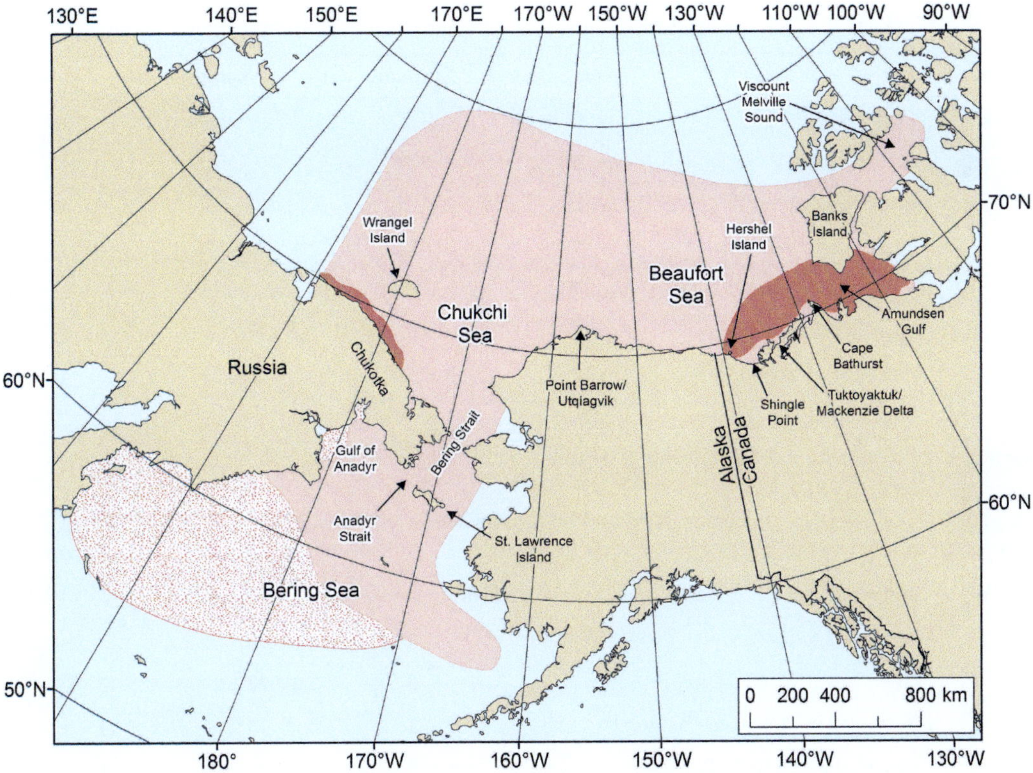

FIGURE 4.2 Range of the Bering-Chukchi-Beaufort (BCB) stock of bowhead whales. Light pink is the total range of frequent occurrences; dark pink is the high-density summer range; speckled pink is the historical range mostly based on Bockstoce et al. (2005). The summer concentration area along the northern coast of Chukotka, Russia, is poorly understood relative to the main summer concentration area in Amundsen Gulf, Canada. Fewer than 1000 whales are thought to summer in the Chukchi Sea (Melnikov and Zeh, 2007).

10 more adult (≥13 m length) female bowhead whales. Adding 10 more adult females to the sample would not significantly alter our results regarding the distribution or behavior of bowhead whales.

Seasonal distribution of tagged Bering-Chukchi-Beaufort bowhead whales

Spring migration

We learned that bowhead whales typically begin migrating northward from wintering grounds in the western Bering and southern Chukchi seas in April and that the migration begins before the Bering Sea is ice-free (Citta et al., 2012; see also Figures 9 and 10 in Citta et al., 2015). Whales typically exited the Bering Sea by the third week of April ($\bar{x} = 17$ April; Table 4.1) and then passed Point Barrow approximately 12 days later (Table 4.2). For whales migrating to the Canadian Beaufort Sea, which is something almost

TABLE 4.1 Dates at which bowhead whales arrive at different boundaries and geographic features.

Season	Event	Average date	Minimum date	Maximum date	*n* whales
Spring	Leave Bering Sea	Apr 17	Apr 1	May 26	15
	Pass Pt. Barrow	Apr 29	Apr 16	May 25	17
	Pass Demarcation Pt.	May 10	Apr 26	May 30	17
	Arrive Cape Bathurst	May 12	May 2	May 26	14
Autumn	Pass Demarcation Pt.	Sep 17	Jul 13	Oct 26	24
	Pass Pt. Barrow	Sep 25	Jul 21	Nov 2	53
	Arrive at Chukotka Coast	Oct 27	Jul 28	Dec 21	40
	Arrive in Bering Sea	Dec 2	Nov 5	Dec 29	24

Demarcation Point marks the boundary between the Alaskan and Canadian Beaufort Seas.

TABLE 4.2 Travel times in days for individual bowhead whales.

Season	Route	Average no. of days	Minimum no. of days	Maximum no. of days	No. of whales
Spring	Bering Sea to Pt. Barrow	12.0	8	16	12
	Pt. Barrow to Demarcation Pt.	8.5	5	22	13
	Demarcation Pt. to Cape Bathurst	5.8	3	13	14
Autumn	Demarcation Pt. to Pt. Barrow	19.0	5	54	21
	Pt. Barrow to Chukotka Coast	32.6	6	82	39
	Pt. Barrow to Bering Sea	69.6	32	94	21

Demarcation Point marks the boundary between the Alaskan and Canadian Beaufort seas.

all whales do (76 of 77 tagged whales), most of the migration through the Chukchi Sea occurs within 50 km of the Alaskan coastline. Once past Point Barrow, migrating whales travel farther from shore, mostly between 80 and 250 km of the Alaskan coastline in the Beaufort Sea. Whales take a generally direct path from Point Barrow to the recurrent polynya at Cape Bathurst (Arrigo and van Dijken, 2004), requiring an average of 9 days to reach the Canadian border at Demarcation Point and 6 more days to reach Cape Bathurst (Table 4.2). Whales tend to linger near the polynya until sea ice leaves Amundsen Gulf.

Most whales migrate from the wintering grounds in the Bering and southern Chukchi seas to Amundsen Gulf, Canada. Only 1 of 26 (~4%) whales that provided location data during the spring migration did not migrate to Amundsen Gulf, Canada. This whale, an adult of unknown sex, was tagged near Point Barrow the prior autumn in 2009, did not leave the Bering Sea until 26 May 2010, at which time it migrated up the coast of Chukotka, Russia, and spent the entire summer in the Chukchi Sea (Citta et al., 2012). This is consistent with the findings of Melnikov and Zeh (2007), which indicated that perhaps 500 whales out of a population of over 16,000 (~3%) may migrate up the Russian coast each spring and do not go to Canada.

Summer range

The main summering area for BCB bowhead whales occurs in the Canadian Beaufort Sea. The distribution of bowhead whales is centered in Amundsen Gulf in May, June, and July (Figs. 4.3–4.4). Bowheads begin moving westward to shelf waters near Tuktoyaktuk and the Mackenzie Delta in July and August (Fig. 4.4). There is currently limited use of waters north of Amundsen Gulf in summer and early autumn. Two tagged whales (both adult males) traveled north along Banks Island in August, one of which spent significant time in Viscount Melville Sound, in the vicinity of a tagged bowhead whale from the East Canada-West Greenland stock (Heide-Jørgensen et al., 2011). All three tagged whales later returned to their respective stock ranges.

The summer distribution of bowhead whales is much more dispersed with more complicated movements than originally thought. Although most whales summer in the Canadian Beaufort Sea, eight of 26 (31%) tagged whales left there in June and July, well before the usual autumn migration in late August and September (Fig. 4.5). Five whales traveled west into the Alaskan Beaufort Sea, three of which passed west of Point Barrow, and then returned to the Canadian Beaufort prior to the autumn migration. Four of these were immature males; one was an immature whale of unknown sex. Two whales, an adult female and an adult of unknown sex, migrated to Russia in midsummer. One whale, an adult male, left Amundsen Gulf and entered the Northwest Passage (Viscount Melville Sound) where it remained until late August. Hence, of the eight whales that left the Canadian Beaufort in June and July, three (38%) were adults and, of the whales of known sex ($n = 6$), five (83%) were male. This suggests that males, perhaps adult males, may be more likely to range outside of typical summering areas. These unexpected movements, especially movements of an adult whale of unknown sex that was tagged near Point Barrow in 2009 and migrated to the Russian coast the following spring, and the fact that

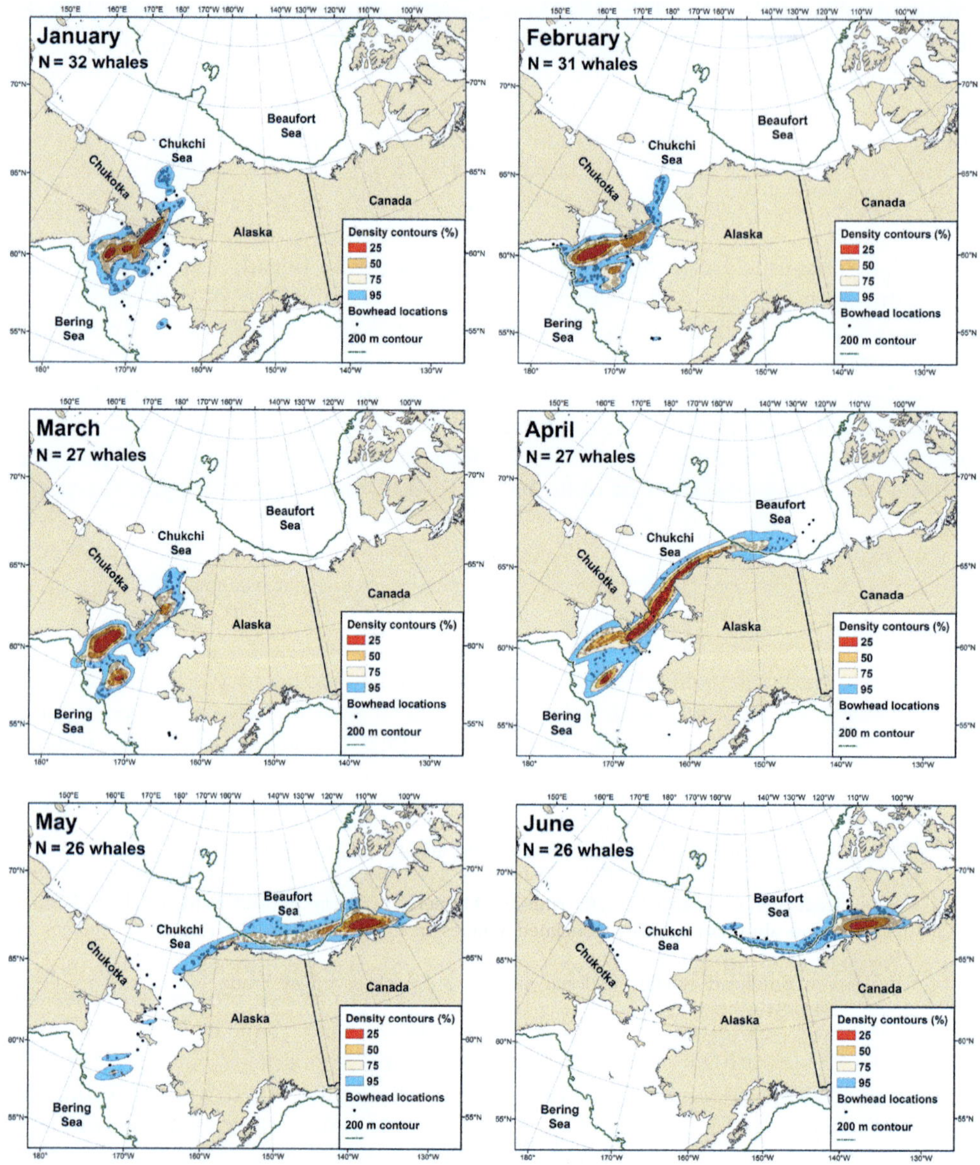

FIGURE 4.3 Kernel densities of bowhead whale locations during January–June, 2006–19. Kernel densities were calculated using smoothed cross-validation as described by Duong and Hazelton (2005) using daily locations estimated with a continuous time Correlated Random Walk model developed by Johnson et al. (2008). Raw location data were first passed through a velocity filter (Freitas et al., 2008). Both the daily locations and the density contours are provided, allowing the reader to directly assess how well the kernel density fits the data.

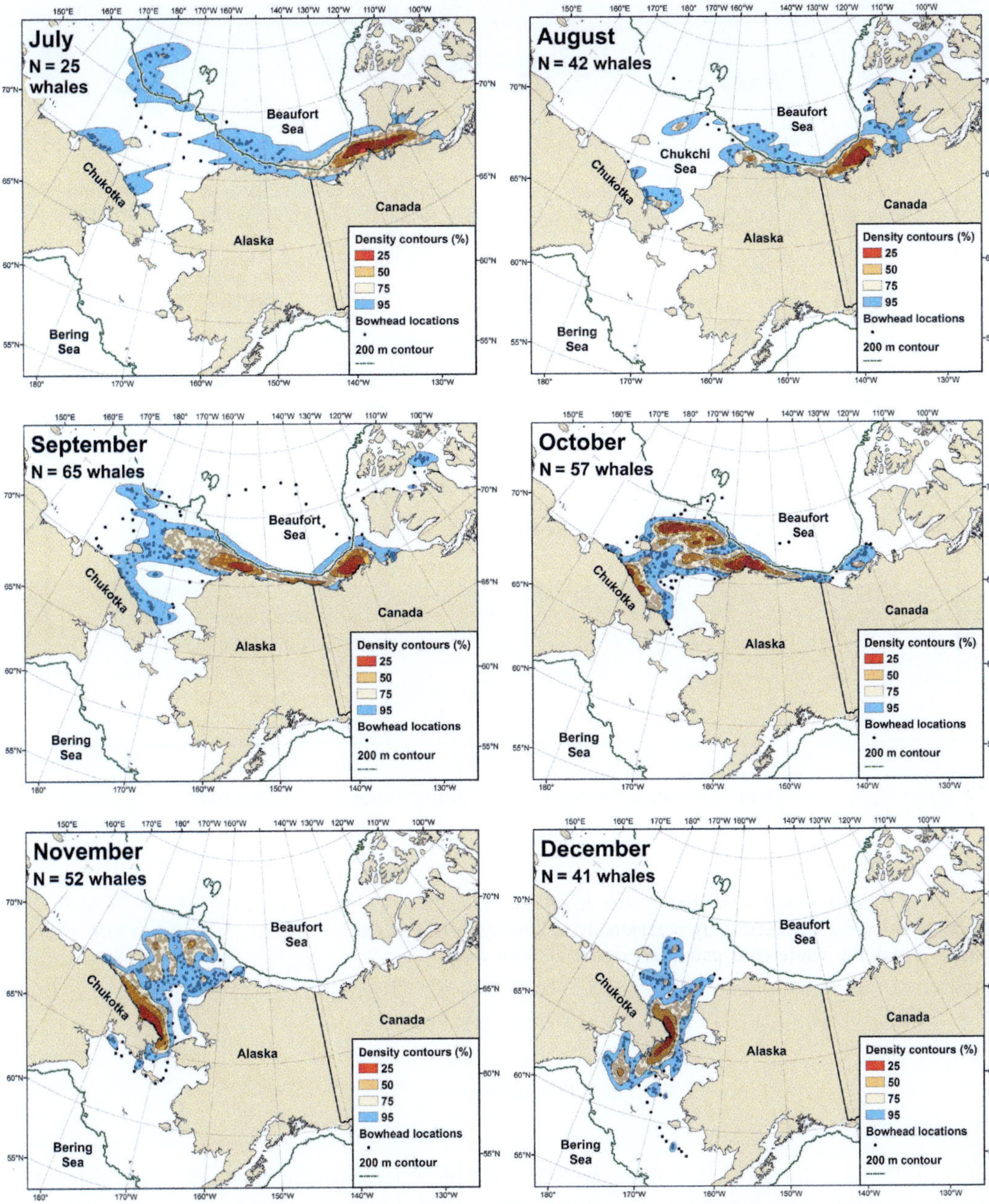

FIGURE 4.4 Kernel densities of bowhead whale locations during July–December, 2006–19. Kernel densities were calculated using smoothed cross-validation as described by Duong and Hazelton (2005) using daily locations estimated with a continuous time Correlated Random Walk model developed by Johnson et al. (2008). Raw location data were first passed through a velocity filter (Freitas et al., 2008). Both the daily locations and the density contours are provided, allowing the reader to directly assess how well the kernel density fits the data.

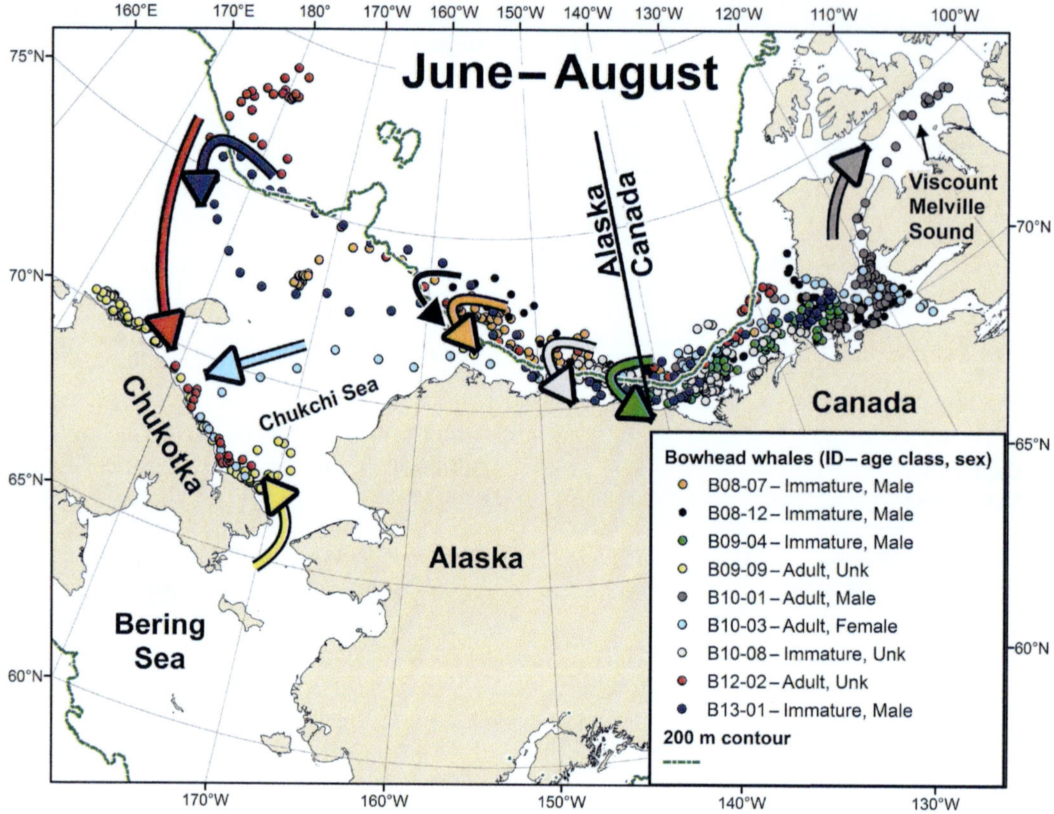

FIGURE 4.5 Anomalous movements of bowhead whales during summer (June–August).

all tagged whales were in the same general area in March when breeding is assumed to occur, provided evidence that there was a single stock of bowhead whales in the Bering, Chukchi, and Beaufort seas (IWC, 2012). The presence of a single stock of BCB bowhead whales was also supported by genetic studies (e.g., Givens et al., 2010; IWC, 2012; Baird et al., 2018), which arguably provide stronger evidence.

Autumn migration

The autumn migration is not as directed as the spring migration. For example, travel time in days between Point Barrow and Demarcation Point averaged 9 days in the spring and 19 days in the autumn (Table 4.2). Likewise, the average travel time between Bering Strait and Point Barrow was 12 days in spring and 70 days in autumn. On average, tagged whales passed into Alaskan waters on 17 September, passed Point Barrow on 25 September, and entered the Bering Sea on 2 December (Table 4.1).

After passing Point Barrow, most use of the Chukchi Sea occurs south of the shelf break and along the northern coast of Chukotka, with little use of the Alaskan coast. Movements generally proceed southward, toward the southern Chukchi and northern Bering seas. At this time, movements through the Chukchi Sea are prolonged enough that referring to them as a "migration" becomes debatable.

Winter range

Bowhead whales typically enter the Bering Sea in early December and exit the following spring in mid-April (Table 4.1). However, there is great variability in when whales enter the Bering Sea and individuals may enter and exit Bering Strait multiple times before moving south toward Anadyr Strait (Figs. 4.3 and 4.4). Whales may also reenter the southern Chukchi at any time in the winter. The winter distribution of bowheads is centered in Russian waters from Bering Strait, south through Anadyr Strait, to the southern end of the Gulf of Anadyr (Fig. 4.3). Almost all use of the Bering Sea by tagged whales has been north of the ice edge (see "The role of sea ice" section). From January to April, whales spend approximately half of their time at or near the seafloor and approximately two-thirds of their dives are square-shaped (where the majority of the dive duration is spent at the maximum depth of the dive; see "Dive behavior" section), suggesting that whales are likely feeding on overwintering zooplankton in diapause near the seafloor. Bowhead whales leave the Bering Sea while it is still ice-covered. Thus, the signal to migrate is

TABLE 4.3 Dive statistics summarized by month.

Month	Mean dive duration in minutes (SE)	Mean dive depth in meters (SE)	Max. dive depth in meters	Proportion of dive shapes (SE)		
				Square-shaped	U-shaped	V-shaped
1	7.7 (1.1)	15.5 (3.9)	89	0.60 (0.03)	0.32 (0.03)	0.07 (0.01)
2	11.0 (1.3)	27.7 (4.1)	115	0.66 (0.03)	0.27 (0.03)	0.06 (0.01)
3	9.6 (1.3)	20.8 (4.1)	115	0.62 (0.03)	0.30 (0.03)	0.08 (0.01)
4	9.4 (1.3)	19.6 (4.0)	324	0.58 (0.03)	0.34 (0.03)	0.07 (0.01)
5	10.1 (1.3)	30.5 (4.0)	420	0.61 (0.03)	0.24 (0.02)	0.13 (0.02)
6	10.1 (1.2)	17.8 (3.9)	428	0.69 (0.03)	0.18 (0.02)	0.12 (0.01)
7	13.6 (1.2)	80.5 (4.0)	316	0.84 (0.02)	0.10 (0.01)	0.06 (0.01)
8	11.7 (1.2)	19.3 (3.9)	238	0.81 (0.02)	0.11 (0.01)	0.08 (0.01)
9	10.9 (1.1)	30.5 (3.8)	252	0.83 (0.02)	0.12 (0.01)	0.05 (0.01)
10	10.5 (1.1)	50.3 (3.8)	246	0.84 (0.02)	0.11 (0.01)	0.04 (0.01)
11	10.4 (1.1)	36.9 (3.9)	134	0.78 (0.02)	0.16 (0.02)	0.05 (0.01)
12	8.5 (1.1)	16.6 (3.9)	101	0.53 (0.03)	0.36 (0.03)	0.09 (0.01)

Descriptions of dive shapes are given in the text.

TABLE 4.4 Proportion of 6-hour summary periods spent by whales at or near the seafloor.

Month	Mean proportion of 6-hour intervals near the seafloor (SE)	Mean water depth at bowhead locations (m)	Number of observations	Number of whales
1	0.57 (0.01)	33.7	1094	21
2	0.54 (0.02)	44.1	840	19
3	0.46 (0.02)	49.3	742	15
4	0.49 (0.02)	168.1	909	14
5	0.20 (0.02)	279.1	1014	17
6	0.13 (0.01)	225.8	1516	19
7	0.50 (0.01)	170.0	1457	19
8	0.67 (0.01)	53.9	1479	29
9	0.67 (0.01)	51.6	2839	40
10	0.72 (0.01)	28.6	2664	34
11	0.67 (0.01)	17.2	2139	33
12	0.60 (0.01)	21.8	1523	24

probably not breakup; rather, it is thought to correspond with zooplankton ending diapause and rising from the seafloor, thereby becoming more dispersed (Citta et al., 2015).

Dive behavior

The average duration of bowhead dives is remarkably consistent throughout the year, ranging from approximately 8 minutes in January to as long as 14 minutes in July (Table 4.3). The duration of dives is positively correlated with depth; longer dives occur in deeper water ($r = 0.76$), likely because bowheads spend approximately half their time at or near the seafloor when located in shelf waters <200 m deep (Table 4.4). The proportions in Table 4.4 are averages; individual whales will sometimes spend >75% of 6-hour intervals at or near the seafloor. The proportion of 6-hour intervals spent at or near the seafloor is much lower in May and June when whales are in the vicinity of the Cape Bathurst polynya and when there is still sea ice in Amundsen Gulf. The polynya contains deep oceanic water (>200 m) and depth temperature data show that there is a thermocline at approximately 100 m where colder water of Pacific origin transitions into deeper, warmer water of Atlantic origin. Almost all use of the polynya occurs above 100 m and we assume, but cannot confirm, this is because whales are feeding on zooplankton rising to feed on the spring phytoplankton bloom.

Some tags that provided dive data also classified dive shape as a function of how much time is spent at the bottom of a dive (Table 4.3). If >50% of a dive's duration is spent at the bottom of a dive (i.e., at the deepest depth), it is classified as "square-shaped." If >20% and <50% of a dive's duration is spent at the bottom of a dive, it is classified as "U-shaped." If <20% of a dive is spent at the bottom of a dive, then it is classified as

"V-shaped." Both square- and U-shaped dives are considered indicative of feeding behavior. Most bowhead dives are classified as square-shaped in all months (Table 4.3). The percentage of U-shaped dives ranges from 10% in July to 36% in December (Table 4.3). V-shaped dives typically make up less than 10% of all dives; the proportion of V-shaped dives is highest during the spring migration in April and May.

We do not report on surface intervals here because most tags were programmed to estimate the duration of sounding dives. Tags do not register a dive until the tag has descended 10 m, at which point the duration of the dive is calculated as the time the tag descends below 2 m and then ascends above 2 m. Hence, near-surface rolling behavior in between sounding dives is pooled into surface intervals. We caution that this measure of surface interval should not be used when estimating availability correction factors, as it does not accurately represent the time whales are visible to aircraft. We optimized all tags deployed since 2018, to collect data that can be used to calculate availability correction factors.

Proximate mechanisms driving distribution

Our current hypothesis is that the main mechanism driving the seasonal distribution of BCB bowhead whales is food availability. Seasonal movements can be largely explained by known or predicted changes in the availability of zooplankton prey and the colocation of whales and oceanographic features such as fronts and stratified layers that are known to help aggregate zooplankton. Zooplankton often aggregate on the seafloor in relatively shallow shelf waters (e.g., Walkusz et al., 2012); this is significant because, based upon dive records, BCB bowhead whales spend at least half of each day at or near the seafloor year-round, with the exception of when they are migrating in the spring or foraging within the Cape Bathurst Polynya where depth is >200 m. The aggregation of zooplankton is important because bowhead whales are thought to require relatively high densities of zooplankton to meet their energy requirements (Lowry, 1993). Bowhead whales appear to know generally where food resources can be found and move from one potential feeding area to another, similar to North Atlantic right whales (*Eubalaena glacialis*; Kenney et al., 2001). However, unlike most other mysticetes (Costa and Williams, 1999), bowhead whales appear to feed year-round and sample the water column as they travel; when food is located, whales pause directed movements to feed, regardless of time-of-year. When whales fail to find food at a known potential feeding area, some make surprising exploratory movements "circling back" to where they started, apparently to check again for food (Fig. 4.5).

Moore et al. (2019) questioned whether the concept of "resource waves" (Armstrong et al., 2016) is a useful model for understanding the movements of bowhead whales. Animals take advantage of resource waves when they track phenological variation in resources across space and time. By following resource waves, animals can take advantage of ephemeral resources longer than if animals were stationary. For example, if bowhead whales moved northward in conjunction with the spring bloom of phytoplankton (to feed upon the rising zooplankton) they would be following a resource wave. Although BCB bowhead whales certainly respond to resource availability, we suggest that memory of important feeding areas has a larger role in predicting movements than following changes in resource availability at a local scale, something also observed in blue whales

(*Balaenoptera musculus*) (Abrahms et al., 2019). We know that when whales migrate from the Bering Sea to the Cape Bathurst polynya, they are not following a resource wave, as little foraging typically occurs during the spring migration. Neither does it appear that whales are following a resource wave when they abandon the Canadian Beaufort Sea and migrate to the coast of Chukotka, Russia, in midsummer, before the typical autumn migration. In between core use areas, movements are generally much faster than the currents that advect zooplankton in shelf waters. Hence, even when whales are traveling in the same direction as local currents, such as during the autumn migration in the Alaskan Beaufort, bowhead whales will typically be moving faster than a resource wave could travel. In autumn, westward currents on the Beaufort Sea shelf near Point Barrow are expected to advect zooplankton approximately 8–13 km/day (Steve Okkonen, pers. comm.). On average, bowhead whales travel from Demarcation Point and Point Barrow in 19 days (Table 4.2), traveling approximately 33 km/day. Hence, it is much more likely that whales simply remember where to locate food within their home range.

While food availability is probably the main mechanism behind movements, these movements are likely tempered by sea ice, which may at times pose a risk of entrapment but may at other times provide protection from killer whales (*Orcinus orca*). These hypotheses are partly speculative; feeding is well documented in some core use areas but not others (see below) and to determine why whales make seemingly exploratory movements requires a great deal of data.

The main core use areas of BCB bowhead whales and the oceanographic factors that help aggregate zooplankton prey are presented in Citta et al. (2015), which considered data collected through 2012. Here, we summarize and refine the descriptions given in that manuscript with new data collected since 2012. Citta et al. (2015) used location data to describe six core use areas: (1) Cape Bathurst; (2) shelf waters near Tuktoyaktuk; (3) Point Barrow; (4) the northern coast of Chukotka, Russia; (5) Anadyr Strait; and (6) the Gulf of Anadyr. Because of its importance to petroleum development and Alaska Native subsistence harvest, we also discuss the eastern Alaskan Beaufort Sea.

Cape Bathurst

The polynya is maintained by wind and wind-driven upwelling of warm water (Williams and Carmack, 2008; Arrigo and van Dijken, 2004). While there is no direct evidence that whales are feeding within the polynya in May and June, the colocation of copepods and bowhead whales at similar depths, in combination with the fact that bowheads leave the Bering Sea area and migrate directly to the Cape Bathurst polynya, is highly suggestive of feeding. Tagged bowhead whales were present within the Cape Bathurst core use area in May and June, and spent the majority of time at depths <100 m, although some use extends to 300 m (see Figure 3c in Citta et al., 2015). Paired depth and temperature profiles indicate a thermocline between colder Pacific water, which is above 100 m, and warmer Atlantic Water, which occurs below 300 m. At this time of year, the highest abundance of large calanoid copepods, *Calanus glacialis* and *C. hyperboreus*, is typically found in the upper 25–50 m of water (Daase et al., 2013; Darnis and Fortier, 2014; Wold et al., 2011) and the temperature stratifications noted above may help aggregate them. As ice breaks up in Amundsen Gulf, whales typically proceed into the gulf. We have yet to make a detailed examination of bowhead behavior in Amundsen Gulf after whales leave the Cape Bathurst polynya.

Tuktoyaktuk shelf

Wind-driven upwelling occurs at Cape Bathurst and along the shelf break, lifting and advecting calanoid copepods onto the shelf near Tuktoyaktuk (Walkusz et al., 2012). The Mackenzie plume may provide a boundary to the westward transport of zooplankton on the shelf and we suspect that whales may leave shelf waters when copepods descend too deep for upwelling to lift them onto the shelf (>100 m depth). Whales will aggregate at other locations along the Mackenzie plume and also near Mackenzie Canyon (Harwood et al., 2010, 2017). However, the main aggregation area for whales is on the shelf, near Tuktoyaktuk. Feeding near Tuktoyaktuk, along the Mackenzie plume, and near Mackenzie Canyon is well documented (e.g., Würsig et al., 1984; Pomerleau et al., 2011; Walkusz et al., 2012).

Eastern Alaskan Beaufort Sea

After leaving the Canadian Beaufort Sea, most whales migrate along the continental shelf westward toward Point Barrow. In the Alaskan Beaufort, whales will sometimes stop to feed prior to reaching Point Barrow if conditions are favorable. Okkonen et al. (2018) describe whales stopping to feed when high river discharges and prior upwelling create feeding opportunities; in short, prior upwelling makes zooplankton available on the shelf and a large river discharge then creates a salinity front that aggregates zooplankton. However, these stops are typically of short duration. Of 23 tagged whales migrating west through the Alaskan Beaufort Sea, only 4 (17%) paused their migration prior to reaching the vicinity of Point Barrow for long enough to be detected in the satellite records; stops shorter than a day are likely not detected. Detected stops ranged from 1 to 9 days ($\bar{x} = 3.4$) in duration. In contrast, tagged bowheads lingered within 100 km of Point Barrow for up to 34 days ($\bar{x} = 8.4$) and within 100 km of Atkinson Point (~90 km north of Tuktoyaktuk) for up to 62 days ($\bar{x} = 17.4$). Hence, although a high proportion (83%) of harvested whales have prey in their stomachs (Lowry et al., 2004; Chapter 28) and many aerial surveys have observed feeding behavior in the eastern Alaskan Beaufort Sea (e.g., Ljungblad et al., 1986; Clarke et al., 2018), the satellite tag data suggests that pauses in migration are not typical and are of short duration when they occur. As such, the eastern Alaskan Beaufort Sea probably does not currently contribute much to the annual energy needs of bowhead whales in most years (Olnes et al., in press). This agrees with the findings of feeding studies conducted during 1997–2000 (Richardson and Thomson, 2002). As part of that study, Thomson et al. (2002) concluded that although approximately half of all bowhead whales were observed feeding in the eastern Alaskan Beaufort Sea, whales apparently only derived a small amount (2.4%) of their annual energy budget from this area on average. Interestingly, Yankee whalers harvested relatively few bowhead whales in the eastern Alaskan Beaufort Sea (Bockstoce et al., 2005), partly because of heavy ice conditions, but also likely because bowhead whales spent little time there.

Point Barrow

Bowhead feeding near Point Barrow is well documented (e.g., Lowry et al., 2004; Moore et al., 2010) and the mechanisms that aggregate zooplankton near Point Barrow in autumn are well understood (Ashjian et al., 2010; Okkonen et al., 2011). East winds promote upwelling and advect zooplankton onto the shelf. When east winds (upwelling winds) are followed by west or south winds, the Alaskan Coastal Current then prevents westward transport of zooplankton and traps them on the shelf. This is known as the "krill trap" (Ashjian et al., 2010; Chapter 26) because it is predominantly euphausiids (krill) from the Bering Sea that are trapped. Whales generally leave the area as sea ice forms; ice may block winds, preventing upwelling near Point Barrow.

Chukotka coast

After leaving Point Barrow, whales are known to follow water of Pacific origin characterized by temperatures <0°C and salinities between 31.5 and 34.25 psu (Citta et al., 2018). Bowhead whales avoid Alaskan Coastal Water and Siberian Shelf Water, which are both composed of relatively warm, fresh river discharge and contain low densities of zooplankton prey (Eisner et al., 2013; Ershova et al., 2015). The location of Siberian Shelf Water, west of Wrangel Island, currently defines the western limit of their range, likely due to its lower intrinsic density of zooplankton prey (Ershova et al., 2015). Chambault et al. (2018) raised the possibility that the avoidance of relatively warm water (>2°C) may be due to thermal stress and not the availability of zooplankton. Although Alaskan Coastal Water is relatively warm (>2°C), we suspect the avoidance of Alaskan Coastal Water is not due to thermal stress because tag data show whales commonly feed in waters much warmer (4°C–6°C) near Tuktoyaktuk.

Tagged bowhead whales show great affinity for the northern coast of Chukotka, Russia, during October–December (Fig. 4.4). Freshwater exiting Siberian and Chukotkan rivers creates a frontal feature that extends along the northern coast of Chukotka from the vicinity of Wrangel Island to Bering Strait (see Fig. 7e in Citta et al., 2015). Krill from the Bering Sea are thought to aggregate along this frontal zone (Moore et al., 1995; Berline et al., 2008) and bowhead whales are known to feed there (Burns, 1993; Moore et al., 1995). Movement into the Bering Sea is correlated with the breakdown of this front and the formation of sea ice. As the rivers freeze in autumn, the salinity front diminishes and stops concentrating zooplankton; therefore, it is likely not the formation of sea ice that precipitates movement into the Bering Sea. Indeed, when whales winter in the southern Chukchi, they are typically found due north of Being Strait and not along the Russian coast, likely because there is no salinity front to aggregate zooplankton.

In recent years, tagged whales often have not traveled all the way to this front to forage, instead lingering in the north-central Chukchi. Our assumption that these whales are feeding is based upon two lines of evidence. First, if whales were not finding food in the north-central Chukchi, it is likely that they would proceed to the coast of Chukotka where zooplankton are known to aggregate (Moore et al., 1995). Second, whales are spending approximately 70% of each 6-hour period at or near the seafloor, which is where zooplankton are expected to aggregate in autumn. Preliminary analyses indicate the shift in distribution toward the north-central Chukchi may be wind-driven.

Anadyr Strait and Gulf of Anadyr

Both Anadyr Strait and the Gulf of Anadyr core use areas are ice covered while whales are present. Citta et al. (2018) found that Anadyr Strait is colocated with an intrusion of relatively high salinity water (>32.5 psu), which is mainly found near the seafloor. Bowhead whales used the entire water column but spent more time on the seafloor than other depths. Likewise, the southernmost core use area at the Gulf of Anadyr lies in the middle of an area where salinity is relatively high (>32.5). This area is also bounded on the north and west by cold Anadyr Water ($<0°C$). In this area, whales spent 26–69% of their time at 75–100 m, which also corresponds to the depth of a strong thermocline (Citta et al., 2015). Furthermore, whales spent more time at the seafloor than at other depths in 82% of 6-hour histograms.

The large amount of time spent near the seafloor ($\sim 50\%$) and the colocation of dive depths and a strong thermocline suggests that bowhead whales are probably feeding on zooplankton in diapause. There are no villages harvesting whales near the southernmost core use area and there is no easy way to observe whale behavior under the sea ice in winter. However, there is ample evidence from the stomach contents of harvested whales that feeding sometimes occurs in both late fall (November) and early spring (April and May) near St. Lawrence Island. Common prey includes amphipods, copepods, and euphausiids (Hazard and Lowry, 1984; Chapter 28). One whale harvested in January of 2017 had consumed euphausiids (Chapter 28) and Noongwook et al. (2007) describes feeding behavior in December. The telemetry data suggest that winter feeding is not anomalous; all tagged bowhead whales showed evidence of winter feeding. Although it is unclear how much energy is obtained in each core use area, winter feeding has major implications for the annual energy budget of BCB bowhead whales (Chapter 16). Unlike other mysticetes (e.g., Costa and Williams, 1999) it may not be necessary for BCB bowhead whales to meet annual energetic requirements during a short summer feeding season.

Role of sea ice

Prior to this study, BCB bowheads were thought to winter in the Bering Sea in broken ice, but with an affinity for the southern ice margin and polynyas (Brueggeman et al., 1987; Bogoslovskaya et al., 1982; Zelensky et al., 1997). We found that while bowhead whales typically do not range south of the ice margin in winter, they are capable of wintering in almost 100% ice concentration, far from the ice margin, and are not reliant on polynyas (Citta et al., 2012). Indeed, whales migrate through heavier ice in the Chukchi and Beaufort seas each spring. Although whales migrate to the Cape Bathurst polynya each spring, their affinity for the polynya is more likely due to feeding conditions associated with upwelling than the presence of open water. This is not to say that bowhead whales are immune from the risk of ice entrapment. Indeed, whales are known to migrate farther offshore during the autumn migration in the Beaufort Sea when ice concentration is higher, possibly due to the risk that wind-driven ice may pile up onshore and entrap whales (Blackwell et al., 2007; Druckenmiller et al., 2018). Responses to heavy ice conditions are discussed in Chapter 7.

That the tagged whales did not go south of the ice margin in winter is curious because the distribution of BCB bowheads is largely independent of sea ice in the Beaufort and Chukchi seas in summer. It is not until whales are near Bering Strait that they show an affinity for ice cover. Staying north of the ice margin may be related to the presence of killer whales, which are known to frequent the ice edge in winter but are rarely observed in bowhead summer ranges. Killer whales are present in the Chukchi between June and November and leave when sea ice forms (Melnikov et al., 2007; Stafford, 2018), and are known to be present along the ice edge in winter (Lowry et al., 1987). That killer whales harass and sometimes kill bowhead whales is indisputable. Approximately 8% of all bowhead whales harvested near Utqiaġvik (and >50% of whales >17 m) have tooth rake marks consistent with killer whale attacks (George et al., 2017; Chapter 29). Willoughby et al. (in review) analyzed photos of bowhead carcasses taken during aerial surveys in the Chukchi and Beaufort seas and found that 18 of 33 (55%) had injuries consistent with killer whale predation, mostly in the Chukchi Sea. By analyzing photos collected during aerial surveys of the Chukchi and Beaufort seas, Willoughby et al. (in review) found that 18 of 33 (55%) bowhead carcasses had injuries consistent with killer whale predation, mostly in the Chukchi Sea. Furthermore, killer whales are known to affect the distribution of bowhead whales in the Eastern Canada-Western Greenland stock (Matthews et al., 2020; Chapter 29).

Recent changes in distribution

It may be tempting to think that recent lighter ice conditions allow BCB bowhead whales to range farther north in summer and autumn than they did when Yankee whalers arrived in the Chukchi and Beaufort seas; however, that conclusion is speculative. The farthest north a bowhead was harvested by the Yankee whalers was ~74°N (Bockstoce et al., 2005) and we have documented a tagged whale approximately 555 km farther north than this (79°N). However, bowhead whales are better adapted to sea ice than whale ships, thus whales may have ranged much farther north than Yankee whalers knew. In general, the distribution of bowhead whales based upon satellite tags mirrors that based upon the harvest locations of Yankee whalers (Bockstoce and Botkin, 1983; Bockstoce et al., 2005). The largest discrepancy is that bowhead whales are no longer found in the southern extent of their historical range (Fig. 4.2).

In recent years, the winter (January–March) range of BCB bowhead whales has shifted northward. The traditional winter range was typically centered southwest of St. Lawrence Island in the Russian Bering Sea (Fig. 4.6). Warm conditions during the winters of 2017, 2018, and 2019 resulted in this area being largely ice-free. No tags were transmitting during January–March of 2017; however, tagged whales shifted their distribution northward during the winters of 2018 and 2019, never moving south of the marginal ice edge. In March of 2019, ice reformed south of St. Lawrence Island and tagged bowheads followed the ice edge southward toward their typical wintering area. The reason why bowhead winter range shifted northward is currently unknown. Although bowhead whales almost certainly avoid killer whales, the lack of sea ice in the Bering Sea is correlated with south winds and enhanced northward currents through Bering Strait. This may have resulted in more advection of zooplankton through Bering Strait, making better foraging opportunities in the Chukchi Sea (Moore et al., 2019). As such, good feeding conditions north of the ice edge may have negated the need for bowhead whales to expose themselves to the risk of predation within their traditional winter range. Regardless, we expect that less sea ice in the Bering Sea will result in more bowhead whales wintering in the Chukchi Sea (Druckenmiller et al., 2018).

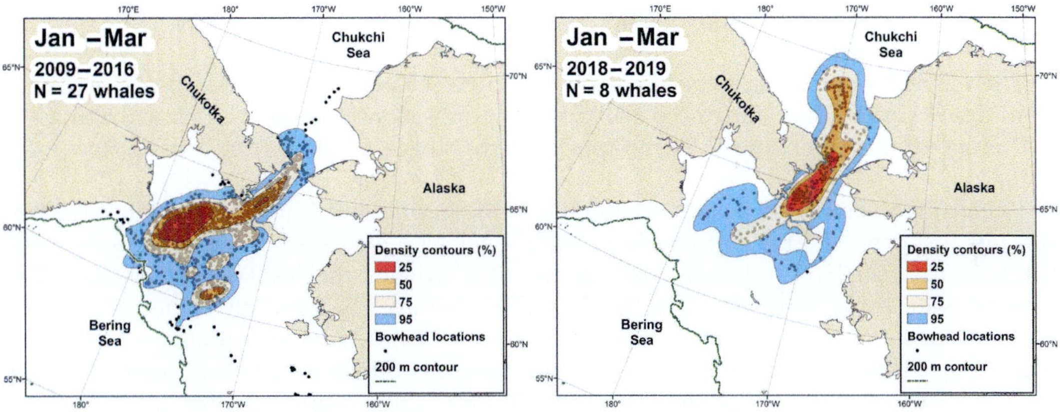

FIGURE 4.6 Northward shift of the winter (January–March) distribution of BCB bowhead whales. The southern core use area observed during 2009–16 was largely abandoned during 2018–19.

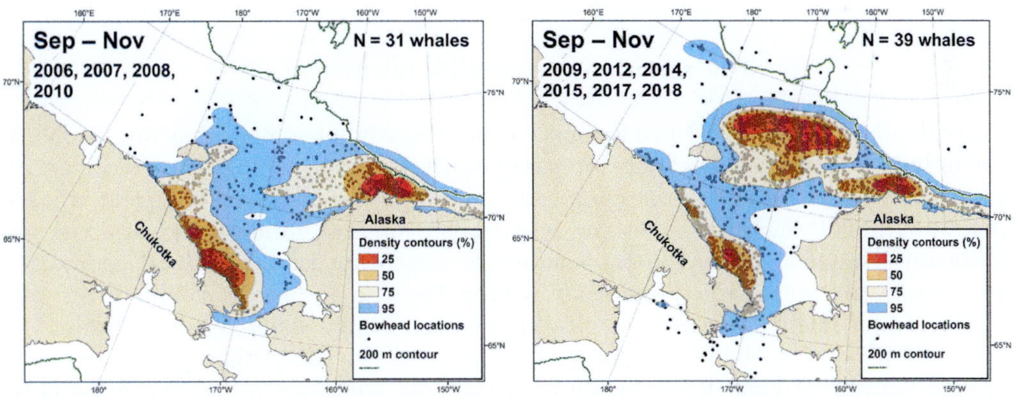

FIGURE 4.7 During most of the initial years of the bowhead tagging study (2006, 2007, 2008, and 2010), there was little use of the north-central Chukchi Sea during the autumn migration. However, in most recent years (2009, 2012, 2014, 2015, 2017, and 2018), there has been extensive use of the north-central Chukchi Sea.

Bowhead whales also sometimes shift their distribution in the autumn away from the coast of Chukotka, Russia, and toward the north-central Chukchi Sea (Fig. 4.7). Whales rarely lingered in the central Chukchi in 2006, 2007, 2008, or 2010; however, use of the north-central Chukchi was extensive in 2009, 2012, 2014, 2015, 2017, and 2018. Results were equivocal for 2011 and 2016 due to too few tagged whales in those years. In most of the early years of the study (4 of 5 years prior to 2011), whales did not linger in the north-central Chukchi. In contrast, the north-central Chukchi received extensive use in almost all years after 2011 (5 of 6). Hence, it appears that use of the north-central Chukchi as an autumn feeding area increased during the study. Preliminary analyses suggest that this shift is wind-driven and that east winds may result in more use of the Russian coast; as such, the shift toward the central Chukchi is likely not permanent. Although we know that bowhead whales followed water of Pacific origin, expected to have higher concentrations of krill, across the Chukchi Sea in autumn (Citta et al., 2018), we have not completely explained what environmental factors predict when whales stop to feed in the north-central Chukchi Sea.

There is not enough tag data to corroborate other changes that we know are occurring from aerial survey data and the observations of subsistence whalers, such as changes in the timing of migration and the timing of the subsistence whaling (e.g., Huntington et al., 2016).

Interestingly, bowhead whales summered in the Bering and Chukchi seas when Yankee whalers first arrived. One hypothesis for why whales no longer summer in the Bering and Chukchi seas (in large numbers) is that the only whales the Yankee whalers failed to remove were those summering in the Canadian Beaufort Sea and that migration to the Canadian Beaufort is the "ghost of whaling past." The modern distribution of bowhead whale is probably very similar to what it was when commercial whaling ceased in 1914. However, it seems likely that bowhead distribution will continue to shift as the climate warms and alters the distribution of sea ice, bowhead prey, and bowhead predators. In effect, climate change will likely obscure the role of commercial whaling on the seasonal distribution of bowhead whales.

Limitations of satellite telemetry

As with all methods (see "Introduction" section), satellite telemetry has limitations. The main technical limitations are that the sample size is small in any given year and the deployment of transmitters is not completely random. This is because tagging whales is logistically difficult due to weather and whale behavior, and because community support is sometimes limited. Deploying transmitters is clearly invasive and local communities often do not want to tag large numbers of whales. Because of this, the sample in any given year is typically small and tags are typically deployed from a single location within a relatively short window of time (e.g., during a month-long period). Although we can develop an understanding of what the majority of whales do, we likely miss the finer details of whale movements in any given year. For example, we can locate the main wintering grounds, but cannot discern other things that are known to occur, such as pulses in migration past Point Barrow or how the migration is segregated by age or reproductive status (e.g., George et al., 2004). Likewise, the earliest a tagged whale has migrated past Point Barrow in the spring is 16 April, while acoustic buoys have detected whales near Point Barrow as early as the end of March (Hannay et al., 2013).

The age and sex distribution of the tagged sample deserves special mention. As stated earlier, to achieve an equal sex ratio and a representative age ratio, we would have to tag approximately 10 more adult (≥13 m length) female bowhead whales. However, adult females are more consistent in their migratory behavior than immature and adult males, which commonly made wide-ranging, exploratory movements. Hence, the total annual range is certainly larger than indicated by the tag data and many of the whales outside of the distribution we describe here are likely immature and adult males.

Another limitation is that some behaviors, such as feeding, are not measured directly and must be inferred from tag data. We often assume that whales are feeding when they linger in one area for many days and spend a large proportion of time near the seafloor (we noted such assumptions earlier). We may also miss pauses in migration if they are too short to be visible in the location data. In general, we think a whale must pause for at least a day for that pause to be distinguishable from location error. As such, short bouts of feeding, be it at the surface or at-depth, will not be detected in the tag data.

Last, as mentioned earlier, tagging is inherently invasive and the effects of tagging on bowhead whales are largely unknown. Although we do not think that tagging directly affects whale behavior for more than a few days, the possibility that tagging may negatively affect the health of tagged whales raises serious concerns among Alaska Natives and researchers alike (e.g., Robbins et al., 2013; Andrews et al., 2019). Indeed, the Alaska Eskimo Whaling Commission passed resolutions to develop less invasive tags in 2005 and 2018, urging researchers to develop and deploy the most humane tags possible. Such concerns will likely continue to limit the feasibility and design of satellite telemetry studies.

Research needs

1. Due to changing environmental conditions, much of the information we describe may already be out of date. The winter distribution of tagged bowhead whales is shifting northward, into the Chukchi Sea. Almost no whales were observed by whalers or during aerial surveys in September or October near Utqiaġvik in 2019 (Chapter 24). Russian observers documented many bowhead whales in mid to late October along the northern coast of Chukotka; hence, it appears that most of the autumn migration occurred offshore of Utqiaġvik in 2019, too far offshore for aerial surveys to detect them (~80 km). Unfortunately, no satellite tags were transmitting at this time. There are also reports that bowhead whales were observed in Amundsen Gulf, near Ulukhaktok, during January and February of 2019, suggesting that some whales may not have migrated to their traditional wintering range in 2019. These patterns have not been observed before. Hence, the most pressing need is to deploy some tags annually. Tags that can monitor environmental covariates, such as temperature and salinity, will prove especially useful because bowhead distribution is tightly linked to water masses with elevated densities of zooplankton and oceanographic features that aggregate zooplankton.

2. Future work on bowhead distribution needs to be more closely linked with research on physical oceanography and zooplankton throughout the range of bowhead whales. To understand how the distribution of bowhead whales may change with a warming climate, we must know how bowhead prey will be affected and how mechanisms that help aggregate prey, such as wind, upwelling, and currents, will change.
3. As sea ice declines, killer whales will likely penetrate farther into the autumn and summer ranges of bowhead whales (Stafford, 2018; Willoughby et al., in review) and have the potential to alter the distribution of the BCB stock. To fully understand the mechanisms behind future changes in the distribution and behavior of BCB bowhead whales, the seasonal distribution of killer whales should also be studied.
4. As always, better technology is needed. Tags that monitor temperature and salinity drain batteries quickly (more so when monitoring salinity) and these tags rarely last more than six months. Tags with longer battery life will be critical to gaining a better understanding of the physical environment year-round.
5. Last, the potential effects of tagging on whales should not be ignored (e.g., Robbins et al., 2013; Andrews et al., 2019). Although subsistence whalers have yet to harvest a tagged whale and we cannot directly comment on how tagging may or may not affect bowhead whales, the Alaska Eskimo Whaling Commission passed resolutions to develop less invasive tags in 2005 and 2018, urging researchers to develop and deploy the most humane tags possible.

Acknowledgments

This study would not have been possible without the support of the Native subsistence whalers of Utqiaġvik, Tuktoyaktuk, Aklavik, Gambell, and Savoonga. Special thanks are reserved for Arnold Brower, Sr., Billy Adams, Harry and Lewis Brower, James and Charles Pokiak, Lois Harwood, Steve Okkonen, Mads Peter Heide-Jørgensen, Mikkel Jensen, and Ellen Lea. Funding was provided by the Bureau of Ocean Energy Management, the U.S. agency responsible for oil and gas leasing, with additional support from Fisheries and Oceans Canada and the U.S. Office of Naval Research. Permits required for research were secured annually in the U.S. and Canada.

References

Abrahms, B., Hazen, E.L., Aikens, E.O., Savoca, M.S., Goldbogen, J.A., Bograd, S.J., et al., 2019. Memory and resource tracking drive blue whale migrations. Proc. Natl. Acad. Sci. U. S. A. 116 (12), 5582–5587.

Andrews, R.D., Baird, R.W., Calambokidis, J., Goertz, C.E.C., Gulland, F.M.D., Heide-Jørgensen, M.P., et al., 2019. Best practices guidelines for cetacean tagging. J. Cetacean Res. Manage. 20, 26–66.

Armstrong, J.B., Takimoto, G., Schindler, D.E., Hayes, M.M., Kauffman, M.J., 2016. Resource waves: phenological diversity enhances foraging opportunities for mobile consumers. Ecology 97 (5), 1099–1112.

Arrigo, K.R., van Dijken, G.L., 2004. Annual cycles of sea ice and phytoplankton in Cape Bathurst polynya, south-eastern Beaufort Sea, Canadian Arctic. Geophys. Res. Lett. 31, L08304. Available from: https://doi.org/10.1029/2003GL018978.

Ashjian, C.J., Braund, S.R., Campbell, R.G., George, J.C., Kruse, J., Maslowski, W., et al., 2010. Climate variability, oceanography, bowhead whale distribution, and Inupiat subsistence whaling near Barrow, Alaska. Arctic 63, 179–194.

Baird, A.B., Givens, G.H., George, J.C., Suydam, R.S., Bickham, J.W., 2018. Stock Structure of Bowhead Whales Inferred from mtDNA and SNP Data. Report to the International Whaling Commission SC/67B/SDDNA/01.

Berline, L., Spitz, Y.H., Ashjian, C.J., Campbell, R.G., Maslowski, W., Moore, S.E., 2008. Euphausiid transport in the western Arctic Ocean. Mar. Ecol. Prog. Ser. 360, 163–178.

Blackwell, S.B., Richardson, W.J., Green, C.R., Streever, B., 2007. Bowhead whale (*Balaena mysticetus*) migration and calling behaviour in the Alaskan Beaufort Sea, Autumn 2001–04: an acoustic localization study. Arctic 60 (3), 255–270.

Bockstoce, J.R., Botkin, D.B., 1983. The Historical Status and Reduction of the Western Arctic Bowhead Whale (*Balaena mysticetus*) Population by the Pelagic Whaling Industry, 1848–1914. Report to the International Whaling Commission, Special Issue 5, pp. 107–141, SC/32/PS16.

Bockstoce, J.R., Botkin, D.B., Philp, A., Collins, B.W., George, J.C., 2005. The geographic distribution of bowhead whales, *Balaena mysticetus*, in the Bering, Chukchi, and Beaufort seas: evidence from whaleship records, 1849–1914. Mar. Fish. Rev. 67, 1–43.

Bogoslovskaya, L.S., Votrogov, L.M., Krupnik, I.I., 1982. The bowhead whale off Chukotka: migrations and aboriginal whaling. Rep. Int. Whal. Comm. 32, 391–399.

Brueggeman, J.J., Webster, B.D., Grotefendt, R., Chapman, D.G., 1987. Monitoring of the Winter Presence of Bowhead Whales in the Navarin Basin Through Association with Sea Ice. Report for the U.S. Minerals Management Service by Envirosphere Company. OCS Study MMS 87-0028.

Burns, J.J., 1993. Epilogue. In: Burns, J.J., Montague, J.H., Cowles, C.J. (Eds.), The Bowhead Whale. Allen Press Inc., Lawrence, KS, pp. 745–764, Special Publication Number 2, The Society for Marine Mammalogy.

Chambault, P., Albertsen, C.M., Patterson, T.A., Hansen, R.G., Tervo, O., Laidre, K.L., et al., 2018. Sea surface temperature predicts the movements of an Arctic cetacean: the bowhead whale. Sci. Rep. 8 (9658). Available from: https://doi.org/10.1038/s41598-018-27966-1.

Citta, J.J., Okkonen, S.R., Quakenbush, L.T., Maslowski, W., Osinski, R., George, J.C., et al., 2018. Oceanographic characteristics associated with bowhead whale movements in the Chukchi Sea in autumn. Deep Sea Res. II 152, 121–131. Available from: https://doi.org/10.1016/j.dsr2.2017.03.009.

Citta, J.J., Quakenbush, L.T., George, J.C., Small, R.J., Heide-Jørgensen, M.P., Brower, H., et al., 2012. Winter movements of bowhead whales (*Balaena mysticetus*) in the Bering Sea. Arctic 65 (1), 13–34.

Citta, J.J., Quakenbush, L.T., Okkonen, S.R., Druckenmiller, M.L., Maslowski, W., Clement-Kinney, J., et al., 2015. Ecological characteristics of core areas used by western Arctic bowhead whales, 2006–2012. Prog. Oceanogr. 136, 201–222. Available from: https://doi.org/10.1016/j.pocean.2014.08.012.

Clark, C.W., Berchok, C.L., Blackwell, S.B., Hannay, D.E., Jones, J., Ponirakis, D., et al., 2015. A year in the acoustic world of bowhead whales in the Bering, Chukchi and Beaufort seas. Prog. Oceanogr. 136, 223–240.

Clark, C.W., Johnson, J.H., 1984. The sounds of the bowhead whale, *Balaena mysticetus*, during the spring migrations of 1979 and 1980. Can. J. Zool. 62 (7), 1436–1441.

Clarke, J.T., Ferguson, M.C., Willoughby, A.L., Brower, A.A., 2018. Bowhead and beluga whale distributions, sighting rates, and habitat associations in the western Beaufort Sea in summer and fall 2009–16, with comparison to 1982–91. Arctic 71, 115–138.

Costa, D.P., Williams, T.M., 1999. Marine mammal energetics. In: Reynolds III, J.E., Rommel, S.A. (Eds.), Biology of Marine Mammals. Smithsonian Institution, p. 578.

Daase, M., Falk-Petersen, S., Varpe, O., Darnis, G., Soreide, J.E., Wold, A., et al., 2013. Timing of reproductive events in the marine copepod *Calanus glacialis*: a pan-Arctic perspective. Can. J. Fish. Aquat. Sci. 70, 871–884.

Darnis, D., Fortier, L., 2014. Temperature, food and the seasonal vertical migration of key arctic copepods in the thermally stratified Amundsen Gulf (Beaufort Sea, Arctic Ocean). J. Plankton Res. 36, 1092–1108. Available from: https://doi.org/10.1093/plankt/fbu035.

Druckenmiller, M.L., Citta, J.J., Ferguson, M.C., Clarke, J.T., George, J.C., Quakenbush, L.T., 2018. Trends in sea ice cover within bowhead whale use areas in the Pacific Arctic. Deep Sea Res. II 152, 95–107. Available from: https://doi.org/10.1016/j.dsr2.2017.10.017.

Duong, T., Hazelton, M.L., 2005. Cross-validation bandwidth matrices for multivariate kernel density estimation. Scand. J. Stat. 32, 485–506.

Eisner, L., Hillgruber, N., Martinson, E., Maselko, J., 2013. Pelagic fish and zooplankton species assemblages in relation to water mass characteristics in the northern Bering and southeast Chukchi seas. Polar Biol. 36, 87–113.

Ershova, E.A., Hopcroft, R.R., Kosobokova, K.N., Matsuno, K., Nelson, R.J., Yamaguchi, A., et al., 2015. Long-term changes in summer zooplankton communities of the western Chukchi Sea, 1945–2012. Oceanography 28 (3), 100–115. Available from: https://doi.org/10.5670/oceanog.2015.60.

Fancy, S.G., Pank, L.F., Douglas, D.C., Curby, C.H., Garner, G.W., Amstrup, S.C., et al., 1988. Satellite Telemetry: A New Tool for Wildlife Research and Management. U.S. Fish and Wildlife Service, Washington DC, p. 54, Res. Pub. No. 172.

Freitas, C., Lydersen, C., Fedak, M.A., Kovacs, K.M., 2008. A simple new algorithm to filter marine mammal Argos locations. Mar. Mammal. Sci. 24, 315–325.

George, J.C., Moore, S.E., Suydam, R.S., 2007. Summary of Stock Structure Research on the Bering-Chukchi-Beaufort Seas Stock of Bowhead Whales 2003–2007. Report to the International Whaling Commission Scientific Committee, SC/59/BRG3.

George, J.C., Sheffield, G., Reed, D.J., Tudor, B., Stimmelmayr, R., Person, B.T., 2017. Frequency of injuries from line entanglements, killer whales, and ship strikes on Bering-Chukchi-Beaufort Seas bowhead whales. Arctic 70 (1), 37–46.

George, J.C., Zeh, J., Suydam, R., Clark, C., 2004. Abundance and population trend (1978–2001) of western Arctic bowhead whales surveyed near Barrow, Alaska. Mar. Mammal. Sci. 20 (4), 755–773.

Givens, G.H., Edmonson, S.L., George, J.C., Suydam, R., Charif, R.A., Rahaman, A., et al., 2016. Horvitz–Thompson whale abundance estimation adjusting for uncertain recapture, temporal availability variation, and intermittent effort. Environmetrics 27 (3), 134–146. Available from: https://doi.org/10.1002/env.2379.

Givens, G.H., Huebinger, R.M., Patton, J.C., Postma, L.D., Lindsay, M., Suydam, R.S., 2010. Population genetics of bowhead whales (*Balaena mysticetus*) in the western Arctic. Arctic 63 (1), 1–12.

Hannay, D.E., Delarue, J., Mouy, X., Martin, B.S., Leary, D., Oswald, J.N., et al., 2013. Marine mammal acoustic detections in the northeastern Chukchi Sea, September 2007–July 2011. Cont. Shelf Res. 67, 127–146.

Harwood, L.A., Auld, J., Joynt, A., Moore, S.E., 2010. Distribution of Bowhead Whales in the SE Beaufort Sea during late summer, 2007–2009. Department of Fisheries and Oceans Canada, Canadian Science Advisory Secretariat, Research Document 2009 (111), 22 p.

Harwood, L.A., Quakenbush, L.T., Small, R.J., George, J.C., Pokiak, J., Pokiak, C., et al., 2017. Movements and inferred foraging by bowhead whales in the Canadian Beaufort Sea during August and September, 2006–12. Arctic 70 (2), 161–176.

Hazard, K.W., Lowry, L.F., 1984. Benthic prey in a bowhead whale from the northern Bering Sea. Arctic 37 (2), 166–168.

Heide-Jørgensen, M.P., Laidre, K.L., Quakenbush, L.T., Citta, J.J., 2011. The northwest passage opens for bowhead whales. Biol. Lett. 8, 270–273. Available from: https://doi.org/10.1098/rsbl.2011.0731.

Hobbs, L.J., Goebel, M.E., 1982. Bowhead Whale Radio Tagging Feasibility Study and Review of Large Cetacean Tagging. NOAA Technical Memorandum NMFS F/NWC-21.

Huntington, H.P., Quakenbush, L.T., Nelson, M., 2016. Effects of changing sea ice on marine mammals and subsistence hunters in northern Alaska from traditional knowledge interviews. Biol. Lett. 12, 20160198. Available from: https://doi.org/10.1098/rsbl.2016.0198.

IWC, 2007. Report of the Scientific Committee, IWC/59/Rep 1.

IWC, 2012. Report of the Scientific Committee, IWC/64/Rep 1, Annex E.

Johnson, D.S., London, J.M., Lea, M.-A., Durban, J.W., 2008. Continuous-time correlated random walk model for animal telemetry data. Ecology 89 (5), 1208–1215.

Kenney, R.D., Mayo, C.A., Winn, H.E., 2001. Migration and foraging strategies at varying spatial scales in western North Atlantic right whales: a review of hypotheses. J. Cetacean Res. Manage. (Spec. Issue) 2, 251–260.

Koski, W.R., Rugh, D.J., Punt, A.E., Zeh, J., 2006. An approach to minimize bias in estimation of the length-frequency distribution of bowhead whales (*Balaena mysticetus*) from aerial photogrammetric data. J. Cetacean Res. Manage. 8 (1), 45–54.

Krutzikowsky, G.K., Mate, B.R., 2000. Dive and surfacing characteristics of bowhead whales (*Balaena mysticetus*) in the Beaufort and Chukchi seas. Can. J. Zool. 78, 1182–1198.

Ljungblad, D.K., Moore, S.E., Clarke, J.T., 1986. Assessment of bowhead whale (*Balaena mysticetus*) feeding pattern in the Alaskan Beaufort and northeastern Chukchi Seas via aerial surveys, fall 1979-1984. Rep. Int. Whal. Comm. 36, 265–272.

Ljungblad, D.K., Thompson, P.O., Moore, S.E., 1982. Underwater sounds recorded from migrating bowhead whales, *Balaena mysticetus* in 1979. J. Acoust. Soc. Am. 71, 477–482.

Lowry, L.F., 1993. Foods and feeding ecology. In: Burns, J.J., Montague, J.H., Cowles, C.J. (Eds.), The Bowhead Whale. Allen Press Inc., Lawrence, KS, pp. 201–238, Special Publication Number 2, The Society for Marine Mammalogy.

Lowry, L.F., Nelson, R.R., Frost, K.J., 1987. Observations of killer whales, *Orcinus orca*, in western Alaska: sightings, strandings, and predation on other marine mammals. Can. Field Nat. 101, 6–12.

Lowry, L.F., Sheffield, G., George, J.C., 2004. Bowhead whale feeding in the Alaskan Beaufort Sea, based on stomach contents analyses. J. Cetacean Res. Manage. 6, 215–223.

Mate, B.R., Krutzikowsky, G.K., Winsor, M.H., 2000. Satellite-monitored movements of radio-tagged bowhead whales in the Beaufort and Chukchi seas during the late-summer feeding season and fall migration. Can. J. Zool. 78, 1168–1181.

Matthews, C.J.D., Breed, G.A., LeBlanc, B., Ferguson, S.H., 2020. Killer whale presence drives bowhead whale preference for sea ice in Arctic seascapes of fear. Proc. Natl. Acad. Sci. 117, 6590–6598. Available from: https://doi.org/10.1073/pnas.1911761117.

Melnikov, V.V., Zagrebin, I.A., Zelensky, G.M., Ainana, L.I., 2007. Killer whales (*Orcinus orca*) in waters adjacent to the Chukotka Peninsula, Russia. J. Cetacean Res. Manage. 9, 53–63.

Melnikov, V.V., Zeh, J.E., 2007. Chukotka Peninsula counts and estimates of the number of migrating bowhead whales. J. Cetacean Res. Manage. 9, 29–35.

Moore, S.E., George, J.C., Coyle, K.O., Weingartner, T.J., 1995. Bowhead whales along the Chukotka coast in Autumn. Arctic 48, 155–160.

Moore, S.E., George, J.C., Sheffield, G., Bacon, J., Ashjian, C.J., 2010. Bowhead whale distribution and feeding near Barrow, Alaska, in late summer 2005–06. Arctic 63, 195–205.

Moore, S.E., Haug, T., Vikingsson, G.A., Stenson, G.B., 2019. Baleen whale ecology in arctic and subarctic seas in an era of rapid habitat alteration. Prog. Oceanogr. 176, 102118. Available from: https://doi.org/10.1016/j.pocean.2019.05.010.

Moore, S.E., Reeves, R.R., 1993. Distribution and movement. In: Burns, J.J., Montague, J.H., Cowles, C.J. (Eds.), The Bowhead Whale. Allen Press Inc., Lawrence, KS, pp. 313–386., Special Publication Number 2, The Society for Marine Mammalogy.

Noongwook, G., The Native Village of Savoonga, The Native Village of Gambell, Huntington, H.P., George, J.C., 2007. Traditional knowledge of the bowhead whale (*Balaena mysticetus*) around St. Lawrence Island, Alaska. Arctic 60 (1), 47–54.

Okkonen, S.R., Ashjian, C.J., Campbell, R.G., Clarke, J.T., Moore, S.E., Taylor, K.D., 2011. Satellite observations of circulation features associated with a bowhead whale feeding 'hotspot' near Barrow, Alaska. Remote Sens. Environ. 115, 2168–2174.

Okkonen, S.R., Clarke, J.T., Potter, R.A., 2018. Relationships among high river discharges, upwelling events, and bowhead whale (*Balaena mysticetus*) occurrence in the central Alaskan Beaufort Sea. Deep Sea Res. II 152, 195–202.

Olnes, J., Citta, J., Quakenbush, L., George, C., Harwood, L., Adams, B., et al., in press. Use of the Alaskan Beaufort Sea by bowhead whales (*Balaena mysticetus*) tagged with satellite transmitters, 2006–2018. Arctic.

Pomerleau, C., Ferguson, S.H., Walkusz, W., 2011. Stomach contents of bowhead whales (*Balaena mysticetus*) from four locations in the Canadian Arctic. Polar Biol. 34, 615–620.

Quakenbush, L.T., Huntington, H.P., 2010. Traditional Knowledge Regarding Bowhead Whales in the Chukchi Sea near Wainwright, Alaska. Final Report OCS Study MMS 2009-063, U.S. Department of the Interior, Bureau of Ocean Energy Management, Regulation and Enforcement, Alaska Outer Continental Shelf Region, and the University of Alaska Fairbanks, Fairbanks.

Richardson, J.W., 1987. Importance of the Eastern Alaskan Beaufort Sea to Feeding Bowhead Whales, 1985–1986. OCS Study MMS 87-0037 U.S. Minerals Management Service, Anchorage, AK.

Richardson, W.J., Thomson, D.H., 2002. Bowhead Whale Feeding in the Eastern Alaskan Beaufort Sea: Update of Scientific and Traditional Information. OCS Study MMS 2002-012; LGL Rep. TA2196-7. Rep. from LGL Ltd., King City, Ont., for U.S. Minerals Management Service, Anchorage, AK, and Herndon, VA. Vol. 1, xliv + 420 p; Vol. 2, 277 p.

Robbins, J., Zerbini, A.N., Gales, N., Gulland, F.M.D., et al., 2013. Satellite Tag Effectiveness and Impacts on Large Whales: Preliminary Results of a Case Study with Gulf of Maine Humpback Whales. Report to the International Whaling Commission SC/65a/SH05.

Smultea, M., Fertl, D., Rugh, D.J., Bacon, C.E., 2012. Summary of Systematic Bowhead Surveys Conducted in the U.S. Beaufort and Chukchi Seas, 1975–2009. U.S. Department of Commerce, NOAA Technical Memorandum NMFS-AFSC-237, 48 p.

Stafford, K.M., 2018. Increasing detections of killer whales (*Orcinus orca*) in the Pacific Arctic. Mar. Mammal. Sci. 35, 696–706.

Stoker, S.W., Krupnik, I.I., 1993. Subsistence whaling. In: Burns, J.J., Montague, J.H., Cowles, C.J. (Eds.), The Bowhead Whale. Allen Press Inc., Lawrence, KS, pp. 579–629., Special Publication Number 2, The Society for Marine Mammalogy.

Suydam, R., George, J.C., 2018. Subsistence Harvest of Bowhead Whales (*Balaena mysticetus*) by Alaskan Natives, 1974 to 2016. Report to the International Whaling Commission SC/67b/AWMP.

Thomson, D.H., Koski, W.R., Richardson, W.J., 2002. Integration and conclusions. In: Richardson, W.J., Thomson, D.H. (Eds.), Bowhead Whale Feeding in the Eastern Alaskan Beaufort Sea: Update of Scientific and Traditional Information, vol. 1. OCS Study MMS 2002-012; LGL Rep. TA2196-7. Rep. from LGL Ltd., King City, Ont., for U.S. Minerals Manage. Serv., Anchorage, AK and Herndon, VA, pp. 23-1 to 23-36.

Walkusz, W., Williams, W.J., Harwood, L.A., Moore, S.E., Stewart, B.E., Kwasniewski, S., 2012. Composition, biomass and energetic content of biota in the vicinity of feeding bowhead whales (*Balaena mysticetus*) in the Cape Bathurst upwelling region (south eastern Beaufort Sea). Deep Sea Res. I 69, 25–35.

Wartzok, D., Watkins, W.A., Würsig, B., Guerrero, J., Schoenherr, J., 1990. Movements and Behaviors of Bowhead Whales. Rep. from Purdue Univ., Fort Wayne, IN, for AMOCO Prod. Co., Anchorage, AK. 197 p.

Wartzok, D., Watkins, W.A., Würsig, B., Malme, C.I., 1989. Movements and Behaviors of Bowhead Whales in Response to Repeated Exposures to Noises Associated with Industrial Activities in the Beaufort Sea. Rep. from Purdue Univ., Fort Wayne, IN, for AMOCO Prod. Co., Anchorage, AK. 228 p.

Williams, W.J., Carmack, E.C., 2008. Combined effect of wind-forcing and isobaths divergence on upwelling at Cape Bathurst Beaufort Sea. J. Mar. Res. 66, 645–663.

Willoughby, A.L., Ferguson, M.C., Stimmelmayr, R., Clarke, J.T., Brower, A.A., in review. Killer whale predation on bowhead whales in the Pacific Arctic—on the rise? Polar Biol.

Wold, A., Darnis, G., Soreide, J., Leu, E., Philippe, B., Fortier, L., et al., 2011. Life strategy and diet of *Calanus glacialis* during the winter–spring transition in Amundsen Gulf, southeastern Beaufort Sea. Polar Biol. 34, 1929–1946.

Würsig, B., Dorsey, E.M., Fraker, M.A., Payne, R.S., Richardson, W.J., 1984. Behavior of bowhead whales, *Balaena mysticetus*, summering in the Beaufort Sea: a description. Fish. Bull. 83, 357–377.

Zelensky, M.A., Melnikov, V., Zagrebin, I., Bychkov, V., 1997. The Role of the Naukan Native Company in Encouraging Subsistence Use of Wildlife Resources by the Chukotka Native People and in Conducting Shore-Based Observations of the Distribution of Bowhead Whale, *Balaena mysticetus*, in the Waters of the Bering Sea and the Chukchi Sea Adjacent to the Chukotka Peninsula (Russia) During 1996. Naukan Production Cooperative, Lavrentiya, Chukotka Autonomous District, p. 152.

CHAPTER

5

Distribution, migrations, and ecology of the Atlantic and the Okhotsk Sea Populations

Mads Peter Heide-Jørgensen[1], *R.G. Hansen*[1] *and O.V. Shpak*[2]

[1]Greenland Institute of Natural Resources, Copenhagen, Denmark [2]A.N. Severtsov Institute of Ecology and Evolution of Russian Academy of Sciences, Moscow, Russia

Introduction

This chapter discusses the distribution, migration patterns, diving and dietary differences of three of the four populations of bowhead whales, the East Canada-West Greenland (ECWG), the East Greenland-Svalbard-Barents Sea (EGSB), and the Okhotsk (OKH) populations (Figs. 5.1 and 5.2). The fourth population, Bering-Chukchi-Beaufort Seas (BCB), is discussed in Chapter 4, abundance estimates are presented in Chapter 6, and the genetic differences between the populations are discussed in Chapter 3.

The East Canada-West Greenland population

Based on catch reports from mostly 19th century whaling captains, Southwell (1898) provided the first overview of the migrations of bowhead whales in the North Atlantic. It took another 100 years before the emerging technology of satellite tracking of whales provided the opportunities for verifying the migratory patterns of the populations east and west of Greenland that Southwell (1898) detailed.

Initial pioneering efforts with satellite tracking of bowhead whales took place in Alaska in 1992 (Mate et al., 2000). In 1999, new methods for tagging and tracking of baleen whales were developed (Heide-Jørgensen et al., 2001, 2003) and they were used to study bowhead whales in West Greenland and subsequently in Canada, Alaska and East Greenland (Ferguson et al., 2010; Quakenbush et al., 2010; Lydersen et al., 2012). The new methods

DOI: https://doi.org/10.1016/B978-0-12-818969-6.00005-4

FIGURE 5.1 Bowhead whale of the East Canada-West Greenland population in spring 2009 in Disko Bay with Disko Island in the background. *Source: Photo by M.P. Heide-Jørgensen.*

FIGURE 5.2 Main summering grounds (*dark pink*), total range (*pink*), and historical range before commercial exploitation (*dotted*) of the East Canada-West Greenland population. *Source: Map by John Citta.*

FIGURE 5.3 Tagging of a bowhead whale in Disko Bay in 2017. *Source: Copyright M.P. Heide-Jørgensen.*

included the use of a modified air gun as a remote tag delivery system and a pole system for deploying instruments at a short range (Fig. 5.3). The duration of data from tracks thus studied was considerably improved with the new methods and yielded surprising new insights on the movement of whales when large numbers of whales were studied. Despite a large number of satellite tags ($n > 100$) and the occasional good duration of the tracks (>400 days), satellite tracking alone does not adequately describe the seasonal ranges of the ECWG population. Information on observations and catches should be included to obtain a more complete picture of the distribution. The tagging of whales has been limited to three localities (Disko Bay, Cumberland Sound, and Foxe Basin) and is for the same reason biased in terms of spatial coverage of the movements of the whales.

Most tagging of whales has been conducted in Disko Bay in winter and spring (Figs. 5.4 and 5.5). When the whales abandon the bay in May–June they either head straight west to Baffin Island, north into Baffin Bay, or even south toward Hudson Strait (Chambault et al., 2018; Heide-Jørgensen et al., 2006). Their departure from Disko Bay follows the recession of sea ice but their movements are also largely influenced by the circulation of cold polar water in Baffin Bay (Chambault et al., 2018). The main summering grounds are all located in the Canadian Arctic Archipelago (along the east coast of Baffin Island, Cumberland Sound, Isabella Bay, northern Foxe Basin, Admiralty Inlet, Prince Regent Inlet, Gulf of Boothia, Doniol-Valcroze et al., 2015). The fall movement occurs ahead of the sea ice formation in Canada and targets the Hudson Strait and areas along the east coast of Baffin

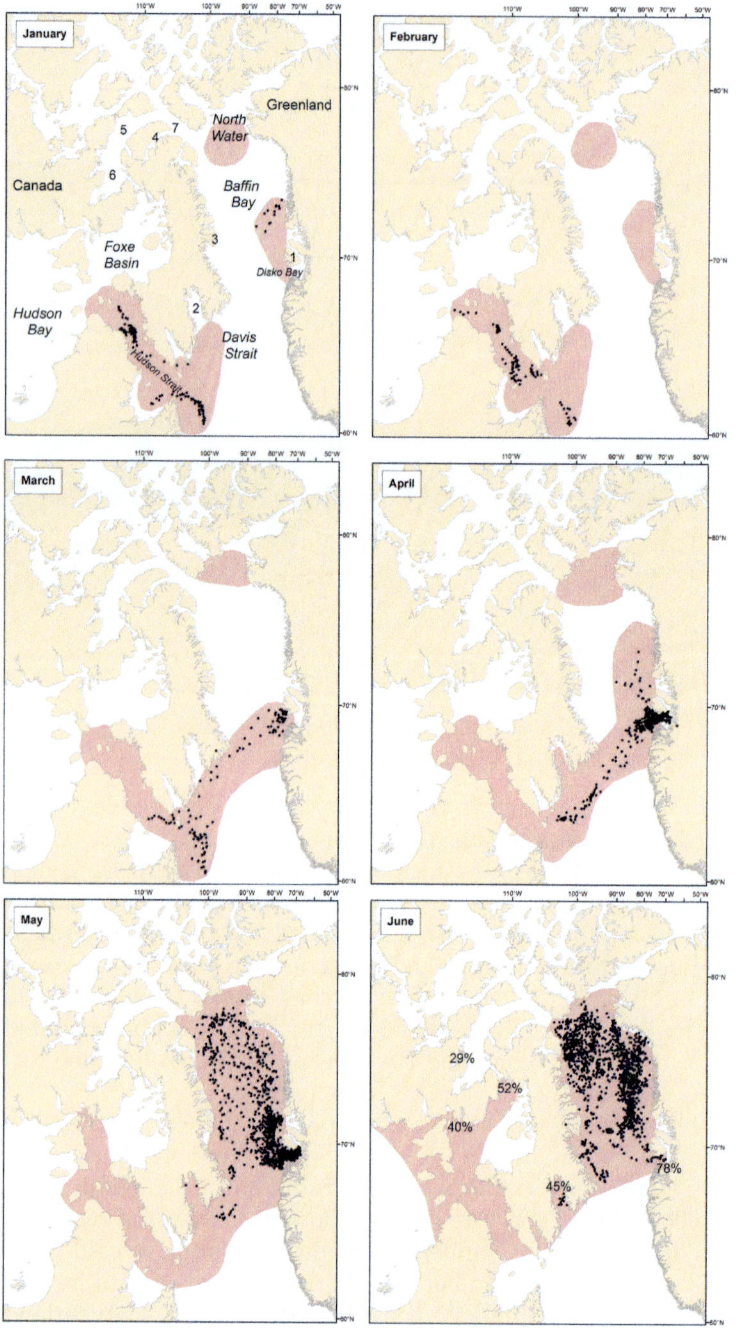

FIGURE 5.4 Seasonal ranges (January–June) of the ECWG bowhead whales compiled from observational and tracking data. The monthly distribution patterns are supplemented with positions from whales tracked from Disko Bay, Cumberland Sound, and Foxe Basin (see Chambault et al., 2018). The following numbers refer to place names mentioned in the text: 1. Disko Island; 2. Cumberland Sound; 3. Isabella Bay; 4. Admiralty Inlet; 5. Prince Regent Inlet; 6. Gulf of Boothia; 7. Lancaster Sound. The percentage of females in samples from five localities is shown on the map for June.

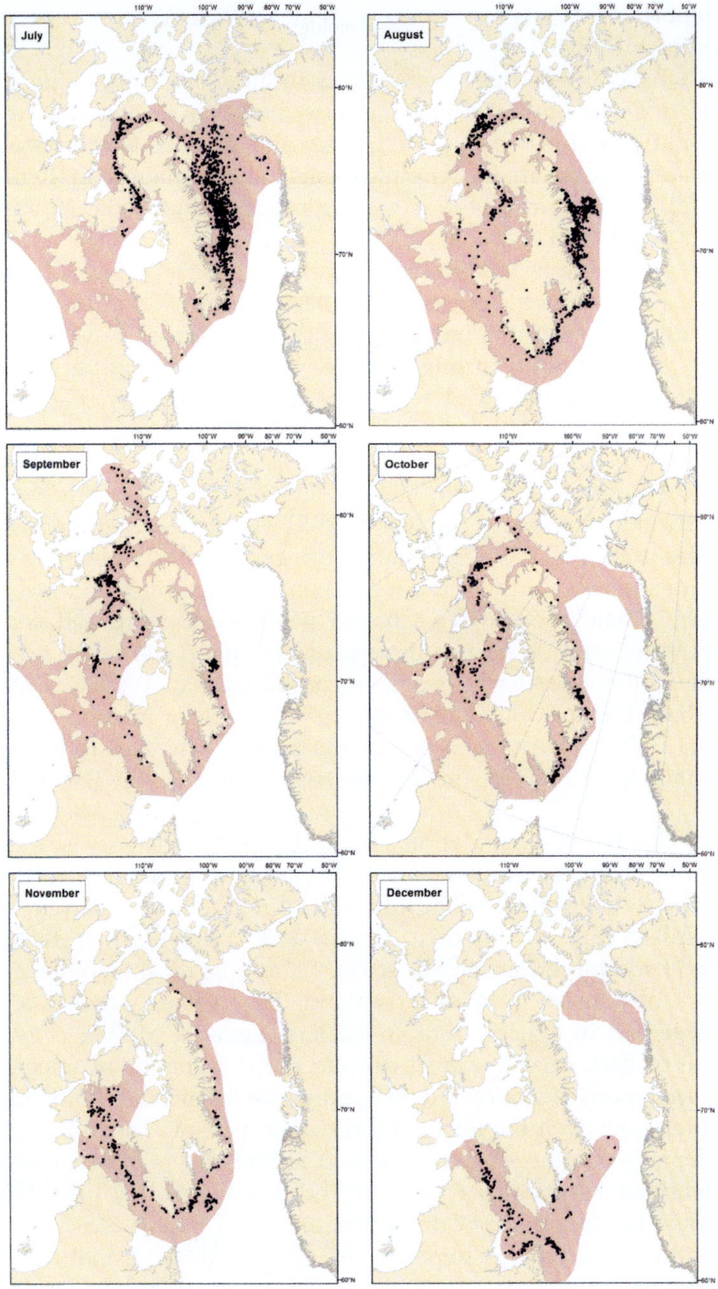

FIGURE 5.5 Seasonal ranges (July–December) of the ECWG bowhead whales compiled from observational and tracking data. See caption of Fig. 5.4 for details.

Island. Some reports also suggest a movement toward West Greenland in the fall (Kapel, 1985; Southwell, 1898).

The bowhead whales utilize open water areas in Hudson Strait, along the east coast of Baffin Island and in Disko Bay for wintering. Low numbers of bowhead whales winter in the North Water in the northern part of Baffin Bay and perhaps at other scattered localities in small open-water refugia in the complex pack ice formation in Baffin Bay (Heide-Jørgensen et al., 2016). Historically the wintering range included the Labrador coast as far south as Newfoundland (McLeod et al., 2008).

Whales that are wintering in Hudson Strait or along the east coast of Baffin Island either remain in the same areas, move west into Foxe Basin, or move north into Lancaster Sound (Ferguson et al., 2010). In addition, movements of whales across Davis Strait from Eastern Canada to West Greenland occur throughout the winter (Chambault et al., 2018).

In August 2010, an adult bowhead whale tagged in Disko Bay, West Greenland, ventured into the Northwest Passage as far west as Viscount Melville Sound. In the same month, an adult whale tagged in Alaska entered this area too. They stayed in Viscount Melville Sound for 10 days confirming that, in light ice years, bowhead whales from the Pacific and the Atlantic may meet in the Northwest Passage (Heide-Jørgensen et al., 2011).

Bowhead whales from the ECWG population are generally not following the strict migratory schedules with seasonal north–south movements as seen for BCB bowhead whales and many other baleen whale species (see Chapter 4). The topographical complexity of the Canadian Arctic Archipelago is part of the reason for this pattern, but oceanographic features like the influx of warm Atlantic water to coastal areas of West Greenland also play a role in shaping the migration routes (Chapter 25). The spring intrusion of warm water in the eastern part of Baffin Bay probably forces the bowhead whales to make a longitudinal shift to cooler summering habitats on the western side of Baffin Bay (Chambault et al., 2018).

The whales of the ECWG population are highly segregated by sex and age classes (Fig. 5.4, June). It is primarily adult males and resting and/or pregnant females that are found in Baffin Bay, while calves, subadults, and nursing females stay in Eastern Canada (Prince Regent Inlet, Gulf of Boothia, Foxe Basin, and northwestern Hudson Bay; Southwell, 1898; Heide-Jørgensen et al., 2011). Sampling of skin biopsies has shown a greater abundance of females in West Greenland (Disko Bay) whereas a more even sex ratio has been observed in aggregations in Eastern Canada. Mating activities have been observed offshore of Disko Bay in April despite the dominance of females, and singing whales are also frequently encountered in spring (Eschricht and Reinhardt, 1866; Tervo et al., 2009; Chapter 23). Given the low proportions of males in Disko Bay, it has been suggested that the females are the singers and field studies have shown that three localized singers were identified as females (Tervo, 2011). It is, however, uncertain whether the vocal activity should be classified as calls or singing (see Chapter 22).

The body length of the whales suggests that primarily large mature whales without calves occupy Disko Bay, whereas primarily mother–calf pairs are found in Foxe Basin (Heide-Jørgensen et al., 2010). The most parsimonious explanation for the large-scale spatial segregation of sexes in bowhead whales in Baffin Bay is that mature females without calves utilize Disko Bay as a foraging ground during certain parts of their reproductive cycle. The main calving areas seem to be in the Canadian Arctic Archipelago and especially the shallow waters of Foxe Basin, where sheltered areas may offer a refuge for young calves, minimizing the predation risk from killer whales as well as reducing the risk of ice entrapment (Finley, 2001; Ferguson et al., 2010; Higdon and Ferguson, 2010).

The East Greenland-Svalbard-Barents Sea population

Southwell (1898) presented the first overview of the migrations of parts of the EGSB population of bowhead whales based on records from the extensive whaling around Svalbard between 1611 and the early 19th century (Allen and Keay, 2006). During the whaling period, bowhead whales were first hunted at the "Northern Whaling Ground" just west of Svalbard in spring (Fig. 5.6), and later in the summer they were found further south in the Greenland Sea at the "Southern Whaling Ground" off the East Greenland pack ice belt. The first satellite tracking of a bowhead whale from 2010 confirmed the historical route of north–south movements described by the whalers (Lydersen et al., 2012). The track revealed that the whale spent most of its time in waters close to the ice edge with little ice coverage, over areas where the bottom slope was relatively steep.

The occurrence of bowhead whales on the East Greenland shelf has been overlooked for centuries. The whaling vessels operating in the 18th and 19th centuries could not

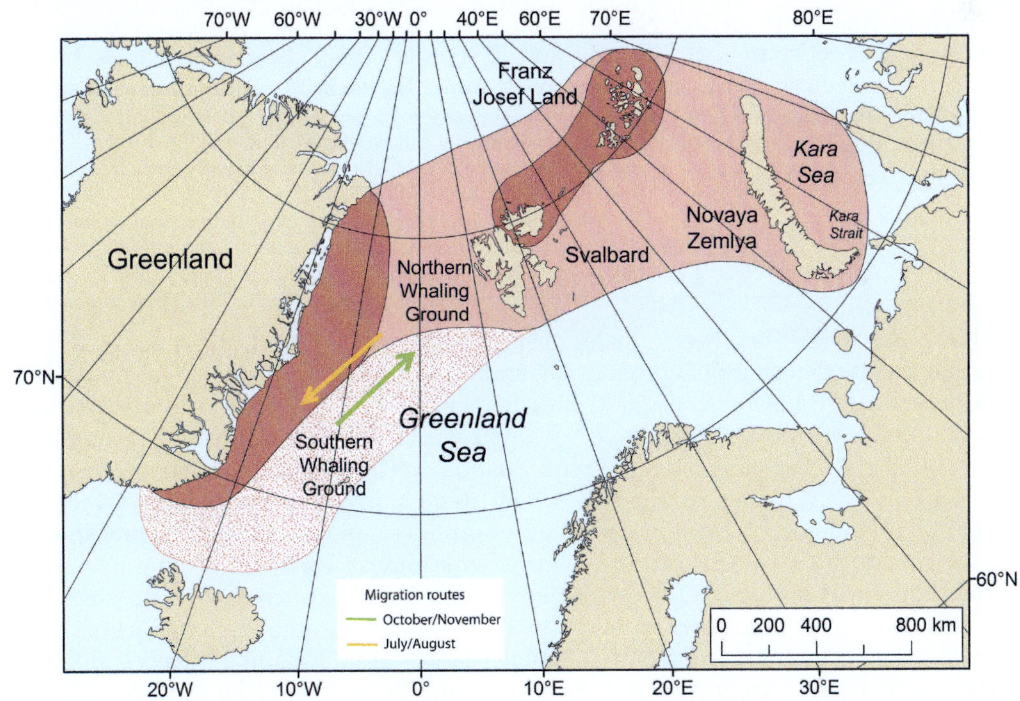

FIGURE 5.6 Main summering grounds (*dark pink*), total range (*pink*), and historical range before commercial exploitation (*dotted*) of the East Greenland-Svalbard-Barents Sea population. *Source: Map by John Citta.*

penetrate the East Greenland pack ice and only in years with low ice concentration did they occasionally reach the southern part of the coast of East Greenland (Scoresby, 1823). By the beginning of the 20th century, the bowhead whale population was severely depleted and expeditions rarely encountered whales, but recent surveys of the shelf and coastal areas of East Greenland have confirmed an increasing number of sightings (Gilg and Born, 2005; Boertmann et al., 2009; de Boer et al. 2019). They are present in surprisingly large abundances in the Northeast Water polynya at the northeastern corner of Greenland (Boertmann et al., 2015; Hansen et al., 2018). Observations of calves in this area provide hope for the recovery of the population (Boertmann et al., 2009; Boertmann and Nielsen, 2010). Intensive and highly diverse singing during winter months recorded off Northeast Greenland suggest that this area is a mating ground for bowhead whales (Stafford et al., 2012; Ahonen et al., 2017) as is also confirmed by aerial observations (Hansen et al., 2018).

Most bowhead whale sightings in western Russian Arctic waters come from the region of Franz Josef Land, where whales have been observed year-round with the majority of summer encounters occurring south of the archipelago (Belikov et al., 1989; Gavrilo, 2015). Observations during the last decade suggest that the waters around Franz Josef Land may be occupied by at least a hundred bowhead whales. Several video- or photographically documented sightings come from different parts of the Kara Sea, with the southernmost in Kara Gates Strait (Chaadaeva et al., 2018a,b; Petrov et al., 2018). Further east, bowhead whales have been observed in the western Laptev Sea (Petrov et al., 2018), and (at least 10 whales) to the north of the Novosibirskie Islands Archipelago, which separates the Laptev and East Siberian seas (Gavrilo and Tretiakov, 2008). Bowhead whales that are believed to come from the Bering-Chukchi-Beaufort Seas (BCB) population occasionally visit the East Siberian Sea. The presence of bowhead whales across all Russian Arctic seas indicates that it is possible individuals from the BCB and EGSB populations may meet.

The Okhotsk Sea population

Okhotsk Sea (OKH) bowhead whales (Fig. 5.7) are geographically and genetically isolated from the neighboring BCB population, and are confined to the waters of the Okhotsk Sea (Fedoseev, 1984; LeDuc et al., 2005; Mescherskiy et al., 2014). Historical distribution of bowhead whales in the ice-free period included both coastal and offshore waters of the northern Okhotsk Sea according to information retrieved from whalers' log-books (Townsend, 1935). Whaling resulted in severe depletion of the population, which may have led to a reduction of the summer distribution range. In recent decades, few sightings have been made in the deep offshore waters, with almost none in late summer months (Ivashchenko and Clapham, 2010; Myasnikov et al., 2016; Gushcherov et al., 2017, 2018, 2019). At present, the larger part of the population summers in the western part of the sea, mostly in the so-called Shantar region (Shantar Islands and all mainland bays south to the archipelago: Udskaya, Tugursky, and Akademii, which includes Konstantina, Ulbansky, and Nikolaya bays) as far south as 53°N. Whales are found in the Shantar region from late May (Doroshenko, 1996) through November (Shpak and Paramonov, 2018).

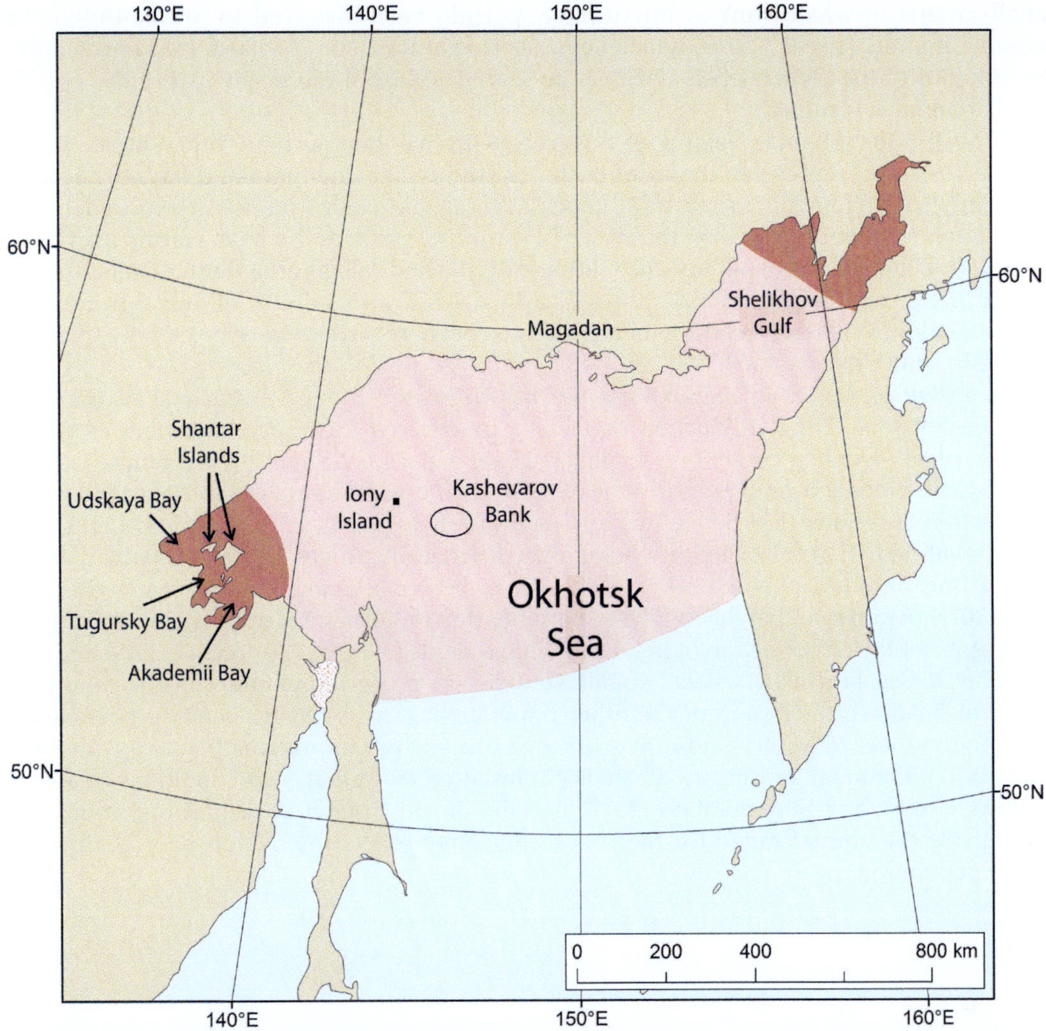

FIGURE 5.7 Main summering grounds (*dark pink*), total range (*pink*), and historical range before commercial exploitation (*dotted*) of the Okhotsk Sea population. *Source: Map by John Citta.*

Wintering areas for Okhotsk Sea bowhead whales remain to be identified, but it is presumed that bowhead whales utilize the leads along the ice edge in the area of Kashevarov Bank (Fedoseev, 1984; Doroshenko, 1996). Lindholm (1863) mentioned that early in spring whales pass Iony Island (i-o-n-y). Doroshenko (1996) believed that in spring whales segregate: immature individuals and mother–calf pairs migrate to the Shantar Islands, while large whales make a northward loop to Shelikhov Gulf, but later in summer join the rest of the population in the Shantar region. This hypothesis is yet to be tested. Local people from the Magadan area observe bowhead whales near the coast in spring (Shpak, unpubl. data), but the directions of their movements have not been reported. Single animals and small groups of whales (up to three) have recently been observed in Shelikhov Gulf in summer months (Shpak, 2016), which suggests that at least some whales feed in the north-eastern part of the Okhotsk Sea and may be segregated from the major part of the population throughout summer.

Although the Shantar region is covered with ice longer than other areas of the Okhotsk Sea and represents the local center of cold water, the mainland bays are ice-free from July to November. Large tidal zones several kilometers wide are exposed to sun and high air temperatures, and the water in estuarine parts of the bays warms up significantly. In Ulbansky Bay, for instance, immature individuals aggregate in waters as shallow as 2.3 m, with surface temperature of up to 16.5°C and salinity of only 8 ppt (Shpak and Paramonov, 2018). Feeding behavior has been recorded in the shallow parts of Ulbansky and Udskaya bays.

The extent of sea ice in the Okhotsk Sea has decreased by 22% between 1980 and 2010 (Radchenko et al., 2010). Reduction of the sea ice-covered area and shortening of the ice period create favorable conditions for killer whale expansion. Since 2011, multiple cases of killer whale predation on bowhead whales—calves and juveniles of up to 10 m long—were recorded in the region (Fig. 5.8). Observations from photographs in 2015 showed that six out of six identified whales had killer whale rake marks on the flukes (Shpak, unpubl. data). In comparison, only 10% of bowhead whales in the ECWG population had rake marks (Reinhart et al., 2013). At present, bowhead whale mortality due to killer whale predation likely equals the surplus of the reproduction of bowhead whales in the Okhotsk Sea (Shpak, 2018).

Okhotsk Sea bowhead whales spend summer in more temperate climate conditions than bowhead whales in any of the other populations, and intensive molting observed in most individuals may serve adaptive functions to such environment (Fig. 5.9; Chernova et al., 2017). Frequent and heavy whale lice infestation as well as skin conditions related to molting (Shpak and Stimmelmayr, 2017) may be linked to environmental conditions that are outside the normal range for the species and may reflect the overall poor health state of the population.

FIGURE 5.8 Killer whale attacking a bowhead whale in the Okhotsk Sea. *Source: Copyright O. Shpak.*

FIGURE 5.9 Molting bowhead whale of the Okhotsk Sea population. *Source: Copyright O. Shpak.*

Diving activity

Information on diving patterns of bowhead whales is only available for the ECWG population where whales instrumented with archival and satellite relayed data loggers have provided detailed short-term diving information or summarized dive data over a longer term (see Chapters 4 and 24).

The maximum diving depth recorded for bowhead whales is 582 m from Disko Bay but the whales rarely (<0.2% of the dives) dive below 300 m (Heide-Jørgensen et al., 2013). The maximum dive duration recorded is 48 minutes and is probably close to their aerobic dive limit (Laidre et al., 2007). Diving activity of bowhead whales in Disko Bay can be split between periods with U-shaped dives targeting a specific depth for an extended period and the occasional and deeper ranging V-shaped dives that probably serve exploratory purposes (Heide-Jørgensen et al., 2013). From estimates of the speed during the bottom phase of U-dives, Simon et al. (2009) calculated that bowhead whales spent 29% of their time on filter feeding in Disko Bay. Depending on the bathymetry and the seasonal depth of the concentration of zooplankton, the U-shaped dives can either focus on the upper part of the water column (<60 m) or target deeper waters with epibenthic plankton layers. In Disko Bay the target depth of dives changed between deep dives in winter to more shallow dives in spring, probably in response to vertical dispersal of zooplankton. A similar deep (>250 m) diving activity in winter has been observed in Hudson Strait (Matthews and Ferguson, 2015).

One indicator of foraging activity is the dive rate (number of dives per hour) where a low rate represents longer dive durations related to longer periods of filtering activity. Nielsen et al. (2015) found that Disko Bay, the east coast of Baffin Island (e.g., Isabella Bay) and the Hudson Strait appeared to be important foraging areas whereas Davis Strait, central and northern parts of Baffin Bay, and Foxe Basin were less important.

Comparison of diet among stocks

Examination of stomach contents of bowhead whales from Eastern Canada indicated feeding on benthic and epibenthic calanoid copepods and mysids while isotope and fatty acid analyses indicated feeding on more pelagic prey and larger calanoid copepods (Pomerleau et al., 2011a, 2012, 2014; Chapter 28). Stomach contents from West Greenland showed feeding on pelagic copepods like *Calanus hyperboreus* (Heide-Jørgensen et al., 2012). It has been suggested that a major part of the feeding of the adult whales takes place in the productive waters of Disko Bay (Laidre et al., 2007), but apparently the bowhead whales abandon Disko Bay and move to less productive areas before the peak density of copepods in the water column is reached. This is similar to the Bering Sea where the whales move into the Beaufort Sea before optimal feeding conditions are reached (Lowry, 1993).

Calanoid copepods are an important prey item of bowhead whales. They are aggregated near the seabed during their winter diapause. In spring they migrate upward in the water column and their abundance peaks there in summer. This likely explains the shallow (<60 m) diving activity during summer for bowhead whales in ECWG (Pomerleau et al., 2011b; Heide-Jørgensen et al., 2013; Laidre et al., 2007; Nielsen et al., 2015). As the copepods reach the upper layers (<50 m) they become more scattered as a result of variability in hydrography (e.g., wind and currents), and it may be more efficient for bowhead whales to target swarms of preascending copepods at lower depths during winter months. The mean depth of U-shaped dives during winter was 382 m and it seems unlikely that they can target epibenthic layers at greater depths (Heide-Jørgensen et al., 2013).

Both east and west of Greenland bowhead whales are frequently found over deep water (>1000 m) where the deep layer of hibernating copepods is out of reach for the whales (Hirche, 1991; Nielsen et al., 2015; de Boer et al., 2019). In these areas the whales are either feeding primarily during summer months and/or they are targeting zooplankton species that can be located closer to the surface (e.g., krill, Euphausiacea). There is little evidence of bowhead whales feeding on euphausiids in the Canadian Arctic, and west and east of Greenland (Pomerleau et al., 2012) but there is ample evidence that euphausiids is a common prey for the BCB population, which leaves the possibility that euphausiids could also be an important prey in deepwater areas of the North Atlantic.

No stomachs from bowhead whales from the Okhotsk Sea have ever been examined for diet studies. Local whalers believed that one of the major food items was the pteropod mollusk *Clione limacina* (*Clio borealis* of Lindholm, 1863), but this information is unconfirmed.

The productive Akademii Bay, one of known concentration and feeding areas for Okhotsk Sea bowhead whales, has high densities of zooplankton species (e.g., *Calanus glacialis*, *Pseudocalanus* sp., *Limacina helicina*) of which at least the *Calanus* species are known to be important prey to bowhead whales (Rogachev et al., 2008). Simultaneous zooplankton sampling and whale observations showed that the plankton biomass was significantly higher at stations located in the immediate proximity to feeding whales (Melnikov and Fedorets, 2016). Especially *C. glacialis* has been found in densities that are similar to those from the bowhead whale feeding ground in Disko Bay, West Greenland (Laidre et al., 2007). Tidal currents and estuarine circulation patterns are driving the advection of copepods into the shallow waters of Akademii Bay and its tributaries where the whales are targeting zooplankton concentrations at depths below 10 m.

Discussion

Bowhead whales share few ecological similarities with other baleen whales. This is likely a result of being restricted to a unique Arctic habitat where cold water seems to be driving the habitat selection. The exception is the OKH bowhead whales that in summer is found in temperate waters, but little is known about their migratory patterns.

Most other baleen whales show seasonal migrations between high latitude feeding grounds in summer and low latitude winter grounds. Baleen whales also often make predictable migrations linked to concentrations of prey resources whereas bowhead whale migrations are less predictable and some move from northern winter grounds to more southern summer grounds. The link between bowhead whale concentration areas and their prey is still not clear although bowhead whales target a limited selection of prey.

Compared to the two other cetaceans that are endemic to the Arctic, the beluga and the narwhal, the bowhead whale clearly has a much larger degree of plasticity in its movement patterns. For instance, whales tagged in West Greenland may move straight across Baffin Bay or instead enter summer grounds in the Canadian Arctic Archipelago through Lancaster Sound or Hudson Strait. Even though bowhead whales show the same level of seasonal site fidelity as observed for belugas and narwhals, they have much wider distributions at all seasons and even show sexual segregation within a population over distances of thousands of kilometers, something unseen for the two other Arctic whales.

It has always been assumed that the bowhead whale is year-round associated with the ice margin and that their migrations are governed by the retreat and formation of sea ice. However, large areas of the Arctic with bowhead whale habitats are ice-free during summer and fall and the seasonal changes in sea temperatures seem to have a greater impact on bowhead whale migrations than sea ice (Chambault et al., 2018). It is an intriguing question if the bowhead whale is essentially coastal, shallow water (<100 m) species that capitalizes on prey advection from deepwater and is only forced into offshore areas with the recent ocean warming or when fast ice prevents entry into coastal areas (Treacy et al., 2006).

Before the whaling era, thousands of bowhead whales were present in coastal areas of Svalbard where it has been speculated that they had a significant impact on zooplankton abundance and composition (Weslawski et al., 2000). Today with receding sea ice and warming temperatures the small EGSB population is found further offshore in more remote areas (Vacquié-Garcia et al., 2017; Storrie et al. 2018). The other extreme is the isolated Okhotsk Sea population that in summer occurs in sea temperatures above 10°C (Shpak and Paramonov, 2018) and where the whales apparently feed almost on the beach in areas where rocks and promontories create eddies that trap zooplankton advected into the shallows (Melnikov and Fedorets, 2016; Rogachev et al., 2008). In the Okhotsk Sea the affinity to beaches may also be driven by the escape response to the presence of killer whales (Fig. 5.10).

Coastal feeding is well known from the BCB population (Citta et al., 2015) and in Baffin Bay the so-called rocknosing fishery involved coastal whaling along the east coast of Baffin Island in areas where the whales were also feeding (Finley, 1990). Rocknosing whales seasonally travel close to shore, occasionally rubbing rocks along the way. Rocknosing in August in Cumberland sound may facilitate the molting process (Fortune et al., 2017). Rocknosing is also known from the Shantar region in the Okhotsk Sea but it seems to be unrelated to feeding (Shpak, unpubl. data).

Pomerleau et al. (2011b) found that bowhead whales in the Gulf of Boothia spent most of their time feeding in shallow depths <16 m. Whether benthic feeding is also part of the

FIGURE 5.10 Bowhead whale close to shore in the Okhotsk Sea. *Source: Copyright O. Shpak.*

FIGURE 5.11 Bowhead whale in the Okhotsk Sea with mud on its head, probably from feeding on benthic prey. *Source: Copyright O. Shpak.*

coastal affinity remains to be studied but gouges on sea floor, mud on the jaws and large benthopelagic fauna at whale locations are consistent with benthic feeding (Fig. 5.11; Hein and Syvitski, 1989; Andersen et al., 2014; Shpak and Paramonov, 2018). The coastal affinity in summer has, however, also been suggested to be important for molting in the warmer shallows with freshwater input and presumed protection from killer whales (Chernova et al., 2017; Fortune et al., 2017; Shpak and Paramonov, 2018).

The development of satellite tracking technology after 2000 (Heide-Jørgensen et al., 2003) has changed our understanding of bowhead whale migrations and distribution, especially for the BCB and ECWG populations, but new satellite tracking information is also being generated for the EGSB population and it will soon broaden our knowledge about this population that inhabits the most remote and inaccessible waters. The only population for which migratory routes and winter areas remain unknown is the OKH. Information on its movements seems particularly important since this is a small population that in many ways is different from bowhead whales in other areas.

While distribution and movements are well understood for some populations, our understanding of the basic ecology of bowhead whales still has major unresolved questions for all populations. Do the whales feed when they are in deep waters? Why do they leave productive areas before the zooplankton abundance peaks at the surface? Do they prefer to target zooplankton swarms at the bottom instead of in the water column? How often do they need to feed and to what extent are their migrations driven by nutritional needs? These are important questions to be solved not just to understand bowhead whales but to understand Arctic pelagic ecosystems at large.

Acknowledgments

We would like to express our gratitude to the hunters of Qeqertarsuaq that for four decades patiently have assisted with studies of whales in Disko Bay. They have in particular been instrumental in developing methods for approaching and tagging bowhead whales in West Greenland in icy winter months. In addition to the hunters also our long-term technician Mikkel Villum Jensen has been critical for the initial design of both pole-systems and the Air Rocket Transmitter System now widely used for tagging of large whales. Without their persistence and knowledge tagging of large whales would not have been developed this far. In Russia we would like to thank the National Park "Russian Arctic" and Maria Gavrilo for data collection in the Barents Sea, and Alexey Paramonov for many years of dedicated work in the Okhotsk Sea.

References

Ahonen, H., Stafford, K.M., de Steur, L., Lydersen, C., Wiig, Ø., Kovacs, K.M., 2017. The underwater soundscape in western Fram Strait: breeding ground of Spitsbergen's endangered bowhead whales. Mar. Pollut. Bull. 123, 97–112.

Allen, R.C., Keay, I., 2006. Bowhead whales in the Eastern Arctic, 1611–1911: population reconstruction with historical whaling records. Environ. Hist. 12, 89–113.

Andersen, O.G.N., Laidre, K., Heide-Jørgensen, M.P., 2014. Benthopelagic shrimp and associated pelagic benthopelagic fauna on a bowhead whale foraging ground in Disko Bay, West Greenland. Mar. Sci. 4 (1), 10–20.

Belikov, S.E., Gorbunov, Y.A., Shilnikov, V.I., 1989. Distribution of pinnipeds and cetaceans in the Soviet Arctic seas and the Bering Sea in winter. Biol. Morya 4, 33–41 (in Russian) (Беликов С. Е., Ю. А. Горбунов, В. И. Шильников. 1989. Распространение ластоногих и китообразных в морях Советской Арктики и Беринговом море в зимний период. Биология моря 4:33-41).

Boertmann, D., Nielsen, R.D., 2010. A bowhead whale calf observed in northeast Greenland waters. Polar Rec. 46, 373–375.

Boertmann, D., Merkel, F., Durinck, J., 2009. Bowhead whales in East Greenland, summers 2006–2008. Polar Biol. 32, 1805–1809.

Boertmann, D., Kyhn, L.A., Witting, L., Heide-Jørgensen, M.P., 2015. A hidden getaway for bowhead whales in the Greenland Sea. Polar Biol. 38, 1315–1319. Available from: https://doi.org/10.1007/s00300-015-1695-y.

Chaadaeva, E.V., Starkov, A.I., Bordukov, K.Y., Danilov, M.B., 2018a. Sightings of bowhead whales (*Balaena mysticetus*) in the Kara Sea. In: The 9th International Conference on "Marine Mammals of Holarctic". Astrakhan 2016. Marine Mammal Council, 2, pp. 230–235. <http://marmam.ru/upload/conf-documents/mmc2016_full.pdf>.

Chaadaeva, E.V., Voyta, L.L., Afanasyeva, G.A., Baleeva, N.V., Starkov, A.I., Danilov, M.B., 2018b. Marine mammals of the Kara Sea: summer season. In: The 9th International Conference on "Marine Mammals of Holarctic". Astrakhan 2016. Marine Mammal Council, 2, pp. 223–230. <http://marmam.ru/upload/conf-documents/mmc2016_full.pdf>.

Chambault, P., Albertsen, C.M., Patterson, T., Hansen, R.G., Tervo, O., Laidre, K.L., et al., 2018. Sea surface temperature predicts the movements of an Arctic cetacean: the bowhead whale. Sci. Rep. 8, 9658. Available from: https://doi.org/10.1038/s41598-018-27966-1.

Chernova, O.F., Shpak, O.V., Kiladze, A.B., Rozhnov, V.V., 2017. Epidermal molting in the bowhead whale *Balaena mysticetus*. Biol. Bull. 44 (6), 591–602. Available from: https://doi.org/10.1134/S1062359017050065.

Citta, J.J., Quakenbush, L.T., Okkonen, S.R., Druckenmiller, M.L., Maslowski, W., Clement-Kinney, J., et al., 2015. Ecological characteristics of core-use areas used by Bering-Chukchi-Beaufort (BCB) bowhead whales, 2006–2012. Progress. Oceanogr. 136, 201–222. Available from: http://dx.doi.org/10.1016/j.pocean.2014.08.012.

de Boer, M.N., Janinhoff, N., Nijs, G., Verdaat, H., 2019. Encouraging encounters: unusual aggregations of bowhead whales *Balaena mysticetus* in the western Fram Strait. Endang. Species Res. 39, 51–62. Available from: https://doi.org/10.3354/esr00948.

Doniol-Valcroze, T., Gosselin, J.-F., Pike, D., Lawson, J., Asselin, N., Hedges, K., et al., 2015. Abundance estimate of the Eastern Canada-West Greenland bowhead whale population based on the 2013 High Arctic Cetacean Survey. DFO Can. Sci. Advis. Sec. Res. Doc. 2015/058. v + 27 p.

Doroshenko, N.V., 1996. Polar whales of the Sea of Okhotsk. Izv. Tikhookean. Nauchno–Issled. Inst. Rybn. Khoz. Okeanogr. 121, 14–25 (in Russian).

Eschricht, F., Reinhardt, J., 1866. The Greenland right-whale with especial reference to its geographical distribution and migrations in times past and present, and to its external and internal characteristics. Ray Soc. Lond. 192, 1–49.

Fedoseev, G.A., 1984. Whale encounters in the ice massifs of the Sea of Okhotsk. Ekologiya (3), 81–83 (in Russian).

Ferguson, S.H., Dueck, L., Loseto, L.L., Luque, S.P., 2010. Bowhead whale *Balaena mysticetus* seasonal selection of sea ice. Mar. Ecol. Prog. Ser. 411, 285–297.

Finley, K.J., 1990. Isabella Bay, Baffin Island: an important historical and present-day concentration area for the endangered bowhead whale (*Balaena mysticetus*) of the Eastern Canadian Arctic. Arctic 43, 137–152.

Finley, K., 2001. Natural history and conservation of the Greenland whale, or bowhead, in the Northwest Atlantic. Arctic 54, 55–76.

Fortune, S.M.E., Koski, W.R., Higdon, J.W., Trites, A.W., Baumgartner, M.F., Ferguson, S.H., 2017. Evidence of molting and the function of 'rock-nosing' behavior in bowhead whales in the eastern Canadian Arctic. PLoS One 12 (11), e0186156. Available from: https://doi.org/10.1371/journal.pone.0186156.

Gavrilo, M.V., 2015. Status of the bowhead whale *Balaena mysticetus* in the waters of Franz Josef Land Archipelago. Paper SC/66a/BRG20 Presented to the IWC Scientific Committee, May 2015, San Diego, USA, 10pp.

Gavrilo, M.V., Tretiakov, V.Y., 2008. Observation of bowhead whales (*Balaena mysticetus*) in the East-Siberian Sea during 2007 season with record-low ice cover. In: The 5th International Conference on "Marine Mammals of Holarctic", Odessa, 2008, pp. 191–193. https://marmam.ru/upload/conf-documents/mmc2008_full.pdf.

Gilg, O., Born, E.W., 2005. Recent sightings of the bowhead whale (*Balaena mysticetus*) in northeast Greenland and the Greenland Sea. Polar Biol. 28, 796–801.

Gushcherov, P.S., Tyupeleev, P.A., Blokhin, S.A., Shkarupa, M.A., Samanov, V.I., Miyashita, T., 2017. Cruise report of the cetacean sighting survey in the northern part of the Sea of Okhotsk in 2016. Paper SC/67a/ASI/03 Presented to the IWC Scientific Committee, May 2017, Bled, Slovenia (unpublished), 26pp.

Gushcherov, P.S., Tyupeleev, P.A., Shkarupa, M.A., Makrak, S.V., Samanov, V.I., Miyashita, T., 2018. Cruise report of the cetacean sighting survey in the eastern part of the Sea of Okhotsk in 2017. Paper SC/67b/ASI/17 Presented to the IWC Scientific Committee, April–May 2018, Bled, Slovenia (unpublished), 25pp.

Gushcherov, P.S., Tyupeleev, P.A., Naberezhnykh, I.A., Makrak, S.V., Samanov, V.I., Miyashita, T., 2019. Cruise report of the cetacean sighting survey in the northwest part of the Sea of Okhotsk in 2018. Paper SC/68a/ASI12 Presented to the IWC Scientific Committee, May 2019, Nairobi, Kenya (unpublished), 20pp.

Hansen, R.G., Borchers, D., Heide-Jørgensen, M.P., 2018. Summer surveys of marine mammals in the Greenland Sea and the Northeast Water and winter survey of marine mammals in the Northeast Water—preliminary report from field work in 2017 and 2018. Greenland Institute of Natural Resources.

Heide-Jørgensen, M.P., Øien, N., Kleivane, L., Jensen, M.V., 2001. A new technique for deploying satellite transmitters on baleen whales: tracking a blue whale (*Balaenoptera musculus*) in the North Atlantic. Mar. Mammal. Sci. 17, 949–954.

Heide-Jørgensen, M.P., Laidre, K.L., Wiig, Ø., Jensen, M.V., Dueck, L., Schmidt, H.C., et al., 2003. From Greenland to Canada in ten days: tracks of bowhead whales, *Balaena mysticetus*, across Baffin Bay. Arctic 56, 21–31.

Heide-Jørgensen, M.P., Laidre, K.L., Jensen, M.V., Dueck, L., Postma, L.D., 2006. Dissolving stock discreteness with satellite tracking: bowhead whales in Baffin Bay. Mar. Mamm. Sci. 22, 34–45.

Heide-Jørgensen, M.P., Laidre, K.L., Wiig, Ø., Postma, L., Dueck, L., Bachmann, L., 2010. Large scale sexual segregation of bowhead whales. Endang. Species Res. 13, 73–78. Available from: https://doi.org/10.3354/esr00315.

Heide-Jørgensen, M.P., Laidre, K.L., Quakenbush, L.T., Citta, J.J., 2011. The Northwest Passage opens for bowhead whales. Biol. Lett. Available from: https://doi.org/10.1098/rsbl.2011.0731.

Heide-Jørgensen, M.P., Garde, E., Nielsen, N.H., Andersen, O.N., 2012. Biological data from the hunt of bowhead whales in West Greenland 2009 and 2010. J. Cetacean Res. Manag. 12, 329–333.

Heide-Jørgensen, M.P., Laidre, K.L., Nielsen, N.H., Hansen, R.G., Røstad, A., 2013. Winter and spring feeding behavior of bowhead whales. Anim. Biotelem. 1–15.

Heide-Jørgensen, M.P., Sinding, M.H., Nielsen, N.H., Rosing-Asvid, A., Hansen, R.G., 2016. Large numbers of marine mammals winter in the North Water polynya. Polar Biol. 39, 1605–1614. Available from: https://doi.org/10.1007/s00300-015-1885-7.

Hein, F.J., Syvitski, P.J.M., 1989. Sea floor gouges and pits in deep fjords, Baffin Island: possible mammalian feeding traces. Goe Mar. Lett. 9, 91–94.

Higdon, J.W., Ferguson, S.H., 2010. Past, present, and future for bowhead whales (*Balaena mysticetus*) in Northwest Hudson Bay. In: Ferguson, S.H., Loseto, L.L., Mallory, M.L. (Eds.), A Little Less Arctic: Top Predators in the World's 159 Largest Northern Inland Sea, Hudson Bay. Springer, pp. 159–177. Available from: https://doi.org/10.1007/978-90-481-9121-5_8.

Hirche, H.J., 1991. Distribution of dominant calanoid copepod species in the Greenland Sea during late fall. Polar Biol. 11, 351–362.

Ivashchenko, Y., Clapham, P., 2010. Bowhead whales *Balaena mysticetus* in the Okhotsk Sea. Mammal. Rev. 40 (1), 65–89. Available from: https://doi.org/10.1111/j.1365-2907.2009.00152.x.

Kapel, F.O., 1985. A note on the net-entanglement of a bowhead whale (*Balaena mysticetus*) in northwest Greenland, November 1980. Rep. Int. Whal. Commn. 35, 377–378.

Laidre, K.L., Heide-Jørgensen, M.P., Nielsen, T.G., 2007. Role of bowhead whale as a predator in West Greenland. Mar. Ecol. Prog. Ser. 346, 285–297.

LeDuc, R.G., Dizon, A.E., Burdin, A.M., et al., 2005. Genetic analyses (mtDNA and microsatellites) of Okhotsk and Bering/Chukchi/Beaufort Seas populations of bowhead whales. J. Cetacean Res. Manage. 7 (2), 107–111.

Lindholm, O., 1863. Whales and How Tides and Currents in the Okhotsk Sea Affect Them. Translated from a Russian unidentified journal by Lydia A. Hutchinson in 1965. UC San Diego: Scripps Institution of Oceanography. <http://escholarship.org/uc/item/8j93s3pv>.

Lowry, L.F., 1993. Foods and feeding ecology. In: Burns, J.J., Montague, J.J., Cowles, C.J. (Eds.), The Bowhead Whale. Society for Marine Mammalogy, Lawrence, KS, pp. 201–238. Spec. Publ. 2.

Lydersen, C., Freitas, C., Wiig, Ø., Bachmann, L., Heide-Jørgensen, M.P., Swift, R., et al., 2012. Lost highway not forgotten: satellite tracking of a bowhead whale (*Balaena mysticetus*) from the critically endangered Spitsbergen stock. Arctic 65, 76–86.

Mate, B.R., Krutzikowsky, G.K., Winsor, M., 2000. Satellite-monitored movements of radio-tagged bowhead whales in the Beaufort and Chukchi Seas during the late-summer feeding season and fall migration. Can. J. Zool. 78, 1168–1181.

Matthews, C.D., Ferguson, S.H., 2015. Seasonal foraging behaviour of Eastern Canada-West Greenland bowhead whales: as assessment of isotopic cycles along baleen. Mar. Ecol. Prog. Ser. 522, 269–286.

McLeod, B.A., Brown, M.W., Moore, M.J., Stevens, W., Barkham, S.H., Barkham, M., et al., 2008. Bowhead whales, and not right whales, were the primary target of 16th and 17th century Basque whalers in the western North Atlantic. Arctic 61, 61–75.

Melnikov, V.V., Fedorets, Y.V., 2016. The distribution of zooplankton and bowhead whales, *Balaena mysticetus* Linnaeus, 1758, in Akademiya Bay, Sea of Okhotsk. Russ. J. Mar. Biol. 42 (3), 216–221.

Mescherskiy, I.G., Chichkina, A.N., Shpak, O.V., Rozhnov, V.V., 2014. Molecular genetic analysis of the Shantar summer group of bowhead whales (*Balaena mysticetus* L.) in the Okhotsk Sea. Russ. J. Genet. 50 (4), 395–405.

Myasnikov, V.G., Vinnikov, A.V., Ryabov, A.A., Tyupeleev, P.A., Gushcherov, P.S., Samanov, V.I., et al., 2016. Cruise report of the cetacean sighting survey in the northern part of the Sea of Okhotsk in 2015. Paper SC/66b/IA/17 Presented to the IWC Scientific Committee, June 2016, Bled, Slovenia (unpublished), 25pp.

Nielsen, N.H., Laidre, K., Larsen, R.S., Heide-Jørgensen, M.P., 2015. Identification of potential foraging areas for bowhead whales in Baffin Bay and adjacent waters. Arctic 68, 169–179.

Petrov, S.A., Isachenko, A.I., Glebova, M.A., Gavrilov, Y.G., Fedotov, S.A., Ponomartsev, N.V., et al., 2018. The results of marine mammal countins during the four expeditions in the Arctic in 2014 and 2015. In: 9th International Conference on "Marine Mammals of Holarctic". Astrakhan 2016. Marine Mammal Council, 2, pp. 91–102. <http://marmam.ru/upload/conf-documents/mmc2016_full.pdf>.

Pomerleau, C., Ferguson, S.H., Walkusz, W., 2011a. Stomach contents of bowhead whales (*Balaena mysticetus*) from four locations in the Canadian Arctic. Polar Biol. 34, 615–620. Available from: https://doi.org/10.1007/s00300-010-0914-9.

Pomerleau, C., Patterson, T.A., Luque, S., Lesage, V., Heide-Jørgensen, M.P., Dueck, L.L., et al., 2011b. Bowhead whale *Balaena mysticetus* diving and movement patterns in the eastern Canadian Arctic: implications for foraging ecology. Endang. Species Res. 15, 167–177. Available from: https://doi.org/10.3354/esr00373.

Pomerleau, C., Lesage, V., Ferguson, S.H., Winkler, G., Petersen, S., Higdon, J., 2012. Prey assemblage isotopic variability as a tool for assessing diet and the spatial distribution of bowhead whale foraging in the Canadian eastern Arctic. Mar. Ecol. Prog. Ser. 469, 161–174.

Pomerleau, C., Lesage, V., Winkler, G., Rosenberg, B., Ferguson, S.H., 2014. Contemporary diet of bowhead whales (*Balaena mysticetus*) from the Eastern Canadian Arctic inferred from fatty acid biomarkers. Arctic 67, 84–92.

Quakenbush, L.T., Citta, J.J., George, J.C., Small, R.J., Heide-Jørgensen, M.P., 2010. Fall and winter movements of bowhead whales (*Balaena mysticetus*) in the Chukchi Sea and within a potential petroleum development area. Arctic 63, 289–307.

Radchenko, V.I., Dulepova, E.P., Figurkin, A.L., Katugin, O.N., Ohshima, K., Nishioka, J., et al., 2010. Status and trends of the Sea of Okhotsk region, 2003–2008. In: McKinnell, S.M., Dagg, M.J. (Eds.), Marine Ecosystems of the North Pacific Ocean, 2003–2008. PICES Special Publication 4, pp. 268–299. 393 pp.

Reinhart, N.R., Ferguson, S.H., Koski, W.R., Higdon, J.W., LeBlanc, B., Tervo, O., et al., 2013. Occurrence of killer whale *Orcinus orca* rake marks on Eastern Canada-West Greenland bowhead whales *Balaena mysticetus*. Polar Biol. 36, 1133–1146.

Rogachev, K.A., Carmack, E.C., Foreman, M.G.G., 2008. Bowhead whales feed on plankton concentrated by estuarine and tidal currents in Academy Bay, Sea of Okhotsk. Cont. Shelf Res. 28, 1811–1826.

Scoresby, W., 1823. Journal of a Voyage to the Northern Whale-Fishery; Including Researches and Discoveries on the Eastern Coast of West Greenland, Made in the Summer of 1822, in the Ship Baffin of Liverpool. Caedmon Reprints, Whitby, UK.

Shpak, O.V., 2016. Update on studies of bowhead whales (*Balaena mysticetus*) in the Okhotsk Sea in 2014–2015. Paper SC/66b/BRG/05 presented to the IWC Scientific Committee, June 2016, Bled, Slovenia (unpublished), 4 pp.

Shpak, O.V., 2018. Okhotsk Sea bowhead whales as a live model on how climate change may affect the species in the polar regions. The Arctic Biodiversity Congress. CAFF, Arctic Council. October 9–11, 2018, Rovaniemi, Finland. <https://www.arcticbiodiversity.is/index.php/iab7-bowhead-whale-conservation-and-future-research-cooperation>.

Shpak, O.V., Stimmelmayr, R., 2017. Preliminary image analysis of acute and chronic injuries, parasites, and skin conditions in the Okhotsk Bowhead whale (*Balaena mysticetus*) stock in the western Okhotsk Sea. Paper SC/67A/E/01 presented to the IWC Scientific Committee, May 2017, Bled, Slovenia (unpublished), 15 pp.

Shpak, O.V., Paramonov, A.Y., 2018. The bowhead whale, *Balaena mysticetus* Linnaeus, 1758, in the western Sea of Okhotsk (2009–2016): distribution pattern, behavior, and threats. Russ. J. Mar. Biol. 44 (3), 210–218.

Simon, M., Johnson, M., Tyack, P., Madsen, P.T., 2009. Behaviour and kinematics of continuous ram filtration in bowhead whales (*Balaena mysticetus*). Proc. R. Soc. B. Available from: https://doi.org/10.1098/rspb.2009.1135.

Southwell, T., 1898. The migration of the right whale (*Balaena mysticetus*). Nat. Sci. 12, 397–414.

Stafford, K.M., Moore, S.E., Berchok, C.L., Wiig, Ø., Lydersen, C., Hansen, E., et al., 2012. Spitsbergen's endangered bowhead whales sing through the polar night. Endang. Species Res. 18, 95–103.

Storrie, L., Lydersen, C., Andersen, M., Wynn, R.B., Kovacs, K.M., 2018. Determining the species assemblage and habitat use of cetaceans in the Svalbard Archipelago, based on observations from 2002 to 2014. Polar Res 37, 2018. Available from: https://doi.org/10.1080/17518369.2018.1463065.

Tervo, O., 2011. Acoustic Behaviour of Bowhead Whales *Balaena mysticetus* in Disko Bay, Western Greenland (Ph. D. thesis). Arctic Station Natural History Museum of Denmark. The PhD School of Science, Faculty of Science, University of Copenhagen, Denmark.

Tervo, O.M., Parks, S.E., Miller, L.A., 2009. Seasonal changes in the vocal behavior of bowhead whales (*Balaena mysticetus*) in Disko Bay, Western-Greenland. J. Acoust. Soc. Am. 126, 1570–1580.

Townsend, C.H., 1935. The distribution of certain whales as shown by logbook records of American whaleships. Zoologica 19 (1), 1–50. with charts.

Treacy, S.D., Gleason, J.S., Cowles, C.J., 2006. Offshore distances of bowhead whales (*Balaena mysticetus*) observed during fall in the Beaufort Sea, 1982–2000. An alternative interpretation. Arctic 59, 83–90.

Vacquié-Garcia, J., Lydersen, C., Marques, T.A., Aars, J., Ahonen, H., Skern-Mauritzen, M., et al., 2017. Late summer distribution and abundance of ice-associated whales in the Norwegian High Arctic. Endang. Species Res. 32, 59–70. Available from: https://doi.org/10.3354/esr00791.

Weslawski, J.M., Hacquebord, L., Stempniewicz, L., Malinga, M., 2000. Greenland whales and walruses in the Svalbard food web before and after exploitation. Oceanologia 2, 37–56.

CHAPTER

6

Abundance

Geof H. Givens[1] *and Mads Peter Heide-Jørgensen*[2]

[1]Givens Statistical Solutions LLC, Fort Collins, CO, United States [2]Greenland Institute of Natural Resources, Copenhagen, Denmark

Introduction

Four distinct populations (stocks) of bowhead whales are recognized, as described in Chapter 3. All bowhead whale stocks were severely depleted from commercial whaling during the last several centuries (Chapter 33). All bowhead stocks were at severe risk of extirpation, and commercial hunting ceased in part because bowheads became too scarce as markets for baleen and whale oil collapsed. Severe depletion led to initial protection from commercial whaling through the 1931 League of Nations Convention, continued in 1946 with the International Convention for the Regulation of Whaling (Montague, 1993). Since then, the bowhead story is—with one exception—a hopeful one: most stocks seem to be recovering and two of them are now quite abundant. The abundance of whale stocks is usually estimated from aerial or ship-based line-transect surveys, shore-based counts of migrating populations (Fig. 6.1), or photographic, genetic, or close-kin capture–recapture analyses, although in some cases only opportunistic sightings are available. For bowhead whales worldwide, all of these types of information have been important.

The Bering-Chukchi-Beaufort Seas stock

Woodby and Botkin (1993) reviewed early estimates of Bering-Chukchi-Beaufort Seas (BCB) stock abundance prior to the start of commercial whaling in 1848 (see their Table 10.2). These estimates ranged from 6100 to 47,000, and the authors presented their own estimate of 10,400–23,000. More recent, population model-based estimates include those of Raftery et al. (1995) (15,394–22,951), and Brandon and Wade (2006) (9466–28,475). Givens et al. (2015) elected not to use catch data from the first few decades of commercial whaling and, consequently, found that the time series of abundance estimates (1978–2011,

The Bowhead Whale
DOI: https://doi.org/10.1016/B978-0-12-818969-6.00006-6

FIGURE 6.1 Bowhead whales of the Bering-Chukchi-Beaufort stock migrate past Utqiaġvik (Barrow) in Spring by swimming through the "lead," an open water channel roughly parallel to the coast in the frozen ocean. Researchers can observe, locate, and count the whales from ice perches (shown here) for the duration of the migration, approximately 2 months. Such data allow for relatively accurate estimates of the stock size. *Source: Photo by Kate Stafford (2011).*

Table 6.1) alone contained little evidence of a recent inflection in the growth trajectory or an upper bound for carrying capacity.

The BCB bowhead whale was listed as endangered under the Endangered Species Act in 1973 and was believed to be severely depleted when the IWC briefly declined to establish a subsistence hunting quota in 1977. The lack of abundance estimates together with an increased number of strikes and landings were both important factors in this decision (Chapters 32 and 38). Subsequently, improved surveys were guided by Iñupiat hunters' knowledge that bowheads could migrate without being visually detected, either under the sea ice or beyond visual range from the ice edge (Albert, 2001). A series of abundance estimates since 1977 has shown that the population numbered in the thousands in the late 1970s and has grown robustly in the past several decades; hence early concerns about stock depletion or even extinction were probably unfounded.

The BCB stock is the most intensively studied population of bowhead whales. From 1978 to 2001, 11 successful ice-based surveys were conducted. In each year, observers stood on elevated perches at the edge of the shore fast ice near the lead, a channel of water near Utqiaġvik (formerly Barrow), Alaska, through which the whales migrate (Fig. 6.1). The observers sighted whales as they migrated northeast during April and May. Only whales within 4 km of the perch could be reliably detected visually. In some years, independent observers operated a second perch; from such data it is possible to estimate detection probabilities, that is, the probability of sighting a bowhead given that it is present within 4 km. These detection probabilities are used to correct raw counts of sightings to

TABLE 6.1 Abundance estimates for the Bering-Chukchi-Beaufort Seas stock of bowhead whales.

Year	Abundance	CV	Reference
1978	4765	0.305	Zeh and Punt (2005)
1980	3885	0.343	Zeh and Punt (2005)
1981	4467	0.273	Zeh and Punt (2005)
1982	7395	0.281	Zeh and Punt (2005)
1983	6573	0.345	Zeh and Punt (2005)
1985	5762	0.253	Zeh and Punt (2005)
1986	8917	0.215	Zeh and Punt (2005)
1987	5298	0.327	Zeh and Punt (2005)
1988	6928	0.12	Zeh and Punt (2005)
1993	8167	0.071	Zeh and Punt (2005)
2001	10,545	0.128	Zeh and Punt (2005)
2004	12,631	0.244	Koski et al. (2010)
2011	16,820	0.052	Givens et al. (2016)

Notes: The estimates from Zeh and Punt (2005) are correlated and differ slightly from originally published estimates. The estimates in 2004 and 2011 are independent of each other and all prior estimates.

estimate the total number of whales that passed within 4 km during the survey period, denoted N_4. In addition, a correction for periods of missed visual watch is also made. The estimate of total abundance is then N_4/P_4 where P_4 is the estimated proportion of bowheads that swim within 4 km of the perch. In some years, P_4 has been estimated from acoustic locations derived from an array of underwater recording devices. In other years, aerial surveys were used to estimate P_4. For years when neither was available, a weighted average value from other years was used. More technical details about survey protocols and analysis are provided by Givens et al. (2016) and the references therein.

Zeh and Punt (2005) summarized these surveys, noted that the abundance estimates are correlated because they use shared information about P_4, and proposed a statistical model to consolidate all the available information. The result is a revised set of abundance estimates that are slightly modified from the originally published values due to the statistical model applied, along with estimates of variance and correlation (Table 6.1).

Koski et al. (2010) provided an estimate of abundance from an entirely independent source: photo-identification data collected (from planes) during the spring migrations in 2003, 2004, and 2005. Using a capture–recapture analysis, they estimated the 2004 population size to be 12,631 (CV = 0.24), excluding calves.

The best, current estimate (2011) of BCB bowhead abundance is 16,820 whales with a 95% confidence interval of (15,176, 18,643) and a CV of 0.052 (Table 6.2), resulting from an ice-based survey completed in 2011 (Givens et al., 2016). Furthermore, these results indicate that the stock has grown at an estimated annual rate of 3.7% (2.9%–4.6%) and nearly quadrupled in size since surveys began in 1978 (Table 6.1 and Fig. 6.2).

TABLE 6.2 Estimated current stock sizes and historical precommercial whaling stock size estimates for all stocks.

Stock	Abundance	CV	95% CI	Year	Current abundance reference	IUCN status	Precommercial whaling abundance ($\times 10^3$); reference
BCB	16,820	0.052	15,176–18,643	2011	Givens et al. (2016)	Least concern	10.4–23.0; Woodby and Botkin (1993); 13.85, Brandon and Wade (2006)
OKS	218	0.22		2016	Cooke et al. (2017)	Endangered	3.0–6.5, Woodby and Botkin (1993)
ECWG	6446	–	3876–10,721	2013	Doniol-Valcroze et al. (2015)	Least concern	11.0, Woodby and Botkin (1993), Rugh et al. (2003)
EGSB	318	–	110–956	2017	Hansen et al. (2018)	Endangered	25.0, Woodby and Botkin (1993); 52.5, Allen and Keay (2006)

Notes: The EGSB abundance estimate pertains to an unknown portion of the overall population and does not represent an estimate of total EGSB bowhead abundance.

Givens et al. (2017) applied an untested capture–recapture analysis technique to photo-id data from three decades to obtain a considerably larger 2011 abundance estimate; however, this has not yet been published. In 2019 two new surveys were conducted. An ice-based survey (with no simultaneous estimation of P_4) was mostly successful, and a corresponding abundance estimate will be attempted. An aerial line transect survey of a huge portion of the Beaufort Sea was also flown. This effort was groundbreaking in its scope and ambition. Estimates of 2019 bowhead abundance from both surveys should be forthcoming in the next several years.

The East Canada-West Greenland stock

In 1977 the IWC adopted a two-stock model for bowhead whales that occupy eastern Canadian and western Greenlandic waters (IWC, 1978). The designation of two stocks was based on the summer distribution of the whales and it was considered the most conservative approach for management. The stocks came to be identified as the Baffin Bay-Davis Strait (BB-DS) stock and the Hudson Bay-Foxe Basin (HB-FB) stock and were treated as separate populations (Mitchell and Reeves, 1981; Cosens et al., 1997; IWC, 1999; Cosens and Innes, 2000; Finley, 1990, 2001) until Heide-Jørgensen et al. (2006) suggested that, based on satellite telemetry, bowhead whales summering in eastern Canadian Arctic and wintering in West Greenland comprise a single stock. This is followed in this volume and tentatively endorsed by IWC (2008).

There are no estimates of the preexploitation population size of the East Canada-West Greenland (ECWG) bowhead whale stock. The most intensive exploitation occurred during the period of commercial whaling (1530–1915) where it is estimated that about 70,000

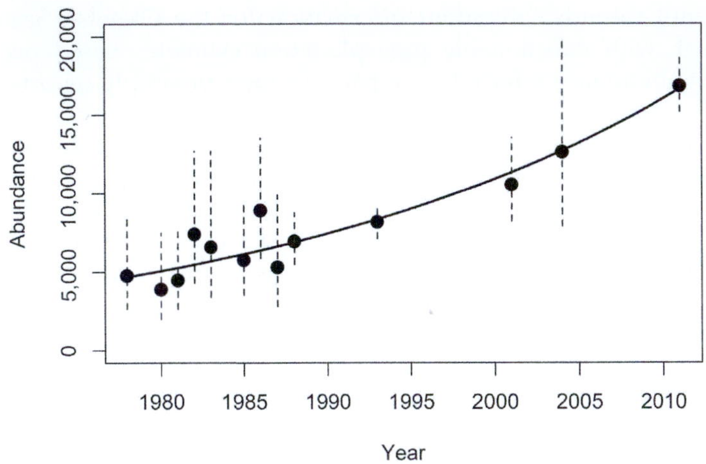

FIGURE 6.2 Estimated abundance and trend of BCB bowhead whales, 1978–2011. *Dashed lines* are 95% confidence intervals.

whales were taken (Higdon, 2010). Considering the magnitude of the accumulated historical catches the pristine population must have been in the tens of thousands.

Local observations and surveys covering parts of the population range showed low numbers of bowhead whales (Reeves and Heide-Jørgensen, 1996; Koski et al., 2006) until around 2000. Then, in 2006, a surprising larger abundance was found in Disko Bay, one of the wintering grounds for bowhead whales in Baffin Bay (Heide-Jørgensen et al., 2007). Also, a cetacean survey covering large parts of the Canadian Arctic Archipelago conducted in 2002 provided an abundance estimate of the summer population of bowhead whales of 6344 whales (95% CI: 3119–12,906), which was accepted by IWC in 2009 (Givens et al., 2009). That estimate was expected to be negatively biased because the full range of the population was not covered in the same year. Later surveys in West Greenland and Canada have confirmed the higher overall abundance (Rekdal et al., 2014; Doniol-Valcroze et al., 2015). The most recent abundance estimate (Table 6.2) for a specific year is 6,446 from an aerial survey in 2013 (95% CI 3,876–10,721) (Doniol-Valcroze et al. 2015). Frasier et al. (2020) used samples collected across 19 years (1995–2013) to apply genetic mark-recapture methods to estimate abundance during the sampling period to be 11,747 (95% CI 8169–20,043). Although the genetic approach was more geographically inclusive than the aerial survey of Doniol-Valcroze et al. (2015), it is difficult to quantify how the use of a broad time period for genetic sampling affects the abundance estimate, particularly if the population has grown during that period. Currently, this stock is listed as "least concern" in the IUCN Red List due to the relatively large abundance estimates that have been obtained in the past two decades (Cooke and Reeves, 2018).

The Okhotsk Sea stock

The Okhotsk Sea (OKS) bowhead stock is listed in the IUCN Red List (Cooke et al., 2018) as endangered and in the Red Data Book of the Russian Federation (Danilov-Danilian, 2001) as Category 1 (endangered). Until recently, abundant information for the Okhotsk Sea stock has been mostly anecdotal, with questionable preexploitation estimates based on incomplete data and more recent abundance inferred from partial counts or simply reports of sightings.

Ivashchenko and Clapham (2010) reviewed a variety of estimates of preexploitation abundance. Mitchell (1977) estimated that Okhotsk Sea bowheads once numbered at least 6600 whales, in order to have sustained the estimated historical catch during 1848–57 when the stock was heavily whaled. Other guesses range from less than half (Berzin and Doroshenko, 1981) to triple (Lindholm, 1888) this number. All these estimates are highly uncertain due to issues of data quality and questionable assumptions.

A comprehensive table of sightings from surveys and other sources from 1967 to 2004 is provided by Ivashchenko and Clapham (2010). Observations range from a few animals to many dozens. Sightings are also reported by Moore and Reeves (1993), Rugh et al. (2003), Meschersky et al. (2014), Melnikov and Fedorets (2016), and Shpak and Paramonov (2018).

Ivashchenko and Clapham (2010) also reviewed some early abundance estimates that were in the low hundreds. Like those authors, we consider these results to be of questionable reliability since they were not "derived from a statistical analysis of data from any systematically designed survey" (Ivashchenko and Clapham, 2010, p. 72) and/or have a huge variance estimate.

The most recent and probably more reliable abundance estimate is given by Cooke et al. (2017). The authors used genetic capture–recapture data from 184 whales obtained during 1995–96, 1999–2000, and 2011–16, from Ulbanskiy Bay in the western Okhotsk Sea near the Shantar Archipelago, with some additional sampling in nearby Konstantina and Udskaya Bays. J. G. Cooke (pers. comm.) indicates that his analysis replaces that of Shpak et al. (2017) which has somewhat higher stock sizes due to the misinterpretation of the "superpopulation abundance" parameter as an actual abundance. Cooke et al. (2017) incorporated a simple population growth model, to estimate a population decline from 643 (CV 0.66) whales in 1995 to 218 (CV 0.22) in 2016 (Table 6.2). A constant population size model yielded an estimate of 258 (CV 0.20) individuals, although this model fits the data slightly less well. The description of the statistical approach used in these analyses is incomplete.

Considering that bowheads have been sighted elsewhere in the Okhotsk Sea during the same (and different) months as in the dataset used in this analysis (Ivashchenko and Clapham, 2010; Cooke et al., 2017), the estimates by Cooke et al. (2017) may represent abundance of a localized feeding aggregation. If so, they underestimate the total number of bowheads in the Okhotsk Sea. Regardless, it is clear that this stock remains severely depleted from prewhaling levels with no clear sign of recovery.

The East Greenland-Svalbard-Barents Sea stock

The East Greenland-Svalbard-Barents Sea (EGSB) stock of bowhead whales was among the first to be exploited commercially and the first bowhead whale population to be depleted. The exploitation was focused on the west coast of Svalbard and the Greenland

Sea and it is estimated that 120,507 whales were killed and that the preexploitation population size was 52,500 whales (Allen and Keay, 2006). The modern range of the EGSB Stock is not fully known but it presumably extends from East Greenland, around Svalbard, east to Frantz Josef Land and Severnaya Zemlya in the Russian Arctic, involving three countries: Denmark/Greenland, Norway, and Russia (Chapter 5).

It has been suggested that fewer than 1000 bowhead whales were left in this stock when the whaling stopped in the Greenland Sea in 1911 (Allen and Keay, 2006). Only 46 sightings of bowhead whales have been reported between 1950 and 2009 in their former center of distribution around Svalbard (Wiig et al., 2010). Dedicated searches for bowhead whales in 2006 failed to detect whales, both visually and acoustically, in their traditional habitat west of Svalbard but did detect bowheads further west in the Fram Strait (Wiig et al., 2007). An increasing number of observations have been made in the cooler and more heavily ice infested waters of East Greenland (Gilg and Born, 2005; Boertmann et al., 2009) and at Franz Josef Land after 1990 (Wiig, 1991; de Korte and Belikov, 1994).

The first sign that the EGSB stock is recovering came from an aerial survey in 2009 of the polynya at the northeastern corner of Greenland (the so-called Northeast Water, see Stirling, 1997; Fig. 5.6, this volume). A fully corrected estimate of 102 whales (95% CI: 32–329) was developed from a systematic aerial survey in an area that is rarely visited and had ice conditions impenetrable for the commercial whaling operations in historical times (Boertman et al., 2015). The next indication that the main distribution of bowhead whales has changed came from a survey in 2015 in the pack ice north of Svalbard where a fully corrected abundance of 343 whales (95% CI: 136 – 862) was obtained, again in a rarely visited area (Vacquié-Garcia et al., 2017). Opportunistic sightings from a cruise ship in the Greenland Sea provided a record of 84 whales in June 2015 in another restricted area centrally in the Greenland Sea (de Boer et al., 2019). Repeated surveys in the Northeast Water in East Greenland gave new abundance estimates of 301 whales (95% CI: 127–769) in March–April 2017 and 318 individuals (95% CI: 110–956) in August–September of 2017 (Hansen et al., 2018). It is important to emphasize that these are estimates for specific and limited regions. The wide and partially unknown distribution far north in inaccessible areas makes it difficult to survey the entire potential range of the EGSB stock and no complete abundance estimates exist.

The new and relatively large abundance estimates from parts of the range of the EGSB stock indicate both that the whales have shifted their distribution from the areas where the commercial whalers found them and that the stock may be recovering, although the stock remains endangered in the IUCN listing (Cooke and Reeves, 2018).

References

Albert, T.F., 2001. The influence of Harry Brower, Sr., an Inupiaq Eskimo Hunter, on the bowhead whale research program conducted at the UIC-NARL faculty by the North Slope Borough. In: Norton, D. (Ed.), Fifty More Years Below Zero: Tributes and Mediations for the Naval Arctic Research Laboratory's First Half Century at Barrow, Alaska. University of Alaska Fairbankst, Fairbanks, AK, pp. 265–278.

Allen, R.C., Keay, I., 2006. Bowhead whales in the Eastern Arctic, 1611–1911: population reconstruction with historical whaling records. Environ. Hist. 12, 89–113.

Berzin, A.A., Doroshenko, N.V., 1981. Right whales of the Okhotsk Sea. Rep. Int. Whal. Comm. 32, 451–455.

Boertmann, D., Merkel, F., Durinck, J., 2009. Bowhead whales in East Greenland, summers 2006–2008. Polar Biol. 32, 1805–1809.

Boertman, D., Kyhn, L.A., Witting, L., Heide-Jørgensen, M.P., 2015. A hidden getaway for bowhead whales in the Greenland Sea. Polar Biol. 38, 1315–1319.

Brandon, J., Wade, P., 2006. Assessment of the Bering-Chukchi-Beaufort Seas stock of bowhead whales using Bayesian model averaging. J. Cetacean Res. Manage. 8, 225–239.

Cooke, J.G., Brownell Jr., R.L., Shpak, O.V., 2018. *Balaena mysticetus* Okhotsk Sea subpopulation. The IUCN Red List of Threatened Species 2018: e.T2469A50345920. <http://www.iucnredlist.org/species/2469/50345920> (Accessed 21 February 2019).

Cooke, J., Reeves, R., 2018. *Balaena mysticetus* East Greenland-Svalbard-Barents Sea subpopulation. The IUCN Red List of Threatened Species 2018: e.T2472A50348144. <https://doi.org/10.2305/IUCN.UK.2018-1.RLTS.T2472A50348144.en> (Accessed 25 May 2019).

Cooke, J.G., Shpak, O.V., Meschersky, I.G., Burdin, A.M., MacLean, S.A., Chichkina, A.N., et al., 2017. Updated estimates of population and trend for Okhotsk Sea bowhead whales. Paper SC/67a/NH10 presented to the Scientific Committee of the International Whaling Commission. <http://www.iwc.int> (Accessed 4 September 2019).

Cosens, S.E., Innes, S., 2000. Distribution and numbers of bowhead whales (*Balaena mysticetus*) in northwestern Hudson Bay in August 1995. Arctic 53, 36–41.

Cosens, S.E., Qamukaq, T., Parker, B., Dueck, L.P., Anardjuak, B., 1997. The distribution and numbers of bowhead whales, *Balaena mysticetus*, in northern Foxe Basin in 1994. Can. Field Nat. 111, 381–388.

Danilov-Danilian, V.I., 2001. Red Data Book of the Russian Federation: Animals. AST & Astrel Publisher, St. Petersburg, Russia (in Russian).

de Boer, M.N., Janinhoff, N., Nijs, G., Verdaat, H., 2019. Encouraging encounters: unusual aggregations of bowhead whales *Balaena mysticetus* in the western Fram Strait. Endanger. Species Res. 39, 51–62.

de Korte, J., Belikov, S.E., 1994. Observations of Greenland whales (*Balaena mysticetus*), Zemlya Frantsa-Iosifa. Polar Rec. 30, 135–136.

Doniol-Valcroze, T., Gosselin, J.-F., Pike, D., Lawson, J., Asselin, N., Hedges, K., et al. (2015). Abundance estimate of the Eastern Canada – West Greenland bowhead whale population based on the 2013 High Arctic Cetacean Survey. DFO Can. Sci. Advis. Sec. Res. Doc. 2015/058.

Finley, K.J., 1990. Isabella Bay, Baffin Island: an important historical and present-day concentration area for the endangered bowhead whale (*Balaena mysticetus*) of the Eastern Canadian Arctic. Arctic 43, 137–152.

Finley, K.J., 2001. Natural history and conservation of the Greenland whale, or bowhead, in the Northwest Atlantic. Arctic 54, 55–76.

Frasier, T.R., Petersen, S.D., Postma, L., Johnson, L., Heide-Jørgensen, M.P., Ferguson, S.H., 2020. Abundance estimation from genetic mark-recapture data when not all sites are sampled: an example with the bowhead whale. Glob. Ecol. Conserv. 22, e00903.

Gilg, O., Born, E.W., 2005. Recent sightings of the bowhead whale (*Balaena mysticetus*) in Northeast Greenland and the Greenland Sea. Polar Biol. 28, 796–801.

Givens, G.H., George, J.C., Suydam, R., 2015. A population dynamics model and assessment of Bering-Chukchi-Beaufort Seas bowhead whales. Paper SC/66a/BRG9 presented to the Scientific Committee of the International Whaling Commission. <http://www.iwc.int> (Accessed 4 September 2019).

Givens, G.H., Edmondson, S.L., George, J.C., Suydam, R., Charif, R.A., Rahaman, A., et al., 2016. Horvitz-Thompson whale abundance estimation adjusting for uncertain recapture, temporal availability variation, and intermittent effort. Environmetrics 26, 1–16.

Givens, G., Koski, B., da Silva, V., Dueck, L., Witting, L., Heide-Jørgensen, M.P., et al., 2009. Report of the working group on abundance estimates for Eastern Canada West Greenland bowhead whales. Annex F (Appendix 3). Rep. Int. Whal. Comm. 11, 188–190 (Suppl.).

Givens, G.H., Mocklin, J.A., Vate Brattström, L., Tudor, B.J., Koski, W.R., George, J.C., et al. (2017). Survival rate and 2011 abundance of Bering-Chukchi-Beaufort Seas bowhead whales from photo-identification data over three decades. Paper SC/67a/AWMP09 presented to the Scientific Committee of the International Whaling Commission. <http://www.iwc.int> (Accessed 4 September 2019).

Hansen, R.G., Borchers, D., Heide-Jørgensen, M.P., 2018. Summer surveys of marine mammals in the Greenland Sea and the Northeast Water and winter survey of marine mammals in the Northeast Water—preliminary report from field work in 2017 and 2018. Greenland Institute of Natural Resources.

Heide-Jørgensen, M.P., Laidre, K.L., Jensen, M.V., Dueck, L., Postma, L.D., 2006. Dissolving stock discreteness with satellite tracking: Bowhead whales in Baffin Bay. Mar. Mammal. Sci. 22, 34–45.

Heide-Jørgensen, M.P., Laidre, K., Borchers, D., Samarra, F., Stern, H., 2007. Increasing abundance of bowhead whales in West Greenland. Biol. Lett. 3, 577–580.

Higdon, J.W., 2010. Commercial and subsistence harvests of bowhead whales (*Balaena mysticetus*) in eastern Canada and West Greenland. J. Cetacean Res. Manage. 11 (2), 185–216.

Ivashchenko, Y., Clapham, P., 2010. Bowhead whales *Balaena mysticetus* in the Okhotsk Sea. Mammal. Rev. 40, 65–89.

IWC, 1978. Report of the scientific committee. Rep. Int. Whal. Comm. 28, 38–92.

IWC, 1999. Report of the scientific committee. J. Cetacean Res. Manage. 1 (Suppl.), 1–284.

IWC, 2008. Report of the scientific committee. J. Cetacean Res. Manage. 10 (Suppl.), 1–74.

Koski, W.R., Heide-Jørgensen, M.P., Laidre, K.L., 2006. Winter abundance of bowhead whales, *Balaena mysticetus*, in the Hudson Strait, March 1981. J. Cetacean Res. Manage. 8 (2), 139–144.

Koski, W.R., Zeh, J., Mocklin, J., Davis, A.R., Rugh, D.J., George, J.C., et al., 2010. Abundance of Bering-Chukchi-Beaufort bowhead whales (*Balaena mysticetus*) in 2004 estimated from photo-identification data. J. Cetacean Res. Manage. 11 (2), 89–99.

Lindholm, O.V., 1888. Whaling. Russkoe sudokhodstvo 34, 11–28 (in Russian).

Melnikov, V.V., Fedorets, Y.V., 2016. The distribution of zooplankton and bowhead whales, *Balaena mysticetus* Linnaeus, 1758, in Akademiya Bay, Sea of Okhotsk. Russ. J. Mar. Biol. 42, 216–221.

Meschersky, I.G., Chichkina, A.N., Shpak, O.V., Rozhnov, V.V., 2014. Molecular genetic analysis of the Shantar summer group of bowhead whales (*Balaena mysticetus* L.) in the Okhotsk Sea. Russ. J. Genet. 50, 395–405.

Mitchell, E., 1977. Initial population size of bowhead whale (*Balaena mysticetus*) stocks: cumulative catch estimates. Paper SC/56/RMP1 presented to the Scientific Committee of the International Whaling Commission. <http://www.iwc.int> (Accessed 4 September 2019).

Mitchell, E., Reeves, R.R., 1981. Catch history and cumulative catch estimates of initial population size of cetaceans in the eastern Canadian Arctic. Rep. Int. Whal. Comm. 31, 645–682.

Montague, J.J., 1993. Introduction. In: Burns, J.J., Montague, J.J., Cowles, C.J. (Eds.), The Bowhead Whale. Society for Marine Mammalogy, Lawrence, KS, Special Publication No. 2, pp. 1–17.

Moore, S.E., Reeves, R.R., 1993. Distribution and movement. In: Burns, J.J., Montague, J.J., Cowles, C.J. (Eds.), The Bowhead Whale. Society for Marine Mammalogy, Lawrence, KS, Special Publication No. 2, pp. 313–386.

Raftery, A.E., Givens, G.H., Zeh, J.E., 1995. Inference from a deterministic population dynamics model for bowhead whales. J. Am. Stat. Assoc. 90, 402–416.

Rekdal, S.L., Hansen, R.G., Borchers, D., Bachmann, L., Laidre, K.L., Wiig, Ø., et al., 2014. Trends in bowhead whales in West Greenland; Aerial surveys vs. genetic capture-recapture. Mar. Mammal. Sci. 31 (1), 133–154.

Reeves, R.R., Heide-Jørgensen, M.P., 1996. Recent status of bowhead whales, *Balaena mysticetus*, in the wintering grounds off West Greenland. Polar Res. 15 (2), 115–125.

Rugh, D., DeMaster, D., Rooney, A., Breiwick, J., Shelden, K., Moore, S., 2003. A review of bowhead whale (*Balaena mysticetus*) stock identity. J. Cetacean Res. Manage. 5, 267–279.

Shpak, O.V., Meschersky, I.G., Kuznetsova, D.M., Chichkina, A.N., Paramonov, A.Y., Rozhnov, V.V., 2017. Abundance estimate of Okhotsk Sea population of the bowhead whale (*Balaena mysticetus* Linnaeus, 1758). Doklady Biol. Sci. 477, 236–238.

Shpak, O.V., Paramonov, A.Y., 2018. The bowhead whale, *Balaena mysticetus* Linnaeus, 1758, in the western Sea of Okhotsk (2009–2016): distribution pattern, behavior, and threats. Russ. J. Mar. Biol. 44, 210–218.

Stirling, I., 1997. The importance of polynyas, ice edges, and leads to marine mammals and birds. J. Mar. Syst. 10 (1–4), 9–21.

Vacquié-Garcia, J., Lydersen, C., Marques, T.A., Aars, J., Ahonen, H., Skern-Mauritzen, M., et al., 2017. Late summer distribution and abundance of ice-associated whales in the Norwegian High Arctic. Endanger. Species Res. 32, 59–70.

Wiig, Ø., 1991. Seven bowhead whales (*Balaena mysticetus* L.) observed at Franz Josef Land in 1990. Mar. Mammal. Sci. 7, 316–319.

Wiig, Ø., Bachmann, L., Janik, V.M., Kovacs, K.M., Lydersen, C., 2007. Spitsbergen bowhead whales revisited. Mar. Mammal. Sci. 23, 688–693.

Wiig, Ø., Bachmann, L., Øien, N., Kovacs, K.M., Lydersen, C., 2010. Observations of bowhead whales (*Balaena mysticetus*) in the Svalbard area 1940–2009. Polar Biol. 33, 979–984.

Woodby, D.A., Botkin, D.B., 1993. Stock sizes prior to commercial whaling. In: Burns, J.J., Montague, J.J., Cowles, C.J. (Eds.), The Bowhead Whale. Society for Marine Mammalogy, Lawrence, KS, Special Publication No. 2, pp. 387–407.

Zeh, J.E., Punt, A.E., 2005. Updated 1978–2001 abundance estimates and their correlations for the Bering-Chukchi-Beaufort Seas stock of bowhead whales. J. Cetacean Res. Manage. 7, 169–175.

CHAPTER

7

Life history, growth, and form

J.C. George[1], J.G.M. Thewissen[2], A. Von Duyke[1], Greg A. Breed[3], Robert Suydam[1], Todd L. Sformo[1], Brian T. Person[1] and H. K. Brower Jr.[4]

[1]Department of Wildlife Management, North Slope Borough, Utqiaġvik, AK, United States [2]Department of Anatomy and Neurobiology, Northeast Ohio Medical University, Rootstown, OH, United States [3]Institute of Arctic Biology, University of Alaska, Fairbanks, Fairbanks, AK, United States [4]North Slope Borough, Mayors Office, Utqiaġvik, AK, United States

Introduction

Bowhead whales (*Balaena mysticetus*) have many unusual and extreme traits that set them apart from all other baleen whales (mysticetes). These specializations include exceptionally long (>4 m) and numerous baleen plates, latest age of sexual maturity, the longest life span, low core body temperatures, a strong association with sea ice, and extreme blubber thickness. The proportionally huge head (one-third body length) has cranial features that allow it to stow the massive baleen rack as well as break through ice to breathe (Finley, 2001; Thewissen et al., 2009; George et al., 2016, Chapter 13). Three of the four bowhead stocks spend their entire lives in sub-Arctic and Arctic seas, collectively giving them a near-continuous circumpolar distribution. The extreme environmental conditions they evolved in likely imposed strong selection pressure leading to the bowhead's highly specialized behavioral, ecological, and physiological characteristics, as well as a life history strategy that is unique among whales and mammals.

This chapter is divided into two sections: (1) growth and form and (2) life history. In growth and form, we describe the general morphology of the bowhead whale and how it changes (allometry) over the life of the animal. In life history, we discuss the life of a typical bowhead whale from birth through its approximately 200-year life span, describing its ecology, and articulating a set of hypotheses that relate the ecological and evolutionary aspects of their life history strategy to their unusual specializations. The details from this chapter are largely based on studies of the Bering–Chukchi–Beaufort (BCB) stock of

The Bowhead Whale
DOI: https://doi.org/10.1016/B978-0-12-818969-6.00007-8

bowhead whales, but are broadly applicable to all stocks (see Chapter 3 for a discussion of stocks).

Growth and form

General description

The bowhead whale is the only extant species in the Genus *Balaena*, and is closely related to right whales (*Eubalaena*) (Chapter 3). They are extremely large mysticetes that live in the sub-Arctic and Arctic seas. By mass, only blue whales are greater. Bowheads are characterized by a highly arched upper jaw (or rostrum), extremely long and generally black baleen plates, black skin, and a rotund profile (Fig. 7.1). As adults, bowheads have proportionally larger heads than all other cetaceans and longer life spans than any other mammal (Eschricht and Reinhardt, 1866; Tomilin, 1957; George et al., 1999). The flippers are relatively large and paddle-like. Their flukes are broad, triangular, taper distally to a narrow blunt apex, and are divided by a broad medial notch.

FIGURE 7.1 A bowhead whale breaches in the Arctic waters near Utqiaġvik. Such aerial displays are common in humpback whales but occur only occasionally in bowheads. The white eyepatch only develops in older individuals, whereas the white chin patch is present at birth and does not change with age. *Source: Photo by Kate Stafford.*

The head is very large, with dorsal nares (i.e., blow holes) that are set at the most superior elevation of the arched rostrum. The lower lips form a dorsally convex arch that encloses the baleen rack and seats with the ventral portion of the rostrum when the mouth is closed. The eye is located just ventral to the lateral midline above the commissure of the mouth gape and approximately a third of the distance from the anterior tip of the rostrum to the fluke notch.

The skin is black, except for a number of white patches. Some white patches develop and expand with age, specifically, around the eye, on the tail stock (peduncle), and flukes (Figs. 7.2 and 7.3), whereas the chin patch is present at birth and only grows in proportion to the animal. Immature whales, even some 10 m in length, may show some small (mm-scale) white areas around the eye. These coalesce to give a gray appearance (peppering) as the whales age. When bowheads reach 13.5 m (45 ft) in length (approximately around sexual maturity), nearly all show some patchy white area around the eye. By 14 m, most whales have continuous white around the eye (Figs. 7.2 and 7.3). In addition to these congenital and age-related white patches, wounds almost always scar bright white, accumulate with age, and remain visible for the life of the animal (Chapters 29 and 36). Unlike the closely related right whales (*Eubalaena* spp.), bowheads lack callosities on the face and lower jaw.

Loss of pigmentation on the peduncle develops with age on the lateral surface. Similar to the areas around the eyes, whales may show a white speckling on the peduncle at 10–11 m, and by a length of ~14.5–15.0 m, essentially all bowheads have obvious white on the peduncle. With advanced age, the peduncle patch may spread to include most, and sometimes the entire surface of the flukes (Fig. 7.2C).

The chin patch (Fig. 7.2) covers the lateral and ventral surfaces of the face. The size varies, but averages about 60 cm on the lateral surface. On the ventral surface, the patch can be longer and reach 4.5 m in a few individuals. About five percent of BCB bowheads lack a chin patch entirely (Fig. 7.2D). On some rare individuals (about 2% of the population), an additional spectacular white patch occurs on the belly. Both the chin and belly patches expand as the animal grows in proportion to size, but it is not known if the pattern changes as they age. As a comparison, white-belly patches occur on about 32% of carefully examined North Atlantic right whales (Hamilton, personal communication, 2019).

Bowhead growth phases

Bowhead growth and development is characterized by several phases. In this regard, bowheads differ from most mammals, including other cetaceans, in that these age cohorts show marked allometric or proportional differences in body shape and morphology. Our current understanding of these age-related differences comes from the examination of bowheads captured for subsistence in northern Alaska by Iñupiat and Yupik hunters, who long ago recognized these distinct age-related morphotypes (Fig. 7.4). Subsistence bowhead hunts in Alaska typically occur in the spring (April–June), fall (August–October), and occasionally winter at Saint Lawrence Island (Chapter 32). Hence, our understanding of morphological change that occurs in the intervening time periods is limited. The transition between morphotypes is gradual and varies among individuals. Though indigenous

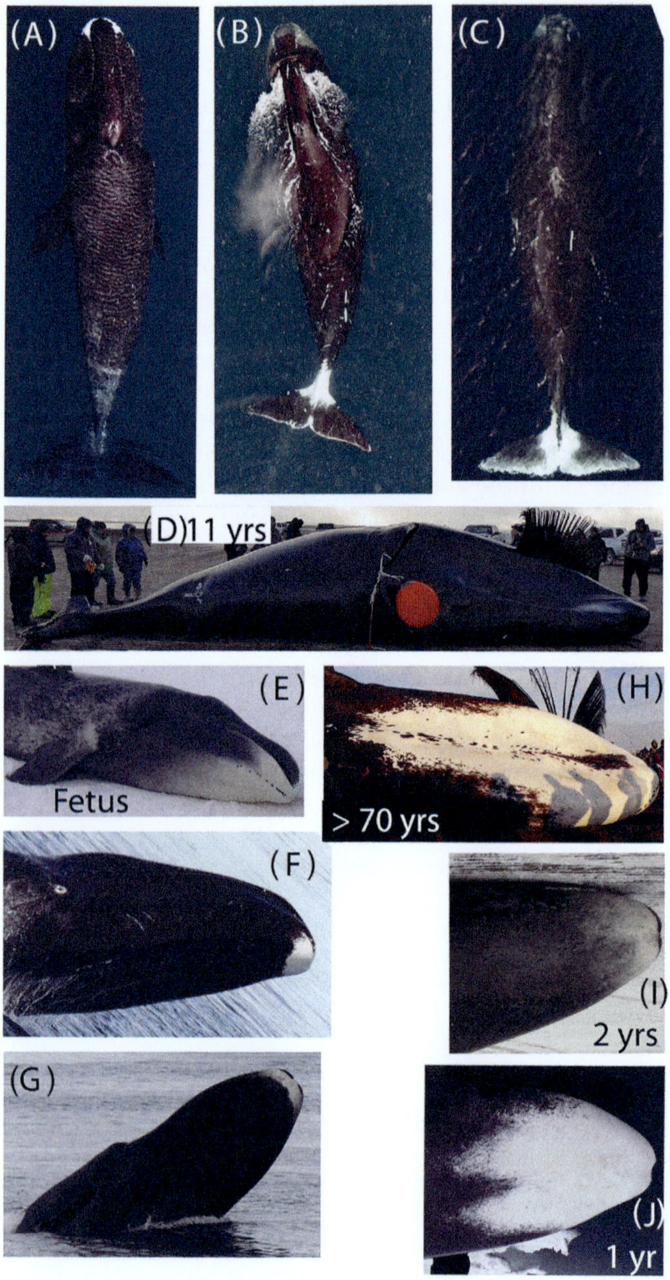

FIGURE 7.2 (A–C) The coloration of peduncle, fluke, and eye in bowhead whales is initially black (A), and turns white with increasing age (B and C), as seen here arranged from A to C. Note also the increasing number of "ice" scars on the back. (D–J) White coloration on chin is variable among individuals; the chin the patch expands proportionally as whale grows in length (D and H, oblique ventral view; E and F, lateral view; G, I, and J, ventral view of chin). D shows a completely black individual. A and C by Vicki Beaver; B by Brenda Rone; D, NSB-DWM 2018B12, by J.G.M. Thewissen; E, NSB-DWM 2007B9F; F–G two images of a whale during the same breach by Kate Stafford, this whale is an adult as evidenced by the white eyepatch; H, NSB-DWM 2013B1, by J.G.M. Thewissen; I, NSB-DWM 2015B13; J, NSB-DWM 2016B7. Images not to scale. A–C, NOAA/ North Slope Borough, remainder North Slope Borough, NMFS Permit No. 14245.

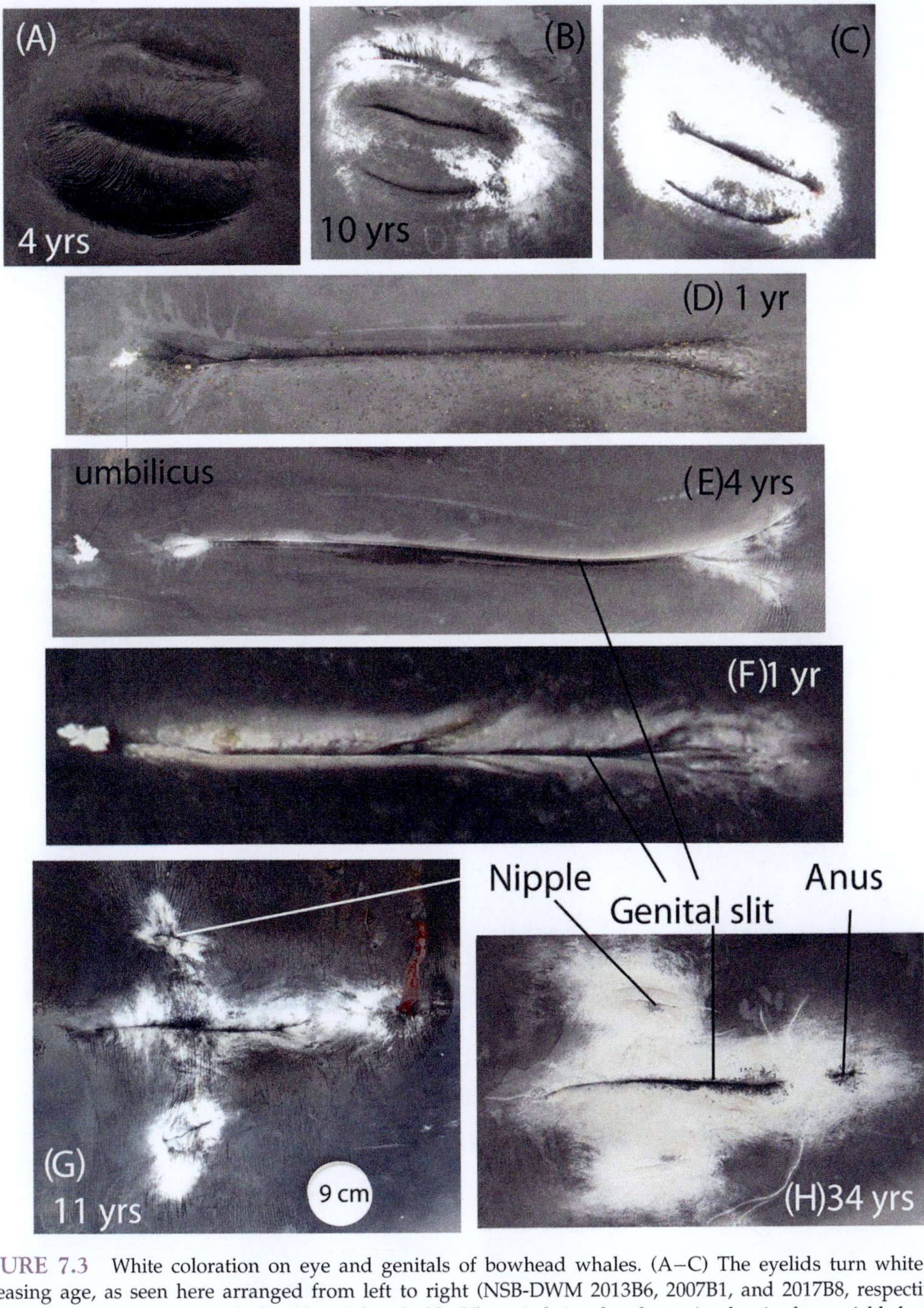

FIGURE 7.3 White coloration on eye and genitals of bowhead whales. (A–C) The eyelids turn white with increasing age, as seen here arranged from left to right (NSB-DWM 2013B6, 2007B1, and 2017B8, respectively). (D–H) White coloration on male (D–F) and female (G–H) genitals is related to aging but is also variable between individuals (NSB-DWM 2018B6, 2015B23, 2018B6, 2018B12, and 2015B24, respectively). Anterior to left in D-H. White disc in G has a diameter of 9 cm. A, C, and E by L. Pierce, G by J.G.M. Thewissen, NMFS Permit No. 14245.

FIGURE 7.4 Four stages in the life of the bowhead whale. (A) Large sexually mature female. Note the numerous scars, white around the eye and peduncle, which is typical of large old adults. (B) A typical yearling or *ingutuq*; note the rotund body and relatively small head. (C) A bowhead in the first decade of life, a body type referred to as *qairaliq*. The *ingutuq* and *qairaliq* can be similar in body length, but differ significantly in the relative size of the head, baleen length, and body profile. (D–E) A newborn calf. All drawings are to the same scale, except E, which is an enlarged version of D. *Source: Illustration by Uko Gorter.*

hunters recognize finer scale gradations between successive growth phases, our description will focus upon the most basic developmental morphotypes. Below, we discuss the growth and form of these developmental morphotypes and their Iñupiat names.

Age: birth to year 1 (neonate or *aġvaaq*)

Neonates, referred to as *aġvaaq* by the Iñupiat (MacLean, 2014), are relatively thin at birth but rapidly increase in girth with nursing, which continues over the next 6–12 months (Figs. 7.4–7.6). The skin is mottled gray. The chin patch is present but is less distinct than for yearlings and adults. The epidermis is very thick (4.5 cm) in neonates and is a distinguishing characteristic recognized by Iñupiat whale hunters.

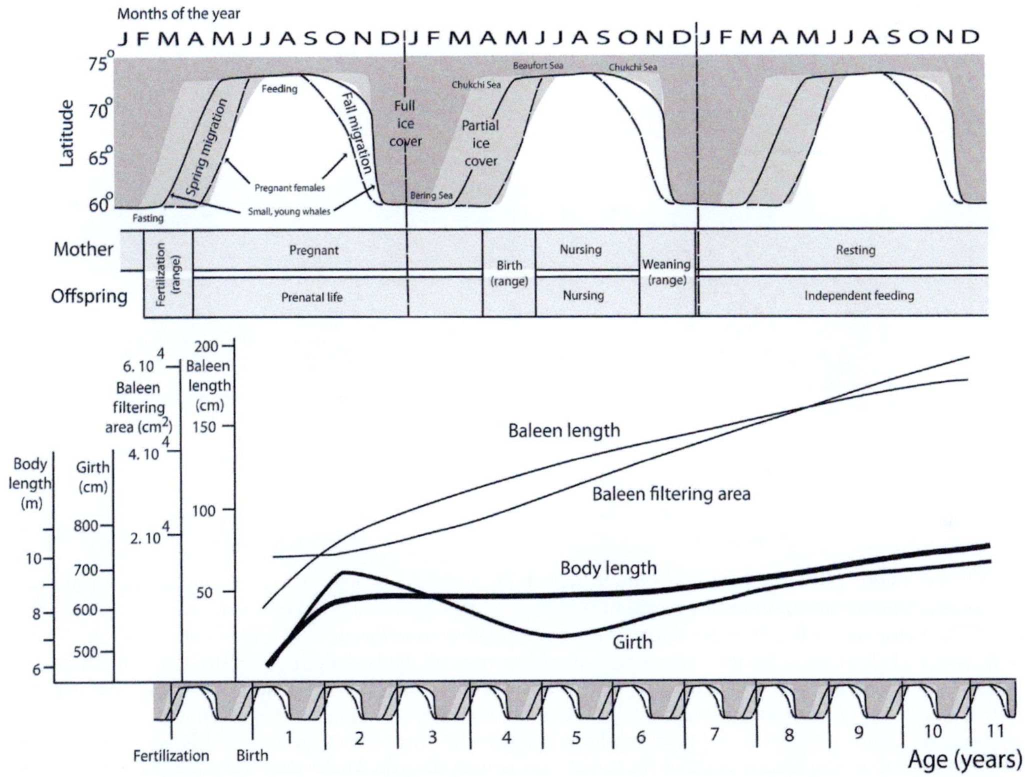

FIGURE 7.5 Generalized figure showing changes in body length, girth, and baleen length for whales age 0–11 years. Calendar at top shows migration, feeding pattern, ice cover, and reproductive cycle for females and calves over a 3-year period. Note the hiatus in growth in body length and the drop in girth (and body mass) following weaning at age 1 while baleen continues to increase. Long baleen is thought to be necessary to enable efficient feeding and is thus a "prerequisite for body length growth" in bowhead whales (Lubetkin et al., 2012). *Source: From George, J.C., Stimmelmayr, R., Suydam, R., Usip, S., Givens, G., Sformo, T., 2016. Severe bone loss as part of the life history strategy of bowhead whales. PLoS One 11 (6), e0156753. Available from: https://doi.org/10.1371/journal.pone.0156753.*

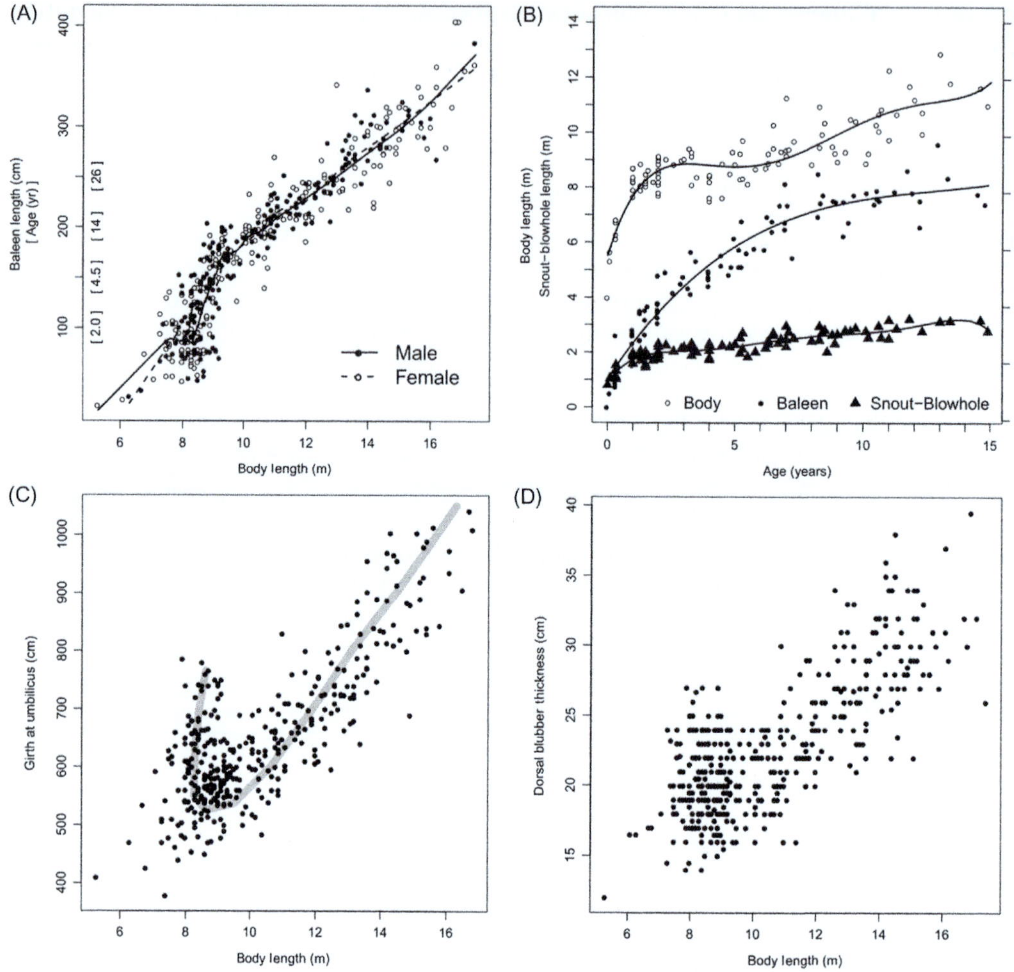

FIGURE 7.6 Four plots documenting growth with body length and age. (A) Baleen length as a function of body length. Note the strong inflection at about 8 to 9 m, which shows emphasis on baleen growth in the first 4 years of life. Lines were fit to the data using the super-smoother function in R (R Core Team, 2014). Males have slightly shorter baleen for a given body length than females. (B) Body length, snout-to-blowhole length, and (greatest) baleen length for whales age 0–15. Note the hiatus in growth in body length following weaning at age 1 while baleen length continues to increase. (C) Girth at the umbilicus and the axilla both show a "J" shaped function with body length whereby *ingutuqs* or yearlings are quite fat from recent nursing and have the highest girth to body length ratios. (D) Dorsal blubber thickness versus body length. There were no statistical differences by sex for the girths measurements of nonpregnant whales.

Several lines of evidence from surveys, postmortem exams, and indigenous knowledge (IK) indicate that parturition peaks in mid-May (see Chapters 8 and 13). Near-term fetuses have only been recovered during the spring hunt, suggesting that conception is strongly seasonal and controlled (Reese et al., 2001). Standard length (i.e., tip of rostrum to fluke notch) measurements of term fetuses harvested in May at Utqiaġvik ranged from 390 to 440 cm (mean = 413, SD = 15.7, n = 10) and probably represent individuals close to birth. Consistent with direct measurements of term fetuses, aerial photogrammetric estimates of newborn's vary between 360 and 450 cm (12.7–14.7 ft; Koski et al., 1993). Durham (1980) reported the length of two neonates harvested in the 1970s at Barrow at 450 and 460 cm. Another one, harvested at Utqiaġvik in 1998, was unusual in being much larger (533 cm) and swimming alone at the time of capture. Together, these data suggest the length at birth between 360 and 530, with most about 420 cm or 13.4 ft in length (Durham, 1980; Koski et al., 1993, NSB data). Their weight is approximately 1,000 kg.

At birth, the bowhead skull is approximately 27% of its body length (Fig. 9.5, this volume; Fig. 7.7). Short baleen plates are visibly erupted and are between 5 and 15 cm in length. The baleen plates are marked by a deviation in the direction of growth, which is visible as a "neonatal notch" in the lateral side of the plates that remains visible for several years. Over their first year of life, baleen grows relatively quickly but slows markedly in subsequent years (see below; Lubetkin et al., 2008).

Bowhead whales nurse for at least 6 months while staying close to their mother, although some nurse longer, similar to other baleen whales (Costa and Williams, 1999; Chapter 13). Growth is rapid during this period. Fat reserves also increase markedly in the first 6–9 months of life. In fall calves, which are 4–5 months old, dorsal blubber thickness (measured 1 m behind the blowhole) is about 13 cm, and after 1 year, blubber makes up approximately 50% of their body mass. Fall-harvested calves average 6.7 m in body length (n = 10), their longest baleen plate averages 37 cm, and the umbilical circumference is about 500 cm. By the end of the first year, the whales are approximately 10,000–12,000 kg.

Calves begin weaning in late autumn with some swimming more independently from their mother by that time based on observations from hunters and aerial surveys. Except for nursing, feeding is probably limited in winter during their first 6–12 months of life. While most weaned individuals may not feed substantially until they reach their summer feeding grounds, a few yearlings (age 1) harvested in spring have had food in their stomach and a couple have had milk suggesting they had recently nursed (NSB data; see Chapter 28). Relatively few spring yearlings remain closely associated with their (presumed) mother during the migration to their summer feeding grounds in the Eastern Beaufort Sea for BCB bowheads, based on aerial observations (Koski et al., 1993).

Age: 1–2 years (*ingutuq*)

The term *ingutuq* is used by Inuit peoples across the Arctic to describe young, fat whales with short baleen. These features indeed characterize bowheads at weaning, and this morphology is retained for approximately 1–2 years. The rich milk and rapid fat accumulation result in a stout, rotund body profile (Figs. 7.4 and 7.6). In the first year, bowheads grow to an average body length of 833 cm (n = 23; SD = 50.1; range = 745–919 cm), or about 1.1 cm/day. They gain roughly 1000 kg in weight per month.

The baleen length of an *ingutuq* averages 77 cm (n = 34). The growth rate of baleen is fastest in the first year of life, growing about 60 cm (Lubetkin et al., 2008). *Ingutuqs* of the BCB stock often migrate before the older, reproductively mature whales. This behavioral

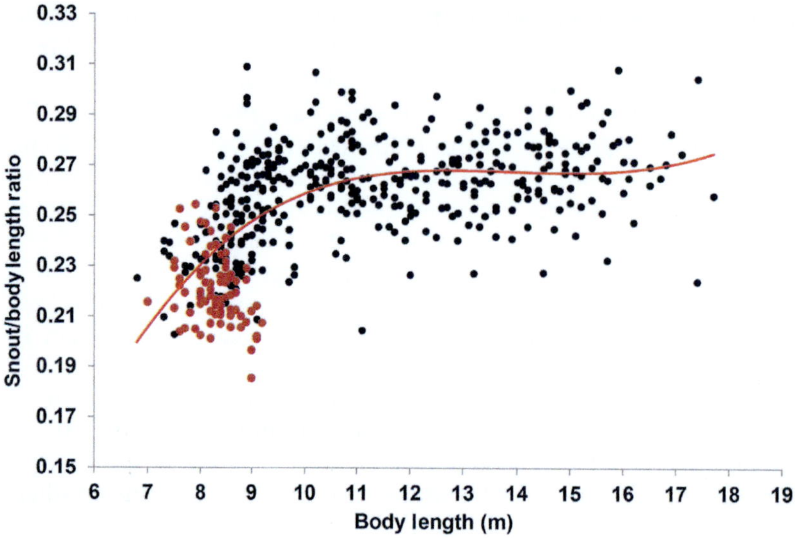

FIGURE 7.7 The relative rostrum size (snout-to-blowhole/body length) plotted against body length. Note the rapid increase in the size of the rostrum relative to body length for whales ~7–9 m long. *Ingutuqs* (yearlings) have relatively small heads (*red dots*). For whales over 10 m, the relative size of the head increases slightly relative to body length throughout its life. A polynomial fit is shown.

difference, combined with the allometric differences, led some hunters and biologists to initially suggest that they represented a different species (Braham et al., 1980; Rooney et al., 2002). Commercial whalers also recognized these growth patterns in both bowhead and right whales, calling *ingutuqs* "suckers" or "short-heads" (Scoresby, 1820; Best and Schell, 1996).

George and Suydam (2014) provide a detailed description of the morphological differences between calves and *ingutuqs*. The body length of *ingutuqs* ranges between 6.8 and 9.2 m (mean = 8.1 m; SD = 0.58; $n = 35$), which overlaps somewhat with calves, and greatly overlaps with the next age class, the *qairaliq*. The length of the longest baleen plates differs between calves (range 29–52 cm) and *ingutuqs* (range 74–105 cm), and clearly distinguishes *ingutuqs* from older whales (George and Suydam, 2014; Fig. 7.6). The rostrum length, as described by the snout-to-blowhole index, is about 22% of body length for *ingutuqs* (Fig. 7.7). The condylobasal skull length of the *ingutuq* is relatively small at about 31% of the body length, as compared to large adults, whose skull can reach 37% of their length (see Chapter 9).

Bone density of *ingutuqs*, especially their ribs, is greater than all other age classes (calves, *qairaliq's*, or adults; George et al., 2016), and fat makes up a larger percentage of body weight. Denser bone is thought to store minerals that are mobilized for later growth and to offset the buoyancy of the large fat reserves (George et al., 2016).

Age: ~2–6 years (*qairiliq*)

After weaning, bowheads lose body mass and girth for several years. The *ingutuq* morphotype transitions into the *qairaliq* type over several years while not growing in body length (Figs. 7.4–7.6). Iñupiat hunters use the term *qairaliq* to refer to the distinctive thin "tapered" bowhead whales that have a larger head and longer baleen than an *ingutuq*. The lateral profile of *ingutuqs* is convex behind the head, whereas it is concave in all older age classes. The *qairaliq* can be similar in body length to *ingutuqs*, but differs morphologically and is older. The *qairaliq* body type may develop at different ages in different individuals. While designated here at 2–6 years, the onset and duration are variable.

The Yankee Whalers had some awareness of these morphotypes, and some recognized larger whales as the *qairaliq* type or "dry skins" whose blubber yielded relatively little oil in some cases (Bockstoce, 1986). Yankee whalers also referred to *qairaliq* as "stunts" (Scoresby, 1820), a reference to their emaciated appearance. The implied inference, that growth was stunted, is correct (George et al., 2016).

During the postweaning stage, bowheads do not grow appreciably in length (George et al., 1999; Lubetkin et al., 2008; George, 2009; Figs. 7.6 and 7.7), but their proportions change. It appears that their torso length changes little while the head keeps growing to become an increasingly larger fraction of body length. Both the rostrum and baleen grow disproportionately faster as a function of body length for the *qairaliq* as compared to neonates and *ingutuqs*. This growth pattern accommodates rapid baleen growth in younger animals (Figs. 7.5 and 7.6). The net growth of the baleen (growth minus wear) of a whale aged 4–6 years is about 20–24 cm/year. The baleen growth rate slows as the years advance to about 15.6 and 17.3 cm/year for adult males and females, respectively (Lubetkin et al., 2008; Chapter 21).

This proportionally larger baleen rack allows more efficient feeding per unit mass. We suspect this is of considerable importance for these arctic whales given they are often feeding in areas with relatively low prey densities compared to most other whale species (Thomson, 1987, 2002; Lubetkin et al., 2012; George et al., 2016; Chapter 14). At the same time, girth decreases for the first years of the *qairaliq* stage and rib bone mineral content drops as whales mobilize stored fat and minerals to increase the size of the baleen rack and head (Chapter 34).

The key event in the postweaning stage is the hiatus in growth in body length. We believe this is due to the relatively small and inefficient baleen rack at the beginning of this phase (Lubetkin et al., 2008). A similar, but less dramatic, growth hiatus occurs in right whales (Best and Schell, 1996), and Ohmura et al. (1969) reported that rapid baleen growth follows weaning in right whales, and continues until the animals are about 13.5 m in length. With their markedly different feeding ecology, the baleen of gray whales (*Eschrichtius robustus*), by contrast, is short and can reach adult size within a year or two, consistent with the apparent lack of a developmental growth hiatus in this species (Rice and Wolman, 1971).

Age: ~6–25 years (sexually immature subadults)

At this stage, most bowheads transition away from the thinner *qairaliq* stage to a more rotund body shape typical of most bowheads. Our data suggest that most bowheads begin to increase in body length, girth, and body mass by 5 or 6 years. Hence, we use this age to indicate the beginning of the subadult stage. Subadults are characterized by slow growth, increasing in body length by about 20–40 cm/year, eventually slowing to ~10 cm/year for individuals >12 m (Koski et al., 1992; George et al., 1999). As the whale's head becomes larger relative to body length, it can accommodate baleen plates of greater length (Chapter 9). This larger rack of baleen facilitates more efficient foraging, which fuels the increase in body length in subadult bowheads. As such, an increase in blubber thickness and bone mineral content also typifies this phase. Meanwhile, older juveniles begin to accumulate wounds that result in white scars that remain visible for the life of the animal (Chapters 29 and 36).

Age: ~26 to 200+ years (sexually mature adults)

Both sexes reach sexual maturity in their mid-20s at about 13.0–13.5 m body length (Chapter 13). After maturity, growth continues up to about 19 m in body length. Various Iñupiat terms apply to the largest whales such as *tiptalaayuk*, which translates approximately as "a 60+ foot whale that makes a fluttering noise when it exhales" (MacLean, 2014; D. Edwardsen, NSB IHLC, 2020). Obtaining precise body lengths for the largest bowheads is difficult because most whales are measured when they are out of the water, yet large whales often cannot be hauled ashore in one piece, and stretching likely occurs in the process (George et al., 2004a). Historical Yankee whaling records include an astonishingly large bowhead 80.5 ft (24.5 m) taken c.1850, but it is impossible to validate (Bockstoce and Burns, 1993).

The length at age for large whales can vary considerably. For example, whale NSB-DWM 1995WW5, a 14.60 m male, was estimated at over 200 years of age, while other whales ~17 m were estimated to be just 40–50 years (Fig. 7.8; Rosa et al., 2013; Wetzel et al., 2017; George et al., 1999). Growth slow's markedly at about 50 years. However, it is not clear when physical maturity occurs or even if growth ceases. There is evidence that, like other large whales, growth is indeterminate and they grow throughout life (George et al., 1999; Bryden, 1972). In sexually mature individuals, the skull continues growing and can reach 37% of body length in the oldest individuals (Chapter 9).

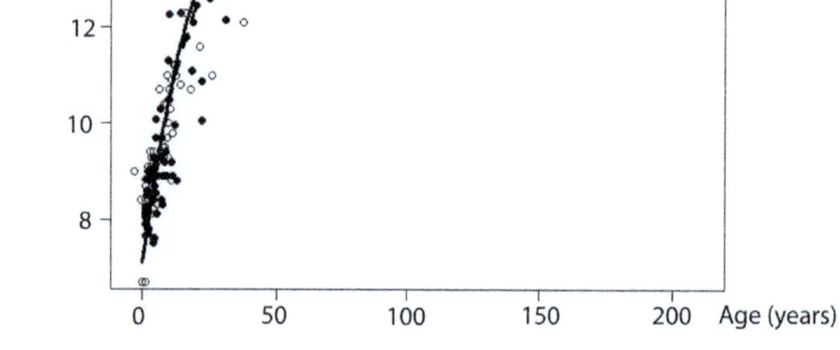

FIGURE 7.8 Plot of body length against age for bowhead whales. Ages were estimated using a variety of methods (Chapter 21). Von Bertalanffy growth models are fit to all age and length data, not taking into account the growth hiatus of immature whales. *Open dots* and *dashed line* are males, *closed dots* and *solid line* are females.

Female bowheads grow longer than males as in other mysticetes (see Fig. 7.8). The longest standard body lengths (tip of rostrum to fluke notch) of whales we measured were an 18.7 m female harvested at Gambell (NSB-DWM 2019G2) and three females harvested and measured by hunters at Utqiaġvik, Wainwright, and Point Hope were 19.2 m in length; however, it is not known exactly how they were measured. The longest male was 17.4 m (NSB-DWM 1995B9), though it may have been slightly shorter, since it was measured to its chin which extends past the maximum extent of the rostrum. The longest "standard" measurement for a male in our NSB records is 16.6 m (NSB-DWM 2007B13). Among whales harvested since 1972, the ten longest were all females. Of the 120 harvested whales over 16 m in length, 96 (80%) were female and 24 were male.

Interestingly, while the longest harvested whales reach 19 m, the greatest length of a living bowhead measured using aerial photogrammetry is 17.1 m (Bill Koski, personal communication, NSB-LGL-NMFS photogrammetry database). This discrepancy, along with some measurements made before and after hauling, suggests that some stretching (up to 9%) occurs when whales are hauled ashore (George et al., 2004a). While this may not entirely explain the 2 m difference, stretching should be considered when comparing landed whale lengths to those estimated from living individuals using aerial photogrammetry.

Specific morphological characteristics

Baleen

Bowheads have, by far, the longest baleen of any whale, with over 600 plates suspended from the rostrum, each having fringe "hairs" along the lingual edge (see Chapter 14). Baleen length is positively correlated with body length. However, the relationship between baleen and body length differs for whales above and below about 9 m in length (see Fig. 7.6). The longest confirmed baleen measured was 409 cm long, with fringe hairs reaching 50 + cm (Chapter 14, NSB-DWM data; George, 2009). Although longer baleen (4.87 m) has been recorded in Yankee whaler logbooks, such records are difficult to confirm (Bockstoce and Burns, 1993). No significant sex differences in baleen length have been found, although the data suggest that baleen length increases slightly faster with body length for males (Figs. 7.6 and 7.8). A few bowheads have baleen plates that are cream white with black edges, and even fewer individuals have some white plates due to gum injury. Overall, about 25% of examined bowheads were scored as having subtle but detectable "white streaks" in the baleen (NSB data).

Blubber

The term blubber is only loosely defined in the older cetacean literature as the thick layer of fat beneath the skin of sea mammals. More strict histological study in the bowhead has demonstrated that the extensive adipose layer generally referred to as "blubber" is actually a fat-filled expansion of the collagenous dermal layer of the skin, exclusive of the thick, pigmented epidermis and the variable hypodermis beneath it (Haldiman and Tarpley, 1993). However, blubber thickness, as we report it here, includes both the dermis and black epidermis or *maktak*. We note that bowhead blubber thickness is somewhat difficult to precisely measure due to its great thickness and minimal rigidity. A fatty hypodermis of variable thickness and less connective tissue support may be present, depending on body location and bowhead stage of development; care must be taken to distinguish it from the overlying dermis (Mau, 2004; Willetto et al., 2002; Fig. 19.1, this volume).

Bowhead blubber is considerably thicker than most other cetaceans, although the blubber of North Pacific right whales (NPRW) is nearly as thick. Two ~14 m NPRW females both had a blubber thickness of 23 cm, while the blubber thickness of 14 m female bowheads averages about 27–28 cm at the same anatomical location (ventral midline near the umbilicus). The largest right whales have blubber to ~38 cm (Ohmura et al., 1969), matching the thickest blubber measurements in our BCB database. Blubber thickness tends to be proportionally thicker for the younger morphotypes (i.e., *ingutuq*) but scales with body length for older whales (Fig. 7.6).

Dorsal blubber thickness (1 m behind of the blowhole) reaches a maximum of 38.5 cm in the largest females (NSB data). Slightly greater blubber thicknesses may occur on other parts of the body. Pregnant females tend to have somewhat thicker blubber, but the difference is marginal. Blubber thicknesses reported for BCB bowheads are consistent with those for similar-sized bowheads harvested in Greenland (Heide-Jørgensen et al., 2012). The average dorsal blubber thickness is slightly thicker than similarly positioned ventral blubber ($\overline{x}_{\text{dorsal}} = 22.6$ cm, SD 5.06, $n = 61$; $\overline{x}_{\text{ventral}} = 20.9$, SD = 4.10, $n = 61$). The proportion of blubber and muscle mass to total body mass is reported in Chapter 16.

The blubber of whales harvested in spring is slightly thicker, but the difference is not significant (George et al., 2015). However, girth is significantly greater in autumn (see "Body length/girth relationships" section). The increase in girth is driven by increases in hypodermal thickness, visceral fat, and possibly increases in muscle mass (Chapter 16). This seasonal pattern is consistent with the findings of Rice and Wolman (1971) for Eastern Pacific gray whales, who also noted that girth changed seasonally while blubber thickness did not.

Body length/girth relationships

Girth at the axilla, umbilicus, and anus shows a "J-shaped" pattern with body length (Fig. 7.6). Yearlings and some young whales carry a large amount of maternal fat from nursing and have a very high girth to body length ratio (mean = 0.79). Axillary girths for some exceptionally rotund *ingutuq* whales were nearly equal to their body length (Fig. 7.6). After individuals grow longer than 10 m, girth increases linearly with body length.

As a comparison, Woodward et al. (2006) defined a "maximum girth statistic" for right whales as (Girthmax/body length). Although methods were somewhat different, the mean statistic for right whales of 0.686 (SE = 0.036) was virtually identical to bowheads 0.680 (SE = 0.004) when all bowhead age groups are pooled. However, *ingutuqs* had a much higher ratio (0.79) than right whales or older bowheads.

Peduncle girth is also highly correlated with body length (Table 7.1; $R^2 = 0.90$; George, 2009) and to some extent age (Archer et al., 2010; Chapter 21). The relationship does not differ by sex. The peduncle increases in diameter through life, possibly increasing

TABLE 7.1 Linear regression model results for selected morphometric characters as a function of body length (BL) and SEX (=1 if male; =0 if female). Body length in meters, all other measurements in cm.

Morphometric character	Linear regression model	R^2Adj	*df*	Sex diff?
Fluke width	−36.09 + 37.31 • (BL) − 30.53 • (SEX) + 3.10 • (BL • SEX)	0.92	315	Yes
Peduncle circumference	23.78 + 10.0 • (BL) − 2.35 • (SEX) + 0.24 • (BL • SEX)	0.9	476	No
Snout-to-blowhole all	−52.18 + 30.49 • (BL) − 18.30 • (SEX) + 1.78 • (BL • SEX)	0.94	503	Yes
Snout-to-blowhole ≤11.5 m	−114.85 + 37.48 • (BL)	0.85	305	No
Snout-to-blowhole >11.5 m	−58.02 + 31.03 • (BL)	0.79	136	No
Anterior flipper length	−29.10 + 18.34 • (BL) − 17.04 • (SEX) + 2.28 • (BL • SEX)	0.94	498	Yes
Longest baleen length ≤9 m	−193.53 + 36.22 • (BL)	0.35	149	No
Longest baleen length >9 m	−52.57 + 23.63 • (BL)	0.83	211	No

propulsive power to allow more effective feeding for older/larger whales that have massive peduncles up to 2.3 m in circumference (NSB-DWM 1995B9). For stranded whales, where peduncle girth is the only measurement available, peduncle circumference can be used to roughly estimate body length.

Pectoral fin

The pectoral fins (flippers) of the bowhead are blunt and paddle shaped. The blubber is relatively thin (2 cm) on the flippers. Flippers insert at about 39% of the body length posteriorly from the tip of the rostrum. Generally, bowhead morphometric characters do not differ between sexes; however, the pectoral fin of males is about 10% larger than females (Fig. 7.9). The proportional length of the limb changes through life but averages about 20% of the body length in males and 18% in females, similar to NPRW (Omura et al., 1969). Differences by sex are significant ($P < 0.001$) for all three flipper measurements: anterior length, posterior length, and width. The largest flipper we measured was 325 cm (anterior length) for a large 17.4 m male with an estimated age of 174 years (NSB-DWM 1995B9). We speculate that the larger pectoral fins may be associated with mating activity.

Flukes

Balaenids have among the broadest flukes of all cetaceans (Fig. 7.9), which may be important for ram filter feeding and for thermoregulation in these large stout animals (Chapter 16). Woodward et al. (2006) noted that, similar to other balaenids, *the rotund right whale has large, high aspect ratio flukes for efficient slow speed cruising that is optimal for their continuous filter feeding technique.* The vasculature of the flukes is highly developed, using counter-current exchange to effectively shed or conserve heat (Elsner et al., 2004).

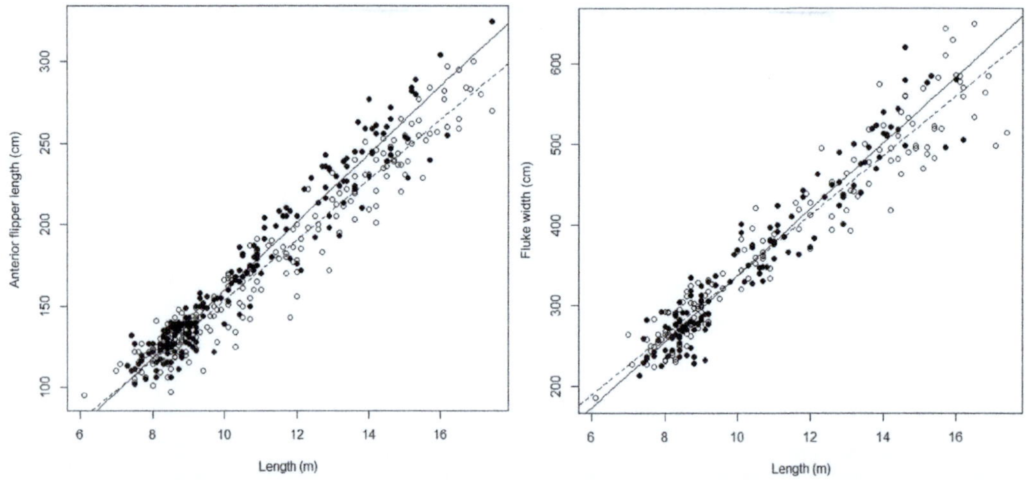

FIGURE 7.9 (Left) Plot of anterior flipper length versus body length. The *solid dots* and *solid line* are male and the *open dots* and *dashed line* are female. The pectoral fin on male bowheads is significantly greater than females and flipper size is highly correlated with body length for both sexes. (Right) Plot of fluke width as a function of body length. The relationship between fluke width and total body length is highly significant; therefore, fluke width can be used to approximate the body length of a bowhead.

The fluke width or span averages about 34% (SD = 0.025, range 0.26–0.41, $n = 182$) of the body length and this ratio increases slightly as the animal grows (Table 7.1; Fig. 7.9). The fluke span ratio is nearly identical to that of right (*Eubalaena* spp.) and humpback whales (*Megaptera novaeangliae*) (Woodward et al., 2006; Moore et al., 2004). In neonates, the caudal edge is somewhat pleated but becomes nearly straight in most adults with a slight concavity lateral to the fluke notch (Fig. 7.2). The fluke to body length ratio has a slight but significant interaction term between sexes ($P = 0.01$) suggesting the flukes of adult males have a slightly faster growth rate with body length (Fig. 7.9). The largest fluke span measured for BCB bowheads was 716 cm on a massive 19 m female (NSB-DWM 2002B3).

Body mass

Bowheads have a rotund body shape and a higher body weight-at-length than all other large whales (Fig. 7.10). Large baleen whales can forage at lower trophic levels (i.e., closer to primary production), which affords a larger and more reliable food source (Goldbogen et al., 2017). Baleen whales also develop huge bodies, which permits massive energy storage in the form of blubber. This, along with their extremely low metabolic rate per unit mass, buffers against disruptions in the availability of prey that would ordinarily cause smaller marine mammals such as dolphins and seals to starve (Chapter 16).

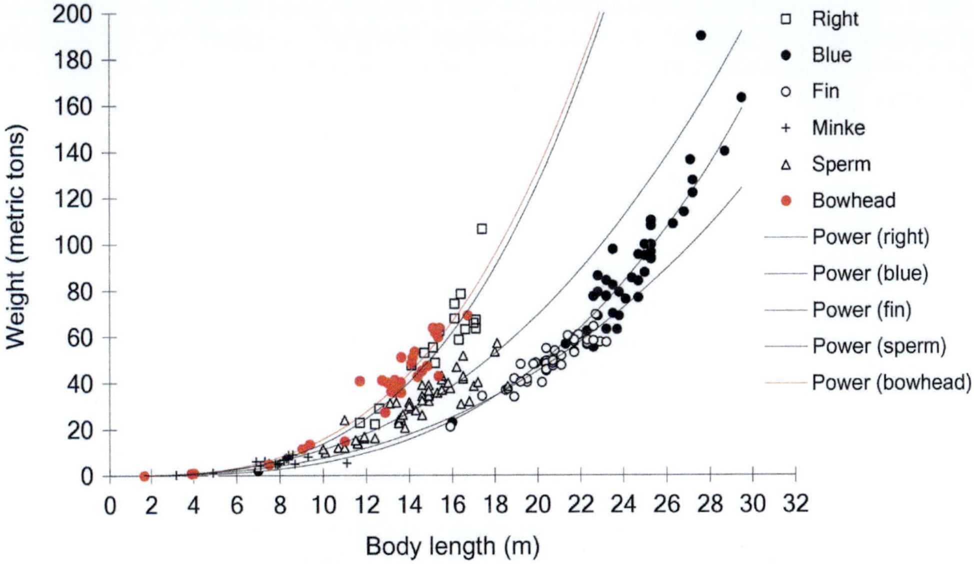

FIGURE 7.10 Plot of comparative body mass for five species of cetaceans. Note that the bowhead and North Pacific right whales are nearly identical in their weight-at-length. The bowhead data are a combination of direct weights and estimated weights based on the body weight equation in the text. *Source: Data for species other than bowhead are from Lockyer, C., 1976, Body weights of some large whales. J. Cons. Int. Explor. Mer. 36 (3), 259–273.*

George (2009) reported on eight harvested bowheads that were weighed either whole or in pieces. A body mass model developed by Rice and Wolman (1971) was fit to the data:

$$W = aLG^2$$

where a is a fitted constant, $a = 38.53$ (95% CI = 35.85–41.21); W = weight (kg); L = length (m); G = girth (m) ($R^2 = 0.987$).

Applying the body mass model to the largest harvested bowheads with reliable girth measurements suggests weights approaching 100 metric tons (mt; 1000,000 kg). For example, whale NSB-DWM 13B1, a 16.5 m female with an axillary girth of 12.6 m, has an estimated mass of about 101 mt (Fig. 7.10), the approximate weight of about 18 adult male African elephants (*Loxodonta africana*, MacDonald, 2010).

The body masses estimated using this model correlated extremely well ($R^2 = 0.983$) with body masses estimated independently using a series of girth measurements to estimate the whales' volume. That method divides the whale into four truncated cones based on the proportion, length, and girth of each section (George et al., 2015). The subsequent mass was then estimated from these volumes assuming a density of ~ 1.0 gm/cm^3, since bowheads are only slightly positively buoyant (George, 2009; George et al., 2015). This suggests either model is useful for estimating bowhead whale mass.

Bowheads and the closely related right whale have very similar weight-at-length relationships that are much higher than other species. Balaenopterids (rorquals), including fin (*Balaenoptera physalus*), blue (*Balaenoptera musculus*), sei (*Balaenoptera borealis*), minke

(*Balaenoptera acutorostrata*), and humpback whales (*Megaptera novaeangliae*), have a lower body mass per unit of body length than bowheads. Rorquals are powerful, fast swimming whales with slender bodies, are heavily muscled presumably as an adaptation for lunge feeding (Chapter 14), and migrate quickly across vast ocean expanses. These large allometric differences are visualized in Fig. 7.10, where we have plotted the mass versus length relationship of bowheads alongside other large cetaceans across four families.

Blue whale-sized bowheads?

The largest bowheads were taken in the early years of the Yankee whale fishery in Alaska, and a few taken late in the fishery. A bowhead taken by Yankee whalers around 1850 yielded 375 barrels of oil or 11,812.5 US gallons from this single animal (44,715.2 L; barrel volume is about 31.5 gallons; Bockstoce and Burns 1993). This whale challenges our understanding of the maximum body mass this species can attain. In a simple experiment, we estimated that 6.4 kg of blubber is required to render 1 gallon of oil (NSB unpublished data). We applied this oil conversion to the oil volume reported for this bowhead and assumed its body mass was 44% blubber (Chapter 16). The result suggests that the whale weighed about 172 metric tons when alive, which is similar to the mass of a large blue whale (Lockyer, 1976). No bowheads similar in size are currently known to exist in any of the bowhead stocks; however, there is little reason to doubt the whaling records of the valuable oil yields. This suggests that bowheads reached masses similar to blue whales prior to commercial whaling.

Longevity

Several lines of evidence suggest that bowheads are the longest lived mammals (Chapter 21). Age analyses indicated that they live over 150 years, and two were estimated near 200 years (George et al., 1999; George and Bockstoce, 2008; Wetzel et al., 2017). The age for whale NSB-DWM 1995WW5 was estimated at 211 years (SE = 35) using aspartic acid racemization of the eye lens nucleus. This whale remains the oldest known bowhead (George et al., 1999) and the oldest known mammal. The whale hunters reported that the animal appeared old, scarred, and had tough meat and blubber. Several of the oldest whales also had an amber colored lens nucleus, which is associated with advanced age (Chapter 30).

Longevity reported for bowheads is greater than for other cetaceans for which age data exist (Ohsumi, 1979; Foote, 2008). Ohsumi (1979) reported ages of 110 and 114 years for a southern hemisphere blue and fin whale respectively based on a count of ear plug laminae. A maximum age among narwhal was estimated at 115 years (Garde et al., 2007).

The recovery of several stone harpoon points found embedded in whales harvested in Wainwright and Utqiaġvik, and a Yankee whaling projectile found in a whale taken at Utqiaġvik in 2007 (George and Bockstoce, 2008) are consistent with estimated ages. Murdoch (1892) noted that stone weapons were being phased out of use in the 1880s at Utqiaġvik (see Chapter 21). Very few of the nearly 1500 bowheads that have been examined by hunters and biologists since 1972 have shown advanced disease (Philo et al., 1993;

Chapter 30), which brings into question the age of reproductive and somatic senescence (Chapter 13). Taken together, the evidence strongly supports extreme longevity in bowheads, consistent with Iñupiat IK that states that bowheads live "two human lifetimes" (Arnold Brower, Jr., personal communication).

Morphometric regressions

Several of the standard cetacean measurements for bowhead whales show strong linear correlations with body length (George, 2009). Exceptions include the girth measurements, which show nonlinear "J" shaped functions with body length and greater scatter in the data. While highly correlated, some relationships such as the snout-to-blowhole and baleen length show inflections with body length which are associated with allometric growth of the head. These morphometric data should be sufficiently robust for investigating regional morphological differences between populations or stocks of whales. We include key regression models for comparative purposes (Table 7.1).

Life history

Life history theory

In this section, we discuss some technical aspects of life history theory, apply them to the bowhead as a case study, and offer a plain-language summary at end. Here, we draw upon ecological and physiological principles to explain the selective forces that favored extreme longevity, slow reproduction, and other distinctive characteristics of bowheads.

An organism's life history is its set of physiological and behavioral phenotypes—constrained by ecology, morphology, and phylogeny—that control life span, age at maturity, growth rate, reproductive rate, fecundity, size at birth, and levels of parental investment and care. To maximize relative fitness, any given organism should strive to produce as many surviving offspring as possible over the course of its life. But given limited resources and natural variability, there are fitness tradeoffs between immediate and future reproduction. Future or "residual reproductive value" is the fitness value of all future reproductive attempts an animal will have (Williams, 1966; Pianka and Parker, 1975; Kirkwood and Rose, 1991). Allocating resources to immediate reproduction may decrease survival and/or come at the expense of one or several future reproductive attempts. Individuals may instead choose to invest resources into growth and somatic (body) maintenance, which should increase the likelihood of success in later reproductive attempts; particularly if their current body condition is poor or if access to critical resources in a given year is low due to scarcity or natural variability (Williams, 1966; Pianka and Parker, 1975). However, somatic investment may be risky if extrinsic mortality, such as predation, prevents survival and later reproduction irrespective of an individual's condition (Reznick and Endler, 1982). Ultimately, Darwinian fitness is the currency that determines which strategies are favorable under varying ecological conditions.

Given the diverse range of ecological niches occupied by all life forms, it is no surprise that animals have evolved a great variety of life histories to maximize this tradeoff. These include such highly successful short-lived species, which typically reproduce only once, live for only a few weeks, and produce hundreds or thousands of offspring that receive essentially no parental investment (Denno and Dingle, 1981). Such life histories have evolved to invest exclusively and intensely on immediate reproduction. Mammals are an example of class of animals that have evolved an iteroparous life history strategy, whereby they produce multiple litters that consist of fewer offspring. For young individuals, forgoing reproduction in the current year and diverting those resources to body maintenance should help to preserve their large residual reproductive value, and would be favorable if long-term fitness increases as a result. Ungulates typify this strategy, often giving birth to a single offspring once per year, investing heavily in that single offspring via nursing and parental defense (Promislow and Harvey, 1990). Certain ecological and phylogenetic constraints influence which strategy is best. Predation (and/or other sources of extrinsic mortality) is an important determinant of the patterns of iteroparity in mammals. For example, mammals that are subject to extremely high predation pressure (e.g., rodents) face selection for early sexual maturation, large litter size, frequent reproduction (often several litters per year), but also a *short* life span. In other words, they "live fast and die young" (Promislow and Harvey, 1990). Alternatively, if extrinsic mortality rates are relatively low, then fitness is maximized when individuals are better competitors and/or survive even when important resources are low or highly variable. These conditions preserve residual reproductive value and select for longer life span and increased parental investment (Wilson and MacArthur, 1967; Mangel et al., 2007). As a clade of the ungulates (Shimamura et al., 1997), Cetacea express this general life history pattern (Thewissen, 1994).

Here, we consider how the unique evolutionary forces and ecological niche of bowhead whales have shaped the development of a life history strategy characterized by extraordinary life spans, exceptionally slow maturation rates, and slow reproductive rates.

Reproduction

Bowhead reproduction is covered in detail in Chapter 13; however, we highlight some aspects here since it is highly relevant to their life history. Gestation lasts a little more than a year, calves nurse for less than a year, and the calving interval is 3–4 years (George et al., 2011). Thus, replenishment of energy stores and/or repair of body wear associated with reproductive investment would appear to take females 1–2 years, and thus reproductive investment in individual offspring is exceptionally large. Finally, unlike some other long-lived whales, there is no clear evidence of reproductive senescence (i.e., menopause), with individuals continuing to reproduce into their second century of life (Foote, 2008; Johnstone and Cant, 2010) (Chapter 13).

Calving areas

In the BCB stock, calving occurs along Alaska's west and north coasts during the spring migration and within a complex network of open channels in the sea ice referred to as "leads." Neonates would seemingly be vulnerable because they are too small to break through sea ice, and the lead systems can rapidly close. Mother–calf pairs tend to migrate later in the spring when leads are wider, possibly to reduce the threat of ice entrapment. Calving within the sea ice and migrating to the Beaufort Sea almost certainly reduces predation risk from arctic killer whales, which are highly ice averse, unlike their Antarctic counterparts (Corkeron and Connor, 1999; Matthews et al., 2020; Chapters 5 and 29). East Canada-West Greenland (ECWG) bowheads show similar calving behavior and migratory timing. Chapter 5 describes their main calving area as "the Canadian Arctic Archipelago and especially the shallow waters of Foxe Basin, where sheltered areas may offer a refuge for young calves, minimizing the predation risk from killer whales as well as reducing the risk of ice entrapment" (Matthews et al., 2020).

Koski et al. (1993) suggested that most calves are born in April through early June, which aligns well with fetal growth predictions. However, they estimated that about 11% of bowhead whale calves were born after mother–calf pairs had passed Point Barrow by June, and noted that a few neonates have been observed in Amundsen Gulf in summer months (see maps in Chapters 3 and 4; Koski et al., 1993). Neonates have been observed by hunters at Saint Lawrence Island in the Bering Sea and by hunters and in surveys over 1000 km north at Utqiaġvik, Alaska. During ice-based whale censuses from 1978 to 2001, observers noted the peak mother–calf passage occurred in late May at Utqiaġvik (George et al., 1995, 2019). Under the assumption that these were neonates, this suggests that most calves were born in the Chukchi Sea, at least in that time period. Taken together, the data collectively indicate BCB whales calve in sea ice leads along the northwest Alaskan coast, with a relatively small fraction being born east of Point Barrow.

While calving events have not been photographed or seen during aerial surveys, a Point Hope hunter (J. Koonook, personal communication) described a likely calving event on May 20, 1986, near Point Hope in which he said "the female had three attendants and blood was noted in the water" in association with the appearance of the calf.

Migration

Many cetaceans and pinnipeds express some degree of migratory behavior. Gray and humpback whales express long, complete migrations that are among the longest of any mammal (Dawbin, 1966; Rice and Wolman, 1971; Wells et al., 1999). These mysticete whales migrate to the seasonally rich, cold waters of the North Pacific, Bering, Chukchi, and even Beaufort Seas in the summer to forage, but return to the mild tropical and subtropical Pacific waters to calve and breed. These warm southern oligotrophic waters offer so little food resources that gray whales typically fast through the winter, giving birth and nursing calves using fat stored during summer foraging.

Bowheads, unlike all other mysticetes, remain in the sub-Arctic and Arctic year-round, forgoing long migrations to temperate and tropical latitudes typical of other species (Chapters 4 and 5). This adaptation is rare among cetaceans, and is only present in three species—bowheads, belugas, and narwhals. Wintering in the sub-Arctic sea ice (e.g., Bering Sea and Davis Strait) allows an opportunity for winter feeding and refuge from predators.

It is classically thought that the principle reason for long migrations to subtropical waters is to minimize thermal stress to their calves and to reduce their metabolic costs.

However, Corkeron and Connor (1999) reviewed the evidence of various hypotheses, and suggested that these warm tropical and subtropical waters are also poor habitat for killer whales, the primary predators of baleen whales and their calves, which are particularly vulnerable. Thus, migrations to warmer waters may also be a predator-avoidance mechanism for these species. Bowheads, by wintering and calving in the dense Arctic sea ice, may have evolved an alternative strategy to migrating to temperate and subtropical areas as a predator-avoidance strategy. Resolving the drivers of large whale migration remains an open question and a rich area for future research.

With regard to bowhead migration timing, annual movements begin with the departure from their wintering areas, that is, the Bering Sea (BCB stock) and Davis Strait (ECWG stock), between late March and early May. Departures are usually, but not always, timed with the onset of spring sea ice breakup. Individuals travel through the lead systems to higher latitude feeding and nursery areas.

The spring migration of BCB bowheads, during the 1970s through the early 2000s, passed Point Barrow in distinct age structured pulses over a 6- to 8-week period (Fig. 7.11). These pulses occurred in roughly the third week of April, early May, and late May (George et al., 2004b). The early pulse included a high proportion of subadult whales, the midpulse was a mixed group with the highest daily passage rates, and the last pulse was mostly large adults of both sexes, pregnant females, and mothers with calves (Koski et al., 2006; George et al., 1995). More contemporary observations by whale hunters at Utqiaġvik and the ice-based surveys conducted since 2010 near Point Barrow indicate that: (1) the migration starts 2 to 3 weeks earlier in spring than in the 1970s and (2) the demographically segregated pulses (Braham et al., 1980; George et al., 2012) are less distinct, possibly because population size has increased. Additional details of the BCB bowhead migratory behavior are given in Chapter 4.

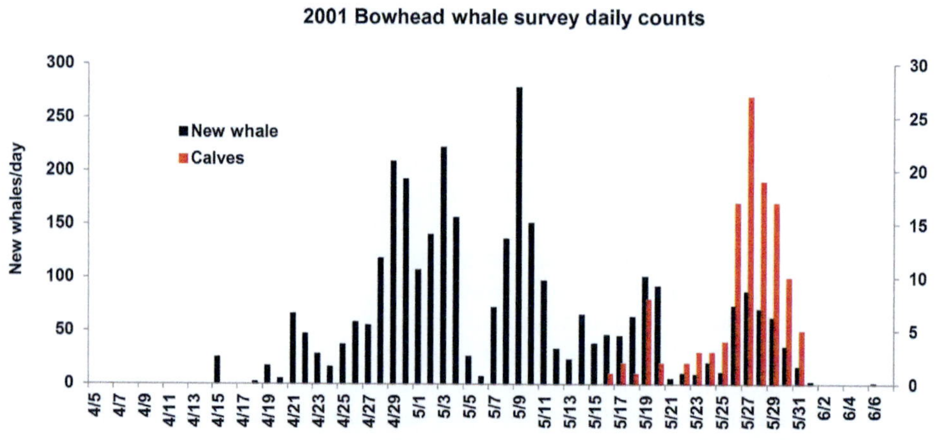

FIGURE 7.11 Counts of bowhead whales during the 2001 bowhead whale census at Point Barrow, Alaska. New whales indicate first time sightings of an individual, excluding calves. Note the three fairly distinct pulses of migrating whales around late-April, the second week in May, and large adults including mothers and calves in late May. The three-peaked pattern of migration in recent years is not as obvious based on hunter observations and more recent surveys.

Ice navigation

One of the bowheads' most remarkable adaptations is their ability to navigate beneath and break through dense sea ice to breathe (George et al., 1989; Fig. 7.12). This is critically important in winter and during the spring migration. While long known to indigenous hunters, ice navigation was not well studied by scientists until the 1970s. These navigational skills are especially evident where, under heavy ice conditions, BCB whales annually transit from Point Barrow to Amundsen Gulf, often through essentially closed leads. Furthermore, in some years, they contend with heavy ice cover even through the summer in the Eastern Beaufort Sea (Harwood et al., 2015).

There are limits to the ice thickness that a bowhead can break through. Migrating whales traversing from the Bering Sea to the Eastern Beaufort Sea in spring are frequently forced to use closed leads. This behavior is not restricted to migratory periods, as analyses of telemetry data by Citta et al. (2012) indicate that some bowheads spend the entire winter within "100% ice cover" based on satellite imagery. While the winter ice cover is almost certainly fractured, one would expect numerous records of bowheads drowning under sea ice. However, such records are exceedingly rare within the Inuit traditional knowledge and in the scientific literature (Nerini et al., 1984; George et al., 1989). Therefore, bowheads are clearly able to locate small cracks in the sea ice to breathe and have behavioral and sensory mechanisms for assessing and avoiding solid, potentially dangerous ice floes. How they accomplish this is unclear, although learning, acoustic/visual assessment, and accumulated experience (memory) accrued over their long life spans likely play a critical

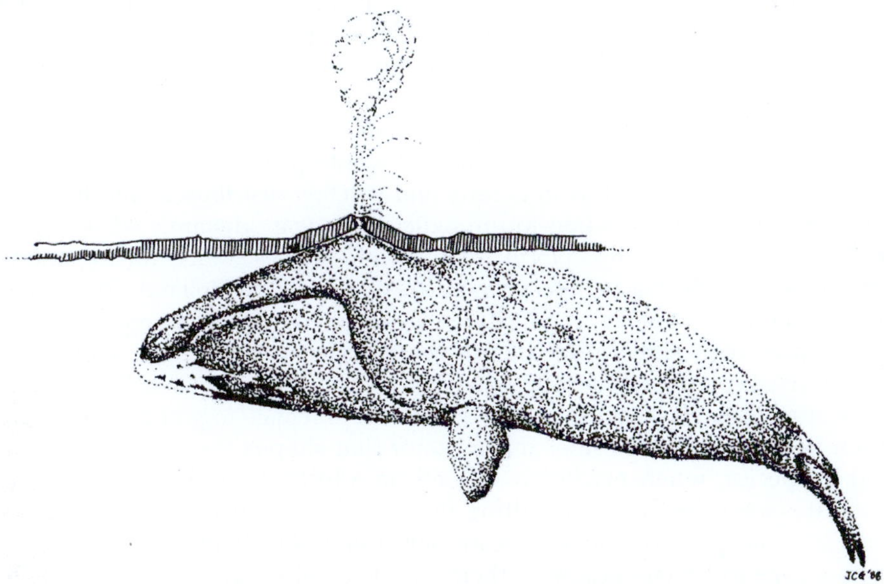

FIGURE 7.12 Sketch of a bowhead whale using its rostrum to break through sea ice to breathe. The ability to break sea ice was a critical morphological modification to life in the arctic seas. Bowheads appear to be able to readily break ice to 20 cm and some Iñupiat hunters describe events in which they broke through ice of 60 cm or more.

role (George et al., 1989; Fagan et al., 2013). Briefly stated, bowheads are highly specialized for living in the dynamic sea ice of the arctic seas.

Evolution of exceptional longevity and delayed sexual maturity

The extraordinary physiological and behavioral traits of bowheads can be understood as life history adaptations that maximize fitness in the cold and ice-covered Arctic and sub-Arctic seas they inhabit (Promislow and Harvey, 1991). As noted earlier, bowheads do not migrate to temperate or tropical waters to calve, and they live among the sea ice nearly year-round. They are pagophyllic or "ice loving." In this high-latitude environment, the prey field is sparse, patchy, and intensely seasonal (Neibauer and Schell, 1993; Chapter 26). Such extreme Arctic conditions very likely imposed strong selective forces that shaped the evolutionary development of bowhead life history.

Longevity in excess of 200 years (George et al., 1999; Wetzel et al., 2017) stands out as perhaps the most dramatic of the many extreme bowhead life history attributes. Longevity in vertebrates is associated with slow growth, long periods of parental dependence, low reproductive, and/or low juvenile survival rates (Stearns, 1992). Life history theory predicts that two primary conditions should select for long life: (1) strong intraspecific competition over evolutionary time scales and/or (2) unpredictable resource availability (Wilson and MacArthur, 1967; van Noordwijk and de Jong, 1986). The first condition selects for long dependent periods and/or high maternal investment because well-prepared offspring better compete among conspecifics. The second condition regarding the unpredictability of resources would likely select for a capital breeding strategy (Jönsson, 1997) in females, which allows for reproduction to occur only upon reaching a certain energetic threshold. Presumably, females that are unable to achieve this energetic threshold will sequester their energetic resources, investing them into somatic growth and maintenance (Lecomte et al., 2010) and/or future reproductive efforts until conditions are amenable to successful reproduction. The pattern of high maternal investment, extended longevity, and low reproductive rate in bowheads is consistent with conditions that would promote strong intraspecific competition (e.g., resources that are low in density and patchily distributed) and highly variable seasonal resource availability leading to unpredictability (e.g., dynamic sea ice conditions). That bowheads and closely related North Atlantic right whales exhibit calving intervals that are longer than 4 years during periods of food stress (Kraus and Rolland, 2007; Miller et al., 1992) further indicates a strategy that includes the facultative delay in reproduction, which would help to preserve residual reproductive value during years of poor resource availability. Moreover, it takes a typical female bowhead about 25 years to reach sexual maturity. One interpretation is that it takes about two decades for a female to prepare for her first calf.

Predation is known to be another major factor that shapes the evolution of life histories. In other species, when predation is high on adults, shorter life spans and faster maturation rates are selected for, resulting in individuals that reproduce early—that is, before they become prey (Reznick and Endler, 1982). Adult bowheads, however, are generally too large to be vulnerable to their only natural predator, and are further protected by sea ice most of the year (Matthews et al., 2020). Calves and juveniles, however, appear acutely vulnerable to predation, further decreasing calf and juvenile survival (Chapter 29).

In theory, over evolutionary time spans, highly variable resource availability and predation should select for high maternal investment in each calf. Even so, it is unclear how many calves survive to reach adulthood, although it is likely to be few as in other mammals. Note however that in the last 100 years following heavy commercial bowhead whaling, calf survival *may* have been higher because the population was below carrying capacity. At the same time, individuals that reach the subadult stage appear to have high survival rates, and survival rates are even higher in adults. Data from harvested bowheads indicate that they maintain a relatively high plane of nutrition and body condition as well as health status (Chapter 30; Philo et al., 1993; George et al., 2015). Tumors are rare (Chapter 30). Consistent with Bradley et al. (2005), who documented relatively low disease prevalence in the Arctic compared with temperate and tropical regions, bowhead health studies indicate few infectious agents are present that could potentially shorten life span, and parasite diversity and loads are low (Chapter 30). Greenland sharks are another good example of an arctic vertebrate that evolved extreme longevity under high-arctic conditions (Nielsen et al., 2016). Moreover, adult bowhead survival rate has been estimated using interyear photo-recapture at 0.996 (Givens et al., 2017). This exceptional survival rate includes hunting mortality, suggesting it may have been slightly higher prior to anthropogenic mortality.

Extremely low mortality in adults, higher mortality in calves, and a 3- to 4-year investment required to produce a single viable offspring create the conditions in which great longevity is favored as a life history strategy. With long intervals between reproduction and low probability that offspring survive, it may take 100 years for an average female bowhead to produce five calves that survive to reproduce themselves. Under such conditions, having exceptionally long life spans would be essential for maximizing lifetime fitness.

As noted earlier, a key adaption that may have led to immense longevity in bowheads was their year-round association with sea ice. This may have also allowed the evolution of biochemical and genetic mechanisms, necessary for their life style (Keane et al., 2015; Chapter 20); that is the evolution of genetic repair mechanisms that ultimately reduce disease and aging attributed to DNA damage (Nussey et al., 2009). Sea ice clearly affords protection from killer whales, and this protection may be at least partially responsible for the year-round sea ice association of bowheads. Never having to leave the protective sea ice refugia probably lowered bowhead mortality (Austad, 1993; Promislow and Harvey, 1990) as compared to closely related right whales.

The thick blubber layers of all whale species buffer against disruptions in prey availability and, together with the low predation rates of adult bowheads by killer whales, would have reduced the major sources of mortality close to zero. In the absence of other sources of mortality, the physiological changes noted earlier would have eventually interacted to lengthen bowhead life span to the extreme we observe today (Nussey et al., 2009; Keane et al., 2015). By delaying death, lengthening reproduction, and selecting for longer life spans, these adaptations would come with commensurate increases in lifetime reproductive success.

Bowheads are well adapted to live in frigid icy Arctic seas and, over a few million years, evolved into their current form as a mammalian outlier. Neither the Inuit people who hunt them, nor scientists who study them, would argue with Charles Darwin's assessment that *The Greenland Whale* [bowhead] *is one of the most wonderful animals in the world...* (Darwin, 1859).

Acknowledgments

We like to thank the whale hunters of the Alaska Eskimo Whaling Commission, and in particular the captains and crews of the village Whaling Captain's Associations for allowing us to examine their whales. The Barrow Whaling Captains Association was particularly helpful in advocating and supporting our research. We are grateful to the hunters for sharing their knowledge, hunting stories, and coffee, assistance, on many long, cold nights. We thank the scientists and whaling crew members who examined whales in subzero weather and unstable ice conditions, including: Tom Albert, Rita Acker, Perry Anashugak, Joe Burgener, Geoff Carroll, Jim Cubbage, Jason Herreman, Gordon Jarrell, Cyd Hanns, Kathy Hazard, Todd O'Hara, Mike Pedersen, Cheryl Rosa, Leslie Pierce, Mike Philo, Paul Nader, J.R. Patee, Dave Ramey, Raymond Tarpley, Victoria Woshner, and Raphaela Stimmelmayr. Geof Givens and Judith Zeh gave statistical advice and analysis. We thank Seiji Ohsumi for providing data and suggestions. Many at the North Slope Borough Department of Wildlife Management assisted with data checking and technical editing including Barbara Tudor and Leslie Pierce.

References

Archer, F.I., Martien, K.K., Taylor, B.L., LeDuc, R.G., Ripley, B.J., Givens, G.H., et al., 2010. A simulation-based approach to evaluating population structure in non-equilibrial populations. J. Cetacean Res. Manage. 11, 101–113.

Austad, S.N., 1993. Retarded senescence in an insular population of Virginia opossums (*Didelphis virginiana*). J. Zool. 229 (4), 695–708. Available from: https://doi.org/10.1111/j.1469-7998.1993.tb02665.x.

Best, P., Schell, D., 1996. Stable isotopes in southern right whales (*Eubalaena australis*) baleen as indicators of seasonal movements, feeding and growth. Mar. Biol. 124, 283–294.

Bockstoce, J.R., 1986. Whales, Ice and Men. University of Washington Press, Seattle, WA, 400 pp.

Bockstoce, J.R., Burns, J.J., 1993. Commercial whaling in the North Pacific sector. In: Burns, J.J., Montague, J.J., Cowles, C.J. (Eds.), The Bowhead Whale. Allen Press, Lawrence, KS, pp. 563–576.

Bradley, M.J., Kutz, S.J., Jenkins, E., O'hara, T.M., 2005. The potential impact of climate change on infectious diseases of Arctic fauna. Int. J. Circumpolar Health 64 (5), 468–477.

Bryden, M.M., 1972. Growth and development of marine mammals. In: Harrison, R.J. (Ed.), Functional Anatomy of Marine Mammals. Academic Press, London, pp. 1–80.

Braham, H.W., Durham, F.E., Jarrell, G.H., Leatherwood, S., 1980. Ingutuk: a morphological variant of the bowhead whale, *Balaena mysticetus*. Mar. Fish. Rev. 42, 70–73.

Citta, J.J., Quakenbush, L.T., George J.C., Small, R.J., Heide-Jorgensen M.P., Brower H., et al., 2012. Winter movements of bowhead whales (*Balaena mysticetus*) in the Bering Sea. Arctic, 65, 1.

Corkeron, P.J., Connor, R.C., 1999. Why do baleen whales migrate? Mar. Mamm. Sci. 15 (4), 1228–1245. Available from: https://doi.org/10.1111/j.1748-7692.1999.tb00887.x.

Costa, D.P., Williams, T.M., 1999. Marine mammal energetics. In: John, E., Reynolds, I.I.I., Rommel, S.A. (Eds.), Biology of Marine Mammals. Smithsonian Institution, p. 578.

Darwin, C., 1859. On the Origin of Species by Means of Natural Selection, or the Preservation of Favoured Races in the Struggle for Life. Murray, London. Reprinted 1909 by Harvard Classics, C.W. Eliot (Ed.), New York, 552 p.

Dawbin, W.H., 1966. The seasonal migratory cycle of humpback whales. In: Whales, Dolphins, and Porpoises, pp. 145–170.

Denno, R.F., Dingle, H. (Eds.), 1981. Insect Life History Patterns: Habitat and Geographic Variation. Springer-Verlag, New York.

Durham, F.E., 1980. External morphology of bowhead fetuses and calves. Mar. Fish. Rev. 42 (9–10), 74–80.

Elsner, R., George, C., O'Hara, T., 2004. Vasomotor responses of isolated peripheral blood vessels from bowhead whales: thermoregulatory implications. Mar. Mamm. Sci. 20 (3), 546–553.

Eschricht, D.F., Reinhardt, J., 1866. On the Greenland right-whale (*Balaena mysticetus*), with special reference to its geographic distribution and migration in times past and present, and to its external and internal characteristics. In: Flower, W.H. (Ed.), Recent Memoirs on the Cetacea. The Ray Society, London, pp. 1–150.

Fagan, W.F., Lewis, M.A., Auger-Méthé, M., Avgar, T., Benhamou, S., Breed, G., et al., 2013. Spatial memory and animal movement. Ecol. Lett. 16 (10), 1316–1329.

Finley, K.J., 2001. Natural history and conservation of the Greenland whale, or bowhead, in the Northwest Atlantic. Arctic 54, 55–76.

Foote, A.D., 2008. Mortality rate acceleration and post-reproductive lifespan in matrilineal whale species. Biol. Lett. 4 (2), 189–191.

Garde, E., Heide-Jørgensen, M.P., Hansen, S.H., Nachman, G., Forchhammer, M.C., 2007. Age-specific growth and remarkable longevity in narwhals (*Monodon monoceros*) from West Greenland as estimated by aspartic acid racemization. J. Mammal. 88 (1), 49–58.

George, J.C., 2009. Growth, Morphology, and Energetics of Bowhead Whales (*Balaena mysticetus*) (Ph.D. thesis). University of Alaska Fairbanks, Alaska, December.

George, J.C., Bockstoce, J.R., 2008. Two historical weapon fragments as an aid to estimating the longevity and movements of bowhead whales. Polar Biol. (31), 751–754.

George, J.C., Suydam, R.S., 2014. Update on characteristics of bowhead whale (*Balaena mysticetus*) calves. In: Paper SC/66b/BRG20 presented to the IWC Scientific Committee.

George, J.C., Clark, C., Carroll, G.M., Ellison, W.T., 1989. Observations on the ice-breaking and ice navigation behavior of migrating bowhead whales (*Balaena mysticetus*) near Point Barrow, Alaska, spring 1985. Arctic 42, 24–30.

George, J.C., Suydam, R.S., Philo, L.M., Albert, T.F., Zeh, J.E., Carroll, G.M., et al., 1995. Report of the spring 1993 census of bowhead whales, *Balaena mysticetus*, off Point Barrow, Alaska with observations on the 1993 subsistence hunt of bowhead whales by Alaska Eskimos. Rep Int. Whal. Commn. 45, 371–386.

George, J.C., Bada, J., Zeh, J., Scott, L., Brown, S.E., O'Hara, T., et al., 1999. Age and growth estimates of bowhead whales (*Balaena mysticetus*) via aspartic acid racemization. Can. J. Zool. 77, 571–580.

George, J.C., Koski, W.R., Suydam, R., Rugh, D., 2004a. Body stretching of bowhead whales during hauling and butchering during the subsistence hunt. In: Paper SC/56/BRG9 presented to the Scientific Committee of the International Whaling Commission.

George, J.C., Zeh, J., Suydam, R., Clark, C., 2004b. Abundance and population trend (1978–2001) of western arctic bowhead whales surveyed near Barrow, Alaska. Mar. Mamm. Sci. 20 (4), 755–773.

George, J.C., Follmann, E., Zeh, J., Sousa, M., Tarpley, R.J., Suydam, R., 2011. A new way to estimate whale age using ovarian corpora counts. Can. J. Zool. 89, 840–852.

George, J.C., Herreman, J., Givens, G. H., Suydam, R., Mocklin, J., Clark, C.W., et al., 2012. Brief review of the 2010 and 2011 bowhead whale abundance surveys near Point Barrow, Alaska. Paper SC/64/AWMP7 presented to the IWC Scientific Committee.

George, J.C., Druckenmiller, M.L., Laidre, K.L., Suydam, R., 2015. Bowhead whale body condition and links to summer sea ice and upwelling in the Beaufort Sea. Prog. Oceanogr. 136 (2015), 250–262.

George, J.C., Stimmelmayr, R., Suydam, R., Usip, S., Givens, G., Sformo, T., 2016. Severe bone loss as part of the life history strategy of bowhead whales. PLoS One 11 (6), e0156753. Available from: https://doi.org/10.1371/journal.pone.0156753.

George, J.C., Tudor, B., Givens, G.H., Mocklin, J., Vate Brattström, L., 2019. Entanglement-scar acquisition rates and scar frequency for Bering-Chukchi-Beaufort Seas bowhead whales using aerial photography. Mar. Mammal. Sci., 35 (4), 1304–1321.

Givens, G.H., Mocklin, J.A., Vate Brattström, L., Tudor, B.J., Koski, W.R., George, J.C., et al., 2017. Survival rate and 2011 abundance of Bering-Chukchi-Beaufort Seas bowhead whales from photo-identification data over three decades. Paper SC/67a/AWMP09.

Goldbogen, J.A., Cade, D.E., Calambokidis, J., Friedlaender, A.S., Potvin, J., Segre, P.S., et al., 2017. How baleen whales feed: the biomechanics of engulfment and filtration. Annu. Rev. Mar. Sci. 9, 367–386.

Haldiman, J.T., Tarpley, R.J., 1993. Anatomy and physiology. In: Burns, J.J., Montague, J.J., Cowles, C.J. (Eds.), The Bowhead Whale. Allen Press, Lawrence, KS.

Hamilton, P., 2019. Personal Communication. Kraus Marine Mammal Conservation Program, Anderson Cabot Center for Ocean Life New England Aquarium Central Wharf Boston, MA.

Harwood, L.A., et al., 2015. Change in the Beaufort Sea ecosystem: diverging trends in body condition and/or production in five marine vertebrate species. Prog. Oceanogr. 136, 263–273. Available from: https://doi.org/10.1016/j.pocean.2015.05.003.

Heide-Jørgensen, M.P., Garde, E., Nielsen, N.H., Andersen, O.N., Hansen, S.H., 2012. A note on biological data from the hunt of bowhead whales in West Greenland 2009–2011. J. Cetacean Res. Manage. 12, 329–333.

Jönsson, K.I., 1997. Capital and income breeding as alternative tactics of resource use in reproduction. Oikos, pp. 57–66. <https://www.jstor.org/stable/3545800>.

Johnstone, R.A., Cant, M.A., 2010. The evolution of menopause in cetaceans and humans: the role of demography. Proc. R. Soc. B: Biol. Sci. 277 (1701), 3765–3771.

Keane, M., Semeiks, J., Webb, A.E., Li, Y.I., Quesada, V., Craig, T., et al., 2015. Insights into the evolution of longevity from the bowhead whale genome. Cell Rep. 10, 112–122.
Kirkwood, T.B.L., Rose, M.R., 1991. Evolution of senescence: late survival sacrificed for reproduction. Philos. Trans. Biol. Sci. 332 (1262), 15–24.
Koski, W.R., Davis, R.A., Miller, G.W., 1992. Growth rates of bowhead whales as determined from low-level aerial photogrammetry. Rep. Int. Whal. Commun. 42, 491–499.
Koski, W.R., Davis, R.A., Miller, G.W., Withrow, D.E., 1993. Reproduction. In: Burns, J.J., Montague, J., Cowles, C.J. (Eds.), The Bowhead Whale. Society for Marine Mammalogy, p. 787.
Koski, W.R., Rugh, D.J., Punt, A.E., Zeh, J., 2006. An approach to minimize bias in estimation of the length frequency distribution of bowhead whales (*Balaena mysticetus*) from aerial photogrammetric data. J. Cetacean Res. Manage. 8 (1), 45–54.
Kraus, S.D., Rolland, R.M., 2007. The Urban Whale. Harvard University Press, Cambridge, MA.
Lecomte, V.J., Sorci, G., Cornet, S., Jaeger, A., Faivre, B., Arnoux, E., et al., 2010. Patterns of aging in the long-lived wandering albatross. Proc. Natl. Acad. Sci. U. S. A. 107 (14), 6370–6375.
Lockyer, C., 1976. Body weights of some large whales. J. Cons. Int. Explor. Mer. 36 (3), 259–273.
Lubetkin, S.C., Zeh, J.E., Rosa, C., George, J.C., 2008. Age estimation for young bowhead whales (*Balaena mysticetus*) using annual baleen growth increments. Can. J. Zool. 86 (6), 525–538.
Lubetkin, S.C., Zeh, J.E., George, J.C., 2012. Statistical modeling of baleen and body length at age in bowhead whales (*Balaena mysticetus*). Can. J. Zool. 90, 915–931.
MacDonald, D.W., 2010. The Encyclopedia of Mammals. Oxford University Press, pp. 128–167.
MacLean, E. 2014. Iñupiatun Uqaluit Taniktun Sivuninit/Iñupiaq to English Dictionary. University of Alaska Press.
Mangel, M., Kindsvater, H.K., Bonsall, M.B., 2007. Evolutionary analysis of life span, competition, and adaptive radiation, motivated by the Pacific rockfishes (Sebastes). Evol. Int. J. Org. Evol. 61 (5), 1208–1224.
Matthews, C.J., Breed, G.A., LeBlanc, B., Ferguson, S.H., 2020. Killer whale presence drives bowhead whale selection for sea ice in Arctic seascapes of fear. Proc. Natl. Acad. Sci. U. S. A. 117(12), 6590–6598.
Mau, T., 2004. Investigations of the Role of Lipids in Marine Mammal Diets, Health and Ecology (Ph.D. dissertation). University of Alaska Fairbanks. Fairbanks, Alaska.
Miller, G.W., Davis, R.A., Koski, W.R., Crone, M.J., Rugh, D.J., Withrow, D.E., et al., 1992. Calving intervals of bowhead whales: an analysis of photographic data. Rep. Int. Whal. Commn. 42, 501–506.
Moore, M.J., Knowlton, A.R., Kraus, S.D., Mclellan, W.A., Bonde, R.K., 2004. Morphometry, gross morphology and available histopathology in North Atlantic right whale (*Eubalaena glacialis*) mortalities (1970–2002). J. Cetacean Res. Manage. 6 (3), 199–214.
Murdoch, J., 1892. Ethnological Results of the Point Barrow Expedition. Smithsonian Institution Press, p. 441.
Neibauer, H.J., Schell, D.M., 1993. Physical environment of the Bering Sea population. In: Burns, J.J., Montague, J.J., Cowles, C.J. (Eds.), The Bowhead Whale. Society for Marine Mammalogy, Special publication No. 2. i-xxxvi + 787 pp.
Nerini, M.K., Braham, H.W., Marquette, W.M., Rugh, D.J., 1984. Life history of the bowhead whale, *Balaena mysticetus* (Mammalia: Cetacea). J. Lond. Zool. Soc. 204, 443–468.
Nielsen, J., Hedeholm, R.B., Heinemeier, J., Bushnell, P.G., Christiansen, J.S., Olsen, J., et al., 2016. Eye lens radiocarbon reveals centuries of longevity in the Greenland shark (*Somniosus microcephalus*). Science 353 (6300), 702–704. Available from: https://doi.org/10.1126/science.aaf1703.
Nussey, D.H., Pemberton, J.M., Pilkington, J.G., Blount, J.D., 2009. Life history correlates of oxidative damage in a free-living mammal population. Funct. Ecol. 23 (4), 809–817. Available from: https://doi.org/10.1111/j.1365-2435.2009.01555.x.
Ohsumi, S., 1979. Interspecies relationships among some biological parameters in cetaceans and estimation of the natural mortality coefficient of the Southern Hemisphere minke whales. Rep. Int. Whal. Commn. 29, 397–406.
Omura, H., Ohsumi, S., Neomoto, T., Nasu, K., Kasuya, T., 1969. Black right whales in the North Pacific. Sci. Rep. Whales Res. Inst. 21, 1–78 + appendices.
Philo, L.M., Shotts, E.B., Jr., George, J.C., 1993. Morbidity and mortality. In: Burns, J.J., Montague, J.J., Cowles, C.J. (Eds.), The Bowhead Whale. Allen Press, Lawrence, KS, 787 pp.
Pianka, E.R., Parker, W.S., 1975. Age-specific reproductive tactics. Am. Nat. 109 (968), 453–464.
Promislow, D.E., Harvey, P.H., 1990. Living fast and dying young: a comparative analysis of life-history variation among mammals. J. Zool. 220 (3), 417–437.
Promislow, D.L., Harvey, P.H., 1991. Mortality rates and the evolution of mammal life histories. Acta Oecol. 12 (1), 119–137.

R Core Team, 2014. R: A Language and Environment for Statistical Computing. R Foundation for Statistical Computing, Vienna, Austria. <http://www.R-project.org/>.

Reese, S., Calvin, J.A., George, J.C., Tarpley, R.J., 2001. Estimation of fetal growth and gestation in bowhead whales. JASA 96 (455), 915–928.

Reznick, D., Endler, J.A., 1982. The impact of predation on life history evolution in Trinidadian guppies (*Poecilia reticulata*). Evolution 36 (1), 160–177.

Rice, D.W., Wolman, A.A., 1971. The life history and ecology of the gray whale (*Eschrichtius robustus*). Special Publication No. 3. U.S. Fish & Wildlife Service, Bureau of Commercial Fisheries, Marine Mammal Biological Laboratory, Seattle, WA. Copies available from: the Secretary-Treasurer of the Society, Dr. Bryan P. Glass, Department of Zoology, Oklahoma State University, Stillwater, OK.

Rooney, A.P., George, J.C., Tarpley, R.J., 2002. Phylogenetic and morphometric analyses of the ingutuk, a morphological variant of the bowhead whale (*Balaena mysticetus*). In: Paper SC/54/BRG19 presented to the IWC Scientific Committee, Shimonoseki, Japan.

Rosa, C., Zeh, J., George, J.C., Botta, O., Zauscher, M., Bada, J., et al., 2013. Age estimates based on aspartic acid racemization for bowhead whales (*Balaena mysticetus*) harvested in 1998–2000 and the relationship between racemization rate and body temperature. Mar. Mamm. Sci. 29 (3), 424–445.

Scoresby, W., Jr., 1820. An Account of the Arctic Regions With a History of the Northern Whale Fishery Volume 1: The Arctic. David and Charles Reprints (1969), Newton Abbot, Devon, England, 551 pp.

Shimamura, M., Yasue, H., Ohshima, K., Abe, H., Kato, H., Kishiro, T., et al., 1997. Molecular evidence from retroposons that whales form a clade within even-toed ungulates. Nature 388 (6643), 666–670.

Stearns, S., 1992. The Evolution of Life Histories. Oxford University Press, Oxford, p. 249.

Thewissen, J.G.M., 1994. Phylogenetic aspects of cetacean origins: a morphological perspective. J. Mammal. Evol. 2 (3), 157–184.

Thewissen, J.G.M., Cooper, L.N., George, J.C., Bajpai, S., 2009. From land to water: the origin of whales, dolphins, and porpoises. Evol. Educ. Outreach 2 (2), 272.

van Noordwijk, A.J., de Jong, G., 1986. Acquisition and allocation of resources: their influence on variation in life history tactics. Am. Nat. 128(1), 137–142.

Wells, R.S., Boness, D.J., Rathbun, G.B., 1999. Behavior. In: Reynolds III, J.E., Rommel, S.A. (Eds.), The Biology of Marine Mammals. Smithsonian Institution Press, p. 578.

Wetzel, D.L., Reynolds III, J.E., Mercurio, P., Givens, G.H., Pulster, E.L., George, J.C., 2017. Age estimation for bowhead whales, *Balaena mysticetus*, using aspartic acid racemization with enhanced hydrolysis and derivatization procedures. J. Cetacean Res. Manage. 17, 9–14.

Willetto, C.E., O'Hara, T.M., Rowles, T., 2002. Bowhead whale health and physiology workshop, October 2001. North Slope Borough, Department of Wildlife Management, Barrow, Alaska, 129 pp.

Williams, G.C., 1966. Natural selection, the costs of reproduction, and a refinement of Lack's principle. Am. Nat. 100 (916), 687–690.

Wilson, E.O., MacArthur, R.H., 1967. The Theory of Island Biogeography. Princeton University Press.

Woodward, B.L., Winn, J.P., Fish, F.E., 2006. Morphological specializations of baleen whales associated with hydrodynamic performance and ecological niche. J. Morphol. 267, 1284–1294.

CHAPTER

8

Prenatal development

J.G.M. Thewissen[1], *D.J. Hillmann*[2], *J.C. George*[3], *R. Stimmelmayr*[3], *Raymond J. Tarpley*[4], *Gay Sheffield*[5] *and Robert Suydam*[3]

[1]Department of Anatomy and Neurobiology, Northeast Ohio Medical University, Rootstown, OH, United States [2]Department of Comparative Biomedical Sciences, School of Veterinary Medicine, Louisiana State University, Baton Rouge, LA, United States [3]Department of Wildlife Management, North Slope Borough, Utqiaġvik, AK, United States [4]Department of Veterinary Integrative Biosciences, College of Veterinary Medicine and Biomedical Sciences, Texas A&M University, College Station, TX, United States [5]Marine Advisory Program, University of Alaska Fairbanks, Nome, AK, United States

Introduction

As in all mammals, remarkable changes in bowhead whale morphology accompany prenatal growth (Fig. 8.1). The most dramatic of these changes take place in the embryonic period, which we estimate to last 8–10 weeks. In the fetal period that follows, the fetus is clearly recognizable as a balaenid and distinct from odontocetes (Stěrba et al., 2000; Thewissen and Heyning, 2007) and balaenopterids (Kükenthal, 1914; Hampe et al., 2015). Even so, dramatic morphological changes continue late into the fetal period, including the loss of teeth and the development of baleen (Thewissen et al., 2017). Koski et al. (1993) and Reese et al. (2001) estimated the length of gestation in bowhead, and Tarpley et al. (Chapter 13) further analyzed and summarized this data, concluding that the gestation length in bowheads is 14 months.

Iñupiat and Yupik people of Alaska harvest bowhead whales of the Bering-Chukchi-Beaufort Seas (BCB) stock twice yearly (Chapters 3 and 32). This subsistence hunt is the source for prenatal specimens of the species studied here. Current evidence suggests that breeding and calving times of bowhead whales are synchronized in the BCB stock (Reese et al., 2001). Breeding may take place in March and thus prenatal specimens cover three periods in the gestational timeline; the period that includes the late embryonic and early fetal period (around

The Bowhead Whale
DOI: https://doi.org/10.1016/B978-0-12-818969-6.00008-X

FIGURE 8.1 A bowhead whale mother swims along her calf, shortly after it was born. The two puffs of water vapor mark their breath. At birth, bowhead whales are more than 4 m long. *Source: Photo by Brenda Rone (NOAA/ North Slope Borough, NMFS Permit 14245).*

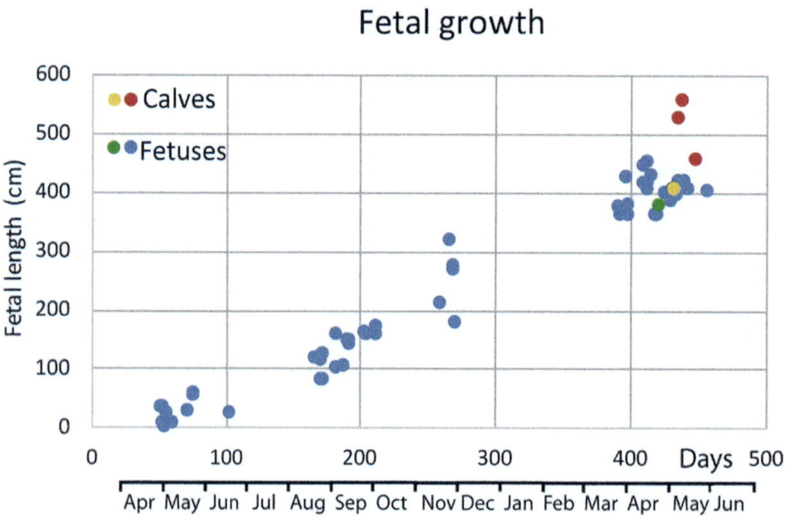

FIGURE 8.2 Size distribution of embryos, fetuses, and neonates of the BCB stock (*red* and *blue*) for a year starting at March 15, the approximate time of start of zygote formation (Reese et al., 2001). Data points for a fetus from the EGSB stock (East Greenland-Svalbard-Barents Sea, *green*; Eschricht and Reinhardt, 1866) and an ECWG calf (East Canada-West Greenland, *yellow*; Heide-Jørgensen et al., 2012) are also shown.

5–10 weeks), a period one third through gestation, and the period just before birth (Fig. 8.2). Thus, collections based on the Iñupiat subsistence hunt do not cover the entire prenatal period, and only limited observations are possible based on these specimens. The purpose of this paper is to characterize the known stages of bowhead whale embryology. Chapter 13, discusses length of gestation and mating periods in detail.

The specimens we studied were collected for scientific study by the North Slope Borough Department of Wildlife Management (NSB-DWM). Their numbering system reflects the year they were caught, the Alaskan village that caught them, the sequence number of the whale in that year, and an appended F that indicates that it is a prenatal specimen, whereas the same number without the F refers to the mother. Hence, NSB-DWM 2007B12F indicates the fetus of the 12th whale caught in 2007. The B indicates that the whale was caught in Utqiaġvik (Barrow). Bowheads are caught in a number of villages (Chapter 32), WW indicates Wainwright, and G, Gambell. Prenatal stages of these embryos are determined following criteria described for odontocetes by Thewissen and Heyning (2007).

Description and comparisons

Embryos of 8.7–9.1 cm

The youngest bowhead whale specimen in our collection (NSB-DWM 1999B7F; Fig. 8.3A) is characterized by unpigmented eyes, incomplete eyelids, and a physiological gut herniation. The total length (TL) of this embryo is 8.7 cm, as measured across the curved back from tip of snout to end of tail, and it pertains to Stage 19 (Thewissen and Heyning, 2007). It is approximately 7 weeks old. Embryo NSB-DWM 2018G3F is somewhat larger (TL = 9.05 cm; 4.6 g after formalin fixation), and somewhat further in development, as it shows the beginnings of eye pigmentation.

These specimens are similar in stage of development as a 85 mm humpback whale specimen (Kükenthal, 1914; Hampe et al., 2015). In the bowhead whales, digit 3 is the longest

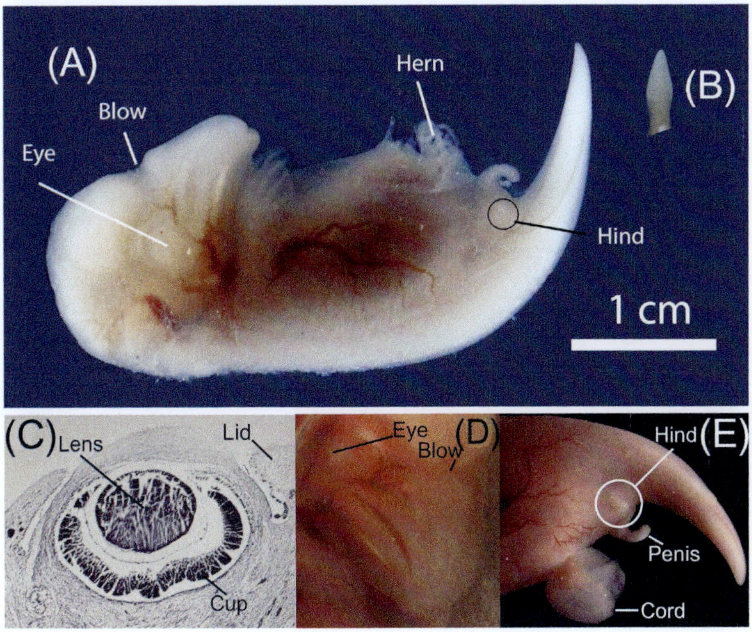

FIGURE 8.3 (A) Bowhead whale embryo of 8.7 cm (NSB-DWM 1999B7F). Note lack of pigmentation of the eye, physiological herniation (Hern) of gut into umbilical cord, position of blowhole (Blow), and remnants of hind limb buds (Hind) in this male fetus. (B) Tail tip in ventral view, showing lateral outgrowths. (C) Histological section of the eye, showing optic cup (Cup), lens vesicle (Lens), and eyelid (Lid) not covering eye. (D) Detail of face and hand, showing position of blowhole. (E) Caudal part of embryo NSB-DWM 2018GF showing penis, hind limb bud (Hind), and cut-off umbilical cord (Cord).

in the handplate and it remains like that after birth. In contrast, the second finger of dolphins outgrows the third in length at stage 17 (Stěrba et al., 2000; Thewissen and Heyning, 2007). Unlike these odontocetes, finger length cannot be used to characterize developmental stage in bowhead whales.

Bowhead whale embryos of this stage show some incipient features of mysticetes, including the elongated rostrum with the maxilla arching over the mandible. Left and right blowholes are located immediately rostral to the braincase, as they are postnatal. This position is similar to that described for odontocetes of this stage (Stěrba et al., 2000). Histological sections show that the dental lamina is present, but no tooth buds have developed (Thewissen et al., 2017). Paired low elevations on either side of the penis represent the remnants of the hind limb bud. Anlagen for nipples and the last remnant of the hind limb bud cooccur in similar-staged embryos of *Megaptera novaeangliae* (Kükenthal, 1914), and *Stenella attenuata* (Sedmera et al., 1997; Thewissen et al., 2006); however, the location of the observed structures in our bowhead whale embryos implies that they are hind limb buds (Anderssen, 1917–1918), not nipples. The tail is circular on cross-section, and has slight lateral outgrowths that are the precursors for the fluke, similar to delphinids of this age (Thewissen, 2018).

Fetuses of 12.8–16.6 cm

NSB-DWM 2016B9F shows partially pigmented eyes, eyelids that are closed over the eye, and a retracted physiological gut herniation (Fig. 8.4). The total length (TL) of this specimen is

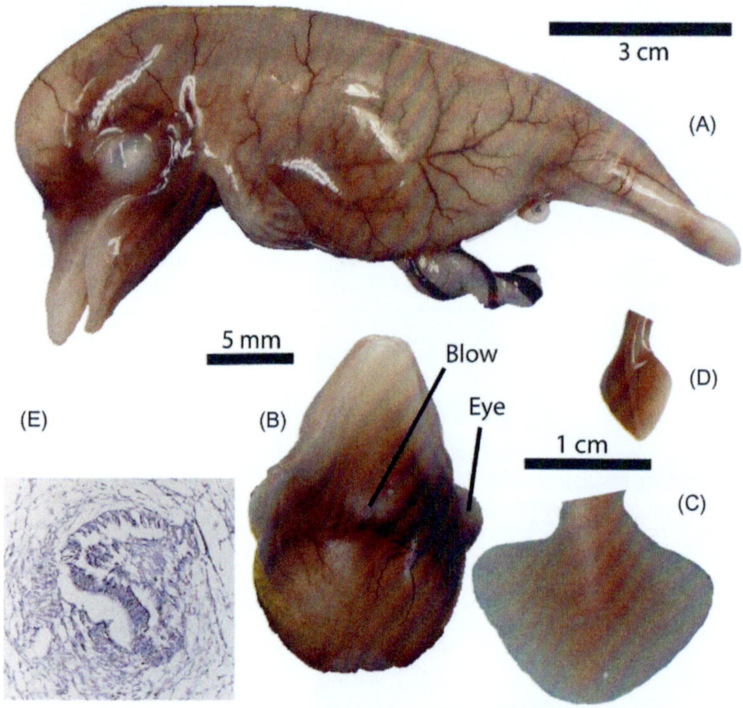

FIGURE 8.4 Bowhead whale fetuses from 12.8 to 16.6 cm. (A) NSB-DWM 2016B9F with dorsal view of head, blowhole and eyes (B), and flukes (C). (D) Tail flukes of NSB-DWM 1999B6F. (E) Histological section of upper tooth cap 1999B6F (slide 121, 20 ×). Scale bar near (B) goes with (B); scale bar near (C) and (D) goes with both.

12.8 cm and it weighs 25 g. It represents Stage 20 (Thewissen and Heyning, 2007), a similar stage to a bowhead whale fetus described by Durham (1980). NSB-DWM 1999B6F (TL = 16.6 cm) is slightly more developed, eye pigmentation has been completed and the eyelids are separated partially by the palpebral fissure. In both specimens, the blowholes are paired and slit-like. Haldiman and Tarpley (1993) report a 13.4 cm fetus that weighed 0.29 kg.

The upper jaw of fetuses of this stage is longer and straighter than the lower jaw, and the lower lip has grown dorsally and is strongly convex. Kükenthal (1914) described the allometric growth of the mouth in similar staged *Megaptera* fetuses, and Armfield et al. (2011) described allometric growth during bowhead whale ontogeny. More than 30 developing teeth occur in both upper and lower jaws and these are in the cap stage (Thewissen et al., 2017). There is no hair on any part of the face.

The tail fluke in the smaller bowhead whale specimen is diamond shaped, and it is more or less triangular in the larger specimen. Phalanges can be seen through the translucent hand plate, and tentative counts of phalanges made. The third finger is longest and has five phalanges, while the second and fourth have three. The fifth finger has a single phalanx, and the only evidence of a first digit is a small outgrowth from the carpus. Cooper et al. (2017) described flipper development and its gene control in cetaceans.

Fetuses of 27.4–40.3 cm

Several of these fetuses were described and measured by Durham (1980), and additional specimens have been recovered for this study (NSB-DWM 2000B3F, 2007B12F, 2013B1F, 2016B12F; Fig. 8.5). The developmental stage for these fetuses is 21/22, and NSB-DWM 2013B1F weighs 0.46 kg. Skin pigmentation is absent in these fetuses. Kükenthal (1914) described the external morphology of a bowhead whale fetus with total length of 520 mm. The flukes are triangular, although they have not taken the full shape of the adult bowhead whale flukes. Penis and clitoris are well developed and not hidden in folds of skin as they are in the adult.

This is the earliest stage at which hairs appear on the face, a similar timing to that of humpback whales (Kükenthal, 1914). Drake et al. (2015) described three clusters of hairs in postnatal individuals (Chapter 14), but an additional region with hairs is present in fetuses of this stage. Two to four hairs, much longer and thicker than hairs on rostrum and mandible tip, occur in a row between the tip of the rostrum and the lateral edge of the blowhole. This placement is similar to the only hair found in fetal dolphins (Drake et al., 2015). No other hair clusters will develop on the body.

The soft tissues of NSB-DWM 2000B3F were cleared and the cartilage was stained green and the bone purple (Fig. 8.5B). The skull shows ossification cores for most bones, including the lacrimal, and none of the bones or their cartilage precursors overlap (which would indicate telescoping, Roston and Roth, 2019). The supraoccipital consists of cartilage, but the interparietal is ossified, and makes up the entire dorsal side of the posterior side of the skull (see discussion by Mead and Fordyce, 2009). In *Stenella attenuata*, the supraoccipital ossifies before the interparietal (Moran et al., 2011). Several small ossified areas occur just anterior to the interparietal.

Meckel's cartilage is present, extending from the ear region into the mandible, which is ossified. Kükenthal (1914) described a trace of the external auditory meatus in a fetus, and

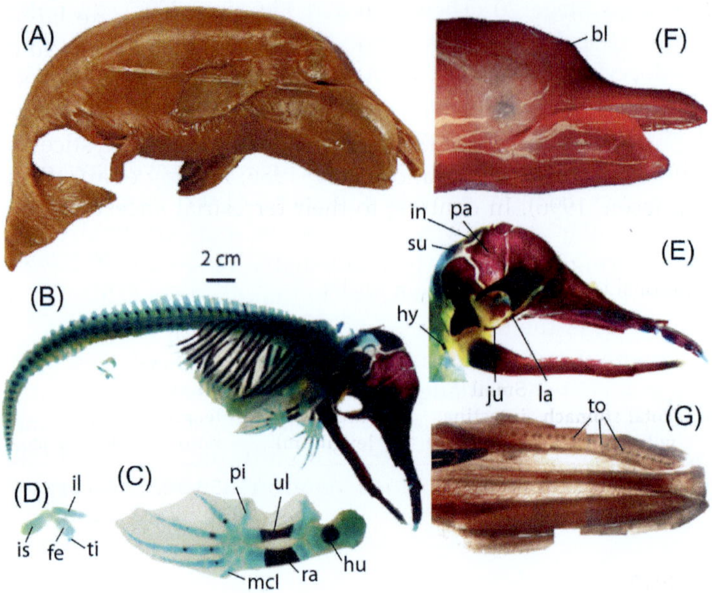

FIGURE 8.5 Bowhead whale fetuses of 27.4–40.3 cm. (A) Fetus NSB-DWM 2000B3F, and in clear-and-stain preparation (B), with forelimb (C), (internal) hind limb cartilages (D), and head (E) enlarged. (F) Head of fetus NSB-DWM 2013B1F. (G) Palate of NSB-DWM 2000B3F, backlit to show many tooth germs. *bl*, blowhole; *fe*, femur; *hu*, humerus; *hy*, hyoid; *il*, ilium; *in*, interparietal; *is*, ischium; *ju*, jugal; *la*, lacrimal; *mcI*, first metacarpal; *pa*, parietal; *pi*, pisiform; *ra*, radius; *su*, supraoccipital; *ti*, tibia; *to*, tooth germs; *ul*, ulna. Scale bar pertains to (A) and (B).

that structure is visible in NSB-DWM 2016B12F too. It is located 7.5 mm caudal to the eye as a small pigmented area that does not indent or pierce the skin.

Four carpals are located distal to the radius and ulna in the forelimb, and a fifth carpal, the pisiform, projects caudally. It is larger than the first metacarpal. The first digit does not have phalanges, and the digital formula for the remaining digits is 3.5.3.2. This matches the digital formula of the adult (Cooper, 2018). Small ossifications can be recognized in metacarpal 2, 3, and 4, and in the proximal phalanges of digits 2 and 3.

In NSB-DWM 2000B3F, four paired cartilage anlagen representing hind extremity elements are visible (Fig. 8.5B–D), with the left and right side group far from the midline. Three of these anlagen meet and form a T-shape; we interpret them as ilium, ischium, and femur. The femur anlagen is directed medially, and connected to the fourth anlagen, the tibia.

Fetuses of this stage have a band of thickened epithelium on their palate where the baleen will form, and protein signaling in this band suggests that this tissue is already patterned to initiate baleen plate formation. Embedded in the palate are more than 25 tooth germs on each side of the upper and lower jaw. These tooth germs are in the bell stage (Thewissen et al., 2017).

Fetuses of 84–175 cm

Between 1992 and 2018, 14 fetuses were recovered in September and October, ranging in length from 84 to 175 cm. We estimate the gestational age of these fetuses to be between 170 and 217 days. The head, and especially the oral cavity, has grown disproportionally in these fetuses, and the mouth opening makes up 85% of the length of the skull. This trend of allometric growth continues in the remainder of prenatal and much of postnatal life

TABLE 8.1 Organ weights of bowhead whale fetuses.

Fetus number	2000B3F	2015B24F	1989B2F	2016B5F	2018WW2F
Total length (cm)	40.3	175	401	432	419
Brain (g)	–	–	–	–	1450
Heart (g)	7.4	1930	9080	14,968	–
Lung (g, single)	11.0	880	–	39,462	–
Liver (g)	41.6	–	–	22,226	–
Kidney (g, mean of two)	9.2	1285	–	–	–

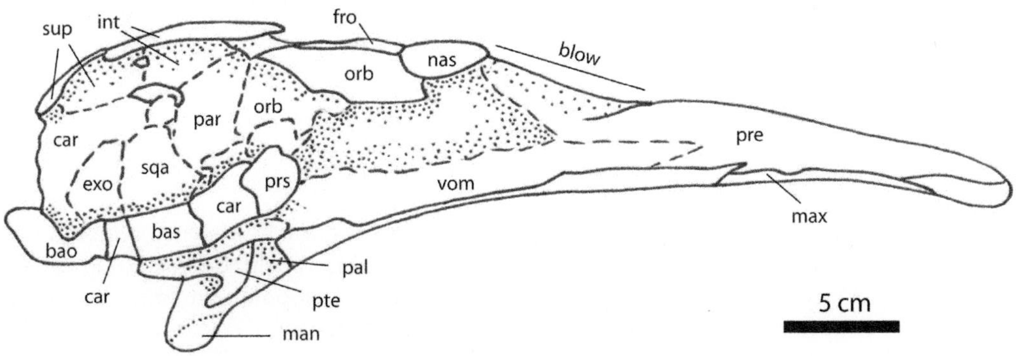

FIGURE 8.6 Sagittal section of bowhead whale fetal skull (NSB-DWM 2007B16F). *bao*, basioccipital; *bas*, basisphenoid; *blow*, extent of blowhole; *car*, cartilage; *exo*, exoccipital; *fro*, frontal; *int*, interparietal; *man*, mandibular fossa; *max*, maxilla; *nas*, nasal; *orb*, orbitosphenoid; *pal*, palatine; *par*, parietal; *pre*, premaxilla; *prs*, presphenoid; *pte*, pterygoid; *sqa*, squamosal; *sup*, supraoccipital; *vom*, vomer.

(Armfield et al., 2011; George et al., 2016). Table 8.1 provides data on relative growth of some internal organs. Baleen is not present in these fetuses, and no remnants of tooth germs remain (Thewissen et al., 2017). The skin of these specimens is pigmented and light gray in color.

The skull of another fetus of this age (NSB-DWM 2007B16F) was skeletonized (Fig. 8.6). Telescoping has started with the interparietal overriding frontal in the midline. Lateral to the nasal opening, the premaxilla and maxilla extend further posterior than the nasal, and the vomer covers most of the basisphenoid ventrally. The nasal opening is well rostral to the braincase, whereas these structures were adjacent in earlier fetuses. Fig. 7.2E (this volume) shows images of a head of an older fetus.

CT scans of one specimen (NSB-DWM 1992B7F) allow the study of development of internal structures such as muscles (Chapter 10). As in other mammals, the fetal period mostly encompasses growth of organs that were initially formed in the embryonic period. The absence of baleen is an exception to this. Another exception is the absence of skinfolds that surround a recess that holds the external genitals. As a result, the penis is exposed in this fetus.

Full-term fetus

At birth, a bowhead whale 1989B2F was 401 cm long, weighed 1050 kg, of which 443 was blubber, consistent with measurements provided by Koski et al. (1993) and George (2009; Chapter 7). The growth that it has undergone was powered by its mother, which, when she reached sexual maturity, had a body more than 20 times as heavy as the calf. She weighed a minimum of 27,000 kg and was 13 m long (George, 2009). While this seems remarkable, the ratio of sizes between full-term fetal and maternal size of approximately 4% is not very different from that of humans or other mammals. Tarpley et al. (1997) described the heart of a full-term fetus, finding that it weighs 1% of the full-term fetal body weight, also similar to other mammals. Mean baleen length for these fetuses is 11.2 cm ($n = 3$).

The body profile of a full-term bowhead whale fetus (or a newborn calf; see Fig. 7.4, this volume) is narrow and slender, unlike that of a whale that is being nursed. This is partially the result of the thin layers of blubber in the calf. Mean dorsal blubber thickness of these fetuses is 6.0 cm ($n = 5$).

Discussion

Heterochrony, the process of differential rates of development of different organs, during prenatal development is a major source of variation that enables evolution, including in cetaceans (Tsai and Fordyce, 2014). Examples of heterochronous development in cetaceans include the formation of supernumerary teeth in toothed whales and prenatal baleen whales (Armfield et al., 2011; Thewissen et al., 2017), and supernumerary phalanges of the forelimb (Richardson and Oelschläger, 2002; Cooper et al., 2017). In general, the timing of initiation of such organs follows that of other mammals, but their later development deviates. This even goes for organs that strongly deviate from traditional terrestrial mammal shapes later in ontogeny, such as hind limb buds (which disappear in cetaceans) and tooth germs (which disappear in mysticetes).

The order Cetacea evolved some structures not present in other mammals, such as tail flukes and baleen. Initiation of the flukes starts when the hind limb buds are still present, and it is possible that signaling leading to fluke formation mirrors that of limb formation, given that both are paired outgrowths partly controlled by axial skeleton development. Baleen development is closely linked to tooth development and uses similar signaling pathways (Thewissen et al., 2017). Both flukes and baleen may represent exaptations, where signaling pathways initially used to build one organ have been exapted to build another.

References

Anderssen, M.S., 1917–1918. Studier over mammarorganernes utvikling hos *Phocaena communis*. Bergen. Mus. Aarb. Naturvidensk. R. 3, 1–45.

Armfield, B.A., George, J.C., Vinyard, C.J., Thewissen, J.G.M., 2011. Allometric patterns of fetal head growth in mysticetes and odontocetes: comparison of *Balaena mysticetus* and *Stenella attenuata*. Mar. Mamm. Sci. 27, 819–827.

Cooper, L.N., 2018. Forelimb anatomy. In: Würsig, B., Thewissen, J.G.M., Kovacs, K. (Eds.), Encyclopedia of Marine Mammals, third ed. London, Elsevier, pp. 385–388.

Cooper, L.N., Sears, K.E., Armfield, B.A., Kala, B., Hubler, M., Thewissen, J.G.M., 2017. Review and experimental evaluation of the embryonic development and evolutionary history of flipper development and hyperphalangy in dolphins (Cetacea: Mammalia). Genesis 56 (1), e23076. Available from: https://doi.org/10.1002/dvg.23076.

Drake, S., Crish, S., George, J.C., Stimmelmayr, R., Thewissen, J.G.M., 2015. Sensory hairs in the bowhead whale, *Balaena mysticetus* (Cetacea, Mammalia). Anat. Rec. 298, 1327–1335.

Durham, F.E., 1980. External morphology of bowhead fetuses and calves. Mar. Fish. Rev. 42, 74–80.

Eschricht, D.F., Reinhardt, J., 1866. On the Greenland Right-Whale (*Balaena mysticetus*). The Ray Society, London, pp. 1–150, 6 pl.

George, J.C., 2009. Growth, Morphology, and Energetics of the Bowhead Whales (*Balaena mysticetus*) (Ph.D. thesis). University of Alaska, Fairbanks, 169pp.

George, J.C., Stimmelmayr, R., Suydam, R., Usip, S., Givens, G., Sformo, T., et al., 2016. Severe bone loss as part of the life history strategy of bowhead whales. PLoS ONE 11 (6), e0156753.

Haldiman, J.T., Tarpley, R.J., 1993. Anatomy and physiology. In: Burns, J.J., Montague, J.J., Cowles, C.J. (Eds.), The Bowhead Whale. Society for Marine Mammalogy, Lawrence, KS, pp. 71–156, Spec. Publ. 2.

Hampe, O., Franke, H., Hipsley, C.A., Kardjilov, N., Müller, J., 2015. Prenatal cranial ossification of the humpback whale. J. Morphol. 276, 564–582.

Heide-Jørgensen, M.P., Garde, E., Nielsen, N.H., Andersen, O.N., 2012. Biological data from the hunt of bowhead whales in West Greenland 2009 and 2010. J. Cetacean Res. Manage. 12, 329–333.

Koski, W.R., Davis, R.A., Miller, G.W., Withrow, D.E., 1993. Reproduction. In: Burns, J.J., Montague, J.J., Cowles, C.J. (Eds.), The Bowhead Whale. Society for Marine Mammalogy, Lawrence, KS, pp. 239–274. Spec. Publ. 2.

Kükenthal, W., 1914. Untersuchungen an Walen, zweiter Teil. Jenaischer Z. Naturw. 51, 1–122.

Mead, J.G., Fordyce, R.E., 2009. The therian skull: a lexicon with emphasis on the odontocetes. Smithson. Contr. Zool. 627, 1–248.

Moran, M.M., Nummela, S., Thewissen, J.G.M., 2011. Development of the skull of the pantropical spotted dolphin (*Stenella attenuata*). Anat. Rec. 294, 1743–1756.

Reese, C.S., Calvin, J.A., George, J.C., Tarpley, R.J., 2001. Estimation of fetal growth and gestation in bowhead whales. J. Am. Stat. Assoc. 96, 915–928.

Richardson, M.K., Oelschläger, H.H.A., 2002. Time, pattern, and heterochrony: a study of hyperphalangy in the dolphin embryo flipper. Evol. Dev. 4, 435–444.

Roston, R.A., Roth, L., 2019. Cetacean skull telescoping brings evolution of cranial sutures into focus. Anat. Rec. 302, 1055–1073. Available from: https://doi.org/10.1002/ar.24079.

Sedmera, D., Míšek, I., Klima, M., 1997. On the development of cetacean extremities. II. Morphogenesis and histogenesis of the flippers in the spotted dolphin (*Stenella attenuata*). Eur. J. Morphol. 35, 117–123.

Stěrba, O., Klima, M., Schlidger, B., 2000. Embryology of dolphins, staging and ageing of embryos and fetuses of some cetaceans. Adv. Anat. Embryol. Cell Biol. 157, 1–133.

Tarpley, R.J., Hillmann, D.J., Henk, W.G., George, J.C., 1997. Observations on the external morphology and vasculature of a fetal heart of the bowhead whale, *Balaena mysticetus*. Anat. Rec. 247, 556–581.

Thewissen, J.G.M., 2018. Highlights of cetacean embryology. Aquat. Mamm. 44, 591–602.

Thewissen, J.G.M., Heyning, J., 2007. Chapter 15, Embryogenesis and development in *Stenella attenuata* and other cetaceans. In: Miller, D. (Ed.), Reproductive Biology and Phylogeny of Cetacea. Science Publishers, Enfield, NH, pp. 307–329.

Thewissen, J.G.M., Cohn, M.J., Stevens, L.S., Bajpai, S., Heyning, J., Horton Jr., W.E., 2006. Developmental basis for hind limb loss in dolphins and the origin of the cetacean bodyplan. Proc. Natl. Acad. Sci. U. S. A. 103, 8414–8418.

Thewissen, J.G.M., Hieronymus, T.L., George, J.C., Suydam, R., Stimmelmayr, R., McBurney, D., 2017. Evolutionary aspects of the development of teeth and baleen in the bowhead whale. J. Anat. 230, 549–566. Available from: https://doi.org/10.1111/joa.12579.

Tsai, C.-H., Fordyce, R.E., 2014. Disparate heterochronic processes in baleen whale evolution. Evol. Biol. 41, 299–307.

species comparisons, the intestinal tract of the whale [illegible] comparisons for new research, the method relies on [illegible] differences in prey [illegible] per feeding event, and this may not necessarily be the case, and can therefore underestimate the number of distinct feeding episodes and gut passage times.

CHAPTER

9

Anatomy of skull and mandible

D.J. Hillmann[1], Raymond J. Tarpley[2], J.C. George[3], P.B. Nader[4] and J.G.M. Thewissen[5]

[1]Department of Comparative Biomedical Sciences, School of Veterinary Medicine, Louisiana State University, Baton Rouge, LA, United States [2]Department of Veterinary Integrative Biosciences, College of Veterinary Medicine and Biomedical Sciences, Texas A&M University, College Station, TX, United States [3]Department of Wildlife Management, North Slope Borough, Utqiaġvik, AK, United States [4]Department of Anatomy, College of Veterinary Medicine, Lincoln Memorial University, Harrogate, TN, United States [5]Department of Anatomy and Neurobiology, Northeast Ohio Medical University, Rootstown, OH, United States

Introduction

The chapter reports on the boney structure of the head of the bowhead whale. The information is presented primarily in the form of illustrations and identifies the bones that form the structural basis of the whale's skull and mandibles. Terminology follows that of Mead and Fordyce (2009), or common comparative anatomical and veterinary terms and some of this material is presented by Hillmann et al. (1999).

The illustrated material is primarily from examinations of seven bowhead whale skulls prepared and mounted for display at various locations in Utqiaġvik, Alaska (Hillmann et al., 1997). These animals were harvested by subsistence whale hunters in Utqiaġvik. Bowhead whales are central to the culture of the indigenous people of northern Alaska (Chapter 31), and local communities are proud of their relation with the whales (Fig. 9.1).

The bones of the skull

The occipital bone is an unpaired bone that makes up much of the caudal aspect of the skull as well as the dorsal surface of the braincase (Figs. 9.2A, B and 9.3A, D). In the fetal

The Bowhead Whale
DOI: https://doi.org/10.1016/B978-0-12-818969-6.00009-1

FIGURE 9.1 Iñupiat hunters of Cross Island, Alaska, line up skulls of the bowhead whales they have harvested. This is a testament of the close connection between people and whales. Skulls of harvested animals form the basis for the information in this chapter. *Source: Photo by Craig George.*

whale, much of the occipital bone is cartilaginous and not yet ossified, and its parts can be recognized (supraoccipital, exoccipital, and basioccipital in Fig. 9.4A–C). Midsagittally, the flattened supraoccipital part overlaps the frontal bones. As in all mammals, the occipital condyle is paired and articulates with the first cervical vertebra, while surrounding most of the foramen magnum. Through the foramen magnum pass the spinal cord (and its coverings) and many blood vessels. The spinal cord of the bowhead in this area is less than 2 cm in diameter and most of the foramen is filled with many small blood vessels (a vascular rete) and the covering (meninges) of the spinal cord (see Fig. 15.6, this volume). The atlanto-occipital joint is formed by the occipital condyles and the first cervical vertebra (the atlas). Neck vertebrae are fused to varying extents in bowheads (Chapter 10), but the atlanto-occipital joint is a true synovial articulation and never fused.

The basioccipital, basisphenoid, and presphenoid are in the median plane and form the floor of the braincase (Fig. 9.4A–C). Ventrally, the presphenoid is cradled by the caudal part of the vomer and palatine bones (Fig. 9.3B). The pterygoids define the lateral walls of the nasopharynx.

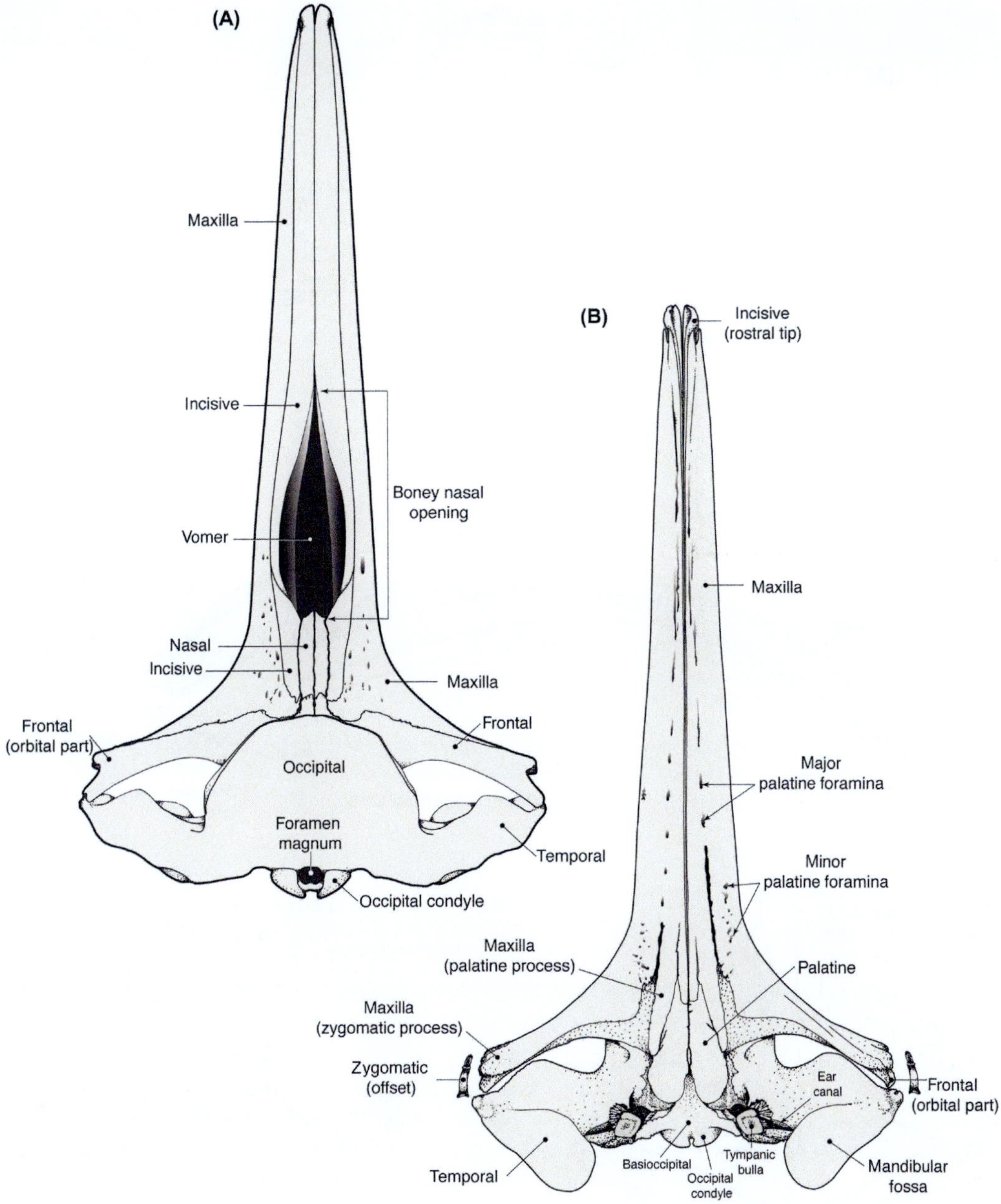

FIGURE 9.2 Dorsal (A) and ventral view (B) of the skull of the bowhead whale. *Source: Drawings by D. Hillmann.*

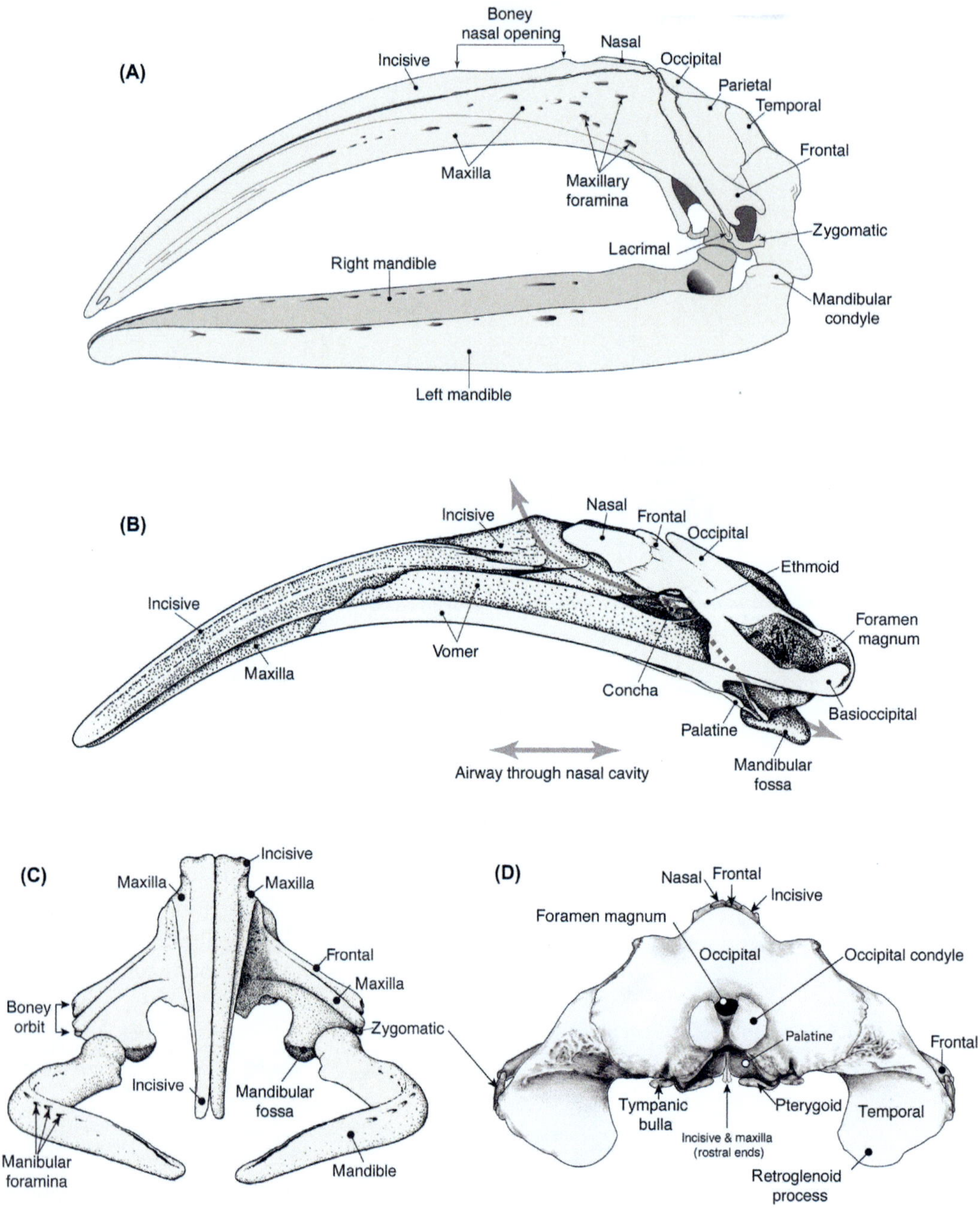

FIGURE 9.3 Lateral (A), midsagittal (B), rostral (C), and caudal (D) views of the skull of the bowhead whale. The mandibles are included in A and C. *Source: Drawings by D. Hillmann.*

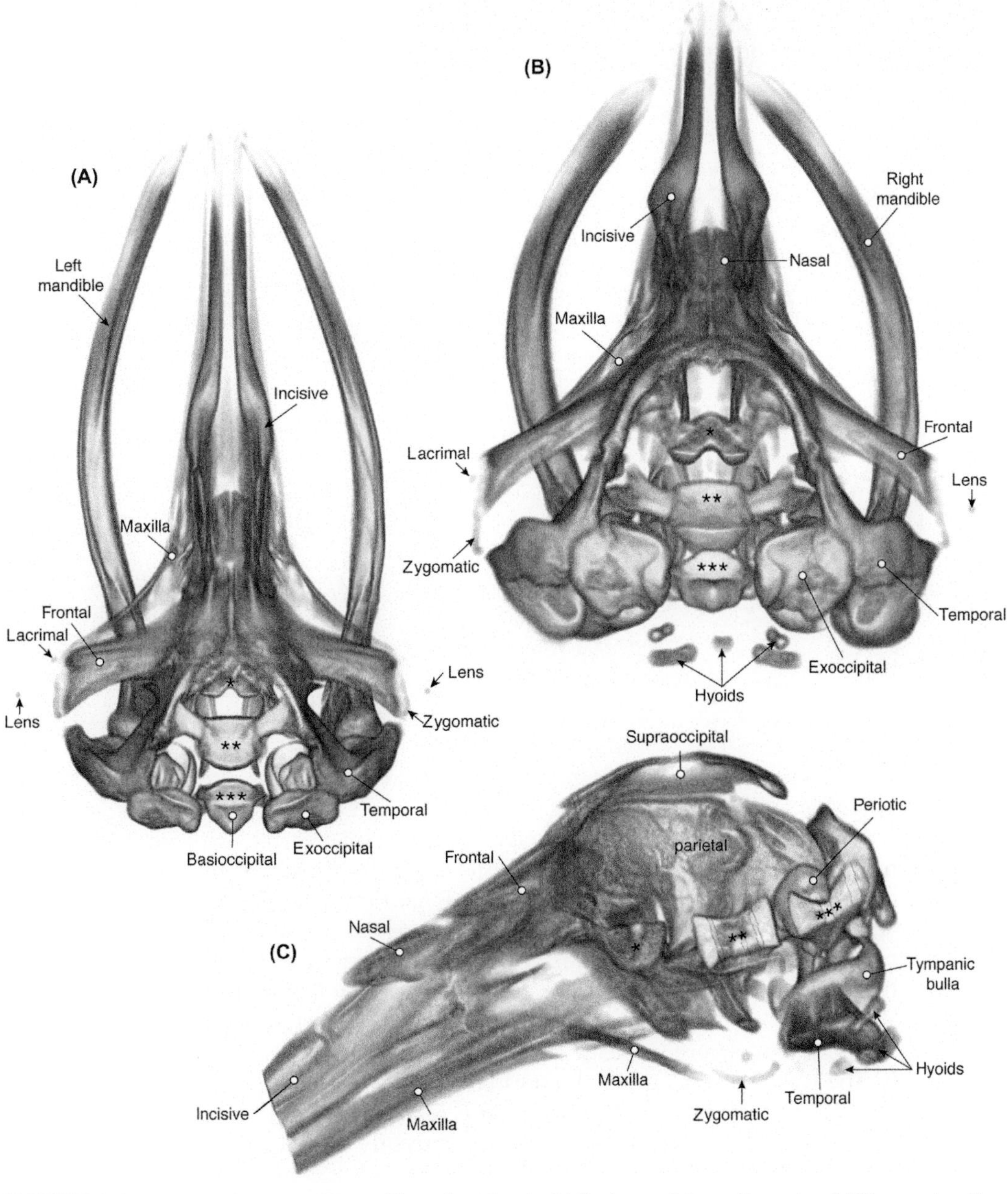

FIGURE 9.4 Dorsal (A), dorso-oblique (B), and midsagittal (C) views of three-dimensional CT reconstructions of a fetal bowhead whale (NSB-DWM 1992B13F, total body length, 145 cm). The supraoccipital part of the occipital is removed in A and B. In A–C, * = presphenoid, ** = basisphenoid, and *** = basioccipital bones. *Source: Images by D. Hillmann.*

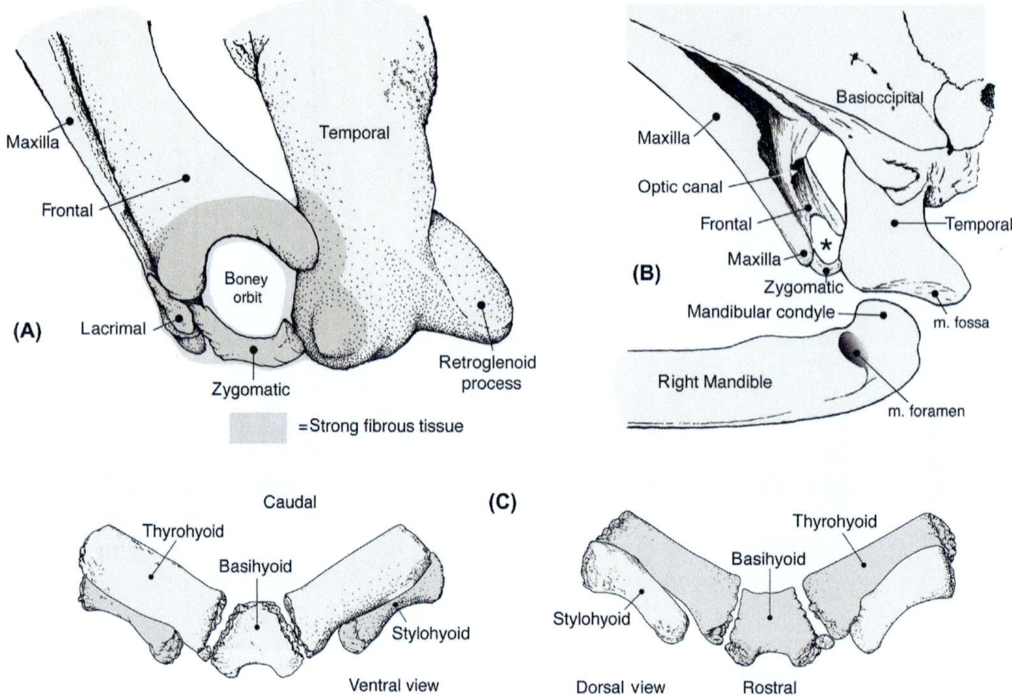

FIGURE 9.5 Lateral view of left orbital region (A), ventromedial view of right caudal palate and temporomandibular joint (B), and hyoid bones in ventral (C) and dorsal (D) views. *Source: Drawings by Hillmann.*

The temporal bones (Figs. 9.2A, B, 9.3A–D, 9.4A–C, and 9.5A, B) form much of the lateral wall of the cranial vault, the caudal aspect of the boney orbit, and the mandibular fossa. The mandibular fossa is located on a large, ventrally projecting process (Figs. 9.2B, 9.3A, C, and 9.5B). Caudal to the fossa is the retroglenoid process (Figs. 9.3D–9.5A), which projects caudally more than ventrally. The articular surface of the mandibular fossa faces somewhat cranioventrally and articulates with the rounded mandibular condyle to form the temporomandibular joint. Auditory ossicles and the bony labyrinth of the petrous part of the temporal bone are described in Chapter 18. Details on the anatomy of the petrous part of the temporal bone were published by Eckdale and Racicot (2015).

Externally, the right and left parietal bones contribute to the lateral surface of the temporal fossa (Fig. 9.3A). Internally, they form the rostrolateral walls of the cranial vault (Fig. 9.4C). The right and left frontal bones project strongly lateral to form the dorsal margin of the boney orbit (Figs. 9.2A, B, 9.3A, C, D, 9.4A, and 9.5A, B), as well as the rostral aspect of the temporal fossa. Internally, the frontals form part of the roof of the nasal cavity (Fig. 9.3B).

The ethmoid bone (Fig. 9.3B) forms the most rostral aspect of the cranial vault and thereby separates the cranial cavity from the nasal cavity. It consists of the cribriform plate and ethmoturbinates (Chapter 18). The vomer, incisive (premaxilla), and maxilla are greatly elongated to form the curved rostrum that supports the baleen rack (Figs. 9.2A and 9.3B). The vomer is V-shaped in cross section and cradles the massive mesorostral cartilage (Mead and Fordyce, 2009) where it forms the foundation for the nasal septum on the midline. The paired nasal bones form the caudal aspect of the boney nasal opening (Figs. 9.2A and 9.3A, B, D), and, with the internal parts of the frontal bones, also form the rostral part of the roof of the nasal cavity (Fig. 9.3B). Frontal, nasal, and the supraoccipital part of the occipital bone form a greatly thickened part of the bowhead whale skull (Figs. 9.2A and 9.3A, B, D) that provides the structural basis for the animal's ability to break through the thickened ice to create breathing holes. The vomer and mesorostral cartilage separate the nasal cavity into right and left halves (Klima, 1999).

The lacrimal bone forms a small part of the rostral edge of the bony orbit (Figs. 9.3A, 9.4A, B, and 9.5A). In older individuals, this bone is fused with the maxilla and frontal bones. Most of the baleen rack is anchored in a deep groove in the maxilla (Lambertsen et al., 1989) pierced by numerous major palatine foramina (Fig. 9.2B). The left and right incisive bones lie dorsal and medial to the two maxillary bones, and all four bones together form most of the rostrum of the bowhead whale head (Figs. 9.2A and 9.3A–C). The incisive bones contribute part of the lateral walls of the boney nasal opening. The left and right palatine bones together make up most of the caudal aspect of the hard palate, and thus the caudal aspect of the roof of the month, extending as far caudal as the ear region (Figs. 9.2B and 9.3B). The most caudal baleen plates are anchored on the palatine bones at the rear of the mouth. The zygomatic bone (also called jugal) (Figs. 9.2B, 9.3A, C, D, 9.4A–C, and 9.5A) connects the zygomatic process of the maxilla to the temporal bone and completes the ventral aspect of the boney orbital margin.

Mandible and hyoid apparatus

The left and right mandibles are curved in the mediolateral plane and are essentially straight in the dorsoventral plane (Figs. 9.3A, C and 9.4A, B). The coronoid process is very small, and rostrally there is no boney contact between the left and right mandibles. The mandibular foramen is large but does not cover the depth of the lower jaw (Fig. 9.5B). The large mandibular foramen leads into the mandibular canal which extends rostrally, carrying the inferior alveolar nerve and numerous blood vessels. Small foramina connect the mandibular canal to the dorsal surface of the mandible and transmit nerves and vessels to the massive lower lips (Fig. 9.3C). The mandibular condyle is located on the condylar process and articulates with the concave mandibular fossa of the temporal bone. A thick meniscus (fibrocartilaginous pad) divides the temporomandibular joint (TMJ) into dorsal and ventral compartments.

The hyoid bones consist of paired stylohyoids and thyrohyoids, and an unpaired basihyoid (Eschricht and Reinhardt, 1866). Attachment of the stylohyoid to the skull is fibrous whereas the attachment of the thyrohyoid to the basihyoid is cartilaginous. The paired thyrohyoids project caudally to attach and cradle the laryngeal apparatus (Fig. 9.5A–C) (Schoenfuss et al., 2014).

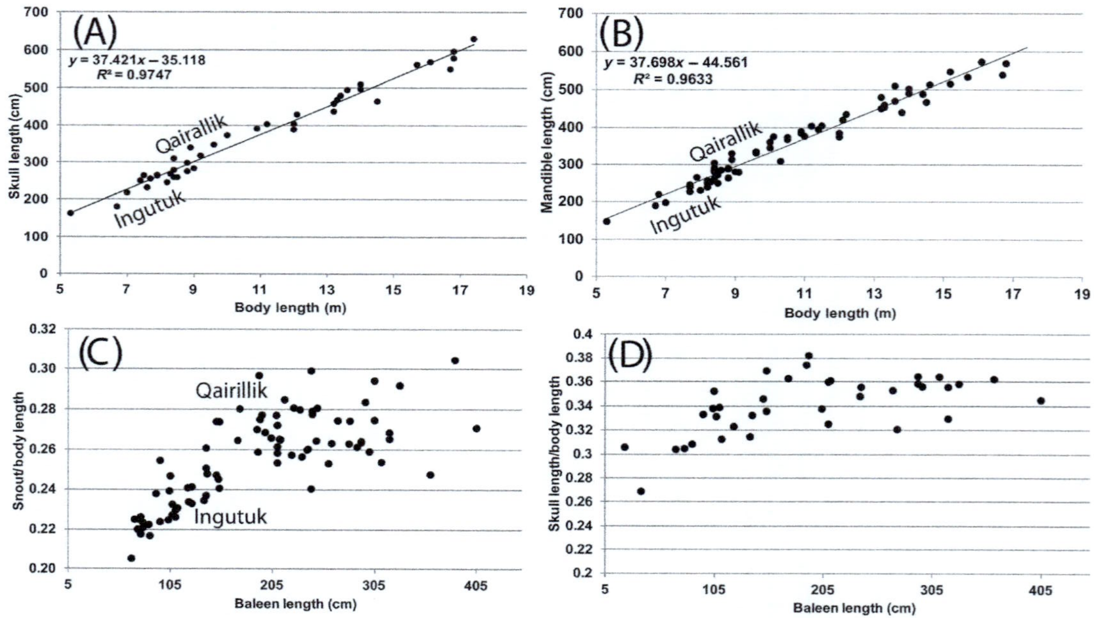

FIGURE 9.6 Skull length (A, condylobasal length, in cm) and mandible length (B) plotted against body length for specimens landed in Utqiaġvik. Ratios of snout length divided by body length (C) and skull length divided by body length (D), plotted against longest baleen length. Longest baleen length correlates well with age (Chapter 21). Ingutuks and qairilliks are stages in the life of bowhead whales (Chapter 7).

Bones of the cranial vault

The cranial vault (also called the cranial cavity or braincase) houses the brain, its meninges, cranial nerves, and blood vessels (Figs. 9.3B and 9.4C). The roof and most caudal aspect of the cranial vault is formed by the (unpaired) supraoccipital part of the occipital bone. The unpaired ethmoid bone forms the rostral aspect of the cranial vault. The (unpaired) basisphenoid and presphenoid bones form the middle and rostral portions of the floor of the cranial vault. The massive paired (right and left) temporal bones contribute much of the lateral walls of the cranial vault. The paired (right and left) parietal bones make up the rostrolateral walls of the cranial vault.

Boney orbit and position of the eye

The boney orbit forms a ring around the eye and is comprised of five bones in the bowhead (Fig. 9.5A), the frontal dorsally, the temporal caudally, the zygomatic ventrally, and the maxillary and lacrimal rostrally. The bowhead whale eye is located 10–15 cm lateral to the boney orbit through which blood vessels, optic nerve, and extrinsic eye muscles pass to reach the eye (Zhu et al., 2000). The eyes are placed laterally on the head, and there is likely very little overlap of the right and left fields of vision (Zhu et al., 2000).

The boney nasal opening and nasal cavity

The boney nasal opening is located just rostral to the highest point of the skull and lies beneath the (soft tissue) blowhole (external nares or nostrils). The margin of the boney nasal opening is formed laterally by the incisive bones and caudally by the nasal bones (Figs. 9.2A and 9.3A).

The nasal cavity is divided into left and right halves by a cartilaginous nasal septum (the mesorostral cartilage of Mead and Fordyce, 2009), which is cradled by the vomer. The nasal cavity is centrally located within the skull where it is the passageway for air during respiration and houses the conchal (turbinate) bones (Fig. 9.3B) (Chapter 18). The nasal cavity is bounded dorsally by the thick, paired nasal bones and laterally by the paired incisive and maxillary bones.

Skull growth

Skull length and mandible length correlate well with body length (Fig. 9.6A and B). George et al. (2015) documented the presence of a growth hiatus between year 1 and 6 in bowhead whales, when the head grows faster than the body (Fig. 9.6C and D; Chapter 7). The short relative size of the rostrum and skull is obvious in 1-year-old whales (ingutuks, Fig. 9.6C), which have short baleen.

References

Eckdale, E.G., Racicot, R.A., 2015. Anatomical evidence for low frequency sensitivity in an archaeocete whale: comparison of the inner ear of the inner ear of *Zygorhiza kochii* with that of crown *Mysticeti*. J. Anat. 226, 1–57.

Eschricht, D.F., Reinhardt, J., 1866. On the Greenland right whale (*Balaena mysticetus*). In: Flower, W.H. (Ed.), Recent Memoirs on the Cetacea. The Ray Society, London, pp. 1–150.

George, J.C., Druckenmiller, M.L., Laidre, K.L., Suydam, R., Person, B., 2015. Bowhead whale body condition and links to summer sea ice and upwelling in the Beaufort Sea. Prog. Oceanogr. 136, 250–262.

Hillmann, D.J., Henk, W.G., Nader, P.B., 1997. Initial observations on the skull and mandibles from Eskimo harvested bowhead whales (*Balaena mysticetus*). Report submitted to the North Slope Borough Department of Wildlife Management, Box 69, Barrow, AK 99723. By the: Department of Veterinary Anatomy & Cell Biology, School of Veterinary Medicine, Louisiana State University, South Stadium Road, Baton Rouge, LA.

Hillmann, D.J., Henk, W.G., Nader, P.B., Zhu, Q., 1999. Anatomy of the Bowhead Whale Skull and Mandible. A brochure for use by the general public to be distributed by the Department of Wildlife Management. North Slope Borough, Barrow, Alaska.

Klima, M., 1999. Development of the cetacean nasal skull. Adv. Anat. Embryol. Cell Biol. 149, 1–143.

Lambertsen, R., Hintz, R., Hirons, A., Kreiton, K., Moor, C., 1989. Characterization of the functional morphology of the mouth of the bowhead whale, *Balaena mysticetus*, with special emphasis on feeding and filtration mechanisms. Report to the Department of Wildlife Management, North Slope Borough, Barrow, Alaska from Ecosystems, Inc., Institute for Environmental Medicine, University of Pennsylvania Medical Center, Philadelphia, PA, 134 pp.

Mead, J.G., Fordyce, R.E., 2009. The therian skull: a lexicon with emphasis on the odontocetes. Smithson. Contrib. Zool. 627, 248.

Schoenfuss, H.L., Bragulla, H.H., Schumacher, J., Henk, W.G., George, J.C., Hillmann, D.J., 2014. The anatomy of the larynx of the bowhead whale, *Balaena mysticetus*, and its sound-producing functions. Anat. Rec. 297, 1316–1330.

Zhu, Q., Henk, W.G., Hillmann, D.J., 2000. Observations on the muscles of the eye of the bowhead whale, *Balaena mysticetus*. Anat. Rec. 259, 189–204.

CHAPTER

10

Postcranial skeleton and musculature

J.G.M. Thewissen[1], *D.J. Hillmann*[2], *J.C. George*[3], *Raymond J. Tarpley*[4], *Gay Sheffield*[5], *R. Stimmelmayr*[3] *and Robert Suydam*[3]

[1]Department of Anatomy and Neurobiology, Northeast Ohio Medical University, Rootstown, OH, United States [2]Department of Comparative Biomedical Sciences, School of Veterinary Medicine, Louisiana State University, Baton Rouge, LA, United States [3]Department of Wildlife Management, North Slope Borough, Utqiaġvik, AK, United States [4]Department of Veterinary Integrative Biosciences, College of Veterinary Medicine and Biomedical Sciences, Texas A&M University, College Station, TX, United States [5]Marine Advisory Program, University of Alaska Fairbanks, Nome, AK, United States

Introduction

The bowhead skeleton (Figs. 10.1 and 10.2) and its musculature have all the hallmarks of a typical cetacean: a streamlined mammal with large muscles to propel the horizontal fluke, with stiff, nonmuscular paddles as forelimbs, no external hind limbs, and no ability to support its weight on land. Within that constraining and generalized cetacean design scheme, bowheads stand out among the largest, fattest, and slowest cetaceans. The specific adaptations of their musculoskeletal system are for efficient feeding, for storage of resources, for minimizing energy expenditure, and for living in water with near-freezing temperatures. To this effect, they have a large head with a large baleen rack, long baleen plates, thick blubber layers, low body temperature, and a locomotor skeleton that emphasizes slow and steady movements.

From a physiological perspective, the most impressive musculoskeletal feature of bowheads is their slow maturation. Bowhead whales become sexually mature around 20–25 years of age, but the growth of their skeleton does not cease until long thereafter. Additionally, the species can reach ages up to 200 years (Chapter 7).

The most complete description of the anatomy of the skeletal system of the species is by Eschricht and Reinhardt (1866), and specific parts of the musculoskeletal systems have been described in detail (such as the larynx, Schoenfuss et al., 2014). Haldiman and

DOI: https://doi.org/10.1016/B978-0-12-818969-6.00010-8

FIGURE 10.1 A brown bear feasts on the carcass of a bowhead whale on the beach of the Okhotsk Sea, Russia. *Source: Photo by Olga Shpak.*

Tarpley (1993) provided citations to a number of unpublished reports about bowhead musculoskeletal anatomy and summarized earlier work on the skeleton. We provide only a summary description here, but we relate some of the anatomies to age and function. The specimens we studied were collected by the North Slope Borough Department of Wildlife Management (NSB-DWM, see Chapter 8).

Muscles of the head and neck

Skull osteology and its ontogeny are described in Chapters 8 and 9. The tongue is the largest organ in the head, making up approximately 5% of its weight (Chapter 7). Much of this weight is adipose tissue, not muscle (Fig. 14.5, this volume). In an 84 cm fetus that was CT-scanned (Figs. 10.3–10.5), only three muscles of the head could be distinguished individually. The temporal and masseter muscles have a continuous insertion on the posterior mandible, with the absence of a strong coronoid process. The jugal arch is weak, and the masseter is small. Temporalis originates from the small temporal fossa. Digastricus is a large muscle because it opens the mouth against the water pressure as the whale is swimming. The temporomandibular joint consists, as in other mammals of two separate synovial cavities separated by a fibrous disc. Left and right mandibles articulate via a fibrous pad and there is no bony contact. Pyenson et al. (2012) described the intermandibular joint in balaenopterids as consisting of dense connective tissue that surrounds a spheroid cavity filled with a viscous, gel-like matrix into which connective tissue papillae project. A similar morphology is found in bowheads, but synovial fluid was not observed and the detailed innervation found in balaenopterids may be lacking.

As bowheads age and grow, their body increases in length until long after sexual maturity. Of note, the head grows disproportionally faster: in a 1-year-old bowhead, the head

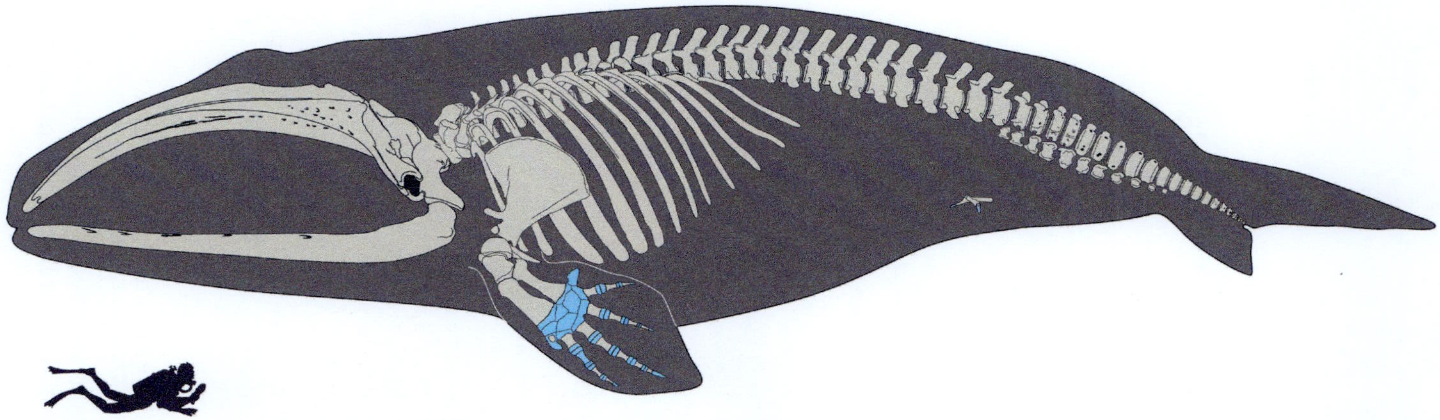

FIGURE 10.2 Skeleton of a sexually mature bowhead whale. Drawing by Daniel Hillmann, with thanks to Eline Lorenzen and Daniel Klingberg Johansson, who shared photographs of a skeleton in the Natural History Museum of Denmark.

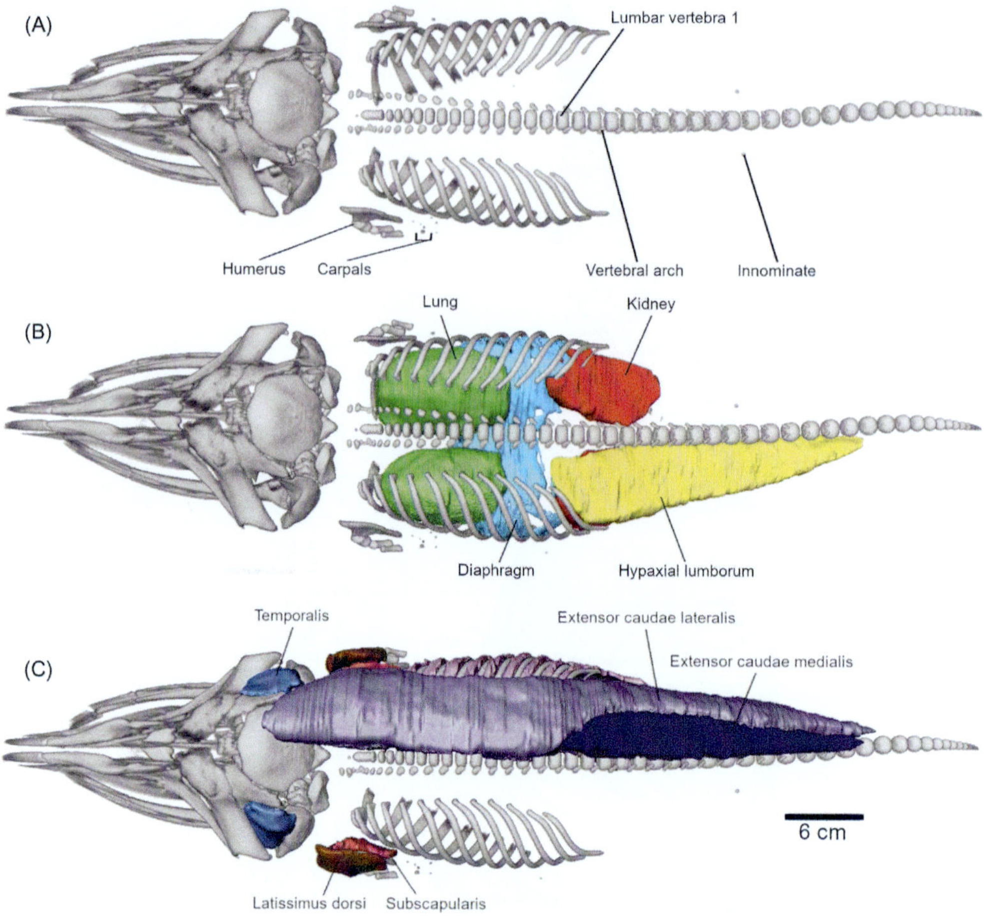

FIGURE 10.3 Dorsal view of skeleton (A), some organs (B) and some muscles (C) of bowhead whale fetus (NSB-DWM 1992B7F, total length, 84 cm). Only some organs and muscle groups could be identified with confidence in these scans, and smaller structures or parts thereof (such as tendons) cannot be seen. This fetus was scanned at Baton Rouge General Medical Center under the direction of D. Hillmann. Identification and 3D reconstruction by J. G. M. Thewissen and K. Mars. NOAA-NMFS permits 314 and 519 to NSB-DWM to T. Albert.

makes up 25% of the body, and in a 30-year-old it makes up approximately 33%. These proportional shifts commence early in ontogeny (Armfield et al., 2011).

Axial skeleton and musculature

Eschricht and Reinhardt (1866) report that bowhead whales commonly have 7 cervical, 13 thoracic, 35 postthoracic vertebrae. The first hemal arch articulates with the caudal part of the 13th postthoracic vertebra, but these numbers can vary somewhat between

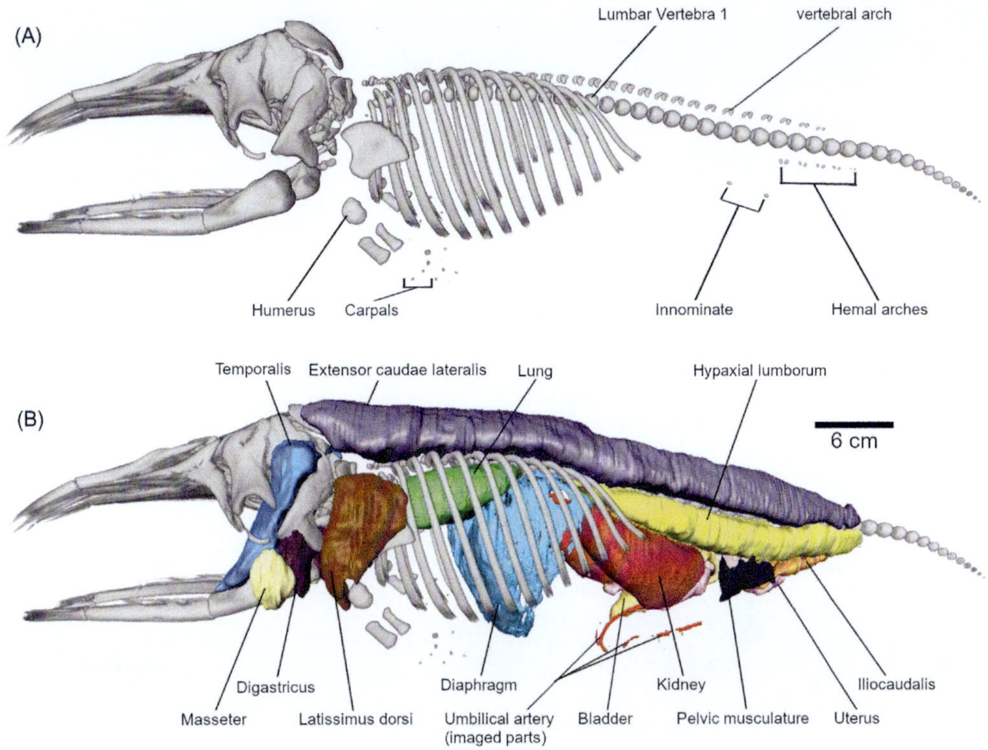

FIGURE 10.4 Left lateral view of skeleton, some muscles, and some organs of fetal bowhead. See Fig. 10.3 for details.

individuals (Fig. 10.2; see discussion of ribs, below). Ten bony hemal arches were recovered by Eschricht and Reinhardt (1866) in their sexually mature individuals, but they noticed the presence of four more cartilaginous ones in the fetus. It is likely that these ossify late in life.

As a rule, the epiphyses of mammal vertebrae fuse with the bodies of their vertebra and length growth ceases in youth. In most cetaceans, this happens long after sexual maturity, meaning that much of the population consists of individuals that are not fully grown. This is also the case in bowhead whales (Fig. 10.6).

In some cetaceans, epiphyses of subsequent cervical vertebrae fuse (here called vertebral fusion) around the same time that the epiphysis fuses to the body of the vertebra (here called epiphyseal fusion). Delphinid cervical vertebrae fuse to different degrees across species (Cozzi et al., 2017), but atlas and axis always display vertebral fusion. Cervical vertebral fusion occurs in balaenids, but not in balaenopterids (Slijper, 1936; VanBuren and Evans, 2016). However, the balaenid pattern differs from that in delphinids. Vertebral fusion of the center of the body of cervical vertebrae two to five is already present in 1-year-old whales (Eschricht and Reinhardt, 1866; Moran et al., 2015), although it is

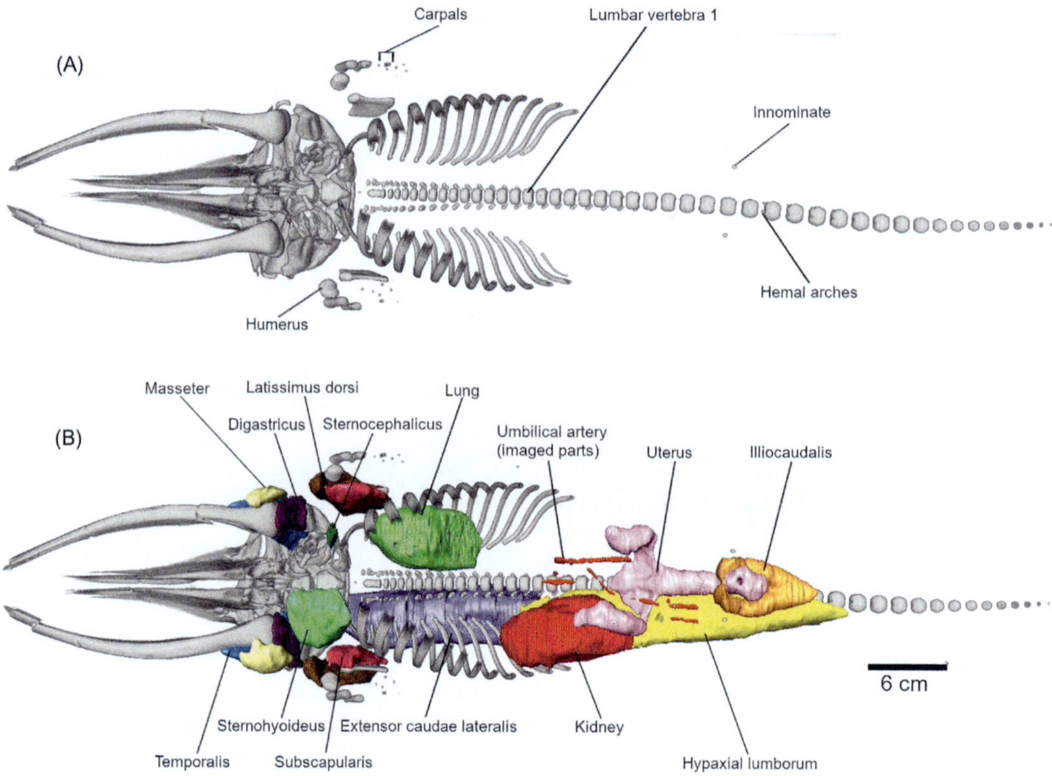

FIGURE 10.5 Ventral view of skeleton (A), some muscles, and some organs (B) of fetal bowhead whale fetus. See Fig. 10.3 for details.

not complete. Even in the fetus, cervical vertebral bodies appear as fused on the CT-scan (Fig. 10.3). The first thoracic vertebra may also be fused to the cervical vertebrae. Peripherally, there is neither vertebral nor epiphyseal fusion in young bowheads. Vertebral fusion of Ce6–7 does not occur until after sexual maturity, and the timing of vertebral fusion of Ce1–2 and Ce5–6 varies greatly, apparently independent of age. Even though vertebral and epiphyseal fusion for all cervical vertebrae is complete in old whales (NSB-DWM 2011B3), there is no sign of either vertebral or epiphyseal fusion in Ce5–7 in a 32-year-old whale (NSB-DWM 2010B15).

Moran et al. (2015) described a 21-year-old bowhead whale (NSB-DWM 2011B9) in which the epiphyses of T1 were fused to their centrum, but all other thoracic epiphyses, as well as postthoracic vertebrae 1–20 remained unfused. Fused epiphyses occurred in postthoracic vertebrae 21 and 22. These vertebrae are located immediately anterior to the fluke and have hemal arches. They are commonly considered caudal vertebrae (although it is controversial whether the first caudal is immediately anterior or posterior to the first hemal arch, Crovetto, 1991). In a much younger whale (NSB-DWM 2009B11), postthoracic vertebrae 15 and 16 had fused epiphyses. Apparently, the degree of epiphyseal fusion

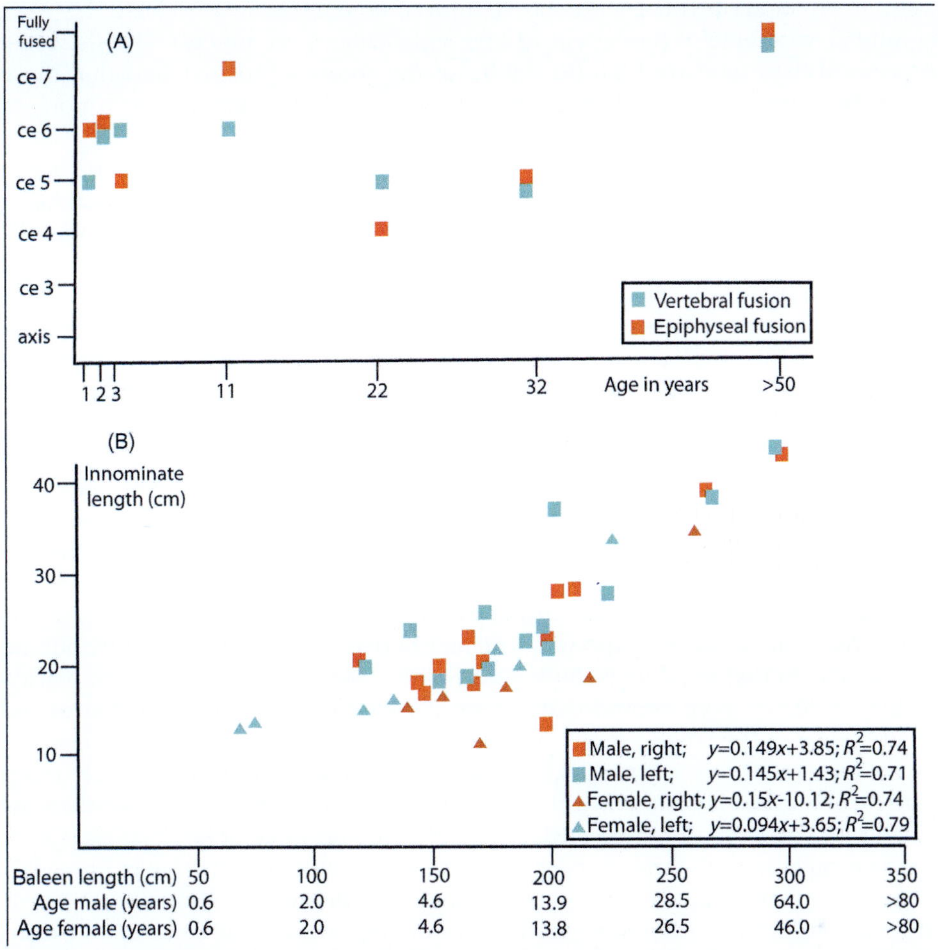

FIGURE 10.6 Osteological measures that change with age in bowhead whales. (A) Fusion of epiphyses (*red*) of the cervical vertebral column and vertebral bodies (vertebral fusion, *blue*) to each other. Plot for epiphyseal fusion indicates the most anterior vertebra with at least one fully unfused epiphysis, where all vertebrae anterior to it have fused epiphyses. Fusion of adjacent vertebrae shows the most anterior vertebra which lacks fusion with the vertebra caudal to it. Neither pattern appears linearly related to age, except that the oldest individual has fully fused centra and epiphysis. (B) Maximum length for male and female, left and right innominates (pelves). Innominates grow with age, female innominates are shorter than male ones, and right innominates are shorter than left ones. Regressions are based on baleen length and ages are calculated based on baleen length (Chapter 21). *Source: (A) Data from Moran, M.M., Bajpai, S., George, J.C., Suydam, R., Usip, S., Thewissen, J.G.M., 2015. Intervertebral and epiphyseal fusion in the postnatal ontogeny of cetaceans and terrestrial mammals. J. Mamm. Evol. 22, 93–109.*

with age is variable, but it implies that the vertebral column will grow for much or all of the life of a bowhead whale, consistent with body length measurements (Chapter 7). Bryden (1972) used epiphyseal fusion as the indicator for physical maturity.

Crovetto (1991) described the vertebral column of large cetaceans in detail. He observed that the lumbar vertebrae of the bowhead and right whales are much wider (across transverse processes) than they are high (to the tip of the spinous process), and that these two dimensions are more similar in thoracic and caudal vertebrae. They are also more similar in the lumbar vertebrae of balaenopterids.

Two large bundles of epaxial muscles extend along the dorsal surface of the transverse processes of the thoracic and postthoracic vertebrae. These axial muscles are enclosed in a tight connective tissue sheet, and derive part of their origin from it as they do in all cetaceans (e.g., Pabst, 1990).

On our CT-scans, two muscle bundles can be clearly recognized. They both converge on the tail and form separate tendons that will insert on the flukes. The larger of these has a broad origin on the skull, as well as the entire thoracic and postthoracic vertebral column. It lays lateral to a smaller muscle mass. The tendon of this large muscle mass is lateral at the peduncle and is usually called the extensor caudae lateralis (Slijper, 1936; Crovetto, 1991).

The bony origin of the smaller muscle mass is located on the postthoracic vertebrae, and the bulk of this muscle is mediodorsal to that of the larger muscle mass. The tendon of the smaller muscle mass enters the peduncle medial to that of the larger muscle mass and is often called extensor caudae medialis (Slijper, 1936). Strickler (1980) identified the muscle mass attaching to extensor caudae lateralis in *Pontoporia* as multifidus. However, the attachment of the two muscle masses appears reversed between bowheads and this odontocete, and more similar to that in *Kogia* (Schulte and Smith, 1918), whereas in balaenopterids and *Physeter*, the tendon is more ventral than lateral (Crovetto, 1991). Until detailed dissections can be undertaken in the bowhead, we prefer to use the terms extensor caudae lateralis and medialis for the entire muscle group. Hypotheses of homology were extensively discussed by Slijper (1936) and are notoriously difficult in epaxial muscles in general. They are particularly difficult for cetaceans. Our nomenclature is of heuristic value only.

On the ventral side of the transverse processes of the lumbar vertebrae is hypaxial lumborum and a similar arrangement was described in odontocetes (Strickler, 1980; Pabst, 1990). It is the largest depressor of the tail. From the relative size of these main propulsive muscles, it is clear that the muscle mass that raises the tail is much larger than the mass that lowers it.

Ribs and sternum

Thirteen pairs of ribs are commonly reported for bowhead whales. Our fetal specimen (Figs. 10.3–10.5) has 14 on the right side and 13 on the left. Eschricht and Reinhardt (1866) described the bowhead skeleton based on 6 individuals; 2 of these had the 14th rib on the right side, and a third individual only had 12 pairs of ribs. Cervical ribs are uncommon in bowheads, unlike in other balaenids (Slijper, 1936).

As in *Caperea* (Buchholtz, 2011), the circumference of the ventral part of the anterior ribs is greater than the dorsal part, and this is not the case for posterior ribs. George et al. (2016) found that bone density and circumference of the anterior ribs change dramatically during the first decade of life. While being nursed, the circumference of rib 2 is great and bone density, in both dorsal and ventral segments of the rib, is high. After weaning, both variables decrease. It is likely that the ribs are a calcium source for the growing baleen rack, as well as a buoyancy control mechanism (George et al., 2016). Locating the main weight of the ribs ventrally will also assist the whale in balance control. Iñupiat hunters were aware of the great weight of 1-year-old bowhead whale ribs and used fragments of these ribs as net weights in fishing (Chapter 34).

Only the first rib is connected to the sternum, similar to other mysticetes. Eschricht and Reinhardt (1866) describe a cartilaginous "sternal rib" that intervenes between the first rib and sternum in a fetus they studied. This fetus was preserved in "spirits of wine" (p. 118). None of the remaining ribs have costal cartilages or are connected to the sternum, a common pattern among cetaceans (Slijper, 1936; Cozzi et al., 2017).

Forelimb muscles and skeleton

Eschricht and Reinhardt (1866) described the osteology of the forelimb. The scapula is large with an infraspinous fossa much larger than the supraspinous fossa, as is common in cetaceans. The humerus has a globular head, with an epiphyseal plate. In a specimen that is older than 60 years (NSB-DWM 1987B6), the head is fully fused to the diaphysis, but the distal epiphysis is not fully fused. Bony epiphyses are present in the distal ulna and radius, and these are not fused to the diaphysis. The entire upper extremity from the humerus down is embedded in a tough fibrous flipper with minimal movement between its parts.

The carpus of a fetus (NSB-DWM 2000B3F, Fig. 10.7) contains cartilaginous anlagen for five distinct carpals. Of these, two can be identified with certainty, the lunate, articulating with radius and ulna, and the pisiform, projecting laterally from the carpus. Two carpals flank the lunate on the medial and lateral sides, articulating also with the radius and ulna, respectively. We consider these proximal carpals, scaphoid, and cuneiform, respectively.

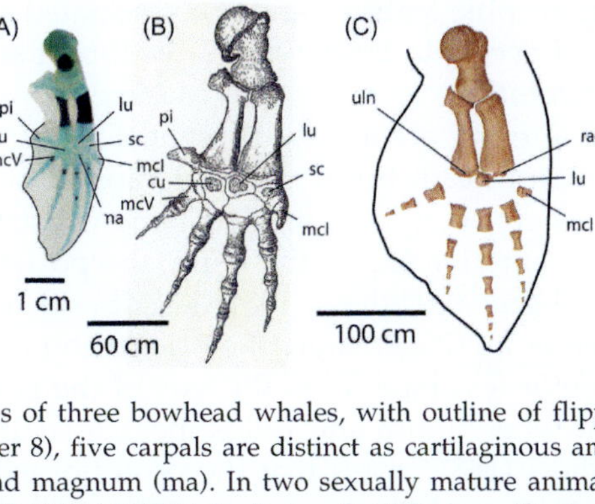

FIGURE 10.7 Forelimb bones of three bowhead whales, with outline of flipper for (A) and (C). In the fetus (A) (NSB-DWM 2000B3F, Chapter 8), five carpals are distinct as cartilaginous anlagen, scaphoid (sc), lunate (lu), cuneiform (cu), pisiform (pi), and magnum (ma). In two sexually mature animals (B) (Eschricht and Reinhardt, 1866, body length, 13.6 m; C, NSB-DWM 1987B6, body length 15.7 m), ossification of the carpus varies, with four partially ossified carpals in (B), and only one in (C). Given its size and baleen length, (C) is older than 60 years and certainly older than (B), but its distal epiphyses of radius (rad) and ulna (uln) remain unfused.

A fifth carpal is located distal to the lunate and also articulates with metacarpals II and III, and we consider it a magnum. This arrangement of carpals in a fetus (Fig. 10.7A) also occurs in a 13.6 m bowhead described by Eschricht and Reinhardt (1866). Its forelimb has unfused epiphyses on radius and ulna, and five carpals consisting mostly of cartilage, but with a bony core in all except the distal row bone. In contrast, NSB-DWM 1987B6, a specimen older than 60 years, only had a single bony carpal: the lunate. Eschricht and Reinhardt (1866) describe in some detail their inferences about the slow ossification of the carpals in ontogeny.

There are five metacarpals, and the digital formula of fetus NSB-DWM 2000B3F is 0.4.5.3.2, with some of its phalanges having a bone core (dark purple in Fig. 10.7A). This digital formula is retained in NSB-DWM 1987B6, whereas there are 0.3.4.3.2 ossified phalanges in the specimen described by Eschricht and Reinhardt (1866). However, none have bony epiphyses, and much of the bulk of these structures consists of cartilage. These authors also note that the absence of phalanges on digit I characterizes *Balaena*, whereas two occur in *Eubalaena*. Cooper et al. (2007a, 2018) summarized patterns of hyperphalangy in cetaceans. Cooper et al. (2007b) described the intrinsic forelimb musculature of the bowhead. These authors only noted a few muscles: a single head of triceps brachii, extensor and flexor digitorum communis (both to digits II–V), flexor digitorum radialis, and flexor carpi ulnaris. Cooper et al. (2017) described gene control of development in the flipper of cetaceans.

Hindlimb

In normal development, there are no external signs of hindlimbs in any extant cetacean, including bowheads. However, bowheads and other cetaceans have hind limb buds as embryos (Chapter 9), and occasionally, a bowhead has external processes on its abdomen that represent the hindlimbs (Fig. 10.8).

All bowhead whales have internal skeletal elements that are remnants of the hindlimb bones. These bones are variable in shape. Struthers (1881) described the shape of the bones in detail, noting sexually dimorphic features and the associated muscles. His study is particularly important as he had access to large whales, well past sexual maturity, which are rarely available for study currently.

An os coxae (pelvis) and a femur are always present (Fig. 10.9). There is not usually an acetabulum or globular femoral head, but a synovial capsule with synovial fluid always connects these bones. The os coxae is an elongated bone consisting of two straight parts that connect at an angle. The anterior part is the ilium, the posterior the pubis, the obturator foramen nearly always absent (Struthers, 1881 describes a foramen "scarcely as large as a crow-quill"). Thewissen et al. (2009) show the orientation of these bones in the body, the axis of the pubis is more or less parallel to the axis of the body. On the lateral surface of the os coxae, at the point where ilium and pubis meet, a shelf occurs and this is where the femur articulates. Struthers (1881) notes that the posterior extremity of the os coxae is thicker in males than in females, and its tip is rugose. The crura of the corpora cavernosa of the penis and clitoris attach in this area (in addition to the interpelvic ligament of Struthers, 1881). No part of the os coxae articulates with its counterpart across the midline.

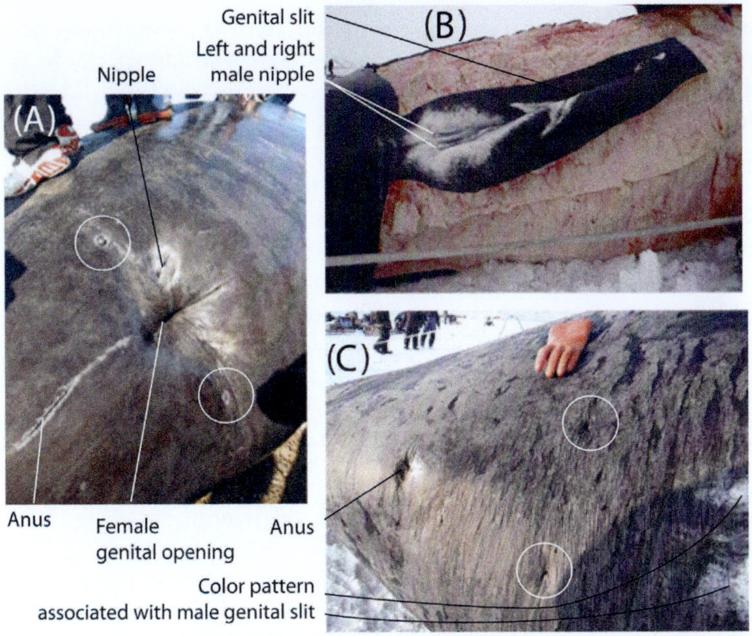

FIGURE 10.8 Hindlimbs in bowhead whales. (A) Genital region of female (NSB-DWM 2007KK3) showing the position of anomalous external hindlimb buds (*circled*) as lateral and slightly caudal to the nipples. (B) Male bowheads usually have nipples positioned just caudal and lateral to the genital slit (NSB-DWM 2010G3, blubber partly removed). (C) Hindlimb remnants in males are far lateral and caudal to the genital slit (NSB-DWM 2011S2), as in females. Cranial is to the right in all photos. *Source:All photographs by Gay Sheffield.*

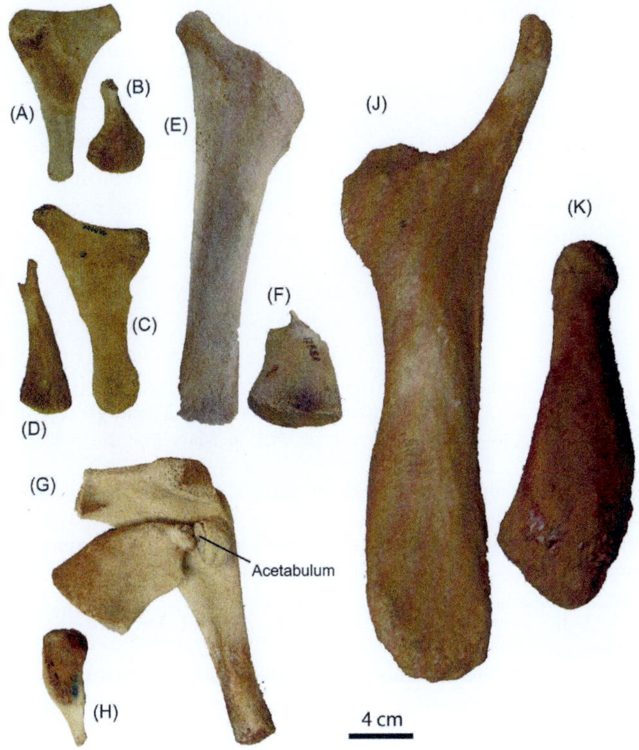

FIGURE 10.9 Os coxae (also called pelvis, A, C, E, J), femur (B, D, F, K), and tibia (H) of bowhead whales. G, os coxae and femur in anatomical position, with poorly formed acetabulum. For ossa coxarum, rostral to top of page; for femora, proximal to top of page (except for G). A and B: 1992B17, female, 1.3 years; C and D: NSB-DWM 1992B20, female, 4 years; E and F: 2017B18, male, 13 years; G, H, and I: 1992B21, male, 7 years; J and K: male, 60 years. Shown to the same scale.

The femur is a triangular bone, with a distal end much wider than the proximal end. Occasionally, an acetabulum and femoral head are present. Distal to the femur, there is usually a triangular piece of cartilage (sometimes ossified) that represents the tibia. Rarely, an additional bone occurs distal to the tibia, which is a presumed metatarsal.

In general, males have larger ossa coxarum than females of the same age (Fig. 10.6), which is probably a reflection of the function of these bones in anchoring the genitals. As in sperm whales (Deimer, 1977), left ossa coxarum are larger than right ossa coxarum of the same individual. Interestingly, similar asymmetry of the pelvis occurs in stickleback fish where this element is under selection for smaller size (Shapiro et al., 2006), which allowed Thewissen et al. (2012) to speculate about its genomic underpinnings.

Conclusions

The most remarkable aspect of the skeletal system of bowhead whales is its slow maturation and its inherent variation. Energy conservation and improving feeding efficiency appear to be important features of bowhead life history (Chapter 7). Delay in skeletal maturation would save energy, and increasing filtering capacity of the baleen rack would improve feeding efficiency. George et al. (2016) proposed that the ribs play a role in this, as a storage site for calcium needed to make baleen, and as part of a buoyancy control mechanism to counteract loss of blubber after weaning. Kim et al. (2014) inferred rapid ossification of odontocete and mysticete cetacean ear bones based on their biochemical composition. It is possible that bowheads expanded these unusual pathways to other parts of their skeleton, making bone a source of resources that can be mobilized quickly.

The muscular system of bowhead whales has been poorly studied. It is likely that muscular adaptation, on a physiological level, is similar to those of other cetaceans. They may be related to diving behaviors and migration. However, it is also likely that physiological adaptations of muscle tissue (e.g., fiber type composition; enzyme activities) to their particular life history strategy occur. With the ability to sample bowhead tissue from harvested individuals, such questions can be answered.

References

Armfield, B.A., George, J.C., Vinyard, C.J., Thewissen, J.G.M., 2011. Allometric patterns of fetal head growth in mysticetes and odontocetes: comparison of *Balaena mysticetus* and *Stenella attenuata*. Mar. Mamm. Sci. 27, 819–827.

Bryden, M.M., 1972. Growth and development of marine mammals. In: Harrison, R.J. (Ed.), Functional Anatomy of Marine Mammals. Academic Press., London, pp. 1–80.

Buchholtz, E., 2011. Vertebral and rib anatomy in *Caperea marginata*: implications for evolutionary patterning of the mammalian vertebral column. Mar. Mam. Sci. 27 (2), 382–397. Available from: https://doi.org/10.1111/j.1748-7692.2010.00411.x.

Cooper, L.N., 2018. Forelimb anatomy. In: Würsig, B., Thewissen, J.G.M., Kovacs, K. (Eds.), Encyclopedia of Marine Mammals, third ed. Elsevier, Amsterdam, pp. 385–388.

Cooper, L.N., Berta, A., Dawson, S.D., Reidenberg, J.S., 2007a. Evolution of hyperphalangy and digit reduction in the cetacean manus. Anat. Rec. 290, 654–672.

Cooper, L.N., Dawson, S.D., Reidenberg, J.S., Berta, A., 2007b. Neuromuscular anatomy and evolution of the cetacean forelimb. Anat. Rec. 290, 1121–1137.

Cooper, L.N., Sears, K.E., Armfield, B.A., Kala, B., Hubler, M., Thewissen, J.G.M., 2017. Review and experimental evaluation of the embryonic development and evolutionary history of flipper development and hyperphalangy in dolphins (Cetacea: Mammalia). Genesis 56 (1). Available from: https://doi.org/10.1002/dvg.23076.

Cozzi, B., Huggenberger, S., Oelschläger, H., 2017. Anatomy of Dolphins: Insights into Body Structure and Function. Academic Press, London, 438 pp.

Crovetto, A., 1991. Etude osteometrique et anatomo-functionelle de la colonne vertebrale chez les grands cetacés. Investig. Cetacea 23, 7–189.

Deimer, P., 1977. Der rudimentäre hintere Extremitätengurdel des Potwals (*Physeter macrocephalus* Linnaeus, 1758), seine Variabilität und Wachstumallometrie. Z. Säugetierk 42, 88–101.

Eschricht, D.F., Reinhardt, J., 1866. On the Greenland Right Whale (*Balaena mysticetus*). The Ray Society, London, 150 pp.

George, J.C., Stimmelmayr, R., Suydam, R., Usip, S., Givens, G., Sformo, T., et al., 2016. Severe bone loss as part of the life history strategy of bowhead whales. PLoS ONE 11 (6), e0156753.

Haldiman, J.T., Tarpley, R.J., 1993. Anatomy and physiology. In: Burns, J.J., Montague, J.J., Cowles, C.J. (Eds.), The Bowhead Whale. Society for Marine Mammalogy, Lawrence, KS, Spec. Publ. 2, pp. 71–156.

Kim, S., Thewissen, J.G.M., Churchill, M.M., Suydam, R.S., Ketten, D.R., Clementz, M.T., 2014. Unique biochemical and mineral composition of whale ear bones. Physiol. Biochem. Zool. 87 (4), 576–584. Available from: https://doi.org/10.1086/676309.

Moran, M.M., Bajpai, S., George, J.C., Suydam, R., Usip, S., Thewissen, J.G.M., 2015. Intervertebral and epiphyseal fusion in the postnatal ontogeny of cetaceans and terrestrial mammals. J. Mamm. Evol. 22, 93–109.

Pabst, D.A., 1990. Axial muscles and connective tissues of the bottlenose dolphin. In: Leatherwood, S., Reeves, R.R. (Eds.), The Bottlenose Dolphin. Academic Press, San Diego, pp. 51–68.

Pyenson, N.D., Goldbogen, J.A., Vogl, A.W., Szathmary, G., Drake, R.L., Shadwick, R.E., 2012. Discovery of a sensory organ that coordinates lunge feeding in rorqual whales. Nature 485, 498–501.

Schoenfuss, H.L., Bragulla, H. H., Schumacher, J., Henk, W. G., George, J.C., Hillmann, D.J., 2014. The anatomy of the larynx of the bowhead whale, *Balaena mysticetus*, and its cound-producing functions. Anat. Rec. 297, 1316–1330.

Schulte, H.V.W., Smith, M.D.F., 1918. The external characters, skeletal muscles, and peripheral nerves of *Kogia breviceps* (Blainville). Bull. Am. Mus. Nat. Hist. 38, 1–72.

Shapiro, M.D., Bell, M.A., Kingsley, D.M., 2006. Parallel genetic origins of pelvic reduction in vertebrates. Proc. Natl. Acad. Sci. U. S. A. 103, 13753–13758.

Slijper, E.J., 1936. Die Cetaceen, vergleichend-anatomisch und systematisch. Ein Beitrag zur vergleichenden Anatomie des Blutgefass, Nerven, und Muskelsystems, sowie des Rumpfskelettes der Säugetiere, mit Studien über die Theorie des Aussterbens und der Foetalisation. Capit. Zool. VI–VII, 590 pp.

Strickler, T.L., 1980. The axial musculature of *Pontoporia blainvillei*, with comments on the organization of this system and its effect on fluke-stroke dynamics in the Cetacea. Am. J. Anat. 157, 49–59.

Struthers, M.D., 1881. The bones, articulations, and muscles of the rudimentary hind-limb of the Greenland right whale (*Balaena mysticetus*). J. Anat. Phys. 15, 142–321.

Thewissen, J.G.M., Cooper, L.N., George, J.C., Bajpai, S., 2009. From land to water: the origin of whales, dolphins and porpoises. Evo. Edu. Outreach 2, 272–288.

Thewissen, J.G.M., Cooper, L.N., Behringer, R.R., 2012. Developmental biology enriches paleontology. J. Vertebr. Paleont. 32, 1223–1234.

VanBuren, C.S., Evans, D.C., 2016. Evolution and function of anterior cervical vertebral fusion in tetrapods. Biol. Rev. 92, 608–626.

CHAPTER

11

Hematology, serum, and urine composition

R. Stimmelmayr[1,2], Lara Horstmann[3], Brian T. Person[1] and J.C. George[1]

[1]Department of Wildlife Management, North Slope Borough, Utqiaġvik, AK, United States
[2]Institute of Arctic Biology, University of Alaska Fairbanks, Fairbanks, AK, United States
[3]College of Fisheries and Ocean Sciences, University of Alaska Fairbanks, Fairbanks, AK, United States

Introduction

Over the past decade, marine mammal clinical medicine has continued to advance, and the use of serum chemistry and hematology as useful diagnostic tools for monitoring the health and physiological processes in captive and free-ranging small cetaceans has been well established (CRC Handbook of Marine Mammal Medicine, 2018). For free-ranging, large whales, health assessments have mainly focused on visual assessment in combination with monitoring of endocrinological aspects using various biomatrices obtained remotely (i.e., biopsies, fecal samples, blow samples, Rolland and Moore, 2018). This chapter synthesizes published and new data about hematology, clinical chemistry, and urine composition of the bowhead whale (Fig. 11.1).

Hematology

The morphology of blood cells and their differential distribution in bowhead whales ($n = 4$) were first described by Medway (1980, 1983). Medway (1983) observed a leucocyte distribution characterized by a marked absence or low detection (<1%) of monocytes, basophils, eosinophils, and a variable neutrophil to lymphocyte ratio among the different individuals. Medway (1983) speculated that this reflected a possible stress response to the hunt (eosinopenia), infectious processes, and/or postmortem body temperature-related

The Bowhead Whale
DOI: https://doi.org/10.1016/B978-0-12-818969-6.00011-X

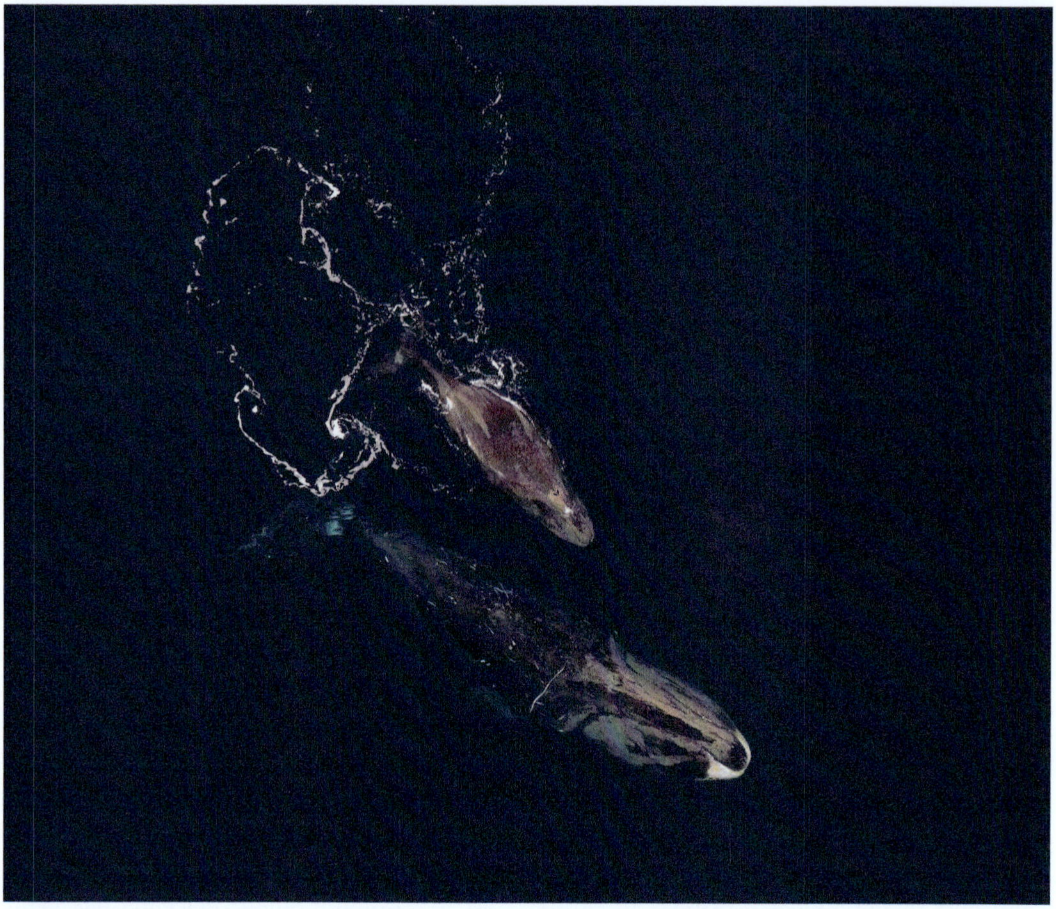

FIGURE 11.1 A bowhead whale mother, her face covered in mud, swims next to her calf. Calves of this size are near the age of weaning. *Source: Photo by Amelia Brower (NOAA/North Slope Borough, NMFS Permit No. 14245).*

differential white cell viability. Given the limited clinical laboratory data available for bowhead whales, our recent health assessment efforts have focused on expanding the clinical pathology data for bowhead whales.

We acquired normal ranges for hematology for 21 immature, healthy bowhead whales. All whales had a total body length of less than <40 feet (12.2 m), and our sample included 11 females and 10 males. We examined these whales after they were landed during the Iñupiat subsistence harvest (see Chapters 34 and 36). Blood samples were collected during spring and fall hunts from 2017 to 2019 in Utqiaġvik, Alaska. We collected blood in a blood vacutainer (EDTA) from an incision into the palatal corpus cavernosum maxillaris (CCM) (Heidel et al., 1996; Ford et al., 2013) within 5–10 hours after the whales died. In the field, specimens were kept from freezing and later on refrigerated at the lab until analyzed (<24 hours from the time of collection; Fig. 11.2). Blood hematologic analyses were performed using an automated analyzer (Abaxis Vetscan HM5). Not all blood variables were

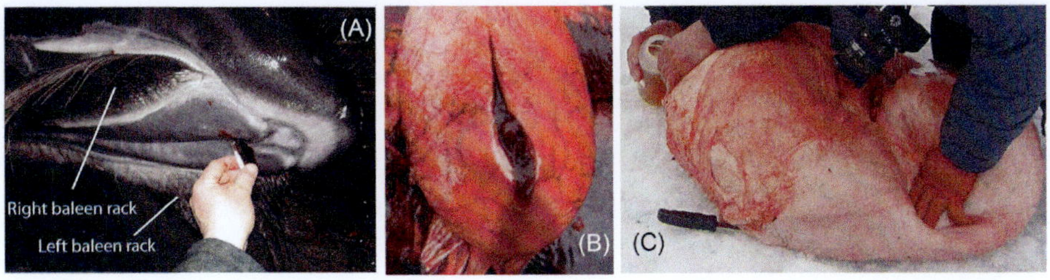

FIGURE 11.2 (A) The palate of a dead bowhead whale is incised to collect the free-flowing blood. Bladder of a bowhead whale with bloody urine (B) and yellow urine, collected in jar on left (C).

assessed in every sample; thus, the total number of samples evaluated for each variable varied. Samples were divided by sex, but further separation (by the season of harvest or age) was not possible due to small sample sizes.

Two single outlying values (hematocrit (HCT) and white blood cell count (WBC)) from two animals (one each) were removed from the analysis. Minimum and maximum observed values are reported (Table 11.1); the latter are considered the best estimate of the reference limits, given the limited sample size (less than 40, Schwacke et al., 2009). For the majority of analytes measured, reference ranges (min to max) are wide in both sexes. It is likely that these findings are partially related to individual variability, but the effect of season (spring vs fall) cannot be excluded. An effect of season on various blood analytes has been shown in both free-ranging and ex situ (captive) cetaceans (Nabi et al., 2019; Nollens et al., 2019). This is similar to observations in humans, reptiles, and other wildlife species (Fröhlich et al., 1997; Græsli et al., 2015; Yang et al., 2019).

Hematological data for other baleen whales is lacking, with the exception of rare studies of live stranded baleen whale calves (gray whale (*Eschrichtius robustus*): Dailey et al., 2000; Reidarson et al., 2001; right whale (*Eubalaena glacialis)*: Harr et al., 2005), thus interpretation of available bowhead whale data in the context of baleen whale physiology currently remains limited. Comparison with data for free-ranging beluga whales (*Delphinapterus leucas*) (CRC, 2002), a toothed whale that is sympatric with bowheads, suggests noted differences in red blood cell indices (red blood cell counts (RBC), hemoglobin (HBG), hematocrit (HCT) and generally lower WBC (lymphocytes, monocytes, eosinophils). Basophils, in contrast, are comparable between bowhead and beluga. Data from other baleen whales (e.g., fin whale (*Balaenoptera physalus*), sei whale (*Balaenoptera borealis*), minke whale (*Balaenoptera acutorostrata*)) are needed to determine if our observed hemo/leucogram are characteristic of all mysticetes, or are unique to bowhead whales, and thus a specific adaptation to their longevity, dive capacity, and general immune response (Keane et al., 2015; Seim et al., 2014; Chapter 22).

Serum electrolytes

Medway (1980, 1983) analyzed serum electrolytes of bowhead whales: sodium (148–170 meq/L (milliequivalents of solute per liter)), potassium (6.1–13.7 meq/L),

TABLE 11.1 Normal hematology ranges for immature healthy bowhead whales caught in Utqiaġvik, Alaska (2017–2019).

		Males		Females	
Cell type	Units	Sample size	Range	Sample size	Range
RBC	10^{12} cell/L	10	0.84–2.33	11	0.04–2.73
HCB	g/L	10	7.9–23.3	11	9.1–24.2
HCT	%	10	13.17–37.66	10	17.73–41.9
MCV	Fl	10	149–165	11	146–165
MCH	Pg	6	76.9–99	6	71.3–98.9
MCHC	g/L	10	51.6–70	9	48.1–74.6
Platelets	10^9 cell/L	10	30,407	10	20,363
WBC					
Leucocytes	10^9 cell/L	9	1.54–7.6	11	1.79–5.57
Neutrophil	10^9 cell/L	10	0.05–3.2	11	0.49–3.01
Lymphocyte	10^9 cell/L	10	0.38–3.43	11	0.82–3.43
Monocyte	10^9 cell/L	10	0.01–0.24	11	0.02–0.2
Eosinophil	10^9 cell/L	10	0–0.51	11	0.01–0.34
Basophil	10^9 cell/L	10	0–0.02	11	0-0.01

Abbreviations: *RBC*, red blood cell counts; *HGB*, hemoglobin; *HCT*, hematocrit; *MCV*, mean corpuscular volume; *MCH*, mean corpuscular hemoglobin; *MCHC*, mean corpuscular hemoglobin concentration; *WBC*, white blood cell counts.

chloride (104–149 meq/L), phosphate/PO_4 (6.7–12.6 mg/dL), calcium (10.03–12.4 mg/dL), magnesium (2.6–4.2 mg/dL). Given the small sample size and probable effects of hunt and hemolysis on these values (especially potassium, sodium), Medway (1983) was hesitant to attribute too much clinical diagnostic value to the electrolyte data. More electrolyte data on harpooned baleen whales have become available recently, and this allows for some comparisons. With the exception of potassium and calcium, both are higher in bowhead whales, serum electrolytes data compares well with available plasma electrolyte values for three other baleen whales, including common minke, sei, and Bryde's whale (*Balaenoptera brydei*) (Birukawa et al., 2005). Comparison with fin whale serum data by Kjeld (2003) suggests similarity for sodium concentrations, but chloride values are lower and magnesium higher in bowhead whales, although Kjeld (2003) did not determine all values. Given the known effect of hemolysis of blood samples on levels of potassium, calcium, and magnesium, it is likely that these bowhead values are also affected. Observed differences in chloride levels between bowhead whale and fin whale most likely reflect individual variability in hydration level. Further, more rigorous studies on plasma and urine electrolytes in bowhead whales will be of interest, as they may shed light on their osmoregulatory ability in addition to providing valuable diagnostic information.

Serum chemistry

Medway (1980, 1983) described serum chemistry for six bowhead whales. This initial study was followed by Heidel et al. (1996), who analyzed specimens from 19 bowhead whales (11 females; 8 males; 6 immature; 13 mature) during the 1992 Inupiat fall harvest. For the majority of analytes measured in serum, reference ranges (min to max) are wide (Table 11.2). It is likely that these findings are partially related to individual variability and age composition, but hunt-related effects (i.e., chase, blast-associated tissue injuries) probably play a role as well. Several analytes commonly associated with significant muscle damage, blood hypovolemia, prerenal azotemia, etc., including aspartate aminotransferase (AST), creatine kinase (CK), potassium, blood urea nitrogen (BUN), and creatinine (CRE) are all elevated in bowhead whales compared to captive toothed whales (Medway, 1983; Heidel et al., 1996). Similar trends are present in serum chemistry values for harpooned fin whales (Lambertsen et al., 1986; Kjeld, 2001) and minke whales (Tryland and Brun, 2001). For the most part, serum chemistry values of bowhead whales are comparable to the fin and minke whales, both members of the family Balaenopteridae; however, differences in individual parameters are evident and probably reflect their sex and age composition, and phylogeny. For bowhead whales, values for serum creatinine (♂ > ♀), sodium (♂ > ♀), and glucose (♂ < ♀) show marked gender differences. An age class effect is present for alkaline phosphatase (ALP) and calcium (immature > mature). This is consistent with the growth dynamics of immature bowhead whales (see Chapter 7) and neonatal gray whales (Reidarson et al., 2001). An effect of reproductive status (pregnant vs

TABLE 11.2 Normal serum chemistry ranges for 19 bowhead whales (11 females; 8 males) ranging from 7.5 to 16.2 m (24′6″–53′1″), caught in Utqiaġvik, Alaska.

	TBL	BUN (mg/dL)	Creatinine (mg/dL)	Glucose (mg/dL)	Total protein (g/dL)	Albumin (g/dL)	Total bilirubin (mg/dL)	ALP (IU/L)	Creatine kinase (IU/L)
Female, *n* = 11	7.5–16.2	60.4–79.9	2.5–7.0	64–116	6.4–11.1	3.6–5.7	0.03–1.8	94–693	113–4544
Males, *n* = 8	8.8–15.0	21.2–73.4	2.3–8.1	9–109	6.2–9.7	4.5–5.5	0.3–1.0	61–693	157–2291

	GGT (IU/L)	AST (IU/L)	Sodium (meq/L)	Potassium (meq/L)	Chloride (meq/L)	Calcium (meq/L)	Phosphorus (meq/L)	Cholesterol (mg/dL)	Triglycerides (mg/dL)	Uric acid (mg/dL)
Females, *n* = 11	43,481.00	112–680	162–188	6.9–14.9	107–133	8.7–12.7	7.6–14.4	285–665	186–628	1.3–8.3
Males, *n* = 8	43,511.00	36–482	179–198	6.8–11	125–137	5.7–13.1	7.8–11.9	352–442	166–425	1.6–6.8

Abbreviations: *TBL*, total bilirubin; *BUN*, blood urea nitrogen; *ALP*, alkaline phosphatase; *GGT*, gamma glutamyl transferase; *AST*, aspartate aminotransferase.

*Data from Heidel, J.R., Philo, L.M., Albert, T.F., Andreasen, C.B., Stang, B.V., 1996. Serum chemistry of bowhead whales (*Balaena mysticetus*). J. Wildl. Dis. 32 (1), 75–79.*

nonpregnant females) is notable for total protein, albumin (ALB), and triglycerides, which are elevated in pregnant females. The effect of age, sex, and reproductive status on serum chemistry values of bowhead whales (Heidel et al., 1996) has not been reported for other baleen whales but has been shown for several toothed whales (Yangtze finless porpoise (*Neophocaena phocaenoides asiaorientalis*), Nabi et al., 2019; killer whale (*Orcinus orca*), Nollens et al., 2019).

Serum chemistry and feeding status

Reference ranges for serum chemistry parameters can provide important insights into seasonal and physiological variabilities of marine mammals. Serum chemistry means and ranges for 52 free-ranging bowheads harvested in Utqiaġvik in 2009–12 are given in Table 11.3. These values are well within the range reported by Heidel et al. (1996) for bowheads harvested during fall season 1992 near Utqiaġvik, as well as minke whales (Tryland and Brun, 2001). However, total body length (as an indicator of age) of the whales included in the study by Heidel et al. (1996) was higher. Variables, such as ALP, have been associated with age (i.e., bone metabolism) and other physiological parameters including pregnancy (Cornell et al., 1988; Schweigert, 1993; Heidel et al., 1996). In a principal component analysis (PCA), our serum chemistry data separated feeding and nonfeeding (empty stomach) whales and was co-correlated with harvest season (i.e., nonfeeding during spring and feeding in fall, Fig. 11.3). Length/age was similar between the feeding and nonfeeding whales (Table 11.3). Variables driving the separation between the two groups in the PCA are higher levels of ALP, amylase (AMY), CRE, and glucose (GLU) in fasting whales, while ALB, alanine aminotransferase (ALT), AST, gamma-glutamyl transferase (GGT), magnesium, and BUN are higher in feeding whales (Fig. 11.3). Although somewhat counterintuitive, this indicates that fasting bowhead whales tend to have higher levels of glucose and enzymes associated with the production of bile and bone metabolism (ALP) and pancreatic juices (AMY). This is similar to dolphins, which also exhibit sustained hyperglycemia during fasting (Ridgway, 2013; Venn-Watson et al., 2013). Sustained hyperglycemia during extended seasonal fast (up to 3 months) is also characteristic of northern elephant seals (*Mirounga angustirostris*), and glucose recycling via the Cori cycle has been suggested for elephant seals (Champagne et al., 2005). During this cycle, gluconeogenesis uses lactate produced by glucose-consuming tissues, such as red blood cells and the brain, and is fueled by ATP generated during fat oxidation in the liver. Glucose is then cycled back through the bloodstream to tissues that need it (Champagne et al., 2005). This scenario is reasonable in fasting whales relying on fat oxidation of their blubber stores.

Creatinine is a waste product of muscle breakdown and higher levels could be indicative of muscle catabolism. However, fasting adapted marine mammals, such as bowhead whales will avoid muscle catabolism (Castellini and Rea, 1992), and other associated indicators of muscle breakdown (e.g., BUN, CK, and AST) do not support muscle utilization in fasting bowhead whales. This contrasts with dolphins who will utilize muscle and other body protein sources along with body fat to sustain gluconeogenesis during prolonged fasts (Ridgway, 2013). Alternatively, dehydration could increase CRE levels (reviewed in

TABLE 11.3 Serum chemistry (mean ± 1 standard deviation) for 54 bowhead whales harvested in Utqiagvik during spring and fall migration, 2009–12 (n = sample size).

Season	Whales Feeding	Length [cm]	ALB [g/dL]	ALP [U/L]	ALT [U/L]	AMY [U/L]	AST [U/L]	BUN [mg/dL]	Ca^{2+} [mg/dL]	CK [U/L]	CRE [mg/dL]	GGT [U/L]	GLOB [g/dL]	GLU [mg/dL]	K^+ [mmol/L]	PHOS [mg/dL]	Mg [mg/dL]	Na^+ [mmol/L]	TBIL [mg/dL]	TP [g/dL]
Fall 2009	7 of 8	945 ± 107	5.1 ± 0.5	211 ± 48	346 ± 543	31 ± 5	254 ± 235	75 ± 13	12.1 ± 1.2	2,922 ± 3,228	1.8 ± 0.4	12 ± 5.7	2.2 ±1.0	94 ± 35	7.6 ± 1.2	10.9 ± 1.7	4.9 ± 1.0	161 ± 5	0.3 ± 0.1	7.3 ± 1.0
Range		800–1,130	4.1–6.0	146–289	7.0–1,517	25–39	0–652	59–94	10.8–14.7	0–7,461	1.1–2.4	2.5–18	1.3–4.4	49–161	5.3–8.9	7.6–12.7	3.2–6.6	153–166	0.3–0.4	6.2–8.9
n		8	8	8	7	8	8	8	8	8	8	8	8	8	7	8	8	7	8	8
Spring 2010	0 of 10	898 ± 112	4.5 ± 1.7	266 ± 90	20 ± 38	36 ± 12	87 ± 131	58 ± 12	10.3 ± 2.8	1,063 ± 1,373	5.6 ± 2.0	4.1 ± 3.1	1.3 ± 0.8	160 ± 91	7.5 ± 1.7	9.1 ± 2.3	3.0 ± 0.8	145 ± 18	0.2 ± 0.1	6.5 ± 1.5
Range		750–1,090	0.0–5.9	136–376	2.5–124	12–53	21–452	34–71	5.9–15.6	0–3,961	2.5–9.4	0.0–11.0	0.0–2.4	77–397	6.0–10.6	5.4–13.3	1.8–4.1	128–163	0.0–0.4	4.3–8.1
n		10	10	10	10	10	10	10	10	10	10	10	10	10	7	10	9	4	10	10
Fall 2010	7 of 7	914 ± 205	4.5 ± 1.1	172 ± 42	18 ± 30	23 ± 20	169 ± 155	50 ± 13	10.4 ± 2.3	2,659 ± 3,280	3.2 ± 1.5	4.6 ± 3.1	1.0 ± 0.3	118 ± 46	6.4 ± 1.6	9.2 ± 2.5	4.3 ± 1.7	143 ± 24	0.3 ± 0.1	5.5 ± 1.1
Range		730–1,250	2.8–5.4	118–241	2.5–84	9–64	38–418	34–70	6.7–13.2	0–9,204	1.9–6.0	2.5–10.0	0.6–1.4	72–180	4.3–8.4	6.0–13.0	2.5–7.2	110–164	0.2–0.4	3.9–6.5
n		7	7	7	7	7	7	7	7	7	7	7	7	7	5	7	6	4	7	7
Fall 2011	9 of 10	1,067 ± 238	4.7 ± 0.5	262 ± 107	50 ± 50	28 ± 8	223 ± 188	66 ± 11	12.2 ± 2.0	1,892 ± 2,486	3.9 ± 1.4	8.3 ± 7.6	2.1 ± 0.6	114 ± 30	7.1 ± 1.1	10.7 ± 2.2	4.5 ± 1.4	146 ± 9	0.3 ± 0.1	7.4 ± 0.7
Range	(1 unknown)	820–1,450	4.0–5.5	82–495	11–155	22–47	40–664	51–83	9.3–15.6	143–8,353	1.4–6.3	2.0–28.0	1.4–2.8	71–167	5.2–8.3	7.6–14.9	2.7–7.4	134–153	0.2–0.4	6.3–8.6
n		10	10	10	10	10	10	10	10	10	10	10	10	10	6	10	9	4	10	10
Spring 2012	2 of 8	898 ± 96	5.0 ± 0.6	399 ± 145	35 ± 54	31 ± 12	151 ± 177	64 ± 10	11.8 ± 0.7	2,817 ± 2,226	5.1 ± 1.0	3.6 ± 2.6	4.0 ± 0.9	126 ± 32	7.3 ± 1.8	11.0 ± 2.9	3.9 ± 1.0	150 ± 6	0.4 ± 0.1	7.8 ± 0.5
Range		775–1,011	4.1–6.2	196–575	2.5–156	16–51	30–532	53–77	11.0–13.4	0–5,058	3.5–6.1	2.0–10.0	2.4–5.3	85–183	5.2–10.2	8.6–17.5	2.8–5.8	139 0 156	0.3–0.4	7.0–8.5
n		8	8	8	8	8	8	8	8	8	7	8	8	8	5	8	8	6	5	8
Fall 2012	9 of 10	984 ± 146	5.2 ± 0.6	300 ± 90	318 ± 458	30 ± 8	490 ± 653	71 ± 9	12.1 ± 1.3	1,509 ± 1,724	3.4 ± 1.6	6.6 ± 2.9	3.7 ± 0.7	147 ± 48	6.7 ± 1.7	10.9 ± 2.1	4.7 ± 0.8	149 ± 2	0.4 ± 0.1	7.5 ± 1.1
Range		838–1,334	4.5–6.1	163–450	2.5–1,337	17–40	38–1,842	56–84	10.9–14.2	0–5,761	1.6–7.4	2.5–13.0	2.9–4.7	86–233	3.1–8.2	8.8–15.3	3.7–6.0	146–150	0.3–0.4	6.0–9.7
n		10	10	10	10	10	10	10	10	10	10	10	10	10	7	10	10	5	10	10
All Feeding	–	974 ± 170	4.9 ± 0.7	257 ± 96	134 ± 302	29 ± 11	297 ± 394	66 ± 14	11.8 ± 1.8	2,243–2,694	3.2 ± 1.6	7.9 ± 5.9	2.5 ± 1.3	122 ± 44	7.0 ± 1.5	10.7 ± 2.5	4.6 ± 1.1	152 ± 13	0.3 ± 0.1	7.1 ± 1.3
Range		730–1,450	2.8–6.1	118–515	2.5–1,517	9–64	0–1,842	34–94	6.7–15.6	0–9,204	1.1–7.4	2.0–28.0	0.6–4.7	49–233	3.1–8.9	6.0–17.5	2.5–7.4	110–166	0.2–0.4	3.9–9.7
n		34	34	34	33	34	34	34	34	34	34	34	34	34	23	34	32	19	33	34
All Non-Feeding	–	895 ± 107	4.7 ± 1.4	309 ± 130	125 ± 327	34 ± 11.8	118 ± 174	62 ± 11	11.0 ± 2.2	1,708 ± 1,798	5.1 ± 1.9	4.3 ± 2.8	2.3 ± 1.5	143 ± 72	7.4 ± 1.6	9.6 ± 2.0	3.6 ± 1.2	148 ± 12	0.3 ± 0.1	6.9 ± 1.3
Range		750–1,090	0.0–6.2	136–575	2.5–1,337	12–53	21–652	34–77	5.9–15.6	0–5,058	2.4–9.4	0.0–11.0	0.0–5.3	77–397	5.2–10.6	5.4–13.3	1.8–6.6	128–163	0.0–0.4	4.3–8.5
n		18	18	18	18	18	18	18	18	18	17	18	18	18	13	18	17	10	15	18

Notes: Feeding and nonfeeding whales were characterized by the presence or absence of stomach contents.

Trumble et al., 2006). Trumble et al. (2006) also note that a proportional increase in muscle mass will yield elevated CRE, thus a loss of blubber mass and a relative increase in muscle mass on the wintering grounds could account for increases in CRE. For feeding whales, ALB, ALT, AST, and GGT levels might be related to season rather than feeding, but these variables were also elevated in captive harbor seals (*Phoca vitulina*) in an experimental feeding trial when fed pollock (*Theragra chalcogramma*, low fat fish) compared to herring (*Clupea pallasii*, high fat fish) or mixed fish diet (Trumble et al., 2006). Similarly, Tryland and Brun (2001) found elevated values of AST and ALT in lipemic (i.e., recent meal) minke whale serum. These variables may therefore be associated with digestion, but specifically a high protein diet (Trumble et al., 2006). That said, lipemia interferences in routine clinical biochemical tests (i.e., ALT) are known, thus we cannot completely exclude an additive lipemic effect on measured ALT values (Walker and Crook, 2013). Blood (or serum) urea nitrogen has also been associated with both an increased intake of dietary nitrogen/protein, but also with stage III fasting (Castellini and Rea, 1992). As BUN in bowhead whales is a variable with greater loading in the PCA in feeding whales (Fig. 11.3), it is likely that BUN is also an indicator of recent food intake, specifically high protein food. Euphausiids have overall higher crude protein content than copepods (6.9% ± 0.6% and 3.3% ± 0.3% dry weight, respectively, see Chapter 12), thus, these variables may indicate prey items to some degree.

Immunoglobulins

Medway (1983) published limited data on the composition of immunoglobulins of bowhead whale blood. Fractions (in g/dL) ranged as follows: albumin, 3.0–4.0; alpha-globulin, 0.5–0.8; beta-globulin, 0.8–1.4; and gamma-globulin, 0.6–1.5 with an albumin/globulin ratio of 1.16–1.67. In general, bowhead whale serum protein electrophoretic separations are comparable in many aspects to those of smaller toothed whales and data collected on two gray whale calves (unpublished data, see Medway, 1983; Reidarson et al., 2001). Reference values for these markers in common bottlenose dolphins (*Tursiops truncatus*) were established rigorously (Schwacke et al., 2009). Because reference values have not been established for bowhead whales, the usefulness of these parameters for the interpretation of pathological patterns in bowhead whales remains limited.

Urine analysis

Harvested bowheads often have voided their bladders (78 out of 113 or 70% of landed whales in Utqiaġvik, Alaska). Urine chemistry in the bladder occurs temporarily before an animal is hunted and is thus less likely to change in composition (Medway, 1983). Thus, postmortem urinalysis remains a potentially valuable tool for assessing renal and metabolic health, biotoxin exposure, and feeding status in bowhead whales. Due to the difficulty and the opportunistic nature of obtaining samples during postmortem examination, few studies are available that have described urine composition and characteristics in baleen whales (Medway, 1980; Kjeld, 2001). The first

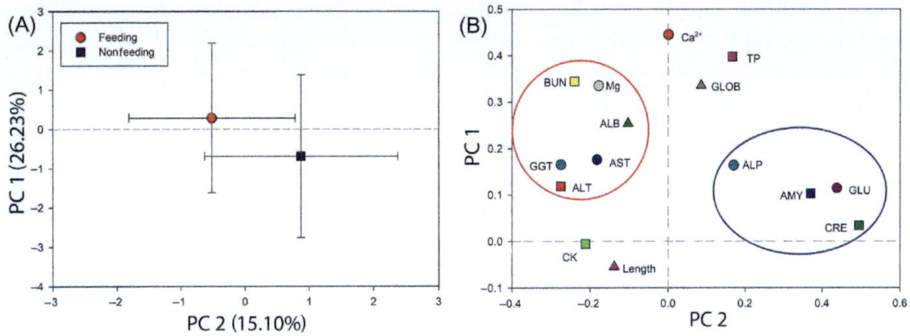

FIGURE 11.3 (A) Principal component analysis for blood chemistry parameters measured in bowhead whale serum. Bowheads were separated by feeding status (i.e., feeding vs nonfeeding based on presence/absence of stomach contents). The first two principal components (PC1 and PC2) explained 41.33% of the variability between feeding and nonfeeding bowhead whales. (B) A loading plot of the variables is shown in the upper right. Variables driving the separation by a positive loading in PC2 (*blueoval*) are alkaline phosphatase (ALP), amylase (AMY), creatinine (CRE), and glucose (GLU), while albumin (ALB), alanine aminotransferase (ALT), aspartate aminotransferase (AST), gamma-glutamyl transpeptidase (GGT), magnesium (Mg), and blood urea nitrogen (BUN) are driving the separation by a negative loading of PC2 (*red circle*).

description of bowhead whale urine was based on a single specimen collected by cystocentesis (urine collected by needle puncture of the urinary bladder) in spring 1978 from an immature (8.4 m) male bowhead. Medway (1980) described the urine as "dark amber, clear, and had no odor. The pH was 5.5, and the specific gravity (SG) was 1.032; there was a trace of protein; and the tests for ketones, glucose, reducing substances, bile pigments, hemoglobin, and urobilinogen were negative. Sediment analysis indicated "few red and white cells, but no casts [urinary casts definition tiny tube-shaped particles made up of white blood cells, red blood cells, or kidney cells]; however, there were myriads of epithelial cells (some were cornified), and there were some bladder transitional cells and caudate cells. There were many unidentified spheroid crystals (probably urates), and there were also occasional oxalate crystals as well as a few triple phosphate crystals."

Medway (1983) characterized urine samples collected during 1979–80 spring and fall hunt from two immature males, one mature male, and one immature female bowhead. Sample outcomes were comparable to the earlier specimen but showed variability in color, clarity, pH, specific gravity, and the following urine analytes: protein, glucose, and hemoglobin. Urinary crystals, as described previously, and sperm were detected on microscopic examination of the sediments of the mature male. Our assessments have not been able to confirm the presence of urinary crystals on microscopic examination of sediment in fresh bowhead urine samples ($n = 6$) (Larshe Hoffland pers. commun), and this may be due to known freeze–thaw cycle effects on enhanced urine crystal formation in cetacean urine (Venn-Watson et al., 2010a,b).

For the establishment of baseline reference values of standard urine analytes, we analyzed 38 urine specimens from landed bowhead whales (2014–18) in Utqiaġvik, Alaska. Urine collection by cystocentesis and occasionally free-catch during butchering occurred within 5–10 hours after the whales died, and specimens were stored at −20°C to −40°C until analyzed. Chemstrip 10UA reagent pack for urine chemistry was read manually and/or by machine (Urisys 1100 analyzer). Urine samples were evaluated for color, odor, and clarity. Urine color was categorized as yellow (1), dark yellow (2), or burgundy to brown (3). Transparency was categorized as clear (clear), slightly cloudy (1) or cloudy (2). A summary of the urinalysis baseline values for bowhead whales is provided in Table 11.4.

Bowhead urine has a mild, but distinct urine odor. Urine color varied between yellow (53%; 20/38), dark yellow (32%; 12/38) burgundy (10%; 4/38) to brown (5%; 2/38). Burgundy color, indicating frank hematuria, was most likely associated with blast-associated trauma. Bowhead urine color has previously been described as dark amber (Medway, 1980). In the subsequent study (Medway, 1983) greater variation in urine color (pale yellow to dark amber) was noted. In fish-eating aquatic mammals, urine color is often reported as deep amber color (Medway and Geraci, 1986). While fish has been identified as an occasional prey species during bowhead whale stomach content analysis (Lowry et al., 2004; Chapter 28), it is by no means representative of the bowhead whale diet. Thus, individual variation in urine color of bowhead whale probably reflects variability in diet, feeding and hydration status. Description of urine color in other baleen whales, where urine has been analyzed in support of osmoregulatory studies, is not available. Most bowhead urine specimens were clear (59%; 20/34), but some were slightly cloudy (35%; 12/34) or cloudy (6%; 2/34). Urine specific gravity ranged between 1.01 and 1.03. Observed values were lower than what was previously reported (SG 1.028–1.035; Medway, 1980). In general, urine dipsticks are too inaccurate to be clinically reliable for the measurement of specific gravity.

Urinary pH ranged from 5 to 6, with most animals having more acidic urine (pH = 5; 79%; 27/34), and consistent with published values (Medway, 1983). The freeze–thaw cycle is known to lower pH measurements of cetacean urine (Daniels et al., 2013), thus, we cannot exclude an effect on our measurements. Fin whales show a greater range in pH (5.5–9.0) and higher mean than bowhead whales (6.6) (Kjeld, 2001). Dipstick evaluation for bilirubin, urobilinogen, and nitrites were negative in all bowhead samples, while 6% (2/34) were positive for glucose, and 3% (1/34) for ketones. Of the examined urine samples, only 32% (11/34) were negative for blood. The remainder of the samples had variable degrees of positive blood dipstick reactions ranging between trace (9%) to 250 (53%). This finding is likely associated with cystocentesis, but blast-associated tissue damage, leading to minor bleeding, cannot be excluded. Protein dipstick evaluation was negative in 29% (10/34) of samples. Protein positive dipsticks (mg/dL) varied, being trace in 32%, 30 in 3%, 100 in 15%, and 500 in 21%. Observed positive protein dipstick reactions were most likely associated with the presence of red blood

TABLE 11.4 Ranges of urine analytes of 34 bowhead whales (collection period 2014–18) caught in Utqiagvik, Alaska.

	LEU (μL)	Nitrite	URO (mg/dL)	PRO (mg/dL)	BLO (Ery/μL)	KET (mg/dL)	BIL (mg/dL)	GLU (mg/dL)	pH
Range	25–500	ND	ND	TR.–500	50–250	15	ND	100–250	5–6
n	3	0	0	10	23	1	0	2	34

Notes: Results for individual values based on a subset of our sample of 34 individuals (*n*). Absence of analyte is indicted as ND (nondetect), and trace amounts as TRACE.
Acronyms: *LEU*, leukocyte; *URO*, urobilinogen; *PRO*, protein; *BLO*, blood; *KET*, ketones; *BIL*, bilirubin; *GLU*, glucose.

cells, as 7 out of 11 urine samples negative for blood were also negative for protein. For the remaining four samples, lower urinary tract contamination or proteinuria cannot be excluded and was not further confirmed. Leucocyte dipstick evaluation was negative for 91% (31/34) of whales. Of the three leucocyte positive samples (Leu/μL) (25;75), one (500) was from a pregnant female with a full-term fetus. No comparative urine analysis data are available for other baleen whales.

Urine electrolytes and aminograms

Medway (1980, 1983) determined the urinary concentrations for sodium (183–310 meq/L), potassium (14.9–87.3 meq/L), chloride (ND–383 meq/L), urea N (560–5000 mg/dL), creatinine (124–540 mg/dL), and osmolality (186–1440 MO/kgH_2O) and characterized urinary free-aminoacids (urine aminogram). Urinary electrolyte concentrations of bowhead whales and creatinine are comparable to data reported for fin and sei whale (Kjeld, 2001, 2003). Urinary osmolality, chloride, sodium, and potassium concentrations fall within the range reported for common minke, sei, Bryde's, and sperm whales (*Physeter macrocephalus*) (Birukawa et al., 2005). Urinary urea N is generally higher in bowhead whales than reported for the other baleen whales (i.e., common minke, sei, Bryde's whale). This could reflect dietary differences, as high urea concentration in urine is thought to be related to high-protein food (Ridgway and Venn-Watson, 2010). However, high urea concentration could be related to phylogenetic differences in osmoregulation mechanisms (Birukawa et al., 2008; Guo et al., 2014; Xu et al., 2013; Wang et al., 2015). Urinary aminograms recently have been described for several small cetaceans, and we suggest that urine aminograms may be useful for the assessment of cetacean health (Miyaji et al., 2010).

Conclusions

The basic knowledge of clinical pathology in bowhead whales has continued to expand, and it provides a useful tool for ad hoc postmortem health assessment of bowhead whales. Furthermore, given the rapid advancement in the development of various clinical metabolomics biomarkers, the importance of routine sampling of bowhead whale blood and urine, as part of the field examination, cannot be overstated for future health assessment efforts of this baleen whale. Such sampling presents an opportunity to examine the multiple associations between host physiology, life history, parasite/toxin burden, seasonal and environmental effects, and pathology in a large whale species.

Acknowledgments

The authors wish to acknowledge the longtime contribution of the many NSB staff in tissue and data collection: Dave Ramey, Cyd Hanns, Brian Person, Todd Sformo, Leslie Pierce, Leandra Sousa, Jason Herrman, Andy Von Duyke, Rita Acker, Craig George and Robert Suydam. We thank all of those who have collected blood and urine of the whales in this study, including R. Acker, P. Detwiler, C. Hanns, J. Herreman, R. Klimstra, L. Pierce, D. Ramey, T. Sformo, L. de Sousa, R. Suydam, and A. Von Duyke.

References

Birukawa, N., Ando, H., Goto, M., Kanda, N., Pastene, L.A., Nakatsuji, H., et al., 2005. Plasma and urine levels of electrolytes, urea and steroid hormones involved in osmoregulation of cetaceans. Zool. Sci. 22 (11), 1245–1257.

Birukawa, N., Ando, H., Goto, M., Kanda, N., Pastene, L.A., Urano, A., 2008. Molecular cloning of urea transporters from the kidneys of baleen and toothed whales. Comp. Biochem. Physiol. B: Biochem. Mol. Biol. 149 (2), 227–235.

Castellini, M.A., Rea, L.D., 1992. The biochemistry of natural fasting at its limits. Experientia 48, 575–582.

Champagne, C.D., Houser, D.S., Crocker, D.E., 2005. Glucose production and substrate cycle activity in a fasting adapted animal, the northern elephant seal. J. Exp. Biol. 208, 859–868.

Cornell, L.H., Duffield, D.S., Joseph, B.E., Stark, B., 1988. Hematology and serum chemistry values in the beluga (*Delphinapterus leucas*). J. Wildl. Dis. 24, 220–224.

CRC Handbook of Marine Mammal Medicine, 2018. 3rd Edition. In Gulland, F.M.D., Dierauf, L.A., Whitman, K. L., (Eds.), CRC Press (Taylor & Francis), Boca Raton, FL, pp. 1, 124 ISBN 9781498796873.

Dailey, M.D., Gulland, F.M., Lowenstine, L.J., Silvagni, P., Howard, D., 2000. Prey, parasites and pathology associated with the mortality of a juvenile gray whale (*Eschrichtius robustus*) stranded along the northern California coast. Dis. Aquat. Organ. 42 (2), 111–117.

Daniels, R., Smith, C., Venn-Watson, S., 2013. Effects of freeze–thaw cycle on urine values from bottlenose dolphins (*Tursiops truncatus*). Aquat. Mamm. 39, 330–334.

Ford, T.J. Jr, Werth, A.J., George, J.C., 2013. An intraoral thermoregulatory organ in the bowhead whale (*Balaena mysticetus*), the corpus cavernosum maxillaris. Anat. Rec. (Hoboken) 296 (4), 701–708.

Fröhlich, M., Sund, M., Russ, S., Hoffmeister, A., Fischer, H.G., Hombach, V., et al., 1997. Seasonal variations of rheological and hemostatic parameters and acute-phase reactants in young, healthy subjects. Arterioscler. Thromb. Vasc. Biol. 17 (11), 2692–2697.

Græsli, A.R., Evans, A.L., Fahlman, Å., Bertelsen, M.F., Blanc, S., Arnemo, J.M., 2015. Seasonal variation in haematological and biochemical variables in free-ranging subadult brown bears (*Ursus arctos*) in Sweden. BMC Vet. Res. 8 (11), 301.

Guo, A., Hao, Y., Wang, J., et al., 2014. Concentrations of osmotically related constituents in plasma and urine of finless porpoise (*Neophocaena asiaeorientalis*): implications for osmoregulatory strategies for marine mammals living in freshwater. Zool. Stud. 53, 10. Available from: https://doi.org/10.1186/1810-522X-53-10.

Harr K.E.; Maners J., Roberts J., Troutman M., Chittick E., Walsh M.T., et al., 2005 Lack of passive transfer in a neonatal right whale calf (*Eubalaena glacialis*). IAAAM Archives 2005.

Heidel, J.R., Philo, L.M., Albert, T.F., Andreasen, C.B., Stang, B.V., 1996. Serum Chemistry of bowhead whales (*Balaena mysticetus*). J. Wildl. Dis. 32 (1), 75–79.

Keane, M., Semeiks, J., Webb, A.E., Li, Y.I., Quesada, V., Craig, T., et al., 2015. Insights into the evolution of longevity from the bowhead whale genome. Cell Rep. 10 (1), 112–122.

Kjeld, M., 2001. Concentrations of electrolytes, hormones, and other constituents in fresh postmortem blood and urine of fin whales (*Balaenoptera physalus*). Can. J. Zool. 79, 438–446.

Kjeld, M., 2003. Salt and water balance of modern baleen whales: rate of urine production and food intake. Can. J. Zool. 81, 606–616.

Lambertsen, R.H., Birnir, B., Bauer, J.E., 1986. Serum chemistry and evidence of renal failure in the North Atlantic fin whale population. J. Wildl. Dis. 22 (3), 389–396.

Lowry, L., Sheffield, G., George, J.C., 2004. Bowhead whale feeding in the Alaskan Beaufort Sea, based on stomach content analysis. J. Cetacean Res. Manage. 6 (3), 215–223.

Medway, W., 1980. Some observations on urine from a bowhead whale. Mar. Fish. Rev. 42, 91–92.

Medway, W., 1983. Examination of blood and urine from Eskimo killed bowhead whales (*Balaena mysticetus*). Aquat. Mamm. 10 (1), 1–8.

Medway, W., Geraci, J.R., 1986. Clinical pathology of marine mammals. In: Young, N. (Ed.), Zoo and Wild Animal Medicine. WB Saunders, Hawaii.

Miyaji, K., Nagao, K., Bannai, M., Asakawa, H., Kohyama, K., Ohtsu, D., et al., 2010. Characteristic metabolism of free amino acids in cetacean plasma: cluster analysis and comparison with mice. PLoS ONE 5 (11), e13808.

Nabi, G., Robeck, T.R., Hao, Y., Wang, D., 2019. Hematologic and biochemical reference interval development and the effect of age, sex, season, and location on hematologic analyte concentrations in critically endangered Yangtze finless porpoise (*Neophocaena asiaeorientalis* ssp.). Front. Physiol. 11 (10), 792.

Nollens, H.H., Robeck, T.R., Schmitt, T.L., Croft, L.L., Osborn, S., McBain, J.F., 2019. Effect of age, sex, and season on the variation in blood analytes of a clinically normal ex situ population of killer whales (*Orcinus orca*). Vet. Clin. Pathol. 48 (1), 100–113.

Reidarson, T.H., McBain, J.F., Yochem, P.K., 2001. Medical and nutritional aspects of a rehabilitating California gray whale calf. Aquat. Mamm. 27 (3), 215–221.

Ridgway, S., 2013. A mini review of dolphin carbohydrate metabolism and suggestions for future research using exhaled air. Front. Endocrinol (Lausanne). 4, 152.

Ridgway, S., Venn-Watson, S., 2010. Effects of fresh and seawater ingestion on osmoregulation in Atlantic bottlenose dolphins (*Tursiops truncatus*). J. Comp. Physiol. B 180 (4), 563–576.

Rolland R.M. and Moore M.J., Health assessment of large whales, In: Gulland, F.M.D., Dierauf L.A., Whitman K. L (Eds.), *CRC Handbook of Marine Mammal Medicine: Health Disease and Rehabilitation*, third ed., 2018, Taylor and Francis Publishers, Boca Raton, FL, 835–848

Schwacke, L.H., Hall, A.J., Townsend, F.I., Wells, R.S., Hansen, L.J., Hohn, A.A., et al., 2009. Hematologic and serum biochemical reference intervals for free-ranging common bottlenose dolphins (*Tursiops truncatus*) and variation in the distributions of clinicopathologic values related to geographic sampling site. Am. J. Vet. Res. 70, 973–985.

Schweigert, F.J., 1993. Effects of fasting and lactation on blood chemistry and urine composition in the grey seal (*Halichoerus grypus*). Comp. Biochem. Physiol. 105A, 353–357.

Seim, I., Ma, S., Zhou, X., Gerashchenko, M.V., Lee, S.G., Suydam, R., et al., 2014. The transcriptome of the bowhead whale *Balaena mysticetus* reveals adaptations of the longest-lived mammal. Aging (Albany NY) 6 (10), 879–899.

Trumble, S.J., Castellini, M.A., Mau, T.L., Castellini, J.M., 2006. Dietary and seasonal influences on blood chemistry and hematology in captive harbor seals. Mar. Mammal. Sci. 22, 104–123.

Tryland, M., Brun, E., 2001. Serum chemistry of the minke whale from the northeastern Atlantic. J. Wildl. Dis. 37 (2), 332–341.

Venn-Watson, S., Smith, C.R., Johnson, A., Daniels, R., Townsend, F., 2010a. Clinical relevance of urate nephrolithiasis in bottlenose dolphins *Tursiops truncatus*. Dis. Aquat. Organ. 2010 (89), 167–177.

Venn-Watson, S., Smith, C.R., Stevenson, S., Parry, C., Daniels, R., Jensen, E., et al., 2013. Blood-based indicators of insulin resistance and metabolic syndrome in bottlenose dolphins (*Tursiops truncatus*). Front Endocrinol (Lausanne). 4, 136.

Venn-Watson, S., Townsend, F.I., Daniels, R.L., Sweeney, J.C., McBain, J.W., Klatsky, L.J., et al., 2010b. Hypocitraturia in common bottlenose dolphins (*Tursiops truncatus*): assessing a potential risk factor for urate nephrolithiasis. Comp. Med. 60, 1–5.

Walker, P.L., Crook, M.A., 2013. Lipaemia: causes, consequences and solutions. Clin. Chim. Acta 418, 30–32.

Wang, J., Yu, X., Hu, B., Zheng, J., Xiao, W., Hao, Y., et al., 2015. Physicochemical evolution and molecular adaptation of the cetacean osmoregulation-related gene UT-A2 and implications for functional studies. Sci. Rep. 12 (5), 8795.

Xu, S., Yang, Y., Zhou, X., Xu, J., Zhou, K., Yang, G., 2013. Adaptive evolution of the osmoregulation-related genes in cetaceans during secondary aquatic adaptation. BMC Evol. Biol. 13, 189. Available from: https://doi.org/10.1186/1471-2148-13-189. PMID: 24015756; PMCID: PMC3848586.

Yang, T., Haas, H.L., Patel, S., Smolowitz, R., James, M.C., Williard, A.S., 2019. Blood biochemistry and haematology of migrating loggerhead turtles (*Caretta caretta*) in the Northwest Atlantic: reference intervals and intra-population comparisons. Conserv. Physiol. 7 (1), coy079. Available from: https://doi.org/10.1093/conphys/coy079.

CHAPTER

12

Anatomy and physiology of the gastrointestinal system

Lara Horstmann

College of Fisheries and Ocean Sciences, University of Alaska Fairbanks, Fairbanks, AK, United States

Introduction

Baleen whales have evolved to exceptionally large body sizes while filter-feeding on mostly small, low-trophic level prey. Whether or not gigantism developed ahead of filter feeding is debated (Fordyce and Marx, 2018; Goldbogen et al., 2019), but Goldbogen (2018) argued that lower mass specific metabolic rate and prolonged search for prey and fasting ability are among the many advantages of large body size. Recently, Goldbogen et al. (2019) demonstrated that foraging efficiency of baleen whales far exceeds that of toothed whales, and the only constraint on baleen whale size is the patchiness (both spatially and seasonally) of their prey. In the terrestrial ecosystem, many ungulates (hoofed mammals) are adapted to forage on low-quality plant material, yet have evolved to very large body sizes. For these mammals, unique adaptions of their gastrointestinal tract, increased retention times, and overall large gut capacity have been credited for attained gigantism (Clauss et al., 2003). Size of terrestrial predators on the other hand appears to be related to prey size, but this relationship is not apparent in aquatic ecosystems (Tucker and Rogers, 2014a). Interestingly, life and search for food in 3D environments is characterized by higher prey consumption rates (Pawar et al., 2012), which in turn require large, voluminous gastrointestinal tracts to accommodate patchy and seasonally abundant prey (Williams et al., 2001). While baleen whales, including bowhead whales are generally considered "gentle giants", it is important to keep in mind, that they are carnivorous and are predominantly feeding on small planktivorous animal prey (see Chapter 28) (Fig. 12.1).

This chapter will review the anatomy of the bowhead whale gastrointestinal tract, and will compare gut structure and physiology of baleen whales to terrestrial relatives. We will review digestive efficiency, lipid and wax ester digestion, gut passage times, and information on the microbiome of bowhead whales.

The Bowhead Whale
DOI: https://doi.org/10.1016/B978-0-12-818969-6.00012-1

FIGURE 12.1 A bowhead whale defecating while feeding in the western Beaufort Sea. *Source: The reddish color of the excrement (bottom, left) is derived from the zooplankton prey it fed on (NOAA/North Slope Borough, NMFS Permit No. 14245).*

Wax ester digestion

Energy transfer from one trophic level to the next is inefficient, known as the 10% rule; therefore, predators feeding on low trophic level zooplankton are profiting from relatively high energy transfer (Tucker and Rogers, 2014b). The diet of bowhead whales is dominated by calanoid copepods (typically trophic level 2; see Chapter 28). Copepods in particular are concentrating phytoplankton-based lipids, for example, long-chain polyunsaturated fatty acids, such as omega-3 fatty acids, and are important links of these essential nutrients to higher trophic levels (Graeve et al., 1994). In addition, copepods are storing lipids in the form of wax esters (Fig. 12.2A), where a long-chain fatty acid is esterified to a long-chain fatty alcohol. While the fatty acids are mainly dietary, the fatty alcohols of copepods are synthesized de novo (Sargent and Falk-Petersen, 1988). Wax ester digestion in fishes and sea birds is generally accomplished by wax ester hydrolase, bile salts, reflux, and increased gut retention times (Place, 1992; Bogevik, 2011). On the other hand, wax ester digestion in cetaceans is not well understood. Minke whales (*Balaenoptera acutorostrata*) and North Atlantic right whales (*Eubalaena glacialis*) are digesting wax esters with high efficiency (Swaim et al., 2009; Nordøy, 1995), although evidence is generally secondary (e.g., presence of fatty alcohols in feces) and no clear mechanisms has been identified. Miller et al. (2019) showed that wax ester digestion in the bowhead whale is highly variable, and two bacterial taxa, *Actinobacillus* and *Cetobacterium*, were correlated with wax ester abundance in the bowhead whale jejunum. These bacterial taxa may therefore be

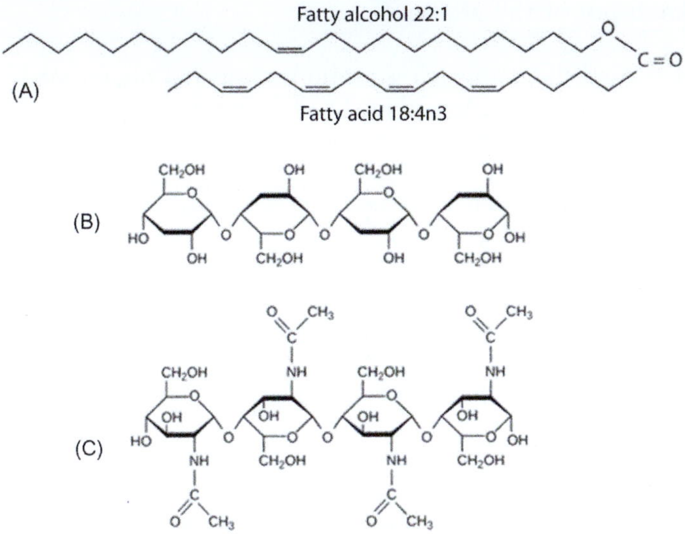

FIGURE 12.2 (A) Typical copepod wax ester containing the fatty alcohol 22:1 and the fatty acid 18:4n3. Naming convention (A:Bn-X) after Budge et al. (2006), where A represents the number of carbon atoms, B the number of double bonds, and X the position of the double bond closest to the terminal methyl group. Chemical structure of cellulose (B) and chitin (C); both molecules build chains by β-1,4 linkage. The hydroxyl group (OH) of cellulose has been replaced by a nitrogen-containing acetyl amine group in chitin.

functionally important in the digestion of wax esters. It remains unclear whether bowhead whales are solely relying on their microbiome to digest wax esters, or if they have additional body-own tools, such as wax ester hydrolase and/or high concentrations of bile salts.

Setting the stage—evolutionary and chemical considerations

To understand the complexities of the gastrointestinal tract of cetaceans, their evolutionary history has to be taken into account. Cetaceans, together with their extant sister family Hippopotamidae, split from a common artiodactyl (even-toed ungulate) ancestor about 67 Ma (e.g., Geisler and Theodor, 2009; Zurano et al., 2019). Both artiodactyls and hippos are mostly herbivorous, consuming a diet rich in the complex carbohydrate, cellulose (Keys et al., 1969; Chritz et al., 2016). Cellulose is often considered the most abundant biopolymer on the planet. The majority of the cellulose biomass is contributed by terrestrial plants due to structural requirements as a result of gravity (Duchesne and Larson, 1989). In contrast, the high density and viscosity of water in marine ecosystems make mechanical support for aquatic plants less important, and chitin, not cellulose, is the most abundant biopolymer in aquatic ecosystems (Cohen-Kupiec and Chet, 1998; Jeuniaux and Voss-Foucart, 1991). Crustacean zooplankton, such as copepods and euphausiids, are using chitin in their exoskeleton, and their chitin biomass production can exceed 2 million metric tons annually (Jeuniaux and Voss-Foucart, 1991). Chemically, chitin and cellulose are remarkably similar molecules, and are only differentiated by an acetyl amine over a hydroxyl group in chitin versus cellulose (Fig. 12.2B and C). With this evolutionary and chemical perspective, we can now put the anatomy and function of the bowhead whale stomach into context.

Anatomy of the stomach

Ruminants are foregut fermenters, and their stomach is a multichambered organ, made up of the rumen, reticulum, omasum, and abomasum (Fig. 12.3A). The rumen is the fermentation vat of the digestive system, and harbors a unique microbial ecosystem that breaks down cellulose and produces volatile fatty acids in the process (Wang et al., 2018). The reticulum is the second chamber, but it is often combined functionally with the rumen to the reticulo-rumen. The reticulum is undergoing complex contraction cycles that appear to function in controlling outflow of digesta to the next chamber and retention of large particles (Okine et al., 1998). Only particles of the correct size are being transferred through the omasum, and the large surface area of this chamber aids in water and salt balance (Clauss et al., 2006). Finally, the abomasum is glandular with chief (pepsinogen and chymosin secreting) and parietal cells (HCl secreting) and functionally similar to the monogastric stomach of carnivores (Hofmann, 1989).

The multichambered stomach of the hippopotamus appears similar to the ruminant stomach (Fig. 12.3B). The first two chambers are the left and right diverticulum (also called parietal and visceral blind sac), large sacs akin to the reticulo-rumen, where volatile fatty

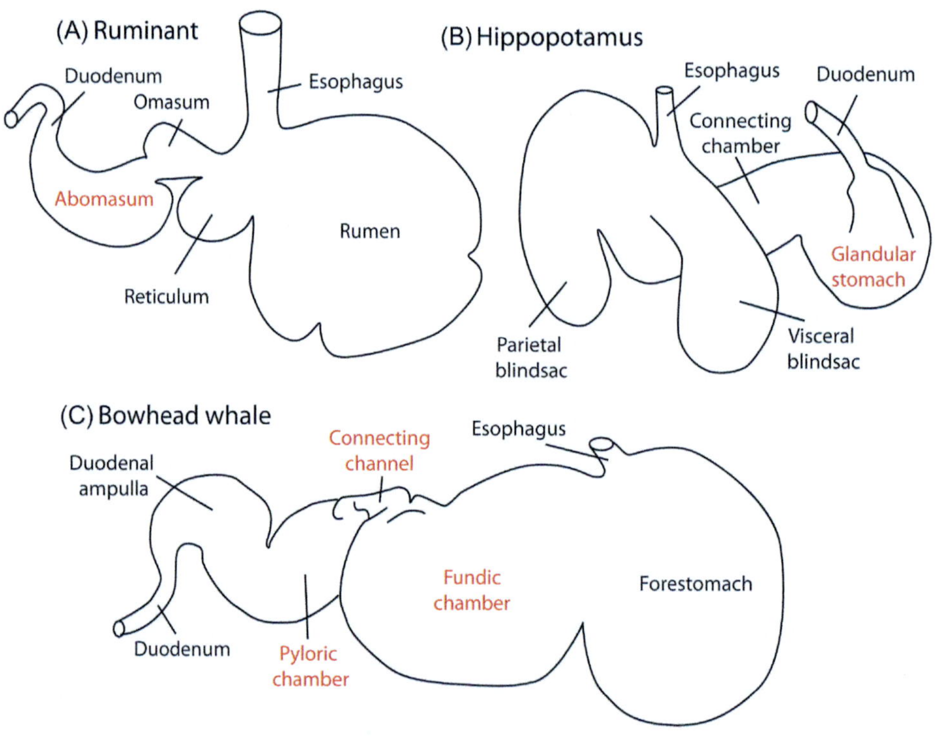

FIGURE 12.3 Comparison of a ruminant stomach (A, after Galloway, 1915), hippopotamus stomach (B, after Clauss et al., 2004), and bowhead whale stomach (C, after Tarpley et al., 1987). The glandular stomach compartments are highlighted in *red*. Drawings are not to scale.

acids and fermentation of ingested food take place (Arman and Field, 1973; Clauss et al., 2004). Selection of particles by size and passage to the final chamber occurs in the median or connecting chamber, while the posterior chamber or glandular stomach is the digestive compartment with chief and parietal cells, similar to the abomasum (Arman and Field, 1973; Clauss et al., 2004).

The stomach of the bowhead whale has been described in great detail by Tarpley et al. (1987). As for ruminants and hippos, the cetacean stomach is multichambered consisting of a forestomach (or esophageal stomach), the fundic chamber (aka main stomach), the connecting channel, and the pyloric chamber (Fig. 12.3C). The forestomach is a large, muscular sac with a white lining thrown in numerous folds (rugae) (Harrison et al., 1970; Tarpley et al., 1987; Olsen et al., 1994). The forestomach is capable of considerable extension, and its volume can be up to 50% of the total stomach volume (Table 12.1; Tarpley et al., 1987; Olsen et al., 1994). The distensible forestomach can thus accommodate large amounts of seasonally abundant prey. The forestomach lining of bowhead whales is white in color (Fig. 12.4), characterized by an absence of glands, and is covered by a keratinized stratified squamous epithelium (Tarpley et al., 1987). Similar to ruminants and hippos, volatile fatty acids are produced in the forestomach, and the unique microbiome of mysticetes includes the presence of chitinolytic bacteria (Herwig and Staley, 1986; Herwig et al., 1984; Olsen et al., 2000; Olsen and Mathiesen, 1996). In contrast to their terrestrial ancestors, the

TABLE 12.1 Stomach volume and intestinal lengths of some baleen whales.

Species	Body length (m)	Forestomach volume (L)	Total stomach volume (L)	Small intestine length (m)	Large intestinal length (m)	Small intestine /body length ratio	References
Bowhead whale	10.4 ± 2.9	–	–	56.5 ± 20.2	–	5.3 ± 0.5	North Slope Borough
Bowhead whale	9	52.7	105.7				Tarpley et al. (1987)
	9.5	–	94.6				
North Pacific right whale	14.8 ± 2.0	–	239.0 ± 60.8 kg	89.9 ± 6.6	4.7 ± 1.5	5.9 ± 0.6	Omura et al. (1969)
Atlantic minke whale	7.1 ± 1.3	156.3 ± 124.7	263.7 ± 170.8[a]	28.0 ± 5.4	2.8 ± 0.5	3.9 ± 0.7	Olsen et al. (1994)
Antarctic minke whale	6.7	70	–	35.4	3.4	5.8	Pérez et al. (2017)
Fin whale	18.2 ± 1.1	372.4 ± 158.2	–	–	–	–	Víkingsson (1997)
Humpback whale	10.3	–	–	57	–	5.5	Besseling et al. (2015)

[a]*Connecting channel not included in Atlantic minke whale.*

Notes: Volumes were derived by filling the stomach (and its components) with water to its maximum capacity.

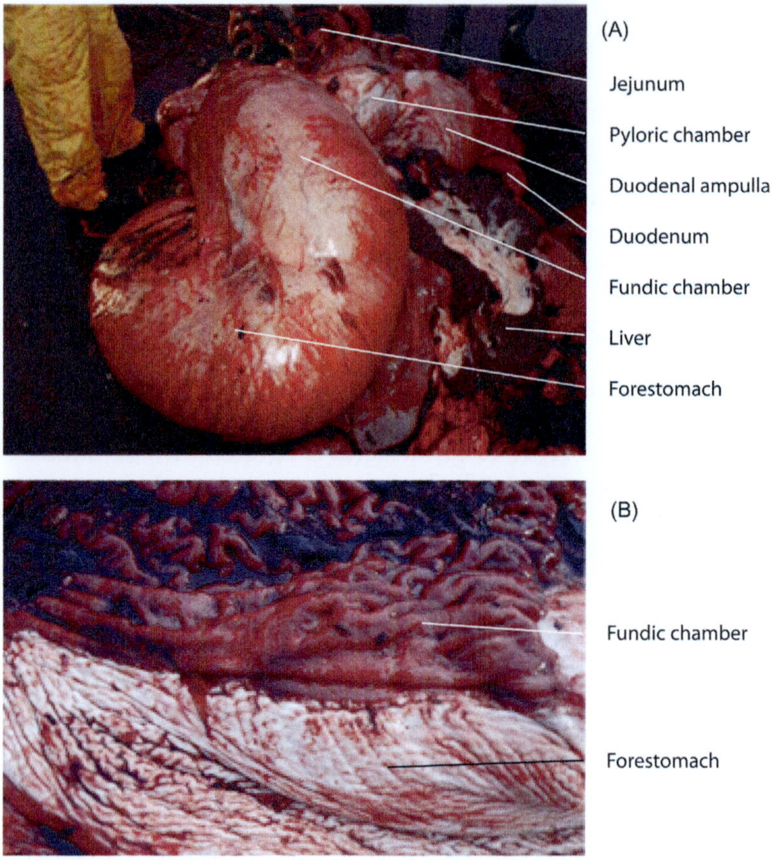

FIGURE 12.4 (A) Overview of a bowhead whale stomach (connecting channel not shown). (B) Stomach lining of a bowhead whale. The nonglandular lining of the forestomach (*white*) can be clearly differentiated from the glandular lining of the fundic chamber (*red*). The demarcation between the two chambers is abrupt. Folds (rugae) are visible in both chambers, but are larger in the fundic chamber than the forestomach. *Source: Photo by (A) C. George, NSB-DWM; (B) R. Stimmelmayr, NSB-DWM.*

forestomach of bowhead whales is not fermenting cellulose, but the chemically similar compound chitin (Fig. 12.2B and C). As much as 70% of prey dry matter disappears during residence in the forestomach, mostly in the form of the chitinous exoskeleton of crustacean prey (Nordøy et al., 1993; Olsen et al., 2000). The forestomach thus serves a similar purpose to the ruminant and hippo fermentation vat by readying the forage and/or prey for chemical digestion in the next chambers.

The forestomach is connected to the fundic chamber by a large orifice, and similar to the reticulo-rumen in ruminants, some mixing occurs. This is supported by the high similarity of microbial communities in the forestomach and fundic chamber (Miller et al., 2019). In contrast to both ruminants and hippos, the fundic chamber is glandular, and chief and parietal cells are present (Tarpley et al., 1987; Olsen et al., 1994). The lining of the fundic chamber appears dark red with many folds (Fig. 12.4). Tarpley et al. (1987) noted that the initial entrance into the fundic chamber has exclusively mucous cells, and is therefore equivalent to the cardiac region of a typical carnivore monogastric stomach.

In many descriptions of cetacean stomachs, the connecting channel has been omitted as an actual stomach chamber. However, Tarpley et al. (1987) argued that the connecting channel is of such functional importance to cetacean digestion, that it should be treated as a distinct stomach compartment. The connecting channel is a tubular structure that, in a 9 m whale, has a diameter of only 5 cm when fully inflated and measures about 17 cm in length (Tarpley et al., 1987). The lining of the connecting channel resembles that of the pyloric chamber and contains mucous glands and a scattering of enteroendocrine cells (e.g., gastrin, a peptide hormone stimulating HCl release) (Tarpley et al., 1987). Most intriguing, the connecting channel is surrounded by smooth muscle, suggesting active regulation and control of the channel diameter (Tarpley et al., 1987). Taken together, the importance of the connecting channel in flow regulation and food retention is apparent. Furthermore, the narrow opening of the channel leaving the fundic chamber may serve as a gatekeeper to pass only foods of appropriate size to the next chamber, thus functionally resembling the omasum of ruminants. This stomach arrangement with the narrow connecting channel and the nonselective filter feeding mechanism of baleen whales makes them particularly vulnerable to plastic pollution in the oceans (Simmonds, 2012). Evidence of plastic ingestion has been found in over 60% of all cetacean species (Fossi et al., 2018), including the bowhead whale, where thus far four cases have been described (see Chapter 30).

Histologically, the pyloric chamber resembles the connecting channel with mucous and enteroendocrine cells. The chamber itself is thin-walled and lined with a red mucosa in longitudinal folds (Tarpley et al., 1987). Some subdivisions in the pyloric chamber are possible, for example, a pyloric septum is present in bowhead whales, and a complete separation leading to 4th and 5th chambers has been described in belugas (*Delphinapterus leucas*), narwhals (*Monodon monoceros*) and some species of beaked whales (Tarpley et al., 1987; Mead, 2007). The pyloric sphincter separates the pyloric chamber from the duodenal ampulla of the small intestine. The arrangement of the sphincter is such that it produces a fold at each end of the orifice, allowing food particles to move from the pyloric chamber to the duodenal ampulla, but not the other way (Tarpley et al., 1987).

While histologically similar to the pyloric chamber with gastric rather than intestinal mucous and enteroendocrine cells, the duodenal ampulla is a dilated sac-like structure that is considered part of the duodenum (Tarpley et al., 1987; Olsen et al., 1994). The presence of the pyloric sphincter separating the two compartments, and not allowing retrograde movement of digesta from the duodenal ampulla to the pyloric chamber, speaks to functionally distinct segments of the bowhead gastrointestinal system (Tarpley et al., 1987). Tarpley et al. (1987) argued that the duodenal ampulla of the bowhead whale functions as a lubrication and pH adjustment chamber for the digesta, before it is being passed to the duodenum proper.

Similar to other mammals, the small intestine of the bowhead whale is divided into the duodenum, jejunum, and ileum. Gastric juices from the liver (bile to emulsify lipids) and pancreas (digestive enzymes, e.g., lipase, trypsinogen) are being secreted into the duodenum shortly after the duodenal ampulla (Olsen et al., 1994; Tarpley et al., 1987) to initiate chemical breakdown of macromolecules (carbohydrates, lipids, proteins, and nucleic acids). As in other cetaceans, a gall bladder is lacking in bowhead whales.

Small intestine to body length ratios in cetaceans are highly variable and can range from short (4 to 6 × body length) in minke whale, right whales (Balaenidae, including bowheads), and humpback whales (*Megaptera novaeangliae*) to ~13 × body length in gray whales (*Eschrichtius robustus*; Yablokov and Bogoslovskaya, 1984) to 25–35 × body length in odontocetes, such as sperm whales (*Physeter macrocephalus*) and La Plata dolphins (*Pontoporia blainvillei*) (Table 12.1; Yamasaki et al., 1975; Williams et al., 2001). For pinnipeds, it has been argued that length of the small intestine is a dive adaptation and related to duration and depth of foraging dives to make up for loss of digestive times during deep or frequent dives when the intestine is not perfused with blood (ischemia; Mårtensson et al., 1998). Known dive durations and myoglobin stores of various cetacean species, particularly when related to body size appear to support this argument (Noren and Williams, 2000). This is further supported when considering that the longest pinniped intestines occur in species consuming difficult to digest foods, for example, chitinous prey, while the opposite is true for cetaceans, where zooplankton consuming mysticetes have much shorter intestines than piscivorous odontocetes.

The presence of a cecum in cetaceans is variable. A cecum is absent in bowheads (Haldiman and Tarpley, 1993), but has been reported in minke whales and river dolphins (Platanistidae) (Olsen et al., 1994; Kelkar et al., 2018). In bowheads, the transition between small and large intestine is marked by an abrupt increase in diameter. Volatile fatty acid-producing bacteria are present in the fin whale (*Balaenoptera physalus*) colon, but their contribution and importance to digestion and nutrient uptake are unknown (Herwig and Staley, 1986).

Gut passage times and fecal isotopes

When working with large cetaceans, logistical constraints make it difficult, if not impossible, to determine actual gut passage times. However, Víkingsson (1997) estimated evacuation rates of forestomach contents to the fundic chamber at 3–6 hours in fin whales based on diurnal differences of prey in the gastrointestinal tract. Passage times from forestomach to fecal evacuation were 15–18 hours. Based on Víkingsson (1997), between 4 and 8 feeding events (i.e., full forestomach) would be possible per day. This is comparable to gut passage times estimated by Besseling et al. (2015) for humpback whales of 14.6 hours. In hippos, retention times of digesta from beginning to end are much longer, between 20 and 107 hours depending on particle size (Clauss et al., 2004), and intermediate in ruminants, between 41 and 65 hours (Udén et al., 1982).

Recently, Arregui et al. (2018) showed that stable isotopes of carbon and nitrogen in fecal material of fin whales are not altered in the digestive process or along the gastrointestinal tract and are indicative of the original prey. This opens new avenues to estimate feeding events and gut passage rates for bowhead whales. To estimate forestomach evacuation rates, we applied stable carbon and nitrogen isotope ratios to the digestive content of different alimentary tract compartments. Generally, stable isotopes are used in ecological research to investigate trophic position, habitat use, and migratory patterns (e.g., Hobson and Schell, 1998; Hoekstra et al., 2002; Horstmann-Dehn et al., 2012). Changes in prey composition can alter stable isotope ratios such that higher trophic level prey will have higher $\delta^{15}N$ values. For fecal matter, $\delta^{13}C$ values may be less useful as an indicator, as changes in $\delta^{13}C$ are also associated with changes in lipid content (DeNiro and Epstein, 1977), that is, increase in lipid and a decrease in $\delta^{13}C$ and vice versa. To evaluate changes in stable isotope ratios in digestive tract compartments of individual bowhead whales, we computed mean and 95% confidence interval for $\delta^{15}N$ and $\delta^{13}C$. The 95% confidence interval length was then applied to the observed stable isotope values of the previous digestive tract compartment (for forestomach the overall mean was used). If the observed value for $\delta^{15}N$ or $\delta^{15}N$ and $\delta^{13}C$ in the following compartment fell outside the 95% confidence interval, a feeding event (i.e., ingestion of different prey) was noted. Within individual whales, we identified on average four distinct stable isotope signature deviations that occur in the forestomach, pyloric chamber/duodenal ampulla, along the small intestine, and along the large intestine (Fig. 12.5). This is in excellent agreement with Víkingsson (1997) based on prey species composition in the intestinal tract of fin whales. While fecal isotopes are an intriguing avenue for new research, the method relies on isotopic differences in prey consumed per feeding event, and this may not necessarily be the case, and can therefore underestimate the number of distinct feeding episodes and gut passage times.

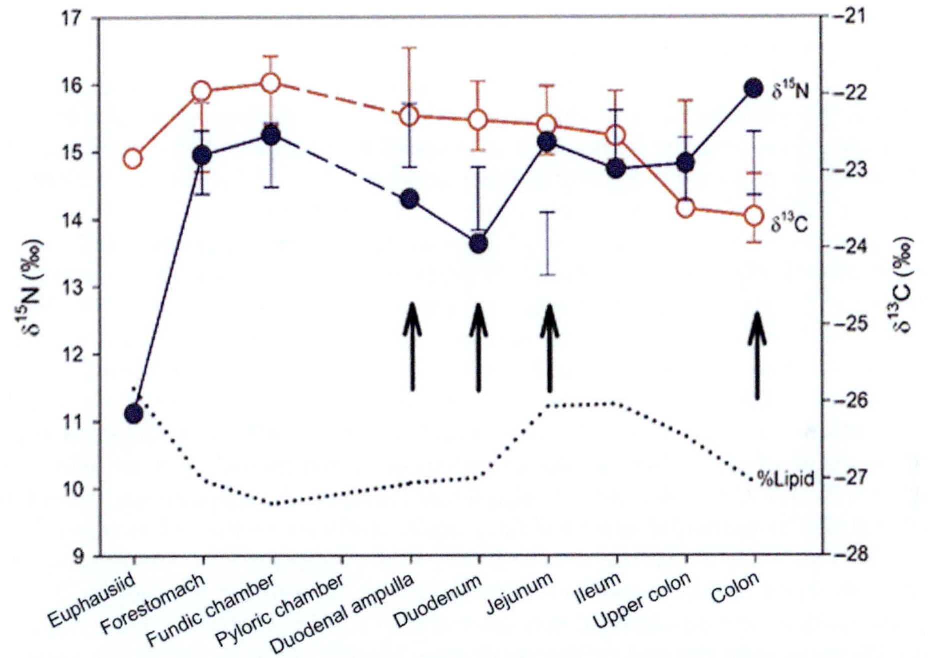

FIGURE 12.5 Stable carbon (*red open symbols*) and nitrogen (*blue solid symbols*) isotopes in the intestinal tract content (sequentially from forestomach to colon) of a male bowhead whale (NSB-DWM 12B15) harvested in fall 2012 in Utqiaġvik (8.4 m total length). Error bars represent the 95% confidence interval of the respective stable isotope mean for the individual based on the stable isotope signature of the previous compartment. The *stippled black line* indicates relative changes in percent lipid in the intestinal tract to provide context for $\delta^{13}C$ values. *Broken blue and red lines* are extrapolated; no sample exists for this gastric compartment. The stomach of this whale contained euphausiids (100% by volume). Stable isotope data are also given for fresh euphausiids (*Thysanoessa* spp.). *Arrows* indicate changes in stable isotope signatures (outside of the confidence interval) that could be evidence of distinct feeding events.

Proximate composition of digesta and fatty acid abundance

Proximate composition analysis provides a quantitative assessment of macronutrients and is most commonly used to evaluate forage quality. Applied to digestive tract content, digestibility and uptake of macronutrients and overall digestive efficiency of prey can be considered (e.g., Fisher et al., 1992; Trumble et al., 2003; Swaim et al., 2009). Proximate composition (percent water, lipid, and crude protein) and gross energy density of bowhead whale intestinal tract content by season and location is provided in Table 12.2. Change in lipid and gross caloric content along the gastrointestinal tract for five bowhead whales is shown in Fig. 12.6. The majority of lipid/gross caloric energy is not taken up until the particles pass the duodenum. This is consistent with typical mammalian digestion under the action of pancreatic lipase in the duodenum (Carey et al., 1983). Miller et al. (2019) also showed that lipid composition significantly changed from the duodenum to the jejunum, specifically, wax esters abundance dropped by about 50%. This is in agreement with our lipid assimilation efficiency ranges (approximately 50%; Table 12.2), and also corresponds with values reported by Nordøy (1995) for minke whales.

Considering that lipid assimilation is only about 50% of available prey lipids, it is important to consider which lipid classes and fatty acids specifically are taken up by bowhead whales. We extracted and quantified 67 fatty acids to determine uptake and modification of fatty acids along the intestinal tract, as well as fatty acid distribution and storage in bowhead whale tissues, specifically blubber and liver. Fig. 12.7 illustrates a multidimensional scaling (MDS) plot of all 67 fatty acids, but highlighting abundance of some example fats, including the essential omega-3 fatty acid 20:5n3 (see Fig. 12.2A, naming convention after Budge et al., 2006). MDS plots visualize relatedness of samples, with those most similar clustering closer together. Fatty acid composition and abundance differed among all tissues, except jejunum and liver (Fig. 12.7) indicating fatty acids taken up in the small intestine are delivered to hepatic short-term storage. Interestingly, in monogastric mammals, digestibility of fatty acids decreases with increasing chain length, but increases with increasing number of double bonds (Carroll, 1958). This does not appear to hold for bowhead whales, where mono- or polyunsaturated long-chain fatty acids, such as 22:1n11, 22:5n3, and 22:6n3 are almost completely taken up. Unsurprisingly, abundance of most fatty acids, in particular essential fatty acids, declined in the colon compared with forestomach (Fig. 12.7), indicating uptake in the small intestine; this was also reported in North Atlantic right whale feces (Swaim et al., 2009). In contrast to triglycerides, which were taken up along the duodenum, wax ester abundance in bowhead whales did not substantially decrease until the end of the small intestine (Miller et al., 2019), suggesting that additional retention may be necessary to break down and absorb these compounds. Some of the most abundant fatty acids in the colon were the long-chain saturated fatty acids, 20:0 and 22:0, as well as microbial fatty acids, for example, anteiso 15:0, which did not occur in the forestomach or other tissues of bowhead whales (Fig. 12.7). Similar increases of some fatty acids, especially odd-chain fatty acids, have been reported for ruminants (Doreau and Ferlay, 1994) and have been attributed to microbial production. Swaim et al. (2009) analyzed fatty acids of copepod prey and North Atlantic right whales feces and showed almost identical increases in long-chain saturated fatty acids and occurrence of

TABLE 12.2 Proximate composition (mean ± 1 standard deviation dry weight) of bowhead whale intestinal tract content, for whales captured during 2009–12.

Location	Season	Compartment	Sample size	%Water	%Lipid	%Nitrogen	%Ash	Crude protein	Caloric content (kJ/g)	Assimilation efficiency
Utqiaġvik	Fall 2009	Forestomach	6	82.7 ± 3.5	48.2 ± 11.0	9.0 ± 0.7	12.9 ± 2.1	4.4 ± 1.3	23.3 ± 0.8	53.6 ± 5.7
		Fundic chamber	5	86.2 ± 2.9	52.1 ± 12.0	8.8 ± 1.0	12.3 ± 2.2	3.3 ± 1.4	23.0 ± 2.3	
		Pyloric chamber	6	88.9 ± 4.9	49.3 ± 10.8	7.2 ± 2.6	11.8 ± 2.2	2.8 ± 1.4	22.8 ± 1.4	
		Duodenum	7	85.1 ± 3.3	54.6 ± 6.8	8.4 ± 1.5	12.6 ± 3.8	3.2 ± 1.3	23.1 ± 2.1	
		Colon	7	78.0 ± 3.5	25.3 ± 8.2	4.1 ± 0.4	45.5 ± 11.4	2.3 ± 0.4	10.6 ± 1.1	
Utqiaġvik	Spring 2010	Colon	14	86.1 ± 4.4	44.4 ± 15.1	6.5 ± 2.7	20.1 ± 17.7	2.6 ± 1.4	19.4 ± 5.3	
St. Lawrence Island	Spring 2010	Duodenum	2	87.7 ± 0.3	50.1 ± 17.7	10. ± 2.8	6.7 ± 0.6	3.9 ± 2.4	23.3 ± 3.0	
		Colon	4	79.0 ± 7.2	43.2 ± 11.5	5.8 ± 1.3	15.4 ± 10.8	3.5 ± 1.1	22.5 ± 3.0	
Wainwright	Spring 2010	Forestomach	1	93.2	65.2	6.4	8.7	0.9	23.0	–
Utqiaġvik	Fall 2010	Forestomach	4	90.3 ± 2.7	50.4 ± 18.1	6.3 ± 2.0	5.1 ± 2.1	1.9 ± 1.1	25.4 ± 2.9	45.3 ± 12.9
		Duodenum	4	88.3 ± 6.1	56.5 ± 10.8	7.6 ± 1.9	6.7 ± 1.5	2.6 ± 2.2	21.7 ± 3.1	
		Colon	5	81.0 ± 4.1	27.0 ± 8.1	5.8 ± 0.8	29.2 ± 10.4	3.5 ± 0.7	15.7 ± 4.4	
Kaktovik	Fall 2010	Colon	2	82.6 ± 1.6	29.7 ± 8.3	6.7 ± 0.6	23.0 ± 2.6	3.9 ± 0.3	18.4 ± 1.0	
St. Lawrence Island	Spring 2011	Colon	2	73.0 ± 0.8	10.6 ± 6.1	4.5 ± 0.5	39.9 ± 2.7	4.0 ± 0.3	11.9 ± 0.3	
Utqiaġvik	Fall 2011	Forestomach	6	81.2 ± 4.9	46.0 ± 21.3	7.0 ± 3.0	5.1 ± 2.6	5.0 ± 4.2	27.8 ± 3.9	43.4 ± 12.2
		Fundic chamber	1	89.0	66.3	5.7	5.2	1.3	26.3	
		Pyloric chamber	2	85.8 ± 9.2	66.3 ± 25.1	5.5 ± 3.5	5.1 ± 4.1	1.2 ± 0.9	20.9 ± 2.5	

(Continued)

TABLE 12.2 (Continued)

Location	Season	Compartment	Sample size	%Water	%Lipid	%Nitrogen	%Ash	Crude protein	Caloric content (kJ/g)	Assimilation efficiency
		Duodenum	6	87.6 ± 1.9	43.8 ± 16.6	6.7 ± 0.9	6.3 ± 0.8	2.8 ± 1.0	22.7 ± 5.2	
		Colon	10	85.3 ± 2.3	27.8 ± 9.0	6.2 ± 1.6	23.0 ± 10.3	3.2 ± 0.9	19.3 ± 4.3	
Kaktovik	Fall 2011	Forestomach	2	77.3 ± 4.3	75.5 ± 17.9	2.2 ± 1.4	1.8 ± 0.7	0.8 ± 0.9	32.7	49.5
		Duodenum	1	89	33.4	8.8	8.3	3.7	18.9	
		Colon	2	84.2 ± 3.0	41.7 ± 23.5	5.1 ± 2.9	24.0 ± 1.6	2.7 ± 2.7	22.8 ± 8.9	
Utqiaġvik	Spring 2012	Colon	7	83.0 ± 3.9	35.9 ± 7.2	8.0 ± 1.3	14.1 ± 8.9	4.5 ± 0.7	24.1 ± 3.8	
St. Lawrence Island	Spring 2012	Forestomach	2	92.8 ± 7.5	7.7 ± 10.9	8.1 ± 8.2	44.8 ± 44.7	4.7 ± 6.6	22.1	–
		Duodenum	4	88.2 ± 1.7	42.5 ± 15.5	7.9 ± 0.8	18.8 ± 9.6	2.7 ± 0.7	20.2 ± 2.9	
		Colon	2	80.8 ± 8.1	30.8 ± 10.6	6.5 ± 0.1	28.4 ± 8.3	3.8 ± 1.3	18.4 ± 4	
Utqiaġvik	Fall 2012	Forestomach	8	84.7 ± 9.2	33.1 ± 22.0	10.4 ± 2.6	14.3 ± 9.8	6.7 ± 6.3	24.0 ± 6.5	41.5 ± 18.8
		Fundic chamber	2	88.5 ± 5.1	22.1 ± 31.3	11.1 ± 1.0	11.3 ± 3.5	6.3 ± 5.4	23.4 ± 0.8	
		Pyloric chamber	3	91.7 ± 1.7	20.5 ± 22.2	10.4 ± 0.3	13.7 ± 6.4	2.6 ± 2.6	23.0 ± 0.6	
		Duodenum		89.0 ± 3.4	40.8 ± 26.8	9.5 ± 2.7	9.3 ± 2.8	4.4 ± 3.7	25.7 ± 2.8	
		Colon	10	82.5 ± 6.1	13.2 ± 10.3	6.5 ± 2.4	35.4 ± 13.4	3.8 ± 2.1	16.0 ± 4.4	
Kaktovik	Fall 2012	Colon	1	85.1	30.9	6.2	29.1	3.0	15.8	

Assimilation efficiency was calculated based on "start" composition of forestomach caloric content to "end" composition of colon caloric content. H_0: no difference in assimilation efficiency among years (Utqiaġvik only), $P = .45$.

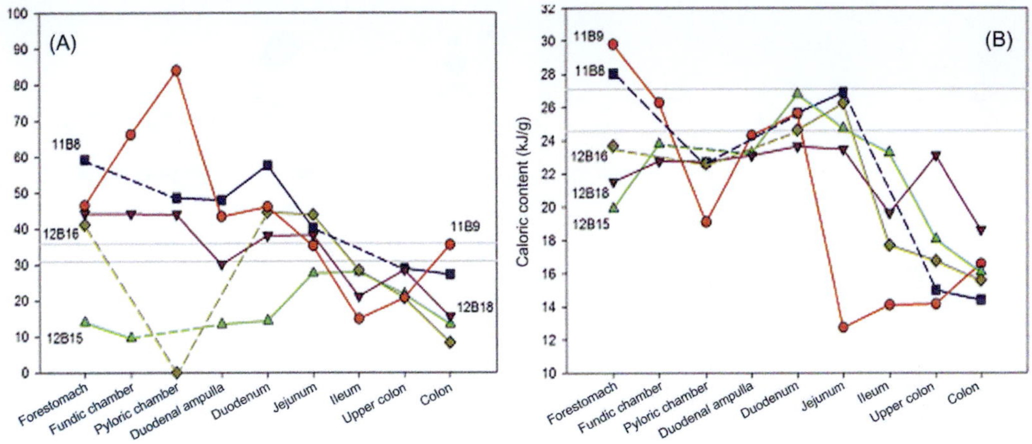

FIGURE 12.6 Changes in (A) lipid (dry weight), and (B) caloric content (kJ/g dry weight) along the digestive tract for five bowhead whales (in different colors and identified by their NSB-DWM numbers) intensively sampled in Fall 2011 and 2012 in Utqiaġvik, Alaska. *Gray lines* reference proximate composition range of euphausiids and copepods. *Broken lines* are extrapolated, and no sample exists for this gastric compartment.

fatty acids that were not identified in the prey. Swaim et al. (2009) contemplated that these long-chain saturated fats may stem from bacterial/microbial production. Recently, Miller et al. (2019) showed that the most common lipid class in the bowhead whale colon are wax esters (between 30% and 40% of colon lipids). However, wax ester abundance in the colon was highly variable (4%–91%), thus prompting Miller et al. (2019) to discuss several factors that may influence wax ester digestibility in individual whales. These authors showed a correlation of wax ester abundance and certain microbial groups in the jejunum, the anatomical location of lipid uptake, and suggested that presence or absence of certain microbes (i.e., *Actinobacillus* and *Cetobacterium*) may influence digestibility (Miller et al., 2019).

Digestive efficiency

Prey and associated composition of macronutrients is an important determinant of digestive efficiency, essentially a measure of energy taken up from prey that is then available for metabolism, growth, and reproduction. Changes in prey quality (e.g., lower amount of lipids, or even specific lipid classes) can substantially alter the energy balance of an organism. For bowhead whales, energy assimilation efficiency was calculated based on "start" composition of forestomach caloric content to "end" composition of colon caloric content (Table 12.2) and ranged between 42% and 54%. Digestive efficiency (or digestible energy) on the other hand was calculated using daily energy intake (daily forestomach volume and energy density of prey in the forestomach, wet weight) and daily fecal energy loss (daily fecal volume and colon energy density, wet weight). For these calculations, we assumed four daily feeding events (see above) and a forestomach volume of

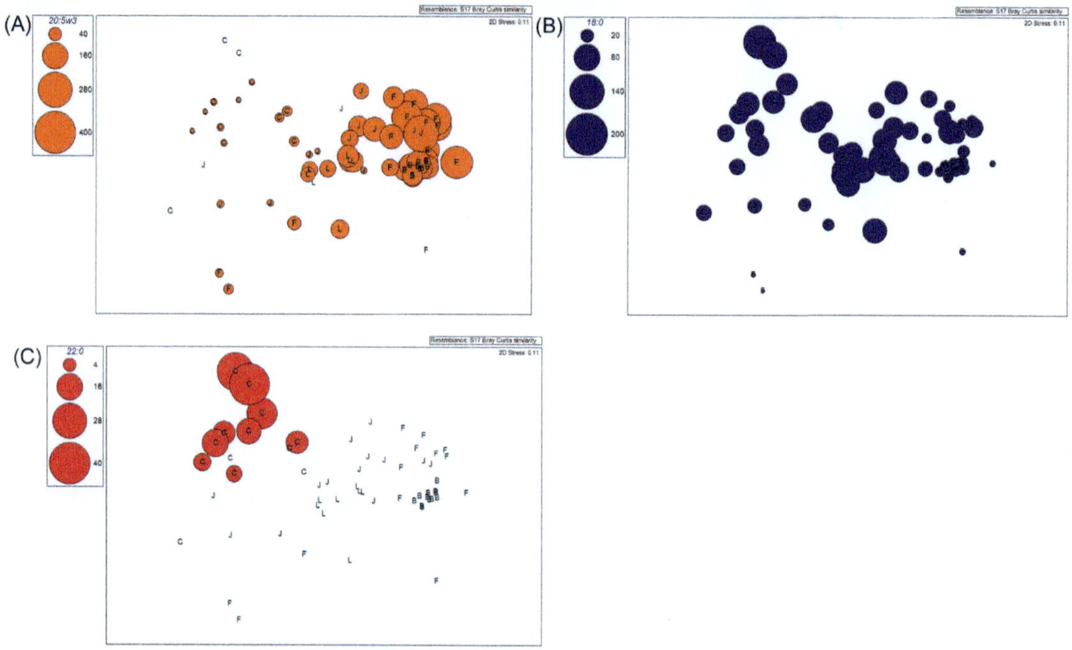

FIGURE 12.7 Multidimensional scaling (MDS) plots of 67 fatty acids, and highlighting abundance of example fatty acids 20:5n3 (A), 18:0 (B), and 22:0 (C) in different gastrointestinal compartments and tissues of bowhead whales sampled in fall 2011–2013 in Utqiaġvik, AK. *B*, Blubber; *C*, Colon; *F*, Forestomach; *J*, Jejunum; *L*, Liver. Size of the bubble reflects concentration of that fatty acid present in the gastric compartment/tissue in mg/g lipid, and clustering reveals similarity of samples. *Letters not enclosed by a bubble* indicate the absence of that fatty acid. The omega-3 fatty acid (20:5n3) is almost completely taken up by the time the digesta reaches the colon. The saturated fatty acid 18:0 shows an interesting pattern, where more of that particular fatty acid is present in the colon compared to the forestomach, or in the case of 22:0, the fatty acid is only present in the colon, but not in the forestomach or any other tissue.

52.7 L for a 9 m subadult bowhead (Table 12.1). Defecation volume and energy loss was estimated by applying the allometric equation implemented by Swaim et al. (2009):

$$FE = 0.85\ BM^{0.63} \tag{12.1}$$

where FE is daily fecal production in gram and BM is body mass in gram. For a subadult bowhead whale (approximately 9 m length and body mass of 12,000 kg), daily defecation mass is about 24.5 kg. Using an average forestomach caloric density of 4.8 kJ wet weight (Table 12.2) and average colon caloric density of 2.2 kJ wet weight (Table 12.2), bowheads are taking in over 1 million kJ per day, and eliminate about 54,000 kJ via defecation per day. This results in a digestive efficiency of 95% and is in excellent agreement with digestive efficiencies reported for North Atlantic right whales (94%, Swaim et al., 2009) and minke whales, for example, 93% and 95% consuming euphausiids of the genus *Thysanoessa* spp. (Mårtensson et al., 1994) and capelin (*Mallotus villosus*), respectively (Mathiesen et al., 1995). Fortune et al. (2013) estimated digestive efficiency of North Atlantic right whales between 74% and 86% depending if the diet contained exclusively Stage 5 *Calanus finmarchicus* or was mixed, respectively. Using in vitro digestibility experiments, Nordøy et al. (1993) confirmed digestive efficiency estimates of approximately 90% in minke whales, eating either krill or fish. Bowhead whale digestive efficiency thus is comparable to those reported for other baleen whales, and variability in assimilation efficiency may be related to differential effectiveness of absorbing lipids from different crustacean prey, or, as suggested by Miller et al. (2019), may be better in some whales than others and related to the microbiome.

Future considerations

Considering the importance of accurate volumetric measurements of the bowhead stomach to estimate energy requirements (Chapter 16), it is surprising that very few of these data actually exist, and only for subadult bowhead whales (~9 m). Measuring forestomach and total stomach volumes for a range of size classes will be an important next step. The bowhead whale ecosystem is undoubtedly changing, and with it food web dynamics, assemblages, and prey quality (e.g., percent lipid, concentrations of essential fatty acids). These changes can have major impacts on the bowhead whale, thus making continued monitoring of proximate composition of prey and nutrient uptake along the gastrointestinal tract critical. This will be particularly important in conjunction with stomach contents analyses to understand digestive efficiency under differing prey compositions. Amino acid uptake along the bowhead whale gut has never been described, and could be essential in growing juvenile and subadult whales, which are living on the edge with regard to their energetic and nutrient requirements. Finally, bowhead whales can undeniably digest wax esters with great efficiency; yet, our understanding of how this is accomplished is lacking. Are bowheads utilizing a specialized microbiome, a body-own mechanism, or a combination of both? Because efficiency of wax ester digestion seems to be highly variable, more research should focus on the microbiome of bowheads in a range of age classes and sexes. A combination of microbiome, diet, and nutrient composition and uptake could shed light on some important digestive mechanisms, and address possible changes and consequences bowheads may have to face in their changing ecosystem.

Acknowledgments

This study was made possible by the willingness of Alaska Native whaling captains, crews, and their families to provide samples and harvest information. A big thank-you goes to C. George and R. Stimmelmayr and the many North Slope Borough Department of Wildlife Management sampling wizards for assistance with sample collections during the harvests.

References

Arman, P., Field, C.R., 1973. Digestion in the hippopotamus. Afr. J. Ecol. 11 (1), 9–17. Available from: https://doi.org/10.1111/j.1365-2028.1973.tb00070.x.

Arregui, M., Borrell, A., Víkingsson, G., Ólafsdóttir, D., Aguilar, A., 2018. Stable isotope analysis of fecal material provides insight into the diet of fin whales. Mar. Mammal. Sci. 34 (4), 1059–1069. Available from: https://doi.org/10.1111/mms.12504.

Besseling, E., Foekema, E.M., Van Franeker, J.A., Leopold, M.F., Kühn, S., Bravo Rebolledo, E.L., et al., 2015. Microplastic in a macro filter feeder: humpback whale *Megaptera novaeangliae*. Mar. Pollut. Bull. 95 (1), 248–252. Available from: https://doi.org/10.1016/j.marpolbul.2015.04.007.

Bogevik, A.S., 2011. Marine wax ester digestion in salmonid fish: a review. Aquac. Res. 42 (11), 1577–1593. Available from: https://doi.org/10.1111/j.1365-2109.2010.02766.x.

Budge, S.M., Iverson, S.J., Koopman, H.N., 2006. Studying trophic ecology in marine ecosystems using fatty acids: a primer on analysis and interpretation. Mar. Mammal. Sci. 22 (4), 759–801. Available from: https://doi.org/10.1111/j.1748-7692.2006.00079.x.

Carey, M.C., Small, D.M., Bliss, C.M., 1983. Lipid digestion and absorption. Annu. Rev. Physiol. 45, 651–677. Available from: https://doi.org/10.1146/annurev.ph.45.030183.003251.

Carroll, K.K., 1958. Digestibility of individual fatty acids in the rat. J. Nutr. 64 (3), 399–410. Available from: https://doi.org/10.1093/jn/64.3.399.

Chritz, K.L., Blumenthal, S.A., Cerling, T.E., Klingel, H., 2016. Hippopotamus (*H. amphibius*) diet change indicates herbaceous plant encroachment following megaherbivore population collapse. Sci. Rep. 6, 32807. Available from: https://doi.org/10.1038/srep32807.

Clauss, M., Frey, R., Kiefer, B., Lechner-Doll, M., Loehlein, W., Polster, C., et al., 2003. The maximum attainable body size of herbivorous mammals: morphophysiological constraints on foregut, and adaptations of hindgut fermenters. Oecologia 136 (1), 14–27. Available from: https://doi.org/10.1007/s00442-003-1254-z.

Clauss, M., Hofmann, R.R., Hummel, J., Adamczewski, J., Nygren, K., Pitra, C., et al., 2006. Macroscopic anatomy of the omasum of free-ranging moose (*Alces alces*) and muskoxen (*Ovibos moschatus*) and a comparison of the omasal laminal surface area in 34 ruminant species. J. Zool. 270 (2), 346–358. Available from: https://doi.org/10.1111/j.1469-7998.2006.00148.x.

Clauss, M., Schwarm, A., Ortmann, S., Alber, D., Flach, E.J., Kühne, R., et al., 2004. Intake, ingesta retention, particle size distribution and digestibility in the Hippopotamidae. Comp. Biochem. Physiol. A: Mol. Integr. Physiol. 139 (4), 449–459. Available from: https://doi.org/10.1016/j.cbpb.2004.10.002.

Cohen-Kupiec, R., Chet, I., 1998. The molecular biology of chitin digestion. Curr. Opin. Biotechnol. 9 (3), 270–277. Available from: https://doi.org/10.1016/S0958-1669(98)80058-X.

DeNiro, M.J., Epstein, S., 1977. Mechanism of carbon isotope fractionation associated with lipid synthesis. Science 197 (4300), 261–263. Available from: https://doi.org/10.1126/science.327543.

Doreau, M., Ferlay, A., 1994. Digestion and utilisation of fatty acids by ruminants. Anim. Feed. Sci. Technol. 45 (3–4), 379–396. Available from: https://doi.org/10.1016/0377-8401(94)90039-6.

Duchesne, L.C., Larson, D.W., 1989. Cellulose and the evolution of plant life. BioScience 39 (4), 238–241. Available from: https://doi.org/10.2307/1311160.

Fisher, K.I., Stewart, R.E.A., Kastelein, R.A., Campbell, L.D., 1992. Apparent digestive efficiency in walruses (*Odobenus rosmarus*) fed herring (*Clupea harengus*) and clams (*Spisula* sp.). Can. J. Zool. 70 (1), 30–36. Available from: https://doi.org/10.1139/z92-005.

Fordyce, R.E., Marx, F.G., 2018. Gigantism precedes filter feeding in baleen whale evolution. Curr. Biol. 28 (10), 1670–1676. Available from: https://doi.org/10.1016/j.cub.2018.04.027.

Fortune, S.M.E., Trites, A.W., Mayo, C.A., Rosen, D.A.S., Hamilton, P.K., 2013. Energetic requirements of North Atlantic right whales and the implications for species recovery. Mar. Ecol. Prog. Ser. 478, 253–272. Available from: https://doi.org/10.3354/meps10000.

Fossi, M.C., Baini, M., Panti, C., Baulch, S., 2018. Impacts of marine litter on cetaceans: a focus on plastic pollution. In: Fossi, M.C., Panti, C. (Eds.), Marine Mammal Ecotoxicology: Impacts of Multiple Stressors on Population Health. Elsevier, pp. 147–184. Available from: http://dx.doi.org/10.1016/B978-0-12-812144-3.00006-1.

Galloway, T.W., 1915. Zoology: A Textbook for Universities, Colleges and Normal Schools, third ed. Blakiston, Philadelphia. Available from: http://hdl.handle.net/2027/hvd.32044072255144.

Geisler, J.H., Theodor, J.M., 2009. Hippopotamus and whale phylogeny. Nature 458 (7236), E1–E4. Available from: https://doi.org/10.1038/nature07776.

Goldbogen, J.A., 2018. Physiological constraints on marine mammal body size. Proc. Natl. Acad. Sci. U. S. A. 115 (16), 3995–3997. Available from: https://doi.org/10.1073/pnas.1804077115.

Goldbogen, J.A., Cade, D.E., Wisniewska, D.M., Potvin, J., Segre, P.S., Savoca, M.S., et al., 2019. Why whales are big but not bigger: physiological drivers and ecological limits in the age of ocean giants. Science 366 (6471), 1367–1372. Available from: https://doi.org/10.1126/science.aax9044.

Graeve, M., Kattner, G., Hagen, W., 1994. Diet-induced changes in the fatty acid composition of Arctic herbivorous copepods: experimental evidence of trophic markers. J. Exp. Mar. Biol. Ecol. 182 (1), 97–110. Available from: https://doi.org/10.1016/0022-0981(94)90213-5.

Haldiman, J.T., Tarpley, R.J., 1993. Anatomy and physiology. In: Burns, J.J., Montague, J.J., Cowles, C.J. (Eds.), The Bowhead Whale. Society for Marine Mammalogy, Lawrence, KS, pp. 71–156. Special Publication Number 2.

Harrison, R.J., Johnson, F.R., Young, B.A., 1970. The oesophagus and stomach of dolphins (*Tursiops, Delphinus, Stenella*). J. Zool. 160, 377–390. Available from: https://doi.org/10.1111/j.1469-7998.1970.tb03088.x.

Herwig, R.P., Staley, J.T., 1986. Anaerobic bacteria from the digestive tract of North Atlantic fin whales (*Balaenoptera physalus*). FEMS Microbiol. Lett. 38 (6), 361–371. Available from: https://doi.org/10.1016/0378-1097(86)90014-5.

Herwig, R.P., Staley, J.T., Nerini, M.K., Braham, H.W., 1984. Baleen whales: preliminary evidence for forestomach microbial fermentation. Appl. Environ. Microbiol. 47 (2), 421–423.

Hobson, K.A., Schell, D.M., 1998. Stable carbon and nitrogen isotope patterns in baleen whales from eastern Arctic bowhead whales (*Balaena mysticetus*). Can. J. Fish. Aquat. Sci. 55 (12), 2601–2607. Available from: https://doi.org/10.1139/cjfas-55-12-2601.

Hoekstra, P.F., Dehn, L.-A., George, J.C., Muir, D.C.G., O'Hara, T.M., 2002. Trophic ecology of bowhead whales (*Balaena mysticetus*) compared with that of other arctic marine biota as interpreted from carbon-, nitrogen-, and sulfur-isotope signatures. Can. J. Zool. 80 (2), 223–231. Available from: https://doi.org/10.1139/z01-229.

Hofmann, R.R., 1989. Evolutionary steps of ecophysiological adaptation and diversification of ruminants: a comparative view of their digestive system. Oecologia 78 (4), 443–457. Available from: https://doi.org/10.1007/BF00378733.

Horstmann-Dehn, L., Follmann, E.H., Rosa, C., Zelensky, G., George, C., 2012. Stable carbon and nitrogen isotope ratios in muscle and epidermis of arctic whales. Mar. Mammal. Sci. 28 (2), E173–E190. Available from: https://doi.org/10.1111/j.1748-7692.2011.00503.x.

Jeuniaux, C., Voss-Foucart, M.F., 1991. Chitin biomass and production in the marine environment. Biochem. Syst. Ecol. 19 (5), 347–356. Available from: https://doi.org/10.1016/0305-1978(91)90051-Z.

Kelkar, N., Dey, S., Deshpande, K., Choudhary, S.K., Dey, S., Morisaka, T., 2018. Foraging and feeding ecology of Platanista: an integrative review. Mammal. Rev. 48 (3), 194–208. Available from: https://doi.org/10.1111/mam.12124.

Keys, J.E., Van Soest, P.J., Young, E.P., 1969. Comparative study of the digestibility of forage cellulose and hemicellulose in ruminants and nonruminants. J. Anim. Sci. 29 (1), 11–15. Available from: https://doi.org/10.2527/jas1969.29111x.

Mårtensson, P.E., Nordøy, E.S., Blix, A.S., 1994. Digestibility of krill (*Euphausia superba* and *Thysanoessa* sp.) in minke whales (*Balaenoptera acutorostrata*) and crabeater seals (*Lobodon carcinophagus*). Br. J. Nutr. 72 (5), 713–716. Available from: https://doi.org/10.1079/bjn19940073.

Mårtensson, P.E., Nordøy, E.S., Messelt, E.B., Blix, A.S., 1998. Gut length, food transit time and diving habit in phocid seals. Polar Biol. 20 (3), 213–217. Available from: https://doi.org/10.1007/s003000050298.

Mathiesen, S.D., Aagnes, T.H., Sørmo, W., Nordøy, E.S., Blix, E.S, Olsen, M.A., 1995. Digestive physiology of minke whales. Dev. Mar. Biol. 4, 351–359. Available from: https://doi.org/10.1016/S0163-6995(06)80036-3.

Mead, J.G., 2007. Stomach anatomy and use in defining systemic relationships of the cetacean family Ziphiidae (beaked whales). Anat. Rec. 290 (6), 581–595. Available from: https://doi.org/10.1002/ar.20536.

Miller, C.A., Holm, H.C., Horstmann, L., George, J.C., Fredricks, H.F., Van Mooy, B.A.S., et al., 2019. Coordinated transformation of the gut microbiome and lipidome of bowhead whales provides novel insights into digestion. ISME J. Available from: https://doi.org/10.1038/s41396-019-0549-y.

Nordøy, E.S., 1995. Do minke whales (*Balaenoptera acutorostrata*) digest wax esters? Br. J. Nutr. 74 (5), 717–722. Available from: https://doi.org/10.1079/bjn19950174.

Nordøy, E.S., Sørmo, W., Blix, A.S., 1993. In vitro digestibility of different prey species of minke whales (*Balaenoptera acutorostrata*). Br. J. Nutr. 70 (2), 485–489. Available from: https://doi.org/10.1079/bjn19930142.

Noren, S.R., Williams, T.M., 2000. Body size and skeletal muscle myoglobin of cetaceans: adaptations for maximizing dive duration. Comp. Biochem. Physiol. A 126 (2), 181–191. Available from: https://doi.org/10.1016/S1095-6433(00)00182-3.

Okine, E.K., Mathison, G.W., Kaske, M., Kennelly, J.J., Christopherson, R.J., 1998. Current understanding of the role of the reticulum and reticulo-omasal orifice in the control of digesta passage from the ruminoreticulum of sheep and cattle. Can. J. Anim. Sci. 78 (1), 15–21. Available from: https://doi.org/10.4141/A97-021.

Olsen, M.A., Blix, A.S., Utsi, T.H.A., Sørmo, W., Mathiesen, S.D., 2000. Chitinolytic bacteria in the minke whale forestomach. Can. J. Microbiol. 46 (1), 85–94. Available from: https://doi.org/10.1139/w99-112.

Olsen, M.A., Nordøy, E.S., Blix, A.S., Mathiesen, S.D., 1994. Functional anatomy of the gastrointestinal system of northeastern Atlantic minke whales (*Balaenoptera acutorostrata*). J. Zool. 234 (1), 55–74. Available from: https://doi.org/10.1111/j.1469-7998.1994.tb06056.x.

Olsen, M.A., Mathiesen, S.D., 1996. Production rates of volatile fatty acids in the minke whale (*Balaenoptera acutorostrata*) forestomach. Br. J. Nutr. 75 (1), 21–31. Available from: https://doi.org/10.1079/bjn19960107.

Omura, H., Ohsumi, S., Nemoto, T., Nasu, K., Kasuya, T., 1969. Black right whales in the North Pacific. Sci. Rep. Whales Res. Inst. 21, 1–78.

Pawar, S., Dell, A.I., Savage, V.M., 2012. Dimensionality of consumer search space drives trophic interaction strengths. Nature 486 (7404), 485–489. Available from: https://doi.org/10.1038/nature11131.

Pérez, W., Lima, M., Büker, M., Clauss, M., 2017. Gross anatomy of the stomach and intestine of an Antarctic minke whale (*Balaenoptera bonaerensis*). Mammalia 81 (1), 111–113. Available from: https://doi.org/10.1515/mammalia-2015-0168.

Place, A.R., 1992. Comparative aspects of lipid digestion and absorption: physiological correlates of wax ester digestion. Am. J. Physiol. Regul. Integr. Comp. Physiol. 263 (3), R464–R471. Available from: https://doi.org/10.1152/ajpregu.1992.263.3.r464.

Sargent, J.R., Falk-Petersen, S., 1988. The lipid biochemistry of calanoid copepods. In: Boxshall, G.A., Schminke, H.K. (Eds.), Biology of Copepods. Springer, Dordrecht, pp. 101–114. Available from: http://dx.doi.org/10.1007/978-94-009-3103-9_9.

Simmonds, M.P., 2012. Cetaceans and marine debris: the great unknown. J. Mar. Biol. 2012, 1–8. Available from: https://doi.org/10.1155/2012/684279.

Swaim, Z.T., Westgate, A.J., Koopman, H.N., Rolland, R.M., Kraus, S.D., 2009. Metabolism of ingested lipids by North Atlantic right whales. Endanger. Species Res. 6 (3), 259–271. Available from: https://doi.org/10.3354/esr00163.

Tarpley, R.J., Sis, R.F., Albert, T.F., Dalton, L.M., George, J.C., 1987. Observations on the anatomy of the stomach and duodenum of the bowhead whale, *Balaena mysticetus*. Am. J. Anat. 180 (3), 295–322. Available from: https://doi.org/10.1002/aja.1001800310.

Trumble, S.J., Barboza, P.S., Castellini, M.A., 2003. Digestive constraints on an aquatic carnivore: effects of feeding frequency and prey composition on harbor seals. J. Comp. Physiol. B 173 (6), 501–509. Available from: https://doi.org/10.1007/s00360-003-0358-4.

Tucker, M.A., Rogers, T.L., 2014a. Examining the prey mass of terrestrial and aquatic carnivorous mammals: minimum, maximum and range. PLoS ONE 9 (8), e106402. Available from: https://doi.org/10.1371/journal.pone.0106402.

Tucker, M.A., Rogers, T.L., 2014b. Examining predator–prey body size, trophic level and body mass across marine and terrestrial mammals. Proc. R. Soc. B 281(1797), 1–9. Available from: https://doi.org/10.1098/rspb.2014.2103.

Udén, P., Rounsaville, T.R., Wiggans, G.R., Van Soest, P.J., 1982. The measurement of liquid and solid digesta retention in ruminants, equines and rabbits given timothy (*Phleum pratense*) hay. Br. J. Nutr. 48 (2), 329–339. Available from: https://doi.org/10.1079/bjn19820117.

Víkingsson, G.A., 1997. Feeding of fin whales (*Balaenoptera physalus*) off Iceland—diurnal and seasonal variation and possible rates. J. Northwest Atl. Fish. Sci. 22, 77–89. Available from: https://doi.org/10.2960/J.v22.a7.

Wang, L., Jiang, X., Xu, H., Zhang, Y., 2018. Metagenomics analyses of cellulose and volatile fatty acid metabolism by microorganisms in the cow rumen. bioRxiv. doi:10.1101/414961.

Williams, T.M., Haun, J., Davis, R.W., Fuiman, L.A., Kohin, S., 2001. A killer appetite: metabolic consequences of carnivory in marine mammals. Comp. Biochem. Physiol. A 129 (4), 785–796. Available from: https://doi.org/10.1016/S1095-6433(01)00347-6.

Yablokov, A.V., Bogoslovskaya, L.S., 1984. A review of Russian research on the biology and commercial whaling of the gray whale. In: Jones, S., Swartz, M.L., Leatherwood, S.L. (Eds.), The Gray Whale: *Eschrichtius robustus*. Academic Press. Available from: https://doi.org/10.1016/B978-0-08-092372-7.50026-1.

Yamasaki, F., Takahashi, K., Kamiya, T., 1975. Digestive tract of La Plata dolphin, *Pontoporia blainvillei*. II. Small and large intestines. Okajimas Folia Anat. Jpn. 52 (1), 1–26. Available from: https://doi.org/10.2535/ofaj1936.52.1_1.

Zurano, J.P., Magalhães, F.M., Asato, A.E., Silva, G., Bidau, C.J., Mesquita, D.O., et al., 2019. Cetartiodactyla: updating a time-calibrated molecular phylogeny. Mol. Phylogenet. Evol. 133, 256–262. Available from: https://doi.org/10.1016/j.ympev.2018.12.015.

CHAPTER

13

Female and male reproduction

Raymond J. Tarpley[1], *D.J. Hillmann*[2], *J.C. George*[3] *and J.G.M. Thewissen*[4]

[1]Department of Veterinary Integrative Biosciences, College of Veterinary Medicine and Biomedical Sciences, Texas A&M University, College Station, TX, United States [2]Department of Comparative Biomedical Sciences, School of Veterinary Medicine, Louisiana State University, Baton Rouge, LA, United States [3]Department of Wildlife Management, North Slope Borough, Utqiaġvik, AK, United States [4]Department of Anatomy and Neurobiology, Northeast Ohio Medical University, Rootstown, OH, United States

Introduction

Cetacean reproductive morphology, even after some 50 million years of genetic isolation from terrestrial relatives (Thewissen and Williams, 2002), has retained the structural strategy typical of eutherian mammals. Despite the novelties of the marine environment (Fig. 13.1), requiring that all reproductive functions from mating to calving and nursing be orchestrated in water, there has been little need to rework basic structural designs. Such evolutionary conservatism is remarkable since aquatic mammals have otherwise invested in a wide array of external and internal body modifications in response to unique habitat demands.

Studies of the reproductive system of bowhead whales have been extensive, as they provide important information for management of populations. In order to help establish sustainable harvest levels for the Bering-Chukchi-Beaufort (BCB) stock of bowhead whales, the National Marine Fisheries Service and later the North Slope Borough science program focused research efforts on female, and later the male, reproductive system. Much of the research reported here is part of a program led by the North Slope Borough Department of Wildlife Management (NSB-DWM), and specimens investigated were harvested as part of the native Alaskan subsistence hunt.

The Bowhead Whale
DOI: https://doi.org/10.1016/B978-0-12-818969-6.00013-3

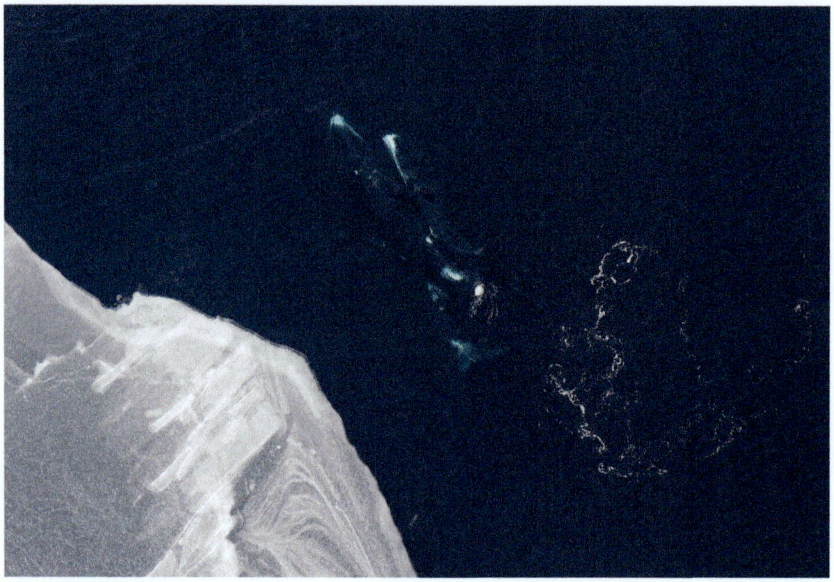

FIGURE 13.1 A female bowhead whale mates with a much smaller male in the Arctic Ocean. The fluke of a third animal is visible underneath them. *Source: Photo by Vicki Beaver (NOAA/North Slope Borough, NMFS Permit No. 14245).*

Reproductive tract morphology

External genitalia

Phenotypic expression of the external genitalia along the ventral midline of the bowhead whale accords with the predominant configuration of other Cetacea. The genital groove is sexually dimorphic: longer in the male and nearly reaching the umbilicus, whereas in the female, it is short and situated just cranial to the anus.

This characteristic is sufficiently pronounced to allow even casual observation to reliably assess sex, and Tarpley and Hillmann (1999) established this quantitatively. Using an expanded dataset of carefully examined whales, we confirmed these results and estimated a mean genital groove/body length ratio for females at 0.029 ($SD = 0.005$; $N = 85$) and for males 0.114 ($SD = 0.014$, $N = 72$), with no sex-based overlap in the range of ratios.

A mammary groove or cleft is positioned on each side of the female genital groove (each cleft housing a single teat) approximately halfway along the genital groove's length (see Chapter 7). The teat is usually enclosed within the cleft, but in lactating whales may protrude externally, possibly under pressure from the caudal most reach of the mammary gland. In the male, the caudal commissure of the genital groove diverges briefly to each side of the midline, ending in the vicinity of short left and right clefts, each generally enclosing a rudimentary teat.

Two bowhead whales have been described with female-like external genitalia, but with undeveloped testes lacking epididymides (Tarpley et al., 1995). Karyotyping of lung fibroblasts from one of these whales (NSB-DWM 82WW1) showed it to be a genetic male. No female structures were identified internally, and both whales were considered to be male pseudohermaphrodites.

Female reproductive tract

The female reproductive tract of the bowhead is bicornuate, comprised of paired, uncoiled uterine horns (cornua) joined caudally to form a single uterine body and cervix (Fig. 13.2A). The entire tract is suspended from the abdominal wall on each side by a well-developed peritoneal bilayer fold, the broad ligament, as in other mammals. The leading edge of the mesovarium is strengthened by the suspensory ligament, a forward extension of the mesovarial bilayer. As the whale grows, the broad ligament, including the suspensory ligament, is increasingly a repository of visceral fat between its peritoneal membranes that may have multiple functional roles.

Ovary

The ovary is an elongate structure, rather blunt at both poles and receiving the mesovarium along a hilus that spans most of the ovary's length, conveying vascular and neural service within the mesovarial bilayer (Figs. 13.2A and B, 13.3A–C, and 13.5A). In the young, sexually immature whale, the ovary tends to be compressed between medial and lateral surfaces (Tarpley et al., 2016). As the whale grows, and particularly as puberty nears, the ovary becomes more robust and thick-bodied, developing an increasing array of grooves and infoldings over its surface (Fig. 13.5A). Ovaries increase in size with whale length and age (Fig. 13.4A and B), although there is considerable variability in ovary size between whales after sexual maturity. Left and right ovaries of the same whale tend to be similar in size (except for bias introduced by a corpus luteum or large corpus albicans). The largest mature ovary was 64 cm long (NSB-DWM 12S7) and the heaviest ovary was 9.6 kg (NSB-DWM 13B1).

Microscopically, the bowhead ovary is typically mammalian. The ovary's surface is covered by a layer of generally cuboidal cells that overlies an investment of irregular collagenous connective tissue, the tunica albuginea. There is evidence of follicular recruitment and atresia in sexually immature ovaries, suggesting an active ovarian cortex even in young whales. Primordial and primary follicles (the latter consisting of cuboidal cells surrounding the oocyte) are most abundant in the outer third of the cortex (Tarpley and Hillmann, 1999). Secondary follicles have a multilayered cellular investment around the oocyte, and the fluid of tertiary (antral) follicles renders many such follicles visible just beneath the ovarian surface. Atretic follicles tend to be more prevalent deeper in the cortex, characterized by a collapsing zona pellucida (Tarpley and Hillmann, 1999).

Sexually mature ovaries are robust and display at least one corpus luteum (CL), whether a CL of the cycle or of pregnancy (Tarpley et al., 2016). Evidence of maturity is provided by the corpus luteum itself or a corpus albicans (CA) as its structural successor. On this basis, 50 females examined prior to 1992 from the BCB bowhead population (for a

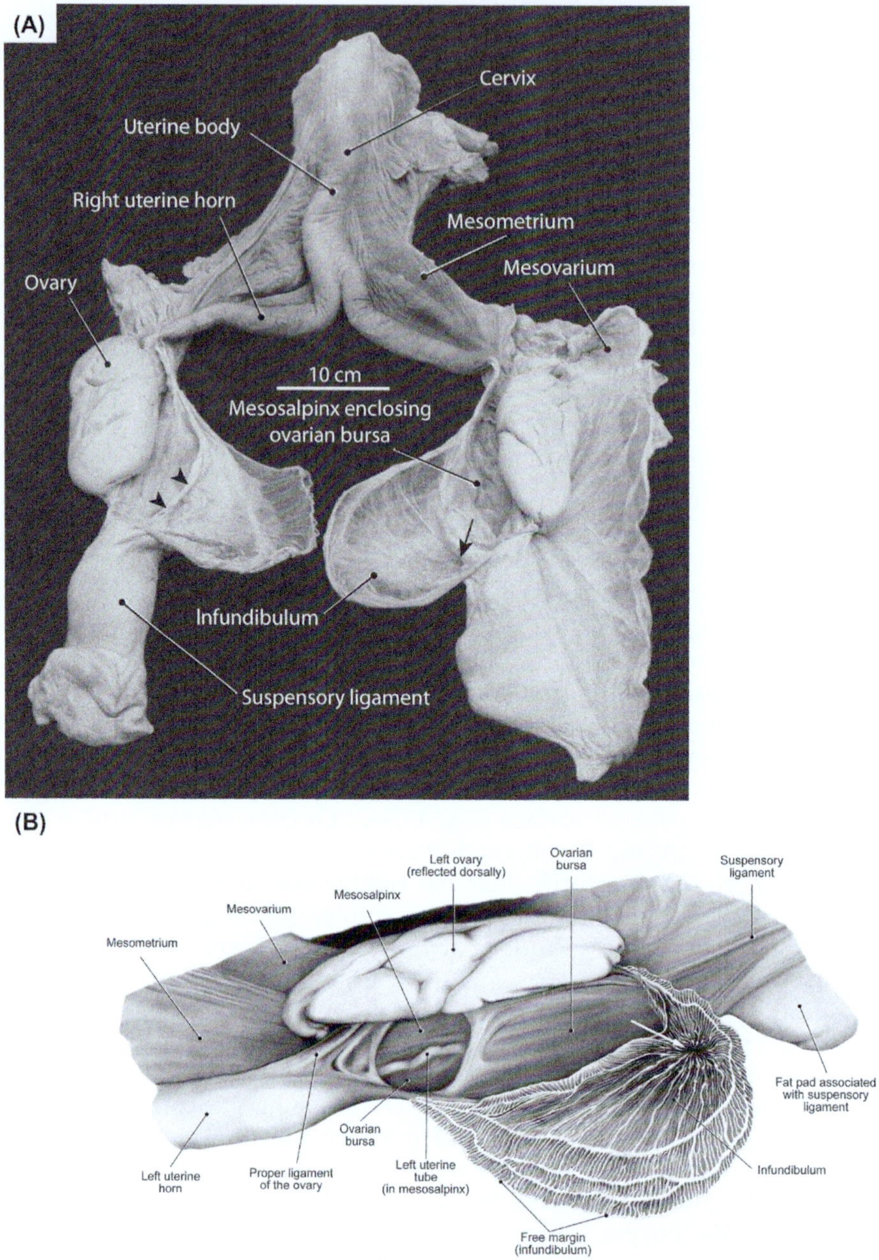

FIGURE 13.2 Dorsal overview of the female reproductive tract (A) and medial view schematic (B) of the left ovary and surrounding structures from a sexually immature spring-caught bowhead (NSB-DWM 96B1; total body length 8.5 m). The bicornuate uterus (A) has elongated horns, and short body, suspended by the broad ligament (mesometrium, mesovarium, and mesosalpinx). An expansive infundibulum extends the reach of the ovarian bursa embracing the ovary. The inner margin of the infundibulum (*black arrowheads* in A) leads into the ampulla of the uterine tube via the abdominal ostium (*black arrow* in A; *white arrow* in B). The uterine tube runs caudally within the mesosalpinx bilayer to reach the uterine horn. The proper ligament of the ovary anchors the caudal aspect of the ovary to the cranial tip of the uterine horn. The suspensory ligament contains considerable fat even in this young whale. *Source: Photo by D.J. Hillmann and R.J. Tarpley. Schematic: D.J. Hillmann.*

description of the population, see Chapter 3), all whales of 14.2 m and over in total body length could be categorized as sexually mature (Tarpley et al., 2016) with the exception of one pubescent 14.2 m whale. Examination of a new, larger dataset of confirmed mature females (based on the presence of a fetus or cycling ovaries) ($N = 185$ through 2018) has identified 12 whales reaching sexual maturity at shorter body lengths (12.6–13.9 m), signaling a possible trend toward sexual maturity occurring at shorter body lengths since 1992.

An irregular surface topography is characteristic of sexually mature ovaries, formed all or in part by protruding corpora (CL and/or CAs), Graafian follicles, and pronounced cortical infoldings (Fig. 13.3A). Rather large, fluid-filled follicles may be present in all ovarian phases, even those with an active CL supporting a pregnancy. With one possible exception (where a single ovary presented two putative CLs) in a careful examination of 70 sexually mature whales, no ovary pair has been found to possess more than a single CL. In contrast, multiple CAs in various stages of involution are commonly identifiable and randomly distributed over the ovarian surface. The former site of ovulation (stigma) is usually apparent (Fig. 13.3A, B, and E). Over time, CAs regress beneath the ovarian surface contour as their size diminishes (Fig. 13.3C–E).

Larger CLs (diameters from 8.7 to 15.0 cm; $N = 8$) have generally supported a fetus, whereas smaller CLs (diameters 6.7–10.6 cm; $N = 4$) may or may not have represented a pregnancy (these four were all spring-caught whales, and an early fetus may have gone undetected; Tarpley et al., 2016). At the macroscopic level, CLs are characterized by a fibrous encapsulated yellowish parenchyma of active luteal cells, often surrounding a gel-filled cavity, a feature that may persist in the younger CAs (Fig. 13.3B and C). Microscopically, the parenchyma is comprised of typical luteal cells. In young CAs following parturition, with or without a central cavity, the luteal tissue takes on a brownish hue as the luteal cells regress. While a central cavity may be found in older CAs, the frequency of this feature diminishes with CA size. Consequently, older CAs are more commonly featured by an irregular central scar, enveloped by the former luteal area (Fig. 13.3D and E).

The former luteal tissue is steadily invaded by a dense network of spiral arteries extending from the rich vasculature of the ovarian medulla, changing the composition of the luteal region to one of vascular dominance surrounding the central scar as the corpus is broken down and absorbed (Fig. 13.3E and F). Smaller CAs are more numerous than larger ones, suggesting accumulation over time due to their permanence or, alternatively, a slowing of the rate of regression prior to absorption. Since CAs below a lower size threshold of a few millimeters have not been detected, even as the spiral arteries work to remove the former luteal cells, the central CA scar may serve as a permanent marker of ovulation (Tarpley et al., 2016). It has not been possible to distinguish CAs associated with pregnancy from those of the cycle, but CA scars can collectively contribute to estimates of productivity and even lifespan of the whale (George et al., 2011). While there is some anecdotal evidence that bowheads are spontaneous ovulators, fertilization efficiency would allow retained CAs to more accurately provide an estimate of pregnancies. On average within the population, ovulations do not appear to preferentially occur in the left or right ovary.

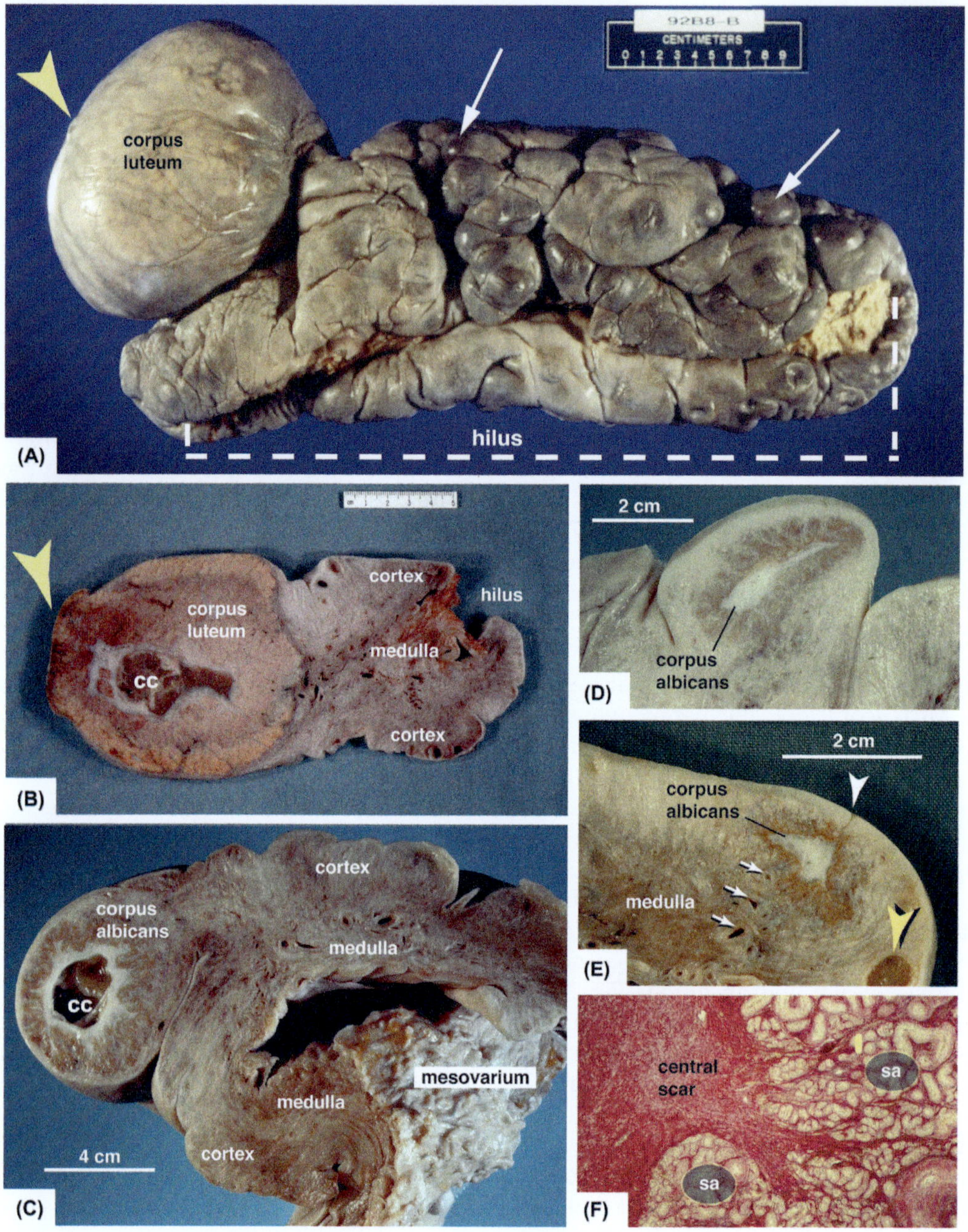

FIGURE 13.3 (A) Ovary of a fall-caught bowhead (NSB-DWM 92B8; total body length 15.7 m), carrying a midterm fetus (total body length 84 cm). The corpus luteum (CL) is approximately 15 cm in diameter, similar to that of NSB-DWM 86WW2, a spring-caught whale (total body length 17.7 m) with a 27.3 cm fetus. Eleven other pregnant whales had smaller CLs. The *yellow arrowhead* indicates the ovulation site. Yellow glandular (luteal) tissue is visible beneath the surface of the CL. Despite the pregnant stage of the estrous cycle, numerous Graafian

Uterine tube

The uterine tube of the bowhead whale has received little attention, but appears to be typical of mammals generally, grossly divisible into an isthmus, ampulla, and infundibulum from distal to proximal. In a brief description of an immature bowhead, the uterine tube took an undulating path forward, initially as the isthmus, from the cranial aspect of the uterine horn, traveling within the bilayered mesosalpinx. About halfway along its length, it continued as the ampulla with a somewhat greater diameter, which in turn transitioned into an expansive, but fragile infundibulum positioned along the length of the ovary (Tarpley and Hillmann, 1999; Fig. 13.2A and B). Kenney et al. (1981) described the mucosa of the uterine tube arranged in thick, longitudinal folds, lined by a simple cuboidal epithelium.

Uterus and cervix

The bowhead whale uterus is bicornuate, lengthy, uncoiled, and suspended within the abdominal cavity by an expansive, fat-filled mesometrium (Figs. 13.2A and 13.5A). The uterine body is short and lined by a mucus-secreting simple columnar epithelium. The uterine body, with a luminal wall gathered into an array of prominent longitudinal folds, bifurcates quickly at its cranial extent into left and right uterine cornua, also replete with longitudinal folds and subfolds in the unexpanded state. The initial uterine bifurcation is not apparent from an external examination of the intact uterus since the uterine horns initially share a median wall within a single external investment until completely diverging cranially. Because of this arrangement, the body of the uterus appears externally to be

follicles are apparent (*white arrows*). The hilus (*dotted line*) extends the full length of the ovary and receives the fatty mesovarium (cut and visible to the right in the hilus). (B) Right ovary in transverse section from a spring-caught whale (NSB-DWM 90B4; total body length 14.9 m) with a 390 cm term fetus. A corpus luteum of 10.6 cm in diameter illustrates a dense field of active luteal tissue with a central cavity (cc) and the stigma (*yellow arrowhead*). The ovarian cortex surrounds the vascular medulla. A portion of the mesovarium enters the hilus of the ovary on the right. (C) Transverse section from a spring-caught lactating whale (NSB-DWM 82WW2; total body length 16.5 m) with a newly formed corpus albicans in the early stages of regression. The former luteal region has lost it yellowish coloration as cells comprising the recent corpus luteum became inactive and degraded. A well-formed central cavity (cc) is bounded by white scar tissue. The ovarian cortex (essentially devoid of antral follicles) and a well-vascularized ovarian medulla surround the hilar region through which a robust, fat-filled mesovarium passes. (D) Transverse section of a regressing corpus albicans from a lactating, spring-caught whale (NSB-DWM 82WW2; total body length 16.5 m). This midsized corpus albicans still protrudes somewhat beyond the ovary's surface. The brownish, former luteal region surrounds a solid central scar. (E) Transverse section of a regressing corpus albicans from a spring-caught, pregnant whale (NSB-DWM 90B4; total body length 14.9 m). The corpus albicans has retracted deeper into the ovarian cortex, no longer projecting above the ovary's surface, even though the stigma is still visible (*white arrowhead*). The former luteal region has diminished in size relative to the central scar as a robust vasculature (*white arrows*) from the ovarian medulla approaches and surrounds the corpus. Spiral arteries from these vessels penetrate and break down the former luteal area over time (see F). A Graafian follicle (*yellow arrowhead*) is visible in the ovarian cortex nearby. (F) Photomicrograph of a midsized corpus albicans from a spring-caught whale (NSB-DWM 82WW2; total body length 16.5 m) illustrating replacement of luteal tissue by an invading network of spiral arteries (sa) surrounding the central scar. *Source: (A) Photo by R.J. Tarpley and D.J. Hillmann. (B–E) Specimen preparation and photo: R.J. Tarpley. (F) Histology preparation by B.A. Merka and photo by R.J. Tarpley.*

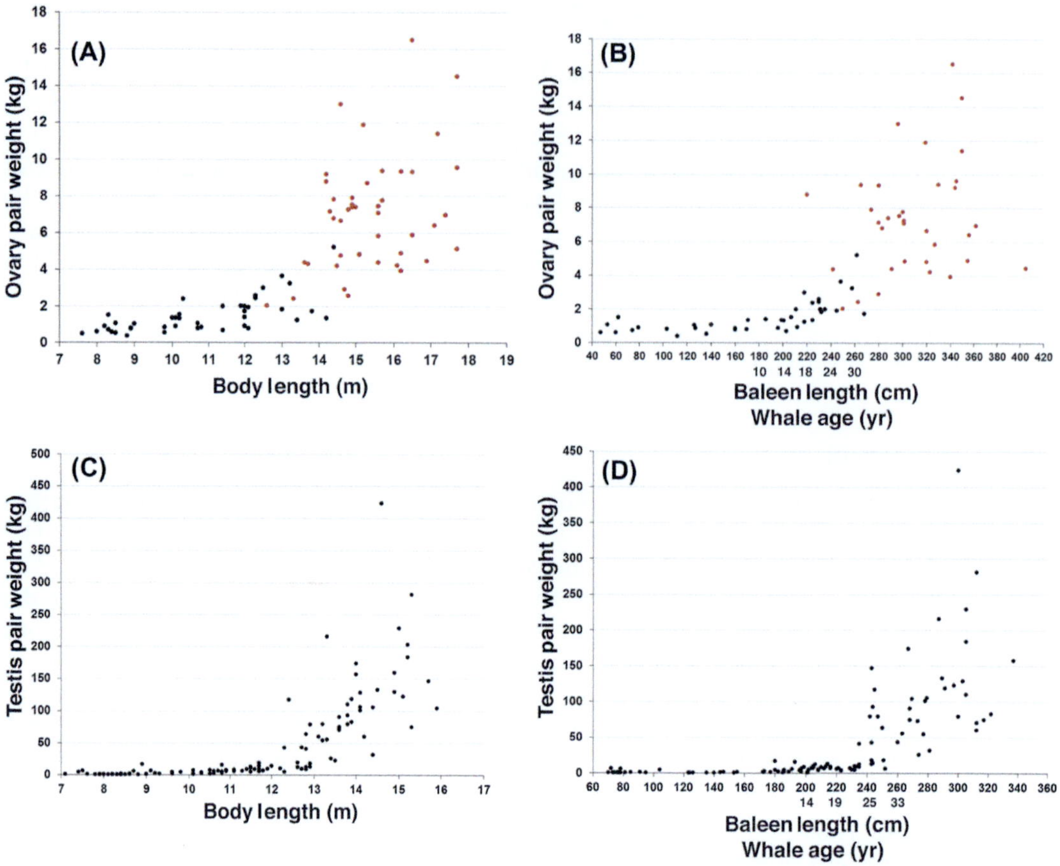

FIGURE 13.4 Scatter plots of bowhead whale reproductive data as a function of body length (m), baleen length (cm) and whale age (yr). Baleen age estimates (based on baleen lengths, Lubetkin et al., 2012) serve as proxies for the whale age values inserted on the X axes (B, D; Chapter 21). (A) Ovary pair weight versus body length. Ovaries increase in weight at a body length of about 12.4–13.0 m. *Red dots* indicate whales confirmed as sexually mature. (B) Ovary pair weight versus baleen length. Ovary weight begins to increase at a baleen length of about 220 cm (an estimated age of 18 years). *Red dots* indicate whales confirmed as sexually mature. (C) Testis pair weight versus body length. Note the inflection in testis weight at a body length of about 12–13 m, indicating onset of sexual maturity. (D) Testis pair weight versus baleen length. Testis weight increases rapidly when baleen length reaches about 240 cm (roughly an age of 25 years).

more extensive than it actually is. Like the uterine body, the mucosa of the uterine horns is arranged in longitudinal folds that are evident even in the fetus (Fig. 13.6A).

Histologically, the uterine horns present the typical mammalian composition with a mucosa consisting of glands embedded in a collagenous matrix and communicating with the lumen lined by a single layer of cuboidal to columnar cells. The degree of endometrial glandular development reflects the stage of whale maturity, as well as the phase of the estrous cycle. In immature whales, the glands have minimal convolution and extend along

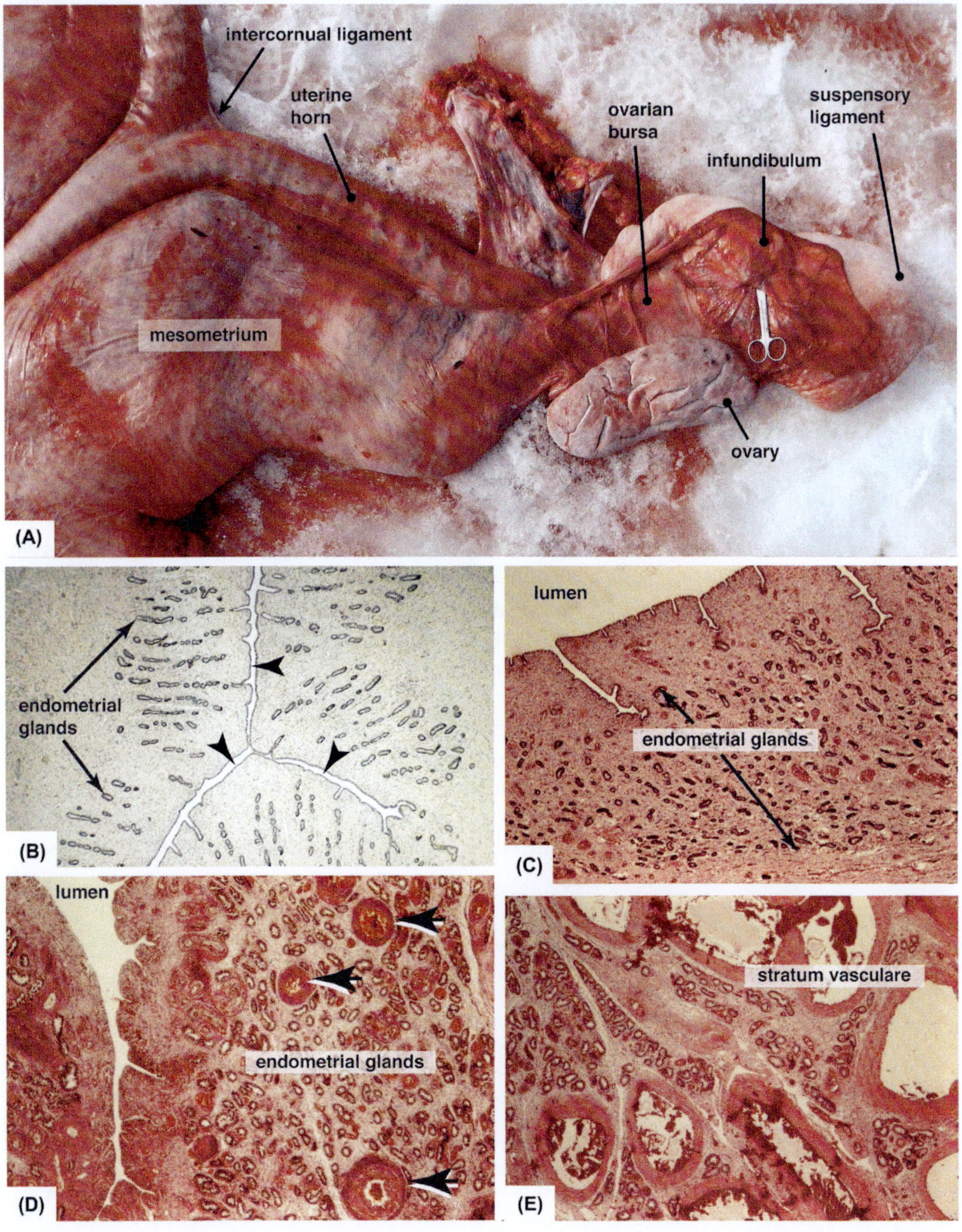

FIGURE 13.5 (A) Reproductive tract of a sexually immature female at the harvest site (13 m spring-caught bowhead, NSB-DWM 16B11). Body length suggests this whale is nearing puberty, and the ovary is more robust than in younger whales, with more extensive cortical infolding and developing Graafian follicles beneath the

linear transects into the mucosa (Fig. 13.5B). In maturing and mature whales, particularly those in a stage of the estrous cycle anticipating pregnancy, the glands are much more developed, more complex, and more obviously branched (Fig. 13.5C–E), reaching deeper into a thickened mucosa wall. Vascular development within the submucosa servicing the glands and mucosa becomes more robust in mature whales (Fig. 13.5E).

The reproductive tract extending from the body of the uterus to the cranial reach of the vaginal cavity is configured into a robust series of caudally directed, transverse, but cone-shaped circular (annular) folds, each of which circumscribes a central orifice (Fig. 13.6A–C). The annular folds become increasingly prominent caudally, as the depth of successive folds increases, along with the blind recesses created between them. The annular folds themselves are permanent and not reducible; they collectively define the passageway of the reproductive tract between the cranial vagina and uterine body. The folds are invested by smooth muscle bundles within their walls and may be capable of strong concentric contractions in the transverse plane. The mucosa overlying the luminal surface of each circular fold features smaller longitudinal folds that may be reducible as luminal expansion demands.

Circular, transverse folds in this region have been reported in many cetaceans, both odontocetes and mysticetes, although the extent of development and number of folds vary among species (Meek, 1918; Ommanney, 1932; Pycraft, 1932; Kleinenberg et al., 1969). Pycraft (1932) referred to the folds collectively as the "pseudocervix," but cetologists have generally counted them as part of the cranial vagina (Slijper, 1962). In an examination of a fetal bowhead whale from Greenland, Meek (1918) suggested the folds were vaginal and had replaced the cervix. Other investigators, however, have considered the folds as an elaborately extended cervix, not unlike the series of folds described in immature domestic bovines (Preuss, 1953), and Fetter and Everitt (1980) chose this approach in their description of the bowhead. Macroscopically, the cone-like appearance of the folds is more robust caudally, and less distinct cranially, allowing the possibility that both the cervical and cranial vaginal regions are represented, a possibility further supported by observations of mucus collection between the more cranial folds, more typical of the cervical region in

ovary's surface (visible as dark spots toward the cranial pole). Fat accumulation in the mesometrium and suspensory ligament is more advanced in this whale than in younger ones (see Fig. 13.2). The ovary has been folded from the ovarian bursa to expose the bursal wall and the margins of the fragile infundibulum. Scissors have been inserted into the abdominal ostium leading into the ampulla of the uterine tube. The external uterine bifurcation is marked by the intercornual ligament. (B) Transverse section through the uterine horn of a sexually immature spring-caught bowhead (NSB-DWM 83B1; total body length 8.5 m). Three longitudinal folds surround a narrow uterine lumen (*black arrowheads*). Endometrial glands are minimally developed, undulating along an otherwise straight, essentially unbranched trajectory as they extend into the uterine wall. Verhoeff van Gieson stain. (C) Uterine endometrial wall of a sexually immature spring-caught bowhead (NSB-DWM 83B2; total body length 14.2 m). This whale had not ovulated, but its body length suggests that it approached puberty, and endometrial glandular development is more advanced relative to that of a smaller whale (B). Hematoxylin & Eosin stain. (D and E) Uterine mucosal wall from a lactating, sexually mature spring-caught bowhead (NSB-DWM 82WW2; total body length 16.5 m). Extensive endometrial glandular development and vascular support (*arrows*) are evident (D). Deeper into the mucosal wall, extensive vascular development (stratum vasculare) is apparent in support of the endometrial glands. Hematoxylin & Eosin stain. *Source: (A) Photo by J.C. George. (B–E) Histology preparation by B.A. Merka and photo by R.J. Tarpley.*

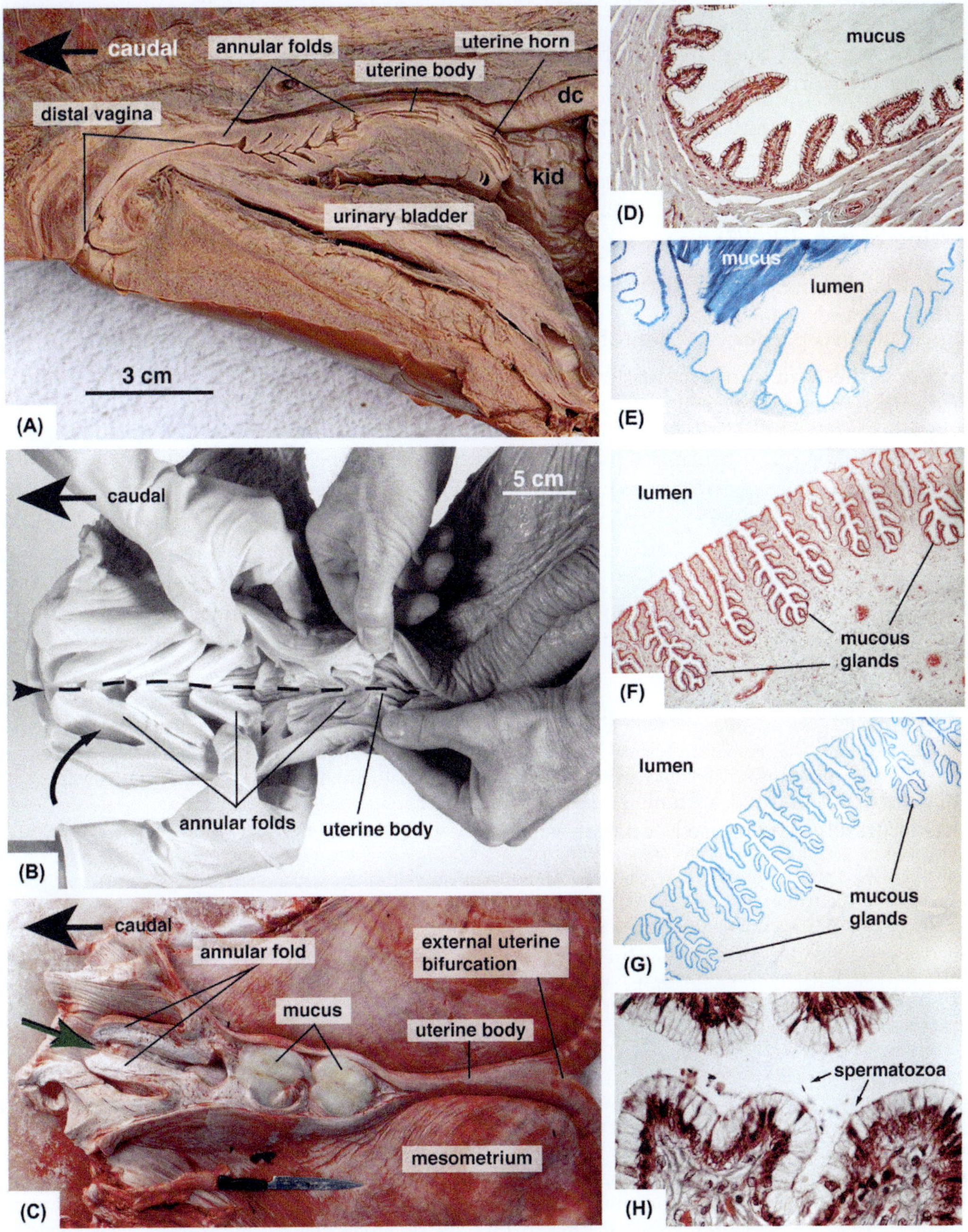

FIGURE 13.6 (A) Female reproductive tract in midsagittal section from a midgestational, 84 cm bowhead fetus (NSB-DWM 92B7F). Six annular folds span the proximal vagina, likely extending into the cervical region. There are no such circular folds in the distal vagina. The right uterine horn is visible cranial to the folds. The reproductive tract extends into the abdominal cavity between the descending colon (dc) dorsally and the urinary

mammals (Fig. 13.6C). Regardless, each fold surrounds a central orifice, which together create a pathway from the vagina to the uterus, while forming a succession of blind "pockets" between folds (Fig. 13.6A–C).

Fetter and Everitt (1980) reported 4–7 annular folds in several bowheads they examined. Ommanney (1932) gathered accounts for several odontocetes and mysticetes where the number of annular folds ranged from 3 to 12, depending on species. An 84 cm fall-collected bowhead fetus (NSB-DWM 92B7F) presented six annular folds at that stage of development (Fig. 13.6A). Tarpley and Hillmann (1999) described six folds in an immature bowhead tract (NSB-DWM 96B1; Fig. 13.6B) although due to the point at which it was transected caudally, it cannot be certain that the specimen included all folds.

The microscopic composition of the mucosa lining the folds needs further clarification since this has important functional implications. Kenney et al. (1981) reported that all annular folds were lined by a stratified squamous epithelium, while Tarpley and Hillmann (1999) found a stratified squamous investment lining the most caudal fold, while others were covered by simple squamous to cuboidal cells, with a transition to low columnar cells in the true cervical region (however, tissue preservation at the microscopic level was not optimum in this specimen). The primary cervical region and uterine body exhibited multiple longitudinal folds projecting into the lumen and lined by mucous cells (Fig. 13.6B). In contrast, the mucosa of at least some of the circular folds consists of columnar mucous cells that continue into numerous branched glandular crypts (Fig. 13.6F–G). This observation is supported at the macroscopic level in a 13 m whale where thick mucus

bladder ventrally. The proximity of the uterine bifurcation to the caudal pole of the left kidney (kid) is evident. (B) Portion of the distal immature female reproductive tract (spring-caught NSB-DWM 96B1; total body length 8.5 m). A series of transverse, annular (circular) folds extends caudally from the uterine body into the vagina. The more distally positioned folds form increasingly prominent elongated cones directed caudally, while those folds located more cranially are less distinct, with more pronounced longitudinal mucosal folds. A centric opening circumscribed by each fold contributes to a collective channel (*dashed line*) from the vagina to the uterus. Blind pockets are created between folds where they meet the primary wall of the reproductive tract (*curved arrow*). (C) Photograph at the harvest site of a sexually immature female reproductive tract from a 13 m spring-caught bowhead (NSB-DWM 16B11). The distal portion of the tract has been opened dorsally to reveal the series of transverse, annular folds that extend caudally from the uterine body. The cone-shaped configuration of the more caudal folds is apparent, each directed rearward and circumscribing a single opening (e.g., *green arrow*) when intact. Mucus is contained within the bounds of two of the cranial folds, but absent in the more caudal ones, perhaps reflecting a functional specialization of the cells lining the mucosal walls. (D–F) Mucosa in the cervical/uterine body region in transverse section (D–E) and from the pitted mucosa of an annular fold (F–G) of a sexually immature spring-caught whale (NSB-DWM 83B2; total body length 14.2 m). The longitudinal folds (D–E) extending into the lumen of the cervical/uterine body region are in contrast to the branched glandular arrangement in the mucosa of an annular fold (F–G). In both regions, however, a single layer of mucous cells lines the folds and glands as evidenced by structural and histochemical stains (Masson Trichrome in D and F and Alcian Blue/Periodic Acid Schiff for glycosaminoglycans in E and G). Mucus is evident within the lumen of the cervical/uterine body region (D–E). In proximity to the columnar mucous cells of the annular fold region in this whale, spermatozoa were lodged within the mucus (Masson Trichrome stain in H). Although this whale had not yet ovulated, its total body length and combined ovary weight suggest it may have been pubescent. *Source: (A) Specimen preparation and photo by D.J. Hillmann. (B) Specimen preparation and photo by R.J. Tarpley and C. A. Curry. (C) Dissection and photo by J.C. George. (D–H) Histology preparation by B.A. Merka, photo by R.J. Tarpley.*

filled the lumen and recesses associated with the more cranial folds (Fig. 13.6C). In a 14.2 m (likely pubescent) whale, spermatozoa were found lodged in the mucus along the surface of the crypts (Fig. 13.6H).

The function of annular folds is unclear. Meek (1918) suggested they may assist rapid ejaculation during intromission by stimulating the terminal cone of the penis during the brief coitus, or their contraction may assist forward movement of spermatozoa. They may also provide sperm capture or storage opportunities to protect from sea water incursion or limited intromission, or allow some degree of sperm competition in the case of multiple matings. Other suggestions have included empowering forceful dilation of this region of the birth canal to facilitate passage of the fetus, or alternatively through constriction, protecting the uterine environment from sea water incursion during mating or parturition (Harrison, 1972). Further histological study of smooth muscle orientation in the walls of the folds may provide information on the capacity of the folds to constrict and/or dilate as needed.

Vagina, vulva, and clitoris

The mucosa of the vaginal cavity proper, apart from any association with the annular folds, is lined by a keratinized stratified squamous epithelium (Kenney et al., 1981). No glands have yet been identified in the vaginal wall proper. The smooth-walled vaginal cavity, positioned beneath the rectum and anal canal, features 10 or so well-developed, but reducible, longitudinal folds (rugae). The vulva is defined by the genital groove, enveloping an elongated clitoris within its cranial angle as the groove folds inward to meet a vestibule receiving the termination of the urethra prior to the vaginal entrance.

Mammary gland

Elongated, dorsoventrally compressed, left and right mammary glands are positioned parasagittally along the midline, each terminating caudally in a single teat, recessed in the nonlactating whale within the mammary cleft on either side of the genital groove. In a likely pubescent whale (NSB-DWM 83B2; total body length 14.2 m), gland and duct development were minimal. The mammary gland of a 15.7 m pregnant female was oblong in shape, 170 cm long, 60 cm wide and 12 cm thick. Kenney et al. (1981) described a teat cistern with complex mucosal wall folding and the presence of smooth muscle bundles between the teat cistern and skin. The teat was covered by a thick stratified squamous epithelium, well-anchored to the underlying collagenous bed.

Male reproductive tract

Testis and epididymis

The testis is an elongated, cylindrical, and smooth-surfaced organ tapering slightly toward equally blunt cranial and caudal poles (Fig. 13.7A). Testis size increases significantly around the time of sexual maturity, but is highly variable among individuals after that (Fig. 13.4C and D). There is some indication that testis size increases in the fall, perhaps related to increased spermatogenesis. The heaviest single testis weighed 211 kg and was 1.5 m in length (NSB-DWM 11B15). Left and right testes, covered by a single-cell layer of peritoneum, rest along the abdominal wall to either side of the midline, with long axes approximating that of the whale. Deep to the peritoneal layer is a robust collagenous tunica albuginea, encapsulating the testis and sending supporting trabeculae into a dense parenchyma of seminiferous tubules (Fig. 13.7B–D). It is noteworthy that, even in sexually mature whales where tubular walls are active, clear evidence of interstitial cells (of Leydig) has not yet been identified (Kenney and Everitt, 1980; Kenney et al., 1981; O'Hara et al., 2002). In prepubertal males, Sertoli cells and spermatogonia have been reported as comprising tubular walls, though tubular lumens are narrow or closed; in sexually mature males, tubular lumens widen and the presence of spermatocytes, spermatids, and spermatozoa have been reported (O'Hara et al., 2002).

Efferent tubules continue the testicular collecting tubules and exit the cranial pole of the testis, penetrating the tunica albuginea, to connect with the ductus epididymidis within the head of the epididymis (Kenney et al., 1981). The body of the epididymis continues linearly and caudally in close apposition to, and along the full length of, the testis until reaching the epididymal tail at the testis' caudal pole, where it continues as the ductus deferens. The epididymal duct itself takes a tortuous course throughout the length of the epididymis, but Kenney et al. (1981) noted that rather than forming a simple duct as in other mammals, it features a complex array of branched glands or crypts that radiate out from the lumen of the primary duct. This unusual feature needs further investigation since mating behaviors in the fall coincident with possible enlargement of testes and spermatogenesis leaves open a consideration of prolonged sperm storage in the epididymis in anticipation of peak ovulations in the late winter and early spring (see "Conception" section). The size of the epididymis in sexually mature whales can rival that of the testis when filled with large quantities of spermatozoa (Haldiman and Tarpley, 1993). The configuration of a single, tortuous duct with radiating crypts continues initially from the tail of the epididymis into the ductus deferens at the caudal pole of the testis. The morphology of spermatozoa can vary with species and warrants further study in the bowhead since abnormalities have been associated in other species with genetic irregularities or environmental insults (Marzec-WróbleWska et al., 2012; Fig. 13.7E).

Ductus deferens, ampulla, and accessory glands

The ductus deferens and its ampulla are similar in arrangement to those of other mammals (Kenney et al., 1981; Kenney and Everitt, 1980). The prostate is the most consistently recognized accessory gland in cetaceans and has been described by Kenney and Everitt (1980) in the bowhead as comprised of serous acini served by ductules leading to ducts lined by a columnar cell bilayer that eventually empty into a transitionally cell-lined urethra. Kenney and Everitt (1980) described a vesicular gland, but this needs further confirmation.

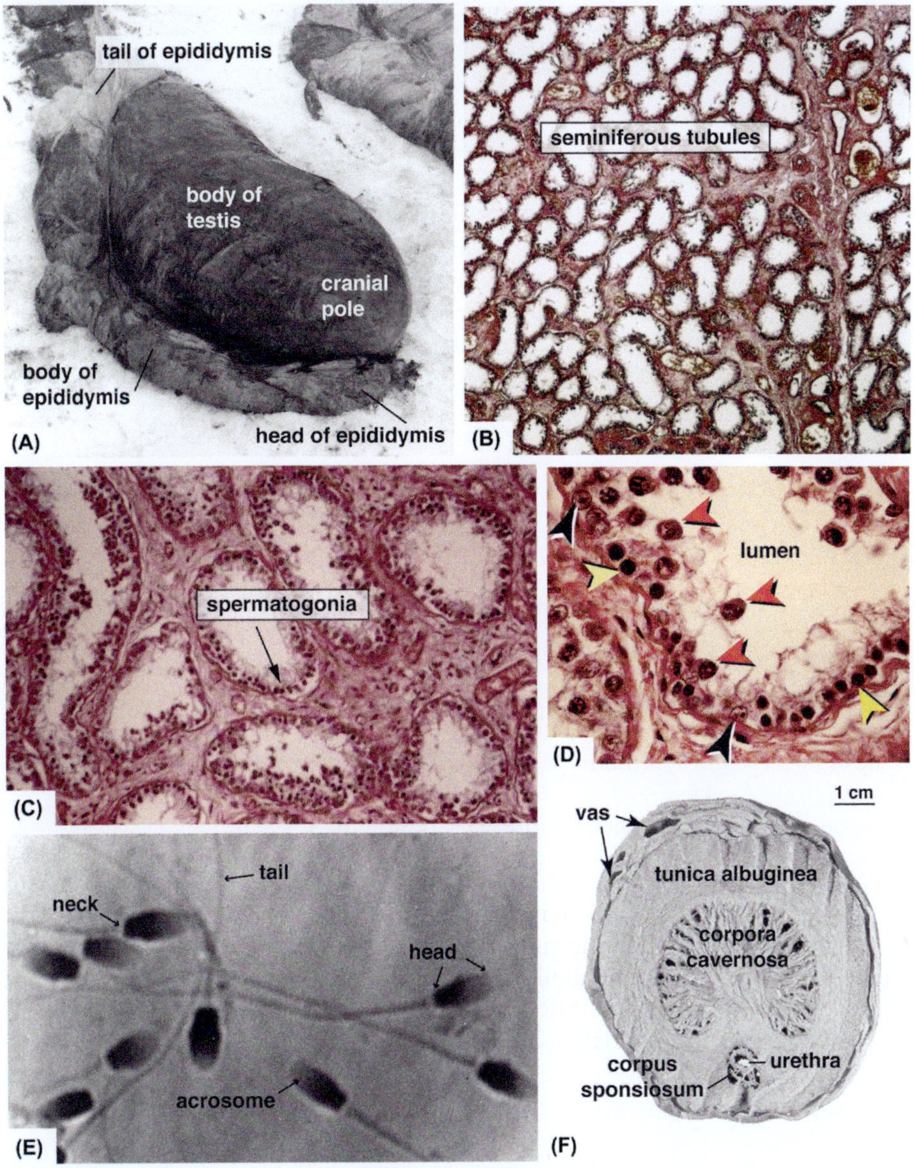

FIGURE 13.7 (A) Testis and epididymis from a sexually mature, spring-caught bowhead (NSB-DWM 82WW2; total body length 16.6 m). The cranial pole of the testis leads into the head of the epididymis. The epididymis body and tail transition to the ductus deferens at the caudal pole of the testis. (B–D) Photomicrographs of well-developed testicular seminiferous tubules from a fall-caught bowhead (NSB-DWM 81KK2; total body length 14 m). Based on body length in relation to testis weight, this whale is in the inflection range for maturity onset (Fig. 13.4C). A relatively dense tubular configuration within the testicular parenchyma and open tubular lumens are evident at lower magnification (B), and at higher magnification (C) the dense nuclei of spermatogonia can be seen lining the tubular walls. Still higher magnification reveals cellular diversity within the tubule (D). While Sertoli cells (*black arrowheads*) and spermatogonia (*yellow arrowheads*) can be observed in prepubertal whales, the

Penis

The fibroelastic penis of the bowhead whale aligns with that described in other cetaceans; it includes a sigmoid flexure and typical supporting musculature, including the ischiocavernosus, bulbospongiosus, and retractor penis (Kenney, 1980; Kenney and Everitt, 1980). There is no baculum (os penis). The terminal cone is smooth and tapered, slightly compressed from side to side at the tip. When retracted, the terminal cone occupies the preputial space provided just inside the genital groove. In transverse section, structures typical of terrestrial artiodactyls are apparent (Fig. 13.7F). The fibrous components are very dense, including a thick tunica albuginea that sends robust collagenous strands in multiple directions throughout the corpora cavernosa. More ventrally, the urethra is surrounded by a well-defined corpus spongiosum. Vascular support to the penis is evident over the dorsum of the terminal cone.

Functional parameters of the bowhead whale reproductive cycle

Female body length and age at sexual maturity

Using a dataset of females with known maturity status, logistic regression was used to assign the probability that a whale of a given length was mature (George et al., 2011). Length at sexual maturity (LSM) was defined as the length at which a female had a 50% probability of being mature. Estimated LSM was 13.4 m (SE: 0.22; 95% CI: 12.9–13.8 m). This estimate is lower but somewhat consistent with our data from earlier examinations of harvested bowhead whales (Tarpley and Hillmann, 1999) and aerial photogrammetric surveys (Koski et al., 1993).

As in the case of most mammals, there is variation in the LSM; however, we suggest that an estimated length of 13.4 m at 50% maturity is the best approximation for LSM in the BCB population, and can be used for comparison with other bowhead populations and other cetaceans. The inflection point for increasing combined (paired) ovary weights correlates well with total body length and generally supports the estimated 13.5 m length for sexual maturity (Fig. 13.4A). Using baleen data, Rosa et al. (2013) estimated an age at sexual maturity for females at 25.8 years (Fig. 13.4B).

Gambell (1968) noted strong correlation between body length and numbers of corpora in sei whales. For bowheads, there is less correlation between total body length and the numbers of corpora, driven in part by the broad range of ages for a given body length (George

presence of primary spermatocytes (*red arrowheads*) along with an open tubular lumen is suggestive of imminent spermatogenesis in this tubule. Hematoxylin & Eosin stain. (E) Photomicrograph of spermatozoa collected from the epididymis of a sexually mature, spring-caught bowhead (NSB-DWM 82WW2; total body length 16.6 m). (F) Cross-sectional photograph of the terminal cone of a bowhead whale penis, demonstrating the robust collagenous composition of the tunica albuginea and corpora cavernosa. Strong collagenous investment restricts expansion of the blood-filled cavities of the corpora cavernosa and corpus spongiosum. The vascular spaces of the corpus spongiosum completely surround the urethra. The terminal cone is contained in a skin-like covering conveying vascular service (vas) over the dorsum of the penis. *Source: (A, E, and F) Photo by R.J. Tarpley. (B–D) Histology preparation by B.A. Merka and photo by R.J. Tarpley.*

et al., 2011; Rosa et al., 2013; Wetzel et al., 2017). In the bowhead, there is a hiatus in growth in body length for young whales, ages 1–4 years (George et al., 2016), and female whales between 14 and 16 meters can be anywhere from 20 to 120 years old.

Ovulation

While sample sizes are not sufficient to precisely quantify ovulation parameters, mating behaviors in bowheads have been observed (though not exclusively) from January through May, coincident with fertilized ovulations and corresponding to increasing day lengths. This is consistent with fin whale observations of Laws (1961) concluding that increasing day length was an exteroceptive factor contributing to ovulation. Fetal length data suggest the peak time for successful ovulations and conception in the bowhead is in late winter and early spring (see "Conception" section). However, since mating behaviors have been observed in other times of the year, such as the fall, and since there is some evidence that testes enlarge in the fall with well-developed seminiferous tubules, the possibility of male or female sperm storage in support of later ovulations might also be considered.

Spontaneous ovulation is anecdotally supported in the bowhead by the occurrence of similar-sized CAs in a single ovary. As an example, one ovary contained three medium-sized CAs with similar diameters (2.2, 2.0, and 1.8 cm). The next largest CA was considerably smaller (0.7 cm). If rapid regression of corpora in the early stages of involution is assumed (as appears likely in the bowhead from the size of CAs in lactating whales), it is possible that all three were formed closely in time (as would be the case where several ovulations occurred in advance of a successful fertilization). However, other (albeit less likely) explanations are also possible for this similarity in size of large CAs (Tarpley and Hillmann, 1999).

Furthermore, only the similar-sized larger CAs can be called upon to support the idea of spontaneous ovulation. This inference is based on the observation that disparities in size will diminish as persistent CAs converge on a minimal size (~5 mm), resulting in many small diameter CAs that are similar in size. Still, the occurrence of spontaneous ovulation in the bowhead would be consistent with findings in other mysticetes. Several authors have argued that spontaneous ovulation occurs in fin, blue, humpback, and sei whales (Mackintosh and Wheeler, 1929; Laurie, 1937; Matthews, 1938; Chittleborough, 1954; Laws, 1961; Gambell, 1968).

Simultaneous multiple ovulations must also be considered in the bowhead, though this is considered rare in other mysticete species. Laws (1961) found simultaneous multiple ovulations to be rare in the fin whale (2.2%) as did Gambell (1968) in the sei whale (2.6%) and Chittleborough (1954) in the humpback whale (0.6%). No single ovary (or pair of ovaries) in the bowhead has been found to contain more than a single corpus luteum (CL) (with one possible exception where two putative CLs were present in a single ovary), suggesting that simultaneous multiple ovulations in this species are also unlikely.

The more important question from a population management perspective may concern ovulatory efficiency relative to conception since the persistence of CAs serve not only as a cumulative record of ovulations but also as markers of successful pregnancies if most ovulations are successfully fertilized. If the bowhead is long-lived as appears to be the case (George et al., 1999), a very large number of CAs would accumulate over a lifetime. Yet, considering the approximate 3-year calving interval in the bowhead with its hiatus in conceptions (see "Pregnancy rate and calving interval" section), the relatively small numbers of CAs in the context of lengthy lifespans give anecdotal support to ovulation efficiency relative to conceptions. If ovulatory efficiency is the norm, CA scars may tend toward serving as records of the number of pregnancies in a given whale.

Conception

A most likely conception time of mid-March through mid-April is best supported by available bowhead fetal ($N = 20$) and neonatal ($N = 3$) lengths using Bayesian statistical methods for small sample sizes (Reese et al., 2001). Typically, a mammalian fetal growth curve is characterized by slow initial growth followed by an accelerated increase later in gestation. This has been the finding for fin whales (Laws, 1959) as well as sei whales (Gambell, 1968). North Slope Borough data from the BCB stock for fetal body lengths ($N = 17$) combined with the most reliable figures of Durham (1980) and Nerini et al. (1984) ($N = 4$) are not conclusive, but are consistent with slow growth early in gestation (see Fig. 10.2).

Anatomical evidence from CLs also supports early March as the beginning of embryonic development. The 15.0 cm CLs from a small fetus (total length of 27.3 cm) of a May-caught whale and an early midterm fetus (total length of 84.0 cm from an early September-caught whale) were the largest CLs in our sample through 2018. Larger fetuses were most commonly associated with smaller CLs (8.3–11.3 cm; $N = 6$). In the humpback whale, Chittleborough (1954) suggested that the corpus luteum of pregnancy increased during the first 2 months of gestation. Similarly, Laws (1961) demonstrated that maximal corpus size is attained approximately 2 months into the gestation period in the fin whale, as did Gambell (1968) in the sei whale. Using these analogies, we can suggest that the corpus luteum associated with the 27.3 cm fetus had attained its maximum size, with the fetus at least 2 months into gestation, aligning with a conception as early as March.

Aerial observations of proposed bowhead mating behaviors, featuring turbulent group interactions near the water's surface (Chapter 23), have been reported throughout a range of months. Presumed sexual activity was documented during March on four occasions in the Bering Sea (Braham et al., 1980) and in one instance in Hudson Strait (Koski et al., 1990). Earlier and later pairings have been suggested by apparent mating activity in January and February (Eschricht and Reinhardt, 1866) and in May during the spring migration past Point Barrow (Everitt and Krogman, 1979; Richardson et al., 1990). Observations of mating-like behavior has also been reported during summer and fall months through extensive aerial surveying (Koski et al., 1993; Chapter 23). A period of conception extending over several months has been shown in other mysticetes (Laws, 1961; Gambell, 1968), and an interval for peak conception running from February through May would not be unreasonable in the bowhead whale. The proposed approximate 14-month gestation period (Reese et al., 2001) and possible peak calving in late spring and early summer (see "Gestation and parturition" section) would align favorably with a late winter to early spring conception period.

In the last decade, Iñupiat hunters and data from bowhead whale surveys have noted that bowheads are passing earlier by Utqiaġvik than when surveys started in 1978 (George et al., 2012). Mother/calf pairs appear to be arriving earlier than in past years as well. For instance, during the 1993 and 2001 seasons, both of which had excellent visibility in late season when mother–calf pairs are more common, the peak number of mother–calf pairs occurred on 27 May (George et al., 1995, 2004). However, in 2011 and 2019, peak mother–calf counts occurred on 11 and 14 May, respectively. Whether or not the conception date is occurring earlier is unknown. That is, while whales are clearly migrating earlier in recent times than in the 1970s, based on the lengths of fetuses by date, it is not clear if there is a trend suggesting earlier conception (George et al., 2012). However, if parturition continued to occur in late spring and early summer, earlier migration arrivals should place the animals farther east of Utqiaġvik with their calves so that hunters would be seeing fewer calves, not more with the early arrivals. Greater numbers of cow/calf sighting supports the possibility that a time shift to earlier conception dates might be occurring.

Gestation and parturition

Using Bayesian statistical methods to model fetal growth from a small bowhead whale dataset confirmed through 1996 (based on body length measurements from 20 fetuses and three neonates), Reese et al. (2001) proposed a gestation length of ~14 months, exceeding the 11-month interval defined for southern right whales (Best, 1981). Mean fetal growth rate, assuming a total body length of 4.3 m at parturition (Koski et al., 1993) and a gestation period of approximately14 months, equals 0.98 cm/day (considerably slower than the 1.83 cm/day estimated by Best (1981) for the southern right whale). A 14-month gestation for the bowhead does not conflict with field observations where spring fetuses are consistently small or large, and fall fetuses are midrange in size. It would be expected that gestations of more than one annual cycle would capture midrange fetuses in the spring fetal population as well, along with multiple size cohorts in the fall, neither of which has been reported.

Out of 47 confirmed pregnant whales where fetal lengths were reliably measured (1969–2018), spring and early summer pregnancies (April–June) were either in early gestation ($N = 15$; fetal lengths 5–60 cm) or nearing parturition ($N = 16$; fetal lengths 366–455 cm), while pregnancies from the fall and early winter (August–December) were in the midgestational range ($N = 16$; fetal lengths 84–216 cm). Using a slightly different dataset, fetal growth through the gestation period is represented in Fig. 8.2. Based on an approximate 14-month gestation period (Reese et al., 2001), this fetal length distribution is compatible with rather tightly timed late spring to early summer births following conception the previous spring in late winter or early spring (March–April).

From a limited dataset, Nerini et al. (1984) estimated term fetal length in the bowhead to range from 400 to 450 cm; Koski et al. (1993) suggested a mean of 430 cm. Applying this to the current dataset of large fetuses ($N = 16$), 12 measured >400 cm and were collected from April 29 to June 15, also compatible with late spring, early summer births.

Davis et al. (1983) and Nerini et al. (1984), however, suggested a less precise bowhead calving interval, extending from March through August. Birth distribution in the southern right whale encompasses a 5- to 6-month period (Best, 1981). Gambell (1968) reported a 4-month range in the sei whale. Most births (77%) in the southern fin whale also occur within a 4-month span (Laws, 1961). Matthews (1937) found that most births in the humpback whale occurred during a 3-month period.

The current bowhead dataset contains two fetuses collected in August, but both were midterm in length (104 and 120 cm). While our limited dataset does not preclude an extended gestational range, it seems to suggest a peak parturition period in late spring and early summer. A narrower calving interval could possibly be more conducive to calf survival if earlier birth and nursing better readies the calf for direct copepod intake on feeding grounds in the fall as a supplement to lactational support (a small sample of stomach contents from young fall-caught whales suggests such feeding may occur).

An expanded dataset of pregnant whales ($N = 62$) from the same time period (1969–2018), confirms that all three fetal size cohorts were well-populated. Small spring and early summer fetuses ($N = 15$), midsize fall and early winter fetuses ($N = 24$), and large spring and early summer fetuses ($N = 23$) were recovered. The small number of young fetuses could reflect the difficulties of detection under field conditions, but the larger midterm and near-term fetuses (not easily missed) might anecdotally suggest that fetal loss during gestation in the bowhead is minimal.

Pregnancy rate and calving interval

The first estimates of bowhead pregnancy rates in the 1980s were compromised by limited sample sizes. Nerini et al. (1984) suggested that pregnancy rate (the number of pregnant females confirmed in a population of mature females accurately examined during a 1-year timeframe) ranged from 0.15 to 0.31 based on an examination of 15 mature females. From a detailed examination of 30 mature females, Tarpley and Hillmann (1999) confirmed 13 were pregnant, yielding a crude pregnancy rate of 0.43. However, they also estimated a pregnancy rate of 0.22 based on the number of pregnant females in the harvest as a proportion of *presumed* sexually mature females over 13.5 m.

Koski et al. (1993) summarized the available evidence on pregnancy rates and calf production from several sources: postmortem examinations, ice-based census calf counts, aerial surveys, and aerial photo-identification data. He concluded that a calving interval of 3–4 years was most plausible and estimated the gross annual recruitment rate at 0.053 calves/noncalf/yr, using aerial photogrammetric data. His calf production estimate is consistent with a calving interval of ~4 years when considered with the estimated proportion of mature females (~20%) in the stock from aerial surveys (Angliss et al., 1995).

Subsequent analyses of pregnancy rates had the advantage of much larger samples of mature females based on an estimated LSM. In George et al. (2011), pregnancy rate (pregnancies/year) was estimated to be the number of reported pregnancies as a function of "potentially" mature females determined by their length and the estimated a LSM as described in George et al. (2011, 2018). This approach is more robust because pregnant females are clearly mature; however, females that are not pregnant are not always examined for maturity status which could positively bias the estimated pregnancy rate. George et al. (2011) had sufficient data to estimate pregnancy rate from fall-harvested whales where pregnancies are far easier to detect with an approximate fetus length of >1 m ($N = 40$); they estimated pregnancy rate at 0.326 per year, suggesting an interbirth interval of 3.1 years (1/0.326).

The most recent estimate of pregnancy rate used a similar approach to George et al. (2011) but used all available mature females from the spring and fall harvests (1976–2016; $N = 208$; George et al., 2018). They estimated a pregnancy rate of 0.317 (95% CI: 0.251–0.385) again suggesting an interbirth interval of just over 3 years.

Rolland et al. (2018) applied progesterone data to estimate calving intervals using two approaches. First, the total timeline (determined by stable isotopes) represented in the baleen of each adult female was divided by the number of progesterone peaks on their baleen plate, yielding calving intervals that ranged from 2.67 to 3.80 years (mean = 3.04 years; $N = 17$ intervals). Second, they estimated the average timeframe between calving events for females with multiple calving cycles. Specifically, they used the time interval from a baseline progesterone data point following a calving event, to the next baseline point following gestation. Using this approach, the estimated duration of calving intervals ($N = 11$) ranged from 2.10 to 5.31 years with a mean of 3.11 years.

The most current analysis (1976–2016) of mature harvested females (BCB stock) supports previous evidence of a 3- to 4-year calving interval with 3-year intervals most common. This is inferred from the estimated pregnancy rate of 0.317 (95% CI: 0.251–0.385) (George et al., 2018). Early literature suggested that 4-year intervals were most common (Rugh et al., 1992), which could have been the case at that time. George et al. (2018) show evidence that pregnancy rates have possibly been increasing over the past 40 years. They also concluded there was no evidence in the reproductive data of density-regulated reproduction or that the BCB stock was approaching carrying capacity. For comparison, fin and sei whales (Laws, 1961; Gambell, 1968), 2-year cycles include pregnancy, lactation, and anestrous. Similarly, a 2-year cycle was reported by Chittleborough (1954) for humpback whales in the southern hemisphere. However, studies of humpback whales in Alaska and Hawaii using benign methods (Baker et al., 1987) suggested a calving interval of 1–5 years. Similar techniques applied to the right whale in the western north Atlantic by Kraus et al. (1986) gave a calving interval range of 2–5 years. Payne (1986) suggested a range of 2–7 years (3 years most common) using benign methods in the southern right whale.

In summary, the bowhead pregnancy rate for BCB mature females is currently about 0.32 suggesting interbirth interval of just over 3 years. The proportion of these pregnancies that go to term is unknown, although fetal cohorts for spring (early and near-term fetuses) and fall (midsized body lengths) are well-represented and not indicative, at least anecdotally, of obvious fetal loss during gestation.

Koski et al. (1993) reviewed information on calf production from a number of data sources that show fairly strong variation from year to year for BCB bowhead whales. Clarke et al. (2018) provided a summary of calf production data from aerial surveys for the period 1982 to 2017. They present total calf counts and the ratio of total calves/total whales seen, which also showed strong variability between years ranging from 0% to over 11%. Variable calf production is consistent with North Atlantic right whales, which ranged from 1 to 31 calves per year (1980–2005) from a population of about 300 animals (Kraus and Rolland, 2007), suggesting a range in the calf/adult ratio by year of about 0.3%–10.0%.

Koski et al. (1993) noted strong variability in annual calf production and evidence of synchrony in calving, which is evident in the North Atlantic right whale as well (Kraus and Rolland, 2007). In a later report, Koski et al. (2008) estimated average annual calf production for BCB bowheads and noted that due to the great variation in calf production several years of surveys were needed to cover the range of annual variation in calving rates. They also noted that migration was protracted in years with high calf production as the females with calves tend to migrate late. Koski et al. (1993) and Zeh et al. (1993) pointed to evidence of synchronization in calf production based on ice-based survey data for years 1978, 1982, and 1986. In those years, high calf counts were intermixed with very low calf counts, suggesting both calving synchrony and a 4-year calving interval.

Lactation

While the length of the lactation period in the bowhead whale is not yet firmly established, Nerini et al. (1984) suggested that a 12-month lactation period was most probable based on evidence from harvested whales. It has been determined in other mysticetes, for example, the fin and sei whales, that weaning typically occurs before or shortly after entry into the feeding area of the southern ocean (Laws, 1961; Gambell, 1968). Although morphologic data from harvested lactating whales may eventually be useful in defining or clarifying the bowhead whale lactation period, they are subject to a wide range of interpretations in small sample sizes and remain inconclusive, being compatible with a 6- to 12-month (or greater) lactation period.

A spring-caught lactating whale had a recently formed CA, measuring 6.3 cm. CL diameters in the bowhead ranged from 6.7 to 15.0 cm in one dataset ($N = 13$), with the smaller CLs found toward the end of gestation or in those whales for which pregnancy was not confirmed, possibly CLs of the cycle (Tarpley and Hillmann, 1999). Based on this, the 6.3 cm CA in the spring-caught lactating whale suggests recent calving and the start of lactation in the spring. A fall-caught (22 October) lactating whale had a CA of 3.4 cm, suggesting that this whale was farther into the lactation period (more advanced CA regression and thus not recently calved).

In fall-caught calves (baleen length < 50 cm), 6 of 9 (66%) had been nursing at the time of capture in September and October, when they are about 5–6 months old (see Chapter 7). Two of these (at Kaktovik and Utqiaġvik) had some invertebrate prey in their stomachs, suggesting a transition to solid food. George and Suydam (2014), reporting on 35 yearlings (~1–1.5 years old) examined in the spring and fall harvests between 1972 and 2013, found only two (~6%) with evidence of nursing (milk in the stomach) at the time of capture. The composite of calf and yearling stomach content findings suggests that bowheads may initiate weaning as early as 5–6 months and that nearly all have completed nursing before their first year.

An investigation of $\delta^{13}C$ and $\delta^{15}N$ isotope oscillations in the baleen plates of young bowheads ($N = 4$) revealed that the nursing period (during which steady nutritional delivery is from the mother's milk) featured minimal isotopic swings relative to that in independently feeding whales. That is, variability in prey intake through time results in oscillatory swings in isotope signals within the baleen (Sensor et al., 2018). The drop in $\delta^{15}N$ in three of the whales was defined as the end of nursing (weaning) along with the initiation of isotopic swings from independent feeding. Baleen length in the three whales at the inflection point for weaning ranged from 62 to 92 cm. Lubetkin et al. (2008) demonstrated that baleen length during the first year is ~75–85 cm, slowing to ~25 cm/yr the following year, and slower yet in subsequent years. Allowing for variability in individual growth, baleen lengths of the three whales in Sensor et al. (2018) suggest that weaning is complete at around a year into the lactation period. This timeframe is consistent with reports in other mysticetes where, depending on species, the lactation period has been estimated in the range of 6–12 months (Slijper, 1962). However, even within a species (both mysticetes and odontocetes), the weaning period can be expected to vary, likely dependent on external food resource availability and, feasibly, the temperament of the mother, as in other mammals.

Male total body length and age at sexual maturity

O'Hara et al. (2002), reporting on a cohort of male bowhead whales ($N = 41$) ranging in total body length from 7.6 to 16.6 m, demonstrated a significant increase in mean testis mass beginning at 12.5–13.0 m in body length (Fig. 13.4). Our most current analysis of testis weights using a larger sample size agrees with an inflection for maturity beginning around a body length of 12.5 m. Lubetkin et al. (2008) determined, from a limited dataset based on baleen growth, that two males 13.0–13.5 m in body length were in their midteens. In a subsequent analysis of a much larger dataset, based on a consideration of baleen length and body length, Lubetkin et al. (2012) estimated the age for maturing males to range from ~20.6 to 24.6 years.

Although Rosa et al. (2013) could more confidently estimate an age at sexual maturity for females (25.9 years, SE: 5.87), data were not sufficient to as precisely determine the age of male sexual maturity. However, using a dataset containing a range of sexually immature and mature whales (20 females and 10 males), they used logistic regression to estimate that 50% of male and female bowheads are sexually mature at an estimated age of 24.8 years. They cautioned, however, that their results "suggest that male and female bowheads may reach sexual maturity at about the same age, but more data from males are required before any firm conclusions can be drawn." Taken together, we suggest from these studies and our most current evidence that the average age at sexual maturity in males is ~25 years, beginning at a body length of about 12.5 m.

Interestingly, while considerable mating activity has been observed in winter and early spring coinciding with the estimated time of conception (see "Conception" section), there is some evidence that testicular activity might be particularly robust in the fall, with testis weights slightly heavier relative to body length in fall-caught whales. Active seminiferous tubules have been observed in mature males from the fall harvest with open lumens and tubular walls comprised of the cellular components of active spermatogenesis. Epididymides can be well-developed, rivaling the testis in size and filled with spermatozoa. Medway (1981) found spermatozoa in the urine of a fall-caught 12.7 m whale—a length that current data (above) suggest to be a newly matured whale. This degree of testicular activity in the fall is intriguing with female ovulation and receptivity many months away. The question is raised not only regarding the possibilities for sperm storage in the epididymis (where tubular structure may be more complex than other mammals—see "Testis and epididymis" section), but also, since mating behaviors have been witnessed in the fall (Chapter 23; Koski et al., 1993), the possibility of insemination and sperm storage in females, awaiting ovulation and conception in late winter and early spring of the following year. The feasibility of mating even in the absence of female receptivity must be considered, since we have evidence of insemination of a female which, though spring-caught and likely pubescent, had not yet ovulated, but nevertheless harbored spermatozoa in the mucus produced by the annular folds caudal to the uterine body (Fig. 13.6H).

Acknowledgements

We would like to express our deep appreciation to the many subsistence captains, crews and villagers of northern Alaska, along with the North Slope Borough Department of Wildlife Management, for their steady support of bowhead whale reproductive studies over many decades.

References

Angliss, R.P., Hobbs, R.C., DeMaster, D.P., 1995. Determination of the sample size necessary to detect changes in length frequency distributions from a recovering population of bowhead whales. Rep. Int. Whal. Commn. 45, 331–334.

Baker, C.S., Perry, A., Herman, L.M., 1987. Reproductive histories of female humpback whales *Megaptera novaeangliae* in the North Pacific. Mar. Ecol. Prog. Ser. 41, 103–114.

Best, P.B., 1981. The Status of Right Whales (Eubalaena glacialis) Off South Africa, 1969–1979. Investigational Report of the Division of Sea Fisheries of South Africa 123, pp. 1–44.

Braham, H.W., Krogman, B., Johnson, J., Marquette, W., Rugh, D., Nerini, M., et al., 1980. Population studies of the bowhead whale (*Balaena mysticetus*): results of the 1979 spring research season. Rep. Int. Whal. Commn. 30, 391–404.

Chittleborough, R.G., 1954. Studies on the ovaries of the humpback whale, *Megaptera nodosa* (Bonnaterre), on the western Australian coast. Aust. J. Mar. Freshw. Res. 5, 35–63.

Clarke, J.T., Ferguson, M.C., Brower, A.A., Willoughby, A.L., 2018. Bowhead whale calves in the western Beaufort Sea, 2012–2017. Paper SC/67b/AWMP03 Presented to the IWC Scientific Committee, Bled, Slovenia.

Davis, R., Koski, W., Miller, G., 1983. Preliminary Assessment of the Length-Frequency Distribution and Gross Annual Reproductive Rate of the Western Arctic Bowhead Whale as Determined with Low-Level Aerial Photography, with Comments on Life History. Report from LGL Ltd., Toronto and Anchorage, to the U.S. Marine Mammal Laboratory, Seattle, Washington, 91 pp.

Durham, F.E., 1980. External morphology of bowhead fetuses and calves. Mar. Fish. Rev. 42, 74–80.

Eschricht, D., Reinhardt, J., 1866. On the Greenland right whale (*Balaena mysticetus*). In: Flower, W.H. (Ed.), Recent Memoirs of the Cetacea. Ray Society, London, 150 pp. + plates.

Everitt, R.D., Krogman, B.D., 1979. Sexual behavior of bowhead whales observed off the north coast of Alaska. Arctic 32, 277–280.

Fetter, A.W., Everitt, J.I., 1980. Tissues (structure/function) (RU 280a). Histological structure of selected organs. In: Kelley, J.J., Laursen, G.I. (Eds.), Investigation of the Occurrence and Behavior Patterns of Whales in the Vicinity of the Beaufort Sea Lease Area. pp. 331–358. Final Report to the Bureau of Land Management from the Naval Arctic Research Laboratory. NTIS No. PB86-153566/AS.

Gambell, R., 1968. Seasonal cycles and reproduction in sei whales of the southern hemisphere. Discov. Rep. 35, 31–134.

George, J.C., Suydam, R.S., 2014. Update on characteristics of bowhead whale (*Balaena mysticetus*) calves. Paper SC/65b/BRG20 Presented to the IWC Scientific Committee.

George, J.C., Suydam, R.S., Philo, L.M., Albert, T.F., Zeh, J.E., Carroll, G.M., 1995. Report of the spring 1993 census of bowhead whales, *Balaena mysticetus*, off Point Barrow, Alaska with observations on the 1993 subsistence hunt of bowhead whales by Alaska Eskimos. Rep. Int. Whal. Commn. 45, 371–386.

George, J.C., Bada, J., Zeh, J., Scott, L., Brown, S.E., O'Hara, T., et al., 1999. Age and growth estimates of bowhead whales (*Balaena mysticetus*) via aspartic racemization. Can. J. Zool. 77, 571–580.

George, J.C., Zeh, J., Suydam, R., Clark, C., 2004. Abundance and population trend (1978–2001) of western arctic bowhead whales surveyed near Barrow, Alaska. Mar. Mamm. Sci. 20 (4), 755–773.

George, J.C., Follman, E., Zeh, J., Sousa, M., Tarpley, R., Suydam, R., et al., 2011. A new way to estimate the age of bowhead whales (*Balaena mysticetus*) using ovarian corpora counts. Can. J. Zool. 89, 840–852.

George, J.C., Herreman, J., Givens, G.H., Suydam, R., Mocklin, J., Clark, C., et al., 2012. Brief overview of the 2010 and 2011 bowhead whale abundance surveys near Point Barrow, Alaska. Paper SC/64/AWMP7 Presented to the IWC Scientific Committee.

George, J.C., Stimmelmayr, R., Suydam, R., Usip, S., Givens, G., Sformo, T., 2016. Severe bone loss as part of the life history strategy of bowhead whales. PLoS One 11 (6), e0156753. Available from: https://doi.org/10.1371/journal.pone.0156753.

George, J.C., Suydam, R., Givens, G., Horstmann, L., Stimmelmayr, R., Sheffield, G., 2018. Length at sexual maturity and pregnancy rates of Bering-Chukchi-Beaufort seas bowhead whales. Paper SC/67b/AWMP07 to the International Whaling Commission Scientific Committee.

Haldiman, J.T., Tarpley, R.J., 1993. Anatomy and physiology. In: Burns, J.J., Montague, J.J., Cowles, C.J. (Eds.), The Bowhead Whale. The Society for Marine Mammalogy, Allen Press, Lawrence, KS, pp. 71–156. Special Publication No. 2.

Harrison, R.J., 1972. Reproduction and reproductive organs in *Platanista indi* and *Platanista gangetica*. Invest. Cetacea 4, 71–82.

Kenney, R.M., 1980. Tissues (structure/function) (RU 280b). In: Kelley, J.J., Laursen, G.L. (Eds.), Investigation of the Occurrence and Behavior Patterns of Whales in the Vicinity of the Beaufort Sea Lease Area. Final Report to the Bureau of Land Management from the Naval Arctic Research Laboratory. pp. 359–393. NTIS No. PB86-153566/AS.

Kenney, R.M., Everitt, J.I., 1980. Tissues (structure/function) (RU 579b). In: Kelley, J.J., Larsen, G.L. (Eds.), Investigation of the Occurrence and Behavior Patterns of Whales in the Vicinity of the Beaufort Sea Lease Area. Final Report to the Bureau of Land Management from the Naval Arctic Research Laboratory. pp. 227–263. NTIS No. PB86-153566/AS.

Kenney, R.M., Garcia, M.C., Everitt, J.I., 1981. The biology of the reproductive and endocrine systems of the bowhead whale, *Balaena mysticetus*, as determined by evaluation of tissues and fluids from subsistence-killed whales. In: Albert, T. (Ed.), Tissue Structural Studies and other Investigations of the Biology of Endangered Whales in the Beaufort Sea. University of Maryland, College Park, MD, pp. 89–156, Final Report to the Bureau of Land Management from the Department of Veterinary Science.

Kleinenberg, S.E., Yablokov, A.V., Bel'kovich, B.M., Tarasevich, M.N., 1969. Beluga (*Delphinapterus leucas*): Investigation of the Species. Translated from the Russian. Israel Program for Scientific Translation, Jerusalem, 376 pp.

Koski, W.R., Miller, G.W., Davis, R.A., 1990. Growth rates and morphology of bowhead whale calves: photogrammetric evidence. In: Extended Abstract Presented at The Fifth Conference on the Biology of the Bowhead Whale, *Balaena mysticetus*, April 1990, Anchorage, Alaska.

Koski, W., Davis, R., Miller, G., Withrow, D., 1993. Reproduction. In: Burns, J.J., Montague, J.J., Cowles, C.J. (Eds.), The Bowhead Whale. The Society for Marine Mammalogy, Allen Press, Lawrence, KS, pp. 239–274, Special Publication No. 2.

Koski, W.R., Zeh, J., George, J.C., 2008. A calf index for monitoring reproductive success in the Bering-Chukchi-Beaufort Seas bowhead whale (*Balaena mysticetus*) population. J. Cetacean Res. Manage. 10 (2), 99–106.

Kraus, S.D., Rolland, R.M., 2007. The Urban Whale. Harvard University Press, Cambridge, MA.

Kraus, S.D., Prescott, J.H., Knowlton, A.R., Stone, G.S., 1986. Migration and Calving of Right Whales (*Eubalaena glacialis*) in Western North Atlantic. Reports of the International Whaling Commission. 10 (special issue), 139–144.

Laurie, A.H., 1937. The age of female blue whales and the effect of whaling on the stock. Discov. Rep. 15, 223–284.

Laws, R.M., 1959. Foetal growth rates of whales with special reference to the fin whale, *Balaenoptera physalus* (L.). Discov. Rep. 29, 281–308.

Laws, R.M., 1961. Reproduction, growth and age of southern fin whales. Discov. Rep. 31, 327–486 + plates 4–7.

Lubetkin, S.C., Zeh, J.E., Rosa, C., George, J.C., 2008. Age estimation for young bowhead whales (*Balaena mysticetus*) using annual baleen growth increments. Can. J. Zool. 86, 525–538.

Lubetkin, S.C., Zeh, J.E., George, J.C., 2012. Statistical modeling of baleen and body length at age in bowhead whales (*Balaena mysticetus*). Can. J. Zool. 90, 915–931.

Mackintosh, N.A., Wheeler, J.F.G., 1929. IV. The reproductive organs. Discov. Rep. 1, 379–471 + plates 25–44.

Marzec-WrÓbleWska, U., KamińsKi, P., ŁaKota, P., 2012. Influence of chemical elements on mammalian spermatozoa. Folia Biol. (Praha) 58, 7–15.

Matthews, L.H., 1937. The humpback whale, *Megaptera nodosa*. Discov. Rep. 17, 7–92.

Matthews, L.H., 1938. The sei whale, *Balaenoptera borealis*. Discov. Rep. 17, 183–290.

Medway, W., 1981. The cytological and clinical evaluation of blood and urine of the bowhead whale, *Balaena mysticetus*. In: Albert, T.F. (Ed.), Tissue Structural Studies and Other Investigations of the Biology of Endangered Whales in the Beaufort Sea. Report to the Bureau of Land Management from the Department of Veterinary Science. University of Maryland, College Park, MD, pp. 201–212. 953pp. NTIS No. PB86-153566. Available from: <http://www.ntis.gov>.

Meek, A., 1918. The reproductive organs of Cetacea. J. Anat. 52, 186–210.

Nerini, M.K., Braham, H.W., Marquette, W.M., Rugh, D.J., 1984. Life history of the bowhead whale, *Balaena mysticetus* (Mammalia: Cetacea). J. Zool. Soc. Lond. 204, 443–468.

O'Hara, T.M., George, J.C., Tarpley, R.J., Burek, K., Suydam, R.S., 2002. Sexual maturation in male bowhead whales (*Balaena mysticetus*) (of the Bering-Chukchi-Beaufort Seas stock. J. Cetacean Res. Manage. 4 (2), 143–148.

Ommanney, F.D., 1932. The urino-genital system of the fin whale (*Balaenoptera physalus*). Discov. Rep. 5, 363–466 + plates 2–3.

Payne, R., 1986. Long Term Behavioral Studies of the Southern Right Whale (*Eubalaena australis*). Reports of the International Whaling Commission. 10 (special issue), pp. 161–167.

Preuss, F., 1953. Beschreibung und Einteilung des Rinderuterus nach funktionellen Gesichtspunkten. Anat. Anz. 100, 59.

Pycraft, W.P., 1932. On the genital organs of a female common dolphin (*Delphinus delphis*). Proc. Zool. Soc. Lond. 192, 807–811 + 3 plates.

Reese, S., Calvin, J.A., George, J.C., Tarpley, R.J., 2001. Estimation of fetal growth and gestation in bowhead whales. J. Am. Stat. Assoc. 96 (455), 915–923.

Richardson, W.J., Greene, Jr., C.R., Koski, W.R., Malme, C.I., Miller, G.W., Smultea, M.A., et al., 1990. Acoustic Effects of Oil Production Activities on Bowhead and White Whales During Spring Migration near Pt. Barrow, Alaska—1989 Phase: Sound Propagation and Whale Responses to Playbacks of Continuous Drilling Noise from an Ice Platform, as Studied in Pack Ice Conditions. Report from LGL Limited, King City, Ontario, to U.S. Minerals Management Service. NTIS No. PB91-105486, 284 pp.

Rolland, R.M., Lysiak, N.S., Graham, K.M., Burgess, E.A., Hunt, K.E., Fuller, R., et al., 2018. Assessing Stress and Reproduction in Bowhead Whales (*Balaena mysticetus*) Using Baleen Hormones. A Final Report for the North Slope Borough/Shell Baseline Studies Program. Contract #2015-102.

Rosa, C., Zeh, J., George, J.C., Botta, O., Zauscher, M., Bada, J., et al., 2013. Age estimates based on aspartic acid racemization for bowhead whales (*Balaena mysticetus*) harvested in 1998–2000 and the relationship between racemization rate and body temperature. Mar. Mamm. Sci. 29 (3), 424–445.

Rugh, D., Miller, G., Withrow, D., Koski, W., 1992. Calving intervals of bowhead whales established through photographic reidentification. J. Mammal. 73, 487–490.

Sensor, J.D., George, J.C., Clementz, M.T., Lovano, D.M., Waugh, D.A., Givens, G.H., et al., 2018. Age estimation in bowhead whales using tympanic bulla histology and baleen isotopes. Mar. Mamm. Sci. 34 (2), 347–364.

Slijper, E.J., 1962. Reproduction. Whales. Cornell University Press, Ithaca, NY, pp. 349–390.

Tarpley, R.J., Hillmann, D.J., 1999. Observations on Ovary Morphology, Fetal Size and Functional Correlates in the Bowhead Whale *Balaena mysticetus*. Final Report to the Department of Wildlife Management, North Slope Borough, Barrow, AK, 201 pp. + appendices.

Tarpley, R.J., Jarrell, G.H., George, J.C., Cubbage, J., Stott, G.G., 1995. Male pseudohermaphroditism in the bowhead whale, *Balaena mysticetus*. J. Mammal. 76 (4), 1267–1275.

Thewissen, J.G.M., Williams, E.M., 2002. The early radiations of Cetacea (Mammalia): Evolutionary pattern and developmental correlations. J. Mammal. 33, 73–90.

Tarpley, R.J., Hillmann, D.J., George, J.C., Zeh, J.E., Suydam, R.S., 2016. Morphometric correlates of the ovary and ovulatory corpora in the bowhead whale, *Balaena mysticetus*. Anat. Rec. 299 (6), 769–797.

Wetzel, D.L., Reynolds III, J.E., Mercurio, P., Givens, G.H., Pulster, E.L., George, J.C., 2017. Age estimation for bowhead whales, *Balaena mysticetus*, using aspartic acid racemization with enhanced hydrolysis and derivatization procedures. J. Cetacean Res. Manage. 17, 9–14.

Zeh, J.E., Clark, C.W., George, J.C., Withrow, D., Carroll, G.M., Koski, W.R., 1993. Current population size and dynamics. In: Burns, J.J., Montague, J.J., Cowles, C.J. (Eds.), The Bowhead Whale. The Society for Marine Mammalogy, Allen Press, Lawrence, KS, pp. 409–489. Special Publication No. 2.

CHAPTER

14

Anatomy and function of feeding

A.J. Werth[1] and Todd L. Sformo[2,3]

[1]Department of Biology, Hampden-Sydney College, Hampden-Sydney, VA, United States

[2]Department of Wildlife Management, North Slope Borough, Utqiaġvik, AK, United States

[3]Institute of Arctic Biology, University of Alaska Fairbanks, Fairbanks, AK, United States

Introduction: baleen and oral morphology

Structural and functional features of bowhead feeding are similar to those of other balaenid mysticetes (right whale species in the sister genus *Eubalaena*), but bowhead feeding morphology can be considered extreme. Darwin himself, in *Origin of Species* (1859), declared that "*the Greenland [bowhead] whale is one of the most wonderful animals in the world, and the baleen, or whalebone, one of its greatest peculiarities.*" The bowhead whale's distinctive form is ideally suited to acquiring and storing energy in a seasonal polar habitat. Its huge, namesake domed head holds the largest filter in the natural world, propelled in ram-driven fashion by its chunky, thickly blubbered body. Compared to other baleen whales, bowheads have longer, finer baleen plates and fringes and a larger, more sharply arched rostrum. Together these produce a filtering system that, in turn, creates high drag that slows movement of the whale. Following a method described by Kawamura (1974), George et al. (2016) calculated a filter area of 40 m^2 for a 16-m adult bowhead. By including not only the medial mat of tangled baleen fringes but also the flattened plate surfaces plus the tiny fringes too, Werth et al. (2018b) calculated the total cumulative baleen area to be over 190 m^2 for a bowhead whale of 16-m body length (Fig. 14.1).

All mysticetes are bulk filter feeders that separate small prey from seawater. Unlike rorquals and gray whales, which ingest and process discrete, single mouthfuls of prey-laden water at a time, bowhead (and right) whales are continuous filter feeders: they ingest a steady stream of water carrying tiny (5–10 mm) copepods, mysids, euphausiids, and other schooling plankton (Chapter 28) (Fig. 14.2). The incurrent stream enters the mouth's center anteriorly through a subrostal gap between paired racks of baleen (Fig. 14.3), each rack with 275–330 (mean 303) plates, suspended from either side of the narrow palate. Lambertsen et al. (1989) documented 320 total plates in a 7.8-m female ingutuk and 310 plates in an 8.6-m ingutuk (see Chapter 7 for life history stages).

The Bowhead Whale

DOI: https://doi.org/10.1016/B978-0-12-818969-6.00014-5

FIGURE 14.1 Bowhead whales are generally observed feeding at or near the sea surface, but the presence of mud plumes near this whale (with seabirds flying overhead) indicates they also feed near the seafloor. *Source: Photo by Vicki Beaver (NOAA/North Slope Borough, NMFS Permit 14245).*

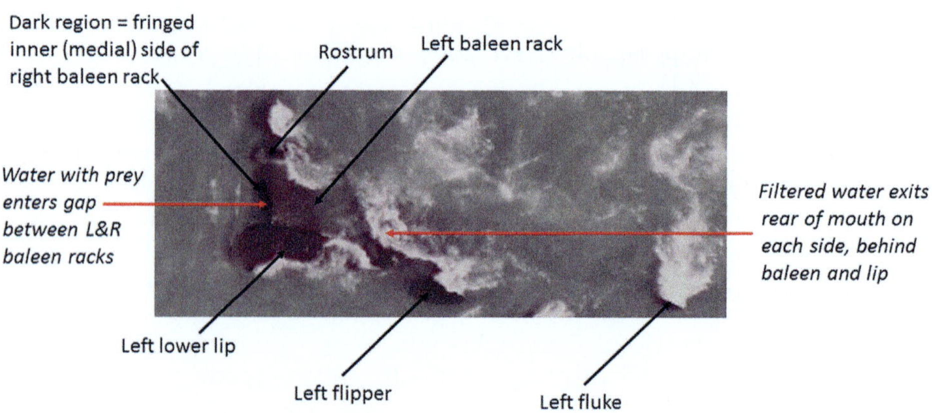

FIGURE 14.2 Like other mysticetes, bowhead whales sometimes turn on their side to capture food at the surface, but prey are not engulfed in a single mouthful. Instead, steady fluking powers ram hydraulic filtration of a steady stream of water. *Source: Photo by Linda Brattstrom, modified from Fish, F.E., Goetz, K.T., Rugh, D.J., Brattström, L.V., 2013. Hydrodynamic patterns associated with echelon formation swimming by feeding bowhead whales* (Balaena mysticetus). *Mar. Mamm. Sci. 29, E498–E507.*

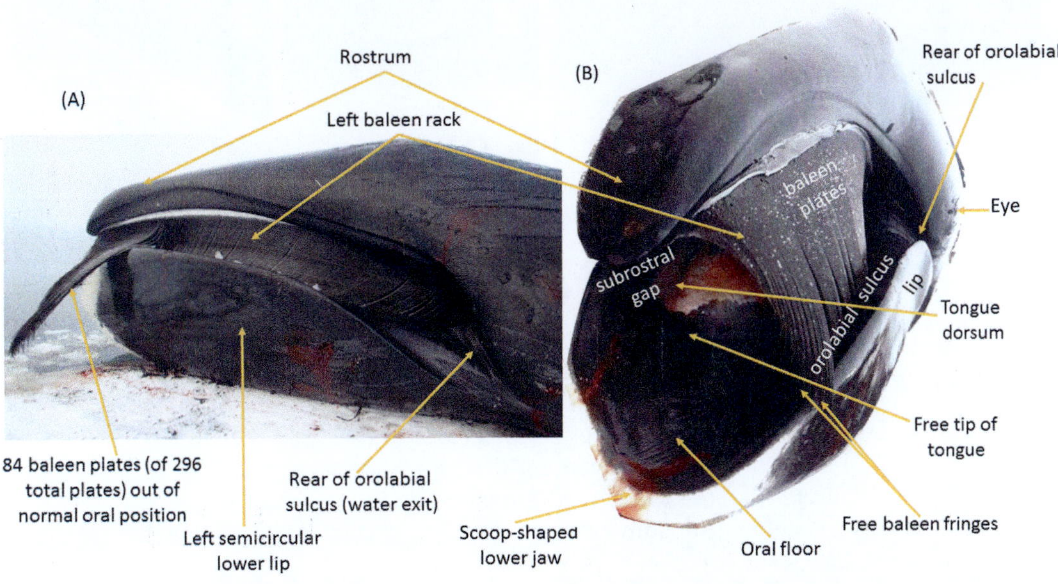

FIGURE 14.3 The bowhead whale's large mouth (seen here in left lateral (A) and oblique views (B)) is dominated by paired racks of keratinous baleen plates serially suspended from the upper jaw. Incurrent water enters under the rostral tip and is channeled between the tongue and lips before exiting posteriorly. *Source: Photos by Department of Wildlife Management North Slope Borough (A) and Todd Sformo (B).*

Unpublished data from Alaska's North Slope Borough suggests there may be as many as 20–30 more plates in the right than in the left baleen rack, though with no sex differences. Water then passes laterally through baleen plates and fringes to a gutter-like longitudinal groove, the orolabial sulcus, on either side of the mouth. This sulcus lies between the lateral (labial) side of each baleen rack and the interior (medial) side of the semicircular lip projecting above the mandible. At the back (posterior) of the mouth on each side is located a roughly oval, jetport-like opening through which filtered excurrent water flows (Fig. 14.3).

Baleen is an ever-growing tissue made entirely of alpha-keratin, with sheet-like cortical faces enclosing a medulla of hollow horn tubules and intertubular keratin. The youngest baleen is found where it emerges from the gums and the oldest at the distal, ventral tip. Like other mammalian integumentary tissues such as hair and nails, baleen grows slowly at its base throughout life, arising from germinal layers in thick (up to 25 cm) gray palatal gingiva (Iñupiat: *mammaq*; Fig. 14.4C) connected via fibrous connective tissue to the maxillary periosteum. Emerging from this matrix are keratin fibers and dead, cornified cells that extend about 25+ cm per year in juveniles and 17.5 cm/year in adults (Lubetkin et al., 2008; Chapter 21). Within the baleen plate itself, tubules emerge as hair-like fringes (sometimes called bristles) when edges of baleen plates erode (Werth et al., 2016b, 2018a) via abrasive friction from prey, water, and oral structures, particularly the tongue, lips, and adjacent baleen (see Fig. 14.4E); with new growth replacing erosive loss (baleen turnover), each adult plate represents approximately 13 years of growth. We estimate that an

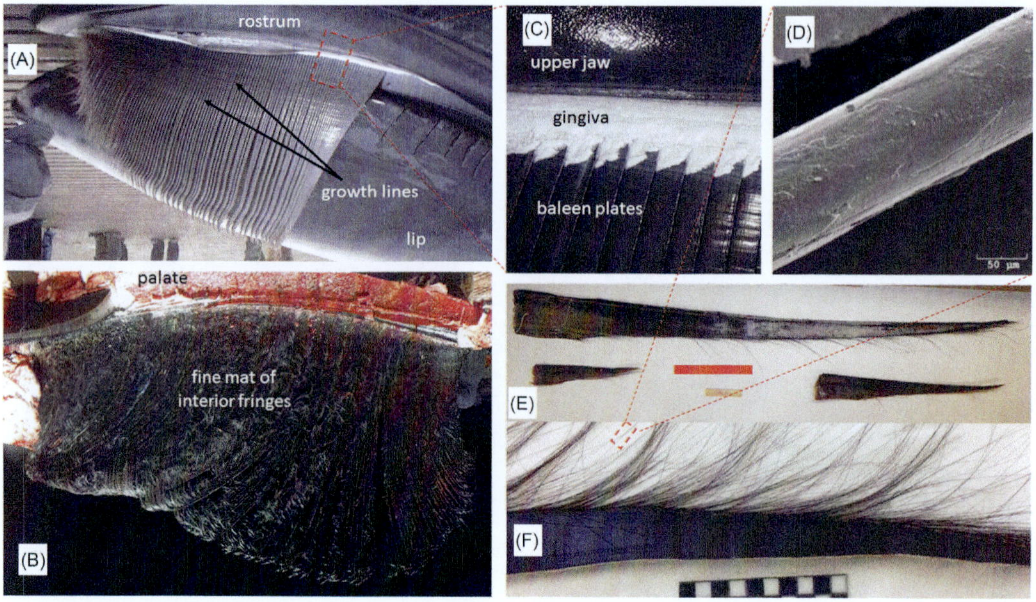

FIGURE 14.4 The bowhead whale's baleen racks comprise over 300 long (3–4 m) plates on each side of the mouth; plates are smooth on their outer (lateral) surface (A; note bent tips and lines indicating growth) but eroded medially to form a dense mat of fine, hair-like bristles or fringes (B, rostral to left), with one fringe seen in close-up view via scanning electron microscope in D (University of Alaska Fairbanks, Advanced Instrumentation Laboratory). Baleen is attached to the upper jaw via tough, rubbery gingiva, with plates emerging like slats of window blinds (*dashed box* in A shows inset C). E shows three plates from whales of varying age (*red scale bar* = 30 cm): 45 cm plate from 0.5-year-old whale (NSB-DWM 2014B12), 85 cm from 1.5-year-old whale (NSB-DWM 2012B15), and 229 cm from 21-year-old whale (NSB-DWM 2017B6). F shows close-up of baleen plate, with *dashed box* indicating single fringe (D). *Source: Photos by Brian Person (A), Hans Thewissen (B, E), Kate Stafford (C), Todd Sformo (D), and Alex Werth (F).*

adult bowhead loses 14 m^2 of baleen area in an average year, the equivalent of 0.05 m^3 (50 L) of baleen volume (Werth et al., 2020).

Plate length varies by position within the rack (Fig. 14.4), with the 120–250th plates (from the rostral tip) being the longest. The maximum reported length of a bowhead baleen plate (sometimes called a lamina) is 427 cm (Scammon, 1874); Bockstoce and Burns (1993) reported a maximum length of 487 cm from old whaling records, but this is unverifiable and probably unreliable. Lengths of 330–380 cm are common in large adults of the Bering-Chukchi-Beaufort Stock of bowhead whales. Each main plate can be thought of as a scalene triangle with the lingual (medial) edge as hypotenuse that is slightly concave due to being eroded into a brush of fine, hair-like fringes. In contrast, the lateral edge is slightly convex except at the distal tip which curves laterally. Plates also curl slightly in the anteroposterior plane, with the lateral (outer) edge directed posteriorly. Because of this triangular shape, plate width varies greatly, but averages $\sim 18 \pm 2$ cm (standard error of mean) in adults; plate thickness averages 0.34 ± 0.04 cm but also varies along its length, with the oldest (ventral) portions being considerably thinner. Plates are suspended vertically in a serial array, spaced (like Venetian window blinds) 0.94 ± 0.12 cm apart. Not only

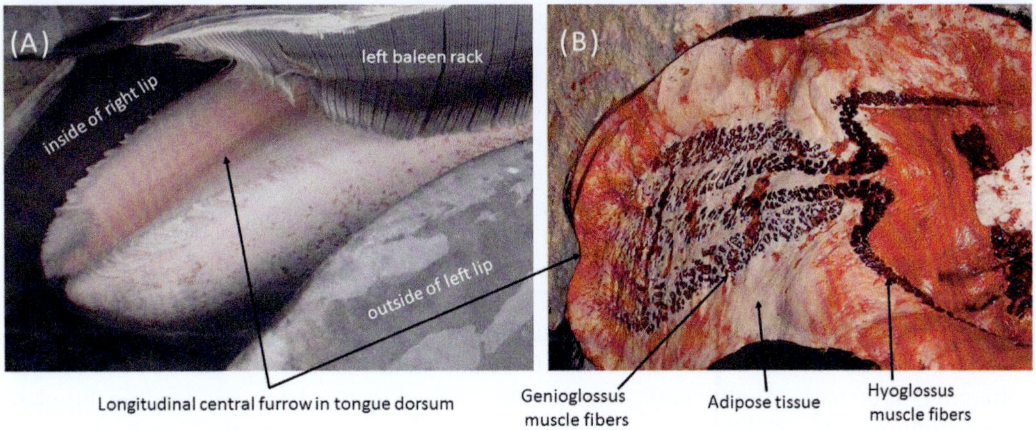

FIGURE 14.5 The bowhead whale's tongue is exceptionally large and firm but lacks internal muscles enabling shape changes (A); it stores energy as fat (B; transverse section) around bundles of muscles that anchor the tongue and enable it to be retracted for swallowing and returned to its usual position. *Source: Photos by J. Craig George (A) and Alex Werth (B).*

are bowhead baleen plates exceptionally long but also fringes are long (194 ± 12.5 mm), fine (0.23 ± 0.02 mm diameter), and dense (27.3 ± 3.9 fringes/cm of plate length, or about 8000 fringes per 3-m long plate; Werth, 2013). As Lambertsen et al. (1989) noted, the density of free fringes per linear centimeter varies widely since fringes emerge from tubules running along each plate's longitudinal axis, whereas a plate's medial (lingual) border meets this axis at varying angles. In some whales there are small, medially located minor plates adjacent to the main baleen plates.

Gape opening occurs via mandibular depression (abduction) from sternomandibularis and depressor mandibulae muscles, and to some extent from cranial elevation. Large temporalis, pterygoid, and masseter muscles serve as jaw adductors, rotating the mandible laterally. This allows the large semicircular lower lips to pivot outward during feeding, aided by the jaw's loose temporomandibular joint with its rounded articular condyle. The tongue (Fig. 14.5) is large: 1 m wide, 1 m high, and 3+ m long in adults (Lambertsen et al., 1989). The tongue is black with scattered white or pink unpigmented spots and is completely attached to the oral floor, apart from a small free tip. A slight central longitudinal furrow runs down in the tongue's dorsum and a spoon-shaped concavity above the free tip is evident. The dorsum is smooth and glandless with keratinized stratified squamous epithelium covering thick muscle bundles and collagen fibers interspersed with abundant adipose deposits (Werth, 2007). About 85% of muscle fibers originate outside the tongue, especially from the mandible (genioglossus) and hyoid (hyoglossus).

Feeding behavior and functional ecology

Bowheads are observed skim feeding at or near the surface, but muddy, abraded snouts and stomach contents suggest they feed at all levels of the water column (Carroll et al., 1987;

Fig. 14.1), although mud plumes may be caused by whales rubbing along the bottom for skin exfoliation during molting (Fortune et al., 2017). Echelon feeding has been reported in which multiple whales skim feed, often while rotated on their sides (Fig. 14.2) in a V-formation similar to that of migrating waterfowl (Würsig et al., 1985); this behavior may improve the efficiency of prey ingestion by limiting drag or taking advantages of hydrodynamic flows that aggregate prey (Fish et al., 2013). However, multiple bowheads may forage together in noncooperative, parallel fashion simply due to prey abundance. Bowheads may also open the mouth to thermoregulate, allowing blood to perfuse the tongue (Chapter 16) and a palatal rete (Ford et al., 2013) to radiate excess heat generated by locomotion.

Bouts of ingestion entering the mouth through the incurrent stream have been observed to occur over several minutes in the case of surface skim feeding (Goldbogen et al., 2017). Biologging data from digital tags temporarily affixed to the backs of feeding bowhead whales (Simon et al., 2009) indicate swim speeds during foraging of <1 m/s, similar to but slightly less than recorded foraging speeds of right whales (van der Hoop et al., 2019). Simon et al. (2009) described bowhead breath-holding foraging dives having a U-shaped pattern: dives averaged 15.2 minutes at 79 m depth, with swim speed during filtration at 0.7–0.8 m/s even with increased (0.12 Hz) fluking and ~10-degree body roll. Brief (2–3 minutes) periods of open gape filtration, indicated by increased fluking, are interspersed with short (~10 second) pauses when fluking slows, during which whales are presumed to close the mouth and swallow accumulated prey (see Chapter 4). In contrast, detailed tag data from a right whale study (van der Hoop et al., 2019) showed repeated pauses of about 3 seconds between 50-second bouts of fluking, likely to swallow accumulated prey during foraging dives of 10–15 minutes. Right whale swim speed decreases by 25% when filtering, from 1.4 to 1.1 m/s, but bowhead locomotion slows more dramatically (40%) due to bowheads' larger cross-sectional mouth area per body size (averaging 4.23 ± 0.66 m^2; Werth, 2004), likely allowing bowheads to save energy by filtering less dense prey aggregations and for shorter foraging bouts relative to right whales (van der Hoop et al., 2019).

Most marine mammals save energy with stroke-and-glide locomotion, but bowhead filtration entails steady, rapid fluking to overcome the enormous drag incurred by the huge oral filter, estimated to increase sixfold when the mouth opens. This is substantially higher than the $3-5\times$ increase in drag above nonforaging swimming (van der Hoop et al., 2019) estimated for right whales. Applying the calculated average 4.23 m^2 anterior oral aperture (for an average 12-m adult bowhead; Werth, 2004) to 0.8 m/s swim speeds during filtration yields a filtration rate of about 3.2 m^3/s, with a consequent estimated daily filtering rate of approximately 80,000 m^3 of water per whale. Previous calculations suggested that, given a copepod density of 0.001 kg/m^3 (Laidre et al., 2007), bowheads would need to filter 800,000 m^3 per day to meet basic energetic demands, indicating a whale foraging for 7 hours per day would need to capture prey at speeds of 7.5 m/s: 10 times their observed speed, with a drag that would be correspondingly 100 times higher. Such a foraging speed is clearly unrealistic. However, far higher copepod densities have been recorded in foraging hot spots (Baumgartner and Mate, 2003), suggesting bowheads can meet energetic demands with their high drag, slow speed strategy. Simon et al. (2009) concluded that bowheads filter 2000 tons of water per dive with a uniquely slow relative swimming speed (<0.07 body length/s), far slower than that of most aquatic predators.

It is apparent that bowhead baleen, like the skull, grows disproportionately faster in young animals (Chapter 7), even at the cost of withdrawing resources from the ribs and other skeletal bones (George et al., 2016). This is likely a life history strategy for creating a large oral filter to gain body mass and volume (to avoid heat loss, among other reasons), leading to age classes with differing body proportions: younger whales have relatively larger baleen racks, particularly when a rack is measured solely by the area its medial, fringed mat covers (Fig. 14.6). However, as body length and girth increases in adults, the surface area of the baleen filter continues its steady pace of growth, with plates being exceptionally long and finely fringed in older bowheads. Cumulative filter area takes into account not only the medial mat but also the planar faces of all plates and the combined area of all free fringes so that total area increases (Fig. 14.6) from 129 m^2 in a 10 m whale to 167 m^2 (14 m whale), 193 m^2 (16 m), and 252 m^2 (18 m). By any measure, these constitute the largest filters in the natural world. Although drag forces limit bowhead filtration (Werth et al., 2018b), the strategy of targeting tiny, concentrated prey with slow foraging speed is clearly effective, yet this makes bowheads susceptible to capturing and ingesting microplastic pollution in seawater (Werth et al., 2019).

A photogrammetric analysis of bowhead baleen led Lambertsen et al. (1989) to speculate that the three-dimensional contours of racks create intraoral flow patterns that might minimize an anterior compressive bow wave. Such a wave could scatter densely aggregated copepods or alert them to the approaching whale, possibly triggering an escape response. Lambertsen et al. (1989) suggested that the progressive narrowing of the oral cavity—as incurrent prey-laden water flows posteriorly through the mouth and along the tapering orolabial sulcus—could create a simple Bernoulli effect, increasing the intraoral flow rate and potentially mitigating any bow wave. Further, Lambertsen et al. (1989) speculated that posteriorly directed intraoral water flow, both incurrent in the center of the mouth dorsal to the tongue and excurrent along each lateral orolabial sulcus, could

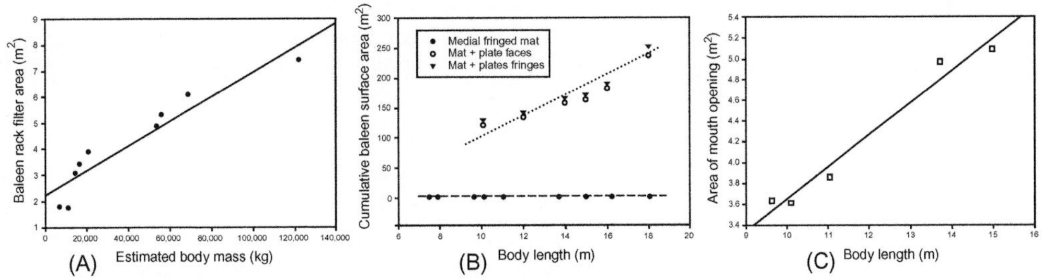

FIGURE 14.6 The bowhead whale's low-porosity, high drag filtration strategy has evolved to exploit seasonal patches of tiny copepod prey. Overall baleen filter surface area can be calculated in various ways but is the largest filter of any whale or other organism. (A) Filter area (following Kawamura, 1974) plotted against estimated body mass for nine individual bowhead whales of known body length. (B) The same filter area for the same nine whales plotted against body length (*solid circles* and *dashed line*), plus cumulative, three-dimensional baleen surface areas (Werth et al., 2018b) taking into account plate faces (*open circles*) and free fringes (*triangles*), with regression trend indicated by a *dotted line.* (C) Measured area of the anterior oral opening plotted against body length for five whales (Werth, 2004).

produce a Venturi effect, with decreased pressure enhancing perpendicular flow from the center of the mouth through each baleen plate, thereby improving filtration efficiency and decreasing drag forces generated by filtration. Laboratory flow experiments by Braithwaite (1983) and Mayo et al. (2001) on sections of bowhead and North Atlantic right whale baleen plates, respectively, demonstrated that balaenid baleen possesses sufficiently small filter porosity to minimize drag yet prove remarkably effective at capturing tiny copepods, often as small as a grain of rice (Fig. 14.4B). Further flow experiments with sections of bowhead baleen in circulating flow tanks (Werth, 2013) and with mathematical and physical models of bowhead bodies (Werth, 2004), corroborated, via kinematics, particle capture, and pressure transducers, the slight Bernoulli and Venturi effects foreseen by Lambertsen et al. (1989, 2005). These hydrodynamic forces are not likely sufficient to pull water into the bowhead mouth but could as predicted preclude a compressive bow wave.

Flow tank experiments simulating oral filtration (Werth, 2013) showed that bowhead baleen exhibits flow-dependent porosity. As flow speed increases, undulating baleen fringes create larger gaps, but because serial plates are so closely spaced, fringes merge with those of adjacent plates to create a tight "mat" with interfringe spaces of about 1 mm. Flow speeds above 1 m/s, the foraging speed observed and recorded from tagged whales, are counterproductive, greatly increasing viscous/friction drag and turbulence while not improving filtration. However, the hydrofoil-like shape of individual baleen plates likely reduces drag and may even generate slight lift forces to ease the energetic burden of bowhead filtration (Werth and Potvin, 2016).

Continued fluid dynamics modeling and experiments (Werth and Potvin, 2016; Potvin and Werth, 2017) strongly support the hypothesis that balaenid oral filtration generally involves tangential or cross-flow filtration rather than, as expected, direct "dead end" or throughput filtration in which water entering the mouth turns perpendicularly and flows directly laterally through the narrow (1 cm) intra-baleen gaps separating adjacent plates. In contrast, cross-flow filtration involves the majority of flow moving along instead of directly through the filter surface, in this case along the medial (lingual) surface on the "inside" of each baleen rack. Not only would such crossflow filtration—which has been demonstrated in continuous filtration schemes of numerous other vertebrates, including sharks, rays, and teleost fishes (Sanderson et al., 2016; Divi et al., 2018; Werth, 2019)—improve overall flow and filtration efficiency, it has the added benefits of moving accumulated prey to the tongue base, where it can be more easily swallowed, and greatly decreasing the risk of filter clogging. This cross (tangential) flow potentially solves the problem of how whales remove or cleanse captured prey from the intricately "sticky" filter (Werth, 2001) and why necropsied whales do not show prey items trapped within baleen fringes. Flow experiments and modeling indicate that although the tongue and lips play a small role, the curved baleen racks are the primary agent of cross-flow. Most particles are captured in the posterior and ventral regions of the baleen, where perpendicular (intra-baleen) flow velocities are highest and intra-baleen pressures lowest (Fig. 14.7; Werth and Potvin, 2016; Potvin and Werth, 2017). Because the majority of flow through the baleen filter (i.e., in gaps between plates) occurs in the rack's posterior and ventral regions, prey accumulates close to the tongue, facilitating swallowing. Evidence for this flow pattern is seen in mechanical wear patterns of baleen plates and fringes (Werth et al., 2016b), which suggest that any prey trapped in baleen can be freed by hydraulic "backflow" flushing via tongue depression (Werth, 2001). Bowhead baleen is highly flexible, resisting breakage and allowing the longest plates to fold posteriorly when the mouth closes (Werth et al., 2018a).

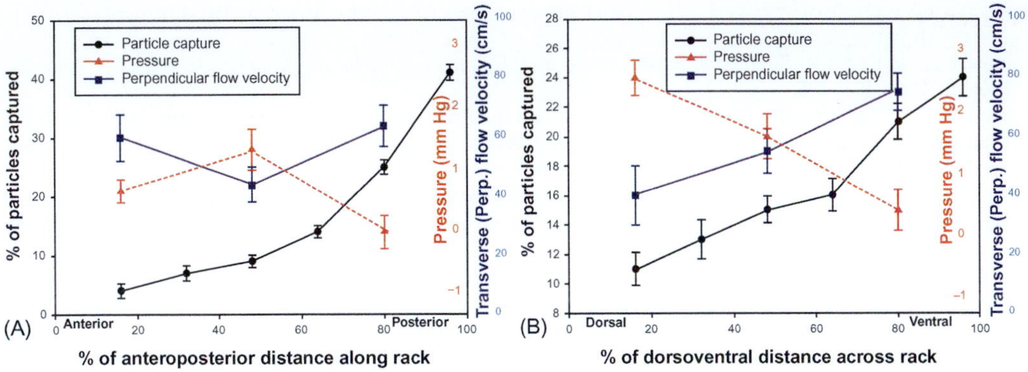

FIGURE 14.7 Laboratory flow experiments indicate that bowhead and right whale filters trap prey by crossflow filtration, with the majority of water flowing through the most posterior and ventral inter-baleen gaps. (A) Changes along the length of a partial baleen rack as water and buoyant particles flow from front to rear in a flow tank. Pressure (in *red*) is lowest in the posterior-most portion of the rack, where inter-baleen flow velocity (in gaps between plates, toward the lateral surface; *blue*) and capture of particles (*black*) is highest. (B) Similar changes from the top to the bottom of a partial bowhead baleen rack in a flow tank. Pressure (*red*) is lowest and flow velocity (*blue*) highest toward the lower-most portion of the rack, leading to steadily greater particle capture (*black*) in the ventral-most portion of the rack. Accumulation of prey in the rack's posterior- and ventral-most regions leads to easier swallowing. *Source: Modified from Werth, A.J., Potvin, J., 2016. Baleen hydrodynamics and morphology of crossflow filtration in balaenid whale suspension feeding. PLoS ONE, 11(2), e0150106.*

As tagging studies continue to provide data on foraging dives, adding knowledge to poorly studied questions including precise depths, durations, and body orientations during feeding bouts (Chapter 4), there is much need for continued functional studies to provide much need for continued functional studies to provide detailed information on movements of the tongue, jaws, and lips during filter feeding, and on how bowhead whales locate and adjust the flow of water into and within the mouth to improve prey capture. Finally, the extent to which climate change and other natural or anthropogenic threats may affect bowhead whale feeding and ultimately bowhead health demands close scrutiny. The good news is that bowhead baleen appears resistant to pH changes caused by ocean acidification (Werth and Whaley, 2019); baleen is also hydrophilic (Werth et al., 2016a) and apparently repels oil (St Aubin et al., 1984; Werth et al., 2019); however, the greatest risks to bowhead feeding may come from climate change that likely involves altered abundance, distribution, and density of planktonic prey. The fine porosity and large area of the oral filter makes bowhead whales particularly susceptible to capture of microplastics or other pollutants (Werth et al., 2019), again, with consequent risk to bowhead health.

Acknowledgments

We thank the Alaska Eskimo Whaling Commission and Barrow Whaling Captain's Association for specimens and access. Specimens were collected under NMFS Permit No. 21386 issued to the North Slope Borough Department of Wildlife Management, Utqiaġvik, Alaska. We thank many colleagues for field and laboratory assistance and fruitful discussions.

References

Baumgartner, M.F., Mate, B.R., 2003. Summertime foraging ecology of North Atlantic right whales. Mar. Ecol. Prog. Ser. 264, 123–135.

Bockstoce, J.R., Burns, J.J., 1993. Commercial whaling in the North Pacific sector. In: Burns, J.J., Montague, J.J., Cowles, C.J. (Eds.), The Bowhead Whale. Society for Marine Mammalogy, Lawrence, KS, pp. 563–577.

Braithwaite, L.F., 1983. Final Report: Effects of Oil on the Feeding Mechanism of the Bowhead Whale, Project RU 679 (Baleen Plate Fouling). Prepared for the US Department of the Interior Under Contract No. AA851-CTO-55. Brigham Young University, Provo, UT, USA.

Carroll, G.M., George, J.C., Lowry, L.F., Coyle, K.O., 1987. Bowhead whale (*Balaena mysticetus*) feeding near Point Barrow, Alaska, during the 1985 spring migrations. Arctic 40, 105–110.

Darwin, C.R., 1859. On the Origin of Species. John Murray, London.

Divi, R.V., Strother, J.A., Paig-Tran, E.W.M., 2018. Manta rays feed using ricochet separation, a novel nonclogging filtration mechanism. Sci. Adv. 4, e9533.

Fish, F.E., Goetz, K.T., Rugh, D.J., Brattström, L.V., 2013. Hydrodynamic patterns associated with echelon formation swimming by feeding bowhead whales (*Balaena mysticetus*). Mar. Mamm. Sci. 29, E498–E507.

Ford, T.J., Werth, A.J., George, J.C., 2013. An intraoral thermoregulatory organ in the bowhead whale (*Balaena mysticetus*), the corpus cavernosum maxillaris. Anat. Rec. 296, 701–708.

Fortune, S.M.E., Koski, W.R., Higdon, J.W., Trites, A.W., Baumgartner, M.F., Ferguson, S.H., 2017. Evidence of molting and the function of "rock-nosing" behavior in bowhead whales in the eastern Canadian Arctic. PLoS ONE 12 (11), e0186156.

George, J.C., Stimmelmayr, R., Suydam, R., Usip, S., Givens, G., Sformo, T., et al., 2016. Severe bone loss as part of the life history strategy of bowhead whales. PLoS ONE 11, e0156753.

Goldbogen, J.A., Cade, D., Calambokidis, J.A., Friedlaender, A.S., Potvin, J., Segre, P.S., et al., 2017. How baleen whales feed: the biomechanics of engulfment and filtration. Ann. Rev. Mar. Sci. 9, 367–386.

Kawamura, A., 1974. Food and feeding ecology in the southern sei whale. Sci. Rep. Whales Res. Inst. 26, 25–144.

Laidre, K.L., Heide-Jørgensen, M.P., Nielsen, T.G., 2007. Role of the bowhead whale as a predator in West Greenland. Mar. Ecol. Prog. Ser. 346, 285–297.

Lambertsen, R.H., Hintz, R.J., Lancaster, W.C., Hirons, A., Kreiton, K.J., Moor, C., 1989. Characterization of the Functional Morphology of the Mouth of the Bowhead Whale, *Balaena mysticetus*, With Special Emphasis on the Feeding and Filtration Mechanisms. Report to the Department of Wildlife Management. North Slope Borough, Barrow.

Lambertsen, R.H., Rasmussen, K.J., Lancaster, W.C., Hintz, R.J., 2005. Functional morphology of the mouth of the bowhead whale and its implications for conservation. J. Mammal. 86, 342–352.

Lubetkin, S.C., Zeh, J.E., Rosa, C., George, J.C., 2008. Age estimation for young bowhead whales (*Balaena mysticetus*) using annual baleen growth increments. Can. J. Zool. 86, 525–538.

Mayo, C.A., Letcher, B.H., Scott, S., 2001. Zooplankton filtering efficiency of the baleen of a North Atlantic right whale, *Eubalaena glacialis*. J. Cetacean Res. Manage. 2, 225–229.

Potvin, J., Werth, A.J., 2017. Oral cavity hydrodynamics and drag production in balaenid whale suspension feeding. PLoS ONE 12 (4), e0175220.

Sanderson, S.L., Roberts, E., Lineburg, J., Brooks, H., 2016. Fish mouths as engineering structures for vortical cross-step filtration. Nat. Commun. 7, e11092.

Scammon, C.M., 1874. The Marine Mammals of the Northwestern Coast of North America. Dover, New York (reprinted 1968).

Simon, M., Johnson, M., Tyack, P., Madsen, P.T., 2009. Behaviour and kinematics of continuous ram filtration in bowhead whales (*Balaena mysticetus*). Proc. R. Soc. B 276, 3819–3828.

St Aubin, D.J., Stinson, R.H., Geraci, J.R., 1984. Aspects of the structure and composition of baleen, and some effects of exposure to petroleum hydrocarbons. Can. J. Zool. 62, 193–198.

van der Hoop, J.M., Nousek-McGregor, A.E., Nowacek, D.P., Parks, S.E., Tyack, P., Madsen, P.T., 2019. Foraging rates of ram-filtering North Atlantic right whales. Funct. Ecol. 33 (7), 1290–1306.

Werth, A.J., 2001. How do mysticetes remove prey trapped in baleen? Bull. Mus. Comp. Zool. 156 (1), 189–203.

Werth, A.J., 2004. Models of hydrodynamic flow in the bowhead whale filter feeding apparatus. J. Exp. Biol. 207 (20), 3569–3580.

Werth, A.J., 2007. Adaptations of the cetacean hyolingual apparatus for aquatic feeding and thermoregulation. Anat. Rec. 290 (6), 546–568.

Werth, A.J., 2013. Flow-dependent porosity of baleen from the bowhead whale (*Balaena mysticetus*). J. Morphol. 216, 1152–1159.

Werth, A.J., Potvin, J., 2016. Baleen hydrodynamics and morphology of crossflow filtration in balaenid whale suspension feeding. PLoS ONE 11 (2), e0150106.

Werth, A.J., Harriss, R.W., Rosario, M.V., George, J.C., Sformo, T.L., 2016a. Hydration affects the physical and mechanical properties of baleen tissue. R. Soc. Open Sci. 3 (10), 160591.

Werth, A.J., Straley, J., Shadwick, R., 2016b. Baleen wear reveals intraoral water flow patterns of mysticete filter feeding. J. Morphol. 277 (4), 453–471.

Werth, A.J., Espada, D.R., Rosario, M.V., Moore, M.J., Sformo, T.L., 2018a. How do baleen whales stow their filter: a comparative biological analysis. J. Exp. Biol. 221, 189233.

Werth, A.J., Potvin, J., Shadwick, R.E., Jensen, M.M., Cade, D.E., Goldbogen, J.A., 2018b. Filtration area scaling and evolution in mysticetes: trophic niche partitioning and the curious cases of the sei and pygmy right whales. Biol. J. Linn. Soc. 125 (2), 264–279.

Werth, A.J., 2019. Variable porosity of throughput and tangential filtration in biological and 3D printed systems. In: Petrova, V.M. (Ed.), Advances in Engineering Research, Vol. 29. Nova Science, New York, pp. 37–88.

Werth, A.J., Blakeney, S.M., Cothren, A.I., 2019. Oil adsorption does not structurally or functionally alter whale baleen. R. Soc. Open Sci. 6 (5), 182194.

Werth, A.J., Whaley, H.R., 2019. Ocean acidification's potential effects on keratin protein in cetacean baleen and other integumentary tissue. Ann. Ecol. Envtl. Sci. 3 (2), 21–28.

Werth, A.J., Sformo, T.L., Lysiak, N.S., Rita, D, George, J.C., 2020. Baleen turnover and gut transit in mysticete whales and its environmental implications. Polar Biol. Available from https://doi.org/10.1007/s00300-020-02673-8.

Würsig, B., Dorsey, E.M., Fraker, M.A., Payne, R.S., Richardson, W.J., 1985. Behavior of bowhead whales, *Balaena mysticetus*, summering in the Beaufort Sea: a description. Fish. Bull. 83, 357–377.

CHAPTER

15

Cardiovascular and pulmonary systems

M.A. Castellini[1] *and P.J. Ponganis*[2]

[1]Graduate School, University of Alaska Fairbanks, Fairbanks, AK, United States
[2]Center for Marine Biotechnology & Biomedicine, Scripps Institution of Oceanography, University of California San Diego, La Jolla, CA, United States

Introduction

We consider the concept that evolutionary adaptation has refined the anatomy and function of the cardiovascular system to best fit the ecological and physiological demands of the bowhead whale. If we design a heart–lung–circulatory system for the whale from first principles, would we derive what we observe in the bowhead, or has evolution taken other paths? (Fig. 15.1).

We explore three drivers that should impact the cardiopulmonary structure and function of the bowhead. First, large body mass and associated low metabolic rate/body mass. Second, diving requirements for oxygen storage and low utilization rates. Finally, a very long life span with implications for cardiac aging and senescence.

In 1993, Haldiman and Tarpley focused on circulatory anatomy and vasculature (Haldiman and Tarpley, 1993). However, at that time, the complex diving patterns of the whales were not well studied (Chapters 4 and 24), diving heart rates were not known, estimates of hydrodynamic requirements and metabolic rate were just beginning (Chapter 16), and their long life spans had not been documented (Chapter 7). We use these new findings to discuss the ecological physiology and comparative biochemistry of the cardiac and pulmonary systems of the bowhead whale.

Cardiovascular and respiratory function

Bowhead whales are slow-swimming continuous filter feeders with low deep body temperatures (regional heterothermy) (Elsner et al., 2004b; Simon et al., 2009). Although gas

The Bowhead Whale
DOI: https://doi.org/10.1016/B978-0-12-818969-6.00015-7

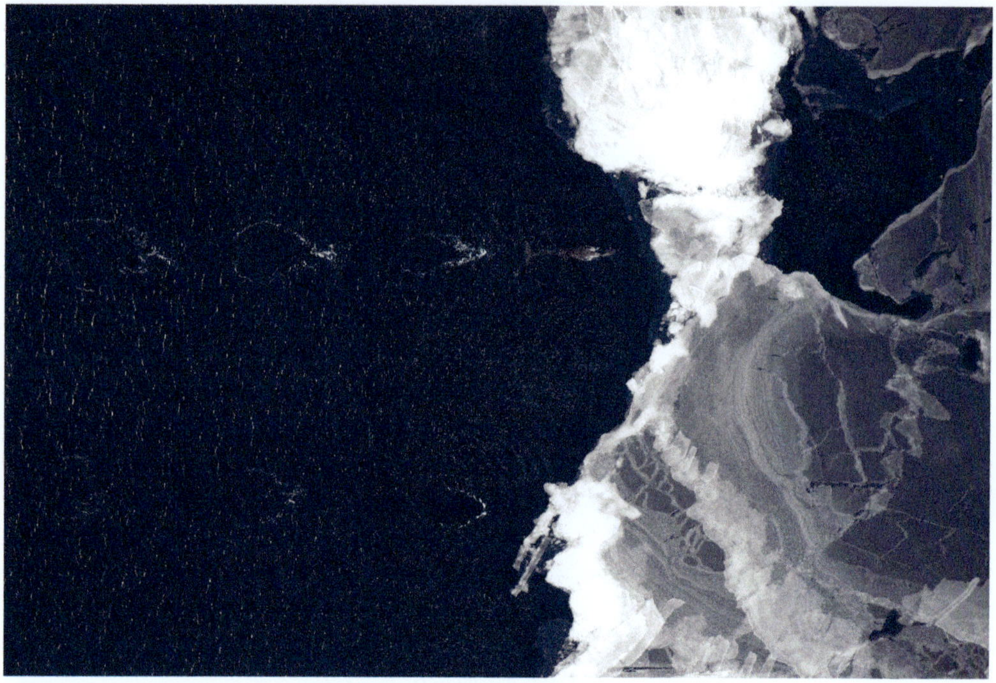

FIGURE 15.1 A bowhead whale is about to disappear underneath the ice, leaving three distinct wakes behind it—evidence of the three breaths it has taken. Wakes lower on the photo indicate the presence of a second whale that has sounded under the ice. *Source: Photo by Vicki Beaver (NOAA/North Slope Borough, NMFS Permit No. 14245).*

exchange must be efficient during surface periods, mass-specific oxygen demand is probably not exceptional and therefore we expect lung volumes to be in the range of 40–60 mL/kg, similar to those of other baleen whales (Kooyman, 1973; Piscitelli et al., 2013). Their wide and short trachea that bifurcates into principal bronchi of equivalent diameter should facilitate airflow into the unicameral (single chamber), rectangular lungs (Henry et al., 1983). Fig. 15.2A–C shows 3D MRI images of a fetal bowhead whale with differing views of the cardiopulmonary assembly. The lungs are highlighted in green.

Presumably, tidal volumes are equivalent to total lung capacity as in other cetaceans (Olsen et al., 1969). The microanatomy of the lung is notable for the apparent absence of bronchioles and the presence of numerous blood vessels beneath the epithelium of the tracheobronchial tree (Henk and Haldiman, 1990). The blood vessels lining the airway are proposed to warm inspired air and/or become engorged at depth to prevent barotrauma (Cozzi et al., 2005; Henk and Haldiman, 1990).

Body mass and diving physiology greatly influence cardiovascular anatomy and function. For an 80,000 kg mammal, allometric equations predict a cardiac mass of 362 kg, and a stroke volume (volume of blood ejected per heartbeat) of 93 L (Holt et al., 1968; Innes et al., 1986). The bowhead heart shares features similar to those found in other whales and marine mammals (Pfeiffer, 1990; Tarpley et al., 1997). These include dorsoventral flattening

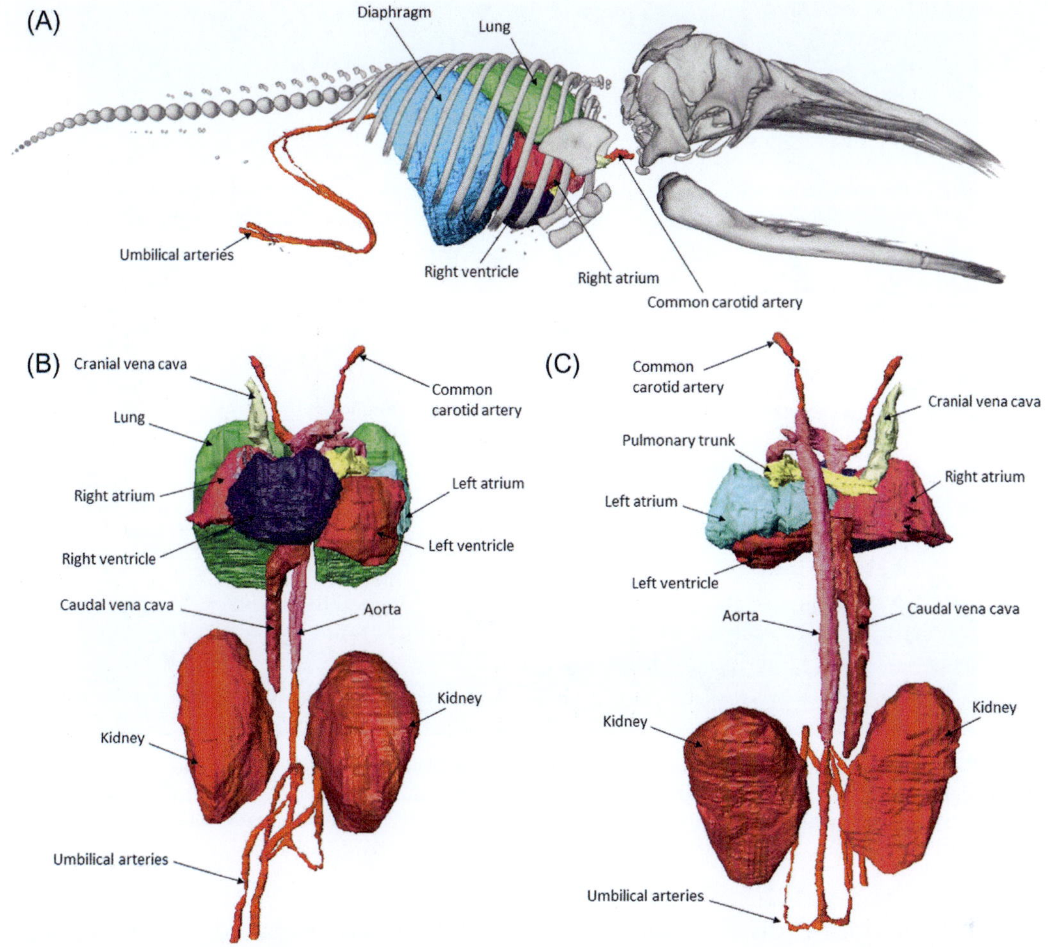

FIGURE 15.2 3D reconstruction of fetal bowhead whale organs (NSB-DWM 1992B7F; 84 cm length; Avizo software). (A) Whole body view, right side. Cardiopulmonary and renal systems in ventral (B) and dorsal (C). CT-scan by D. Hillmann, reconstruction by K. Mars and J. G. M. Thewissen.

of the heart, numerous coronary artery anastomoses (to facilitate blood oxygen delivery to the heart), large Purkinje fibers (for more rapid transmission of cardiac action potentials) to allow efficient contraction of such a large heart (Noujaim et al., 2004), and increased glycogen storage (as an anaerobic energy source) (Pfeiffer, 1990) (Fig. 15.3).

As body mass increases, resting heart rate and metabolic rate decline (Kleiber, 1975; Stahl, 1967) and equations predict a resting heart rate of 14 beats/min (bpm) for an 80,000 kg whale (Stahl, 1967). However, due to the cardiovascular dive response (decrease in heart rate, peripheral vasoconstriction, and redistribution of blood flow), heart rates during dives and surface periods are usually below or above predicted resting levels, respectively (Ponganis et al., 2011). Predicted heart rates in an 80,000 kg bowhead whale

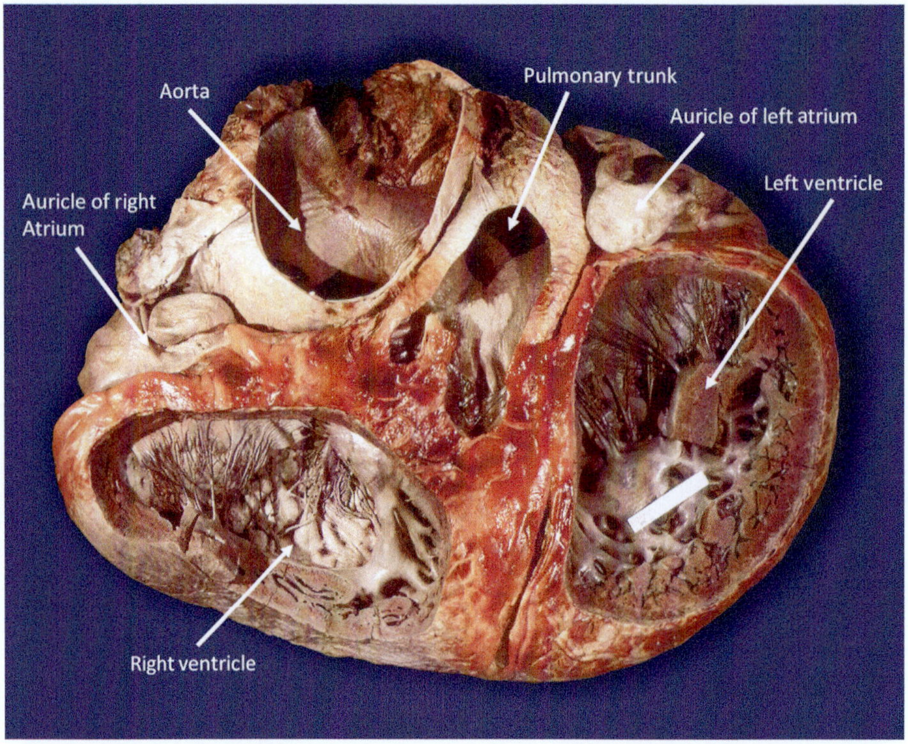

FIGURE 15.3 Cross section of a bowhead whale heart (NSB-DWM 1989B4) in ventral and caudal view with left and right ventricles, right atrium, and pulmonary trunk opened. Scale bar in left ventricle measures 15 cm. Estimated whale body mass of >52,000 kg; measured heart mass of 177 kg. *Source: Dissection by D. Hillman, photo by Geoff Carol (Tarpley et al., 1993). NMFS permits # 314 and #519 to NSB-DWM, T. Albert.*

are <10 bpm during dives, while surface heart rates are expected to be ≥23 bpm (Fig. 15.4). Heart rates in this range have recently been documented during surface and dive intervals in a blue whale (Goldbogen et al., 2019).

We also expect that low muscle metabolic rates due to the slow swim speeds of bowhead whales, combined with a three- to fourfold higher myoglobin concentration than those of large rorqual whales (blue whales (*Balaenoptera musculus*), fin whales (*Balaenoptera physalus*), and sei whales (*Balaenoptera borealis*) (Hochachka and Foreman, 1993; Lawrie, 1953; Noren and Williams, 2000; Tawara, 1950) will provide longer durations of aerobic muscle metabolism. These features minimize the need for adjustments of heart rate and blood flow to provide supplemental blood oxygen to the muscle during a dive (Davis and Williams, 2012; Williams et al., 2015). Furthermore, some data suggest the body temperature of the bowhead is low for a mammal (~34°C) (see Chapter 16). Low temperatures will reduce metabolic rate, and presumably decrease demand for oxygen delivery (cardiac function) to the tissues.

At slow heart rates during a dive, blood pressure and flow must be supported during the long pauses between heartbeats in order to provide perfusion to critical organs. The enlarged, highly compliant, and elastic aortic arch of the fin whale has been postulated to

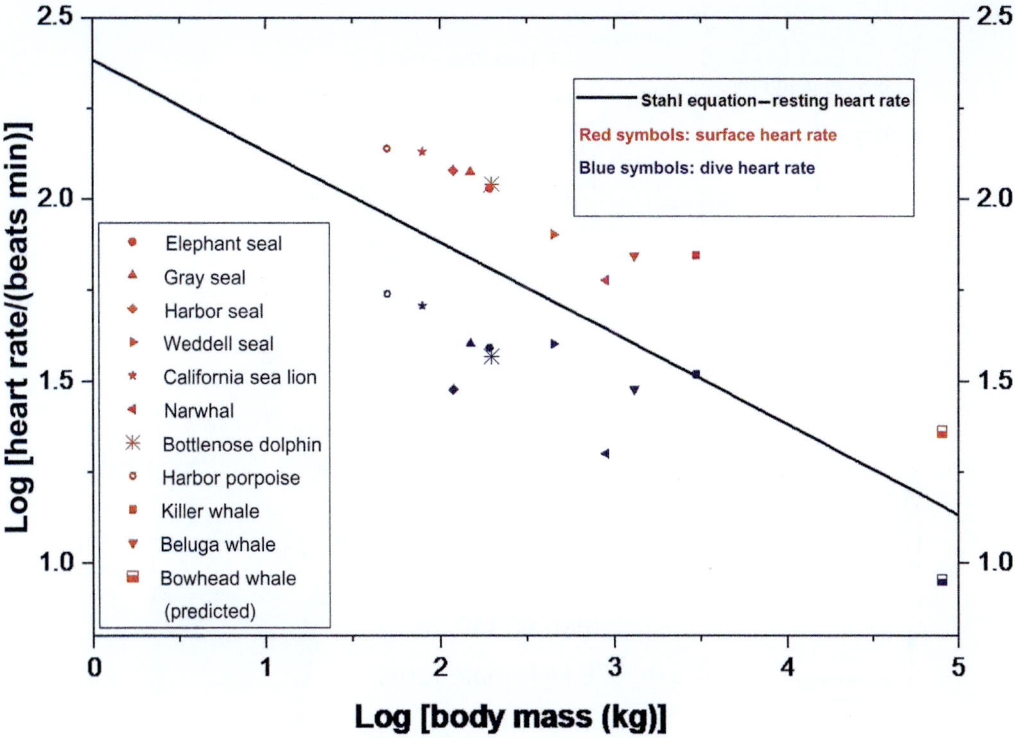

FIGURE 15.4 Log–log plot (base 10) of heart rate relative to body mass using predicted resting heart rate (Stahl, 1967) along with measured surface and dive heart rates of marine mammals (Andrews et al., 1997; Bickett et al., 2019; Fedak, 1986; Hill et al., 1987; McDonald et al., 2018; McDonald and Ponganis, 2014; Thompson and Fedak, 1993; Williams et al., 1999, 2017). Heart rates are from dives of routine duration and considered within the aerobic dive limit of each species. The average surface heart rate was 1.67 × the predicted resting heart rate while that of dive heart rates was 0.64 × the predicted rate. Based on those ratios and Stahl's equation for an 80,000 kg bowhead whale, surface and dive heart rates are predicted to be in the range of 23 and 9 beats/min, respectively.

function as an elastic reservoir (windkessel) to maintain blood pressure and flow during diastole (relaxation phase of the heartbeat) (Shadwick and Gosline, 1994). During systole (ejection phase of the heartbeat), the arch expands to accommodate the stroke volume of the heart, and then, during diastole, it gradually contracts to maintain blood pressure and peripheral flow. Similar to the fin whale, the bowhead whale also has an enlarged aortic arch that is present even in the fetus (Tarpley et al., 1997). Fig. 15.5 shows the large aortic arch system in a fetal bowhead whale. The diameter of the widest section of the fetal aortic arch is 49% greater than that of the descending aorta.

The dimensions and biomechanical properties of the arch and descending aorta of the fin whale have been suggested to decrease aortic impedance and allow for a more efficient circulation during the higher heart rates of the surface period (Shadwick and Gosline, 1994). Aortic impedance is decreased through destructive interference of outgoing and reflected pressure waves at heart rates above 20 bpm. We suspect that similar

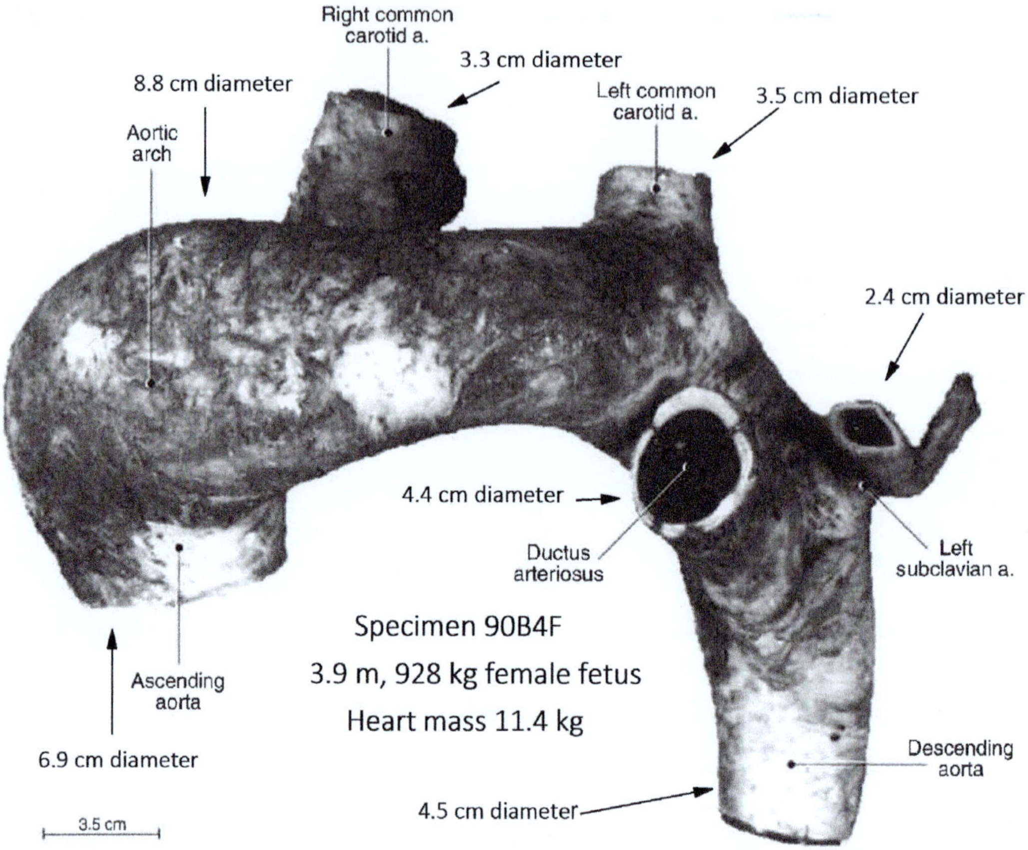

FIGURE 15.5 Enlarged aortic arch in the image from dissected heart specimen of a fetal bowhead whale (NSB-DWM 90B4F) at the level of the ascending aorta distal to the aortic sinuses. *Source: Modified from Tarpley, R. J., Hillmann, D.J., Henk, W.G., 1993. Observations of the external morphology and coronary vasculature of the heart of the bowhead whale* Balaena mysticetus. *Report to the Department of Wildlife Management North Slope Borough, Barrow, AK, p. 245. NMFS permits # 314 and #519 to NSB-DWM, T. Albert.*

characteristics of the bowhead whale aorta provide for decreased cardiac workload and a more efficient circulation during the surface period.

Recent anatomical and modeling studies in cetaceans have investigated the potential significance of elevated intra-abdominal pressures secondary to abdominal compression during movement of the fluke (Lillie et al., 2013, 2017, 2018). Elevations in intra-abdominal pressure could lead to increased inflow from the abdomen into the thorax and overdistention of the heart. In species with more active fluking, diaphragms were stiffer (greater collagen content), but vena caval sphincters (though present in all cetaceans examined) were not more prominent. The authors postulated that extrathoracic vena caval collapse (vascular waterfall) was a more probable mechanism to regulate inflow into the thorax. Secondarily, potential increased retrograde flow from the abdomen toward the heart and brain via the extradural vein due to caval compression may be minimized by the presence

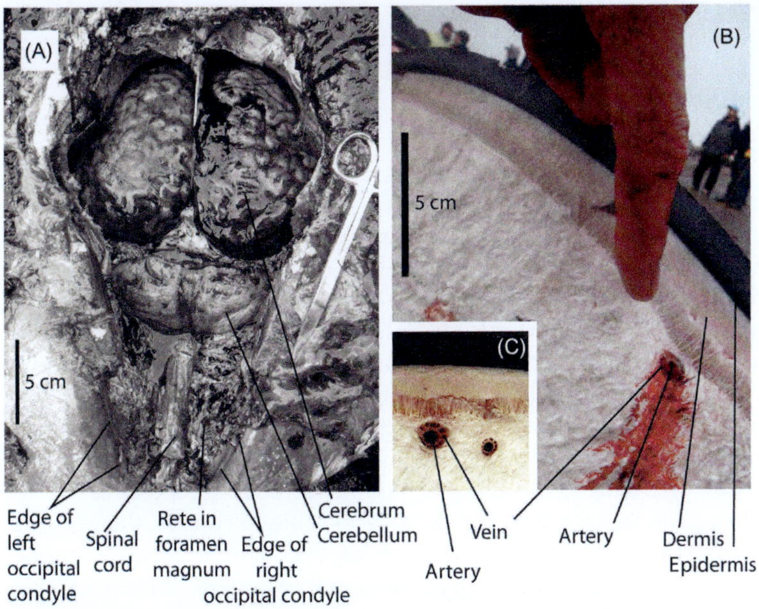

FIGURE 15.6 (A) Dorsal view of bowhead whale brain (with skull roof removed) and foramen magnum, showing the narrow spinal cord located in the enormous foramen magnum, which is otherwise filled by a rete mirabile. (B) Cross section of part of the fluke, showing an artery surrounded by veins, an arrangement that forms a countercurrent heat exchanger. (C) Countercurrent heat exchanger enlarged. *Source: (A) Photo: J. G. M. Thewissen. (B) Photo: J. G. M. Thewissen. (C) Photo: C. George, North Slope Borough.*

of intravertebral retia mirabilia, which are present in the bowhead whale (Pfeiffer and Kinkead, 1990).

Other interesting features of the circulation in the bowhead whale include the spleen and the tail fluke vasculature. The spleen in an adult whale is small (2.5-5 kg in adults) for body mass, and unlikely to serve as a significant blood storage organ (C. George, pers. obs.; Haldiman and Tarpley, 1993). This is an interesting contrast to the large spleen of some deep and long diving seals, where it acts as a reservoir to store viscous blood when red cells are not needed for oxygen delivery during surface periods and can reach 10%–12% of body mass (Castellini and Castellini, 1993; de Silva et al., 2014; Hurford et al., 1996). At that relationship, the spleen for a 40,000 kg whale would weigh 4000 kg! This is an example of a prediction (small spleen = less viscous blood) that is not borne out in bowheads. As in other cetaceans, the tail fluke vasculature is characterized by a deep, heat-conserving, countercurrent exchange arrangement of central arteries surrounded by veins, and a superficial, heat-dissipating arteriovenous anastomotic system (see Fig. 15.6) (Elsner et al., 2004a). Activation of the sympathetic nervous system and constriction of the superficial vessels should allow for heat preservation via the countercurrent exchange system during diving or cold-stress.

Aging and cardiovascular function

Bowhead whales can live up to 200 years of age or older (see Chapters 20 and 21) (George et al., 1999; Keane et al., 2015). There are not enough data to postulate cardiac disease in bowheads, but generally individuals are not found with age-related pathologies (George et al., 1999; Philo et al., 1993; Chapter 30).

Cardiac aging and disease must be either extremely rare or long delayed in bowheads. The oldest whales captured were described as "fast," "vigorous," with "amazing power" while swimming, diving, or foraging and did not show evidence of loss of exercise function (Craig George and Billy Adams, pers. comm., 2019, Utqiaġvik, Alaska).

Structural adaptations

In healthy humans, there are suites of cardiac changes that occur with age. By about 80 year, human hearts show an increase in stiffness and a loss of elasticity in the aorta. This tends to produce left ventricular hypertrophy and increases in systolic blood pressure. Along with a loss of pacemaker cells, the resting heart rate decreases and the ability to increase HR during exercise (scope) is reduced (Cheitlin, 2003) These changes would not be adaptive for a long-lived diving mammal. The bowhead whale demonstrates adaptations that either mitigate these aging problems or strongly delay them. For example, the highly expandable aortic branch/bulb directly counteracts the effects of a relatively rigid, aged aorta. Further, the fine neural control of heart rate during diving and surfacing periods directly runs counter to the age-induced loss of pacemaker cells, and a diving whale must be able to control alterations in blood pressure. Finally, marine mammals as a group exhibit low aerobic scope (animal's capacity to increase its aerobic metabolic rate above maintenance levels) during exercise (Elsner, 1988), and may already be "aged" by comparison with terrestrial mammals.

Molecular adaptations

Evidence suggests that enhanced DNA damage repair mechanisms (telomere replacement) are involved in longevity of some species, though there are inconsistencies in that pattern (see Chapter 20) (Jarman et al., 2015; Olsen et al., 2012). Telomere chemistry impacts heart cellular function as cardiac tissue ages in mammals (Anversa et al., 2005). We predict substantial molecular aging protective mechanisms in bowhead cardiac tissue, especially in telomere biochemistry. However, this has not yet been investigated.

We speculate that the cardiac adaptations that revolve around diving, large mass, and low metabolic rate may counter the impacts of aging. While diving adaptations alone might not be enough to impart "old age tolerance" in bowhead whales, the combination of these along with low metabolic rate/mass and low body temperature may form a suite of adaptations that are consistent with longevity.

Summary

The bowhead cardiopulmonary system exhibits adaptations in structure and function for diving, large mass, low metabolic rate, and longevity. Some cardiopulmonary patterns are expected for a diving mammal (bradycardia), but others are not (small spleen). Some are consistent with all cetaceans (fluking influence on caval blood flow) and others may be unique (cardiac longevity). Some adaptations may be under several evolutionary pressures (e.g., large and compliant aortic arch) that not only affords a benefit to diving (smoothing out pulsatile blood flow from the heart) but may also provide an antiaging benefit against stiffening blood vessels.

While this chapter provides an overview of bowhead whale cardiopulmonary features, there remain interesting questions for future investigations including, diving heart rate, blood volume, neural patterns in the heart through high-resolution ECGs, cardiac disease, and the impacts of aging on heart and blood vessels.

Acknowledgments

We thank the many hunters, local community members and scientific teams that provided information, samples, images and history for this chapter. Special thanks to editors Thewissen and George for their time, interest and help. Support to PJP from ONR grant N00014-19-1-2455.

References

Andrews, R.D., Jones, D.R., Williams, J.D., Thorson, P.H., Oliver, G.W., Costa, D.P., et al., 1997. Heart rates of northern elephant seals diving at sea and resting on the beach. J. Exp. Biol. 200, 2083–2095.

Anversa, P., Rota, M., Urbanek, K., Hosoda, T., Sonnenblick, E.H., Leri, A., et al., 2005. Myocardial aging—a stem cell problem. Basic Res. Cardiol. 100, 482–493.

Bickett, N.J., Tift, M.S., St. Leger, J., Ponganis, P.J., 2019. Heart rates, heart rate profiles, and electrocardiograms in three killer whales, a beluga, and a pilot whale: an exploratory investigation. Mar. Mamm. Sci. 35, 1112–1132.

Castellini, J.M., Castellini, M.A., 1993. Estimation of splenic volume and its relationship to long-duration apnea in seals. Physiol. Zool. 66, 619–627.

Cheitlin, M.D., 2003. Cardiovascular physiology-changes with aging. Am. J. Geriatr. Cardiol. 12, 9–13.

Cozzi, B., Bagnoli, P., Acocella, F., Costantino, M.L., 2005. Structure and biomechanical properties of the trachea of the striped dolphin *Stenella coeruleoalba*: evidence for evolutionary adaptations to diving. Anat. Rec. 284, 500–510.

Davis, R.W., Williams, T.M., 2012. The marine mammal dive response is exercise modulated to maximize aerobic dive duration. J. Comp. Physiol. A 198, 583–591.

e Silva, F.M.D.O., Carvalho, V.L., Guimarães, J.P., Vergara-Parente, J.E., de Meirelles, A.C.O., Marmontel, M., et al., 2014. Accessory spleen in cetaceans and its relevance as a secondary lymphoid organ. Zoomorphology 133, 343–350.

Elsner, R., 1988. Anaerobic contributions to metabolism in diving seals. Can. J. Zool. 66, 142–143.

Elsner, R., George, J.C., O'Hara, T., 2004a. Vasomotor responses of isolated peripheral blood vessels from bowhead whales: thermoregulatory implications. Mar. Mamm. Sci. 20, 546–553.

Elsner, R., Meiselman, H.J., Baskurt, O.K., 2004b. Temperature-viscosity relations of bowhead whale blood: a possible mechanism for maintaining cold blood flow. Mar. Mamm. Sci. 20, 339–344.

Fedak, M.A., 1986. Diving and exercise in seals: a benthic perspective. In: Brubakk, A.O., Kanwisher, J.W., Sudnes, G. (Eds.), Diving in Animals and Man. Tapir Publishers, Trondheim, pp. 11–32.

George, J.C., Bada, J., Zeh, J., Scott, L., Brown, S.E., O'Hara, T., et al., 1999. Age and growth estimates of bowhead whales (*Balaena mysticetus*) via aspartic acid racemization. Can. J. Zool. 77, 571–580.

Goldbogen, J.A., Cade, D.E., Calambokidis, J., Czapanskiy, M.F., Fahlbusch, J., Friedlaender, A.S., et al., 2019. Extreme bradycardia and tachycardia in the world's largest animal. Proc. Natl. Acad. Sci. U. S. A. 116, 25329–25332.

Haldiman, J.T., Tarpley, R.J., 1993. Anatomy and physiology. In: Burns, J.J., Montague, J.J., Cowles, C.J. (Eds.), The Bowhead Whale. Allen Press, Lawrence, KS, pp. 71–150.

Henk, W.G., Haldiman, J.T., 1990. Microanatomy of the lung of the bowhead whale *Balaena mysticetus*. Anat. Rec. 226, 187–197.

Henry, R.W., Haldiman, J.T., Albert, T.F., Henk, W.G., Abdelbaki, Y.Z., Duffield, D.W., 1983. Gross anatomy of the respiratory system of the bowhead whale, *Balaena mysticetus*. Anat. Rec. 207, 435–449.

Hill, R.D., Schneider, R.C., Liggins, G.C., Schuette, A.H., Elliott, R.L., Guppy, M., et al., 1987. Heart rate and body temperature during free diving of Weddell seals. Am. J. Physiol. 253, R344–R351.

Hochachka, P.W., Foreman, R.A., 1993. Phocid and cetacean blueprints of muscle metabolism. Can. J. Zool. 71, 2089–2098.

Holt, J.P., Rhode, E.A., Kines, H., 1968. Ventricular volumes and body weight in mammals. Am. J. Physiol. 215, 704–715.

Hurford, W.E., Hochachka, P.W., Schneider, R.C., Guyton, G.P., Stanek, K.S., Zapol, D.G., et al., 1996. Splenic contraction, catecholamine release, and blood volume redistribution during diving in the Weddell seal. J. Appl. Physiol. 80, 298–306.

Innes, S., Lavigne, D.M., Earle, W.M., Kovacs, K.M., 1986. Estimating feeding rates of marine mammals from heart mass to body mass ratios. Mar. Mamm. Sci. 2, 227–229.

Jarman, S.N., Polanowski, A.M., Faux, C.E., Robbins, J., De Paoli-Iseppi, R., Bravington, M., et al., 2015. Molecular biomarkers for chronological age in animal ecology. Mol. Ecol. 24, 4826–4847.

Keane, M., Semeiks, J., Webb, A.E., Li, Y.I., Quesada, V., Craig, T., et al., 2015. Insights into the evolution of longevity from the bowhead whale genome. Cell Rep. 10, 112–122.

Kleiber, M., 1975. The Fire of Life: An Introduction to Animal Energetics. Kreiger, New York, p. 454.

Kooyman, G.L., 1973. Respiratory adaptations in marine mammals. Am. Zool. 13, 457–468.

Lawrie, R., 1953. The activity of the cytochrome system in muscle and its relation to myoglobin. Biochem. J. 55, 298–305.

Lillie, M.A., Piscitelli, M.A., Vogl, A.W., Gosline, J.M., Shadwick, R.E., 2013. Cardiovascular design in fin whales: high-stiffness arteries protect against adverse pressure gradients at depth. J. Exp. Biol. 216, 2548–2563.

Lillie, M.A., Vogl, A.W., Raverty, S., Haulena, M., McLellan, W.A., Stenson, G.B., et al., 2017. Controlling thoracic pressures in cetaceans during a breath-hold dive: importance of the diaphragm. J. Exp. Biol. 220, 3464–3477.

Lillie, M.A., Vogl, A.W., Raverty, S., Haulena, M., McLellan, W.A., Stenson, G.B., et al., 2018. The caval sphincter in cetaceans and its predicted role in controlling venous flow during a dive. J. Exp. Biol. 221. Available from: https://doi.org/10.1242/jeb.177212.

McDonald, B.I., Johnson, M., Madsen, P.T., 2018. Dive heart rate in harbour porpoises is influenced by exercise and expectations. J. Exp. Biol. 221.

McDonald, B.I., Ponganis, P.J., 2014. Deep-diving sea lions exhibit extreme bradycardia in long-duration dives. J. Exp. Biol. 217, 1525–1534.

Noren, S.R., Williams, T.M., 2000. Body size and skeletal muscle myoglobin of cetaceans: adaptations for maximizing dive duration. Comp. Biochem. Physiol. 126, 181–191.

Noujaim, S.F., Lucca, E., Munoz, V., Persaud, D., Berenfeld, O., Meijler, F.L., et al., 2004. From mouse to whale: a universal scaling relation for the PR interval of the electrocardiogram of mammals. Circulation 110, 2802–2808.

Olsen, C.R., Elsner, R., Hale, F.C., Kenney, D.W., 1969. "Blow" of the pilot whale. Science 163, 953–955.

Olsen, M.T., Berube, M., Robbins, J., Palsboll, P.J., 2012. Empirical evaluation of humpback whale telomere length estimates; quality control and factors causing variability in the singleplex and multiplex qPCR methods. BMC Genet. 13, 77.

Pfeiffer, C.J., 1990. Observations on the ultrastructural morphology of the bowhead whale (*Balaena mysticetus*) heart. J. Zoo Wildl. Med. 48–55.

Pfeiffer, C.J., Kinkead, T.P., 1990. Microanatomy of retia mirabilia of bowhead whale foramen magnum and mandibular foramen. Acta Anat (Basel) 139, 141–150.

Philo, L.M., Shotts, E., George, J.C., 1993. Morbidity and mortality. The bowhead whale. Soc. Mar. Mammal. 275–312. Spec Publ.

Piscitelli, M.A., Raverty, S.A., Lillie, M.A., Shadwick, R.E., 2013. A review of cetacean lung morphology and mechanics. J. Morphol. 274, 1425–1440.

Ponganis, P.J., Meir, J.U., Williams, C.L., 2011. In pursuit of Irving and Scholander: a review of oxygen store management in seals and penguins. J. Exp. Zool. 214, 3325–3339.

Shadwick, R.E., Gosline, J.M., 1994. Arterial mechanics in the fin whale suggest a unique hemodynamic design. Am. J. Physiol. 267, R805–R818.

Simon, M., Johnson, M., Tyack, P., Madsen, P.T., 2009. Behaviour and kinematics of continuous ram filtration in bowhead whales (*Balaena mysticetus*). Proc. R. Soc. Edinb. Biol. 276, 3819–3828.

Stahl, W.R., 1967. Scaling of respiratory variables in mammals. J. Appl. Physiol. 22, 453–460.
Tarpley, R.J., Hillmann, D.J., Henk, W.G., 1993. Observations of the external morphology and coronary vasculature of the heart of the bowhead whale *Balaena mysticetus*. Report to the Department of Wildlife Management North Slope Borough, Barrow, AK, p. 245.
Tarpley, R.J., Hillmann, D.J., Henk, W.G., George, J.C., 1997. Observations on the external morphology and vasculature of a fetal heart of the bowhead whale, *Balaena mysticetus*. Anat. Rec. 247, 556–581.
Tawara, T., 1950. On the respiratory pigments of whale (Studies on whale blood II.). Sci. Rep. Whales Res. Inst. 3, 95–101.
Thompson, D., Fedak, M.A., 1993. Cardiac responses of grey seals during diving at sea. J. Exp. Biol. 174, 139–164.
Williams, T.M., Blackwell, S.B., Richter, B., Sinding, M.-H.S., Heide-Jørgensen, M.P., 2017. Paradoxical escape responses by narwhals (*Monodon monoceros*). Science 358, 1328–1331.
Williams, T.M., Fuiman, L.A., Kendall, T., Berry, P., Richter, B., Noren, S.R., et al., 2015. Exercise at depth alters bradycardia and incidence of cardiac anomalies in deep-diving marine mammals. Nat. Commun. 6, 6055. Available from: https://doi.org/10.1038/ncomms7055.
Williams, T.M., Haun, J.E., Friedl, W.A., 1999. The diving physiology of bottlenose dolphins (*Tursiops truncatus*). I. Balancing the demands of exercise for energy conservation at depth. J. Exp. Biol. 202, 2739–2748.

CHAPTER

16

Thermoregulation and energetics

J.C. George[1], Lara Horstmann[2], S. Fortune[3], Todd L. Sformo[1], Robert Elsner[2,*] and Erich Follmann[4,*]

[1]Department of Wildlife Management, North Slope Borough, Utqiaġvik, AK, United States [2]College of Fisheries and Ocean Sciences, University of Alaska Fairbanks, Fairbanks, AK, United States [3]Institute for Oceans and Fisheries, Marine Mammal Research Unit, University of British Columbia, Vancouver, BC, Canada [4]Institute of Arctic Biology, University of Alaska Fairbanks, Fairbanks, AK, United States

Introduction

The energy balance of any animal is maintained by matching energy intake with energy expenditures over time. For large whales, the general strategy is to optimize energy intake by selecting seasonally high caloric food in large quantities in areas where it is abundant (Lockyer, 1981; Thomson, 2002; Costa and Williams, 1999). Once found, they fine-tune their energy and time expended during foraging, such as by adjusting short-shallow and long-deep dives (Matthews and Ferguson, 2015). Baleen whales are arguably the most efficient foragers among vertebrates by exploiting abundant small prey along with developing extreme body size (Goldbogen et al., 2019). How large mysticetes locate these high-density patches is not fully understood, but oceanographic cues, maternally driven site fidelity to traditional feeding areas are important, and olfaction may play a role as well (Chapters 4, 24, 28, this volume; Moore and Reeves, 1993; Baumgartner and Mate, 2003; Thewissen et al., 2011). While the mechanics of prey ingestion varies amongst mysticetes, balaenids (bowhead and right whales) strain prey-laden water with their baleen as they swim (see Chapter 14). Because filter-feeding results in increased drag and energy expenditure, successful foraging requires that the energetic density of bowhead prey exceeds the costs incurred while feeding (van der Hoop et al., 2019).

Large cetaceans have unique energetic challenges due to their massive size, the high thermal conductivity of water with subsequent heat loss in polar seas, variable and

* Posthumous.

The Bowhead Whale
DOI: https://doi.org/10.1016/B978-0-12-818969-6.00016-9

seasonal food supplies, anthropogenic disturbances, predators, and many other factors (Matthews and Ferguson, 2015; Elsner, 1999; Peters, 1989; Young et al., 2020). Even among other large cetaceans, bowheads are unusual in several physiological and anatomical traits that affect energetics, including greater blubber thickness, low body temperature, and feeding in winter (Hokkanen, 1990; George et al., 2009a; George et al., 1999; Koski et al., 1993; Haldiman and Tarpley, 1993; Matthews and Ferguson, 2015; Pomerleau et al., 2018). Furthermore, three of the four stocks remain in arctic waters with temperatures at or below 0°C much of the year with brief periods of exposure to relatively warm waters to 10°C during summer (Fig.16.1; Chapter 25). All of these factors led to unique energetic adaptations and challenges for bowhead whales.

Understanding the energetic requirements and nutritional status of marine mammal populations has become increasingly important from a conservation perspective (Braithwaite et al., 2015; Villegas-Amtmann et al., 2015). Most estimates of the nutritional requirements of large marine mammals are based on energetic models (Lockyer, 1981; Fortune et al., 2013). State and Federal managers, as well as the Alaska Eskimo Whaling Commission, can use this information to determine where key foraging habitats occur throughout their range and regulate human activities such as commercial fishing, vessel

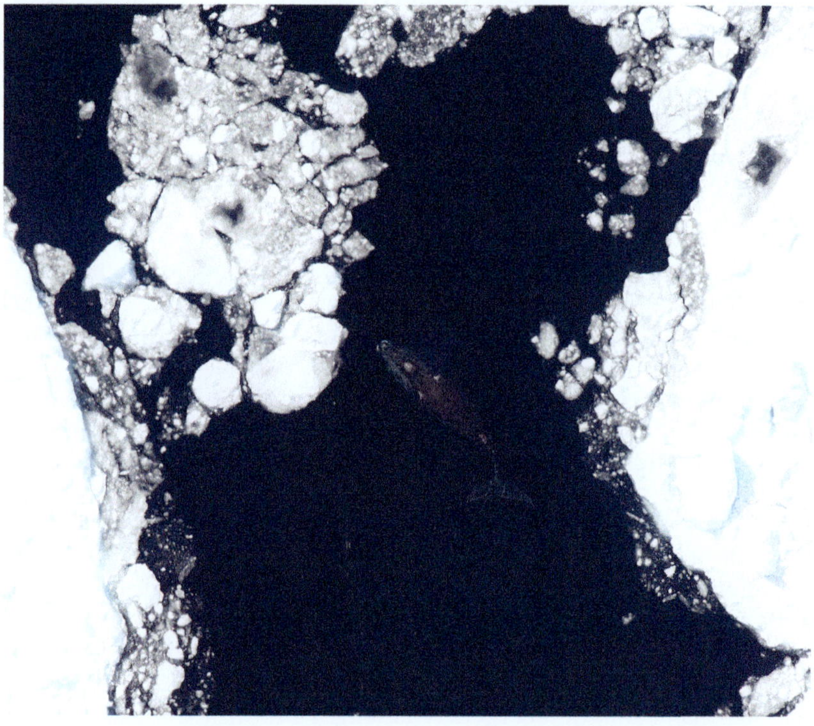

FIGURE 16.1 Aerial photograph of a bowhead whale on migration through ice-leads in the NE Chukchi Sea, off the Alaskan coast. The seawater temperatures in this photograph are near freezing (−1.8°C, 28.7°F), but the whale is not cold-stressed, and in fact may have to shed heat when migrating to maintain normal body temperature. *(Source: North Slope Borough Photo ID Survey, NMFS permit No. 14245).*

traffic, and oil and gas exploration that might affect foraging success. (see Chapter 38; Noren, 2011; Reeves et al., 2012). The energetic implications of climate-induced shifts in prey quality and quantity may be of increasing concern for arctic cetaceans such as the bowhead whale (see Chapter 27).

Direct measurements of metabolic rates of bowheads and other large whales are not available due to logistical constraints, and captive studies are not feasible. Consequently, we use indirect methods of estimating metabolic rates that include allometric models of mass and basal metabolic rate (BMR) (Kleiber, 1975), estimation of ingestion rates, tidal volume and respiration rates, and heat loss models. Two of the most important parameters for estimating energy requirements are body mass and BMR—again, both are logistically difficult to measure in the field for large whales (Fortune et al., 2013).

Collaboration with subsistence hunters has provided a unique opportunity to research the morphometrics, body composition, and thermoregulation of bowhead whales (Chapter 34). For example, with access to fresh, landed animals, we were able to obtain direct measurements of internal and regional body temperatures (George et al., 2009a,b), which would otherwise be impossible to measure. We also include relevant information on bowhead body mass, the thermoconductivity of blubber, lung volumes, thermoregulatory mechanisms in the flukes, and several other aspects of bowhead physiology; all of which are needed for estimating metabolic rates.

A full energetic treatment of a free-ranging bowhead whale, or population of whales, is beyond the scope of this chapter (Costa and Williams, 1999; Noren, 2011; Jeanniard-du-Dot et al., 2017). Instead we review some aspects of bowhead energetics including body temperatures, body mass, thermoregulation, and resting metabolic rate (RMR) predictions by various researchers using several methods (Brodie, 1981; Thomson, 2002; George et al., 2009a,b). Lastly, we identify important areas for future research.

Body temperature

The concept of body temperature is complex in living animals. Temperatures vary by body region, age, activity, and a myriad of interactions between these factors (Schmidt-Nielsen, 1997). However, deep body (core) temperatures may only vary by a few degrees in healthy nonhibernating mammals including humans (Guyton, 1968; Prosser, 1991; Kleiber, 1975; Schmidt-Nielsen, 1997). Nonetheless, body core temperatures vary by species and body mass and have implications for RMR, longevity, and other traits (Conti et al., 2006; Schmidt-Nielsen, 1997).

A cooperative study with Iñupiat whale hunters offered an rare opportunity to measure body temperatures of recently harvested bowhead whales (George et al., 2009a,b; Chapter 34). The core body temperature of harvested bowhead whales was 33.8°C on average ($N = 28$ individuals; $SD = 0.83$; range 32.4°C–35.3°C) (George et al., 2009a). Core temperatures were a few degrees lower for bowhead than other cetaceans, and most other nonhibernating eutherian mammals, with mean core body temperatures of 38°C ± 2°C (Schmidt-Nielsen, 1997; Prosser, 1991). For comparison, postmortem body temperatures of some other large whales include the following: minke whales (*Balaenoptera acutorostrata*), 34.7°C ($SD = 0.8$) (Folkow and Blix, 1992; Vongraven et al., 1990), fin whales (*Balaenoptera physalus*), 36.1°C (Brodie and Paasche, 1985), and humpback whales (*Megaptera novaeangliae*), 34.7°C (Morrison, 1962). Lower body temperatures in mammals and other vertebrates are associated with lower metabolic rates and energy conservation strategies (Schmidt-Nielsen, 1997).

Heat loss measurements

Direct heat loss measurements from large whales are few due to logistical constraints but are useful in theoretical models by providing important insights into cold water adaptations (Marchand, 1996; Kvadsheim et al., 1997).

George et al. (2009a) reported a time series of temperature measurements using a data logger for bowhead whales taken immediately after death. For two whales, with a time series of measurements over 5 hours, the core temperatures were quite stable, where one animal cooled 0.39°C (whale 98B10), while another whale (98BSL) warmed 0.84°C (Fig. 16.2). These data suggest that bowheads lose heat very slowly (0.1°C/h) after death—even in very cold water (~3°C). Similarly, Innes (1986) measured core temperatures of a 13-m beach-cast fin whale 8 days after death and found it had only cooled to 28°C, which could be explained by the slow heat loss and thermal inertia of a large body mass.

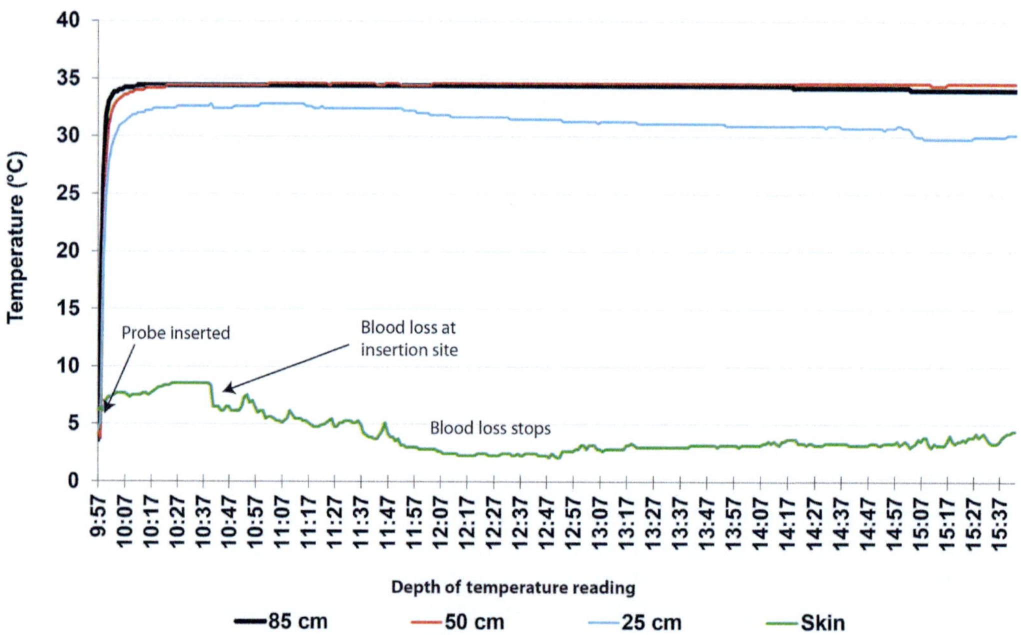

FIGURE 16.2 Time series (5 h) of body temperatures at various depths within a bowhead whale (NSB-DWM 98B10, a 13.0 m, female) started immediately after death. Note that the skin temperature was elevated initially by blood loss through the probe insertion site, and then stabilized at the ambient sea temperature (~3°C–4°C). The body temperature at 85 cm depth and 50 cm differed only slightly and essentially did not change over time. The temperature in the muscle (25 cm depth) was several degrees cooler than core temperature and did exhibit some cooling.

Our observations are consistent with the traditional knowledge of whale hunters (Albert, 2001; Chapter 34). Hunters know that unless a bowhead is butchered and the blubber removed in less than ~12 hours, it will not cool and will decompose internally. As Captain Edward Hopson, Sr., of Utqiaġvik, told us years ago when we conducted this study, *"You will learn that the bowhead whale is a thermos bottle."* The Iñupiat term for a dead bloated whale is *avataayyuniq* (ah-va-Tie-un-nik). Although the muscle and internal organs become necrotic, their *maktak* (skin and blubber) remains cool and is edible.

Regional heterothermy

Heterothermy is a condition in which different body regions are routinely held at different temperatures at the same time. To some extent, all mammals exhibit a degree of heterothermy, but it is more highly developed in arctic mammals including bowheads (Albert and Panuska, 1978; Schmidt-Nielsen, 1997; Elsner, 1999). The core temperature of nonhibernating mammals remains relatively constant; however, an animal in heat balance must always have a thermal gradient away from its core to the surface or it would overheat (Schmidt-Nielsen, 1997).

To document temperature gradients across the body, George et al. (2009b) reported postmortem temperature measurements at 33 different locations across the whale ($n = 6$). They found a wide range of temperatures suggesting extreme regional-heterothermy in bowhead whales (Fig. 16.3). In all mammals, surface and appendage temperatures tend to be lower than core temperatures, but the temperature difference in bowhead whales is marked (Williams et al., 1999). Their skin temperature is held near ambient water temperature which is around 0°C much of the year. If the skin temperature was significantly elevated above ambient, even a large animal like the bowhead would rapidly lose heat (Schmidt-Nielsen, 1997).

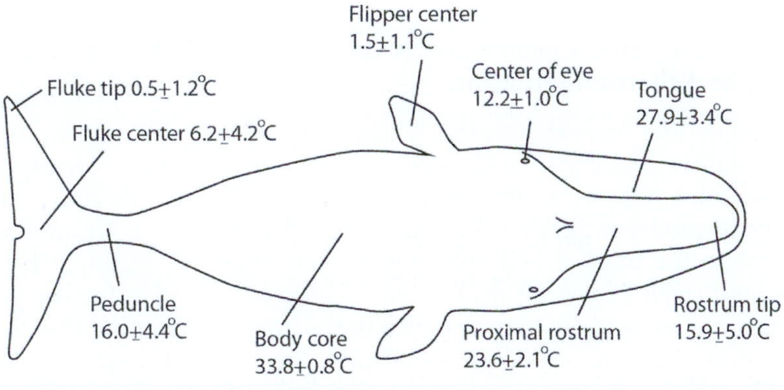

FIGURE 16.3 Sketch of an adult bowhead whale showing the average temperatures found across various parts of the body taken beneath the skin, or against the bone (rostrum, flipper), or deep in the connective tissue (fluke). These temperatures were taken several hours postmortem, hence, the peripheral temperatures are somewhat cooler than in a living whale; however, the internal temperatures are quite stable over time (George et al., 2009b; NSB unpublished data.).

The peripheral temperatures of the flukes and pectoral fins (flipper shown in Fig. 16.3 are affected to some degree by postmortem time. A temperature taken at the center of the fluke was much higher than at the flukes' distal tip. This is higher than those taken a few hours later during the harvest. Nevertheless, these measurements reflect *relative* temperature differences and the data indicate that the deep body temperature changes very little over time. The exact temperatures of these body regions in a living bowhead may differ from those reported here, but they illustrate the wide variation in temperatures across the body.

We found that bowhead whale appendages, flippers, and flukes are maintained at surprisingly low temperatures which is consistent with other cetaceans (Williams et al., 1999). Compared with the rest of their body, bowhead whale flukes have a very thin blubber layer (1.5 cm) sandwiched between the epidermis and a tendinous layer (Haldiman and Tarpley, 1993; Elsner et al., 2004). The thermal conductivity of the blubber of the flukes was higher (0.26 W/m K) than that of the thorax (0.20 W/m K). The flukes also have sophisticated countercurrent heat exchange mechanisms, which are important for conserving or dissipating heat depending on environmental conditions and the behavioral state of the animal (Blix, 2016) (see *Anatomical Specializations* section below; see Chapter 15).

Cooling the appendages (by restricting blood flow) and other regions of the body will reduce overall metabolic demands. On the other hand, bowheads also need effective mechanisms to dissipate heat when highly active, such as during migration and feeding (Hokkanen, 1990). This is accomplished by moving blood to their periphery including the flukes, tongue, and epidermis (see *Anatomical Specializations* section below) through countercurrent heat exchange.

With regard to core body temperatures (T_{bc}), George et al. (2009a) found fairly uniform temperatures among whales; however, the larger and presumably older whales had slightly lower body temperature than smaller whales, which is consistent with humans and other mammals (Waalen and Buxbaum, 2011). Consequently, large and old bowheads may have lower metabolic rates than predicted by Kleiber's law.

Thick blubber and thermoregulation

Bowhead whales live in near-freezing water most of the year, so one might suspect them to be constantly challenged with heat conservation and have correspondingly elevated metabolic rates. However, the opposite may be the case in some circumstances. Hokkanen's (1990) model suggested that bowhead whales could withstand a 200°C thermal gradient across the blubber and still remain thermoneutral, stating, "*The insulation of the bowhead is good enough to enable it to swim in liquid oxygen.*" He further predicted that bowhead whales are more likely to experience issues with overheating due to activity and thus need to find ways to dissipate significant amounts of heat (Hokkanen, 1990). However, Hokkanen's predictions were based on metabolic rates estimated using the standard Kleiber prediction, whereas our data suggest bowhead metabolic rates fall below classic predictions (see *Resting Metabolic Rates* section).

Blubber serves several functions, including buoyancy, streamlining, thermoregulation, energy storage, and as an endocrine organ (Pabst et al., 1999). Strong thermal gradients across the blubber help maintain core temperature (Fig. 16.4). The temperature gradient extends through the muscle and into the body, and while the temperature varies by location on the whale, it was roughly 25°C at the muscle–blubber interface for several whales (Fig. 16.4). In a calibration study of heat-loss energetic models using captive harp seals (*Pagophilus groenlandicus*), Kvadsheim et al. (1997) found that the temperature of the muscle–blubber interface was lower than the body core temperature and significantly affected model performance. Earlier studies assumed the temperature at the muscle–blubber interface to be the same as the core temperature, leading to an overestimate of heat loss and hence metabolic rates. Once this and other adjustments were made, the models performed well against measured metabolic rates of live animals.

As early as the 1960s, researchers suggested that the blubber of large whales serves primarily as energy storage and secondarily for thermoregulation (Kanwisher and Sundnes, 1965; Kanwisher and Ridgway, 1983). Among other things, these authors noted that small cetaceans, such as harbor porpoise (*Phocoena phocoena*) with comparatively thin blubber (approximately 2 cm), successfully inhabit polar seas. Similarly, arctic ringed seals (*Pusa hispida*) inhabit near-freezing seawater year-round with only about 5 cm of blubber (Schmidt-Nielsen, 1997). Burns (1993) suggested the primary function of the bowhead's extremely thick blubber is an energetic buffer against seasons or extended periods with low prey abundance (Chapter 7). Bowhead whales live in a less predictable environment than other large whale species with regard to food availability; hence, it is likely that the primary function of their very thick blubber is energy storage and secondarily for thermoregulation.

The seasonal migration by large whales to low latitudes has classically been proposed as a means to avoid the thermoregulatory stress of polar seas during winter

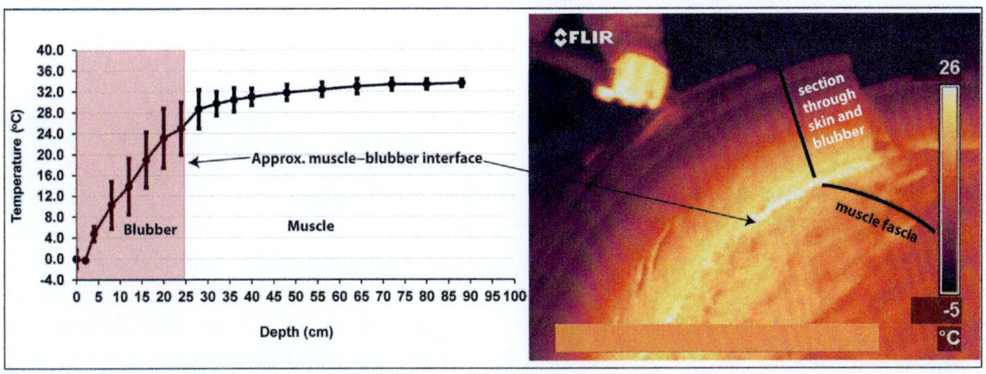

FIGURE 16.4 Left: Plot of the average temperature gradient of measurements taken near the umbilicus for seven whales. Note the temperature gradient extends through the muscle (>~25 cm) and into the body. Right: Image from a "forward-looking infrared radiometer" (FLIR) of a bowhead whale (NSB-DWM 04B10, 7.9 m male) taken as blubber is being removed by subsistence hunters. The temperature of the skin is slightly below 0°C in most cases and the muscle–blubber interface varies by location in the whale. For the umbilicus area, the average temperature at the muscle–blubber interface was 25°C ± 5°C ($n = 7$). The FLIR temperatures agree well with direct blubber measurements taken using a temperature probe. Note the warm gloveless hand (upper left) of the whale hunter flensing the whale.

(Kenney et al., 2001; Dingle, 1996). For example, baleen whales typically migrate considerable distances to warmer, but less productive areas for calving and nursing (Braithwaite et al., 2015). This habitat shift is thought necessary to protect neonates with thin blubber against hypothermia in cold water (Sterns and Friedlander, 2018). However, observers reported occasional births of North Atlantic right whales (*Eubalaena glacialis*) in the Gulf of Maine during the winter—far north of their typical calving grounds off the Florida coast (Patrician et al., 2009). Furthermore, a number of alternative hypotheses have been proposed for seasonal migration to warmer waters, such as predator avoidance (e.g., killer whales, *Orcinus orca*) and facilitating epidermal molt necessary for skin maintenance (Chernova et al., 2016; Durban and Pitman, 2012; Corkeron and Connor, 1999); energetic models suggest that most large whales could tolerate cold seas during winter; however, their neonates may not be as capable. With regard to wintering under arctic conditions, a study underway on the bowhead's skin indicates the presence of "antifreeze" proteins which may protect the skin from freezing when surfacing during periods of extreme cold in winter (T. Sformo, unpubl.).

Basal and resting metabolic rates

Resting metabolic rates and Kleiber law

Most large whale energetic estimates are based on an allometric relationship between body mass and basal metabolic rate, known as the Kleiber law (Kleiber, 1975; Costa and Williams, 1999; Fortune et al., 2013; Laidre et al., 2007; Lockyer, 2007). Basal metabolism represents the minimum amount of energy required to sustain life at rest and assumes the animal is nonreproductive, thermally neutral, and fasting (i.e., postabsorptive) (Kleiber, 1975).

The Kleiber Law derived from studies on captive terrestrial mammals where the BMR was measured, is as follows:

$$\mathrm{BMR} = 3.771M^{0.75} \tag{16.1}$$

where BMR is measured in Watts and M is body mass (kg) (Kleiber, 1975; Lockyer, 2007).

We propose a somewhat revised model using a lipid catabolism value of 5.46 W/L O_2 for fat (see Schmidt-Nielsen, 1997), because healthy marine mammals typically break down blubber and resist muscle catabolism. The exponent (3/4, Kleiber Law vs 2/3, Rubner's Surface Law) in metabolic scaling has been a point of contention and much debate. Predictions by Ballesteros et al. (2018) suggest that body mass should instead be raised to the 2/3 (0.667) power and not 0.75 as in the standard Kleiber prediction, in particular for cold-adapted animals because it incorporates the importance of heat dissipation more accurately. The revised equation is:

$$\mathrm{RMR} = 3.693M^{0.667} \tag{16.2}$$

The relationship between body temperature and metabolic rate illustrates an important way to reduce energetic costs (Marchand, 1996; Elsner, 1999). The "Q_{10} effect" describes this relationship, whereby metabolic rates change by a factor of 2–3 for every 10°C increase or decrease in body temperature. The relationship is described by the following equation:

$$R_2 = R_1 \cdot Q_{10}^{(T_2 - T_1)/10} \tag{16.3}$$

where R_2 and R_1 are the metabolic rates at temperatures T_2 and T_1 (Schmidt-Nielsen, 1997).

Therefore, due to their lower core body temperature, we should expect the metabolic rate of bowhead whales to be lower than mammals that have more typical body temperatures of 37°C–38°C. For example, a bowhead with a body temperature of 33.8°C would have a metabolic rate of roughly 80% (using an average Q_{10} of 2) of a mammal of the same mass with a body temperature of 37°C (Elsner, 1999; Schmidt-Nielsen, 1997). The Q_{10} effect applies to all metabolically active tissues, which are primarily muscle and organs. As shown in Fig. 16.4, the temperature of the appendages and skeletal muscle of bowheads are far lower than the core temperatures and could result in a significant reduction in energy requirements of the significant portions of the body, peripheral tissues, and limbs. Because the Kleiber Law for mammals was developed using species with more typical body temperatures, we suspect that Kleiber-based estimates may overestimate bowhead metabolic rates because their body temperatures are lower.

Bowhead resting metabolic via heat loss models

BMR requires several physiological states to be met (e.g., thermally neutral, postabsorptive, nonreproductive, and inactive (Kleiber, 1975)); however, in this chapter, we define resting metabolic rate (RMR) as primarily inactive. In free-ranging whales, any assumptions about absorptive and reproductive states are almost impossible to validate.

George et al. (2009a) constructed a heat-loss model to estimate RMR for bowheads by dividing the animal into four "truncated cones" using measured body proportions and girth measurements. The model included blubber thickness, (measured) blubber thermal conductivity, girth/surface area, and temperature at the muscle–blubber interface for each cone. The results are shown in Table 16.1. Brodie (1981) made one of the first calculations of bowhead RMR using a heat loss model and obtained a value of 4800 W for a 13.7 m, 46,000 kg whale (Table 16.3), which consistent with our estimates.

TABLE 16.1 Estimated resting metabolic rates in Watts (mean ± standard error) for a 9, 13, and 16 m bowhead whale using the heat loss model that divided the animal into a series of truncated cones.

Body length	9 m Whale	13 m Whale	16 m Whale
Estimated metabolic rate	1764 ± 95	3213 ± 168	5070 ± 244

Notes: The standard deviation was estimated by Monte Carlo sampling with 1000 repetitions using temperatures measured at the muscle–blubber interface, and average blubber thickness. Whale lengths were chosen to represent a: (1) typical subadult; (2) whale at the length of sexual maturity; and (3) relatively large adult.

Metabolic rates estimated by lung volume and respiration

Metabolic rate estimates can be approximated using lung volume and O_2 extraction efficiencies and are useful here for making an independent comparison with other metabolic rate estimates (Folkow and Blix, 1992). This approach requires several assumptions, mainly that blow frequency and lung volume are related to energy demand, and further that tidal volume and O_2 extraction efficiency in the lungs are stable. However, Fahlman et al. (2016) showed that both tidal volume and O_2 extraction can vary substantially in bottlenose dolphins (*Tursiops truncatus*) as a function of exercise that increases the variance of the estimates. Nevertheless, when comparing actual breath-by-breath tidal volumes and O_2 extraction efficiency to assumed stable tidal volumes and O_2 extraction, O_2 consumption can be overestimated by as much as 500% (Fahlman et al., 2016). While these experiments are enlightening, they are performed with trained, aquarium-housed odontocetes, and a comparison to free-ranging mysticetes is logistically unattainable.

We compiled allometric and physiological variables for a bowhead whale and estimated metabolic demands (Table 16.2). Bowhead whale blow frequency during migration was reported as 0.4 blow/min (Krutzikowsky and Mate, 2000) and 0.7 blow/min when feeding (Carroll et al., 1987). During migration, the estimated O_2 consumption was between 2.9 and 3.3 mL O_2/min kg and between 5.1 and 5.8 mL O_2/min kg when feeding (calculations based on Table 16.2). When converting these values to power output or metabolic rate, it is important to recognize that a substantial portion of a bowhead whale's body mass (44%) is blubber—that is more or less metabolically inert—so calculations should arguably be based on lean body mass. The majority of bowhead whale prey consists of lipids (e.g., copepods, Chapter 12); therefore, heat output was estimated for fat oxidation (Table 16.2).

TABLE 16.2 Allometric and physiological variables for a 9 m subadult bowhead whale (12,000 kg) and an adult 13 m whale (30,000 kg) used in estimating metabolic output based on respiration rates. Row four provides references for parameters in the estimates.

Body length (m); mass (BM) (kg)	Blubber mass (kg)	Lung mass (LM) (kg)	Total lung capacity (TLC) (L)	Vital capacity (VC) (L)	Tidal volume (V_T) (L)	O_2 extraction efficiency (L)	Heat produced during fat oxidation (kJ/L O_2)
9 m; 12,000 kg	5280	108	270	216–243	194–219	87–99	19.7
13 m; 30,000 kg	13,200	270	675	540–608	486–547	219–246	19.7
–	44% of BM	0.9% of BM	2–5 mL/g LM[a]	80%–90% of TLC	90% of VC	45% of V_T	–
–	Table 16.4; Chapter 7 (this volume)	Table 16.4	Kooyman and Sinnett (1979)	Fahlman et al. (2016)	Olsen et al. (1969)	Blix and Folkow (1995)	Schmidt-Nielsen (1997)

[a]*Estimate based on lung inflation measurements. Bowhead whales fall on the lower end of total lung capacity estimates for other large whales at around 2.5 mL/g (this chapter; George, 2009a).*

For a 9-m migrating and feeding bowhead whale, the power output is 6400–7300 W and 11,300–12,600 W, respectively. This range agrees well with metabolic rate estimates for minke whales of 80 kJ/h day^{-1} corrected for body size, and arrived at using a somewhat different method (Blix and Folkow, 1995). When applied to an average 9-m bowhead whale (12,000 kg minus 44% blubber) while swimming, the estimated metabolic rate is 6200 W. This is about two times higher than the RMR estimates based on a heat-loss model which is expected for an active animal (George et al., 2009a; Table 16.3).

Summary of metabolic rate estimates

A range of RMR estimates using different methodologies is shown in Table 16.3. The Kleiber prediction (1a) for the RMR of a 13-m, 30,000-kg bowhead is approximately 8600 W. The modified Kleiber predictions that exclude blubber and use the lower metabolic scaling produce much lower RMR estimates (e.g., Table 16.2, models 1b,2a, 2b). An obvious pattern is that RMR estimates based on heat-loss models and oxygen consumption are considerably lower than predicted for whales using the standard Kleiber model (1a). We suspect that differences are in part due to the bowheads' higher proportion of blubber, physiological differences, and somewhat lower body temperature than other large whales.

Other researchers have similar concerns that the Kleiber prediction may not accurately predict the bowhead metabolic rate (Thomson, 2002). The bowheads' blubber constitutes nearly half of their body mass (Table 16.4; Chapter 7). Consequently, because blubber is a mostly nonthermogenic tissue, the Kleiber prediction may overestimate metabolic rate. Thomson (2002) used three different methods to estimate RMR for bowheads and also

TABLE 16.3 Estimates of resting metabolic rates (RMR, Watts) for bowhead whales by various researchers for an approximately 13-m (30,000 kg) bowhead.

Model	RMR (W)	Comment
Kleiber prediction 1a	8596	Eq. 16.1; blubber included
Kleiber prediction 1b	5565	Eq. 16.1; Standard Kleiber prediction; blubber excluded
Kleiber prediction 2a	3578	Eq. (16.2) modified, Kleiber prediction, includes blubber; $Mass^{0.667}$ (Ballesteros et al., 2018)
Kleiber prediction 2b	2430	Eq. (16.2), modified Kleiber prediction, excludes blubber; $Mass^{0.667}$ (Ballesteros et al., 2018)
Brodie (1981)	4600	Estimated RMR using self-constructed heat loss model
Thomson (2002)	3300	RMR using heat loss models, 13.5 m female, 31.5 metric tons
Thomson (2002)	8000	Respiration method, uses lung volumes and observed respiration rates while feeding model
George et al. (2009a)	3213 + 168	13 m, 30,000 kg whale. Heat loss

Notes: The equations for the standard Kleiber law are from Schmidt-Nielsen (1997); modifications are noted in the comments.

estimated total food consumption requirements (kg/prey) on a daily and annual basis (Table 16.3). He estimated food acquisition rates of a bowhead based on swim speed and known prey densities across the Bering-Chukchi-Beaufort (BCB) summer range. What Thomson (2002) found is provocative. He suggested that given the measured prey densities available in the Eastern Beaufort Sea, a bowhead's estimated prey ingestion rates were insufficient to satisfy their nutritional requirements even when using the lowest theoretical energetic estimates. This is particularly true if bowheads are assumed to have metabolic rates similar to other large cetaceans. Thomson concluded, "*The theoretical energy requirements of bowheads appear to be quite low and are in keeping with adaptations that bowheads possess for living in a cold environment where food is relatively scarce compared to some other marine waters*." Nonetheless, he also noted that bowheads are likely better at finding areas with higher prey concentrations than their sampling surveys did.

Thomson (2002) also realized that the estimated length of the summer feeding season, and the possibility of winter feeding, had major implications on the amount of food a bowhead would need to ingest on a daily basis. At the time, there was some controversial analysis based on stable carbon isotopes of bowhead tissues suggesting they fed in winter (Schell and Saupe, 1993). Since then considerable evidence for winter feeding has become available based on direct data from stomach examinations, indirect data from satellite telemetry suggesting feeding-dive behavior during winter, and further isotopic analyses (Citta et al., 2015; Lee et al., 2005; Chapters 4 and 28). Recent studies using stable isotope data from East Canada-West Greenland (ECWG) bowheads suggest they feed continuously throughout the winter but at a reduced rate (Matthews and Ferguson, 2015; Pomerleau et al., 2018). Obviously, winter feeding could offer significant energetic advantages in eliminating the need for long migrations and long periods of fasting (see *Duration of Feeding Season and Winter Feeding* section below). The true RMR for bowhead whales likely falls within the range of estimates given in Table 16.3. While the "best" model choice is not specified, we suggest the standard Kleiber prediction overestimates the bowhead's metabolic rate.

Energetic models for East Canada-West Greenland bowheads

A quantitative bioenergetics model has been generated to predict the food requirements of ECWG bowhead whales residing in Disko Bay, Greenland (Laidre et al., 2007). Predominately, nonlactating, adult females seasonally reside in Disko Bay, whereby 20% of the animals in this area are sexually immature and 80% are mature, and the sex ratio is heavily skewed toward females with 85:15 (female:male) (Heide-Jørgensen et al., 2010). When predicted energy needs were compared with estimates of prey consumption, it was determined that there was sufficient food available to support the 250 animals that seasonally occupy the area (Laidre et al., 2007).

Laidre et al. (2007) predicted higher mean daily energy needs of bowhead whales compared with lactating North Atlantic right whales (NARW) is likely attributed to different assumptions regarding field metabolic rate. For example, bowheads were assumed to expend 2.5 times their predicted BMR (using the Kleiber prediction) during foraging and nonforaging (e.g., migrating) related activities (Kenney et al., 1997; Laidre et al., 2007), which agrees with estimates for right whales. However, biologging studies conducted on North Atlantic right whale using digital acoustic recording tags, quantified the drag incurred by foraging and nonforaging whales (van der Hoop et al., 2019) and found that whales expended ~500 MJ/d (5787 W) while foraging and ~30% less when not feeding (Nousek-McGregor, 2010). Consequently, the right whale model used a lower metabolic scalar of ~1.8 times the Kleiber prediction to account for field metabolic rate of nonpregnant animals (Fortune et al., 2013). Obtaining similar estimates of energy expenditure during different activity states is required for bowhead whales to permit comparisons of behavior-specific energy needs. Furthermore, comparisons of predicted energy needs with other species require knowledge of the variability of daily food requirements of bowhead whales between demographic groups.

Anatomical specializations

A bowhead whale would have difficulty dissipating heat unless the blubber is circumvented. Blood can efficiently transfer considerable heat to peripheral regions, such as the skin and appendages of a whale, where it is subsequently lost to the environment (Hokkanen, 1990; Pabst et al., 1999; Schmidt-Nielsen, 1997). This means bowheads must be able to shunt a significant amount of blood (heat) to their extremities, such as the flukes, skin in the abdominal and lumbar regions, pectoral limbs, as well as the tongue when necessary (Kvadsheim and Folkow, 1997; Scholander and Schevill, 1955).

We suggest the bowhead's tongue is an important thermoregulatory organ due to its great size and mass (2140 kg in a 12.87 m male) and its relatively warm core (~30°C). Werth (2007) and personal observations of the tongue show it has a "periarterial venous rete rosette" or countercurrent vascular system similar to those reported in other species (Scholander and Schevill, 1955; Heyning and Mead, 1997) that can conserve heat when necessary. That is, the large amount of adipose tissue in the bowhead tongue may prevent unnecessary heat loss during filter feeding for instance (Werth, 2007; Heyning, 2001) but can be circumvented to dissipate heat when necessary. The intraoral rete (*corpus cavernosum maxillaris*) is also useful in thermoregulation despite its relatively small size and may be particularly important in protecting the brain from overheating (Ford et al., 2013).

The epidermis of the bowhead is of interest as it is among the thickest of any mammal. Why it is so thick is puzzling, however, T.F. Albert (pers. comm., 2019) suggests that its function may be associated with heat dissipation. Most body heat is ultimately lost through the skin and it is highly vascularized in bowhead whales (Haldiman et al., 1985; Haldiman and Tarpley, 1993). The thermal conductivity (TC) of the skin (0.53 W/m K) is over twice that of the underlying blubber and so would effectively transfer heat to the water (George et al., 2009a), and supports the idea that the thick skin may be associated with thermoregulation. Bowhead blubber in the abdominal and lumbar regions appears to be well vascularized (Haldiman and Tarpley, 1993) and conductivity of the blubber increases when it is perfused with blood, augmenting heat loss when necessary.

Like other large whales, the flukes of the bowhead whale have a complex countercurrent system of blood vessels which can either dissipate (like a radiator fin) or conserve heat (Fig. 16.5). Within the fluke, a central artery is surrounded by veins and an alternate

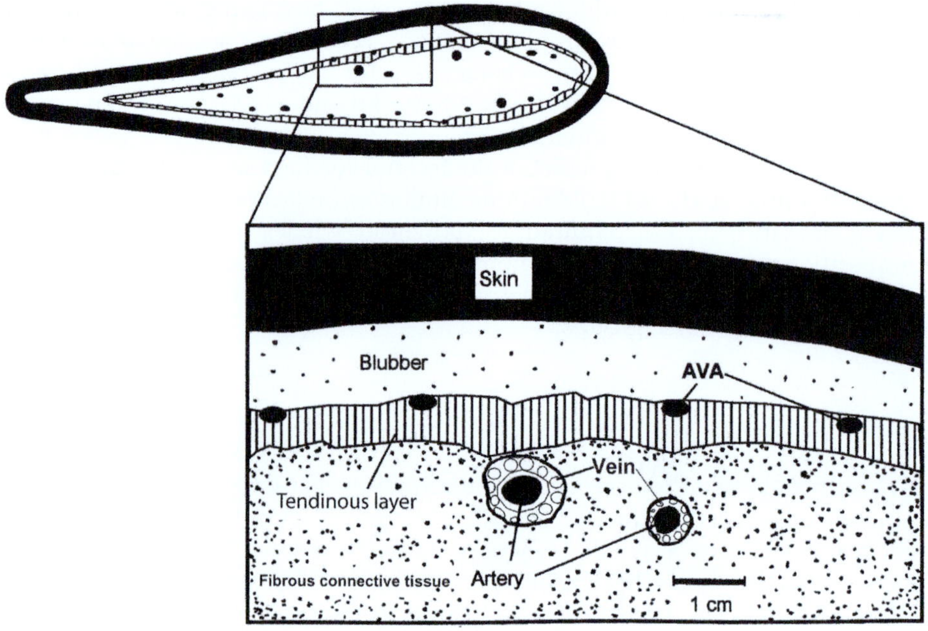

FIGURE 16.5 Sketch of a bowhead whale fluke showing the countercurrent heat exchange system (central artery and surrounding veins), tendinous layer, blubber, and skin (epidermis). Also shown is the arteriovenous anastomoses (AVA) system used for shunting blood close to the skin to dissipate heat when necessary. Note that the blubber layer is only about 1 cm thick on the flukes *(Drawing modified from Elsner et al., 2004).*

arteriovenous anastomoses (AVA) system with vessels near the skin (Elsner et al., 2004; Fig. 16.5). In the vasoconstricted state, the countercurrent system reduces heat loss from the flukes by returning blood through the countercurrent system. Alternatively, in the relaxed state, Elsner et al. (2004) speculated that the AVA system likely functions to dissipate heat when body temperatures are elevated (Fig. 16.5) by shunting blood to vessels close to the skin.

In a recent analysis of satellite telemetry data for ECWG bowhead whales, Chambault et al. (2018) found that bowheads targeted waters with relatively low sea surface temperatures and suggested this was possibly to avoid thermal stress. Furthermore, they speculated that rapid warming in arctic seas could reduce or eliminate important bowhead habitat (Chapter 27). Citta et al. (2018) found similar behavior for satellite-tagged BCB bowheads in which whales targeted relatively cold (<2°C) waters of Pacific origin when crossing the Chukchi Sea in autumn. However, they interpreted the bowheads' associations with these waters as seeking more productive feeding opportunities and higher prey densities—not necessarily avoidance of warmer waters to reduce thermal stress. Okhotsk Sea bowheads feed in warm surface waters (~15°C), and BCB bowheads will feed in relatively warm waters (6°C–8°C) at some locations and times, suggesting that they can tolerate relatively warm conditions. Recently, a bowhead has regularly visited the Gulf of Maine near Cape Cod in summer, suggesting the animal is tolerant of elevated water temperatures (Accardo et al., 2018; see Fig. 1.1). While it is unknown if bowheads are stressed at such temperatures, the question is an intriguing one and warrants further investigation.

Other energetic considerations

Relative organ size, muscle, and blubber proportions

Inferences can be made about the relative energetic levels of various vertebrates based on their anatomy and physiology. The smaller, more athletic whales and porpoises have high muscle mass relative to their body weight (Costa and Williams, 1999). Bowheads appear to have a slightly lower heart/body mass ratio (0.6%) than North Pacific right whales but are consistent with the mammalian average. The heart-to-body mass ratio (0.6%) for terrestrial mammals is fairly consistent across a range of body sizes (Schmidt-Nielsen, 1997; Williams et al., 2015). The lung mass to total body mass ratio (0.7%–0.9%) is similar between right whales and bowheads. The muscle mass of right whales (32%) and blue whales (40%) is considerably higher than bowheads (Omura et al., 1969; North Slope Borough (NSB) unpublished data; Table 16.4). The lower muscle mass of bowhead whales (19%) is surprising and may be associated with the lower metabolic rates and the bowhead's disproportionally large blubber mass.

Duration of feeding season and winter feeding

As noted earlier, the length of the feeding season for large whales has major implications to the rate at which they must ingest prey and accumulate energy on a daily basis. Year-round feeding may offer a significant advantage to this species, despite the presumably higher metabolic demands of wintering in very cold water (see Chapter 28). A recent study conducted on Eastern Canada-West Greenland bowhead whales found evidence of year-round feeding in Cumberland Sound, Nunavut based on analysis of horizontal and vertical movement data using satellite-linked time-depth recorders (Fortune et al., In Press). For example, Pacific gray whales (*Eschrichtius robustus*) must meet their annual energetic needs in a 5- to 6-month feeding season, as well as migrate some 8000 km to

TABLE 16.4 Comparison of mean percent heart, lung, and muscle mass relative to total body mass (blubber included) for three species of large baleen whale: bowhead whale, North Pacific right whale, and blue whale.

Species	Muscle (%)	Blubber (%)	Heart (%)	Lung (%)
Bowhead whale[a]	19	44	0.6	0.9
NP Right whale[b]	32	40	0.8	0.7
Blue whale[c]	40	27	0.5	0.7

[a]*George et al. (2009a,b); NSB unpublished data (muscle* N = 5 *whales; heart* N = 2; *lung* N = 4 *whales; fetuses excluded).*
[b]*Omura et al. (1969).*
[c]*Noncalf; subadult and adult* (N = 4)*; Nishiwaki (1950).*
Notes: Units are given as a percentage (%) of body mass.

wintering areas where little, if any, feeding occurs (Rice and Wolman, 1971). Furthermore, females must nurse their calves while fasting. A disruption in the summer feeding season (due to disturbance, heavy ice cover, etc.) of just 2 weeks may reduce their feeding season by 8%–10% and have energetic consequences (Perryman et al., 2002). In the case of a bowhead, however, a similar disruption would only be 4%–5% of their (year-long) feeding, and a fairly minor energetic loss.

Comparisons with North Atlantic right whales

It is instructive to compare bowhead energetics with their close relative, the North Atlantic right whale (NARW), where much is known about their diet, feeding behavior, and ingestion rates (Baumgartner and Mate, 2003; Baumgartner et al., 2017), body condition (Miller et al., 2012), dive behavior (Baumgartner and Mate, 2003; van der Hoop et al., 2019), and migration (Record et al., 2019). A bioenergetics model constructed for different demographic groups of North Atlantic right whales (calves, juveniles, resting adults, pregnant, and lactating females) found that calves had the lowest mean daily energy needs compared with lactating females, which were 2.2 times higher (Fortune et al., 2013). Furthermore, estimates of energy needs were compared with measurements of prey density and energy content to quantify how much energy whales consumed on a daily basis. Interestingly, modeling suggests that lactating NARW females are living on an "energetic edge" and are not always able to consume sufficient energy to meet their own needs as well as the high cost of nursing a growing calf (Fortune et al., 2013). These findings are consistent with other studies showing that North Atlantic and southern (*E. australis*) right whales experience periods of nutritional stress (Best and Schell, 1996; Miller et al., 2012; Pettis et al., 2004) and that lactating females may experience energetic deficits that could have population level effects, such as low population growth rates and a slow recovery.

Energetics of locomotion

The migration and swim speed of bowheads is about 3–4 km/h, which is lower than other large cetaceans (Wursig and Clark, 1993; Zeh et al., 1993; Woodward et al., 2006; see Chapter 23). Compared to rorquals (e.g., blue, fin, minke, and humpback whales), bowheads are considerably stockier (Woodward et al., 2006; Chapter 7). Consequently, they likely incur more drag when swimming and may compensate for this in part by reducing their swim speed (Simon et al., 2009). This may also explain why bowheads employ a fight strategy when faced with predators, such as killer whales, compared with more streamlined whales that employ a flee strategy (Ford and Reeves, 2008; Chapter 29). Also, slower swim speeds may help offset the energetic costs of high drag during feeding associated with the bowhead's large baleen rack (Simon et al., 2009; Chapter 14).

Blix and Folkow (1995) followed radio-instrumented minke whales and recorded their exhalations over a 24-hour period to estimate respiration rates during various behaviors. They suggest that the cost-of-swimming at slow speeds (3–4 km/h) was "remarkably low" and barely above resting levels. Similarly, the bowhead's relatively slow swim speeds (median speed: 3.5 and 3.2 km/h; 1987 and 1988 seasons respectively) for migrating whales in spring may be an energy conservation strategy (Zeh et al., 1993; Chapter 23). Also, the urgency to quickly reach the summer feeding grounds may be lower for an animal that feeds year-round (Chapter 28).

Growth and reproduction costs

Growth and reproduction induce considerable increases in energetic demands as estimated for North Atlantic right whales (Fortune et al., 2013). In the case of bowheads, females nurse their young from ~1000 kg at birth to 10–12,000 kg in year one (Chapter 7). Hence, the energetic drain on the mother to achieve this weight gain is considerable and may lead to a 25% reduction in her body weight as indicated by recent drone-based studies of southern right whales (*Eubalaena australis*) (Christiansen et al., 2018). Therefore, bowheads use a "capital breeding" strategy, where several years are required to gain sufficient body condition to become pregnant, complete gestation, and nurse a calf (Koski et al., 1993). This pattern is consistent with their calving interval of 3–4 years (Koski et al., 1993; Chapter 13).

Cost of disturbance

Industrial activities and killer whale densities have increased in the Arctic in recent decades, while sea-ice volume has decreased. Braithwaite et al. (2015) and Reeves et al. (2014) succinctly summarized the current situation in the Arctic Seas, "Sea ice losses are also a major stimulant to increased industrial interest [e.g., commercial fishing, shipping, oil and gas, tourism] in the Arctic in previously ice-covered areas." This highlights the need for predictions of bowhead food requirements to make reasonable forecasts about effects from disturbance and responses to climate and ecosystem change (Chapter 27). While the energetic costs of anthropogenic disturbance have been considered for bowheads, we are not aware of any published estimates. Similarly, while quantitative estimates of killer whale avoidance have not been calculated, the impacts of lost foraging opportunities are discussed in Chapter 29, where G.A. Breed suggests that the lost feeding opportunities and reduced calf or juvenile growth could be significant.

Fasting endurance of bowhead whales

Given the bowhead's very thick blubber and low metabolic rates, an obvious question is, how long could a bowhead whale survive when fasting or when food is lacking? Theoretically, bowheads should be able to survive longer than other cetaceans due to their greater blubber reserves and apparently low metabolic rate. To address this, we made some assumptions about assimilation efficiency (see Chapter 12), percent lipid in the blubber, and blubber volume.

We estimate that a 9-m subadult bowhead whale that is relying solely on its lipid reserves can fast up to 5 months based on a "swimming" metabolic rate and up to 8 months at an RMR (Table 16.3). This assumes the body mass is 44% blubber and the blubber lipid is 60%–75% (Horstmann-Dehn and George, 2013). These estimates assume no metabolic rate adjustments during fasting (Rea and Costa, 1992). Adults can likely endure much longer periods (George et al., 2009a) than subadults. A rough calculation for a resting 13-m bowhead, suggests it could survive ~2 years before it utilized 50% of its lipid reserves, and 4 years to use them entirely. However, as the lipids are "used up" in the blubber, its thermal resistance drops. Therefore, a bowhead could not draw down their blubber lipids to very low levels and still survive in icy-water. Nonetheless, these simple calculations support Burns (1993) speculations that the excessively thick blubber of bowheads serves as a buffer against multiyear prey failures. As noted above, juvenile bowhead whales undergoing the postweaning growth hiatus have comparatively little ability to fast and are much more vulnerable to starvation (Chapter 7; Lubetkin et al., 2008; George et al., 2009a).

Zooplankton species composition, abundance, and biomass in the Arctic is highly variable and is dependent on a number of factors including the timing of the spring bloom, which is controlled by the timing of sea ice retreat in the spring and reformation in the fall (Ashjian et al., 2010; Blix, 2016; Okkonen et al., 2011; Questel et al., 2013; see Chapter 26). Atmospheric and oceanographic conditions affect sea ice conditions, which are becoming increasingly variable overtime. For bowheads to maintain a positive energy balance, they need to find prey of sufficient quality and quantity. Consequently, an overaccumulation of lipid stores may serve as a built-in fail-safe for unfavorable prey years (Chapter 7). Migration time from the Bering Strait to the Amundsen Gulf feeding area typically takes less than 30 days over a distance of about 2200 km (Chapter 4)—a considerable journey but far shorter than migrations of other large whale species. Based on our calculations, even sporadic feeding bouts during migrating could be energetically important and is consistent with the characterization of the BCB bowheads' return-migration to the Bering Sea as a "feeding migration" (Chapter 4; Lowry et al., 2004).

Summary

Energetic models are useful for understanding how bowhead whales adapted to the arctic seas. They are also useful for management strategies for bowheads and other marine mammals. Estimates of food requirements and nutritional status can be coupled with biooceanographic studies of a region (Plourde et al., 2019), so managers can identify and predict important foraging habitat for whales and consider mitigation strategies (Gavrilchuk et al., 2019).

Based on several lines of evidence, bowhead whales appear to have metabolic rates that are lower than other similar-sized baleen whales. Low metabolic rates offer some disadvantages, such as slow growth rates, delayed maturation and long intercalf intervals; however, they also offer some advantages. When food is limited, as in winter, or years with low prey densities, it may be possible for a bowhead to persist for long periods at a relatively low metabolic cost. Low metabolic rates may also extend bowhead longevity and allow more opportunities for reproduction. Prey availability in the Arctic is highly variable, including seasons where "food is virtually lacking" (Burns, 1993). Selection may have modified the bowhead to store large amounts of lipid in the form of thick blubber to survive such periods (Burns, 1993; Chapter 7). In nature, however, given the stresses associated with migration, reproduction, and predation, it is unclear how long a bowhead could actually persist on their blubber reserves with limited resources. Kraus and Rolland (2007) nicely summarized the strategy for the closely related North Atlantic right whale as, "survive the lean years, and reproduce in the good ones."

Suggestions for further research

The following are suggestions for future research with regard to bowhead whale energetic studies.

- Continue measurements of core body temperatures and thermal gradients in bowhead whales to better define normal values to detect aberrations, illness, or changes. An important experiment would be to attach a telemetry tag that continuously measures temperatures of the surface of the skin and within the blubber along the tag's anchor). Such data would refine energetic models, and could be used to assess possible effects of warming sea temperatures on bowhead distribution and physiology.
- Make additional measurements of lung volume and heart mass. Empirical data on cetacean lung and heart volume and mass are limited and can be used to refine energetic and physiological models for large whales.
- Take additional measurements of stomach volume, intestinal length and volume, digestive efficiency, lipid and protein digestion and uptake. Conduct further investigations on the duration of feeding bouts and evacuation rates of the gastrointestinal tract. Such information will help inform how large whales process prey, prey volume per feeding bout, and how often they feed and defecate.
- Continue investigations of winter feeding as based on stomach contents of harvested whales in the Bering Sea communities.
- Construct a full bioenergetics model that predicts the mean daily energy needs of different demographic groups of bowhead whales and compare energy inputs with energy outputs to assess nutritional status under current environmental conditions. Such models could provide information useful for conservation and management, such as protecting important feeding areas.

Acknowledgments

We wish to thank the members of the Barrow Whaling Captains' Association for allowing us to take specimens and temperature measurements of their whales. We thank Alaska Eskimo Whaling Commission and North Slope Borough Department of Wildlife Management for their support through the years. The following people assisted with field logistics: Dave Ramey, Joe Burgener Harry Brower, Jr., Charles D.N. Brower, Cyd Hanns, Todd O'Hara, Mike Philo, and Robert Suydam.

Thomas Albert, James Gessaman, John Reynolds III, Matthew Sturm, Doug Goering, Pham Quang, and Judy Zeh offered suggestions and assisted with the heat flow analyses and statistical modeling. We appreciate Edward Hopson Sr.'s comments on whale insulation. We thank the US Army- CRREL laboratory for providing equipment and technical help as well as the UAF Institute of Arctic Biology and UAF Engineering Department. Terrie Williams kindly the manuscript. Finally, we are deeply indebted to Robert Elsner and Erich Follmann (UAF) for their inspiration, friendship, and guidance—without them, the thermoregulation and energetic studies on bowheads would not have been conducted.

References

Accardo, C.M., Ganley, L.C., Brown, M.W., Duley, P.A., George, J.C., Reeves, R.R., et al., 2018. Sightings of a bowhead whale (*Balaena mysticetus*) in the Gulf of Maine and its interactions with other baleen whales. J. Cetacean Res. Manage. 19, 23–30.

Albert, T.F., 2001. The influence of Harry Brower, Sr., an Inupiaq Eskimo Hunter, on the bowhead whale research program conducted at the UIC-NARL faculty by the North Slope Borough. In: Norton, D. (Ed.), Fifty More Years Below Zero: Tributes and Mediations for the Naval Arctic Research Laboratory's First Half Century at Barrow. University of Alaska Fairbanks, Alaska, 2001 xvi. 576 p., ill., maps, 20 cm. ISBN 0-919034098-5.

Albert, T.F., Panuska, J.A., 1978. Regional heterothermy and cardiovascular responses during induced hypothermia in non-hibernated and hibernated woodchucks, *Marmota monax*. Comp. Biochem. Phys. A. Physiol. 60 (1), 1–6.

Ashjian, C.J., Braund, S.R., Campbell, R.G., et al., 2010. Climate variability, oceanography, bowhead whale distribution, and Inupiat subsistence whaling near Barrow, Alaska. Arctic 63, 179–194.

Ballesteros, F.J., Martinez, V.J., Luque, B., Lacasa, L., Valor, E., Moya, A., 2018. On the thermodynamic origin of metabolic scaling. Sci. Rep. 8 (1), 1–10.

Baumgartner, M.F., Mate, B.R., 2003. Summertime foraging ecology of North Atlantic right whales. Mar. Ecol. Prog. Ser. 264, 123–135.

Baumgartner, M.F., Wenzel, F.W., Lysiak, N.S.J., Patrician, M.R., 2017. North Atlantic right whale foraging ecology and its role in human-caused mortality. Mar. Ecol. Prog. Ser. 581, 165–181. Available from: https://doi.org/10.3354/meps12315.

Best, P., Schell, D., 1996. Stable isotopes in southern right whales (*Eubalaena australis*) baleen as indicators of seasonal movements, feeding and growth. Mar. Biol. 124, 283–294.

Blix, A.S., 2016. Adaptations to polar life in mammals and birds. J. Exp. Biol. 219, 1093–1105. Available from: https://doi.org/10.1242/jeb.120477.

Blix, A.S., Folkow, L.P., 1995. Daily energy expenditure in free living minke whales. Acta Physiol. Scand. 153, 61–66.

Braithwaite, J.E., Meeuwig, J.J., Hipsey, M.R., 2015. Optimal migration energetics of humpback whales and the implications of disturbance. Conserv. Physiol. 3. Available from: https://doi.org/10.1093/conphys/cov001.

Brodie, P.F., 1981. A preliminary investigation of the energetics of the bowhead whale (*Balaena mysticetus* L). Rep. Int. Whal. Commn. 31, 501–502.

Brodie, P., Paasche, A., 1985. Thermoregulation and energetics of fin and sei whales based on postmortem, stratified temperature measurements. Can. J. Zool. 63, 2267–2269.

Burns, J.J., 1993. Epilog. In: Burns, J.J., Montague, J., Cowles, C.J. (Eds.), The Bowhead Whale. Society for Marine Mammalogy, 787 pp.

Carroll, G.M., George, J.C., Lowry, L.F., Coyle, K.O., 1987. Bowhead whale (*Balaena mysticetus*) feeding near Point Barrow, Alaska, during the 1985 spring migration. Arctic 40, 105–110.

Chambault, P., Albertsen, C.M., Patterson, T.A., Hansen, R.G., Tervo, O., Laidre, K.L., et al., 2018. Sea surface temperature predicts the movements of an Arctic cetacean: the bowhead whale. Sci. Rep. 8, 9658.

Chernova, O.F., Shpak, O.V., Kiladze, A.B., et al., 2016. Summer molting of bowhead whales *Balaena mysticetus Linnaeus*, 1758, of the Okhotsk Sea population. Dokl. Biol. Sci. 471, 261–265.

Christiansen, F., Vivier, F., Charlton, C., Ward, R., Amerson, A., Burnell, S., et al., 2018. Maternal body size and condition determine calf growth rates in southern right whales. Mar. Ecol. Prog. Ser. 592, 267–281. Available from: https://doi.org/10.3354/meps12522.

Citta, J.J., Okkonen, S.R., Quakenbush, L.T., Maslowski, W., Osinski, R., George, J.C., et al., 2018. Oceanographic characteristics associated with bowhead whale movements in the Chukchi Sea in autumn. Deep Sea Res. II 152, 121–131. Available from: https://doi.org/10.1016/j.dsr2.2017.03.009.

Citta, J.J., Quakenbush, L.T., Okkonen, S.R., Druckenmiller, M.L., Maslowski, W., Clement-Kinney, J., et al., 2015. Ecological characteristics of core areas used by western Arctic bowhead whales, 2006–2012. Prog. Oceanogr. 136, 201–222. Available from: https://doi.org/10.1016/j.pocean.2014.08.012.

Conti, B., Sanchez-Alavez, M., Winsky-Sommerer, R., Morale, M.C., Lucero, J., Brownell, S., et al., 2006. Transgenic mice with a reduced core body temperature have an increased life span. Science 314, 825–828.

Corkeron, P.J., Connor, R.C., 1999. Why do baleen whales migrate. Mar. Mamm. Sci. 15, 1228–1245. Available from: https://doi.org/10.1111/j.1748-7692.1999.tb00887.x.

Costa, D.P., Williams, T.M., 1999. Marine mammal energetics. Biology of Marine Mammals. Smithsonian Institution, Washington, DC, p. 578.

Dingle, H., 1996. Migration. The Biology of Life on the Move. Oxford University Press, New York.

Durban, J.W., Pitman, R.L., 2012. Antarctic killer whales make rapid, round-trip movements to subtropical waters: evidence for physiological maintenance migrations? Biol. Lett. 8, 274–277.

Elsner, R., 1999. Living in water: solutions to physiological problems. In: Reynolds III, J.E., Rommel, S.A. (Eds.), Biology of Marine Mammals. Smithsonian Institution, Washington, DC, pp. 73–116.

Elsner, R., George, J.C., O'Hara, T.M., 2004. Vasomotor responses of isolated peripheral blood vessels from bowhead whales: thermoregulatory implications. Mar. Mamm. Sci. 20, 546–553.

Fahlman, A., van der Hoop, J., Moore, M.J., Rocho-Levine, J., Brodsky, M., 2016. Estimating energetics in cetaceans from respiratory frequency: why we need to understand physiology. Biol. Open 5, 436–442.

Folkow, L.P., Blix, A.S., 1992. Metabolic rates of minke whales (*Balaenoptera acutorostrata*) in cold water. Scand. J. Phys. 146, 141–150.

Ford, J.K., Reeves, R.R., 2008. Fight or flight: antipredator strategies of baleen whales. Mamm. Rev. 38, 50–86.

Ford Jr, T.J., Werth, A.J., George, J.C., 2013. An intraoral thermoregulatory organ in the bowhead whale (*Balaena mysticetus*), the corpus cavernosum maxillaris. Anat. Rec. 296 (4), 701–708.

Fortune, S.M., Trites, A.W., Mayo, C.A., Rosen, D.A., Hamilton, P.K., 2013. Energetic requirements of North Atlantic right whales and the implications for species recovery. Mar. Ecol. Prog. Ser 478, 253–272.

Fortune S.M.E, Ferguson S.H., Trites A.W., LeBlanc, B., LeMay, V., Hudson, J.M., et al. In Press. Seasonal diving and foraging behaviour of Eastern Canada-West Greenland bowhead whales. Mar. Ecol. Prog. Ser. Available from: https://doi.org/10.3354/meps13356.

Gavrilchuk, K., Lesage, V., Fortune, S., Trites, A.W., Plourde, S., 2019. A mechanistic approach to predicting suitable foraging habitat for reproductively mature North Atlantic right whales in the Gulf of St. Lawrence. DFO Can. Sci. Advis. Sec. Res. Doc. 2018/nnn. vi + xx p.

George, J.C., Bada, J., Zeh, J., Scott, L., Brown, S.E., O'Hara, T., et al., 1999. Age and growth estimates of bowhead whales (*Balaena mysticetus*) via aspartic acid racemization. Can. J. Zool. 77, 571–580.

George, J.C., Goering, D., Follmann, E., Elsner, R., Sturm, M., Brower Jr., H., 2009a. Low metabolic rates of bowhead whales (*Balaena mysticetus*). In: George, J.C. (Ed.), Growth, Morphology, and Energetics of Bowhead Whales (*Balaena mysticetus*) (Ph.D. dissertation). University of Alaska Fairbanks, Alaska.

George, J.C., Follmann, E., Elsner, R., Albert, T., 2009b. Regional heterothermy in the bowhead whale (*Balaena mysticetus*). In: George, J.C. (Ed.), Growth, Morphology, and Energetics of Bowhead Whales (*Balaena mysticetus*) (Ph.D. dissertation). University of Alaska Fairbanks, Alaska.

Goldbogen, J.A., Cade, D.E., Wisniewska, D.M., Potvin, J., Segre, P.S., Savoca, M.S., et al., 2019. Why whales are big but not bigger: physiological drivers and ecological limits in the age of ocean giants. Science 366, 1367–1372.

Guyton, A.G., 1968. Textbook of Medical Physiology. Saunders Company, Philadelphia, PA, 1194 p.

Haldiman, J.T., Tarpley, R.T., 1993. Anatomy and physiology. In: Burns, J.J., Montague, J.J., Cowles, C.J. (Eds.), The Bowhead Whale. Special Publication No. 2. Society for Marine Mammalogy, 787 pp.

Haldiman, J.T., Henk, W.G., Henry, R.W., Albert, T.F., Abdelbaki, Y.Z., Duffield, D.W., 1985. Epidermal and papillary dermal characteristics of the bowhead whale (*Balaena mysticetus*). Anat. Rec. 211, 391–402.

Heide-Jørgensen, M.P., Laidre, K.L., Wiig, Ø., Postma, L., Dueck, L., Bachmann, L., 2010. Large-scale sexual segregation of bowhead whales. Endanger. Species Res. 13 (1), 73–78.

Heyning, J.E., 2001. Thermoregulation in feeding baleen whales: morphological and physiological evidence. Aquat. Mamm. 27, 284–288.

Heyning, J.E., Mead, J.G., 1997. Thermoregulation in the mouths of feeding gray whales. Science 278, 1138–1139.

Hokkanen, J.E.I., 1990. Temperature regulation of marine mammals. J. Theor. Biol. 145, 465–485.

Horstmann-Dehn, L., George, C., 2013. Bowhead whale digestive efficiency. In: Shelden, K.E.W., Mocklin, J.A. (Eds.), Bowhead Whale Feeding Ecology Study (BOWFEST) in the western Beaufort Sea. Final Report, OCS Study BOEM 2013-0114. National Marine Mammal Laboratory, Alaska Fisheries Science Center, NMFS, NOAA, Seattle, WA.

Jeanniard-du-Dot, T., Guinet, C., Arnould, J.P.Y., et al., 2017. Accelerometers can measure total and activity-specific energy expenditures in free-ranging marine mammals only if linked to time-activity budgets. Funct. Ecol. 31, 377–386. Available from: https://doi.org/10.1111/1365-2435.12729.

Innes, S., 1986. How fast should a dead whale cool? Can. J. Zool. 64 (9), 2064–2065.

Kanwisher, J., Sundnes, G., 1965. Physiology of small cetacean. Hvalradets Skrifter 48, 45–53.

Kanwisher, J.W., Ridgway, S.H., 1983. The physiological ecology of whales and porpoises. Sci. Am. 248, 111–120.

Kenney, R.D., Mayo, C.A., Winn, H.E., 2001. Migration and foraging strategies at varying spatial scales in western North Atlantic right whales: a review of hypotheses. J. Cetacean Res. Manag. (Special Issue) 2, 251–260.

Kenney, R.D., Scott, G.P., Thompson, T.J., Winn, H.E., 1997. Estimates of prey consumption and trophic impacts of cetaceans in the USA northeast continental shelf ecosystem. J. Northwest Atl. Fish Sci. 22, 155–171.

Kleiber, M., 1975. The Fire of Life: An Introduction to Animal Energetics. Robert E. Kreiger Co, Huntington, NY.

Kooyman, G.L., Sinnett, E.E., 1979. Mechanical properties of the harbor porpoise lung, *Phocoena phocoena*. Respir. Physiol. 36, 287–300.

Koski, W.R., Davis, R.A., Miller, G.W., Withrow, D.E., 1993. Reproduction. In: Burns, J.J., Montague, J.J., Cowles, C.J. (Eds.), The Bowhead Whale. Special Publication No. 2. Society for Marine Mammalogy, 787pp.

Kraus, S.D., Rolland, R.M., 2007. The Urban Whale. Harvard University Press, Cambridge, MA, 543 pp.

Krutzikowsky, G.K., Mate, B.R., 2000. Dive and surfacing characteristics of bowhead whales (*Balaena mysticetus*) in the Beaufort and Chukchi seas. Can. J. Zool. 78, 1182–1198.

Kvadsheim, P.H., Folkow, L.P., 1997. Blubber and flipper heat transfer in harp seals. Acta Physiol. Scand. 161, 385–395.

Kvadsheim, P.H., Gotaas, A.R.L., Folkow, L.P., Blix, A.S., 1997. An experimental validation of heat loss models for marine mammals. J. Theor. Biol. 184, 15–23.

Laidre, K.L., Heide-Jørgensen, M.P., Nielsen, T.G., Gissel Nielsen, T., 2007. Role of the bowhead whale as a predator in West Greenland. Mar. Ecol. Prog. Ser. 346, 285–297. Available from: https://doi.org/10.3354/meps06995.

Lee, S.H., Schell, D.M., McDonald, T.L., Richardson, W.J., 2005. Regional and seasonal feeding by bowhead whales *Balaena mysticetus* as indicated by stable isotope ratios. Mar. Ecol. Prog. Ser. 285, 271–287.

Lockyer, C., 1981. Growth and energy budgets of large baleen whales from the Southern Hemisphere. Food Agric. Organ. 3, 379–487.

Lockyer, C., 2007. All creatures great and smaller: a study in cetacean life history energetics. J. Mar. Biol. Assoc. 87, 1035–1045. Available from: https://doi.org/10.1017/s0025315407054720.

Lowry, L.F., Sheffield, G., George, J.C., 2004. Bowhead whale feeding in the Alaskan Beaufort Sea, based on stomach contents analyses. J. Cetacean Res. Manage. 6, 215–223.

Lubetkin, S., Zeh, J.E., Rosa, C., George, J.C., 2008. Age estimation for young bowhead whales (*Balaena mysticetus*) using annual baleen growth increments. Can. J. Zool. 86, 525–538. Available from: https://doi.org/10.1139/Z08-028.

Marchand, P.J., 1996. Life in the Cold: An Introduction to Winter Ecology. University Press of New England, Hanover, NH, 305 pp.

Matthews, C.J.D., Ferguson, S.H., 2015. Seasonal foraging behavior of Eastern Canada-West Greenland bowhead whales: an assessment of isotopic cycles along baleen. Mar. Ecol. Prog. Ser. 522, 269–286.

Miller, C.A., Best, P.B., Perryman, W.L., Baumgartner, M.F., Moore, M.J., 2012. Body shape changes associated with reproductive status, nutritive condition and growth in right whales *Eubalaena glacialis* and *E. australis*. Mar. Ecol. Prog. Ser. 459, 135–156.

Moore, S.E., Reeves, R.R., 1993. Distribution and movement. In: Burns, J.J., Montague, J.J., Cowles, C.J. (Eds.), The Bowhead Whale. Special Publication No. 2. Society for Marine Mammalogy, 787 pp.

Morrison, P., 1962. Body temperatures in some Australian mammals. III. Cetacea (Megaptera). Biol. Bull. 123, 154–162.

Nishiwaki, M., 1950. On the body weight of whales. Sci. Rep. Whales Res. Inst. 4, 184–209.

Noren, D.P., 2011. Estimated field metabolic rates and prey requirements of resident killer whales. Mar. Mamm. Sci. 27, 60–77.

Nousek-McGregor, A.E., 2010. The cost of locomotion in North Atlantic right whales Eubalaena glacialis. Unpublished doctoral dissertation). Duke University, Durham, NC.

Okkonen, S.R., Ashjian, C.J., Campbell, R.G., Clarke, J.T., Moore, S.E., Taylor, K.D., 2011. Satellite observations of circulation features associated with a bowhead whale feeding 'hotspot' near Barrow, Alaska. Rem. Sens. Environ. 115 (8), 2168–2174.

Olsen, C.R., Hale, F.C., Elsner, R., 1969. Mechanics of ventilation in the pilot whale. Respir. Physiol. 112, 77–85.

Omura, H., Ohsumi, S., Nemoto, T., Nasu, K., Kasuya, T., 1969. Black Right Whales in the North Pacific. Reprinted from the Scientific Reports of the Whales Research Institute, 21, 50 pp.

Pabst, D.A., Rommel, S.A., McLellan, W.A., 1999. The functional morphology of marine mammals. In: Reynolds III, J.E., Rommel, S.A. (Eds.), Biology of Marine Mammals. Smithsonian Institution, Washington, DC, p. 578.

Patrician, M.R., Biedron, I.S., Esch, H.C., et al., 2009. Evidence of a North Atlantic right whale calf (*Eubalaena glacialis*) born in northeastern US waters. Mar. Mamm. Sci. 25, 462–477. Available from: https://doi.org/10.1111/j.1748-7692.2008.00261.x.

Perryman, W.L., Watters, G.M., Schwarz, L.K., 2002. Examination of the relationship between season ice and calf production in the Eastern Pacific population of gray whales. Paper SC/54/BRG4 Presented at the 54th IWC Scientific Committee, Shimonoseki, Japan.

Peters, R.H., 1989. The Ecological Implications of Body Size. Cambridge University Press, Cambridge.

Pettis, H.M., Rolland, R.M., Hamilton, P.K., et al., 2004. Visual health assessment of North Atlantic right whales (*Eubalaena glacialis*) using photographs. Can. J. Zool. 82, 8–19. Available from: https://doi.org/10.1139/z03-207.

Plourde, S., Lehoux, C., Johnson, C.L., Perrin, G., Lesage, V., 2019. North Atlantic right whale (*Eubalaena glacialis*) and its food: (I) A spatial climatology of Calanus biomass and potential foraging habitats in Canadian waters. J. Plankton Res. 41 (5), 667–685.

Pomerleau, C., Matthews, C.J., Gobeil, C., Stern, G.A., Ferguson, S.H., Macdonald, R.W., 2018. Mercury and stable isotope cycles in baleen plates are consistent with year-round feeding in two bowhead whale (*Balaena mysticetus*) populations. Polar Biol. 41, 1881–1893.

Prosser, C.L. (Ed.), 1991. Environmental and Metabolic Animal Physiology: Comparative Animal Physiology. fourth ed. John Wiley and Sons, Hoboken, NJ.

Questel, J.M., Clarke, C., Hopcroft, R.R., 2013. Seasonal and interannual variation in the planktonic communities of the northeastern Chukchi Sea during the summer and early fall. Cont. Shelf Res. 67, 23–41.

Rea, L.D., Costa, D.P., 1992. Changes in standard metabolism during long-term fasting in northern elephant seal pups (*Mirounga angustirostris*). Phys. Zool. 65, 97–111.

Record, N.R., Runge, J.A., Pendleton, D.E., Balch, W.M., Davies, K.T., Pershing, A.J., et al., 2019. Rapid climate-driven circulation changes threaten conservation of endangered North Atlantic right whales. Oceanography 32 (2), 162–169.

Reeves, R., Rosa, C., George, J.C., et al., 2012. Implications of Arctic industrial growth and strategies to mitigate future vessel and fishing gear impacts on bowhead whales. Mar. Pol. 36, 454–462. Available from: https://doi.org/10.1016/j.marpol.2011.08.005.

Reeves, R.R., Ewins, P.J., Agbayani, S., Heide-Jørgensen, M.P., Kovacs, K.M., Lydersen, C., et al., 2014. Distribution of endemic cetaceans in relation to hydrocarbon development and commercial shipping in a warming Arctic. Mar. Pol. 44, 375–389.

Rice, D.W., Wolman, A.A., 1971. The life history and ecology of the gray whale (*Eschrichtius robustus*). Special Publication No. 3. U.S. Fish & Wildlife Service, Bureau of Commercial Fisheries, Marine Mammal Biological Laboratory, Seattle, WA. Copies available from: the Secretary-Treasurer of the Society, Dr. Bryan P. Glass, Department of Zoology, Oklahoma State University, Stillwater, OK.

Schell, D., Saupe, S., 1993. Feeding and growth as indicated by stable isotopes. In: Burns, J.J., Montague, J.J., Cowles, C.J. (Eds.), The Bowhead Whale. Special Publication No. 2. Society for Marine Mammalogy, i-xxxvi + 787pp.

Schmidt-Nielsen, K., 1997. Animal Physiology: Adaptation and Environment. Cambridge University Press, Cambridge, 611 pp.

Scholander, P.F., Schevill, W.E., 1955. Counter-current vascular heat exchange in the fins of whales. J. Appl. Phys. 8, 279–282.

Simon, M., Johnson, M.J., Tyack, P., Madsen, P.T., 2009. Behavior and kinematics of continuous ram filtration in bowhead whales (*Balaena mysticetus*). Proc. R. Soc. Lond. 276, 3819–3828.

Stern, S.J., Friedlander, A.S., 2018. In: Wursig, B., Thewissen, J.G.M., Kovacs, K.M. (Eds.), Encyclopedia of Marine Mammals. Academic Press, Cambridge, MA.

Thewissen, J.G.M., George, J.C., Rosa, C., Kishida, T., 2011. Olfaction and brain size in the bowhead whale (*Balaena mysticetus*). Mar. Mamm. Sci. 27, 282–294.

Thomson, D.H., 2002. Energetics of bowhead whales. In: Richardson, W.J., Thomson, D.H. (Eds.), Bowhead Whale Feeding in the Eastern Alaska Beaufort Sea: Update of Scientific and Traditional Information. For OCS Study MMS 2002-012; LGL Rep. TA 2196-7. Rep. From LGL Ltd., King City, Ontario Department of Interior, Mineral Management Service. Anchorage, AK 99598 and Herdon, VA. p. 22-1 to 22-40. Vol. 1, xliv + 420 p; Vol. 2; 277 p.

van der Hoop, J.M., Nousek-McGregor, A.E., Nowacek, D.P., et al., 2019. Foraging rates of ram-filtering North Atlantic right whales. Funct. Ecol. 33, 1290–1306. Available from: https://doi.org/10.1111/1365-2435.13357.

Villegas-Amtmann, S., Schwarz, L.K., Sumich, J.L., et al., 2015. A bioenergetics model to evaluate demographic consequences of disturbance in marine mammals applied to gray whales. Ecosphere 6, 1–19. Available from: https://doi.org/10.1890/ES15-00146.1.

Vongraven, D., Ekker, M., Espelien, A.R., Aarvik, F.J., 1990. Postmortem body temperatures in the minke whale, *Balaenoptera acutorostrata*. Can. J. Zool. 68, 140–143.

Waalen, J., Buxbaum, J.N., 2011. Is older colder or colder older? The association of age with body temperature in 18,630 individuals. J. Gerontol. Biol. Sci. Med. Sci. 66A, 487–492. Available from: https://doi.org/10.1093/gerona/glr001. PMCID 3107024.

Werth, A.J., 2007. Adaptations of the cetacean hyolingual apparatus for aquatic feeding and thermoregulation. Anat. Rec. 290, 546–568.

Williams, T.M., Bengtson, P., Steller, D.L., Croll, D.A., Davis, R.W., 2015. The healthy heart: lessons from nature's elite athletes. Physiology 30, 349–357.

Williams, T.M., Noren, D., Berry, P., Estes, J.A., Allison, C., Kirtland, J., 1999. The diving physiology of bottlenose dolphins (*Tursiops truncatus*). III. Thermoregulation at depth. J. Exp. Biol. 202, 2763–2769.

Woodward, B.L., Winn, J.P., Fish, F.E., 2006. Morphological specializations of baleen whales associated with hydrodynamic performance and ecological niche. J. Morphol. 267, 1284–1294.

Wursig, B., Clark, C., 1993. Behavior. In: Burns, J.J., Montague, J.J., Cowles, C.J. (Eds.), The Bowhead Whale. Allen Press, Lawrence, KS.

Young, B.G., Fortune, S.M.E., Koski, W.R., Raverty, S.A., Kilabuk, R., Ferguson, S.H., 2020. Evidence of killer whale predation on a yearling bowhead whale in Cumberland Sound, Nunavut. Arct. Sci. 6, 53–61.

Zeh, J.E., Clark, C.W., George, J.C., Withrow, D., Carroll, G.M., Koski, W.R., 1993. Current population size and dynamics. In: Burns, J.J., Montague, J.J., Cowles, C.J. (Eds.), The Bowhead Whale. Special Publication No. 2. Society for Marine Mammalogy, pp. 409–489, i-xxxvi + 787 pp.

CHAPTER

17

Brain

Sam Ridgway[1], *Patrick R. Hof*[2] *and Mary Ann Raghanti*[3]

[1]Department of Pathology, University of California, San Diego and National Marine Mammal Foundation, San Diego, CA, United States [2]Nash Family Department of Neuroscience and Friedman Brain Institute, Icahn School of Medicine at Mount Sinai, New York, NY, United States [3]Department of Anthropology, School of Biomedical Sciences, and Brain Health Research Institute, Kent State University, Kent, OH, United States

Introduction

The evolution of vertebrates was associated with the invention of a "new head" structure, which allowed the transition from a passive to a more active predatory lifestyle (Gans and Northcutt, 1983; Haberland et al., 2009). From zebrafish to humans, early brain development is the primary determinant of skull shape (Wada et al., 2005). As cetacean ancestors moved into the aquatic environment, they needed to breathe while swimming. Thus, the nasal aperture moved from the midface to the top of the skull. This is a common feature of all cetaceans. However, among cetaceans, different prey and modes of feeding have resulted in considerable variation in skull shape and other features such as asymmetry of cranial structures (Mead and Fordyce, 2009). Despite long skulls, cetacean brains appear shortened relative to those of terrestrial mammals, with a forebrain that has been rotated ventrally conforming to the modified braincase. In addition, the anatomical location of functional cortical regions is similar to the arrangement observed in many ungulates, but differs from that of most rodents and primates. For example, the auditory and visual cortex in dolphins and porpoises are located in the parietal region rather than the typical mammalian locations of temporal and occipital lobes (Berns et al., 2015; Sokolov et al., 1972). However, in dolphins, auditory responses have been documented in temporal regions of the cortex (Bullock and Ridgway, 1972) and neural connections from the auditory midbrain to the temporal lobe have been found by magnetic resonance imaging (Berns et al., 2015) (Fig. 17.1).

Our understanding of the cetacean brain has been informed mostly by studies of delphinids and, to a lesser extent, other odontocetes. The structure and function of the mysticete brain are less well known, and our limited knowledge is based mostly on a handful

The Bowhead Whale
DOI: https://doi.org/10.1016/B978-0-12-818969-6.00017-0

FIGURE 17.1 A bowhead mother and neonate calf swim (right) near two beluga whales (left). Bowheads have brains approximately 1.5 times as large as that of belugas, but their body weight is more than 80 times greater. A human brain is smaller than either. *Source: Photo by Corey Accardo under NOAA/North Slope Borough, NMFS Permit No. 14245.*

of studies in balaenopterids (Butti et al., 2015; Dell et al., 2016; Hof and Van der Gucht, 2007; Kraus and Pilleri, 1969; Pilleri, 1966a,b; Ratner et al., 2010). Several studies focused on the bowhead whale brain (Breathnach, 1955; Duffield et al., 1992; Pilleri, 1964; Raghanti et al., 2018; Thewissen et al., 2011), and those results will be summarized here in the context of cetacean neuroanatomy.

Description and comparisons

Shape and size of the bowhead whale brain

In most cetaceans, the cerebrum is wider than long (see Ridgway et al., 2016; Fig. 19). However, in the bowhead, the situation is reversed. The mean length of the bowhead cerebrum measured by Duffield et al. (1992) was 175 mm while the mean width was 135 mm. The opposite is true for most odontocetes, including the beluga (Fig. 17.2).

Overview and surface morphology of the bowhead whale brain

Cetaceans possess the largest brains among living species. The largest measured are found in the adult male sperm whales (*Physeter macrocephalus*) and adult killer whales (*Orcinus orca*) which may weigh 6000–9000 g (Ridgway and Hanson, 2014).

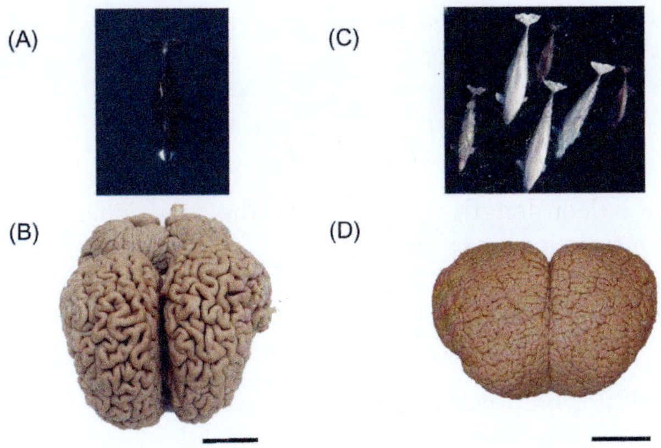

FIGURE 17.2 (A and B) Bowhead whale and the dorsal surface of the brain. (C and D) Beluga whales and the dorsal surface of the cerebrum. Note the micro gyri of the beluga resulting in a much greater gyrification compared to the bowhead. Also, note that, unlike the bowhead, the beluga cerebrum is wider than long. Scale bars (for C and D) are 5 cm. *Source: Photographs by Amelia Brower (A) and Vicki Beaver (C) under NOAA/North Slope Borough, NMFS Permit No. 14245.*

Encephalization refers to the evolutionary increase in brain size, and the encephalization quotient (EQ) goes further in that it takes into account body mass and the expected brain size for an animal of that size (Jerison, 1973, 1977). Cetacean EQs range from very low in large baleen whales (e.g., 0.2 for the blue whale, *Balaenoptera musculus*) to some that are close to the human EQ of 5.72 (e.g., 5.44 for the river dolphin, *Sotalia fluviatilis*) (Boddy et al., 2012). The bowhead EQ for mature length animals is similar to that of the blue whale around 0.2 (see data in Ridgway et al., 2016). Notably, the bowhead whale brain occupies only 35%–41% of the braincase, with the remainder of space being occupied by the dural rete mirabile (Thewissen et al., 2011). The amount of nonbrain tissue in the cranial vault of cetaceans is variable and one must wonder about the function of such extensive vascular networks.

Remarkable features of the bowhead whale brain include a blunted temporal pole, which stands in stark contrast to the extensive temporal regions present in other cetaceans, especially the odontocetes (Fig. 17.3). In fact, the temporal pole of the bowhead whale brain appears so truncated that the insula is exposed on the lateral surface. The bowhead whale brain also does not appear to possess the anteroposterior flexure characteristic of most cetacean brains, although brain extraction procedures may have significantly distorted the morphology of examined specimens (Duffield et al., 1992). The cerebellum is large with a well-developed vermis and lateral expansions of the cerebellar hemispheres. The cerebellum as a percentage of total brain of three brains we measured was 24.8, 21.5, and 20.4, the largest cerebellum percentage we have found among cetaceans. Another notable feature is that the bowhead brain does not have an interthalamic adhesion (Duffield et al., 1992). We concur with this observation and note that dolphins and other cetaceans we have observed do have an interthalamic adhesion connecting the two hemispheres.

The surface morphology of the bowhead whale brain has been well described (Duffield et al., 1992; Raghanti et al., 2018). The cortical gyrification pattern of the bowhead cerebral cortex is consistent with what is typically observed in other cetaceans, including odontocetes and mysticetes. Many features of this pattern are also shared with artiodactyls, perissodactyls, and carnivores and include the concentric organization of gyri surrounding a

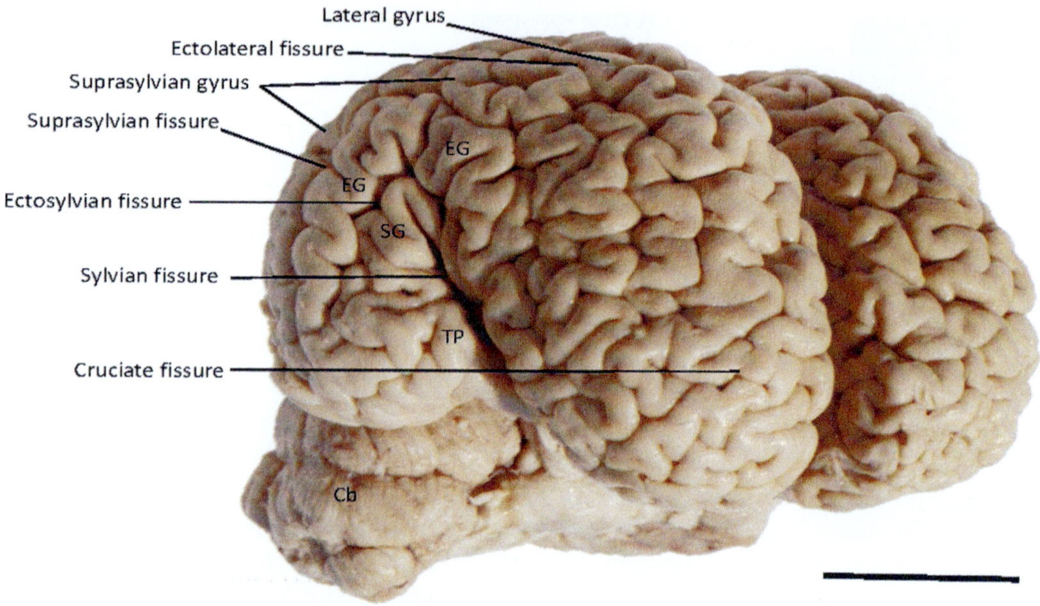

FIGURE 17.3 Lateral view of the bowhead whale brain with labels of some fissures and gyri. Abbreviations: *Cb*, cerebellum; *EG*, ectosylvian gyrus; *SG*, sylvian gyrus; *TP*, temporal pole. Scale bar = 5 cm.

verticalized sylvian cleft on the lateral aspect. A number of concentric gyri and sulci extend from the sylvian fissure within the temporoparietal operculum to include the ectosylvian and suprasylvian fissures that define the ectosylvian, suprasylvian, and entolateral gyri (see Fig. 17.3). Superiorly, the entolateral fissure separates the entolateral and lateral gyri. A well-defined cingulate gyrus is readily visible on the midline lying superior a relatively small and thin corpus callosum.

Gyrencephalic index and corpus callosum size

A highly convoluted cortical surface is characteristic of the cetacean brain. The gyrencephalic index (GI) is a measure of this cortical folding, obtained by dividing the sum of the perimeter of the pial contour, including gyri and sulci, by the sum of the outer contour of the brain (Ridgway and Brownson, 1984; Zilles et al., 1989). The average bowhead GI is 2.32 (Raghanti et al., 2018), which is lower than the typical cetacean GI of >5 (e.g., *Kogia simus*, 5.26; *Tursiops truncatus*, 5.63; *Orcinus orca*, 5.70; *Megaptera novaeangliae*, 5.35), but comparable to that of most artiodactyls (e.g., *Sus scrofa*, 2.16; *Bos taurus indicus*, 2.53; *Odocoileus virginianus*, 2.27) (Manger et al., 2012). As with other mammals, there is an isometric relationship between brain size and gyrencephaly among cetaceans (Ridgway et al., 2016). However, *Balaena* is an outlier having a lower index of folding and less cortex surface area relative to brain size (note differences in gyrencephaly apparent when we compare the brains shown in Fig. 17.2).

The midsagittal surface area of the bowhead corpus callosum relative to brain mass (CCA:BM) is higher than what has been reported for other cetacean species (Manger et al., 2010; Tarpley and Ridgway, 1994). In general, cetaceans possess smaller corpus callosum surface areas relative to terrestrial species. This feature may be related to having smaller-than-predicted cerebral cortex size relative to brain mass or to lateralization of function, particularly for the unique unihemispheric slow wave sleep that is characteristic of cetaceans (Lyamin et al., 2008; Manger, 2006; Ridgway and Brownson, 1984). However, the bowhead whale CCA:BM is consistent with those of both the hippopotamus and pygmy hippopotamus (Butti et al., 2014; Raghanti et al., 2018).

Olfaction

Most cetaceans have dramatically reduced olfaction, with altered olfactory bulb anatomy and a reduced number of olfactory receptor genes in the majority of mysticetes to virtually no olfaction in odontocetes (Dehnhardt, 2002; Kishida et al., 2015; Oelschläger and Oelschläger, 2008; Pihlström, 2008). The bowhead whale, in contrast, possesses an olfactory bulb that makes up approximately 0.13% of brain mass, a nonoccluded cribriform plate, and retains a higher number of functioning olfactory receptors (Thewissen et al., 2011; Chapter 18). In short, it appears that the bowhead has a more developed sense of smell relative to humans, who have a higher percentage of olfactory receptor pseudogenes than the bowhead and whose olfactory bulb makes up a smaller proportion of overall brain size (Thewissen et al., 2011).

Odontocetes have a terminal nerve, a small nerve that runs along with the olfactory nerve in humans and many other animals (Ridgway et al., 1987). Odontocete embryos studied so far do have olfactory bulbs, yet, as adults, they have no olfactory bulbs or nerves. Olfactory structures may play a role in skull asymmetry of odontocetes. Olfactory bulbs in bowheads are surprisingly asymmetrical. In one individual (NSB-DWM 2009B14), the left olfactory bulb weighed 2.06 g, and the right bulb only 1.64 g (Thewissen, pers. comm.).

Sleptsov (1939) studied development of the skull and nasal system of odontocetes—*Delphinus*, *Delphinapterus*, and *Phocoena*. Early embryos have olfactory bulbs, nerves, and tracts (Sleptsov, 1939). As the embryo grows, the left olfactory parts degenerate first, resulting in the cranial asymmetries and a leftward deviation of the nares. This leftward deviation has also been found in three fossil Eocene archaeocete whales (Fahlke et al., 2011). Consistent with the earlier suggestions of others (e.g., Heyning and Mead, 1990), these authors relate the early evolution of asymmetry to the development of directional hearing and echolocation in odontocetes. These ideas are not inconsistent with Sleptsov's suggestion that early reduction of the left olfactory apparatus of odontocetes may trigger asymmetry. Olfactory contrasts between odontocetes and mysticetes suggests that the well-developed olfactory system (Thewissen et al., 2011) is a basis for the bowhead's symmetrical skull (Heyning and Mead, 1990).

Hippocampus

In many species, the hippocampus is connected to virtually all regions of the neocortex and plays a critical role in learning and memory, information retrieval, and spatial memory (Andersen et al., 2007). Neurons within the hippocampus exhibit experience-dependent plasticity, with larger hippocampal volume being associated with increased capacity for learning and memory, as has been reported for a variety of species such as food-caching birds and multilingual humans (Bellander et al., 2015; Cnotka et al., 2008; Tarr et al., 2009). Hippocampus size scales in an exponential, rather than linear, fashion among mammals (Patzke et al., 2015). However, cetaceans are known for having relatively small hippocampi that lack the adult neurogenesis observed in terrestrial mammals (Manger, 2006; Morgane et al., 1980; Patzke et al., 2014, 2015). The bowhead hippocampus is even more reduced in size relative to that of other cetaceans (Patzke et al., 2015), and its location is unique in being found within the lateral ventricle beneath the corpus callosum (Duffield et al., 1992; Raghanti et al., 2018). The presence of an expansive hippocampus in hippopotamids (Butti et al., 2014) suggests that the cetacean hippocampus was reduced early during their return to an aquatic lifestyle. Some have argued that the cetacean hippocampus size, coupled with irregular sleep patterns, is indicative of a diminished cognitive capacity (Manger, 2006; Patzke et al., 2015) while others suggest that the hippocampus is particularly vulnerable to acoustic insults (Wright et al., 2017). The anatomical differences in hippocampus structure between obligatory aquatic and terrestrial mammals are impossible to interpret in terms of cognitive ability, particularly when considering the advanced cognition of corvids and parrots that lack the typical mammalian hippocampus morphology and the exceptionally large size of the hippocampus in hippopotamids.

Cerebral cortex cytoarchitecture

The cytoarchitecture of the cetacean cerebral cortex is variable and complex, with significant differences among species. Consistent with the cytoarchitecture reported for other cetacean species, the bowhead whale cerebral cortex is relatively thin, agranular (i.e., lacking a layer IV), neuron-sparse overall, and with a thick layer I (Fig. 17.4). The lack of a layer IV in cetaceans suggests a change in connectivity, with thalamic input being possibly directed to layers I–III rather than the typical mammalian distribution of layers I, III, and IV (Ferrer and Perera, 1988; Furutani, 2008; Glezer and Morgane, 1990; Hof and Van der Gucht, 2007; Jones, 2007). Bowhead whales lack layer II neuron clusters that are present in other cetaceans, including odontocetes and mysticetes (Butti and Hof, 2010; Hof et al., 2005; Hof and Van der Gucht, 2007; Jacobs et al., 1984; Manger et al., 1998). These neuron clusters may represent a special connectivity that involves thalamic afferent input to more superficial cortical layers in the absence of a granular layer (Hof and Van der Gucht, 2007). Both layer I and the white matter of the bowhead cortex are more cellular than what is observed in terrestrial species. As expected, the cytoarchitecture of the different cortical areas shows regional variation in cell density and composition (Raghanti et al., 2018).

A distinctive feature of the bowhead whale neocortex is the ubiquitous presence of von Economo neurons (VEN) and fork neurons. VEN are considered specialized cortical projection neurons that are easily recognized for their large, spindle-shaped soma and thick basal and apical dendrites and for neurons possess a soma that divides into two thick

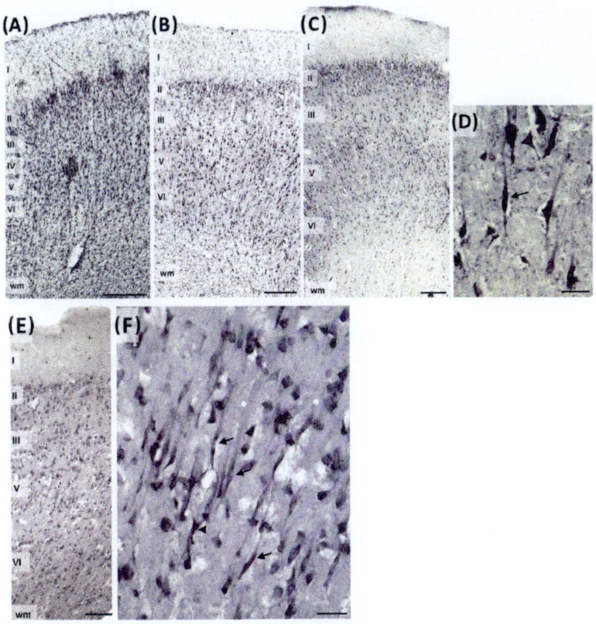

FIGURE 17.4 Photomicrographs showing the cytoarchitecture of the anterior cingulate cortex of Nile Hippopotamus (A), minke whale (B), and bowhead whale (C). Cortical layers are indicated by roman numerals, *wm*, white matter. Note the distinctive layer II neuron clusters present in the hippopotamus. A von Economo neuron from the bowhead whale anterior cingulate cortex is indicated by the *arrow* in (D). The cytoarchitecture of the bowhead insula is shown in (E), with additional von Economo neurons shown in (F). Scale bars for panels A, B, C, and E = 500 μm. Scale bars in D and F = 50 μm.

apical dendrites and one basal dendrite (e.g., Butti et al., 2013; Nimchinsky et al., 1999; Raghanti et al., 2015a,b). VEN and fork neurons are present in other cetaceans, elephants, artiodactyls, carnivores, and primates, but are typically limited to only a few cortical regions. In contrast, VEN and fork neurons were present in every cortical region of the bowhead whale brain, and often in multiple cortical layers (predominantly in layer V, but also in layers II and III). Interestingly, this ubiquitous distribution of VEN and fork neurons is also characteristic of hippopotamids (Butti et al., 2014; Raghanti et al., 2015a,b). In this context, it is worth noting that bowheads are in a phylogenetically basal position among mysticetes and therefore the widespread distribution of these neurons may have been conserved in the bowhead from an ancestral condition shared with modern hippopotamids.

Discussion

The bowhead whale brain displays several characteristics that are consistent with other cetaceans, with its surface morphology being most similar to that of the southern right whale (Duffield et al., 1992; Pilleri, 1966b). Unique characteristics of the bowhead whale include enhanced olfaction relative to other cetaceans (Thewissen et al., 2011), a larger corpus callosum and decreased gyrencephaly, which are more characteristic of artiodactyls (Raghanti et al., 2018), and a dramatically reduced hippocampus located within the lateral ventricle, ventral to the corpus callosum (Duffield et al., 1992; Raghanti et al., 2018).

Unique features of the cerebral cortex cytoarchitecture include the absence of layer II neuron clusters that are observed in other cetaceans, and the ubiquitous distribution of VEN and fork neurons (Raghanti et al., 2018).

The spinal cord and peripheral nervous system have been little studied. The innervation of the skin and sensory vibrissae deserve future study (Drake et al., 2015; Chapter 18). Olfaction is another avenue for additional research as the bowhead whale possesses the largest olfactory bulbs observed in any cetacean species to date (Thewissen et al., 2011).

The development of skull asymmetry in odontocetes in contrast to mysticetes still needs further study. Analysis of symmetry during the embryonic period may be enlightening to determine if the findings of Sleptsov (1939) can be replicated. Likely there are additional mysticete asymmetries. Developmental biology may inform paleontology on these issues and vice versa (Thewissen et al., 2012).

Acknowledgments

We are grateful to Raphaela Stimmelmayr, Hans Thewissen, J. Craig George, Paul Nader, Department of Wildlife Management, North Slope Borough, Barrow, AK, Barrow Whaling Commission and Alaska Eskimo Whaling Commission for bowhead whale brains. Through this and other projects, Iñupiat Eskimos have made significant contributions to our knowledge of cetaceans.

References

Andersen, P., Morris, R., Amaral, D.G., Bliss, T., O'Keefe, J., 2007. The Hippocampus Book. Oxford University Press, New York.

Bellander, M., Berggren, R., Mårtensson, J., Brehmer, Y., Wenger, E., Li, T.Q., et al., 2015. Behavioral correlates of changes in hippocampal gray matter structure during acquisition of foreign vocabulary. Neuroimage 131, 205–213.

Berns, G.F., Cook, P.F., Foxley, S., Jbabdi, S., Miller, K.L., Marino, L., 2015. Diffusion tensor imaging of dolphin brains reveals direct auditory pathway to temporal lobe. Proc. Biol. Sci. 282 (1811), 20151203.

Boddy, A.M., McGowen, M.R., Sherwood, C.C., Grossman, L.I., Goodman, M., Wildman, D.E., 2012. Comparative analysis of encephalization in mammals reveals relaxed constraints on anthropoid and cetacean brain scaling. J. Evol. Biol. 25, 981–994.

Breathnach, A.S., 1955. The surface features of the brain of the humpback whale (*Megaptera novaeangliae*). J. Anat. 89, 343–354.

Bullock, T.H., Ridgway, S.H., 1972. Evoked potentials in the central auditory system of alert porpoises to their own and artificial sounds. How do Brains Work? Birkhäuser, Boston, MA.

Butti, C., Fordyce, R.E., Raghanti, M.A., Gu, X., Bonar, C.J., Wicinski, B., et al., 2014. The cerebral cortex of the pygmy hippopotamus, *Hexaprotodon liberiensis* (Cetartiodactyla, Hippopotamidae). Anat. Rec. 297, 670–700.

Butti, C., Hof, P.R., 2010. The insular cortex: a comparative perspective. Brain Struct. Funct. 214, 477–493.

Butti, C., Janeway, C.M., Townshend, C., Ridgway, S.H., Manger, P.R., Sherwood, C.C., et al., 2015. Neuronal morphology in cetartiodactyls. I. A comparative Golgi analysis of neuronal morphology in the bottlenose dolphin (*Tursiops truncatus*), the minke whale (*Balaenoptera acutorostrata*), and the humpback whale (*Megaptera novaeangliae*). Brain Struct. Funct. 220, 3339–3368.

Butti, C., Santos, M., Uppal, N., Hof, P.R., 2013. Von Economo neurons: clinical and evolutionary perspectives. Cortex 49, 312–326.

Cnotka, J., Mohle, M., Rehkamper, G., 2008. Navigational experience affects hippocampus size in homing pigeons. Brain Behav. Evol. 72, 233–238.

Dehnhardt, G., 2002. Sensory systems. In: Hoelzel, R. (Ed.), Marine Mammal Biology. Blackwell Publishing Company, Oxford, UK.

Dell, L.A., Karlsson, K.A., Patzke, N., Spocter, M.A., Siegel, J.M., Manger, P.R., 2016. Organization of the sleep related neural systems in the brain of the minke whale (*Balaenoptera acutorostrata*). J. Comp. Neurol. 524, 2018–2035.

Drake, S.E., Crish, S.D., George, J.C., Stimmelmayr, R., Thewissen, J.G.M., 2015. Sensory hairs in the bowhead whale *Balaena mysticetus* (Cetacea, Mammalia). Anat. Rec. 298, 1327–1335.

Duffield, D.W., Haldiman, J.T., Henk, W.G., 1992. Surface morphology of the forebrain of the bowhead whale, *Balaena mysticetus*. Mar. Mammal. Sci. 8, 354–378.

Fahlke, J.M., Gingerich, P.D., Welsh, R.C., Wood, A.R., 2011. Cranial asymmetry in Eocene archaeocete whales and the evolution of directional hearing in water. Proc. Natl. Acad. Sci. U. S. A. 108, 14545–14548.

Ferrer, I., Perera, M., 1988. Structure and nerve cell organisation in the cerebral cortex of the dolphin *Stenella coeruleoalba* a Golgi study. With special attention to the primary auditory area. Anat. Embryol. (Berl.) 178, 161–173.

Furutani, R., 2008. Laminar and cytoarchitectonic features of the cerebral cortex in the Risso's dolphin (*Grampus griseus*), striped dolphin (*Stenella coeruleoalba*), and bottlenose dolphin (*Tursiops truncatus*). J. Anat. 213, 241–248.

Gans, C., Northcutt, R.G., 1983. Neural crest and the origin of vertebrates: a new head. Science. 220, 268–273.

Glezer, I.I., Morgane, P.J., 1990. Ultrastructure of synapses and golgi analysis of neurons in neocortex of the lateral gyrus (visual cortex) of the dolphin and pilot whale [published erratum appears in Brain Res Bull 1990 Apr;24 (4):639–40]. Brain Res. Bull. 24, 401–427.

Haberland, M., Mokalled, M.H., Montgomery, R.L., Olson, E.N., 2009. Epigenetic control of skull morphogenesis by histone deacetylase 8. Genes Dev. 23, 1625–1630.

Heyning, J.E., Mead, J.G., 1990. Evolution of the nasal anatomy of cetaceans. In: Thomas, J., Kastelein, R. (Eds.), Sensory Abilities of Cetaceans. Springer, Boston, MA.

Hof, P.R., Chanis, R., Marino, L., 2005. Cortical complexity in cetacean brains. Anat. Rec. A Discov. Mol. Cell Evol. Biol. 287, 1142–1152.

Hof, P.R., Van der Gucht, E., 2007. Structure of the cerebral cortex of the humpback whale, *Megaptera novaeangliae* (Cetacea, Mysticeti, Balaenopteridae). Anat. Rec. 290, 1–31.

Jacobs, M.S., Galaburda, A.M., McFarland, W.L., Morgane, P.J., 1984. The insular formations of the dolphin brain: quantitative cytoarchitectonic studies of the insular component of the limbic lobe. J. Comp. Neurol. 225, 396–432.

Jerison, H.J., 1973. Evolution of the Brain and Intelligence. Academic Press, New York.

Jerison, H.J., 1977. The theory of encephalization. Ann. N. Y. Acad. Sci. 30, 146–160.

Jones, E., 2007. The Thalamus, second ed. Cambridge University Press, Cambridge.

Kishida, T., Thewissen, J.G.M., Usip, S., Suydam, R.S., George, J.C., 2015. Organization and distribution of glomeruli in the bowhead whale olfactory bulb. Peer J. e897.

Kraus, C., Pilleri, G., 1969. Zur Histologie der Grosshirnrinde von *Balaenoptera borealis* Lesson (Cetacea, Mysticeti). Investig. Cetacea 1, 151–170.

Lyamin, O.I., Manger, P.R., Ridgway, S.H., Mukhametov, L.M., Siegel, J.M., 2008. Cetacean sleep: an unusual form of mammalian sleep. Neurosci. Biobehav. Rev. 32, 1451–1484.

Manger, P., Sum, M., Szymanski, M., Ridgway, S.H., Krubitzer, L., 1998. Modular subdivisions of dolphin insular cortex: does evolutionary history repeat itself? J. Cogn. Neurosci. 10, 153–166.

Manger, P.R., 2006. An examination of cetacean brain structure with a novel hypothesis correlating thermogenesis to the evolution of a big brain. Biol. Rev. 81, 293–338.

Manger, P.R., Hemingway, J., Haagensen, M., Gilissen, E., 2010. Cross-sectional area of the elephant corpus callosum: comparison to other Eutherian mammals. Neuoscience 167, 815–824.

Manger, P.R., Prowse, M., Haagensen, M., Hemingway, J., 2012. Quantitative analysis of neocortical gyrencephaly in African elephants (*Loxodonta africana*) and six species of cetaceans: comparisons with other mammals. J. Comp. Neurol. 520, 2430–2439.

Mead, J.G., Fordyce, R.E., 2009. The therian skull: a lexicon with emphasis on the odontocetes. Smithson. Contrib. Zool. 1–249. Available from: https://doi.org/10.5479/si.00810282.627.

Morgane, P., Jacobs, M., McFarland, W., 1980. The anatomy of the brain of the bottlenose dolphin (*Tursiops truncatus*). Surface configuration of the telencephalin of the bottlenose dolphin with comparative anatomical observations in four other cetacean species. Brain Res. Bull. 5, 1–107.

Nimchinsky, E.A., Gilissen, E., Allman, J.M., Perl, D.P., Erwin, J.M., Hof, P.R., 1999. A neuronal morphologic type unique to humans and great apes. Proc. Natl. Acad. Sci. U. S. A. 96, 5268–5273.

Oelschläger, H.H.A., Oelschläger, J.S., 2008. Brain. In: Perrin, W.F., Würsig, B., Thewissen, J.G.M.H. (Eds.), Encyclopedia of Marine Mammals. Elsevier, Amsterdam, Netherlands.

Patzke, N., Olaleye, O., Haagensen, M., Hof, P.R., Ihunwo, A.O., Manger, P.R., 2014. Organization and chemical neuroanatomy of the African elephant (*Loxodonta africana*) hippocampus. Brain Struct. Funct. 219, 1587–1601.

Patzke, N., Spocter, M.A., Karlsson, K.E., Bertelsen, M.F., Haagensen, M., Chawana, R., et al., 2015. In contrast to many other mammals, cetaceans have relatively small hippocampi that appear to lack adult neurogenesis. Brain Struct. Funct. 220, 361–383.

Pihlström, H., 2008. Comparative anatomy and physiology of chemical senses in aquatic mammals. In: Thewissen, J.M., Nummela, S. (Eds.), Sensory Evolution on the Threshold: Adaptations in Secondarily Aquatic Vertebrates. University of California Press, Berkeley, CA.

Pilleri, G., 1964. Morphologie des Gehirnes des "Southern right whale", *Eubalaena australis* Desmoulins 1822 (*Cetacea, Mysticeti, Balaenidae*). Acta Zool. 46, 246–272.

Pilleri, G., 1966a. Morphologie des Gehirnes des Buckelwasl, *Megaptera novaeangliae* Borowski (*Cetacea, Mysticeti, Balaenopteridae*). J. Hirnforsch. 8, 437–491.

Pilleri, G., 1966b. Morphologie des Gehirnes des Seiwals, *Balaenoptera borealis* Lesson (*Cetacea, Mysticeti, Balaenopteridae*). J. Hirnforsch. 8, 221–267.

Raghanti, M.A., Spurlock, L.B., Treichler, F.R., Weigel, S.E., Stimmelmayr, R., Butti, C., et al., 2015a. An analysis of von Economo neurons in the cerebral cortex of cetaceans, artiodactyls, and perissodactyls. Brain Struct. Funct. 220, 2303–2314.

Raghanti, M.A., Spurlock, L.B., Uppal, N., Sherwood, C.C., Butti, C., Hof, P.R., 2015b. Von Economo neurons. In: Zilles, K., Amunts, K. (Eds.), Brain Mapping: An Encyclopedic Reference. Elsevier Science & Technology Books.

Raghanti, M.A., Wicinski, B., Meierovich, R., Warda, T., Dickstein, D.L., Reidenberg, J.S., et al., 2018. A comparison of the cortical structure of the bowhead whale (Balaena Mysticetus), a basal mysticete, with other cetceans. Anat. Rec. 302, 745–760.

Ratner, C., Riise, J., Eriksen, N., Pakkenberg, B., 2010. A postmortem study of the corpus callosum in the common minke whale (*Balaenoptera acutorostrata*). Mar. Mammal. Sci. 27, 688–700.

Ridgway, S.H., Brownson, R.H., 1984. Relative brain sizes and cortical surface areas of odontocetes. Acta Zool. Fen. 172, 149–152.

Ridgway, S.H., Carlin, K.P., Van Alstyne, K.R., Hanson, A.C., Tarpley, R.J., 2016. Comparison of dolphin's body and brain measurements with four other groups of cetaceans reveals great diversity. Brain Behav. Evol. 88, 235–257.

Ridgway, S.H., Demski, L.S., Bullock, T.H., Schwanzel-Fukuda, M., 1987. The terminal nerve in odontocete cetaceans. Ann. N. Y. Acad. Sci. 519, 201–212.

Ridgway, S.H., Hanson, A.C., 2014. Sperm whales and killer whales with the largest brains of all toothed whales show extreme differences in cerebellum. Brain Behav. Evol. 83, 266–274.

Sleptsov, M.M., 1939. On the problem of the asymmetry of the skull in odontoceti. Zoologichcskiy Zhumal 18, 367–386.

Sokolov, V.E., Ladygina, T.F., Supin, A., 1972. Localization of sensory zones in the dolphin cerebra cortex. Dokl. Akad. Nauk. SSSR 202, 490–493.

Tarpley, R.J., Ridgway, S.H., 1994. Corpus callosum size in delphinid cetaceans. Brain Behav. Evol. 44, 156–165.

Tarr, B.A., Rabinowitz, J.S., Ali Imtiaz, M., De Voogd, T.J., 2009. Captivity reduces hippocampal volume but not survival of new cells in a food-storing bird. Dev. Neurobiol. 69, 972–981.

Thewissen, J.G.M., Cooper, L.N., Behringer, R.R., 2012. Developmental biology enriches paleontology. J. Vert. Paleontol. 32, 1223–1234.

Thewissen, J.G.M., George, J., Rosa, C., Kishida, T., 2011. Olfaction and brain size in the bowhead whale (*Balaena mysticetus*). Mar. Mammal. Sci. 27, 282–294.

Wada, N., Javidan, Y., Nelson, S., Carney, T.J., Kelsh, R.N., Schilling, T.F., 2005. Hedgehog signaling is required for cranial neural crest morphogenesis and chondrogenesis at the midline in the zebrafish skull. Development 132, 3977–3988.

Wright, A., Scadeng, M., Stec, D., Dubowitz, R., Ridgway, S.H., Leger, J.S., 2017. Neuroanatomy of the killer whale (*Orcinus orca*): a magnetic resonance imaging investigation of structure with insights on function and evolution. Brain Struct. Funct. 222, 417–436. Available from: https://doi.org/10.1007/s00429-016-1225-x.

Zilles, K., Armstrong, E., Moser, K.H., Schleicher, A., Stephan, H., 1989. Gyrification in the cerebral cortex of primates. Brain Behav. Evol. 34, 143–150.

CHAPTER

18

Sensory systems

J.G.M. Thewissen[1], *J.C. George*[2], *Robert Suydam*[2] *and Todd L. Sformo*[2]

[1]Department of Anatomy and Neurobiology, Northeast Ohio Medical University, Rootstown, OH, United States [2]Department of Wildlife Management, North Slope Borough, Utqiaġvik, AK, United States

Introduction

All aquatic mammals had terrestrial ancestors whose sensory organs were adapted for functioning with air as a medium and a firm substrate as a platform. As they evolved to live in water, the sense organs changed. Even after millions of years of evolution, some of these taxa (e.g., pinnipeds, otters) live critical parts of their lives on land, and therefore their sense organs need to function in both air and water. Cetaceans and sirenians are obligately aquatic but the legacy of their terrestrially adapted sense organs is still present in their functional morphology. Additional changes in the sense organs were needed for specific clades. Bowhead whales live in cold water among the ice (Fig. 18.1), and their sensory organs underwent changes related to the Arctic environment.

In general, the secondarily aquatic sense organs of marine mammals have been studied well (Watkins and Wartzok, 1985; Wartzok and Ketten, 1999; Dehnhardt, 2002; Thewissen and Nummela, 2008). The current chapter describes sensory anatomy and function in bowhead whales. The sensory anatomy retains aspects of the sensory function of its terrestrial ancestors, its cetacean relatives, as well as specific features that are adaptations to its environment.

Most information on the sensory system of bowhead whales for this review was gathered by studying samples collected by the North Slope Borough, Department of Wildlife Management from Iñupiat subsistence–hunted. Chapter 8 explains collecting and sample curation of these specimens.

The Bowhead Whale
DOI: https://doi.org/10.1016/B978-0-12-818969-6.00018-2

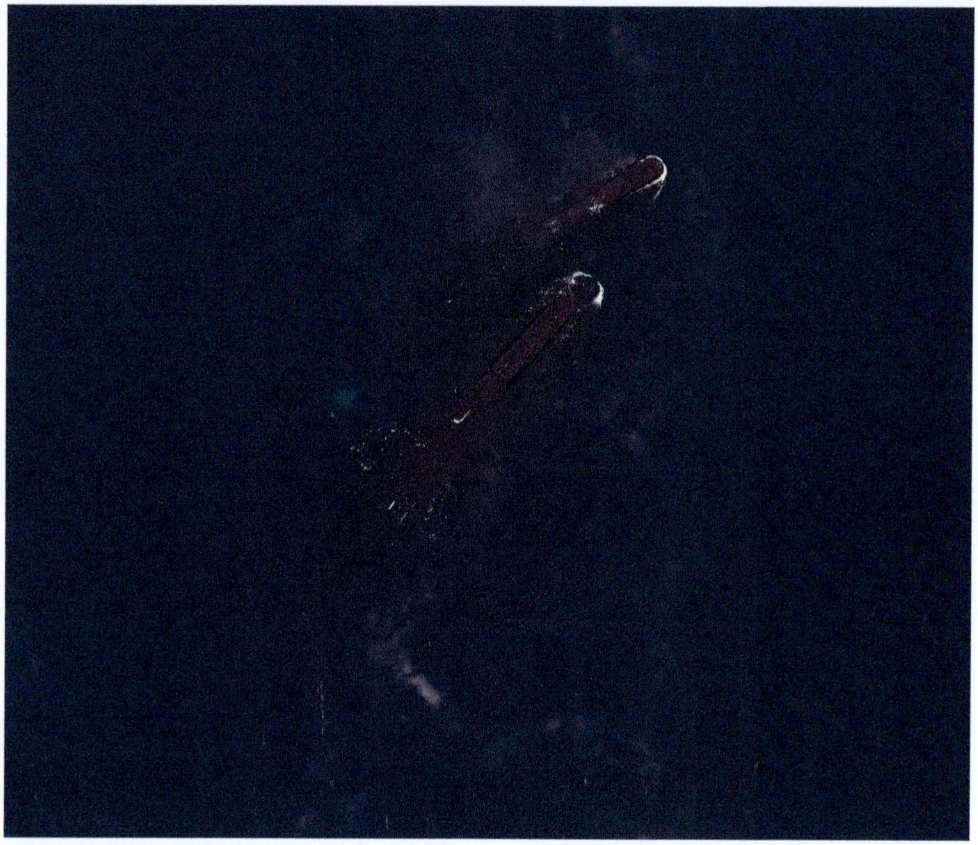

FIGURE 18.1 Two bowhead whales in the Chukchi Sea covered with a thin layer of young ice. The lower animal has opened its blowholes. When inhaling, bowheads can detect airborne odors. Source: *Photo by Amelia Brower (NOAA/North Slope Borough, NMFS Permit 14245).*

Olfaction and gustation

Thewissen et al. (2011) showed that bowhead whales retain an olfactory bulb, the part of the brain dedicated to processing smells. The bowhead olfactory bulb is located at the end of a long recess of the cranial cavity in adult individuals (Fig. 18.2). As a result, the olfactory bulb is immediately adjacent to the nasal chamber, an expanded part of the nasal passages located ventral and caudal to the blowhole. Nerve fibers pass from the olfactory epithelium of the nasal chamber to the olfactory bulb through foramina in the cribriform plate. The bowhead bony anatomy related to olfaction was described by Flower (Eschricht and Reinhardt, 1866) and is similar to that of other mysticetes (Cave, 1988; Godfrey et al., 2013). In bowheads, the olfactory bulb makes up approximately 0.13% of the total brain weight (Thewissen et al., 2011). This percentage is similar to that in many monkeys and considerably larger than that in apes (0.06%) and humans (0.008%). This suggests, by analogy, that olfaction is functional in bowheads. In one of our bowhead specimens

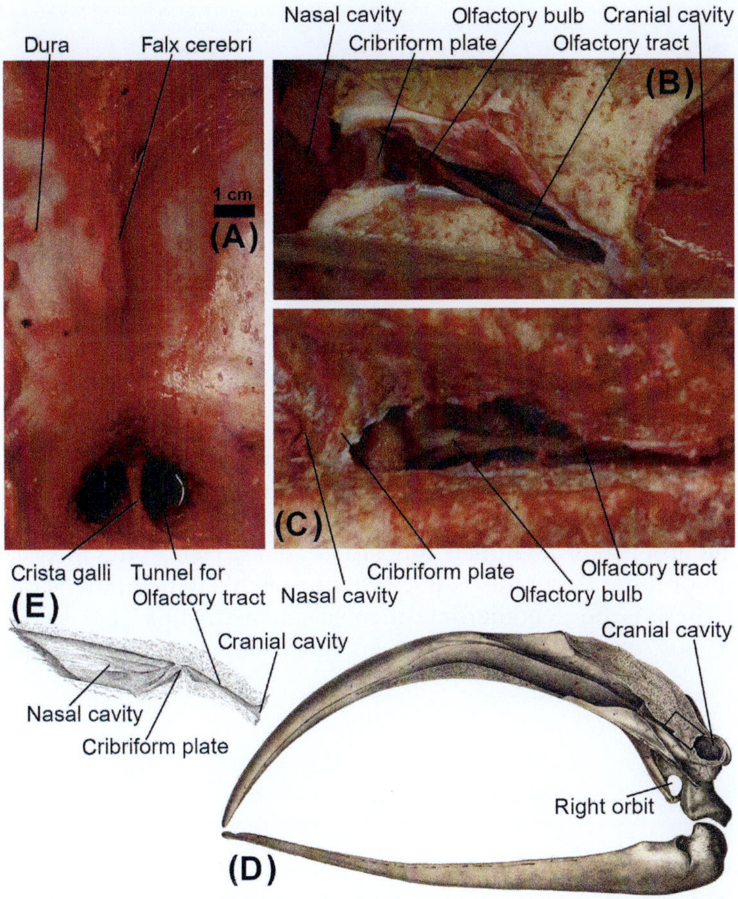

FIGURE 18.2 Olfaction-related structures in the bowhead whale. (A) Anterior wall of the cranial cavity showing the tunnels for left and right olfactory tract in the ethmoid bone (NSB-DWM 2008B11). (B) and (C) Cribriform plate and adjacent areas dissected from lateral [(B) NSB-DWM 2009B11; (C) NSB-DWM 2014B18]. Note that olfactory bulb and tract are much smaller than the cavities they are located in. (D) Sagittal cut of bowhead whale skull, with rectangle showing approximate location of (B) and (C). (E) Bony olfactory morphology in left lateral view. Source: *From (D) Eschricht, D.F., Reinhardt, J. 1866. On the Greenland Right-Whale (*Balaena mysticetus*). The Ray Society, London, pp. 1–150, 6pl; (E) Flower in Eschricht, D.F., Reinhardt, J. 1866. On the Greenland Right-Whale (*Balaena mysticetus*). The Ray Society, London, pp. 1–150, 6pl.*

(NSB-DWM 2009B14), we weighed the left and right olfactory bulb, finding that these are surprisingly different in size (Chapter 17). Histologically, the olfactory bulb is as complex as that of terrestrial mammals (Fig. 18.3). This is in contrast with modern odontocetes, where no olfactory bulb is present (Oelschläger and Oelschläger, 2002).

Genomic data indicate that 49% of bowhead olfactory receptor genes are mutated and nonfunctional (Thewissen et al., 2011). As such, bowheads have more functional olfactory receptor genes than minke whales (58% nonfunctional) and much more than odontocetes (more than 75% nonfunctional; Kishida et al., 2007; McGowen et al., 2008). Taken together, this suggests that olfaction is of importance to bowheads, and it has been proposed that mysticetes detect airborne odorants of their prey (Cave, 1988; Thewissen et al., 2011), in a fashion similar to procellariiform birds (Nevitt et al., 1995).

Bowheads, or any other modern cetacean for that matter, lack the vomeronasal organ and its neural associate the accessory olfactory bulb (Kishida et al., 2015). The status of the sense of taste is difficult to determine in bowheads. Their tongue is smooth and appears to lack taste buds (Haldiman and Tarpley, 1993; Chapter 14). The genome of all

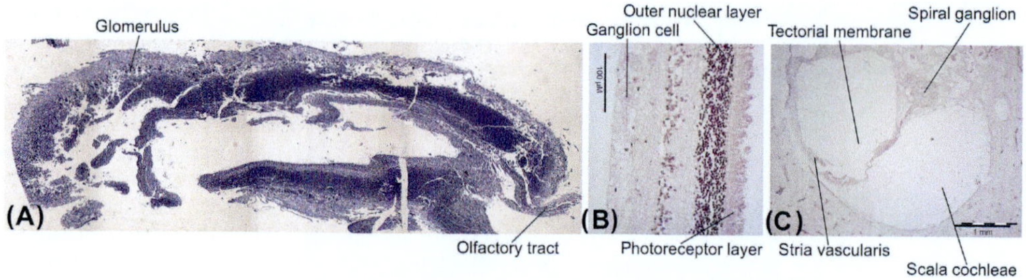

FIGURE 18.3 Histology of bowhead sensory organs. (A) Olfactory bulb, in horizontal section, showing the layers typical of other mammals (NSB-DWM 2009B14), lateral to top of photo, caudal to right. (B) Section of the retina (NSB-DWM 2018B1), showing unusually large ganglion cells. (C) cross section through cochlear whorl (NSB-DWM 2009B6, slide 540).

cetaceans lacks the genes responsible for detecting sweet, bitter, umami, and sour flavorants (Feng et al, 2014; Zhu et al., 2014; Kishida et al., 2015). However, behavioral experiments have noted that some cetaceans respond to bitter and sour flavorants (Nachtigall and Hall, 1984), in spite of the absence of identified functional genes. Hence, the mechanism for the possible detection of these flavorants is not known. It is likely that bowheads can detect salty flavors, and there is no genome data inconsistent with this. It is likely that bowheads retain a nervous terminalis as do other cetaceans, but unlike most mammals (Chapter 17).

Vision and magnetosense

Mayer (1852) and Zhu et al. (2001) provided a detailed description of the gross anatomy of the bowhead eye. The eyes are located at the widest part of the head and their optical axes deviate 150–180 degrees, making it unlikely that bowheads have stereoscopic vision. In life the eyeball protrudes from the orbit, but the deep conjunctival sac and large retractor bulbi muscle allow the eyeball to be retracted far into the orbit. Traditional knowledge confirms this behavior (Dubielzig and Aquirre, 1980), and similar behavior has been documented in other cetaceans (Dawson et al., 1972). Bowhead eyes are surrounded by a ring of glands that includes the homologues of the lacrimal, Harderian, and conjunctival glands (Rehorek et al., 2020) In investigated species, such glands produce "whale tears" that may play a role in refracting light and/or protecting the eye from abrasion (Kröger and Katzir, 2008). Rehorek et al. (2020) proposed that these glands have a function as a defense mechanism against bacteria.

Since the bowhead head is large and the eyes placed far laterally, the optic nerve is long; more than 1 m in large individuals (Zhu et al., 2001). This affects the response time to visual information, and nerve fibers in the optic nerve have large diameters (Smith et al., 2018, In prep), in order to increase transmission speeds. The optic nerve of bowhead whales contains approximately 245,000 axons, comparable to 347,000 in humpback and 252,000 in fin whales (Smith et al., 2018).

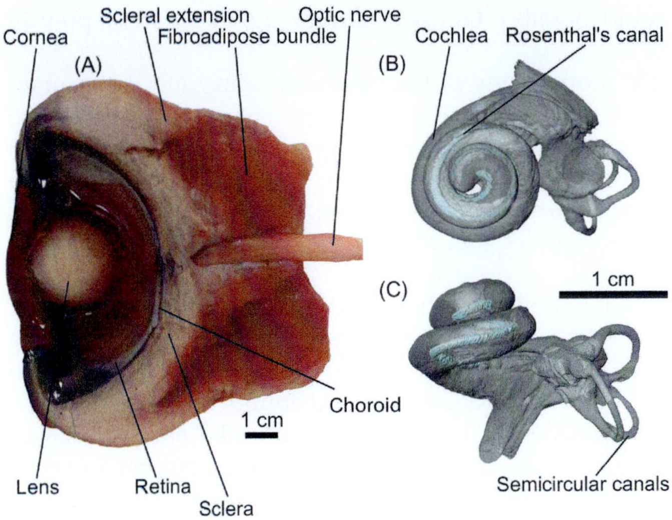

FIGURE 18.4 (A) Section through bowhead left eye (NSB-DWM 2019B1), showing oval shape, thick sclera, and globular lens. Note that the eyeball is flattened (somewhat exaggerated in this dissection). (B–C) 3D reconstructions of CT scans of bowhead bony labyrinth of the inner ear (NSB-DWM 2011B9).

The eyeball is oval in cross section (Fig. 18.4A) and flattened in the direction of its optical axis. The cornea and sclera are thick, the latter three times as thick as in other cetaceans (Zhu et al., 2001). The cornea has limited refractive abilities underwater; its thickness probably is related to the need to withstand abrasion from (ice and sediment) particles. The lens is spherical with no or limited ciliary musculature to change its shape (Mayer, 1852). These are adaptations for seeing under water, where the convex lens is the main refractive structure (Kröger and Katzir, 2008), in spite of being less effective than in air. A rete mirabile is located deep to the eyeball and surrounds the optic nerve. As in all cetaceans (Slijper, 1962), a tapetum lucidum is present. The pupil is oval, with the average rostrocaudal length 1.4 times that of the dorsoventral height (Zhu et al., 2001).

Zhu et al. (2001) observed that the bowhead retina lacked a macroscopically observable fovea and macula, similar to other cetaceans (Pilleri and Wandeler, 1970). However, a streak of high ganglion cell density occurs in investigated mysticetes (Mass and Supin, 2018), but its presence has not been studied in bowhead. Histologically, the bowhead retina consists of the same layers as in other mammals (Fig. 18.3B), and ganglion cells are enormous (Smith et al., In prep), as in other cetaceans. This is possibly an adaptation for low-light conditions at depth (Reuter and Peichl, 2008). The photosensitive layer of the mammalian retina consists of rod and cone cells that contain opsins as their visual pigments. Absorption maxima for the rod opsin of bowheads have not been determined, but Bischoff et al. (2012) made inferences based on a study of the genome of several mysticete species. They estimated that the bowhead maximum absorption frequency is 493 nm, which is below that of land mammals but above that of odontocetes. These authors characterized this frequency as optimal for surface waters.

Meredith et al. (2013) found that bowhead whales have no functional working cone opsin genes, which implies that they only have a single functional visual pigment (rod opsin). That would indicate that bowheads are color-blind. Meredith et al. (2013) considered the absence of working cone opsins as an adaptation to limit interference with the function of rod opsin and that it could possibly be used to track bioluminescent prey in the darkness of the deep ocean.

Schweikert et al. (2016) studied the morphology of the bowhead retina and determined that cone cells were present, but that they lack the areas where cone opsins would be located: the outer segments. They proposed that cone cells were retained in bowhead in order to support the processing of visual information collected by rod cells. Cones have been documented in the retina of fin whales (Mass and Supin, 1997) and minke whales (Murayama et al., 1992), which are also species that are rod monochromats according to Meredith et al. (2013).

Smith et al. (In prep) studied regional variations in bowhead retina histology and found that cones maintain outer segments in some, but not all regions of the retina. This begs the question of what function these cones have, since they must not contain functional cone opsins. Smith et al. (In prep) used immunohistochemistry and found that two proteins known to be involved with magnetoreception in birds, CRY (Wiltschko and Wiltschko, 2002; Watari et al., 2012) and MAGR (Qin et al., 2016), are found in rods, but not in cones of the bowhead retina. While this evidence is only suggestive, it is possible that cone cells of the bowhead retina are engaged in detecting the magnetic field. Nießner et al. (2016) did not detect the presence of CRY1a in the retina of two other cetaceans (*Globicephala* and *Balaenoptera acutorostrata*) using immunohistochemistry, although they only investigated a small part of the retina. The presence of magnetosense in bowhead is consistent with behavioral studies on other cetaceans (reviewed by Hoffman and Wilkens, 2008; Kremers et al., 2016). An alternate hypothesis is that the location of the organ detecting the magnetic field in cetaceans is in the dura mater (Cozzi et al., 2017). This does not resolve what the function of cones lacking an outer segment is.

Audition

The mammalian ear consists of outer, middle, and inner ear. Functionally, the cetacean outer and middle ear differ greatly from that of land mammals, but the inner ear does not. As in all cetaceans, the bowhead external auditory meatus, a part of the outer ear, is anatomically patent but functionally closed, and there is no opening to the surface (Chapter 8). The external auditory meatus sheds its epithelium annually and these shed layers accumulate and form "earwax." Usually there are no growth layers in this wax (Rehorek et al., 2018). As in all mysticetes, the eardrum is thick and fibrous, one part of it projects into the external auditory meatus, the "glove finger," whereas another part projects into the middle ear cavity, the tympanic ligament (Fig. 18.5).

The tympanic bone surrounds the middle ear cavity, and it is attached to the petrosal by means of two synostoses dorsolateral to the middle ear cavity (Rehorek et al., 2018), an arrangement similar to that of right whales (Parks et al., 2007). As a result, the medial lip of the tympanic, the involucrum, is surrounded by soft tissue: mostly fat and a connective tissue capsule. The bowhead involucrum shows growth layers that can be used in age estimation (Sensor, 2017, Chapter 21). Parks et al. (2007) note that this structure is present in three of four right whales.

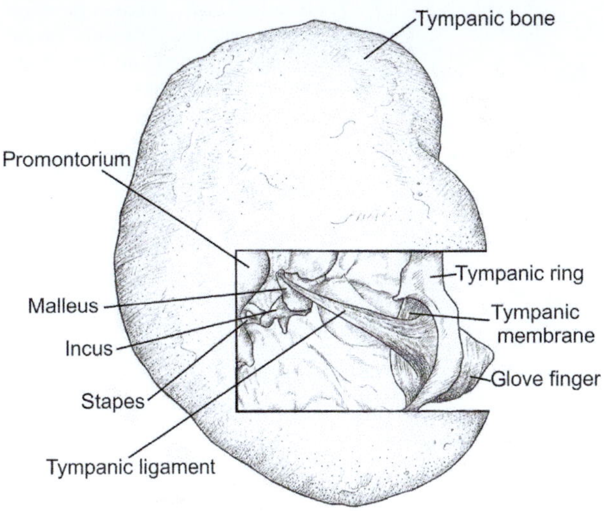

FIGURE 18.5 Diagram of the left middle ear of the bowhead whale in ventral view, with a window cut into the tympanic bone. Lateral to right and rostral to the top of the page. Only bones and tympanic membrane are shown. Iñupiat hunters refer to the tympanic bone as the "eardrum" or siuti, and this structure plays an important role in Iñupiat culture. Source: *Illustration by Kat LoGrande.*

The anterior process of the malleus is fused to the tympanic bone rostral to the tympanic membrane, as it is in all cetaceans. Synovial joints are present between malleus and incus and between incus and stapes. The stapes is firmly attached in the oval window, but it is not synostosed, at least in young whales (as based on histological sections of NSB-DWM 2009B6, 6 years old). External auditory meatus, tympanic membrane, and ear ossicles are involved in sound transmission in land mammals (Lombard and Hetherington, 1993), but that mechanism is unlikely to function in cetaceans (Wartzok and Ketten, 1999). In many odontocetes a fat pad in the mandible conducts sound, and the mandibular foramen of bowhead whales does contain a fat pad (Fig. 3N in Thewissen et al., 2017). Currently, however, the most reasonable hypothesis for sound conduction in mysticetes is that of Cranford and Krysl (2018). They suggested, based on modeling of fin whale CT-data, that bone conduction through the skull is involved. Whereas bone conduction is generally thought to be inconsistent with binaural hearing (Nummela et al., 2007), the models by Cranford and Krysl (2018) show that this is not the case in fin whales. The firm synostosis between the skull and the petrosal bone of bowheads is consistent with the bone conduction hypothesis.

Acoustic data (Chapter 22) indicate that bowhead whales communicate using sound frequencies of 0.02–3.5 kHz, and this implies that this frequency range is included in their hearing range. In addition, there are indications that bowheads listen to the reflections of their sounds to determine the location of objects around them (George et al., 1989), a basic form of echolocation. Parks et al. (2007) modeled hearing range of North Atlantic right whales, a close relative of bowheads, and found it most consistent with a frequency range of 0.01–22 kHz.

Sensor (2017) provided the most detailed description of the bowhead inner ear. The bowhead cochlea consists of 2.25 whorls (Fig. 18.4B), and its basilar membrane has a mean length of 43.4 mm ($n = 4$). Rosenthal's canal (where the spiral ganglion is located) has a mean length of 41.2 mm ($n = 6$) and tapers apically. The mean number of spiral ganglion

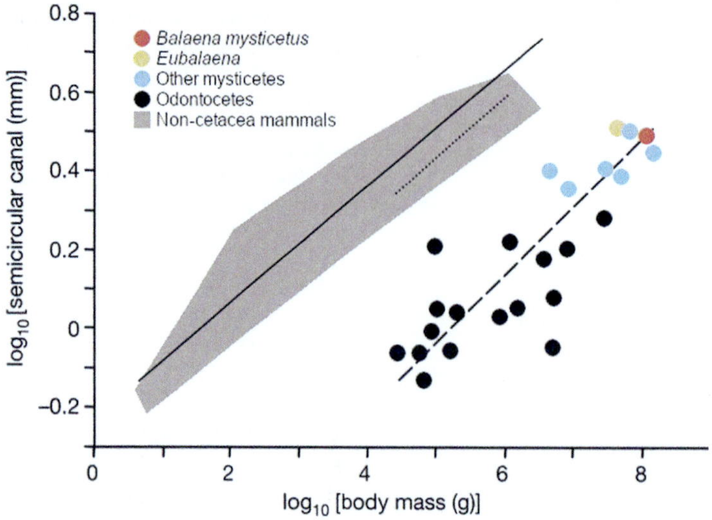

FIGURE 18.6 Radius of semicircular canals, as measured from CT-scans, plotted against body size for cetaceans (individual points with regression line) and all other mammalian orders (in gray envelope with regression line). Dotted line represents regression for modern artiodactyls, the closest land relatives of cetaceans. Source: *Redrawn from Spoor, F., Bajpai, S., Hussain, S.T., Kumar, K., Thewissen, J.G.M., 2002. Vestibular evidence for the evolution of aquatic behaviour in early cetaceans. Nature 417, 163–166.*

cells is 152,672 (s.d. = 45.888, $n = 3$), similar to that of fin whales (134,098, Wartzok and Ketten, 1999) and humpback whales (156,374, Ketten, 1992).

Balance

The organ of balance in mammals consists of two sac-like organs that measure linear acceleration and three semicircular canals that measure angular acceleration (Sipla and Spoor, 2008). Aspects of size and shape of the latter can be studied using CT-scan, as the semicircular canals are a system of cavities in the petrosal bone of the ear. This is the only part of the organ of balance that has been studied in bowhead whales.

The radius of the semicircular canals broadly scales linearly (on a log scale) with body size across all mammals except cetaceans (Fig. 18.6). These canals of cetaceans have a diameter approximately three orders of magnitude smaller than those of other mammals (Spoor et al., 2002). Bowheads have average-sized canals among cetaceans. At this point, there is no satisfactory hypothesis for the small size of cetacean semicircular canals or why there is so much variation among cetacean semicircular canal radii (but see Spoor and Thewissen, 2007; Kandel and Hullar, 2010).

Mechanosense

Haldiman and Tarpley (1993) described patches of large sensory papillae in bowhead skin on the eyelids, tips of the rostrum, and lips and suggested that they function as pressure sensors. Bowheads are one of the few cetacean species that have hairs (Drake et al., 2015). Hairs occur in three clusters in postnatal individuals. Approximately 30 hairs occur on the tip of mandibles in distinct left and right clusters, with a row of 10 or so hairs extending posteriorly from this cluster on the mandible. Skin patches surrounding the mandibular hairs are often white. Approximately 50 hairs occur on left and right side of the tip of the rostrum, forming a single (unpaired) group. Finally, an arc of approximately five hairs occurs immediately posterior to the blowhole on both sides. Hairs in this last cluster are thicker than those in other clusters (Drake et al., 2015) and could serve to detect wind direction as the blowhole is above the water level. That might allow the whale to determine the direction from which perceived odorants originate.

Discussion

Bowhead whales, like other Arctic marine mammals, need their sense organs to detect prey, predators, and conspecifics, find breathing holes and migration routes, and to communicate with each other. Direct measurement of the function of sense organs of bowheads is currently impossible, but inferences can be made based on study of the anatomy and on observations of bowhead behavior. Traditional knowledge contributes independent observations to our understanding of the bowhead sensory landscape.

Bowhead vision is useful over short distances only. Traditional knowledge suggests that bowheads can see humans (such as hunters) standing on ice, when their eyes are submerged or emerged. Hunters also report that bowheads sometime lift their head out of the water as if to get a better look at an object on the ice. Iñupiat hunters in Utqiaġvik avoid wearing red on the ice, as they believe that this color is particularly easily visible to bowheads. Genomic evidence suggests that bowheads cannot see color (Meredith et al., 2013), but it is possible that bowheads use an unknown visual pigment in their cones, or that red contrasts greatly with the surrounding ice, and it is the contrast that the whales are perceiving. In an experiment designed to study interactions with fishing gear, Kraus et al. (2014) found that right whales collide less frequently with red and orange ropes than with green ropes hanging in the water. Right whales, just as bowheads, lack the cone pigments that would allow them to see red, but red objects contrast strongly against the green ocean or white ice and would be perceived by those species as black. It is clear that bowheads have a good sense of vision.

The sense of smell of bowhead is well known to Iñupiat hunters, and they avoid creating airborne odorants when they are hunting. Anatomical evidence confirmed that bowheads have a good sense of smell (Thewissen et al., 2011) and that it is used to detect food sources (Kishida et al., 2015). It is likely that bowheads are able to detect their prey, possibly because of the release of dimethylsulfide (DMS) into the air, as is known for krill-eating birds. Olfaction is a nondirectional sense organ, and only a small part of the whale is exposed to air as it inhales. It is likely that the sensory hairs near a bowhead blowhole are used to detect wind direction indicating where attractive smells are coming from. Right whales have been observed to swim into the wind when tourists on whale watching boats near them smelled boiled cabbage, which is how humans perceive DMS odors.

Hearing is also excellent, and hunters use skin boats that move through the water quietly as they hunt bowhead whales in spring, and attempts are made to limit loud pounding noises even on land, as hunting proceeds in waters near the ice-edge (Murdoch, 1885, Chapter 34). Iñupiat hunters were the first to note the effects of seismic surveys on bowhead migratory paths (Albert, 2001). Bowheads use low-frequency sounds to communicate with each other and these sounds travel for hundreds of miles through the water.

The limited evidence for magnetoreception in bowhead whales presented by Smith et al. (2018, In prep) assists in understanding migration trajectories. Balaenids in general frequent coastal areas where the coastline or ocean floor could be used in navigation. However, in spring bowheads migrate due east from Point Barrow to the eastern Beaufort Sea, passing areas where the sea is too deep (more than 3000 m) to allow navigation based on vision (Chapter 4). It is possible that they use magnetoreception to navigate through these areas.

References

Albert, T.F., 2001. The influence of Harry Brower, Sr., an Inupiaq Eskimo Hunter, on the bowhead whale research program conducted at the UIC-NARL faculty by the North Slope Borough. In: Norton, D. (Ed.), Fifty More Years Below Zero: Tributes and Mediations for the Naval Arctic Research Laboratory's First Half Century at Barrow, Alaska. University of Alaska, Fairbanks, pp. 265–278.

Bischoff, N., Nickle, B., Cronin, T.W., Velasquez, S., Fasick, J.I., 2012. Deep-sea and pelagic rod visual pigments identified in the mysticete whales. Visual Neurosci. 29, 95–103.

Cave, A.J.E., 1988. Note on olfactory ability of mysticetes. J. Zool. (London) 214, 307–311.

Cozzi, B., Huggenberer, S., Oelschläger, H., 2017. Anatomy of Dolphins, Insights Into Body Structure and Function. Elsevier, Amsterdam, 438 p.

Cranford, T., Krysl, P., 2018. Sound paths, cetaceans. In: Würsig, B., Thewissen, J.G.M., Kovacs, K. (Eds.), Encyclopedia of Marine Mammals, third ed. Academic Press, San Diego, CA, pp. 901–904.

Dawson, W., Birndorf, L.A., Perez, J.M., 1972. Gross anatomy of the dolphin eye (*Tursiops truncatus*). Cetology 10, 1–12.

Dehnhardt, G., 2002. Sensory systems. In: Hoelzel, A.R. (Ed.), Marine Mammal Biology, an Evolutionary Approach. Blackwell Science, Hoboken, New Jersey, pp. 116–141.

Drake, S.E., Crish, S.D., George, J.C., Stimmelmayr, R., Thewissen, J.G.M., 2015. Sensory hairs in the bowhead whale, *Balaena mysticetus* (Cetacea, Mammalia). Anat. Rec. 298, 1327–1335.

Dubielzig, R., Aquirre, G., 1980. Tissues (structure/function) RU 5790. In: Kelley, J.J., Larson, G.L. (Eds.), Investigations on the Occurrence and Behavior Patterns of Whales in the Vicinity of the Beaufort Sea Lease Area. Final Rep. Bur. Land Manag. Naval Arctic Res. Lab., pp. 283–296, NTIS No. PB86–1535661AS.

Eschricht, D.F., Reinhardt, J., 1866. On the Greenland Right-Whale (*Balaena mysticetus*). The Ray Society, London, pp. 1–150, 6pl.

Feng, P., Zheng, J., Rossiter, S.J., Wang, D., Zhao, H., 2014. Massive losses of taste receptor genes in toothed and baleen whales. Genom. Biol. Evol. Available from: https://doi.org/10.1093/gbe/evu095.

George, J.C., Clark, C., Carroll, G.M., Ellison, W.T., 1989. Observations on the ice-breaking and ice navigation behavior of migrating bowhead whales (*Balaena mysticetus*), near Point Barrow, Alaskaspring 1985 Arctic 42, 24–30.

Godfrey, S.J., Geisler, J., Fitzgerald, E.M.G., 2013. On the olfactory anatomy in an archaic whale (Protocetidae, Cetacea) and the minke whale *Balaenoptera acutorostrata* (Balaenopteridae, Cetacea). Anat. Rec. 296, 257–272.

Haldiman, J.T., Tarpley, R.J., 1993. Anatomy and physiology. In: Burns, J.J., Montague, J.J., Cowles, C.J. (Eds.), The Bowhead Whale. Soc. Mar. Mamm., Spec. Publ. 2, Lawrence, KS, pp. 71–156.

Hoffman, M.H., Wilkens, L.A., 2008. Magnetoreception. In: Thewissen, J.G.M., Nummela, S. (Eds.), Sensory Biology on the Threshold. Univ. California Pr., Berkeley, CA, pp. 317–324.

Kandel, B.M., Hullar, T.E., 2010. The relationship of head movements to semicircular canal size in cetaceans. J. Exp. Biol. 213, 1175–1181.

Ketten, D.R., 1992. The marine mammal ear: specializations for aquatic audition and echolocation. In: Webster, D.B., Fay, R.R., Popper, A.N. (Eds.), The Evolutionary Biology of Hearing. Springer-Verlag, New York, pp. 717–750.

Kishida, T., Kubota, S., Shirayama, Y., Fukami, H., 2007. The olfactory receptor gene repertoires in secondary-adapted marine vertebrates: evidence for reduction of the functional proportions in cetaceans. Biol. Lett. 3, 428–430.

Kishida, T., Thewissen, J.G.M., Hayakawa, T., Imai, H., Agata, K., 2015. Aquatic adaptation and the evolution of smell and taste in whales. Zool. Lett. Available from: https://doi.org/10.1186/s40851-014-0002-z.

Kraus, S., Fasick, J., Werner, T., McFarron, P., 2014. Enhancing the visibility of fishing ropes to reduce right whale entanglements. BREP 1, 67–75.

Kremers, D., Célérier, A., Schaal, B., Campagna, S., Trabalon, M., Böye, M., et al., 2016. Sensory perception in cetaceans: Part II—Promising experimental approaches to study chemoreception in dolphins. Front. Ecol. Evol. Available from: https://doi.org/10.3389/fevo.2016.00050.

Kröger, R.H.H., Katzir, G., 2008. Comparative anatomy of vision in aquatic tetrapods. In: Thewissen, J.G.M., Nummela, S. (Eds.), Sensory Biology on the Threshold, Adaptations in Secondarily Aquatic Vertebrates. Univ. California Pr., Berkeley, CA, pp. 121–148.

Lombard, R.E., Hetherington, T.E., 1993. Structural basis of hearing and sound transmission. In: Hanken, J., Hall, B.K. (Eds.), The Skull, Volume 3, Functional and Evolutionary Mechanisms. Univ. Chicago Pr, pp. 241–302.

Mass, A.M., Supin, A.Y., 1997. Ocular anatomy, retinal ganglion cell distribution and visual resolution in the grey whale, *Eschrichtius gibbosus*. Aquat. Mammal 23, 17–28.

Mass, A.M., Supin, A.Y., 2018. Vision. In: Würsig, B., Thewissen, J.G.M., Kovacs, K. (Eds.), Encyclopedia of Marine Mammals. Elsevier, London, pp. 1035–1044.

Mayer, E.F.J.K., 1852. Anatomische Untersuchungen über das Auge der Cetaceen nebst Bemerkungen über das Auge des Menschen und der Thiere. Henry & Cohen, Bonn, 55 p.

McGowen, M.R., Clark, C., Gatesy, J., 2008. The vestigial olfactory receptor subgenome of odontocete whales: phylogenetic congruence between gene-tree reconciliation and supermatrix methods. Syst. Biol 57, 574–590.

Meredith, R.W., Gatesy, J., Emerling, C.A., York, V.M., Springer, M.S., 2013. Rod monochromacy and the coevolution of cetacean retinal opsins. PLoS Genet. 9, e1003432. Available from: https://doi.org/10.1371/journal.pgen.1003432.

Murayama, T., Fujise, Y., Aoki, I., Ishii, T., 1992. Histological characteristics and distribution of ganglion cells in the retinae of the Dall's porpoise and minke whale. In: Thomas, J.A., Kastelein, R.A., Supin, A.Y. (Eds.), Marine Mammal Sensory Systems. Springer, Boston, MA, pp. 137–145.

Murdoch, J., 1885. "Natural History," in Patrick Henry Ray, Report of the International Polar Expedition to Point Barrow, Alaska. Government Printing Office, Washington, DC.

Nachtigall, P.E., Hall, R.W., 1984. Taste reception in the bottlenosed dolphin. Acta Zool. Fenn. 172, 147148.

Nevitt, G.A., Veit, R.R., Kareiva, P., 1995. Dimethyl sulphide as a foraging cue for Antarctic procellariiform seabirds. Nature 376, 680–682.

Nießner, C., Denzau, S., Malkemper, E.P., Gross, J.C., Burda, H., Winklhofer, M., et al., 2016. Cryptochrome 1 in retinal cone photoreceptors suggests a novel functional role in mammals. Sci. Rep 6, 21848.48.

Nummela, S., Thewissen, J.G.M., Bajpai, S., Hussain, T., Kumar, K., 2007. Sound transmission in archaic and modern whales: anatomical adaptations for underwater hearing. Anat. Rec. 290, 716–733.

Oelschläger, H.H.A., Oelschläger, J.S., 2002. Brain. In: Perrin, W.P., Würsig, B., Thewissen, J.G.M. (Eds.), The Encyclopedia of Marine Mammals. Academic Pr., San Diego, CA, pp. 133–158.

Parks, S.E., Ketten, D.R., O'Malley, J.T., Arruda, J., 2007. Anatomical predictions of hearing in the north Atlantic right whale. Anat. Rec. 290, 734–744.

Pilleri, G., Wandeler, A., 1970. Ontogeny and functional anatomy of the eye of the fin whale, *Balaenoptera physalus*. Invest. Cetacea 2, 179–229.

Qin, S., Yin, H., Yang, C., et al., 2016. A magnetic biocompass. Nat. Mater. 15, 217–226.

Rehorek, S.J., Stimmelmayr, R., George, J.C., Suydam, R., McBurney, D.M., Thewissen, J.G.M., 2018. Structure of the external auditory meatus of the bowhead whale (*Balaena mysticetus*) and its relation to their seasonal migration. J. Anat. Available from: https://doi10.1111/joa.129908.

Rehorek, S.J., Stimmelmayr, R., George, J.C., Suydam, R., McBurney, D.M., Thewissen, J.G.M., 2020. Whale tears in bowhead and beluga: source and function. J. Morphol. 281, 316–325. Available from: https://doi.org/10.1002/jmor.21099.

Reuter, T., Peichl, L., 2008. Structure and function of the Retina. In: Thewissen, J.G.M., Nummela, S. (Eds.), Sensory Biology on the Threshold, Adaptations in Secondarily Aquatic Vertebrates. Univ. California Pr, Berkeley, CA, pp. 149–173.

Schweikert, L.E., Fasick, J.I., Grace, M.S., 2016. Evolutionary loss of cone photoreception in balaenid whales reveals circuit stability in the mammalian retina. J. Comp. Neurol. 524, 2873–2885.

Sensor, J.D., 2017. Hearing and Age Estimation in Two Species of Arctic Whale (Ph.D. thesis). Kent State Univ., Kent, OH, 132 p.

Sipla, J.S., Spoor, F., 2008. The physics and physiology of balance. In: Thewissen, J.G.M., Nummela, S. (Eds.), Sensory Biology on the Threshold, Adaptations in Secondarily Aquatic Vertebrates. University of California Press, pp. 227–232.

Slijper, E.J., 1962. Whales. Hutchinson, London.

Smith, M.A., Waugh, D., George, J.C., Suydam, R.S., Thewissen, J.G.M., Crish, S.D., In prep. A comparative analysis of cone photoreceptors in bowhead and beluga whales.

Smith, M.A., Plyler, E.S., George, J.C., Thewissen, J., Crish, S.D., 2018. Comparative analysis of the visual system in two whale species. Soc. Neurosci. Abstract. Available from: https://www.abstractsonline.com/pp8/#!/4649/presentation/38268.

Spoor, F., Thewissen, J.G.M., 2007. Comparative and functional anatomy of balance in aquatic mammals. In: Thewissen, J.G.M., Nummela, S. (Eds.), Sensory Biology on the Threshold, Adaptations in Secondarily Aquatic Vertebrates. University of California Press, pp. 257–284.

Spoor, F., Bajpai, S., Hussain, S.T., Kumar, K., Thewissen, J.G.M., 2002. Vestibular evidence for the evolution of aquatic behaviour in early cetaceans. Nature 417, 163–166.

Thewissen, J.G.M., Nummela, S., 2008. Sensory Evolution on the Threshold: Adaptations in Secondarily Aquatic Vertebrates. University of California Press, Berkeley, CA, p. 351.

Thewissen, J.G.M., George, J.C., Rosa, C., Kishida, T., 2011. Olfaction and brain size in the bowhead whale (*Balaena mysticetus*). Mar. Mammal Sci 27, 282–294.

Thewissen, J.G.M., Hieronymus, T.L., George, J.C., Suydam, R., Stimmelmayr, R., McBurney, D., 2017. Evolutionary aspects of the development of teeth and baleen in the bowhead whale. J. Anat. 230, 549–566. Available from: https://doi.org/10.1111/joa.12579.

Wartzok, D., Ketten, D.R., 1999. Marine mammal sensory systems. In: Reynolds III, J.E., Rommel, S.A. (Eds.), Biology of Marine Mammals. Smiths. Inst. Pr., Washington, DC, pp. 117–175.

Watari, R., Yamaguchi, C., Zemba, W., Kubo, Y., Okano, K., Okano, T., 2012. Light-dependent structural change of chicken retinal Cryptochrome4. J. Biol. Chem. 287, 42634–42641.

Watkins, W.A., Wartzok, D., 1985. Sensory biophysics od marine mammals. Mar. Mammal Sci. I, 219–260.

Wiltschko, W., Wiltschko, R., 2002. Magnetic compass orientation in birds and its physiological basis. Naturwiss 89, 445–452.

Zhu, Q., Hillman, D.J., Henk, W.G., 2001. Morphology of the eye and surrounding structures of the bowhead whale, *Balaena mysticetus*. Mar. Mammal. Sci. 17, 729–750.

Zhu, K., Zhou, X., Xu, S., Sun, D., Ren, W., Zhou, K., et al., 2014. The loss of taste genes in cetaceans. BCM Evol. Biol. 14, 218.

CHAPTER

19

Endocrinology and blubber physiology

Rosalind M. Rolland

Anderson Cabot Center for Ocean Life, New England Aquarium, Boston, MA, United States

Introduction

The endocrine system is comprised of organs and tissues which synthesize and secrete hormones. Hormones bind to receptors, exerting actions on target cells to maintain homeostasis and orchestrate responses to internal demands and external factors. While endocrine studies of bowhead whales (*Balaena mysticetus*) are limited to date, research on other baleen whales (suborder Mysticeti) shows multiple hormones can be measured in many different sample types. Hormones have classically been analyzed in blood, but advances in physiological studies of mysticetes have led to new endocrine assessments using alternative biological matrices, including blubber, urine, feces, baleen, earplugs, and exhaled breath. Mysticete endocrine studies encompass research on reproductive status (using estrogens, progestogens, androgens), adrenal corticosteroids as stress indicators (cortisol, corticosterone, aldosterone), and energetic balance (using thyroid hormones and leptin). Analytical techniques involve hormone (or metabolite) extraction from the sample matrix, followed by analysis using immunoassays (enzyme or radioimmmunoassay) or liquid chromatography–mass spectrometry.

This chapter includes research on bowhead whale endocrinology and blubber physiology, and summarizes endocrine studies in other mysticetes with potential applicability to bowheads based on similar physiology and life history strategies (see Chapter 7). For a description of endocrine system anatomy in bowhead whales see Haldiman and Tarpley (1993).

Blubber structure and physiology

Blubber is a dynamic tissue with a key role in energy deposition and mobilization regulated primarily by the hormone leptin. For a detailed description of bowhead blubber histology see Haldiman and Tarpley (1993) and Rosa (2006). The term "blubber"

The Bowhead Whale
DOI: https://doi.org/10.1016/B978-0-12-818969-6.00019-4

refers to subcutaneous adipose tissue that is part of the dermis, which lies just beneath the epidermis, or skin (Fig. 19.1; Haldiman and Tarpley, 1993). Underneath the dermis is the hypodermis, which also contains adipose tissue (especially pronounced in nursing calves and postweaning juveniles, see Fig. 19.1A), but is distinct structurally and histologically from the blubber layer. Blubber consists of adipose cells within a structural matrix of dense collagen fibers. Adipocyte cell size and collagen density vary by depth, and more blood vessels occur in deeper layers (Rosa, 2006; Ball et al., 2015). Ball et al. (2015) found greater fiber density in superficial blubber layers, and the largest adipocytes in deep layers with a gradual trend in structural changes. Although blubber thickness does not vary significantly between fall and spring in adult bowheads (George et al., 2015), Ball et al. (2015) found spring-sampled adults had smaller adipocytes and higher structural collagen density. Rosa (2006) did not detect this seasonal difference, but found a positive correlation between blubber collagen percent and blubber thickness. Ball et al. (2015) found higher levels of collagen in adults, and larger adipocytes in juveniles.

Leptin: the blubber hormone

Bowheads depend upon deposition of energetic reserves in blubber and utilization of those resources during fasting intervals (e.g., winter) to meet the demands of growth, reproduction, migration, and other energetically demanding life history events. Leptin is a

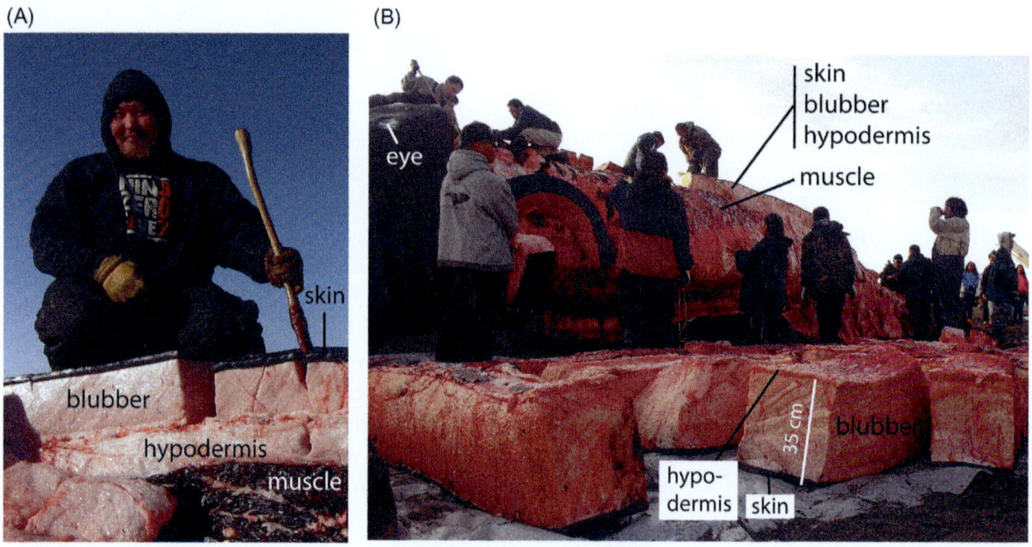

FIGURE 19.1 Blubber of bowhead whales caught as part of the Iñupiat subsistence hunt. (A) Cross-section of skin, blubber, and hypodermis from a yearling ingutuk whale (see Chapter 7 for life history stages), showing underlying muscle and the thickened hypodermis characteristic of young bowhead whales (NSB-DWM 2017B3; pictured with Iñupiaq whale hunter (used with permission), Kyle Bodfish; photo by J. C. George). (B) Blubber strips removed from a pregnant female bowhead showing a very thick blubber layer, with the whale in the background (NSB-DWM 2013B1, photo by J. G. M. Thewissen) (NOAA/North Slope Borough, NMFS Permit 14245).

peptide hormone synthesized by adipocytes, and it controls fat stores and energy balance in mammals (Ahima and Flier, 2000). Leptin binds to receptors in the hypothalamus, triggering upregulation of metabolism, lipolysis, and appetite suppression. Generally, higher leptin levels occur with greater adipose stores, and circulating leptin increases prior to migration (Ball et al., 2017).

Ball et al. (2013, 2017) investigated leptin gene expression, leptin receptors, and lipolysis in blubber, and found they varied markedly by season, blubber depth, and age. Leptin mRNA transcripts were highest in adult and subadult whales in the fall when lipid stores were maximized after summer feeding, compared to spring following a winter of fasting or intermittent feeding. Seasonal differences in leptin gene expression were particularly pronounced in the deepest blubber layers. Lipolytic gene expression was also highest in deeper blubber, and levels increased in spring consistent with lipid mobilization during migration. Mature bowheads sampled in the fall with thick blubber layers had up to 10-fold higher circulating leptin titers compared to other mammals (Ball et al., 2017). Younger whales (<9 m body length, well before sexual maturity) exhibited reduced leptin gene and receptor expression, as well as lower circulating leptin levels and lipolytic gene transcripts. These ontogenetic differences in leptin expression and lipolysis may be related to an adaptive energy-sparing metabolic state during maturation of baleen in young whales, when feeding efficiency is low (George et al., 2016).

Circulating hormones

Because there are currently no methods to collect blood from free-swimming large, whales, samples are collected postmortem (see Chapter 11). Circulating hormone levels reflect real-time endocrine status, and concentrations fluctuate based on many factors including pulsatile secretion, diurnal and seasonal influences, age, reproductive state, and social activity. These potentially confounding factors must be considered when interpreting hormone levels in blood (and other tissues).

Reproductive hormones

Serum hormones have been analyzed in harvested whales, with a particular focus on the gonadal steroid hormones (estrogens, progestogens, androgens) to assess reproductive state and seasonality. Serum progesterone was highly elevated in pregnant bowheads (Kellar et al., 2013), concurring with findings in other mysticetes (Birukawa et al., 2005; Kjeld et al., 1992, 2003, 2004, 2006). Progesterone was also significantly higher in pregnant bowhead urine and blubber, showing it is an accurate biomarker of pregnancy in multiple sample types (Kellar et al., 2013). Histology of reproductive organs coupled with hormone measurement demonstrates the sensitivity of serum hormones for assessment of sexual maturity, and male seasonality in baleen whales (Table 19.1).

TABLE 19.1 Endocrine studies in baleen whales organized by sample type analyzed.

Sample	Species	Hormones measured	References
Blood	*Balaena mysticetus*	P, TH	Kellar et al. (2013), Rosa et al. (2007)
	Eubalaena glacialis	B, F	Rolland et al. (2017)
	Balaenoptera physalus	E, P, T, F, TH, A	Kjeld (2001), Kjeld et al. (1992, 2006)
	Balaenoptera borealis	E, P, T	Kjeld et al., (2003)
		E, P, T, B, F, A	Birukawa et al. (2005)
	Balaenoptera acutorostrata	E, P, T, FSH, LH	Fukui et al. (1996), Kjeld et al. (2004), Mogoe et al. (2000), Suzuki et al. (2001), Watanabe et al. (2004)
		E, P, T, B, F, A	Birukawa et al. (2005)
	Balaenoptera edeni brydei	E, P, T, B, F, A	Birukawa et al. (2005)
	B. edeni edeni	E, T, FSH, LH	Watanabe et al. (2004)
Muscle	*B. acutorostrata*	P	Yoshioka et al. (1994)
Urine	*B. mysticetus*	P	Kellar et al. (2013)
	B. acutorostrata, B. borealis, B. brydei	E, P, T, B, F, A	Birukawa et al. (2005)
Feces	*B. mysticetus*	B	Rolland et al. (2019)
	E. glacialis	E, P, T, B, TH, A	Burgess et al. (2017), Corkeron et al. (2017), Hunt et al. (2006), Rolland et al. (2005, 2012, 2017)
	B. musculus	P, F, B	Valenzuela-Molina et al. (2018)
	Megaptera novaeangliae	E, P, T, B, TH	Hunt et al. (2019)
Blubber	*B. mysticetus*	P, L	Ball et al. (2013, 2017), Kellar et al. (2013)
	B. musculus	P, F	Atkinson et al. (2020)
	B. acutorostrata	P	Mansour et al. (2002)
	B. physalus	P, T	Carone et al., 2019
	M. novaeangliae	P	Clark et al. (2016), Pallin et al. (2018)
		T	Vu et al. (2015), Cates et al. (2019)
		E, P, T	Mello et al. (2017), Mingramm et al. (2020)
		E, P, T, B, F[a]	Dalle Luche et al. (2019, 2020)
	Eschrichtius robustus	P, T, HB	Hayden et al. (2017)
Baleen	*B. mysticetus*	E, P, T, TH, B, F, A	Hirt (2015), Hunt et al. (2014b, 2017b, 2018), Rolland et al. (2019)
	E. glacialis	E, P, T, TH, B, F, A	Fernández Ajó et al. (2018), Hunt et al. (2016, 2017a, 2017b, 2018), Lysiak et al. (2018)

(*Continued*)

TABLE 19.1 (Continued)

Sample	Species	Hormones measured	References
	Eubalaena australis	B, F	Fernández Ajó et al. (2018)
	B. musculus	T, B, F	Hunt et al. (2017b, 2018)
	B. acutorostrata, M. novaeangliae, B. physalus, E. robustus	E, P, T, TH, F, B, A	Hunt et al. (2017b)
Earplug	*B. musculus*	F, T	Trumble et al. (2013, 2018)
	M. novaeangliae, B. physalus	F	Trumble et al. (2018)
Breath	*M. novaeangliae*	P, T, F	Dunstan et al. (2012), Hogg et al. (2009)
Exhaled	*E. glacialis*	E, P, T, B, F, TH	Burgess et al. (2018), Hogg et al. (2009), Hunt et al. (2014a)

[a]*Multiple other hormone metabolites measured.*
Hormones and gonadotropins are abbreviated as: *E*, estrogen; *P*, progesterone; *T*, testosterone; *F*, cortisol; *B*, corticosterone; *HB*, hydrocortisone; *TH*, thyroid hormones; *A*, aldosterone; *L*, leptin; *FSH*, follicle stimulating hormone; *LH*, luteinizing hormone.

Thyroid hormones

Thyroid hormones (THs) are the primary regulator of metabolic rate, responding particularly to energetic challenges and environmental temperature shifts. Generally, thyroid activity decreases with limited food resources or higher environmental temperatures (Behringer et al., 2018). The thyroid gland secretes the prohormone thyroxine (T4) under control of the hypothalamus and pituitary gland, and T4 is converted to tri-iodothyronine (T3), the physiologically active hormone. Circulating forms of THs are either bound to thyroid-binding globulin or "free," which is the biologically active hormone fraction. Thyroid hormones have critical effects on metabolism, thermoregulation, growth, and reproduction (Norris and Carr, 2013).

Rosa et al. (2007) examined serum THs and thyroid gland histology in Bering-Chukchi-Beaufort Seas bowhead whales. Thyroid levels did not differ by age or sex, although pregnant and lactating females had lower free and total T4 levels. While circulating THs did not differ by season, histology showed fall-harvested whales had larger thyroid gland follicles with increased colloid content of the prohormone thyroglobulin suggesting increased thyroid activity. The relative stability in THs in this study is consistent with this hormone's role in maintaining homeostasis in healthy whales, and significant changes in THs could be a valuable biomarker for impacts of nutritional and thermal stressors.

Adrenal glucocorticoid hormones

Because the adrenal gland response to stressors occurs within minutes, interpretation of glucocorticoid (GC) hormones in blood from hunted whales to assess stress is potentially confounded by pursuit and capture. Nevertheless, GCs and aldosterone (another adrenal corticosteroid hormone) have been assayed in serum (and urine) from hunted baleen whales (Birukawa et al., 2005). Both serum and fecal GCs were highly elevated in a chronically injured, stranded North Atlantic right whale (*Eubalaena glacialis*; hereafter referred to as "right whales"; Rolland et al., 2017). Interestingly, Kjeld (2001) found no relationship between chase time in fin whales (*Balaenoptera physalus*) and cortisol or aldosterone, and a positive correlation with THs.

Quantifying hormones in alternative sample types

Alternative samples used for endocrine studies include: feces, urine, blubber, exhaled breath (blow), baleen, and earplugs. Immunoassays using these samples often measure metabolites (or breakdown products) of the parent hormone. It is crucial to conduct hormone assay validations for different species and sample types, and develop baseline values by demographic group and reproductive state before applying these approaches to research questions.

Blubber steroid hormones

Lipophilic (fat-loving) steroid hormones accumulate in blubber. Temporal dynamics of hormone accumulation and mobilization in mysticete blubber are not well studied, and differences in hormone concentration with blubber depth and body location have been reported in humpback whales (*Megaptera novaeangliae*) complicating data interpretation (Mello et al., 2017).

Pregnancy was reliably detected by highly elevated blubber progesterone in bowheads (Kellar et al., 2013), concurring with studies of humpback, fin, blue (*Balaenoptera musculus*) and minke whales (*Balaenoptera acutorostrata*) (Atkinson et al., 2020; Carone et al., 2019; Clark et al., 2016; Mansour et al., 2002; Pallin et al., 2018). Blubber progesterone levels did not correlate with gestational stage in bowheads. Mature bowheads had higher blubber progesterone than immature whales reflecting endocrine changes accompanying sexual maturity, as was also reported in blue whales (Atkinson et al., 2020). In bowheads, progesterone concentrations showed a positive relationship between serum and urine and between urine and blubber (Kellar et al., 2013). Serum and blubber levels had a significant positive correlation only if pregnant whales were included in the analysis. Results from this study suggested a lag time between changes in circulating hormone levels and blubber concentrations of steroid hormones, potentially as long as weeks or months.

Numerous other steroid hormones and gonadotropins have been quantified in blubber from other mysticetes (Table 19.1). Seasonal elevation of blubber testosterone was found in biopsy samples from male humpback and fin whales corresponding with the breeding season (Carone et al., 2019; Cates et al., 2019; Vu et al., 2015). Additionally, several GCs have been measured in baleen whale blubber potentially providing insights into adrenal stress responses and physiological state (Atkinson et al., 2020; Dalle Luche et al., 2019; Hayden et al., 2017). Therefore, research on steroid hormones in bowheads could be conducted using blubber collected at harvest or via remote biopsy darts in free-swimming whales.

Fecal hormones

Analysis of hormone metabolites in fecal samples has been successfully applied to baleen whales, including bowheads (Table 19.1). Circulating steroid hormones are metabolized by the liver and excreted into the gastrointestinal tract with bile (Palme, 2005). Fecal hormone levels integrate average levels of circulating hormones over a period of time previous to sample collection that depends on hormone turnover rates and gastrointestinal transit time. In right whales, this temporal relationship has been estimated at 1–2 days (Rolland et al., 2005), and is likely similar in bowheads.

Assessing reproductive status

Two decades of fecal hormone research in North Atlantic right whales has generated long-term physiological data on reproductive status, sexual maturity, and stress responses (Burgess et al., 2017; Hunt et al., 2006; Rolland et al., 2005, 2012, 2017). Because this population is intensively studied, fecal hormone levels could be compared in whales of independently confirmed sex, age, and calving history to determine if expected mammalian hormone patterns occurred (e.g., high concentrations of progesterone during gestation). In the mysticetes studied thus far, gestation is characterized by fecal progesterone metabolites several orders of magnitude above those of other reproductive states, resulting in very accurate pregnancy detection (Corkeron et al., 2017; Hunt et al., 2019; Rolland et al., 2005; Valenzuela-Molina et al., 2018). Fecal androgen:estrogen ratios differentiated males from nonpregnant females in right and humpback whales (Hunt et al., 2019; Rolland et al., 2005). Lactating female right whales had higher fecal estrogen levels than nonpregnant females, and elevated fecal androgen levels reflected male sexual maturity (Rolland et al., 2005). Similar fecal hormone patterns have been found in Bering-Chukchi-Beaufort Seas bowheads (R. Rolland, unpublished data).

Assessing stress responses

Metabolites of the major GCs involved in the adrenal stress response (e.g., cortisol and corticosterone) have been analyzed in fecal samples from bowheads and other baleen whales (Table 19.1). Increased fecal corticosterone was found in a severely entangled bowhead whale in poor health, along with highly elevated baleen GCs (see below; Rolland et al., 2019). Severe, prolonged entanglement in fishing gear resulted in extreme elevations of fecal GCs in right whales (Rolland et al., 2017). A live-stranded right whale with chronic injuries also had highly elevated fecal GCs and increased serum cortisol and corticosterone (Rolland et al., 2017). Finally, a marked decline in large vessel traffic in the Bay of Fundy, Canada, following the terrorist attacks of September 11, 2001 resulted in decreased low frequency underwater ship noise, and a significant reduction in fecal corticosterone in right whales, providing evidence for a chronic stress response related to underwater noise (Rolland et al., 2012).

Fecal GCs were significantly influenced by sex and reproductive state in right, humpback, and blue whales, with pregnant females and mature males showing physiologically normal elevations (Hunt et al., 2019; Rolland et al., 2017; Valenzuela-Molina et al., 2018). Aldosterone is another corticosteroid secreted in response to adrenal activation, which is less influenced by physiologic state in right whales, and may be helpful in distinguishing normal variation from stress responses to extrinsic factors (Burgess et al., 2017). Hunt et al. (2019)

validated an assay for fecal TH, which may prove useful for assessing nutritional condition and metabolic state.

Baleen hormones

Multiple gonadal, adrenal, and thyroid hormones have been analyzed in baleen plates of bowheads and other mysticetes (Table 19.1). Serial sampling for hormones along baleen plates (Fig. 19.2) generates a longitudinal record of physiological responses spanning up to a decade or more before death, depending on the length of the plate. Timelines have been applied to the hormone data using estimated baleen growth rates or stable isotope analyses. Peaks of baleen progesterone in adult female bowheads suggested gestation events, and mature females also had higher progesterone levels than sexually immature whales (Hirt, 2015; Hunt et al., 2014b). Hirt (2015) reported that baleen progesterone concentration was positively correlated with fetal size. Estimates of pregnancy timing in right whales based on baleen progesterone, coincided with documented calving events, also indicating reproductive history is recorded in baleen (Hunt et al., 2016; Lysiak et al., 2018).

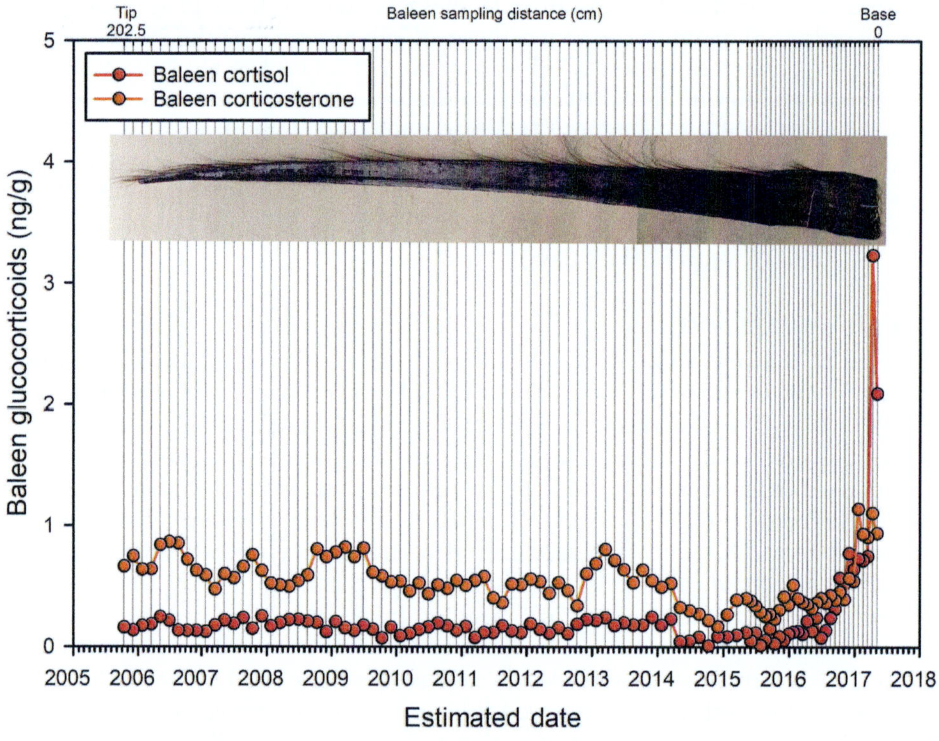

FIGURE 19.2 Graph of baleen cortisol and corticosterone concentrations from a bowhead (NSB-DWM2017B6) that was severely entangled in fishing gear about 6 months before harvest, with a baleen plate to show orientation and approximate areas of samples taken. Peaks in cortisol and, to a lesser extent, corticosterone were detected in baleen following entanglement showing a chronic stress response. *Source: Modified from Rolland, R.M., Graham, K.M., Stimmelmayr, R., Suydam, R.S., George, J.C., 2019. Chronic stress from fishing gear entanglement is recorded in baleen from a bowhead whale* (Balaena mysticetus)*. Mar. Mamm. Sci. 35(4), 1625–1642.*

Cyclic variation in baleen testosterone levels in adult male bowhead, right and blue whales suggested reproductive seasonality (Hunt et al., 2018).

Marked increases in baleen cortisol, especially, and corticosterone were found in a bowhead whale that was severely entangled in fishing gear and in poor body condition at harvest (Rolland et al., 2019; Fig. 19.2). Chronic stress was also indicated by highly elevated fecal corticosterone. Similar increases in baleen corticosterone were seen in response to severe fishing gear entanglement in a right whale (Lysiak et al., 2018). Fernández Ajó et al. (2018) found increased baleen GCs in two southern right whale calves (*Eubalaena australis*) with severe wounds inflicted by kelp gulls. Thus, prolonged adrenal activation indicating chronic stress is recorded in baleen, although the relationship between the magnitude and timing of the stressor and the concentration of baleen hormones is not well understood.

Earplug hormones

Steroid hormones also accumulate in lipid-rich earplugs, which increase in length throughout life in mysticetes via semiannual deposition of growth layer groups, or laminae. Reproductive and GC hormones have been quantified in earplugs from mysticetes (Table 19.1). Sequential analysis of hormones in earplug laminae is thought to represent a lifetime physiological profile with a 6-month temporal resolution (Trumble et al., 2013, 2018). Cortisol doubled in an earplug from a male blue whale over the time series, and testosterone levels were used to infer sexual maturation at c. 10 years (Trumble et al., 2013). Baseline-corrected cortisol profiles in earplugs from blue, humpback and fin whales correlated with whaling in the Northern Hemisphere from 1900 to 1999, and with sea surface temperature anomalies from 1970 to 2016 (Trumble et al., 2018). Thus, lifetime hormone profiles generated by analysis of earplugs show promise for studying mysticete reproductive biology and chronic stress related to intrinsic and extrinsic factors.

Hormones in exhaled breath

Although not yet applied to bowheads, analysis of exhaled breath (blow) from humpback and right whales shows that multiple gonadal, adrenal, and thyroid hormones can be detected, although concentrations are generally quite low (Hogg et al., 2009; Dunstan et al., 2012; Hunt et al., 2014a; Burgess et al., 2018; Table 19.1). Samples are collected using a pole with an attached sampling device positioned over the blowholes during exhalation. The major challenge with blow sampling is variable dilution by seawater and unknown total sample volume, complicating interpretation of hormone concentrations. Burgess et al. (2018) tested urea to normalize blow sample concentration, and showed it has potential to improve the biological relevance of hormone concentrations.

Discussion

The ability to obtain multiple sample types from harvested bowhead whales presents a valuable opportunity to evaluate relationships between hormones measured in different biological matrices. Interpretation of hormones is greatly enhanced by the other morphological, histological, and health-related data collected, including evidence of anthropogenic impacts (George et al., 2019). Studies of other mysticetes indicate that endocrine techniques could be applied to free-swimming bowheads where boat-based research is feasible.

While hormone concentrations vary between tissues, general patterns occur. Thus highly elevated progesterone indicates pregnancy, and androgens can reveal male maturity and reproductive seasonality. Glucocorticoid and thyroid hormones inform about stress, energetic state, and overall health. Hormone analyses of baleen and earplugs generate a longitudinal record of physiological data while the whale was living, including both stress responses and reproductive history.

Bowhead whales are faced with increasing pressures from climate change, seismic exploration, resource extraction, commercial shipping, and fishing activity (Chapter 27). Endocrine studies present a viable approach to understand basic physiology and investigate the implications of a rapidly changing Arctic ecosystem for health, reproduction, and fitness at both individual and population levels.

Acknowledgments

With gratitude to the North Slope Borough Department of Wildlife Management biologists and staff, the whaling captains and crews and the Alaska Eskimo Whaling Commission for their support for sampling bowhead whales for scientific research. Special thanks to Scott D. Kraus for his suggestions, which greatly improved this chapter.

References

Ahima, R.S., Flier, J.S., 2000. Leptin. Annu. Rev. Physiol. 62, 413–437.

Atkinson, S., Gendron, D., Branch, T.A., Mashburn, K.L., Melica, V., Enriquez-Paredes, L.E., et al., 2020. Pregnancy rate and biomarker validations from the blubber of eastern North Pacific blue whales. Mar. Mamm. Sci. 36 (1), 6–28.

Ball, H.C., Holmes, R.K., Londraville, R.L., Thewissen, J.G., Duff, R.J., 2013. Leptin in whales: validation and measurement of mRNA expression by absolute quantitative real-time PCR. PLoS ONE 8, e54277.

Ball, H.C., Stavarz, M., Oldaker, J., Usip, S., Londraville, R.L., George, J.C., et al., 2015. Seasonal and ontogenetic variation in subcutaneous adipose of the bowhead whale (*Balaena mysticetus*). Anat. Rec. 298, 1416–1423.

Ball, H., Londraville, R., Prokop, J., George, J.C., Suydam, R., Vinyard, C., et al., 2017. Beyond thermoregulation: metabolic function of cetacean blubber in migrating bowhead and beluga whales. J. Comp. Physiol. B 187, 235–252.

Behringer, V., Deimel, C., Hohmann, G., Negrey, J., Schaebs, F.S., Deschner, T., 2018. Applications for non-invasive thyroid hormone measurements in mammalian ecology, growth, and maintenance. Horm. Behav. 105, 66–85.

Birukawa, N., Ando, H., Goto, M., Kanda, N., Pastene, L.A., Nakatsuji, H., et al., 2005. Plasma and urine levels of electrolytes, urea and steroid hormones involved in osmoregulation of cetaceans. Zool. Sci. 22, 1245–1258.

Burgess, E.A., Hunt, K.E., Kraus, S.D., Rolland, R.M., 2017. Adrenal responses of large whales: integrating fecal aldosterone as a complementary biomarker to glucocorticoids. Gen. Comp. Endocrinol. 252, 103–110.

Burgess, E.A., Hunt, K.E., Kraus, S.D., Rolland, R.M., 2018. Quantifying hormones in exhaled breath for physiological assessment of large whales at sea. Sci. Rep. 8 (1), 10031.

Carone, E., Pardo, M.A., Atkinson, S., Mashburn, K., Perez-Puig H., Enriquez-Paredes L., et al., Sex steroid hormones and behavior reveal seasonal reproduction in a resident fin whale population. Conserv. Physiol. 7, 2019, coz059.

Cates, K.A., Atkinson, S., Gabriele, C.M., Pack, A.A., Straley, J.M., Yin, S., Testosterone trends within and across seasons in male humpback whales (*Megaptera novaeangliae*) from Hawaii and Alaska,.Gen. Comp. Endocrinol. 279, 164–173.

Clark, C.T., Fleming, A.H., Calambokidis, J., Kellar, N.M., Allen, C.D., Catelani, K.N., et al., 2016. Heavy with child? Pregnancy status and stable isotope ratios as determined from biopsies of humpback whales. Conserv. Physiol. 4 (1), cow050.

Corkeron, P., Rolland, R.M., Hunt, K.E., Kraus, S.D., 2017. A right whale pootree: classification trees of faecal hormones identify reproductive states in North Atlantic right whales (*Eubalaena glacialis*). Conserv. Physiol. 5 (1), cox006.

Dalle Luche, G., Bengtson Nash, S., Kucklick, J.R., Mingramm, F.M., Boggs, A.S., 2019. Liquid chromatography tandem mass spectrometry for the quantification of steroid hormone profiles in blubber from stranded humpback whales (*Megaptera novaeangliae*). Conserv. Physiol. 7 (1), coz030.

Dalle Luche, G., Boggs, A.S.P., Kucklick, J.R., GroB, J., Hawker, D.W., Bengtson Nash, S., 2020. Androstenedione and testosterone but not progesterone are potential biomarkers of pregnancy in Humpback Whales (*Megaptera novaeangliae*) approaching parturition, Sci. Rep. 10, 2954.

Dunstan, J., Gledhell, A., Hall, A., Miller, P., Ramp, C., 2012. Quantification of the Hormones Progesterone and Cortisol in Whale Breath Samples Using Novel, Non-Invasive Sampling and Analysis with Highly-sensitive ACQUITY UPLC and Xevo TQ-S. Waters Corporation, Milford, Massachusetts, p. 8.

Fernández Ajó, A.A., Hunt, K.E., Uhart, M., Rowntree, V., Sironi, M., Marón, C.F., et al., 2018. Lifetime glucocorticoid profiles in baleen of right whale calves: potential relationships to chronic stress of repeated wounding by Kelp Gulls. Conserv. Physiol. 6 (1), coy045.

Fukui, Y., Mogoe, T., Jung, Y., Terawaki, Y., Miyamoto, A., Ishikawa, H., et al., 1996. Relationships among morphological status, steroid hormones, and post-thawing viability of frozen spermatozoa of male minke whales (*Balaenoptera acutorostrata*). Mar. Mamm. Sci. 12, 28–37.

George, J.C., Druckenmiller, M.L., Laidre, K.L., Suydam, R., Person, B., 2015. Bowhead whale body condition and links to summer sea ice and upwelling in the Beaufort Sea. Prog. Oceanogr. 136, 250–262.

George, J.C., Stimmelmayr, R., Suydam, R., Usip, S., Givens, G., Sformo, T., et al., 2016. Severe bone loss as part of the life history strategy of bowhead whales. PLoS ONE 11, e0156753.

George, J.C., Tudor, B., Givens, G.H., Mocklin, J., Vate Brattström, L., 2019. Entanglement-scar acquisition rates and scar frequency for Bering-Chukchi-Beaufort Seas bowhead whales using aerial photography. Mar. Mamm. Sci. 35 (4), 1304–1321.

Haldiman, J.T., Tarpley, R.J., 1993. Anatomy and physiology. In: Burns, J.J., Montague, J.J., Cowles, C.J. (Eds.), The Bowhead Whale. Society for Marine Mammology, Lawrence, KS, pp. 71–156.

Hayden, M., Bhawal, R., Escobedo, J., Harmon, C., O'Hara, T.M., Klein, D., et al., 2017. Nanospray liquid chromatography/tandem mass spectrometry analysis of steroids from gray whale blubber. Rapid Commun. Mass Spectrom. 31, 1088–1094.

Hirt, S.J., 2015. Sexual Dimorphism, the Big-Mother Hypothesis, and Analyses of Hormones, Minerals, and Stable Isotopes to Determine Reproductive Events in Bowhead Whales (*Balaena mysticetus*) (Ph.D. thesis). Auburn University, December 12, 2015.

Hogg, C.J., Rogers, T.L., Shorter, A., Barton, K., Miller, P.J.O., Nowacek, D., 2009. Determination of steroid hormones in whale blow: it is possible. Mar. Mamm. Sci. 25 (3), 605–618.

Hunt, K.E., Rolland, R.M., Kraus, S.D., Wasser, S.K., 2006. Analysis of fecal glucocorticoids in the North Atlantic right whale (*Eubalaena glacialis*). Gen. Comp. Endocrinol. 148, 260–272.

Hunt, K.E., Rolland, R.M., Kraus, S.D., 2014a. Detection of steroid and thyroid hormones via immunoassay of North Atlantic right whale (*Eubalaena glacialis*) respiratory vapor. Mar. Mamm. Sci. 30 (2), 796–809.

Hunt, K.E., Stimmelmayr, R., George, C., Hanns, C., Suydam, R., Brower Jr., H., et al., 2014b. Baleen hormones: a novel tool for retrospective assessment of stress and reproduction in bowhead whales (*Balaena mysticetus*). Conserv. Physiol. 2 (1), cou030.

Hunt, K.E., Lysiak, N.S., Moore, M.J., Rolland, R.M., 2016. Longitudinal progesterone profiles in baleen from female North Atlantic right whales (*Eubalaena glacialis*) match known calving history. Conserv. Physiol. 4 (1), cow014.

Hunt, K.E., Lysiak, N.S., Moore, M., Rolland, R.M., 2017a. Multi-year longitudinal profiles of cortisol and corticosterone recovered from baleen of North Atlantic right whales (*Eubalaena glacialis*). Gen. Comp. Endocrinol. 254, 50–59.

Hunt, K.E., Lysiak, N.S., Robbins, J., Moore, M.J., Seton, R.E., Torres, L., et al., 2017b. Multiple steroid and thyroid hormones detected in baleen from eight whale species. Conserv. Physiol. 5 (1), cox061.

Hunt, K.E., Lysiak, N.S., Matthews, C.J., Lowe, C., Fernández Ajó, A., Dillon, D., et al., 2018. Multi-year patterns in testosterone, cortisol and corticosterone in baleen from adult males of three whale species. Conserv. Physiol. 6 (1), coy049.

Hunt, K.E., Robbins, J., Buck, C.L., Bérubé, M., Rolland, R.M., 2019. Evaluation of fecal hormones for noninvasive research on reproduction and stress in humpback whales (*Megaptera novaeangliae*). Gen. Comp. Endocrinol. 280, 24–34.

Kellar, N.M., Keliher, J., Trego, M.L., Catelani, K.N., Hanns, C., George, J.C., et al., 2013. Variation of bowhead whale progesterone concentrations across demographic groups and sample matrices. Endanger. Species Res. 22, 61–72.

Kjeld, M., 2001. Concentrations of electrolytes, hormones, and other constituents in fresh postmortem blood and urine of fin whales (*Balaenoptera physalus*). Can. J. Zool. 79, 438–446.

Kjeld, J., Sigurjonsson, J., Arnason, A., 1992. Sex hormone concentrations in blood serum from the North Atlantic fin whale (*Balaenoptera physalus*). J. Endocrinol. 134, 405–413.

Kjeld, M., Víkingsson, G.A., Alfredsson, Á., Ólafsson, Ö., Árnason, A., 2003. Salt and water balance of modern baleen whales: rate of urine production and food intake. Can. J. Zool. 81, 606–616.

Kjeld, M., Alfredsson, Á., Ólafsson, Ö., Tryland, M., Christensen, I., Stuen, S., et al., 2004. Changes in blood testosterone and progesterone concentrations of the North Atlantic minke whale (*Balaenoptera acutorostrata*) during the feeding season. Can. J. Fish. Aquat. Sci. 61, 230–237.

Kjeld, M., Ólafsson, Ö., Víkingsson, G.A., Sigurjónsson, J., 2006. Sex hormones and reproductive status of the North Atlantic fin whales (*Balaenoptera physalus*) during the feeding season. Aquat. Mamm. 32, 75.

Lysiak, N.S.J., Trumble, S.J., Knowlton, A.R., Moore, M.J., 2018. Characterizing the duration and severity of fishing gear entanglement on a North Atlantic right whale (*Eubalaena glacialis*) using stable isotopes, steroid and thyroid hormones in baleen. Front. Mar. Sci. 5, 168.

Mansour, A.A.H., Mkay, D.W., Lien, J., Orr, J.C., Banoub, J.H., ØIen, N., et al., 2002. Determination of pregnancy status from blubber samples in minke whales (*Balaenoptera acutorostrata*). Mar. Mamm. Sci. 18 (1), 112–120.

Mello, D., Colosio, A., Marcondes, M., Viau, P., Oliveira, C., 2017. Feasibility of using humpback whale blubber to measure sex hormones. J. Exp. Mar. Biol. Ecol. 486, 32–41.

Mingramm, F.M.J., Keeley, T., Whitworth, D.J., Dunlop, R.A., 2020. The influence of physiological status on the reproductive behavior of humpback whales (*Megaptera novaeangliae*), *Horm. Behav.*, 117, 104606.

Mogoe, T., Suzuki, T., Asada, M., Fukui, Y., Ishikawa, H., Ohsumi, S., 2000. Functional reduction of the southern minke whale (*Balaenoptera acutorostrata*) testis during the feeding season. Mar. Mamm. Sci. 16 (3), 559–569.

Norris, D.O., Carr, J.A., 2013. Vertebrate Endocrinology. Academic Press, New York.

Pallin, L., Robbins, J., Kellar, N., Bérubé, M., Friedlaender, A., 2018. Validation of a blubber-based endocrine pregnancy test for humpback whales. Conserv. Physiol. 6, coy031.

Palme, R., 2005. Measuring fecal steroids: guidelines for practical application. Ann. N. Y. Acad. Sci. 1046, 75–80.

Rolland, R.M., Hunt, K.E., Kraus, S.D., Wasser, S.K., 2005. Assessing reproductive status of right whales (*Eubalaena glacialis*) using fecal hormone metabolites. Gen. Comp. Endocrinol. 142, 308–317.

Rolland, R.M., Parks, S.E., Hunt, K.E., Castellote, M., Corkeron, P.J., Nowacek, D.P., et al., 2012. Evidence that ship noise increases stress in right whales. Proc. R. Soc. B 279, 2363–2368.

Rolland, R.M., McLellan, W.A., Moore, M.J., Harms, C.A., Burgess, E.A., Hunt, K.E., 2017. Fecal glucocorticoids and anthropogenic injury and mortality in North Atlantic right whales (*Eubalaena glacialis*). Endanger. Species Res. 34, 417–429.

Rolland, R.M., Graham, K.M., Stimmelmayr, R., Suydam, R.S., George, J.C., 2019. Chronic stress from fishing gear entanglement is recorded in baleen from a bowhead whale (*Balaena mysticetus*). Mar. Mamm. Sci. 35 (4), 1625–1642.

Rosa, C., 2006. Health Assessment in the Bowhead Whale (Ph.D. thesis) University of Alaska Fairbanks, May 2006.

Rosa, C., O'Hara, T.M., Hoekstra, P.F., Refsal, K.R., Blake, J.E., 2007. Serum thyroid hormone concentrations and thyroid histomorphology as biomarkers in bowhead whales (*Balaena mysticetus*). Can. J. Zool. 85, 609–618.

Suzuki, T., Mogoe, T., Asada, M., Miyamoto, A., Tetsuka, M., Ishikawa, H., et al., 2001. Plasma and pituitary concentrations of gonadotropins (FSH and LH) in minke whales (*Balaenoptera acutorostrata*) during the feeding season. Theriogenology 55, 1127–1141.

Trumble, S.J., Robinson, E.M., Berman-Kowalewskic, M., Potter, C.W., Usenko, S., 2013. Blue whale earplug reveals lifetime contaminant exposure and hormone profiles. Proc. Natl. Acad. Sci. U. S. A. 110, 16922–16926.

Trumble, S.J., Norman, S.A., Crain, D.D., Mansouri, F., Winfield, Z.C., Sabin, R., et al., 2018. Baleen whale cortisol levels reveal a physiological response to 20th century whaling. Nat. Commun. 9, 4587.

Valenzuela-Molina, M., Atkinson, S., Mashburn, K., Gendron, D., Brownell Jr, R.L., 2018. Fecal steroid hormones reveal reproductive state in female blue whales sampled in the Gulf of California, Mexico. Gen. Comp. Endocrinol. 261, 127–135.

Vu, E.T., Clark, C., Catelani, K., Kellar, N.M., Calambokidis, J., 2015. Seasonal blubber testosterone concentrations of male humpback whales (*Megaptera novaeangliae*). Mar. Mamm. Sci. 31 (3), 1258–1264.

Watanabe, H., Mogoe, T., Asada, M., Hayashi, K., Fujise, Y., Ishikawa, H., et al., 2004. Relationship between serum sex hormone concentrations and histology of seminiferous tubules of captured baleen whales in the Western North Pacific during the feeding season. J. Reprod. Dev. 50, 419–427.

Yoshioka, M., Okumura, T., Aida, K., Fujise, Y., 1994. A proposed technique for quantifying muscle progesterone content in minke whales (*Balaenoptera acutorostrata*). Can. J. Zool. 72, 368–370.

CHAPTER

20

Molecular insights into anatomy and physiology

Lisa Noelle Cooper[1] and Vera Gorbunova[2]

[1]Department of Anatomy and Neurobiology, Musculoskeletal Research Group, Northeast Ohio Medical University, Rootstown, OH, United States [2]Department of Biology, University of Rochester, Rochester, NY, United States

Introduction

The bowhead whale is of the largest and longest lived animals on earth (Fig. 20.1). Despite the challenges of living and giving birth in arctic waters, having vastblubber stores, eating a fat-rich diet, and migrations, bowheads are estimated to reach ages of upto 268 years (George et al., 1999; Mayne et al., 2019; Seim et al., 2014; Tacutu et al., 2012). Recently, the bowhead genome and transcriptome libraries (Keane et al., 2015; Seim et al., 2014) were made publicly available for study. Bowheads have probably benefited from both the expansion and decay in different parts of their genome in evolving a long and caompartively healthy life. Here we discuss the expansions (duplication) and losses (inactivation) in genes associated with the derived anatomy of physiology of cetaceans, and when evidence supports it, bowheads. Duplication of genes allows for new copies to be free from selection for their original function and instead evolve novel functions. Gene loss/inactivation can be the result of an evolutionary relaxation in selection for a function that became obsolete but can also be a mechanism for adaptation as gene inactivations in whales may improve replication accuracy (Huelsmann et al., 2019). Molecular biologists are now undertaking laboratory experiments with whale samples that are informed by the bowhead genome and transcriptome to tackle questions that were inaccessible using classical model organisms such as rodents or fish. Biomedical researchers are applying insights gained from research on bowheads into investigations of potential therapies for aging, senescence, and cancer. Moreover, evolutionary biologists are using these results to inform our understanding of how different key traits evolved sequentially in deep time.

The Bowhead Whale
DOI: https://doi.org/10.1016/B978-0-12-818969-6.00020-0

FIGURE 20.1 A large, adult bowhead whale. The white flukes and multiple white scars are evidence of advanced age. Bowhead whales can reach ages of more than 200 years, and their genes display adaptations to minimize the effects of aging. *Source: Photo by Vicki Beaver (NOAA/North Slope Borough, NMFS Permit No. 14245).*

This chapter reviews the molecular basis for bowhead whale longevity and survival in their unique environment and offers insights for further exploration into the molecular mechanisms that shape the extraordinary lives of these animals.

Adapting the genome to extend life span

The bowhead life span probably extends to 268 years (George et al., 1999; Mayne et al., 2019; Seim et al., 2014; Tacutu et al., 2012; Chapter 21) due to genomic modifications that maintain DNA integrity and delay aging and disease. All cetaceans, including bowheads, benefit from slower mutation rates compared to terrestrial mammals; these slower rates contribute to the inhibition of cancer (Tollis et al., 2019). Bowheads show duplications in genes associated with energy metabolism and aging (*DLD*), DNA repair (*PCNA*), controlled cellular growth (*LAMTOR1*), mitosis (*ARPP19*), mitochondrial and cellular stress responses (*STOML2*, *HSBP1*), cellular growth and repair (*SMS*), tumor suppression (*ST13*), ultraviolet sensitivity (*UVRAG*), and apoptosis (*PDCD5*) (Keane et al., 2015; Tollis et al., 2019). Bowheads also show unique mutations and evidence of positive selection in genes associated with DNA repair (*POLE*, *ERCC1*, *ERCC3*) (Keane et al., 2015; Tollis et al., 2019). Cetaceans also display loss of *POLM* that results in correlations with improved tolerance of oxidative DNA damage and increased accuracy in replication of genes (Huelsmann et al., 2019).

Expanding the thermal limits for mammalian life

Compared to other cetaceans, bowheads have exceptionally thick blubber within their integument that insulates them from the cold Arctic waters. Blubber consists of an intricate network of collagen fibers supporting large numbers of adipose cells, and it is the lipids within this blubber that are metabolized seasonally in migrating bowheads (Chapter 19). Uncoupling proteins (*UCP1*) function in nonshivering heat production in most mammals (Zhu et al., 2018), this gene evolved a truncation (Keane et al., 2015) about 52 Mya (Gaudry et al., 2017). Decay of this gene may have been a consequence in thermoregulation that resulted from insulating blubber and is associated with the evolution of extreme cold hardiness (Gaudry et al., 2017; Zhu et al., 2018). Bowheads have an extra copy of an adipogenic gene (*UCHL3*) and show evidence of positive selection in leptin (*LEP*), another gene associated with adipogenesis and lipolysis (Keane et al., 2015). *LEP* expression levels within the blubber of bowheads are ~10 times greater than rodents and humans and ~4 times greater than beluga whales, undergo ~50-fold fluctuations associated with seasonal shifts in lipolysis, and expression levels of adults exceed those of younger bowheads (Ball et al., 2017). Adult bowheads also display higher levels of *LEP* expression in bone compared to young cohorts (Cooper et al., in review). Bowheads therefore evolved genomic modifications that support the generation of exceptionally large blubber stores, and this allows them to tolerate extreme cold and undergo arduous seasonal migrations.

Streamlining the sensory system for dim light and salty seas

Most toothed whales lack the ability to smell, and most of their olfactory receptor (OR) and marker protein genes are dysfunctional (McGowen et al., 2014; Springer and Gatesy, 2017). Bowheads, and possibly other mysticetes, partially rely on airborne scent to locate aggregations of copepods and euphausiids that have distinctive odors (Thewissen et al., 2011). Anatomical structures supporting olfaction are still present in bowheads (olfactory epithelium, olfactory bulb of the brain), and bowheads have some functional ORs (Thewissen et al., 2011). Bowheads lack a dorsal region (domain) of the olfactory bulb that supports avoidance behaviors in response to odors of predators (Kishida et al., 2015). This could have resulted as a response to low predation pressure. Finally, the chemosensitive vomeronasal organ was lost ~45 Mya in ancient fossil whales, and this morphological loss was eventually accompanied by decay in vomeronasal receptors genes (V1Rs) (Kishida et al., 2015).

Humans are broadly able to taste salty, bitter, sour, sweet, and umami flavors, but cetaceans, including bowheads, probably taste only salt (Zhu et al., 2014) and their tongue has few or no taste buds. Retention of salt taste receptor genes may not be linked to taste but instead may facilitate sodium reabsorption and water balance in the kidneys (Venton, 2014; Zhu et al., 2014). Bowheads have one intact bitter gene receptor (*T2R16*) (Feng et al., 2014), but all taste receptor genes associated with sweet/umami (*T1Rs*), other bitter (*T2Rs*) and sour (*PKD2/1*) taste receptor genes lost function probably in ancient whales 36–52 Mya (Feng et al., 2014). The proper function of these genes requires an interaction with gustadin (*GNAT3*) that also lost function during this time frame (Kishida et al., 2015).

Within the mammalian retina, cones support color discrimination in bright light, and rods are sensitive to light in dim conditions. The retina of bowheads contains both rods and cones, but their vision is optimized for rod-based vision. Bowheads display inactivating mutations in cone pigment genes *SWS1*, as in all modern cetaceans that have been genetically screened, and *LWS* in bowheads and right whales (Meredith et al., 2013), and within the cone phototransduction cascade (*PDE6C, CNGA3, CNGB3*) (Springer et al., 2016a), suggesting that they lack color vision. This emphasis on rod-based vision that is optimal for low-light conditions probably evolved about 13 Mya in the common ancestor of bowheads and right whales (Springer et al., 2016a).

Maintaining healthy skin in icy waters

The skin of bowheads is thick, smooth, mostly lacks hair and associated sebaceous glands, does not usually freeze, and is rapidly renewed. This rapid turnover of the skin minimizes the need for elaborate inflammatory responses and mechanical repair. Bowheads, as in all other tested cetaceans, lost function of genes associated with hair follicles (*FGF22*; Nam et al., 2017) and hair keratins (*KRTAP*; Sun et al., 2017), lack some genes associated with sebaceous glands (*MC5R*; Springer and Gatesy, 2018, *DGAT2L6, MOGAT3, AWAT1/2, ELOVL3, FABP9*; Lopes-Marques et al., 2019b), and have lost or diminished function of genes associated with skin inflammatory responses (*IL20, CCL27*) (Lopes-Marques et al., 2018, 2019a). The loss of hair follicles and sebaceous glands reflects adaptations of the skin of bowheads to an aquatic life, while the loss of inflammatory genes may be contributing to overall skin health.

Growing teeth and ever-growing baleen

Bowheads use baleen to strain prey from seawater. Unlike teeth, baleen is composed mostly of keratin proteins and calcium salts. Although adults lack teeth, embryonic and fetal bowheads initially form tooth buds made of dentin that probably lack structurally coherent enamel (Thewissen et al., 2017). Inactivation of enamel-forming and enamel mineralization genes began about ~20 to 30 Mya with *MMP20* and *ACPT* inactive in modern baleen whales and *AMEL, AMBN,* and *C4orf26* inactivate in bowheads and right whales (Deméré et al., 2008; Meredith et al., 2009, 2011; Mu et al., 2019; Springer et al., 2016b).

Incipient tooth buds ultimately disappear in ontogeny, and the oral epithelium thickens to support growth of baleen plates. The oral epithelium of bowheads displays FGF4 protein signals in a pattern that is similar to baleen plates, although this molecule was historically only associated with the development of teeth in other mammals (Thewissen et al., 2017). Bowheads may have therefore exapted FGF4 signaling to support the development and continuous growth of baleen plates (Thewissen et al., 2017).

Unusual growth of the skeleton

Although the skeletal anatomy of bowheads is well described (Eschricht and Reinhardt, 1866), the derived anatomy and physiology of the skeleton remain mostly unknown. The pelvis and hind limbs of bowheads are vestigial and do not function in locomotion (Thewissen et al., 2009). The reduced pelvic bones are asymmetrical (Chapter 10), a shape characteristic of sticklebacks and manatees with modified *PITX-1* expression (Shapiro et al., 2006). Embryonic cetaceans have hindlimb buds that are absorbed by the body before birth. In dolphins, hindlimb buds fail to synthesize SHH, a protein that is essential for growth and patterning of limbs (Thewissen et al., 2006). Hind limbs were lost in ancient cetaceans about 34 Mya ago (Thewissen et al., 2006), and this loss was thought to release developmental constraints on forelimbs, allowing for the evolution of novel morphologies (Wang et al., 2009). The forelimbs of cetaceans are modified into flippers that encase digits with extra phalanges (hyperphalangy), a trait that evolved about 30 Mya (Cooper et al., 2007). In dolphins, hyperphalangy is the result of a prolonged developmental period of outgrowth of digits under the control of FGFs and also increased segmentation of these digits into synovial joints associated with increased WNT signaling (Cooper et al., 2018).

The ribs of fetal and immature bowheads are exceptionally dense with minerals and are eroded internally until the marrow cavity replaces the minerals with adipose and hematopoietic cells in sexually mature adults (George et al., 2016). Throughout this process, *EZH2* probably drives diminished secretion of the bone matrix by bone cells (Cooper et al., in review). Within the liver, high *FGF23* expression may also contribute to this low bone mass phenotype (Nam et al., 2017), although its mechanism of action is not fully understood in whales.

Future work

Bowheads have evolved novel aspects of their biology to optimize health and life span (Chapter 15), generate huge blubber stores, change rib phenotype with age, and reduce functionality in sensory pathways except those associated with hearing and touch. Ongoing studies into the molecular biology of the bowhead whale may further elucidate patterns of gene expression and cellular activity that differentiate bowheads from other mammals. It is still unclear what genes are involved in the production of baleen, how bowheads keep their organs young, and what ultimately causes death in these animals, outside of hunting and predation.

Throughout their extensive life span, bowheads maintain multiple populations of stem cells that are continuously active, and no studies have addressed the molecular regulation of these cells that presumably do not undergo significant age-related declines in function.

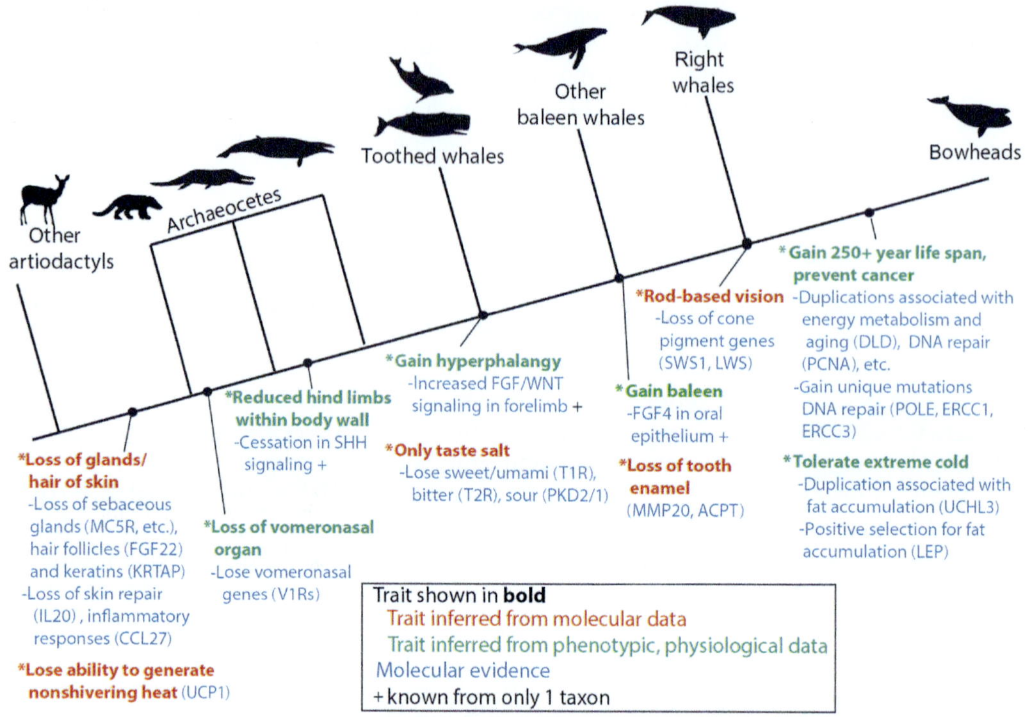

FIGURE 20.2 The evolution of traits leading to unique characteristics of bowhead whales. Traits inferred from molecular data are shown in orange. Traits inferred from phenotypic and/or physiological data are shown in green. Molecular evidence of gene loss or duplication is shown in blue. Evidence from just a single taxon is shown by a +. Placement of trait changes on the evolutionary tree is inferred based on the best available evidence (Chapter 1).

These include keratinocytes that continuously form baleen, epithelial cells that give rise to the rapidly renewed skin, and cells that continuously deposit bone in the postcranial skeleton. It could be that bowheads employ novel molecular mechanisms to retain the integrity of these cell populations throughout their extended life spans (Fig. 20.2).

How bowheads remain essentially cancer free throughout their extended life span also remains a mystery. Bowhead whales are among the largest mammals on Earth. Because cells in whales are not much different in size from other mammals, bowheads have thousands of times greater numbers of cells than terrestrial mammals. As cancer originates from individual cells, statistically one would expect much higher rates of cancer in whales than in smaller mammals, which is referred to as "Peto's paradox" (Nunney, 1999; Tollis et al., 2017). Peto's paradox can be explained by bowheads having evolved additional tumor suppression mechanisms (Seluanov et al., 2018). Although changes in genes involved in DNA repair suggest that bowheads have more efficient ways of maintaining their genomes, other mechanisms remain to be determined. Study of these traits does not guarantee that cures to human health can be found, but it does provide extra incentive to understand the physiological consequences of genomic shifts and understand their evolutionary origins. For any species, body design is a compromise between adaptations to the environment and the specifics of its evolutionary heritage. If patterns of genomic change can be understood in their evolutionary context with links to function and morphology, it will lead to a deeper understanding of the adaptability of the body and the changeability of the genome.

Acknowledgments

We thank the editors and reviewers for insightful comments. We thank the Barrow Whaling Captain's Association, the Wainwright Whaling Captain's Association, and the Alaska Eskimo Whaling Commission for samples of harvested whales, and the North Slope Borough Department of Wildlife Management for logistical support of field studies.

References

Ball, H.C., Londraville, R.L., Prokop, J.W., George, J.C., Suydam, R.S., Vinyard, C., et al., 2017. Beyond thermoregulation: metabolic function of cetacean blubber in migrating bowhead and beluga whales. J. Comp. Physiol. B 187, 235–252.

Cooper, L.N., Ball, H.C., Vinyard, C.J., Safadi, F.F., George, J.C., Thewissen, J.G.M., in review. Linking gene expression and phenotypic changes in the developmental and evolutionary origins of osteosclerosis in the ribs of bowhead whales (*Balaena mysticetus*). J Exp Zool B Mol Dev Evol.

Cooper, L.N., Berta, A., Dawson, S.D., Reidenberg, J.S., 2007. Evolution of hyperphalangy and digit reduction in the cetacean manus. Anat. Rec. 290, 654–672.

Cooper, L.N., Sears, K.E., Armfield, B.A., Kala, B., Hubler, M., Thewissen, J.G.M., 2018. Review and experimental evaluation of the embryonic development and evolutionary history of flipper development and hyperphalangy in dolphins (Cetacea: Mammalia). Genesis 56.

Deméré, T.A., McGowen, M.R., Berta, A., Gatesy, J., 2008. Morphological and molecular evidence for a stepwise evolutionary transition from teeth to baleen in mysticete whales. Syst. Biol. 57, 15–37.

Eschricht, D.F., Reinhardt, J., 1866. On the Greenland right-whale (Balaena mysticetus, Linn.) with special reference to its geographical distribution and migrations in times past and present, and to its external and internal characteristics, In: Hardwick, R. (Ed.). Ray Society, London, p. 150.

Feng, P., Zheng, J., Rossiter, S.J., Wang, D., Zhao, H., 2014. Massive losses of taste receptor genes in toothed and baleen whales. Genome Biol. Evol. 6, 1254–1265.

Gaudry, M.J., Jastroch, M., Treberg, J.R., Hofreiter, M., Paijmans, J.L.A., Starrett, J., et al., 2017. Inactivation of thermogenic UCP1 as a historical contingency in multiple placental mammal clades. Sci. Adv. 3, e1602878.

George, J.C., Bada, J., Zeh, J., Scott, L., Brown, S.E., O'Hara, T., et al., 1999. Age and growth estimates of bowhead whales (*Balaena mysticetus*) via aspartic acid racemization. Can. J. Zool. 77, 517–580.

George, J.C., Stimmelmayr, R., Suydam, R., Usip, S., Givens, G., Sformo, T., et al., 2016. Severe bone loss as part of the life history strategy of bowhead whales. PLoS One 11, e0156753.

Huelsmann, M., Hecker, N., Springer, M.S., Gatesy, J., Sharma, V., Hiller, M., 2019. Genes lost during the transition from land to water in cetaceans highlight genomic changes associated with aquatic adaptations. Sci. Adv. 5, eaaw6671.

Keane, M., Semeiks, J., Webb, A.E., Li, Y.I., Quesada, V., Craig, T., et al., 2015. Insights into the evolution of longevity from the bowhead whale genome. Cell Rep. 10, 112–122.

Kishida, T., Thewissen, J., Hayakawa, T., Imai, H., Agata, K., 2015. Aquatic adaptation and the evolution of smell and taste in whales. Zool. Lett. 1, 9.

Lopes-Marques, M., Alves, L.Q., Fonseca, M.M., Secci-Petretto, G., Machado, A.M., Ruivo, R., et al., 2019a. Convergent inactivation of the skin-specific C-C motif chemokine ligand 27 in mammalian evolution. Immunogenetics 71, 363–372.

Lopes-Marques, M., Machado, A.M., Barbosa, S., Fonseca, M.M., Ruivo, R., Castro, L.F.C., 2018. Cetacea are natural knockouts for IL20. Immunogenetics 70, 681–687.

Lopes-Marques, M., Machado, A.M., Alves, L.Q., Fonseca, M.M., Barbosa, S., Sinding, M.-H.S., et al., 2019b. Complete inactivation of sebum-producing genes parallels the loss of sebaceous glands in Cetacea. Mol. Biol. Evol. 36, 1270–1280.

Mayne, B., Berry, O., Davies, C., Farley, J., Jarman, S., 2019. A genomic predictor of lifespan in vertebrates. Sci. Rep. 9, 17866.

McGowen, M.R., Gatesy, J., Wildman, D.E., 2014. Molecular evolution tracks macroevolutionary transitions in Cetacea. Trends Ecol. Evol. 29, 336–346.

Meredith, R.W., Gatesy, J., Murphy, W.J., Ryder, O.A., Springer, M.S., 2009. Molecular decay of the tooth gene enamelin (ENAM) mirrors the loss of enamel in the fossil record of placental mammals. PLoS Genet. 5, e1000634.

Meredith, R.W., Gatesy, J., Cheng, J., Springer, M.S., 2011. Pseudogenization of the tooth gene enamelysin (*MMP20*) in the common ancestor of extant baleen whales. Proc. Biol. Sci. 278, 993–1002.

Meredith, R.W., Gatesy, J., Emerling, C.A., York, V.M., Springer, M.S., 2013. Rod monochromacy and the coevolution of cetacean retinal opsins. PLoS Genet. 9, e1003432.

Mu, Y., Huang, X., Liu, R., Gai, Y., Liang, N., Yin, D., et al., 2019. ACPT gene is inactivated in mammalian lineages that lack enamel and teeth. bioRxiv 861633.

Nam, K., Lee, K.W., Chung, O., Yim, H.-S., Cha, S.-S., Lee, S.-W., et al., 2017. Analysis of the FGF gene family provides insights into aquatic adaptation in cetaceans. Sci. Rep. 7, 40233.

Nunney, L., 1999. Lineage selection and the evolution of multistage carcinogenesis. Proc. Biol. Sci. 266, 493–498.

Seim, I., Ma, S., Zhou, X., Gerashchenko, M.V., Lee, S.G., Suydam, R., et al., 2014. The transcriptome of the bowhead whale *Balaena mysticetus* reveals adaptations of the longest-lived mammal. Aging (Albany NY) 6, 879–899.

Seluanov, A., Gladyshev, V.N., Vijg, J., Gorbunova, V., 2018. Mechanisms of cancer resistance in long-lived mammals. Nat. Rev. Cancer 18, 433–441.

Shapiro, M.D., Bell, M.A., Kingsley, D.M., 2006. Parallel genetic origins of pelvic reduction in vertebrates. Proc. Natl. Acad. Sci. USA 103, 13753–13758.

Springer, M.S., Gatesy, J., 2017. Inactivation of the olfactory marker protein (OMP) gene in river dolphins and other odontocete cetaceans. Mol. Phylogenet. Evol. 109, 375–387.

Springer, M.S., Gatesy, J., 2018. Evolution of the MC5R gene in placental mammals with evidence for its inactivation in multiple lineages that lack sebaceous glands. Mol. Phylogenet. Evol. 120, 364–374.

Springer, M.S., Emerling, C.A., Fugate, N., Patel, R., Starrett, J., Morin, P.A., et al., 2016a. Inactivation of cone-specific phototransduction genes in rod monochromatic cetaceans. Front. Ecol. Evol. 4, 61.

Springer, M.S., Starrett, J., Morin, P.A., Lanzetti, A., Hayashi, C., Gatesy, J., 2016b. Inactivation of C4orf26 in toothless placental mammals. Mol. Phylogenet. Evol. 95, 34–45.

Sun, X., Zhang, Z., Sun, Y., Li, J., Xu, S., Yang, G., 2017. Comparative genomics analyses of alpha-keratins reveal insights into evolutionary adaptation of marine mammals. Front. Zool. 14, 41.

Tacutu, R., Craig, T., Budovsky, A., Wuttke, D., Lehmann, G., Taranukha, D., et al., 2012. Human ageing genomic resources: integrated databases and tools for the biology and genetics of ageing. Nucleic Acids Res. 41, D1027–D1033.

Thewissen, J.G.M., Cohn, M.J., Stevens, L.S., Bajpai, S., Heyning, J., Horton, W.E., 2006. Developmental basis for hind-limb loss in dolphins and origin of the cetacean bodyplan. Proc. Natl. Acad. Sci. USA 103, 8414–8418.

Thewissen, J.G.M., Cooper, L.N., George, J.C., Bajpai, S., 2009. From land to water: the origin of whales, dolphins, and porpoises. Evol.: Educ. Outreach 2, 272–288.

Thewissen, J.G.M., George, J., Rosa, C., Kishida, T., 2011. Olfaction and brain size in the bowhead whale (*Balaena mysticetus*). Mar. Mammal Sci. 27, 282–294.

Thewissen, J.G., Hieronymus, T.L., George, J.C., Suydam, R., Stimmelmayr, R., McBurney, D., 2017. Evolutionary aspects of the development of teeth and baleen in the bowhead whale. J. Anat. 230, 549–566.

Tollis, M., Boddy, A.M., Maley, C.C., 2017. Peto's Paradox: how has evolution solved the problem of cancer prevention? BMC Biol. 15, 60.

Tollis, M., Robbins, J., Webb, A.E., Kuderna, L.F.K., Caulin, A.F., Garcia, J.D., et al., 2019. Return to the sea, get huge, beat cancer: an analysis of cetacean genomes including an assembly for the humpback whale (*Megaptera novaeangliae*). Mol. Biol. Evol. 36, 1746–1763.

Venton, D., 2014. Highlight: a matter of taste-whales have abandoned their ability to taste food. Genome Biol. Evol. 6, 1266–1267.

Wang, Z., Yuan, L., Rossiter, S.J., Zuo, X., Ru, B., Zhong, H., et al., 2009. Adaptive evolution of 5′HoxD genes in the origin and diversification of the cetacean flipper. Mol. Biol. Evol. 26, 613–622.

Zhu, K., Zhou, X., Xu, S., Sun, D., Ren, W., Zhou, K., et al., 2014. The loss of taste genes in cetaceans. BMC Evol. Biol. 14, 218.

Zhu, K., Ge, D., Wen, Z., Xia, L., Yang, Q., 2018. Evolutionary genetics of hypoxia and cold tolerance in mammals. J. Mol. Evol. 86, 618–634.

CHAPTER

21

Age estimation

J.C. George[1], S.C. Lubetkin[2], Judith E. Zeh[3], J.G.M. Thewissen[4], D. Wetzel[5] and Geof H. Givens[6]

[1]Department of Wildlife Management, North Slope Borough, Utqiaġvik, AK, United States [2]Seattle, WA, United States [3]Department of Statistics, University of Washington, Seattle, WA, United States [4]Department of Anatomy and Neurobiology, Northeast Ohio Medical University, Rootstown, OH, United States [5]Mote Marine Laboratory, Sarasota, FL, United States [6]Givens Statistical Solutions LLC, Fort Collins, CO, United States

Introduction

Knowledge of the age of an individual animal is fundamental to understanding its biology and life history; and knowledge of the age distribution within a population is an essential part of population modeling and many fisheries and wildlife management programs (Dimmick and Pelton, 1997; Schmidt-Nielsen, 1997; Punt and Butterworth, 1999). The age of toothed whales and many terrestrial mammals can be estimated by reading annual bands, called growth layer groups (GLGs), in sectioned teeth (Perrin and Myrick, 1980; Hohn et al., 1989; Klevezal, 1996; Read et al., 2018). Mysticetes are more challenging as they lack teeth, but annual layers in their earwax are often used for age estimation (Purves, 1955; Roe, 1967; Blokhin, 1984; Trumble et al., 2015). Bowheads commonly have earwax that consists of annually shed tissue layers, but the resulting wax plug is difficult to locate and dissect, and layers are commonly indistinct (Perrin and Myrick, 1980; Rosa et al., 2013; Rehorek et al., 2019), although exceptions occur (Trumble et al., 2015). In the 1980s methods that had been developed to estimate ages in other mysticete whales were either ineffective on bowheads (e.g., earplugs) or thought to be ineffective and overestimate age (Nerini, 1983a,b; Fig. 21.1).

Another common method of estimating the age of mysticetes is photographic identification of individuals across years as is done effectively for right whales (Kraus and Rolland, 2007), humpbacks, and other species (Katona et al., 1980; Fig. 21.1). However, bowheads lack dorsal

The Bowhead Whale
DOI: https://doi.org/10.1016/B978-0-12-818969-6.00021-2

FIGURE 21.1 Aerial photo of a heavily scarred "ancient" bowhead whale near Point Barrow, Alaska in spring 2011. The scarring is likely caused by scraping against sea ice, as well as killer whale attacks, fishing gear entanglement, and underwater obstructions. Remarkably, this very old female had recently given birth, note the newborn calf at her left flank. Several harvested females over 100 years old have been found to be pregnant (Chapter 13). NMFS Permit No. 14245.

fins, ridges, and callosities, and nearly all bowhead calves and yearlings lack natural identifying marks or scars. Marks on older bowhead whales are difficult to document from land, sea ice, and boat-based platforms. Bowhead photo reidentification is effective with high-quality vertical aerial photographs, but such photos are difficult and expensive to obtain (Rugh et al., 1992).

When Burns et al. (1993) published the book on the biology of the bowhead whales, only one technique, baleen carbon cycling, was available for estimating a bowhead's age. This method, however, was only effective for whales up through their mid-teens, and it was then unknown that a bowhead could live over 200 years. A number of new techniques for estimating the age of bowhead whales have been developed in the past 30 years, including refinements of the baleen aging technique. We review and compare eight techniques or approaches here and discuss their applicability to different age groups. Essentially, all the data analyzed in this chapter were from collaborative research with Native Alaskan whale hunters.

Age estimation using baleen

Bowhead whale baleen grows throughout life with a growth rate that decreases with age but appears to stabilize by the mid-teens (Withrow et al., 1992; Schell and Saupe, 1993; Lubetkin et al., 2008; Matthews and Ferguson, 2015). The plates wear primarily by rubbing and chafing against the tongue and lip, and the wear rate has been modeled as a function of baleen length and is a key variable in the age models (Lubetkin et al., 2012).

The Bering–Chukchi–Beaufort (BCB) and East Canada–West Greenland (ECWG) stocks of bowhead undertake annual migrations between waters that differ isotopically (Chapters 3–5), resulting in differences in the stable isotope ratios of their prey. Isotopic signals from their prey are recorded in the baleen, and thus the baleen plates retain a signal of the annual migration and can be used for age estimation (Nerini et al., 1987; Schell et al., 1989a,b, Schell and Saupe, 1993; Lubetkin et al., 2008; Matthews and Ferguson, 2015; Sensor et al., 2018). The results from isotope studies on baleen can be used to calibrate growth in baleen plates, allowing age estimation by simply measuring the length of the longest plate (Lubetkin et al., 2012).

Cycles in stable carbon isotope values

The carbon isotope signal ($\delta^{13}C$; Schell et al., 1989a,b) of the bowhead whales from the BCB stock (Chapter 4) generally shows a cyclical pattern in baleen plates, with high $\delta^{13}C$ values when the whales are in the Beaufort Sea, and lower values when they are in the Bering/Chukchi Sea (Fig. 21.2). With the assumption that the whales migrate annually and the baleen plate has limited wear, a whale's age can be estimated by counting the number of migration cycles (Schell and Saupe, 1993; Fig. 21.2), or minimum age estimation for worn plates. While annual migration is by far most common, occasionally a bowhead whale will not complete the full migration or it may change its typical feeding pattern, resulting in a $\delta^{13}C$ plateau across multiple years of baleen growth (Lubetkin et al., 2008; Sensor et al., 2018; Chapter 4).

Age estimates based on stable isotope cycles for BCB bowheads can be extended several years if baleen wear is accounted for. Lubetkin et al. (2008) extended the cycle-counting approach by developing a growth increment model and using it to estimate ages through the late teens. For ECWG bowheads, sulfur, carbon, and nitrogen baleen isotope ratios show considerable variation, but $\delta^{15}N$ ratios appear to provide an annual cyclic signal that can be used for estimating the age (Matthews and Ferguson, 2015).

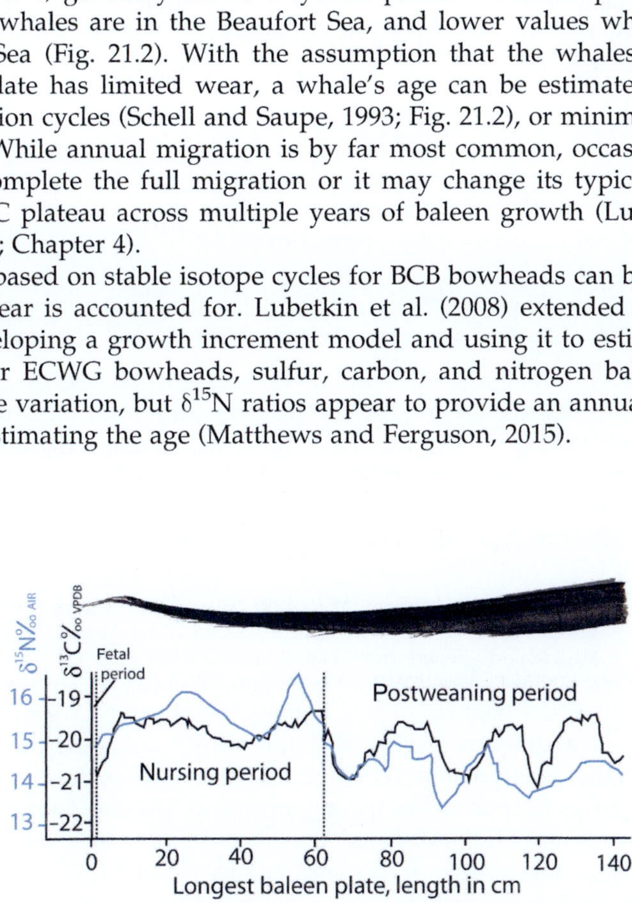

FIGURE 21.2 C (black) and N (blue) isotope trace of the baleen of a bowhead whale from the BCB stock (NSB-DWM 2013B6, from Sensor et al., 2018), with image of the sampled baleen plate lined up above it. The plate shows the distinct values for baleen formed before birth and, during the nursing period (of approximately 9 to 12 months, Chapter 7) and after weaning. Low $\delta^{13}C$ and $\delta^{15}N$ values are consistent with isotope values of the Bering and Chukchi Sea, high values are consistent with time spent in the Beaufort Sea. The trace ends when the whale was harvested while migrating near Utqiaġvik.

Baleen length method

Using isotopic records, Lubetkin et al. (2008, 2012) determined the gross and net (growth minus wear) baleen growth rates for bowheads in the BCB stock. By the end of the first year of life, bowheads grow 70–80 cm of baleen which includes the prenatal baleen (15 cm) (Lubetkin et al., 2008, Table 4). In the second year, bowhead baleen typically grows an additional 35 cm. The growth of baleen gradually slows to about 17.5 cm/year (Lubetkin et al., 2008, Tables 4 and 5). This understanding of growth rate thus allows age estimation based on the rate of change in early annual baleen growth rates.

Lubetkin et al. (2012) found that multistage versions of the von Bertalanffy II model were particularly useful for estimating whale age from baleen length. We extracted the relevant equations for easy access.

The usual form of the von Bertalanffy II equation has the longest baleen plate length [$L(t)$] as a function of age (t):

$$L(t) = L_{max} \times \left[1 - \left(1 - \frac{L_0}{L_{max}}\right) \times \exp\left(\frac{-kt}{L_{max} - L_0}\right)\right]$$

To find the age of a whale based on a baleen length up to 180 cm, the equation can be algebraically rearranged to solve for t as a function of L:

$$t(L) = -\frac{(L_{max1} - L_0)}{k_1} \times \ln\left(\frac{L_{max1} - L}{L_{max1} - L_0}\right)$$

where L_{max1} is the maximum length of initial (preadolescent) baleen, L_0 is the baleen length at birth, and k_1 defines the decline in growth rate over time from birth to adolescence, for both male and female young whales that have not reached the adolescent baleen growth spurt (which occurs around $L_{max1} = 187.82$, $L_0 = 21.26$ cm, $k_1 = 53.6$; see Lubetkin et al., 2012, Table 5, for parameter estimates and standard errors).

For whales with baleen plates >180 cm, a different version of the equation is applied:

$$t(L) = \text{Age}_{gs} - \frac{(L_{max2} - L_{trans})}{k_2} \times \ln\left(\frac{L_{max2} - L}{L_{max2} - L_{trans}}\right)$$

where L_{max2} is the sex-specific maximum baleen length, L_{trans} is the baleen length when the secondary growth spurt begins, and k_2 defines the decline in growth rate over time from the age of the growth spurt (Age_{gs}) through adulthood. When the first equation is used for baleen lengths >180 cm, the resulting age estimates are extremely high because the curve becomes flat as it approaches the asymptote at 187 cm. That is why L_{trans} is less than L_{max1}. Lubetkin et al. (2012) estimated that $\text{Age}_{gs} = 10$ years, with $L_{trans} = 181.15$ cm and $k_2 = 5.17$, with L_{max2} for males = 317.71 and females = 374.26 cm.

As a general rule, the baleen-cycle counting age estimation method is quite accurate for whales younger than about 10 years, or baleen lengths less than about 175 cm, but can be extended to late teens if changing annual baleen growth rates are modeled as noted earlier. However, the growth of baleen for older whales is slow but predictable, such that the total length of baleen plates of known-aged whales could be modeled to estimate age by baleen length alone. Hence, Lubetkin et al. (2012) demonstrated that baleen-length models can be used to estimate ages of whales up to about 60 years old (baleen up to about 325 cm). This eliminates the need for expensive laboratory analysis but has less accuracy and precision for older whales than can be obtained using other methods.

Age estimation using growth layers in the tympanic bone

The involucrum of the tympanic bone (the bulla) of the mysticete skull grows with age and displays little or no remodeling. Growth appears cyclic, slowing down and speeding up in different seasons (Fig. 21.3). This leads to the formation of GLGs that can be used to estimate age (Klevezal and Mitchell, 1971; Christensen, 1981; Christensen, 1995; Hohn, 2009). Nerini (1983a) pioneered this method for bowhead whales, and it was later refined by Sensor et al. (2018). Unlike baleen, which grows at a rate that slows down gradually with age but is less dependent on external factors, the thickness of the bowhead involucrum does vary with external factors, possibly related to feeding or changes in migration. The bowhead involucrum displays three zones that are morphologically and histologically clearly distinct (Sensor et al., 2018). The innermost of these zones represents the fetal period and the middle zone is formed during the period that the whale is nursed or about 6–12 months of the whale's life (Chapter 13). The outermost zone is thinner than the others and displays GLGs, appearing as alternating layers of lighter and darker bone. When polished or on histological thin-section, these layers can be counted and represent an annual cyclical pattern, as confirmed by the study of isotope cycles in baleen plates.

As whales age, involucrum GLGs become thinner and difficult to read in some whales older than 20 years. In whales older than 30 years, they are too thin to yield consistent results or can be undetectable in some cases. The GLGs tend to be clear and match an annual signal.

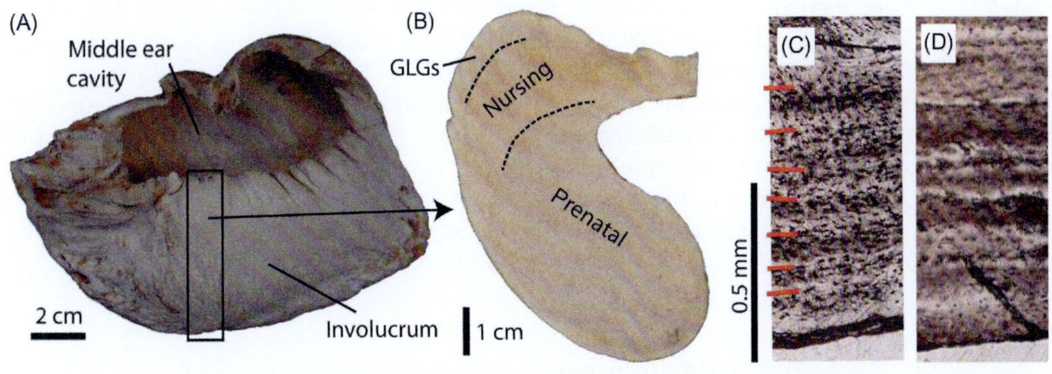

FIGURE 21.3 (A) Tympanic bulla of a bowhead whale (NSB-DWM, 2015B2), with a slice removed for age estimation. (B) Thin section of the slice shows three morphological regions formed in the (1) prenatal, (2) nursing, and (3) the post nursing period. (C) Region 3 forms annual GLGs; GLGs are indicated with red marks. (D) Slide of 2013B18 showing about 10 GLGs or years. *GLG*, Growth layer group. *Source: From Sensor, J.D., George, J.C., Clementz, M.T., Lovano, D.M., Waugh, D.A., Givens, G., et al., 2018. Age estimation in bowhead whales using tympanic bulla histology and baleen isotopes. Mar. Mammal Sci. Available from: https://doi.org/10.1111/mms.12476.*

However, there are additional bands that are often not repeated in a consistent pattern within individual GLGs, which is similar to the bands in the dentin of odontocetes (Perrin and Myrick, 1980). These bands may document important biological events (e.g., fasting and pregnancy) rather than years but sometimes do mimic GLGs. Unlike annual cycles in the baleen, involucrum GLGs are not of a constant width or thickness, making them difficult to interpret in some individuals. A period of stress may slow or stop growth temporarily, leading to the formation of an accessory layer that may interfere with accurate age assessment (Sensor et al., 2018). Hence, growth marks in the tympanic bone should not be considered a calendar of the whales' life, but rather as a "diary," where certain events are emphasized (e.g., feeding) and others not recorded, although the annual GLGs are generally preserved.

Age estimation based on aspartic acid racemization

Measurement of aspartic acid racemization (AAR) in the eye lens has been used to estimate the ages of humans (Masters et al., 1977), bowhead whales (Nerini, 1983b; George et al., 1999; Rosa et al., 2013; Wetzel et al., 2017), minke whales (Olsen and Øien, 2002; Olsen and Sunde, 2002), narwhals (Garde et al., 2007), and other vertebrates. The amino acid, aspartic acid, in the eye lens of vertebrates is always formed as the L enantiomer, and the nucleus of the lens is formed during prenatal life. Over time, aspartic acid racemizes into the D enantiomer at a relatively slow stable rate (termed k_{asp}) if its temperature is constant (George et al., 1999; Rosa et al., 2013; Wetzel et al., 2017) until the D/L ratio reaches 1.0. The ratio of D/L can thus be used to estimate age. Bowhead whales have body temperatures below those of humans and fin whales, the taxa initially used to determine k_{asp} (George et al., 1999). Bowhead k_{asp} has now been directly estimated by regressing D/L ratios with age estimates using other methods such as baleen aging and corpora counts (Rosa et al., 2013). Bowhead k_{asp} is estimated as 0.977×10^{-3}/year. For humans and fin whales, k_{asp} is estimated to be 1.25×10^{-3}/year and 1.16×10^{-3}/year, respectively (Rosa et al., 2013). The bowhead k_{asp} is lower than other whales and humans, possibly because of their lower body temperatures (Chapter 16).

Standard errors of age estimates from the AAR technique increase with whale age, but the coefficient of variation (CV) decreases with whale age, making age estimates of old whales proportionately more accurate than those of young whales. This technique is typically not too useful to estimate the age of bowhead whales that are less than 15 years or about 11–12 m long and is most useful for whales more than 13 m long, which is the length at which they are reaching sexual maturity at around 25 years (Chapter 13).

Age estimation based on ovarian corpora

The ovaries of sexually mature females contain corpora lutea and albicantia (Tarpley et al., 2016; George et al., 2011). A corpus luteum (CL) forms when ovulation results in pregnancy, and this body regresses into a corpus albicans (CA) after the pregnancy ends.

Counts of ovarian corpora can be used in estimating the age of a sexually mature female if assumptions about bowhead reproduction are made. First, it is assumed that the CA persists for the life of the animal, as has been shown in gray whales (Rice and Wolman, 1971). Second, it is assumed that every ovulation results in a CA or ovulation scar.

The model used to estimate age using ovaries also requires knowledge of certain life-history variables. Rosa et al. (2013) found that female bowheads become sexually mature at around 25.9 years, and George et al. (2011) estimated the intercalf interval at about 3.1 years. Aerial photo-recaptures also suggest that the calving interval is approximately 3–4 years (Koski et al., 1993; Chapter 13). The estimated ovulation rate (of mature females) is OR = 0.332 with a standard error of 0.111 (George et al., 2011). Estimated age can then be calculated as:

$$\text{Age} = \text{ASM} + (\text{TC} - 1)/\text{OR} + \text{shift}$$

where ASM = estimated age at sexual maturity = 25.9 (Rosa et al., 2013) and TC = total CL and CA from both ovaries. Because a pregnancy lasts more than a year, the *shift* term, which ranges from 0 to 3 years, is needed as an adjustment for the harvest date of the whale and whether the whale had a CL and/or term fetus (George et al., 2011).

Age estimation using corpora counts is methodologically straightforward for sexually mature females from which ovaries are available. Ovaries are preserved and cut into 5 mm slices, where a trained observer can visually count the CAs and CLs.

Standard errors of the age estimates derived from corpora counts increase with age, but the CVs remain fairly constant (roughly 20%) over the range of ages and whale lengths (13–19.2 m) we analyzed. However, whales estimated to be 85 years and older had slightly higher CVs (George et al., 2011).

Age estimation based on whaling artifacts

Occasionally, old whaling implements have been recovered from the bodies of recently harvested bowhead whales of the BCB population and can be used to estimate their minimum age (Philo et al., 1993; George et al., 1999; Weintraub, 1996). Yankee whaling tools changed relatively quickly as the whaling industry developed, and artifacts are archived in museums. Whalers and whale hunters also engrave identifying marks (termed "owner marks") on their whaling tools; therefore it is occasionally possible to date an earlier (failed) hunting event. This provides a minimum age for a whale caught in a later year. A brass whaling projectile (whale bomb) from a design patented in 1879 was recovered from a whale in 2007. The authors suggested that the device was used close to its manufacture date since whaling equipment inventory was turned over rapidly during the commercial whaling period; they suggested a minimum age for the whale of about 128 years (George and Bockstoce, 2008).

Traditional stone whaling points have been found in bowheads harvested by Iñupiat hunters in the 1980s–90s (Fig. 21.4; George et al., 1999; Weintraub, 1996). Iñupiat hunters switched to hand-thrown Yankee whaling tools in the 1880s, suggesting that these whales

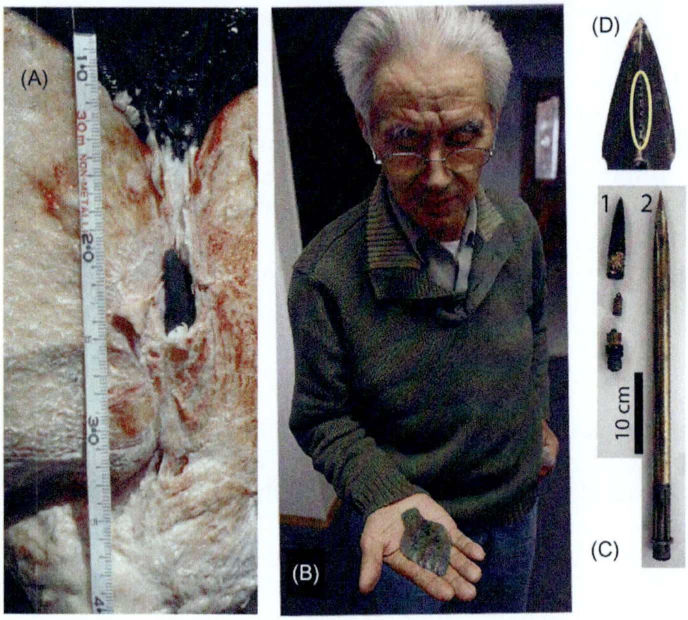

FIGURE 21.4 (A) Stone harpoon point (minimum age around 120 years) embedded in the blubber of whale 92B2, estimated age 133 yr (SE = 40). (B) Captain B. Ahmaogak, Sr. with stone harpoon point recovered from his whale. (C) (1) A brass Yankee "bomb lance" fuse head fragment recovered from whale 07B8 shown with (2). an 1879 patent "bomb lance" with nearly identical design (D) Close-up of the point in panel (C) showing six "owner marks". *Source: Photos Left to right: J.C. George, B. Hess, J. Bockstoce, J.C. George.*

were at least 100 years old at the time of capture (Murdoch, 1892; George et al., 1999). Finally, traditional knowledge of bowhead whale age suggests longevity on the order of "two human lifetimes" (Chapter 34).

Age estimation based on photo-recapture

Photo-identification is a commonly used technique to age whales in studies conducted in all regions of the world (Kraus and Rolland, 2007; Hammond, 2018). Similarly, a number of researchers have used photographic recaptures of well-marked bowheads to estimate abundance, growth rates, and scar acquisition rates (e.g., Rugh et al., 1992; Koski et al., 1993; Koski et al., 2010; Givens et al., 2018, George et al., 2019). However, since calves are rarely marked, estimating the chronological age of a bowhead has to date not been possible and only minimum ages are obtainable. Several photo-recaptures 26 years apart have been found and used with other recaptures in a recent abundance and survival analysis (Givens et al., 2018).

Estimating age using morphometric data

Archer et al. (2010) created a classification and regression tree using bowhead morphometric data, together with ages estimated by the methods described earlier, to classify whales into 10 age bins with cutoffs at 3, 5, 10, 18, 26, 37, 50, 60, 90, and >90 years. The

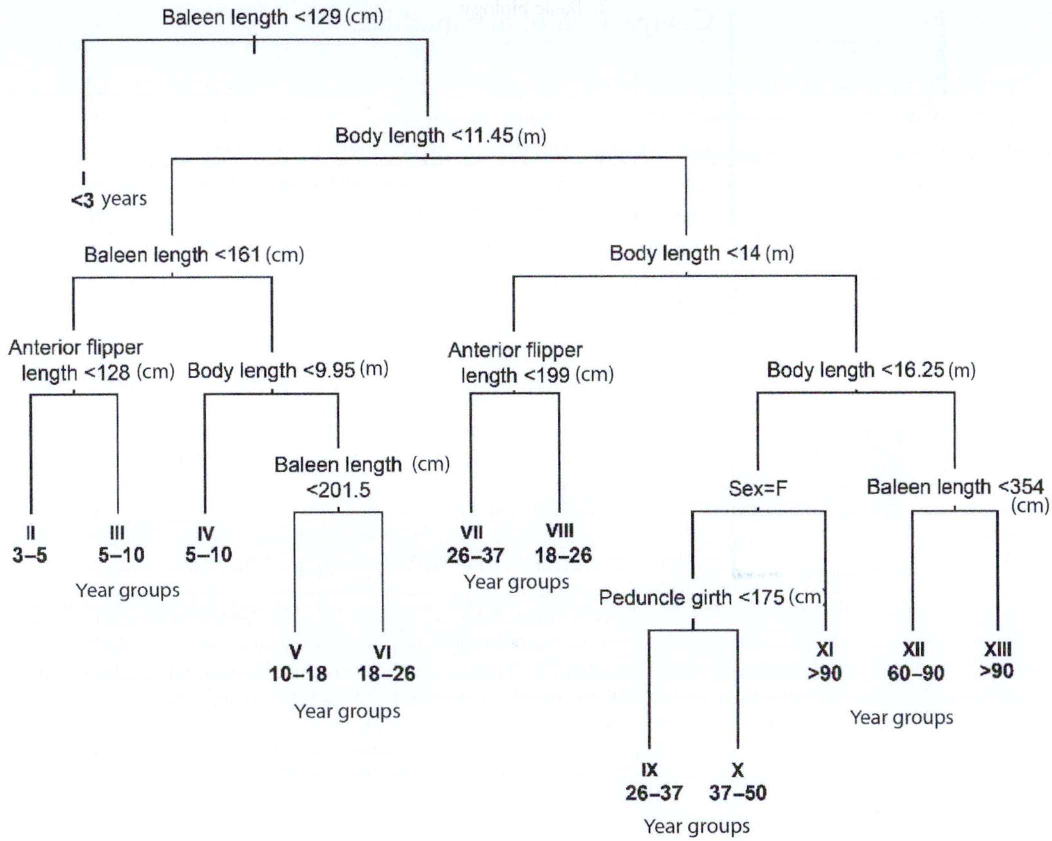

FIGURE 21.5 A classification and regression tree showing the primary splits used for estimating bowhead age. Cases that meet the criteria at each node are sent left and down. The Roman numerals are the leaf identifiers and below them is the estimated age bin for each leaf (Archer et al., 2010); body length in m and morphometric measurements in cm.

variables in the model included age, sex, body length, baleen length, anterior flipper length, peduncle girth, and size of the peduncle white patch (Fig. 21.5). Baleen length and body length were the most used predictors in the tree. Morita and George (2014) applied a similar classification analysis using morphometric data to classify bowheads into three age bins: over 90 years (very old), 60–90 years (old), and under 60 (younger). The analysis indicated that the most useful information for age classification of males was body length and peduncle girth. Females' ages were best predicted by anterior flipper length, body length, baleen length, and peduncle girth.

Classification models of morphometric data to estimate age can be further refined with the larger dataset of known-aged whales available since these publications. An advantage of this and the baleen-length aging methods is that there is little associated cost.

Comparison of age methods

No single method is fully applicable for aging bowheads through their more than 200 year lifespan, but a combination of techniques can be quite effective. As described in Chapter 7 bowheads go through several growth phases, including rapid growth to year 1, followed by a growth hiatus of several years, and then slow growth to maturity with extreme longevity. This complicates age estimation, and each method has strengths and weaknesses depending on life stage of the whale (Figs. 21.6 and 21.7). The recently developed DNA methylation technique may eventually be an accurate method for all age classes, but it has not yet been applied to bowheads (Polanowski et al., 2014).

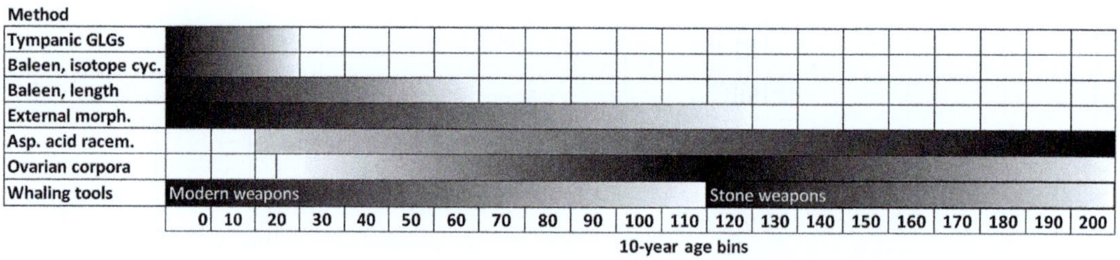

FIGURE 21.6 Gantt chart showing the applicability of various age-estimation techniques for whales in different age/growth stages. The shading roughly reflects the relative accuracy of each technique. The dark colors suggest a higher applicability and accuracy. Whale body length associated with these ages can be approximated from Fig. 21.7.

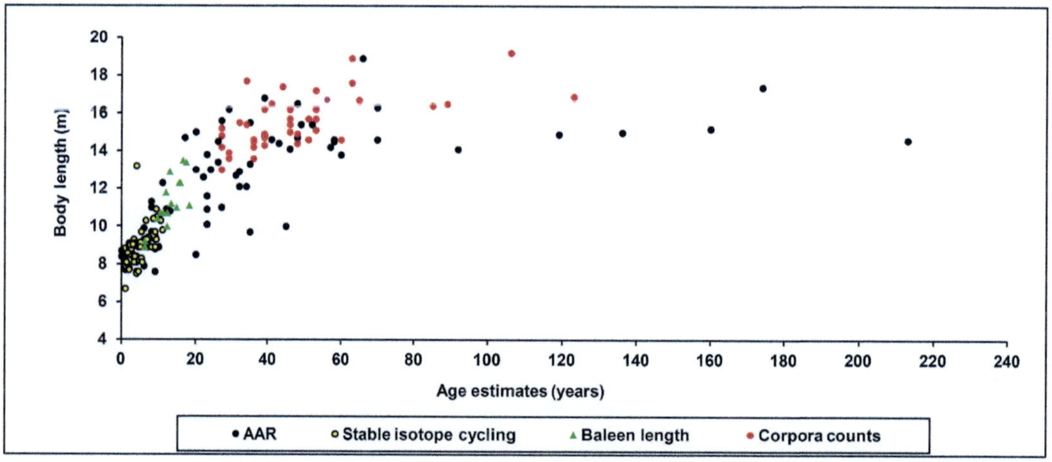

FIGURE 21.7 Plot showing age estimates for bowhead whales using four methods: AAR, stable isotope cycling, baleen length, and corpora counts. *AAR*, Aspartic acid racemization. *Source: Data from Lubetkin, S.C., Zeh, J.E., George, J.C., 2012. Statistical modeling of baleen and body length at age in bowhead whales (Balaena mysticetus). Can. J. Zool. 90, 915–931.*

Age 0 (birth)–20

Whales in this age group generally have baleen lengths less than 220 cm and body lengths less than about 11 m. For this group the baleen $\delta^{13}C$ cycle counting and tympanic bone GLG age methods are the most accurate and precise. We recommend sampling the baleen (for stable carbon isotopes) at 2.5 cm intervals or less along the entire length of the baleen, to best delineate the annual cycles and more accurately estimate whale's age. Baleen length alone can be used to quite accurately age whales less than about age 5, but the precision of the age estimates decreases for whales older than that (Lubetkin et al., 2008). Baleen-cycle counting and growth increment models can be used for whales into their teens (Lubetkin et al., 2008). The "morphometric age" method is quite accurate for this age group; however, it is also useful for all age groups, at varying levels of accuracy (Fig. 21.5; Archer et al., 2010).

Age 20–30

Several age methods are applicable for whales in this group which includes the age at which bowheads become sexually mature (~25 years). Body lengths range from about 12 to 15 m. The tympanic bulla GLG method and the baleen-based methods are the most accurate for this age group. The AAR method is applicable as well, particularly if repeated measures are used, but the *relative* precision or CV of the age estimate can be high (e.g., Rosa et al., 2013; Wetzel et al., 2017). The corpora counting method can be used for females that have reached sexual maturity and ovulated. Baleen- and body-length at age models are also effective in this age range (Lubetkin et al., 2012).

Age 30–150+

The AAR and the corpora counting method are respectively, the first and second best method to estimate the age for older bowheads in this age group. Baleen-length models can be effectively used following Lubetkin et al. (2012) for whales up to around 60 years, but the uncertainty in the age estimate is high and increases as whales' body and baleen length reach their asymptotic values.

References

Archer, F.I., Martien, K., Taylor, B., Leduc, R., Ripley, B., Givens, G., 2010. A simulation-based approach to evaluating population structure in non-equilibrial populations. JCRM 11 (2), 101–113.

Burns, J.J., Montague, J. and C.J. Cowles (eds.). 1993. The Bowhead Whale. Society for Marine Mammalogy. 787 pp.

Blokhin, S.A., 1984. Investigations of gray whales taken in the Chukchi coastal waters, U.S.S.R. In: Jones, M.L., Swartz, S.L., Leatherwood, S. (Eds.), The Gray Whale *Eschrichtius robustus*. Academic Press, Inc, London, pp. 487–509.

Christensen, I., 1981. Age determination of minke whales, *Balaenoptera acutorostrata*, from laminated structure in the tympanic bullae. Rep. Int. Whal. Commn. 31, 245–253.

Christensen, I., 1995. Interpretation of growth layers in the periosteal zone of tympanic bulla from minke whales, *Balaenoptera acutorostrata*. In Developments in Marine Biology (Vol. 4, pp. 413-423).

Clark, S. T., Odell, D. K., and Lacinak, C. T. 2000. Aspects of growth in captive killer whales (Orcinus orca). Marine Mammal Science 16 (1): 110–123.

Dimmick, Pelton, 1997. In: Bookhout, T.A. (Ed.), Research and Management Techniques for Wildlife and Habitats, fifth ed. The Wildlife Society, Bethesda, MD.

Garde, E., Heide-Jørgensen, M.P., Hansen, S.H., Nachman, G., Forchhammer, M.C., 2007. Age-specific growth and remarkable longevity in narwhals (*Monodon monoceros*) from West Greenland as estimated by aspartic acid racemization. J. Mammal. 88, 49–58.

George, J.C., Bockstoce, J.R., 2008. Two historical weapon fragments as an aid to estimating the longevity and movements of bowhead whales. Polar Biol. 31, 751–754.

George, J.C., Bada, J., Zeh, J., Scott, L., Brown, S.E., O'Hara, T., et al., 1999. Age and growth estimates of bowhead whales (*Balaena mysticetus*) via aspartic acid racemization. Can. J. Zool. 77, 571–580.

George, J.C., Follmann, E., Zeh, J., Sousa, M., Tarpley, R., Suydam, R., et al., 2011. A new way to estimate the age of bowhead whales (*Balaena mysticetus*) using ovarian corpora counts. Can. J. Zool. 89, 840–852.

George, J.C., Tudor, B., Givens, G.H., Mocklin, J., Vate Brattström, L., 2019. Entanglement-scar acquisition rates and scar frequency for Bering-Chukchi-Beaufort Seas bowhead whales using aerial photography. Mar. Mammal Sci. Available from: https://doi.org/10.1111/mms.12597.

Givens, G.H., Mocklin, J.A., Vate Brattström, L., Tudor, B.J., Koski, W.R., Zeh, J.E., et al., 2018. Survival rate and 2011 abundance of Bering-Chukchi-Beaufort Seas bowhead whales from photo-identification data over three decades. In: Paper SC/67b/XX Submitted to the Scientific Committee of the International Whaling Commission, Bled.

Hammond, P.S., 2018. Mark-recapture. In: Wursig, B., Thewissen, H., Kovacs, K.M. (Eds.), Encyclopedia of Marine Mammals. Academic Press.

Hohn, A. 2009. Age estimation. Pages 11–17 in W. F. Perrin, B.Wursig and J. G. M. Thewissen, eds. Encyclopedia of marine mammals. Academic Press, Amsterdam, The Netherlands.

Hohn, A., Scott, M.D., Wells, R.S., Sweeney, J.C., Irvine, A.B., 1989. Growth layers in teeth from known-age, free ranging bottlenose dolphins. Mar. Mammal Sci. 5, 315–342, Issue (3):1–50. Available from: https://doi.org/10.1111/j.1748-7692.1989.tb00346.x.

Katona, S.K., Harcourt, P., Perkins, J.S., Kraus, S.D., 1980. Humpback Whales of the Western North Atlantic—A Catalogue of Individuals Identified by Fluke Photographs, second ed. College of the Atlantic, Bar Harbor, ME, p. 169.

Klevezal, G.A., 1996. Recording Structures of Mammals: Determination of Age and Reconstruction of Life History. A. A. Balkema, Rotterdam.

Klevezal, G.A. and Mitchell, E.D., 1971. On the annual layers in the bones of whalebone whales. Zoologicheskii Zhurnal, 50, pp.1114-1116.

Koski, W.R., Davis, R.A., Miller, G.W., Withrow, D.E., 1993. Reproduction. In: Burns, J.J., Montague, J.J., Cowles, C.J. (Eds.), Bowhead Whale. pp. 239–274. Special Publication No. 2 of the Society for Marine Mammalogy. i–xxxvi + 787 p.

Koski, W.R., Zeh, J., Mocklin, J., Davis, A.R., Rugh, D.J., George, J.C., et al., 2010. Abundance of Bering-Chukchi-Beaufort bowhead whales (*Balaena mysticetus*) in 2004 estimated from photo-identification data. J. Cetacean Res. Manage. 11 (2), 89–99.

Kraus, S.D., Rolland, R.M., 2007. The Urban Whale. Harvard University Press, Cambridge, MA.

Lubetkin, S.C., Zeh, J.E., George, J.C., 2012. Statistical modeling of baleen and body length at age in bowhead whales (*Balaena mysticetus*). Can. J. Zool. 90, 915–931.

Lubetkin, S.C., Zeh, J.E., Rosa, C., George, J.C., 2008. Age estimation for young bowhead whales (*Balaena mysticetus*) using annual baleen growth increments. Can. J. Zool. 86 (6), 525–538. Available from: https://doi.org/10.1139/Z08-028.

Masters, P.M., Bada, J.L., Zigler, J.S., 1977. Aspartic acid racemization in the human lens during ageing and in cataract formation. Nature 268, 71–73.

Matthews, C.J.D., Ferguson, S.H., 2015. Seasonal foraging behaviour of Eastern Canada-West Greenland bowhead whales: an assessment of isotopic cycles along baleen. Mar. Ecol. Prog. Ser. 522, 269–286. Available from: https://doi.org/10.3354/meps11145.

Morita, J., George, J.C., 2014. Age classification of bowhead whales using recursive partitioning. J. Cetacean Res. Manage 14, 15–21.

Murdoch, J., 1892. Ethnological Results of the Point Barrow Expedition. Smithsonian Institution Press, p. 441.

Nerini, M.K., 1983a. Age Determination Techniques Applied to Mysticete Whales (M.S. thesis). University of Washington, School of Fisheries, Seattle, WA.

Nerini, M.K., 1983b. Age determination of fin whales (*Balaenoptera physalus*) based upon aspartic acid racemization in the lens nucleus. Rep. Int. Whal. Commn. 34, 447–448.

Nerini, M.K., Braham, H.W., Marquette, W.M., and Rugh, D.J. 1984. Life history of the bowhead whale, Balaena mysticetus (Mammalia: Cetacea). *Journal of the London Zoology Society* 204:443-468.

Nerini, M.K., Brooks, J.M., Pflaum, R., 1987. Preliminary examination of carbon isotope composition of bowhead whale baleen plates. In: Pap. No. SC/39/PS23 Presented to the International Whaling Commission Scientific Committee (unpublished). <http://iwcoffice.org/publications/doclist.htm> (accessed 25.04.08).

Olsen, E., Øien, M., 2002. A comparison of age determination methods when applied to North Atlantic minke whales. In: Paper SC/54/RMP7 Presented to the International Whaling Commission Scientific Committee.

Olsen, E., Sunde, J., 2002. Age determination of mink whales (*Balaenoptera acutorostrata*) using the aspartic acid racemization technique. Sarsia 87, 1–8.

Perrin, W.F., Myrick, Jr., A.C., 1980. Report of the workshop. Age determination of toothed whales and sirenians. In: Report of the International Whaling Commission (Special Issue 3), pp. 1–50.

Philo, L.M., Shotts, E.B., Jr. and George, J.C. 1993. Morbidity and mortality. In: J.J. Burns, J.J. Montague and C.J. Cowles (eds.) The Bowhead Whale. Allen Press, Lawrence, Kansas. 787 pp.

Polanowski, A.M., Robbins, J., Chandler, D., Jarman, S.N., 2014. Epigenetic estimation of age in humpback whales. Mol. Ecol. Resour. 14, 976–987. Available from: https://doi.org/10.1111/1755-0998.12247.

Punt, A.E., Butterworth, D.S., 1999. On assessment of the Bering-Chukchi-Beaufort Seas stock of bowhead whales (*Balaena mysticetus*) using a Bayesian approach. J. Cetacean Res. Manage. 1 (1), 53–711.

Purves, P.E., 1955. The wax plug in the external auditory meatus on the mysticeti. In: Disc Rep 27, 293–302 + plates XIV–XVIII.

Read, F.L., Hohn, A.A., Lockyer, C.H., 2018. A Review of Age Estimation Methods in Marine Mammals With Special Reference to Monodontids. NAMCO.

Rehorek, S.J., Stimmelmayr, R., George, J.C., Suydam, R., McBurney, D.M. and Thewissen, J.G.M., 2019. The role of desmosomes in the ear plug formation in the bowhead whale (Balaena mysticetus). The Anatomical Record.

Rice, D.W., Wolman, A.A., 1971. The Life History and Ecology of the Gray Whale (*Eschrichtius robustus*). U.S. Fish & Wildlife Service, Bureau of Commercial Fisheries, Marine Mammal Biological Laboratory, Seattle, WA, Special Publication No. 3.

Roe, H.S.J., 1967. Seasonal formation of laminae in the ear plug of the fin whale. In: Disc Rep 35, 1–30 + plates I–IV.

Rosa, C., Zeh, J., George, J.C., et al., 2013. Age estimates based on aspartic acid racemization for bowhead whales (*Balaena mysticetus*) harvested in 1998–2000 and the relationship between racemization rate and body temperature. Mar. Mammal Sci. 29 (3), 424–445.

Rugh, D.J., Braham, H.W., Miller, G.W., 1992. Methods for photographic identification of bowhead whales, *Balaena mysticetus*. Can. J. Zool. 70, 617–624.

Schell, D.M., Saupe, S.M., 1993. Feeding and growth as indicated by stable isotopes. In: Burns, J.J., Montague, J.J., Cowles, C.J. (Eds.), The Bowhead Whale. pp. 491–509. Special Publication No. 2 of the Society for Marine Mammalogy. i–xxxvi + 787 p.

Schell, D.M., Saupe, S.M., Haubenstock, N., 1989a. Bowhead (*Balaena mysticetus*) growth and feeding as estimated by $\delta^{13}C$ techniques. Mar. Biol. 103, 433–443.

Schell, D.M., Saupe, S.M., Haubenstock, N., 1989b. Natural isotope abundance in bowhead whale (*Balaena mysticetus*) baleen: markers of aging and habitat usage. In: Rundel, P.W., Ehleringer, J.R., Nagy, K.A. (Eds.), Stable Isotopes in Ecological Research. pp. 260–269.

Schmidt-Nielsen, 1997. Animal Physiology: Adaptation and Environment. Cambridge University Press, Cambridge, p. 611.

Sensor, J.D., George, J.C., Clementz, M.T., Lovano, D.M., Waugh, D.A., Givens, G., et al., 2018. Age estimation in bowhead whales using tympanic bulla histology and baleen isotopes. Mar. Mammal Sci. Available from: https://doi.org/10.1111/mms.12476.

Tarpley, R.J., Hillmann, D.J., George, J.C., Zeh, J.E., Suydam, R.S., 2016. Morphometric correlates of the ovary and ovulatory corpora in the bowhead whale, *Balaena mysticetus*. Anat. Rec. 299 (6), 769–797.

Trumble, S.J., Robinson, E.M., Berman-Kowalewski, M., Potter, C.W. and Usenko, S., 2013. Blue whale earplug reveals lifetime contaminant exposure and hormone profiles. Proceedings of the National Academy of Sciences, 110(42), pp.16922–16926.

Trumble, S.J., Usenko, S., Sanders, R., Stevens, A.N., Thewissen, J.G.M., 2015. Bowhead Whale Earplug Reveals Lifetime Chemical Profiles. Poster. Society for Marine Mammalogy.

Weintraub, B., 1996. Harpoon Blades Point to Long-Lived Whales. National Geographic Society, Washington, DC.

Wetzel, D.L., Reynolds III, J.E., Mercurio, P., Givens, G.H., Pulster, E.L., George, J.C., 2017. Age estimation for bowhead whales, *Balaena mysticetus*, using aspartic acid racemization with enhanced hydrolysis and derivatization procedures. J. Cetacean Res. Manage. 17, 9–14.

Withrow, D., Burke Jr., R., Jones, L., Brooks, J., 1992. Variations in $\delta^{13}C$ carbon isotope ratios in bowhead baleen plates used to estimate age. Rep. Int. Whal. Commn 42, 469–473.

CHAPTER

22

Acoustic behavior

Kathleen M. Stafford[1] and Christopher W. Clark[2]

[1]Applied Physics Laboratory, University of Washington, Seattle, WA, United States

[2]Cornell Lab of Ornithology, Center for Conservation Bioacoustics, Cornell University, Ithaca, NY, United States

Introduction

Hearing is the sense that marine mammals, particularly the large whales, rely most heavily upon. Sound travels efficiently underwater, much faster and/or further than chemical or visual cues. Sound is the primary modality used for basic life functions such as communicating, navigating, finding food, and detecting predators; and the reception of sound can provide information on an animal's environment (Payne and Webb, 1971; Tyack, 1981; Edds-Walton, 1997; Tyack and Clark, 2000; Clark and Ellison, 2004; Johnson et al., 2004). For baleen whales, acoustic signals are generally divided into "calls" and "song" (Clark, 1990). Additionally, a unique signal type, referred to as a gunshot, has been observed primarily in right whales (*Eubalaena* spp., Clark, 1983; Parks et al., 2012), but also in bowheads (*Balaena mysticetus*, Würsig and Clark, 1993). In general, calls are relatively simple, short duration signals produced in a wide variety of social contexts throughout the year and mostly serve as a means by which animals communicate with each other. Songs, in contrast, are composed from multiple sounds (i.e., units or notes) produced in patterns (i.e., phrases) such that one hears both the repetitions and the cadence of the patterns. Songs are produced seasonally, typically between late fall and late spring, and singing is assumed to be a male reproductive display that very likely plays a role in reproduction either by affecting male–male interactions, providing females with information about a male's fitness, or some combination of both (Cholewiak et al., 2018) (Fig. 22.1).

Discoveries about bowhead acoustic behaviors, both calling and singing, have been highly dependent on when and where recordings took place as well as the level of effort dedicated to collecting recordings. While the earliest published accounts of bowhead sounds were those of Yankee whalers who described both cries (calls) and songs of bowhead whales (Aldrich, 1889), the earliest recordings of bowhead calls were collected in the

The Bowhead Whale

DOI: https://doi.org/10.1016/B978-0-12-818969-6.00022-4

FIGURE 22.1 A pair of bowheads surface in a small opening in heavy ice in the Chukchi Sea, exposing only the snout with blowholes. Loud singing was heard from a dipping hydrophone shortly before these animals surfaced. *Source: Photo by Kate Stafford.*

late 1960s (Poulter, 1968). In the late 1970s, more extensive documentation of bowhead calls and the first evidence that bowheads sing emerged during a few spring-summer visual-acoustic aerial surveys in the Chukchi and Beaufort Seas (Ljungblad et al., 1982), and preliminary efforts to survey the Bering–Chukchi–Beaufort (BCB) population during its springtime migration past Point Barrow, Alaska (Cummings and Holliday, 1985, 1987; Clark and Johnson, 1984). Beginning in the early 1980s, the establishment of a springtime visual-acoustic survey effort off Point Barrow provided the first long-term, acoustic dataset documenting the relative complexity of bowhead calling behavior. Analysis of these data also indicated that singing whales recorded during the migration sang songs with relatively similar phrases within a season, but that bowhead song changed from year to year (Clark et al., 1986a,b, 1996; Würsig and Clark, 1993; Zeh et al., 1993).

Because sound production and reception (hearing) is such an important sense for marine mammals, understanding the behavioral context of sound production and the acoustic ecology of these animals is paramount to understanding their life histories. As the only baleen whale to inhabit the Arctic year-round, bowhead whales are faced with unique challenges in communicating, navigating, finding mates, and surviving in an environment that is dominated by ice and the polar night, and which is undergoing rapid physical transformations as a result of a changing climate. Despite these challenges, it is obvious that the acoustic repertoire of bowhead whales, both in terms of calls and songs, is as remarkable as the whales themselves—there is no greater sense of amazement than when one dips a hydrophone into a springtime lead in the Arctic ice and is greeted by a cacophony of bowhead calls and their complex, two-voiced songs.

Bowhead whale sounds

Upon first listening, one might conclude that bowhead whales produce relatively simple calls throughout the year; however, as recording efforts have been extended to collections spanning decades and entire years, and expanded to include data from multiple populations, the conclusion of vocal simplicity in the species is an anthropocentric misunderstanding, to say the least. We have often found ourselves listening to an Arctic acoustic scene composed from the voices of bearded seals (*Erignathus barbatus*), beluga whales (*Delphinapterus leucas*), walrus (*Odobenus rosmarus*), and bowheads, and myriad sounds of ice. While following each of those voices as separate acoustic threads in a recording, we sometimes would become aware that there were sounds that ebbed and flowed in the mix; these started out in the realm of bowhead acoustic simplicity, but evolved into something different than, yet somehow familiar to, normal bowhead sounds. These collections of experiences lead us to conclude that bowheads are not only superb mimics of the ice and other species living in the ice, but are also endowed with an innate ability to create totally novel sounds and combinations of sounds that cannot be described or categorized as some form of a bowhead sound "type." How then is one to derive a dictionary of sorts, some way to categorize and possibly translate the open-ended repertoire of bowhead sounds into meaning? This dilemma was further compounded when we realized that bowhead singers often produce two sounds simultaneously, sounds that are harmonically and temporally unrelated, and are in some ways similar to the suboscine songbirds that sing with two voices (Greenewalt, 1968).

These observations of acoustic display complexities by bowheads present a challenge when it comes to deducing, let alone understanding, the functional significance (i.e., biological meaning) of their different "types" of sounds. Any such attempt is extremely limited given our present inability to associate the sound we record and analyze with a behavioral context. We are left with the task of linking an almost open-ended collection of sounds with a very simplistic set of contextual categories.

In any case, there is certainly a logical suite of behavioral contexts for which bowheads use sounds, and for some of these there have been a few acoustic observations that are at least consistent with a biological function. An obvious conclusion is that bowheads of both sexes and likely all age classes use their sounds as acoustic signals for such basic life functions as communicating between mother–calf pairs and between combinations of individual animals; navigating though and around ice; and coordinating between groups of migrating whales; and that song is a male acoustic display that can provide reliable cues of male attributes attractive to females and threatening to rival males or both (Ljungblad et al., 1980, 1982; Ellison et al., 1987; George et al., 1989; Clark, 1989; Stafford et al., 2012, 2018b).

Sound production

Relatively little is known about how baleen whales produce their remarkable variety of sounds, but it is believed that, like other mammals, the larynx, in particular the laryngeal sac (called the laryngeal diverticulum in Schoenfuss et al., 2014), laryngeal folds, and associated cartilages, in conjunction with the lungs are likely the anatomical structures used for sound production. Because the respiratory tract has to remain closed while an animal is underwater, whales are restricted to using a limited amount of air for sound production (Reidenberg and Laitman, 2008). Air is recycled between the lungs and laryngeal sac via the folds which are found on either side of a slit that connects the sac to the main air passage (Hosokawa, 1950; Reidenberg and Laitman, 2007; Schoenfuss et al., 2014). Detailed descriptions of the larynx and laryngeal sac of the bowhead date back as far as the mid-1800s (Sandifort, 1831 [in Eschricht et al., 1866]; Eschricht et al., 1866). Examining this anatomy in terms of its role in bowhead sound production is somewhat more recent (Haldiman and Tarpley, 1993; Schoenfuss et al., 2014).

The laryngeal folds of baleen whales are thought to serve a similar function as vocal folds in land mammals: sound is produced as the folds vibrate, while air is cycled between the lungs and the laryngeal sac (Reidenberg and Laitman, 2007; Adam et al., 2013). Within the interior of the larynx is a U-shaped fold (the U-fold of Reidenberg and Laitman, 2007) that is anterior to the laryngeal sac and attached to extensions of the arytenoid cartilages. The U-fold may serve as something of a valve between the lungs and laryngeal sac (Reidenberg and Laitman, 2007; Damien et al., 2019). The laryngeal sac of the Balaenidae is smaller than that of Balaenopteridae, and for the bowhead has been described as "rudimentary," but highly muscular and flexible (Beauregard and Boulart, 1882). This flexibility enables the same air to be repeatedly pumped and recycled between the lungs and the sac so that the animal can continue to produce sounds while remaining underwater for extended periods of time (Reidenberg and Laitman, 2008). Presumably, bowhead whales are capable of controlling the internal air pressure in both the lungs and the laryngeal sac to vibrate the vocal folds and laryngeal cartilages (including the arytenoids, cricoids, corniculate flaps, and nasopharynx) to create sounds (Gandilhon et al., 2014; Damien et al., 2019). How exactly these many elements work together to produce the two voices and infinitely variable repertoire of the bowhead is presently unknown, and deserving of greater attention.

Bowhead whale calls

Bowhead whale calls have been classified by their degree of frequency modulation (FM) and/or amplitude modulation (AM) and generally range from ~25 to 500 Hz (Clark and Johnson, 1984; Clark et al., 1986b, 1988; Würsig and Clark, 1993). The majority of bowhead calls are characterized as simple, FM, low-frequency sounds, but within these constraints can nevertheless vary in bandwidth and modulation pattern (Ljungblad et al., 1982; Clark and Johnson, 1984; Thode et al., 2012). Although nomenclature for simple bowhead whale call types has often varied by study, calls can be generally classified as FM downsweeps (decreasing in frequency over time, Fig. 22.2A and B), upsweeps (increasing in frequency over time, Fig. 22.2C), tonal or constant (no change in frequency over time), and inflected (increasing and then decreasing; Clark and Johnson, 1984; Fig. 22.2A). These calls typically last 1–2 seconds in the frequency bandwidth of 120–400 Hz. Exceptions that have been described include higher frequency (400–1000 Hz) tonal calls (Clark and Johnson, 1984; Würsig and Clark, 1993). Other calls, often called complex moans or pulsed tonals (Fig. 22.2B and C; Ljungblad et al., 1982; Clark and Johnson, 1984), can include both FM and AM components in the same signal or be entirely pulsive (Ljungblad et al., 1982; Clark and Johnson, 1984). Such signals can have broader bandwidths (with frequencies as high as 3 kHz) and more variable durations (e.g., less than 1 second to over 7 seconds, Clark and Johnson, 1984).

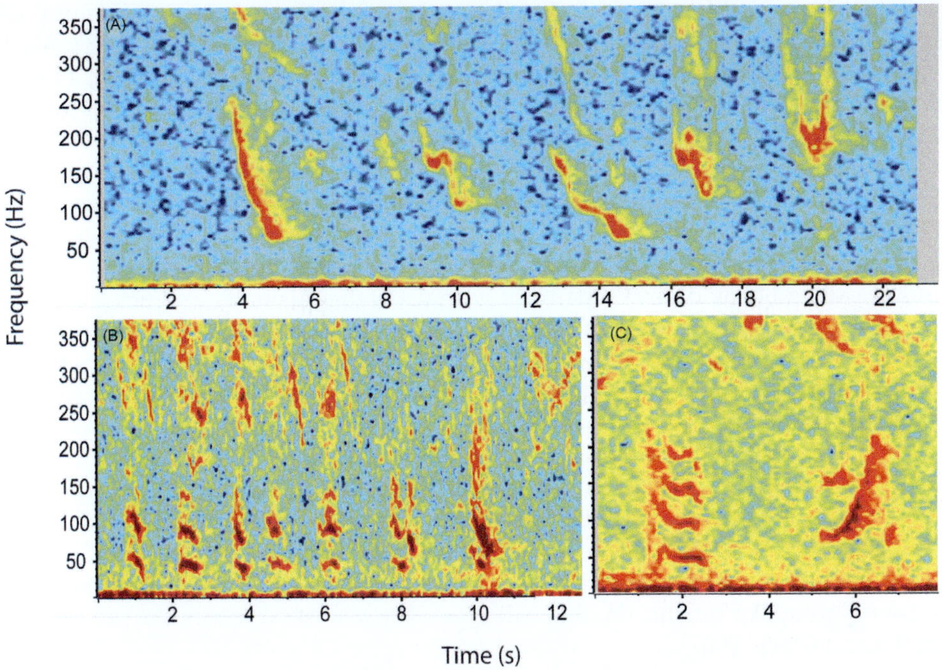

FIGURE 22.2 Spectrogram examples of various bowhead whale calls. (A) Series of similar simple calls showing degrees of frequency modulation; (B) series of complex calls; (C) single complex call followed up simple upsweep. All spectrograms produced with a 0.128 s FFT, 75% overlap, Hann window.

The call repertoire of bowhead whales has been well-described from these early works and was reviewed by Würsig and Clark (1993). Since these earlier studies, year-round monitoring with autonomous, under-ice recording systems has greatly expanded our knowledge of annual variability in bowhead calling and singing as well as the seasonal occurrences and distributions of the different populations throughout much of their home ranges (Moore et al., 2010, 2012; Clark et al., 2015; Ahonen et al., 2017). The vast majority of studies of bowhead whale acoustic behavior and repertoire are from the western Arctic, BCB population. Given the relative similarity of certain call types as typically described for different populations of right whales (Clark, 1982, 1983; Parks et al., 2007; Jacobs et al., 2018), it might be presumed that different bowhead populations also share a common call repertoire. Clearly, a comparative study is needed to address this hypothesis, ideally based on data from all four bowhead populations at similar seasonal scales and under similar behavioral contexts.

From several studies that have occurred in different regions and seasons, there is some evidence that the relative proportions of different call types and call rates produced by bowheads change within and between years (Ljungblad et al., 1982; Blackwell et al., 2007, 2011; Charif et al., 2013). During the spring migration in the 1980's, simple low-frequency FM upsweeps (i.e., upcalls) and downsweeps (i.e., downcalls) represented almost 60% of all nonsong call types, while complex calls were infrequently recorded (Clark et al., 1986b). Further, the proportion of complex calls declined in spring as the migration progressed (Würsig and Clark, 1993). On the spring and fall feeding grounds in the BCB, the most commonly recorded call types were simple calls (Ljungblad et al., 1982; Blackwell et al., 2007, 2011; Charif et al., 2013). In Disko Bay, West Greenland, during 11 hours of recordings from late winter to early spring, constant frequency calls were the most commonly recorded call type followed by complex amplitude-modulated calls. Unlike in other regions, frequency-modulated simple calls were relatively uncommon (Tervo et al., 2009). Can this observation be explained given that Disko Bay is only occupied seasonally and is not necessarily along a migratory corridor (Heide-Jørgensen et al., 2007; Rekdal et al., 2014)? What is lacking from such disparate results between populations is a broad-scale synthesis of calling behavior and repertoire changes both seasonally and geographically.

Behavioral contexts have been inferred for some bowhead calling behaviors, however, the dynamic icy environment in which bowheads live makes it extremely difficult to directly associate call types with specific behavioral contexts. It has been hypothesized that bowhead whales use sound to assess the thickness of sea ice and to avoid areas of very heavy ice during their spring migration (Ellison et al., 1987; Clark, 1989; George et al., 1989). Further, migrating bowheads have been found to use simple, individually distinctive calls in a form of counter-calling to synchronize migratory movements (Clark et al., 1986a; Clark, 1989). Bowhead whales, like right whales produce upcalls; in right whales, the upcall has been shown to function as a contact call often resulting in calling animals joining into a group: in bowheads, this call type may also serve a similar purpose (Clark, 1983; Parks and Clark, 2007).

Some bowhead calls are thought to be directional with louder amplitudes produced forward of the animal, although this has not been shown consistently (Clark et al., 1986b; Blackwell et al., 2011). The strongest support comes from a long-term, acoustic dataset during the westward fall migration along the shallow Beaufort Sea shelf (Blackwell et al., 2011). That study found relatively higher amplitudes for a limited number of FM upsweeps and complex calls when the recording hydrophone was in front (i.e., anterior aspect) of the presumed orientation of migrating animals. In that study, the signal-to-noise ratio of those calls was 1.8 times greater than calls when the recording hydrophone was behind (i.e., posterior aspect) the presumed orientation of migrating animals. Analyses of similar data in deeper water found that while calls in shallow water were strongly directional, those in deeper water were not (Blackwell, pers. comm.). A possible explanation for this difference is that in shallow waters, the lower frequencies of calls (which are likely not directional) are stripped off leaving only higher frequencies which appear to be directional. In deeper water, the lower frequencies are included in the recorded data and the overall call does not appear directional (Blackwell, pers. comm.). A detailed examination of the directionality of the different frequency bands of calls in a range of water depths should provide an increased understanding of frequency-dependent directionality in bowhead signals (both calls and songs). Call directionality, while not an intrinsic characteristic of bowhead calls, may be useful to animals migrating along the narrow Beaufort Sea shelf and in the shallow Chukchi Sea in fall. Likewise, if bowheads are using some calls to navigate in ice, then projecting signals forward would allow the animals to sense their environment over a greater range and thus help them make decisions about their migratory movements. This was evident from acoustic data in the 1980s that showed that bowheads were navigating around a large, deep-keeled multiyear ice floe (George et al., 1989).

Call sequences

Bowhead whales will sometimes produce sequences of simple frequency-modulated calls in bouts of up to 25 similar calls (Ljungblad et al., 1982; Würsig and Clark, 1993; Stafford et al., 2008; Delarue et al., 2009). Because these patterns may repeat numerous times, they have sometimes been labeled "simple song" (e.g., Stafford et al., 2008; Delarue et al., 2009; Johnson et al., 2014). Although they may repeat, these sequences differ from song (see section below) in that they consist of simple calls that are more limited in bandwidth than song units, the time gaps between calls are evenly spaced, and there is never simultaneous production of two distinct sounds. During the winter these sequences can persist for weeks at a time, occupying a frequency band that is just beneath that in which song occurs. Call sequences have been recorded both on wintering grounds, but also during migration and on feeding grounds.

Counter-calling

A different form of calling behavior has been observed during springtime migrations off Point Barrow, Alaska using distributed arrays that enabled locating and tracking multiple calling whales (Clark et al., 1986a; Clark, 1989; Würsig and Clark, 1993). Careful analysis of tracks of calling whales revealed that in the simplest case, each call track was composed of calls with remarkably similar acoustic features (i.e., a call signature, e.g., duration, frequency sweep rate, inflection point), and such calls in a track were subsequently referred to as signature calls (Clark et al., 1986a,b; Clark, 1989). It was further observed that within about 5–10 seconds after the occurrence of one form of a signature call in a call track, a second signature call type occurred in a second call track. This behavior was then repeated throughout the overlapping durations of the two tracks and was referred to as a form of counter-calling. Time gaps between counter-calls were typically on the order of 5–10 minutes (Clark, 1989). At times, there could be as many as 5–10 call tracks, and thus 5–10 different signature call types in the counter-calling episode. The timing and spacing of the different acoustic tracks suggested that a group of animals was maintaining its integrity during migration by counter-calling. These counter-calling episodes could last for many minutes to hours as the group of animals migrated through the early spring ice (Würsig and Clark, 1993). As more of these counter-calling episodes became apparent, it was also observed that some callers in the acoustic herd would switch to the call signature of another member of the herd; based on consecutive locations of the callers, which may potentially be a form of social acoustic mimicry (Clark et al., 1986a).

Bowhead whale gunshots

An unusual sound, observed in all right whale populations and referred to as a "gunshot" (Clark, 1983; Stafford et al., 2010; Parks et al., 2012), has also been recorded and attributed to bowheads during the spring migration off Point Barrow, Alaska. Gunshots are short, loud, broadband percussive signals that are hypothesized to be part of an agonistic display and possibly a form of male song in right whales (Clark, 1983; Parks et al., 2012;

Crance et al. 2019) and have been recorded in surface active groups of both right and bowhead whales (Würsig and Clark, 1993; Parks, 2003). The most unambiguous attribution of gunshot sounds to a bowhead was made under relatively open lead conditions, with little ice in the lead, using a distributed array deployed along the ice edge, such that the acoustic locations of a series of gunshots coincided with the locations of a call track. Other observations of potential bowhead gunshots during the spring migration were confounded by a limited number of sounds that could not be consistently associated with a call track and with ice conditions that could produce broadband impulsive sounds.

Source level, calling depth, and detection distance

The source levels of some simple calls from bowheads in the BCB have been determined to range from ~156–160 dB re 1 μPa up to 190 dB re 1 μPa (Clark and Johnson, 1984; Ljungblad et al., 1982; Cummings and Holliday, 1987; Thode et al., 2016). In the nearshore environment of the Chukchi and Beaufort Seas, these sounds have been heard out to at least 50 km under low ambient noise conditions (Greene et al., 2004; Blackwell et al., 2007). Most calls are likely only localizable to about 10 km under mean ambient noise conditions in shallow water during the summer and fall (Cummings and Holliday, 1985; Bonnel et al., 2014; Abadi et al., 2014). These signals have been modeled to be produced by whales both near the surface but also to depths between 22 and 30 m (Thode et al., 2016). During the spring migration, calling whales were regularly localized to 10 km away from an acoustic array (Clark et al., 1986b, 1988, 1996) but may have been heard from even greater distances. These results contrast with estimates of "song notes" from bowhead whales recorded in deeper water off West Greenland (Tervo et al., 2012). In that study, in which recordings were made under the sea ice, source levels were estimated at 185 dB re 1 μPa with estimates of detection distances from 40 to 130 km, which are both considerably louder and at greater distances, respectively, than reports from western Arctic studies.

Bowhead whale song

Bowhead whales are the only nonrorqual that is known to sing. While bowhead and most balaenopterids are known to produce song, this vocal behavior is not known to occur in right whales (*Balaena* spp.), gray whales (*Eschrichtius robustus*), or pygmy right whales (*Caperea marginata*). Indeed, of all the mysticetes, only bowhead and humpback (*Megaptera novaeangliae*) whales sing complex songs whose structures change over short- and long-time scales. Bowhead song generally consists of one to three phrase types, repeated multiple times and sung in a consistent order. An individual song lasts anywhere from about 45 seconds to over 120 seconds and is repeated for many hours with time gaps of several seconds to several minutes between songs. Our ability to reliably assess how long an individual sings, referred to as song bout duration, is impeded by the inability to determine if a singer swam out of the detection range of the hydrophone-recording system or actually stopped singing. Song phrases are composed of both FM and AM, harmonically rich units that range from 50 to 5000 Hz (Clark and Johnson, 1984; Stafford et al., 2008, 2012).

Furthermore, bowhead whales have the ability to sing "with two voices." That is, a singer can produce two distinct, harmonically unrelated sounds simultaneously; one of which is usually a simple, low-frequency sweep while the other is higher frequency and broaderband (Clark and Johnson, 1984; Würsig and Clark, 1993; Tervo et al., 2011a). Individual units that compose songs as well as the songs themselves, can vary by month and year (Tervo et al., 2009; Stafford et al., 2018b). Often, a song will start with a higher frequency-modulated note, followed by multiple repetitions of complex, harmonically rich FM/AM signals with a simultaneous low-frequency sweep (see Fig. 22.3 for example). The song or phrase will end with a long, low-frequency downsweep (Stafford et al., 2008). Singing begins in late October or early November and continues into April or early May (Delarue et al., 2009; Ahonen et al., 2017; Stafford et al., 2018b).

Prior to 2008, reports of bowhead whale singing had been restricted to recordings from at most a few months during the spring migration off Point Barrow, Alaska. In contrast, recordings from two studies north of Spitsbergen (the East Greenland-Svalbard-Barents Sea, EGSB) and off West Greenland (ECWG), gave some indication that bowheads (or a population of bowhead whales) would sing more than a single song each year (Clark, 1990; Stafford et al., 2008). The development and deployment of instrumentation that could record year-round in the Arctic provided increased evidence that singing was

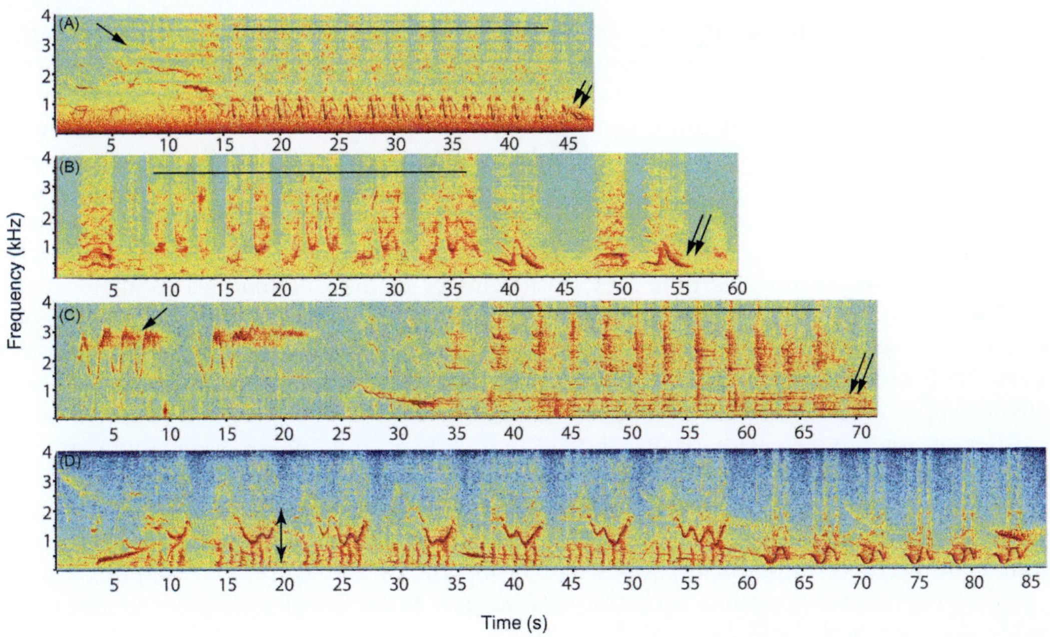

FIGURE 22.3 Exemplar songs showing general song structure and variability therein. Note the relatively higher frequency first note (*single arrow* in A and C) followed by multiple repetitions of similar complex notes (*line* in A–C), terminal lower-frequency tonal note (*double arrow* in A–C), and "two voice" singing (*two-headed arrow* in D). (A) Song from Fram Strait, December 24, 2013; (B) Song from Bering Strait, November 20, 2012; (C) Song from Davis Strait, April 8, 2007; (D) Song from Point Barrow Alaska, April 28, 1988. All spectrograms produced with a 0.128 s FFT, 75% overlap, Hann window.

seasonal (from roughly November through April or early May) and that singers in both western and eastern Arctic populations produced multiple song types during the winter (Delarue et al., 2009; Stafford et al., 2012, 2018b). Year-round recordings in 2008–2014 from Fram Strait, which is thought to be a wintering area for EGSB whales, illustrate the prolific singing ability of bowhead whales. Beginning in late October and continuing through April annually, EGSB bowhead whales sang dozens of novel song types that were not slight variations of previous songs, as occurs within populations of humpback whales (Payne and Payne, 1985), but were distinctly different from one another (Stafford et al., 2012, 2018b). Further, each song type appeared more or less sequentially, and after a few days or weeks, was seldom heard in the following weeks or months. No song type was ever recorded in more than a single November to April singing season based on 4 years of data (Stafford et al., 2018b). While the EGSB is the only population for which year-round studies of song analysis have been undertaken, it appears that both the BCB and the ECWG populations produce multiple song types within a single season with no repetitions of a song type across seasons (Stafford et al., 2008; Delarue et al., 2009; Tervo et al., 2011b; Johnson et al., 2014).

While there have been multiple song types recorded from the same population within a single singing season, the question remains as to *who* is singing the different song types; for example, does an individual sing only one song type or multiple song types? Thus, for example, recordings from Fram Strait demonstrate that a population of singers is capable of singing many dozens of different song types within a season (from November through May, Stafford et al., 2018b). Because those data were from a single omni-directional hydrophone, it could not be determined if individual whales each sing a unique song, if there is song sharing among whales, or if individuals switch songs within a season. To try to determine this, 6 weeks of data from a single hydrophone in an acoustic array deployed during the spring 2011 (12 April–27 May) visual-acoustic census off Point Barrow, Alaska, were examined for different song types (Johnson et al., 2014). BCB bowhead whales migrate north and east along the coast of Alaska in the spring on their way to the eastern Beaufort Sea, and during this time they do not linger in the Chukchi Sea (Braham et al., 1980; Moore and Reeves, 1993; Zeh et al., 1993). Johnson et al., 2014 proposed that different whales were singing the same song type if that song type was recorded at least 24 hours apart. Their study assumed that a singing whale would be migrating through, and not remain within acoustic detection range of the hydrophone for more than 1 day. Similarly, if different song types were recorded with more than 24 hours between them, it was assumed those songs were produced by different whales (Johnson et al., 2014). Analysis of these data suggests that some singers share some song types during the spring migration, but others do not (Johnson et al., 2014). Furthermore, similar to the overwintering data from Fram Strait (a time during which bowhead whales are likely not migrating), song types off Pt. Barrow appeared and disappeared, not to be heard again during the same spring (Johnson et al., 2014; Stafford et al., 2018b).

Song has been recorded from three of the four populations of bowhead whales (there are no data at present from the Okhotsk Sea population), and the songs all share the characteristics of being composed of 1–3 repeated phrases, "two voices," and highly variable combinations of frequency- and amplitude-modulated units (Clark and Johnson, 1984; Würsig and Clark, 1993; Stafford et al., 2008, 2012; Delarue et al., 2009; Tervo et al., 2011a;

Johnson et al., 2014). What has not yet been explored, is whether there are population-level differences in singing behavior or song composition as evidenced by the types of notes and the syntax that characterize songs in different populations. Data from multiple years from both Fram Strait and Disko Bay, West Greenland, suggest that the notes used to compose songs in these two regions change completely between years (Tervo et al., 2011b; Stafford et al., 2018b). At the moment, at least based on data from Fram Strait and Disko Bay, the within-population monthly and annual variability in song note types and song composition is so great that it may prove difficult to distinguish between populations acoustically as has been observed in at least two other baleen whale species (blue whales, e.g., Stafford et al., 2001; and fin whales, e.g., Hatch, 2004; Castellote et al., 2011).

Acoustic ecology of the bowhead whale

Bowhead whale acoustic behavior is varied, complex, and dynamic. As a species that relies on sound and lives in an acoustically fluctuating environment, bowheads must respond to changes in the acoustic environment that surrounds them. The Iñupiat in the Arctic knew that whales could hear well underwater and that silence on the ice while hunting in spring was important (Brower, 1942). This Arctic acoustic environment of the bowhead is influenced by both biotic and abiotic factors. The biotic factors include the sounds of other bowheads, but also other vocal marine mammals including ice seals, walrus, belugas, and narwhals (*Monodon monoceros*) and relatively recently, subarctic species (Stafford et al., 2018a, Stafford, 2019). Abiotic factors are natural noises from ice, wind, and geophony and, increasingly, anthropogenic signals from shipping and resource extraction. These factors, and the acoustic behavior of bowhead whales, vary seasonally.

In some locations during the winter and early spring, there is a nearly continuous chorus of bowhead singers, and the aggregate energy from singers can dominate the 50–5000 Hz frequency band (Stafford et al., 2018a,b). At other times of year, especially during summer and the fall feeding and migration period, the acoustic bandwidth used by bowheads is one-tenth of that used in spring, and calls tend to be spread out more in time with many fewer calls per hour than during periods with singing. The behavioral functions and biological significance of singing and calling are most certainly different. Song is presumed to be a male reproductive display adapted to attract females and repel rival males, or both. This behavior may not initiate an obvious response from female listeners, but might change the singing behavior of other male singers (e.g., Tyack, 1981 for humpback whales). Calls, on the other hand, may require listening specifically for a response from a con-specific (e.g. to coordinate movements) or bistatically, by listening for the reflection of a call off of ice (George et al., 1989).

As with the acoustic behavior of bowheads, the natural background acoustic environment in which they live also changes seasonally. Depending upon the location in the Arctic, in winter and early spring, bowheads sing amid the trills of bearded seals, the knocks of walrus, and the whistles of beluga whales and narwhals. All of these bioacoustic signals can co-occur with the grinding, scraping, and explosions of mobile sea ice. These elements combine to produce arguably the highest natural ambient noise levels in which bowheads live throughout their lives (Clark et al., 2015; Ahonen et al., 2017; Stafford et al., 2018a). And yet, a careful listener can distinguish each of these signals from each other, as surely as the animals themselves must do.

Given that there are now reasonably large and growing collections of recordings from at least the BCB and EGSB populations, it is now feasible to address questions and hypotheses regarding decadal changes in bowhead acoustic behaviors under natural ambient noise conditions and anthropogenically modified noise conditions (see Chapter 35). Some initial indication of changes in acoustic behavior under natural ambient noise conditions has been demonstrated for the BCB population during its spring migration off Point Barrow, Alaska. The estimated trend of this population over the 33-year period from 1978 through 2011 indicates a 3.7% annual rate of increase (95% confidence of 2.9%–4.6%), growing from an abundance estimate of 5000 in 1978 to an estimate of almost 17,000 animals in 2011 (Givens et al., 2016). A comparison of locatable bowhead sounds (calls and song notes) from 1993 through 2011 showed a >500% increase over this 18-year period (Clark et al., 2018). When compared to the 1993 acoustic data, the bioacoustic bowhead scene in 2011 had become extremely complex in terms of the amount of calling and singing and the types of sounds produced. In some cases, the sound density (number per unit time) was so high that it became difficult to clearly identify individual sounds (Clark et al., 2018).

Climate change-driven reductions in Arctic sea ice will change the acoustic environment in which the bowhead whales' elaborate acoustic behavior evolved. Abiotic changes may include increasing wind noise (and perhaps decreasing sea ice noise) in the summer and fall and increasing anthropogenic inputs from shipping (commercial and tourism) and subsea resource exploration and extraction. In the presence of increasing noise, bowhead whales call more often and more loudly, perhaps in response to a decrease in their communication space (Clark et al., 2009, Chapter 35). But at very high noise levels, they cease acoustic communication. This is particularly evident for whales in the presence of noise from seismic exploration (Blackwell et al., 2015; see Chapter 35). This differential response complicates how we can assess and mitigate the impacts of noise on animals. Biotic changes in the acoustic environment of the bowhead include the introduction of soniferous subarctic species such as the humpback whale that may compete for acoustic space with bowheads, or inspire new song behavior, as the former expands their habitat in space and time (Woodgate et al., 2015). Increasing occurrence of killer whales in the Arctic could induce bowheads to "go silent" to avoid detection (see Chapter 29; Ferguson et al., 2010; Stafford, 2019). And of course, the lack of sea ice could allow whales from the BCB, ECWG, and EGSB populations to interbreed (Heide-Jørgensen et al., 2012). How these changes in the Arctic acoustic environment will influence the acoustic ecology of the bowhead, and how this remarkable species will adapt to such changes, will likely strongly influence the next 40 years of bowhead whale acoustic studies.

References

Abadi, S.H., Thode, A.M., Blackwell, S.B., Dowling, D.R., 2014. Ranging bowhead whale calls in a shallow-water dispersive waveguide. J. Acoust. Soc. Am. 136, 130–144.

Adam, O., Cazau, D., Gandilhon, N., Fabre, B., Laitman, J.T., Reidenberg, J.S., 2013. New acoustic model for humpback whale sound production. Appl. Acoust. 74 (10), 1182–1190.

Ahonen, H., Stafford, K.M., de Steur, L., Lydersen, C., Wiig, Ø., Kovacs, K.M., 2017. The underwater soundscape in western Fram Strait: breeding ground of Spitsbergen's endangered bowhead whales. Mar. Pollut. Bull. 123, 97–112.

Aldrich, H.L., 1889. Arctic Alaska and Siberia, or, Eight Months with the Arctic Whalemen. Rand, McNally, Chicago, IL.

Beauregard, M.H., Boulart, R., 1882. Recherches sur le larynx et la trachée des Balaenides. J. Anat. Physiol. Norm. Pat. Homme Anim. 18, 1–26.

Blackwell, S.B., McDonald, T.L., Kim, K.H., Aerts, L.A.M., Richardson, W.J., Greene, C.R., et al., 2011. Directionality of bowhead whale calls measured with multiple sensors. Mar. Mamm. Sci. 28 (1), 200–212.

Blackwell, S.B., Nations, C.S., McDonald, T.L., Thode, A.M., Mathias, D., Kim, K., et al., 2015. Effects of airgun sounds on bowhead whale calling rates: evidence for two behavioral thresholds. PLoS One 10, e0125720.

Blackwell, S.B., Richardson, W.J., Greene, C.R.J., Streever, B., 2007. Bowhead whale (*Balaena mysticetus*) migration and calling behaviour in the Alaskan Beaufort Sea, Autumn 2001–04: an acoustic localization study. Arctic 60, 255–270.

Bonnel, J., Thode, A.M., Blackwell, S.B., Kim, K., Michael Macrander, A., 2014. Range estimation of bowhead whale (*Balaena mysticetus*) calls in the Arctic using a single hydrophone. J. Acoust. Soc. Am. 136, 145–155.

Braham, H., Fraker, M., Krogman, B.D., 1980. Spring migration of the Western Arctic population of bowhead whales. Mar. Fish. Rev. 42, 36–46.

Brower, C.D., 1942. Fifty Years Below Zero: A Lifetime of Adventure in the Far North. Dodd, Mead and Co., New York.

Castellote, M., Clark, C.W., Lammers, M.O., 2011. Fin whale (*Balaenoptera physalus*) population identity in the western Mediterranean Sea. Mar. Mamm. Sci. 28, 325–344.

Charif, R.A., Rahaman, A., Muirhead, C.A., Pitzrick, M.S., Warde, A.M., Hall, J., et al., 2013. Bowhead whale acoustic activity in the southeast Beaufort Sea during late summer 2008–2010. J. Acoust. Soc. Am. 134, 4323–4334.

Cholewiak, D.M., Cerchio, S., Jacobsen, J.K., Urbán, R.J., Clark, C.W., 2018. Songbird dynamics under the sea: acoustic interactions between humpback whales suggest song mediates male interactions. R. Soc. Open Sci. 5 (2), 171298.

Clark, C., Ellison, W.T., Beeman, K., 1986a. Acoustic tracking of migrating bowhead whales. Oceans 86, 341–346.

Clark, C.W., 1982. The acoustic repertoire of the southern right whale, a quantitative analysis. Anim. Behav. 30, 1060–1071.

Clark, C.W., 1983. Acoustic communication and behavior of the southern right whale. In: Payne, R.S. (Ed.), Behavior and Communication of Whales. Westview Press, Boulder, CO, pp. 163–198.

Clark, C.W., 1989. The use of bowhead whale call tracks based on call characteristics as an independent means of determining tracking parameters. Rep. Int. Whal. Commn 39, 111–113.

Clark, C.W., 1990. Acoustic behavior of mysticete whales. In: Thomas, J., Kastelein, R.A. (Eds.), Sensory Abilities of Cetaceans. Plenum Press, New York, pp. 571–583.

Clark, C.W., Berchok, C.L., Blackwell, S.B., Hannay, D.E., Jones, J., Ponirakis, D., et al., 2015. A year in the acoustic world of bowhead whales in the Bering, Chukchi and Beaufort Seas. Prog. Oceanogr. 136, 223–240.

Clark, C.W., Charif, R., Hawthorne, D., Rahaman, A., Givens, G.H., George, J.C., et al., 2018. Acoustic data from the spring 2011 bowhead whale census at Point Barrow, Alaska. J. Cetacean Res. Manage. 19, 31–42.

Clark, C.W., Charif, R., Mitchell, S., Colby, J., 1996. Distribution and behavior of the bowhead whale, *Balaena mysticetus*, based on analysis of acoustic data collected during the 1993 spring migration off Point Barrow, Alaska. Rep. Int. Whal. Commn 46, 541–554.

Clark, C.W., Ellison, W.T., 2004. Potential use of low-frequency sounds by baleen whales for probing the environment: evidence from models and empirical measurements. In: Thomas, J.A., Moss, C.F., Vater, M. (Eds.), Echolocation in Bats and Dolphins. The University of Chicago Press, Chicago, IL, pp. 564–585.

Clark, C.W., Ellison, W.T., Beeman, K., 1986b. An acoustic study of bowhead whales, *Balaena mysticetus*, off Point Barrow, Alaska, during the 1984 spring migration. Report from Marine Acoustics, Clinton, MA, for North Slope Borough, Barrow, AK, 145 pp.

Clark, C.W., Ellison, W.T., Beeman, K., 1988. An acoustic study of bowhead whales, *Balaena mysticetus*, off Point Barrow, Alaska, during the 1985 spring migration. Report from Marine Acoustics, Clinton, MA, for North Slope Borough, Barrow, AK, 143 pp.

Clark, C.W., Ellison, W.T., Southall, B.L., Hatch, L., Van Parijs, S.M., Frankel, A., et al., 2009. Acoustic masking in marine ecosystems: intuitions, analysis, and implication. Mar. Ecol. Prog. Ser. 395, 201–222.

Clark, C.W., Johnson, J.H., 1984. The sounds of the bowhead whale, *Balaena mysticetus*, during the spring migrations of 1979 and 1980. Can. J. Zool. 62, 1436–1441.

Crance, J.L., Berchok, C.L., Wright, D.L., Brewer, A.M., Woodrich, D.F. 2019. Song production by the North Pacific right whale, Eubalaena japonica, J. Acoust. Soc. Am. 145, 3467–3479.

Cummings, W.C., Holliday, D.V., 1985. Passive acoustic location of bowhead whales in a population census off Point Barrow, Alaska. J. Acoust. Soc. Am. 78, 1163–1169.

Cummings, W.C., Holliday, D.V., 1987. Sounds and source levels from bowhead whales off Pt. Barrow, Alaska. J. Acoust. Soc. Am. 82, 814–821.

Damien, J., Adam, O., Cazau, D., White, P., Laitman, J.T., Reidenberg, J.S., 2018. Anatomy and functional morphology of the Mysticete rorqual whale larynx: phonation positions of the U-fold. Anat. Rec. 302 (5), 703–717.

Delarue, J., Laurinolli, M., Martin, B., 2009. Bowhead whale (*Balaena mysticetus*) songs in the Chukchi Sea between October 2007 and May 2008. J. Acoust. Soc. Am. 126, 3319–3328.

Edds-Walton, P.L., 1997. Acoustic communication signals of mysticete whales. Bioacoustics 8 (1–2), 47–60.

Ellison, W.T., Clark, C.W., Bishop, G.C., 1987. Potential use of surface reverberation by bowhead whales, *Balaena mysticetus*, in under-ice navigation: preliminary considerations. Rep. Int. Whal. Commn. 37, 329–332.

Eschricht, D.F., Lilljeborg, W., Reinhardt, J., 1866. In: Flower, W.H. (Ed.), Recent Memoirs on the Cetacea. Royal Society, London, pp. 1–344.

Ferguson, S., Higdon, J., Chmelnitsky, E., 2010. The rise of killer whales as a major Arctic predator. In: Ferguson, S., Loseto, L., Mallory, M. (Eds.), A Little Less Arctic: Top Predators in the World's Largest Northern Inland Sea. Springer, Hudson Bay, pp. 117–136.

Gandilhon, N., Adam, O., Cazau, D., Laitman, J.T., Reidenberg, J.S., 2014. Two new theoretical roles of the laryngeal sac of humpback whales. Mar. Mamm. Sci. 31 (2), 774–781.

George, J.C., Clark, C., Carroll, G.M., Ellison, W.T., 1989. Observations on the ice-breaking and ice navigation behavior of migrating bowhead whales (*Balaena mysticetus*) near Point Barrow, Alaska, Spring 1985. Arctic 42, 24–30.

Givens, G.H., Edmondson, S.L., George, J.C., Suydam, R., Charif, R.A., Rahaman, A., et al., 2016. Horvitz–Thompson whale abundance estimation adjusting for uncertain recapture, temporal availability variation, and intermittent effort. Environmetrics 27 (3), 134–146.

Greene, C., McLennan, M., Mcdonald, T., Jakubczak, R., Richardson, W., 2004. Directional frequency and recording (DIFAR) sensors in seafloor recorders to locate calling bowhead whales during their fall migration. J. Acoust. Soc. Am. 116, 799–813.

Greenewalt, C.H., 1968. The two acoustical sources. Bird Song: Acoustics and Physiology. Smithsonian Institution Press, Washington, DC, pp. 55–78.

Haldiman, J.T., Tarpley, R.J., 1993. Anatomy and physiology. In: Burns, J.J., Montague, J.J., Cowles, C.J. (Eds.), The Bowhead Whale. The Society for Marine Mammalogy, pp. 71–156, Special Publication Number 2.

Hatch, L.T., 2004. Male Genes and Male Songs: Integrating Genetic and Acoustic Data in Defining Fin Whale, *Balaenoptera physalus*, Management Units (Ph.D. dissertation). Cornell University, Ithaca, NY.

Heide-Jørgensen, M.P., Laidre, K., Borchers, D., Samarra, F., Stern, H., 2007. Increasing abundance of bowhead whales in West Greenland. Biol. Lett. 3 (5), 577–580.

Heide-Jørgensen, M.P., Laidre, K.L., Quakenbush, L.T., Citta, J.J., 2012. The Northwest Passage opens for bowhead whales. Biol. Lett. 8 (2), 270–273.

Hosokawa, H., 1950. On the Cetacean larynx, with special remarks on the laryngeal sack of the sei whale and the aryteno-epiglottideal tube of the sperm whale. Sci. Rep. Whales Res. Inst. 3, 23–62.

Jacobs, E., Duffy, M., Magolan, J., Galletti Vernazzania, B., Cabrera, E., Landea, R., et al., 2018. First acoustic recordings of critically endangered eastern South Pacific southern right whales (*Eubalaena australis*). Mar. Mamm. Sci. 35 (1), 284–289.

Johnson, H.D., Stafford, K.M., George, J.C., Ambrose Jr., W.G., Clark, C.W., 2014. Song sharing and diversity in the Bering-Chukchi-Beaufort population of bowhead whales (*Balaena mysticetus*), spring 2011. Mar. Mamm. Sci. 31, 902–922.

Johnson, M., Madsen, P.T., Zimmer, W.T., de Soto, N.A., Tyack, P.L., 2004. Beaked whales echolocate on prey. Proc. R. Soc. B: Biol. Sci. 271 (Suppl_6), S383–S386.

Ljungblad, D., Thompson, P., Moore, S.E., 1982. Underwater sounds recorded from migrating bowhead whales, *Balaena mysticetus*, in 1979. J. Acoust. Soc. Am. 71, 477–482. Available from: https://doi.org/10.1121/1.387419.

Ljungblad, D.K., Leatherwood, S., Dahlheim, M.E., 1980. Sounds recorded in the presence of an adult and calf bowhead whale. Mar. Fish. Rev. 42, 86–87.

Moore, S.E., Reeves, R.R., 1993. Distribution and movement. In: Burns, J.J., Montague, J.J., Cowles, C.J. (Eds.), The Bowhead Whale. The Society for Marine Mammalogy, pp. 313–386, Special Publication Number 2.

Moore, S.E., Stafford, K.M., Melling, H., Berchok, C., Wiig, Ø., Kovacs, K.M., et al., 2012. Comparing marine mammal acoustic habitats in Atlantic and Pacific sectors of the High Arctic: year-long records from Fram Strait and the Chukchi Plateau. Polar Biol. 35, 475–480.

Moore, S.E., Stafford, K.M., Munger, L.M., 2010. Acoustic and visual surveys for bowhead whales in the western Beaufort and far northeastern Chukchi seas. Deep Sea Res. II 57, 153–157.

Parks, S.E., 2003. Response of North Atlantic right whales (*Eubalaena glacialis*) to playback of calls recorded from surface active groups in both the North and South Atlantic. Mar. Mamm. Sci. 19, 563–580.

Parks, S.E., Clark, C.W., 2007. Acoustic communication: social sounds and the potential impacts of noise. In: Kraus, S.D., Rolland, R.M. (Eds.), The Urban Whale: North Atlantic Right Whales at the Crossroads. Harvard University Press, Cambridge, MA, pp. 310–332.

Parks, S.E., Clark, C.W., Tyack, P.L., 2007. Short- and long-term changes in right whale calling behavior: the potential effects of noise on acoustic communication. J. Acoust. Soc. Am. 122 (6), 3725–3731.

Parks, S.E., Hotchkin, C.F., Cortopassi, K.A., Clark, C.W., 2012. Characteristics of gunshot sound displays by North Atlantic right whales in the Bay of Fundy. J. Acoust. Soc. Am. 131 (4), 3173–3179.

Payne, K., Payne, R., 1985. Large-scale changes over 19 years in songs of humpback whales in Bermuda. Z. Tierpsychol. 68 (2), 89–114.

Payne, R., Webb, D., 1971. Orientation by means of long range acoustic signaling in baleen whales. Ann. N. Y. Acad. Sci. 188 (1), 110–141.

Poulter, T.C., 1968. Marine mammals. In: Sebeok, T. (Ed.), Animal Communication. Indiana University Press, Bloomington, pp. 405–465.

Reidenberg, J.S., Laitman, J.T., 2007. Discovery of a low frequency sound source in Mysticeti (baleen whales): anatomical establishment of a vocal fold homolog. Anat. Rec. 290 (6), 745–759.

Reidenberg, J.S., Laitman, J.T., 2008. Sisters of the sinuses: cetacean air sacs. Anat. Rec. 291 (11), 1389–1396.

Rekdal, S.L., Hansen, R.G., Borchers, D., Bachmann, L., Laidre, K.L., Wiig, Ø., et al., 2014. Trends in bowhead whales in West Greenland: aerial surveys vs. genetic capture-recapture analyses. Mar. Mamm. Sci. 31 (1), 133–154.

Richardson, W.J., Finley, K.J., Miller, G.W., Davis, R.A., Koski, W., 1995. Feeding, social and migration behavior of bowhead whales, *Balaena mysticetus*, in Baffin Bay vs. the Beaufort Sea—regions with different amounts of human activity. Mar. Mamm. Sci. 11, 1–45.

Sandifort, G., 1831. Bijdragen tot de ontledkundige Kennis der Walvisschen. Nieuwe Verhandlgr. 1e Kl. Nederl. Inst. 3, 223–270.

Schoenfuss, H.L., Bragulla, H.H., Schumacher, J., Henk, W.G., George, J.C., Hillmann, D.J., 2014. The anatomy of the larynx of the bowhead whale, *Balaena mysticetus*, and its sound-producing functions. Anat. Rec. 297 (7), 1316–1330.

Stafford, K.M., 2019. Increasing detections of killer whales (*Orcinus orca*), in the Pacific Arctic. Mar. Mamm. Sci. 35 (2), 696–706.

Stafford, K.M., Castellote, M., Guerra, M., Berchok, C.L., 2018a. Seasonal acoustic environments of beluga and bowhead whale core-use regions in the Pacific Arctic. Deep Sea Res. II 152, 108–120.

Stafford, K.M., Lydersen, C., Wiig, Ø., Kovacs, K.M., 2018b. Extreme diversity in the songs of Spitsbergen's bowhead whales. Biol. Lett. 14, 20180056. https://doi.org/10.1098/rsbl.2018.0056.

Stafford, K.M., Moore, S.E., Berchok, C.L., Wiig, Ø., Lydersen, C., Hansen, E., et al., 2012. Spitsbergen's endangered bowhead whales sing through the polar night. Endanger. Species Res. 18, 95–103.

Stafford, K.M., Moore, S.E., Laidre, K.L., Heide-Jørgensen, M.P., 2008. Bowhead whale springtime song off West Greenland. J. Acoust. Soc. Am. 124, 3315–3323.

Stafford, K.M., Moore, S.E., Stabeno, P.J., Holliday, D.V., Napp, J.M., Mellinger, D.K., 2010. Biophysical ocean observation in the southeastern Bering Sea. Geophys. Res. Lett. 37 (2), 2009GL040724.

Stafford, K.M., Nieukirk, S.L., Fox, C.G., 2001. Geographic and seasonal variation of blue whale calls in the North Pacific. J. Cetacean Res. Manage. 3, 65–76.

Tervo, O.M., Christoffersen, M.F., Parks, S.E., Kristensen, R.M., Madsen, P.T., 2011a. Evidence for simultaneous sound production in the bowhead whale (*Balaena mysticetus*). J. Acoust. Soc. Am. 130 (4), 2257–2262.

Tervo, O.M., Christoffersen, M.F., Simon, M., Miller, L.A., Jensen, F.H., Parks, S.E., et al., 2012. High source levels and small active space of high-pitched song in bowhead whales (*Balaena mysticetus*). PLoS One 7, e52072.

Tervo, O.M., Parks, S.E., Christoffersen, M.F., Miller, L.A., Kristensen, R.M., 2011b. Annual changes in the winter song of bowhead whales (*Balaena mysticetus*) in Disko Bay, Western Greenland. Mar. Mamm. Sci. 27, E241–E252.

Tervo, O.M., Parks, S.E., Miller, L.A., 2009. Seasonal changes in the vocal behavior of bowhead whales (*Balaena mysticetus*) in Disko Bay, Western-Greenland. J. Acoust. Soc. Am. 126, 1570–1580.

Thode, A.M., Blackwell, S.B., Seger, K.D., Conrad, A.S., Kim, K.H., Macrander, M.A., 2016. Source level and calling depth distributions of migrating bowhead whale calls in the shallow Beaufort Sea. J. Acoust. Soc. Am. 140, 4288–4297.

Thode, A.M., Kim, K.H., Blackwell, S.B., Greene, C.R., Nations, C.S., McDonald, T.L., et al., 2012. Automated detection and localization of bowhead whale sounds in the presence of seismic airgun surveys. J. Acoust. Soc. Am. 131, 3726–3747.

Tyack, P., 1981. Interactions between singing Hawaiian humpback whales and conspecifics nearby. Behav. Ecol. Sociobiol. 8 (2), 105–116.

Tyack, P.L., Clark, C.W., 2000. Communication and acoustic behavior of dolphins and whales. In: Au, W., Popper, A.N., Fay, R.R. (Eds.), Hearing by Whales and Dolphins. Springer Handbook of Auditory Research, pp. 156–224.

Woodgate, R.A., Stafford, K.M., Prahl F.G., 2015. A Synthesis of Year-Round Interdisciplinary Mooring Measurements in the Bering Strait (1990–2014) and the RUSALCA Years (2004–2011). Oceanography 28(3), 46–67. Available from: https://doi.org/10.2307/24861901.

Würsig, B., Clark, C.W., 1993. Behavior. In: Burns, J.J., Montague, J.J., Cowles, C.J. (Eds.), The Bowhead Whale. The Society for Marine Mammalogy, pp. 157–199, Special Publication Number 2.

Zeh, J.E., Clark, C.W., George, J.C., Withrow, D.E., Carroll, G.M., Koski, W.R., 1993. Current population size and dynamics. In: Burns, J.J., Montague, J.J., Cowles, C.J. (Eds.), The Bowhead Whale. The Society for Marine Mammalogy, pp. 409–489, Special Publication Number 2.

CHAPTER

23

Natural and potentially disturbed behavior of bowhead whales

Bernd Würsig[1] *and William R. Koski*[2]

[1]Texas A&M University at Galveston, Galveston, TX, United States

[2]LGL Limited, King City, ON, Canada

Introduction

Bowhead whales (*Balaena mysticetus*, Fig. 23.1) are animals of extremes—they have the thickest blubber layer of any cetacean (George, 2009), longest and most numerous baleen plates of any mysticete (Chapter 14), and greatest longevity of any mammal (George et al., 1999, 2011; Rosa et al., 2013; Wetzel et al., 2017). Males also have among the largest testes on the Earth (George, 2009; O'Hara et al., 2002); and females manage to gestate and nurse a youngster to self-sufficiency in 2 years (Koski et al., 1993, 2010), a remarkable physical and physiological feat explored elsewhere in this compendium (Chapter 13). Then as they grow into adulthood, males and females ascend to reproductive maturity in their late teens to late 20s with males maturing earlier than females (Koski et al., 1993; Rosa et al., 2013). During their long lives, extending in some cases to well over 100 years, females may produce many offspring (Miller et al., 1992). The first part of this chapter explores the behaviors that are dictated by these extremes, and one might argue that the behaviors allow the extremes. In the second part on bowhead whales living with human intrusion, we summarize some of present knowledge of reactions to human-caused disturbance.

Bowhead whales of the far northern reaches of the northern hemisphere are related to right whales (*Eubalaena* spp.) of both hemispheres. Comparisons of the lifestyles and behaviors of the family Balaenidae are likely to lend a more complete understanding of this group of mysticetes that feeds in slow filter-feeding manner and evolved a sexual technique of male–male competition for females at physical and sperm competition levels (Payne, 1995; Brownell and Ralls, 1986). Both of these feeding and sexual techniques are somewhat related to those of another group of mysticetes, the Eschrichtiidae or gray whales (*Eschrichtius robustus*), where most feeding is via suctioning prey from the bottom,

The Bowhead Whale
DOI: https://doi.org/10.1016/B978-0-12-818969-6.00023-6

FIGURE 23.1 Sexual activity in a group of bowhead whales in the Bering Strait region. The female (upside down in the top left corner) is being aggressively pursued by several males. The resting whales on the lower right are probably males "catching their breath." *Source: Photo by W.R. Koski, LGL Limited (NOAA/North Slope Borough, NMFS Permit No. 14245).*

and where male–male competition at behavioral and physiological levels is of paramount importance.

This chapter is divided into two main sections. The first one, on (presumably) undisturbed behavior, relies in large part on behavioral descriptions from circling aircraft, made with real-time descriptions by one or usually two people, video recording by another person for later detailed analyses, and overall descriptions and periodic monitoring of sounds from sonobuoys (devices that receive underwater sound and relay it by radio in real time) by a fourth person. Most, but not all, of these basic descriptions were accomplished in the 1980s and 1990s, as referenced throughout this first section. We do not comment specifically on migrational or simply traveling movements, as these are covered in Chapter 4. However, East Canada–West Greenland (ECWG) and Bering–Chukchi–Beaufort Seas (BCB) bowhead whales are recently comingling (to as yet unknown degree, but likely increasing with time), as Arctic sea ice in summer becomes less and makes connections easier in both directions (Heide-Jørgensen et al., 2011). Genetic differences are marginal between the BCB and ECWG stocks, probably due to historic interchanges, particularly when compared with the Okhotsk Sea (OKH) stock (Chapter 3). The second section is on (potentially) disturbed behavior associated with anthropogenic activities. While many of the early disturbance data were gathered from circling airplanes, other early data came from ice-based observers. In the 2000s a wealth of new disturbance information was acquired, mostly during aerial surveys before, during, and after industry operations, but also with the aid of remote sensing devices. For the potentially disturbed events, we rely largely on these newer data for the summary here.

Undisturbed activities

Feeding behavior

Bowhead whale feeding occurs at the surface, in the water column, and near or at the bottom, surely depending on where dense prey is found. When feeding at the surface, we can see right whales and bowheads with their mouths open, the lower jaw perhaps 3–5 m below the surface. They scull along with some of the largest animal propellers on the Earth—flukes of bowheads can measure, in the extreme, about 7 m across and average about one-third of the body length for physically mature whales (Chapter 7) (BOWPHOT, 2014). Bowheads use their enormous mouths as the equivalent of copepod gathering nets, in a process termed ram feeding (Werth, 2004; Lambertsen et al., 2005; Chapter 14). However, the whales have an advantage over nets, as their baleen does not become occluded with an overabundance of copepods. Instead, the design of the long interlocking baleen plates with their interior fine fringes assures that food is efficiently sluiced to the gullet and does not foul the baleen net on the way (Werth, 2004). Electronically sensed data for bowheads feeding at depth (probably in general similar to those feeding at or near the surface) indicate that they have a very slow fluke rate and speed of <1 m/s (Simon et al., 2009), slower than that of their slightly smaller taxonomic cousin, the northern right whale (*Eubalaena borealis*; van der Hoop et al., 2019).

Feeding at or near the surface is termed skim feeding, although whales are technically only "skimming" when most food is concentrated directly at the surface. As mentioned previously, such feeding usually occurs with the lower jaw dropped down so that the relatively small triangular opening of the front of the mouth, about 2.4 m vertical by 2.2 m horizontal on a 14–15 m long whale (Werth, 2004 and Chapter 14), can be seen by a human observer on shore, ice, or an aerial platform (airplane, helicopter, and drone). The body is somewhat curved upward in front, so that the top of the head is normally higher (nearer the surface or partially out of water) when skim feeding than when a whale is simply lying at the surface or swimming without feeding. Skim feeding usually occurs while bowheads aggregate in a productive food area but are separated from each other by 75 m or more (but see next). The original aggregation in the area probably initiated by vocal communication (Chapter 22) but there often does not seem to be social collaboration while skim feeding.

When food appears concentrated directly at the surface (as sometimes indicated by groups of feeding phalaropes, family Scolopacidae), bowhead whales at times turn onto either left or right side, with mouths just as open as during feeding in dorsum-up position, but now one side of both upper and lower jaw at the surface, as well as one side of the long baleen plates. At the same time the massive fluke tip uppermost in the water column slowly moves back and forth, with perhaps a bit of white water from the tip as it barely touches or overreaches the surface. The humans' view from an airplane reminds of a shark tail slicing back and forth through the water. We assume that such a sidewise position allows especially close-to-the surface feeding without the whale's propulsion system, that massive fluke, coming too close to the surface—as it would on its upward swing in dorsum-up position, creating disadvantageous surface vortex drag (Fig. 24.1).

Much of balaenid whale feeding occurs on clouds of calanoid copepods, the smallest of crustaceans eaten by any whale. However, bowheads also feed on larger prey, including mysid and euphausiid crustaceans (Saupe et al., 1989; Schell and Saupe, 1993; Chapter 28). The latter prey are rapid swimmers, and it is possible that they could outswim the mouth of an approaching bowhead (or right) whale, depending on detection distance. Euphausiid speeds are about 2–35 cm/s (Ignatyev, 1997; De Robertis et al., 2003). This is where a particular coordinated foraging strategy may be most useful; both bowhead (Chapter 24, see Fig. 24.1) and right whales (Fig. 23.2) at times feed in so-called "echelon swimming" (Würsig et al., 1985; Würsig and Clark, 1993) formation. Two to one-half dozen or more whales swim forward together, side-by-side about one-quarter whale length apart but staggered with each whale about one-half whale length ahead or behind the adjacent whale. This coordinated feeding occurs to and over 0.5 hours, with no one whale having apparent leadership, but the front animal changing as the direction of the entire group changes. The formation is reminiscent of geese flying in echelon for aerodynamic reasons, but for the whales the formation is surely to help concentrate prey into their mouths, with prey avoidance perhaps hampered by the barrier created by the adjacent whale's body. Fish et al. (2013) used detailed photographs of echelon feeding bowhead whales, spaced closer than in the earlier observations, and surmised that, while the previous may be true, the vortex

FIGURE 23.2 Right whales and bowhead whales have similar ram feeding techniques, here shown by several southern right whales (*Eubalaena australis*) off Peninsula Valdés, Argentina, in echelon formation as also described next. *Source: Photo by B. Würsig.*

effects (and perhaps a Bernoulli effect of whale walls side-by-side) are also likely to enhance feeding in echelon. The front animal may receive a temporary advantage by being the first to cleave into the euphausiid swarm. Blue and fin whales (*Balaenoptera musculus* and *Balaenoptera physalus*, respectively) feeding in two-whale staggered "simple" echelon may be using similar advantage of each other's "body walls" (Würsig, 1989).

Prey are not often so conveniently (for whales or humans studying them) concentrated at the surface. Bowhead prey are often concentrated well below the surface (Griffiths and Thomson, 2002; Griffiths et al., 2002; Heide-Jørgensen et al., 2013; Citta et al., 2015), and much feeding occurs sufficiently far down that the whales are either invisible or barely visible at a depth of several meters. Surface skim feeding can go on quite uninterrupted by the whales' need to breathe—whales may feed >1 hour without pause, while breathing in regular bouts at the surface. In contrast, bowheads feeding somewhere below the surface, including near the bottom, need to stop feeding, swim to the surface to breathe several times, and then dive and resume feeding. In the 1980s and 1990s, during observations from an airplane circling at about 450 or 600 m above sea level, we generally left an identifiable focal whale's location if we had not resighted it within about 30 minutes. This likely was a mistake, as indicated by accounts from commercial whalers as well as Inuit whale hunters (Bockstoce, 1986).

It is now known from tagging data that bowheads can dive up to 1 hour while feeding in the water column or at the bottom (Krutzikowsky and Mate, 2000; Citta et al., 2015), at depths at times exceeding 400 m (Heide-Jørgensen et al., 2013). Modern DTAG and related electronically gathered data for bowheads have documented depth of dive and swimming speeds during ram feeding, and approximate volume filtered per unit time (Simon et al., 2009; Chapter 24). Earlier aerial-observation data showed that whales in the Beaufort Sea often come to the surface with copious amounts of mud streaming from their bodies, often from the top of the head, suggesting that the whales had been swimming ventrum-up near the bottom (Würsig et al., 1985). However, this does not necessarily mean bottom feeding as had originally been thought; it could also be a part of simply rubbing the skin while exfoliating. Iñupiat whale hunters have described similar behavior along the lead edge whereby bowheads were observed swimming in ventrum-up orientation (Harry Brower, Sr., Pers. Comm.). In the Beaufort Sea, bowheads have been seen with mud streaming from the corner of their mouth (Koski et al., 2009), suggesting that it is common for bowheads to feed at or near the bottom (Mocklin et al., 2011). However, many of the observations in Mocklin et al. (2011) could have been due to mud sticking to whale bodies while they were rubbing on the sea floor or ice during molting, as described for eastern Canadian bowheads (ECWG stock) by Fortune et al. (2017a,b).

Further information on ecology of feeding is summarized in Chapter 24, and particularly valuable accounts of feeding off Pt. Barrow in late summer/early autumn as related to ecology and prey are presented by Ashjian et al. (2010) and Moore et al. (2010). Richardson et al. (1995a) compare feeding in generally deeper waters of the ECWG stock (along the east coast of Baffin Island) with those of the BCB stock, for late summer and early autumn. Heide-Jørgensen et al. (2013) describe ECWG bowheads feeding at depths 100–400 m in early spring (March), and then diving less deeply later in spring as copepods ascended.

Social and sexual behavior

Bowhead whales often aggregate in an area because food is more abundant there (Chapter 24). We assume that such aggregations are not truly social but do not know to what degree bowhead calls and perhaps other feeding-related sounds may be responsible in fostering the aggregation (Chapter 22). When whales feed in echelon formation, they are coordinating activities. There may be more "socializing" by sound than we can definitively state, since there often appears to be an overall synchrony in surfacings and dives of whales foraging below the surface, with "almost all" whales in the area up at one time and "almost all" diving at one time. This observation, although widespread and frequent, has not been put to rigorous analysis (Ljungblad et al., 1980; Würsig et al., 1985; Würsig and Clark, 1993). It is possible that whales are coordinating surface-dive activities because (1) they are coordinating in taking prey at depth, such as the surface-observable echelon formations and (2) synchronized surface-dive cycles may facilitate staying in acoustic communication for various purposes, including avoidance of potential predation, such as by killer whales (*Orcinus orca*) or humans. This latter hypothesis was expressed in more detail by Würsig and Clark (1993). Facilitation of predator avoidance could explain why synchronized surface-dive cycles have been seen even among widely dispersed whales.

Much social activity manifests by whales nudging and chasing each other, rolling over and around each other, or just traveling in close proximity. We do not presently know how much of close-contact social interaction is of a sexual nature and how much might have other functions similar to "rock-nosing" (Reeves et al., 1983; Fortune et al., 2017a,b; Chapter 7). If sexual in nature, it could involve animals attempting to repel a potential rival or attempting to gain mating access (or evade it). If similar to rock-nosing, the close contact may be to stimulate epidermal regeneration or sloughing of dead skin. Fortune et al. (2017a,b) describe social behavior of small immature whales, mostly 7–10 m long, that were rubbing on rocks below the surface. Those same whales while at the surface were engaged in rubbing behavior that appeared similar to some aspects of sexual behavior in spring but was behavior related to skin rubbing during molting.

At times, one male and female occur in an apparent sexual interaction, with both rolling over each other and with the male attempting to insert his penis (Würsig and Clark, 1993), similar to occasional observations of southern right whale, *Eubalaena australis*, interactions (Würsig, 2000). More often, there is a melee of whales, with apparently 1 estrous female the center of attention of more than 2, at times as many as 17, males attempting to gain access to the female. Typically, the female turns her ventrum (belly) toward the surface, while males in dorsal-up position attempt to mate with her, approaching her from each side and attempting to push her down. There is much jostling for position among males during these attempts, and much rapid movement, resulting in white-water splashes at the surface that have so far deterred careful analyses from video recordings (Fig. 23.1). Nevertheless, the surface activity, while rapid and forceful, does not seem equivalent to the male–male interactions seen in surface-active groups (SAGs) of aggressively interacting humpback whales, *Megaptera novaeangliae* (Tyack and Whitehead, 1982), or to the bowhead male–male sexual interactions seen at Isabella Bay, Baffin Island, of what seemed to be large subadult males (Richardson et al., 1995b).

We surmise that sexual interactions of bowhead whales, like those of right and gray whales, are of a less aggressive, more fluid, nature than those of other species. These muted behavioral interactions apparently allow males to pursue a female in estrus with the female making mating with her difficult and therefore exhibiting a form of female choice for the most ardent and most successful suitor. Males presumably also use their large testes to produce large volumes of sperm that can result in "sperm competition" when there are multimate (polygynandrous) matings (Brownell and Ralls, 1986). Payne (1995) hypothesized that male southern right whales (with very similar mating behaviors and testis morphologies as bowhead whales) may be using (1) sperm competition among males and (2) perhaps males even helping each other in mating attempts, as a part of behavioral reciprocal altruism (Trivers, 1971). If so, this would be a very different sexual strategy from that of the taxonomic family Balaenopteridae (rorquals). Rorquals have relatively small testes and, at least for humpback whales, aggressive social/sexual interactions between males that at times lead to gross injuries and even death (Brownell and Ralls, 1986; Pack et al., 1998). As an aside, Accardo et al. (2018) describe a bowhead whale off the northeast coast (well south of normal bowhead whale ranges) of the United States socially (possibly sexually) interacting with groups of North Atlantic right whales, and it is conceivable that such interactions could occur more often with changes in habitats for the two species (see Fig. 1.1).

Quite a bit of apparently sexual behavior in SAGs has been observed in autumn, in the BCB and ECWG populations (Richardson et al., 1995b). This seems odd, as almost all calves are born from late March to late June, after a gestation of slightly longer than 1 year (Koski et al., 1993) and the only observation of extremely large numbers of mating whales was in early April in the Bering Sea (Koski et al., 2005). We surmise that much of autumn sexual behavior is not procreative and may have a learning function. The few documented sessions of sexual activity during late summer/early autumn when lengths of whales have been measured have included only subadult whales (Würsig et al., 2002). Sexual segregation of whales, with nonparturient females, females with young, and males occurring in largely different areas, has been described (from genetic data) in the ECWG (Heide-Jørgensen et al., 2010). Such sexual segregation, perhaps at least in part related to females attempting to avoid sexual advances by males, may be occurring elsewhere, as also described for southern right whales off Peninsula Valdés, Argentina (Payne, 1995).

Mother/calf interactions

Bowhead whales tend to mate in spring, have a calf the next year, also in early spring (Koski et al., 1993), and nurse for at least 6 months but some possibly 10–12 months, so that most yearlings are weaned and on their own by the spring following their birth year. Two of ten calves inadvertently harvested off Pt. Barrow and Kaktovik in autumn, with estimated ages 5–6 months, had milk and invertebrates in the stomach, indicating at least some nonnursing feeding by some calves born that year (Chapters 7, and 28). Mother-yearling separation takes place before or early during the next year's migration to feeding grounds, although a few yearlings are seen with their mothers during spring migration past Barrow; with rare yearlings (1%–2%) harvested with milk in their stomachs

(Chapters 7, and 33). Those yearlings were probably born during late spring of the previous year (Koski et al., 2010). This pattern is rather similar to that of right whales, in which mother–calf separation is generally at about 12 months (Payne, 1995; Brown et al., 2001), and like the strategy of gray or rorqual whales, where separation of mother–calf is usually at or less than 1 year.

Our experience comes from observing newborn calves and yearlings during spring migration and near or on the summer feeding grounds. The physical differences between young calves and yearlings are obvious (Koski et al., 1993; compare Figs. 8.1 and 11.1), with newborns in May (but not afterward) often "riding" on the backs of mothers (Würsig et al., 1999), possibly only during the first 2 months of life. Riding is accompanied by very little or no movement of the calf's fluke, and it appears that the smaller body is being pulled along by the "slipstream" of the mother, possibly especially important for newborn calves during spring migration (Würsig et al., 1993). The slipstream effect is likely very important for the developing young calf, as is postulated for dolphins (Fellner et al., 2006). Yearling bowheads rarely are with their mothers in spring in the Bering Sea in early April and near Pt. Barrow from late April to early June. Most yearlings pass Pt. Barrow during the latter stages of the migration, after separating from their mothers and after the majority of older (age 2 +) juvenile whales have passed (Koski et al., 2010). We presume these yearlings separated from their mothers prior to or during early phases of the spring migration.

When not "riding" (or swimming near the mother as the calf becomes older) the newborn calf orients to the mother's genital and teat area, stays submerged (and apparently nursing) for less than 1 minute at a time, and then generally swims under the mother's body to surface on the other side. It is not known whether alternately left and right teat nursing takes place as a result (Würsig and Clark, 1993). During mid-summer to autumn, mothers often leave their calves at the surface for at least 30 minutes while the mother feeds at depth, with calves as far as 1.6 km from the resurfacing mother (Würsig and Clark, 1993). While mother and calf underwater sounds have not definitely been isolated, it is possible that lower and higher frequency alternating calls received via sonobuoys deployed near a mother–calf pair have come from the mother and calf, respectively, and represented contact calling when the two were separated (Würsig et al., 1985).

Play

Bowhead whales often socialize at or near the surface, with touching by head or flippers, and rolling over each other. It is not known how much of this is simply social play and how much may have sexual connotation or molting functions.

Bowheads also can be aerially acrobatic, with forceful flipper and tail slaps onto the water surface, head-out slow or lunging movements, and breaches. At times, one bowhead engages in all these aerial acrobatics during a bout of activity; as an extreme case, a whale was seen to breach 64 times, tail slap 36 times, and flipper-slap 48 times in 75 minutes. While much aerial activity has been termed play (i.e., occurring for no apparent purpose except possible relief from boredom), aerial activity can probably also be a part of communication among conspecifics. It may also be an outgrowth of social/sexual activities; at times, aerial activity appears to indicate "frustration" by a whale apparently left out of social interactions (Würsig and Clark, 1993). For whales, in general, the loud percussion sounds from flipper and tail slaps and breaching (which often increase with increasing wind speed and wave action) may function in mate advertisement (Whitehead, 1985) and—for right whales at least—long-distance contact (Payne, 1995).

The most obvious indication of play without apparent other function has been of bowhead whales interacting with inanimate objects such as logs at the surface, as well as chemical floating dye markers that indicated sonobuoy positions. Near the Canadian Mackenzie Estuary, large logs (up to 10 m or so in length) commonly float in the Beaufort Sea, and bowhead whales at times nudge, push, attempt to submerge, and balance on their dorsa such logs, described in more detail in studies of Würsig and Clark (1993). Even here though possible functions may be stimulation of skin sensation related to molting [as described for rubbing behavior by Fortune et al. (2017a,b)], or "practicing" for the kinds of activities seen when whales interact in surface-active sexual groupings, multiple males seek to obtain access to a female. Fig. 23.3 shows several common surface-active behaviors, often termed parts of "play," but with perhaps other functions as well.

Predator avoidance/responses

The major predators of bowhead whales are killer whales (Ferguson et al., 2012a,b; Chapter 29) and humans (Mitchell and Reeves, 1982). Much of whale responses to killer whales come from interviews of Inuit peoples, as summarized by the Nunavut Wildlife Management Board (2000), and from whaling logs. When bowheads are confronted by killer whales, they attempt to retreat to thick ice when available, or as close to shore as possible. However, bowheads also strike back at killer whales, especially with their large and powerful flukes, and can maim and even kill the attacking predators. Their fear of killer whales apparently tends to be greater than their fear of humans. There is an account in the whaling logs of bowhead whales moving into cracks in fast ice when killer whales were present, and the whalers killing the bowheads one by one; the bowheads were huddled together and did not disperse or return to open water, apparently for fear that the killer whales might still be there (Reeves et al., 1983). In the eastern Canadian Arctic, bowhead whales swim to the coastline to hide from killer whales. They get as close to shore as possible, at times so close that a human could step on them from shore (Koski, personal observation). Similar near-shore movement has been documented for western Okhotsk Sea (OKH) bowhead whales when killer whales were present, in this case with nonforceful blows, slow and quiet movements, and no leaps or flipper slaps even the day after killer whale presence (Shpak and Paramonov, 2018).

When faced with human predators, bowhead whales tend to "go silent," with very slow movements, barely breaking the surface with single quiet slow breaths (Koski and Johnson, 1987). These breaths are quite unusual and differ from their normally forceful blows. They move from open water into the ice and have been seen hiding on the other side of ice flows from Inuit hunters (Würsig, personal observation). We have seen a bowhead whale slowly circle an ice floe as Inuit hunters circled on the other side, the whale apparently aware of the hunters' boat movements, attempting to always stay on the opposite side of the floe. Bockstoce (1986) included numerous accounts by Yankee whalers of

FIGURE 23.3 (A) A "flipper wave" that can also be part of flipper slapping onto the water surface. (B) A "nose out" or "spy-hop" that can also end in a chin slap onto the surface. (C) A partial breach, with a very percussive in-air and underwater sound upon landing. (D) A tail slap, in this case at the same time as a blow, with strongly arched body. (E and F) Log play: such log manipulation can also be a part of skin stimulation or rubbing. *Source: (A–D) and (F) taken by Olga Shpak in the Okhotsk Sea, Russia; (E) taken by as part of the NMFS survey in the Beaufort Sea, United States.*

bowheads being extremely shy and reacting to the slightest sound when pursued by whale boats. Bowheads became increasingly shy as commercial whaling progressed, the whalers commented, "Where I whaled last voyage now looks like a deserted village" and "The whales appear very shy. They don't like cold iron." (p. 101, Bockstoce, 1986). In fact, after just 7 years of initiation of bowhead whaling in the Northern Bering Sea in 1848, catches dropped from >2000 whales in a season to zero, and possibly also the result of depletion of the stock (Bockstoce, 1986). During recent biopsy sampling programs (Fig. 23.4) in Cumberland Sound, Nunavut, bowhead whales were generally approached to within a few meters without an observable reaction to an 8 m boat, provided that the boat approached at slow speed; but after several days of sampling in the same area, whales become evasive when the boat approached within several hundred meters at the same speed (Koski, personal observation).

Surfacing, diving, respirations

Possibly the best data obtained on surfacing and respiration behaviors for mysticete whales come from long-term studies on bowhead whales (Thomas et al., 2002a,b; Robertson et al., 2013; Zeh et al., 1993). Overall, there are positive correlations among length of surfacing, duration of previous dive, duration of subsequent dive, and number of respirations per surfacing (Robertson et al., 2013). This makes sense, as whales need to respire more (and therefore for longer) as the length of their dives increases. The respiration (="blow") interval is relatively constant, at a mean of about 13–15 seconds between blows while at the surface. It is the surface time, and therefore number of respirations per surfacing, that changes with dive time for full (physiological) respiration of the whale. While it would be helpful to put these parameters onto a true physiological expense regime for (1) resting, (2) socially/sexually active, (3) traveling, and (4) feeding whales, this has not been fully accomplished to date (but see next for DTAG and other

FIGURE 23.4 A small bowhead swims slowly when approached by a boat-conducting biopsy sampling in Cumberland Sound, Nunavut. *Source: Photo collected using a Phantom 3 drone by Thomas Seitz of VDOS Global.*

electronically obtained data related to subsurface feeding). Energy requirements of whales have been assumed to be directly related to the volume of oxygen consumed (Sumich, 1983; Folkow and Blix, 1992), which depends on the number of respirations and the volume of the lungs. Thomson (2002) estimated amount of prey taken in by a bowhead whale per time per swimming speed for whales of various sizes based on measurements (and overall estimates) of density of prey in the BCB seas and concluded that it is probable that bowhead whales have lower energy requirements than other large cetaceans.

The general pattern of movement of a bowhead whale is of the whale surfacing at an angle (generally less than 45 degrees to the surface) from depth, and then slowly moving along the surface while breathing. These breaths tend to be at about 14 seconds apart, with anywhere from 1 to 20 (rarely) respirations during that surfacing, averaging around 5 respirations during the surfacing for a nonmother bowhead (Robertson et al., 2013). Time at surface is about 1–2 minutes, and length of previous and subsequent dives is on average 8 minutes. This surface and dive behavior, noted in summer and fall, is consistent with surface and dive behavior during spring migration (Zeh et al., 1993). However, dives can be shorter or longer depending on behavior state, whale size, and water depth. Whales feeding below the surface in deeper water dive longer and need more "recovery time" at the surface than those diving shallowly or not at all (Würsig et al., 1984). Of course, whales busy with activities (generally feeding) at depth need to come to the surface to breathe and therefore need to stop all other activities for those life-sustaining breaths of air. Simon et al. (2009) showed that the slow ram filtration feeding of bowhead whales at depth maximizes energy efficiency and hypothesized that this is only possible with rather small crustaceans that cannot swim away rapidly from the approaching mouth.

Bowhead whales often show a "predive flex" before the final breath prior to diving, and the predive flex is especially pronounced before they dive steeply. The flex sequence consists of the front one-third (approximately) of the body, mainly the head, and the posterior third, mainly the peduncle and fluke, lifting up relative to the mid one-third, anatomically described as lordosis. At the height of lordosis, the whale gives a final forceful blow, surges forward, often with white water along the flanks, then arches the back high (reverse lordosis), lifts tail out of water, and dives steeply. Predive flexes are most pronounced when whales feed at some depth, over three to four times their own body length, and the entire predive flex and associated behaviors are surely designed for the whale to efficiently (and rapidly) leave the surface and dive steeply to feeding depth. The strong downward curvature of the body, including the fluke lifted out of water, is reminiscent of an experienced human skin diver leaving the surface by bending down at the waist, lifting legs and flippers above the surface to provide in-air weight unencumbered by buoyancy in water, and glide downward before the flippers (and whale tail) can provide propulsive force.

Potential and known disturbance reactions

We have learned much about reactions of bowheads to industry operations since *The Bowhead Whale* (Burns et al., 1993) and *Marine Mammals and Noise* (Richardson et al., 1995c), but much of the new knowledge has been about the presence, absence, or distance of whales near activities rather than reactions of individual whales. Some of the earlier information came from experiments where whales were exposed to projected industrial sounds or small-scale versions of the actual activities, with limited observations of whales near full-scale industrial operations. The newer information provides a different view because it includes the physical presence of whales near operational disturbance sources (e.g., operating seismic surveys or exploratory drilling) and includes other disturbance factors such as support vessels or structures not present during a playback experiment. Also, playback experiments often are "subscale," which results in a sound source being weaker (such as from a sound projector) at a given distance from the source and which does not replicate a full-scale seismic or drilling operation even if received sound level is the same.

One problem in interpreting whale detections (or lack of detections) near an industrial source is that we do not know what distribution of whales would have been in the absence of industrial activities, and it can change markedly within and between seasons (see Chapter 24). Surveys before and after an anthropogenic activity and in years with and without activity can help one to ascertain behavioral and distributional changes. Since the disturbing activities can change behaviors (e.g., calling, surfacing, and dive durations) and make the whales more or less detectable as a result (Richardson et al., 1985, 1995c; Robertson et al., 2016a,b), differences in detectability further confound disturbance reaction data.

Richardson and Malme provide an extensive summary of how various sources of industrial sounds affect bowhead whale distribution and behavior. Here, we discuss newer data to help one to clarify the earlier studies. Chapter 35, presents more detailed information on whale calling behavior with and without potential disturbance.

Traveling whales

Traveling whales near seismic operations

Richardson and Malme summarized reactions and distribution of bowhead whales near seismic operations up to the early 1990s. Behavior state of whales was not always known, but some observations were for whales in autumn (Ljungblad et al., 1988), when most are migrating westward and so probably included travelling. While sounds from seismic operations were audible in the water several tens of kilometers away, bowhead whales rarely showed overt avoidance beyond about 8 km and they sometimes did not show avoidance responses until seismic vessels were about 3 km away (Fig. 23.5). At the avoidance distances, there were highly significant differences in surfacing, respiration, and dive behaviors in comparison to undisturbed whales, and similar changes in behavior occurred in whales at slightly greater distances (5–10 km) before they showed overt avoidance. More subtle changes in behavior were observed out to distances to about 75 km away during the Richardson et al. (1986) study in summering areas, where most whales were probably feeding rather than traveling, but the distances where these changes occurred were variable and not all whales changed behavior when closer. Sound levels with avoidance were 150–180 dB re 1 μPa, and those where subtle changes in behavior were evident varied from 120 to 135 dB. However, avoidance has also been documented at 20–30 km, with received sound levels estimated as low as 120–135 dB re 1 μPa (Miller et al., 1999).

A study was conducted by Miller et al. (1997, 1999) of bowhead whale reactions to seismic operations in the mid-Beaufort in 1996, 1997, and 1998. For all 3 years the aerial

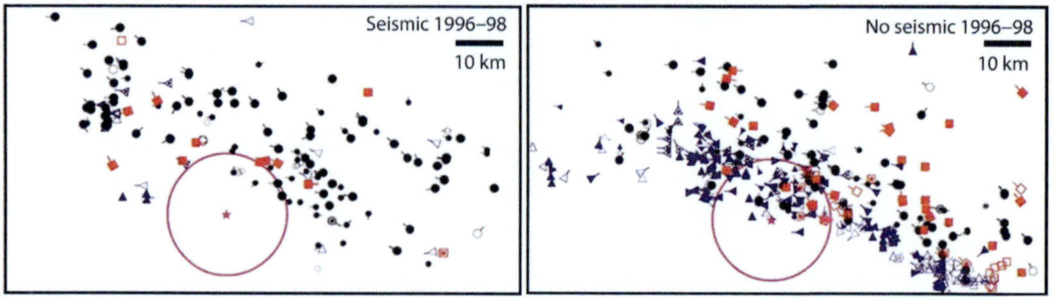

FIGURE 23.5 Distribution of bowhead whales around operating and nonoperating seismic vessel in the mid-Beaufort Sea nearshore waters during 1996 (*black circles*), 1997 (*blue triangles*), and 1998 (*red squares*) ($n = 133$ with-seismic operations, $n = 295$ without seismic operations). Note the greatly reduced detections within about 20 km when seismic vessels were operating. Bowheads in this study were largely migrating (traveling); these whales are known to be more sensitive to noise than feeding whales. Small symbols indicate shallower water, and large symbols deeper water. See Miller et al. (1997, 1999) for details.

survey resulted in a very high number of bowhead detections (428) on transect, of which 133 occurred with no seismic, and 295 with seismic. The whales in this study were primarily traveling and showed avoidance of an area within about 20 km of active seismic operations where received levels of sounds from the seismic operations would have been about 120–130 dB re 1 μPa (rms) (Fig. 23.5). Ship board observers saw no bowheads in 1998. The activity state of those whales was predominantly traveling and other considerations, including being closer to shore and in a shallower water than the 2006–08 studies, may partially explain why the conclusions differed. Also, there were a number of vessels associated with moving the seismic array and the bottom-founded recorders. Other explanatory factors included that whales have been found to react differently when migrating than when feeding (Robertson et al., 2016b). Possible changes in sightability when exposed to seismic sounds were not taken into account in the analysis of Miller et al. (1999).

Traveling whales near drilling operations

During autumn migration in the Beaufort Sea, traveling bowheads appeared to be more responsive to drilling operations than during summer. Most bowheads avoided a drillship by about 10 km during a 1986 study, with one observation of a whale circumventing a drilling operation by about 20 km (Koski and Johnson, 1987). Another study using data from 1993 observed bowheads avoiding a 20 km region around an active drilling platform (Davies, 1997). The difference in distances where whales avoided drilling operations between summer and autumn may have been due to different behavior states during autumn (primarily migrating) than summer (primarily feeding). Koski and Johnson (1987) found that durations of surfacing and mean blow interval were positively correlated with distance to the drillship or nearest active support vessel, but overall behavior was only weakly influenced by the nearest active ship.

Richardson et al. and McDonald et al. summarized the results of several years (2001–04) of intensive acoustic-location study of the man-made Northstar production island NW of

Prudhoe Bay, Alaska (see more detail Chapter 35). This remains the only major study of the bowhead's response to an "offshore" production platform in Arctic Alaska. Their analysis indicated that acoustic locations of whales nearest the island tended to be farther offshore (0.7–2.2 km), when certain industrial sounds emanating from the Island were higher. However, it is not known if there was a change in acoustic behavior or if the whales actually moved slightly offshore. The finding was surprising since the estimated received sound levels where the whales were migrating were very low and near ambient levels.

Traveling whales near ice management and anchor handling

Two of the noisier drilling operations by the oil and gas industry in Arctic waters are managing ice to prevent encroachment on drilling and production facilities and anchor handling associated with mobile drilling structures. While the sounds produced by these two activities vary, they are especially noisy, and are here discussed together.

Richardson et al. (1995a) conducted playbacks of icebreaker sounds to bowhead whales migrating through spring leads in sea ice near Barrow, Alaska, and recorded behavior of whales as they approached and passed the sound projectors. The projectors could not produce the sounds at the intensity of a real icebreaker, or produce the low-frequency components of icebreaker sounds, or provide the physical stimuli of the actual icebreaker. Nonetheless, they provided an approximate simulation of icebreaker sounds. Bowheads typically reacted to icebreaker sounds at signal-to-noise ratios similar to those at which they reacted to other industrial sounds, that is at about 20 dB above the level of the ambient sounds in the dominant one-third octave band. Therefore bowheads might respond to icebreaker sounds when up to 95 km from a real icebreaker (managing ice), with 10–50 km the usual distances where reactions are to be expected. Subtle changes in behavior might occur at much longer distances.

Feeding whales

From 1996 to 2010, aerial surveys were conducted near Prudhoe Bay and from Camden to Harrison Bay, to monitor bowhead whale distribution and movement before, during, and after industrial activities (Miller et al., 1997, 1999; Funk et al., 2007; Thomas et al., 2007; Koski et al., 2009, 2011). Behaviors used to suggest feeding included a non-WNW direction of movement (WNW is indicative of autumn migration along the Alaskan North Slope), slow and sometimes medium speed of movement, reduced speed of movement during a surfacing, turns during a surfacing, open mouths, mud streaming from the mouth or body of the whale, and steep fluke-out dives.

Feeding whales near seismic operations

Large numbers of feeding bowheads, including mothers and calves, lingered about 10–50 km southwest of an active seismic operation during 2007 and 2008 (Koski et al., 2009). About 2500 bowheads were estimated to be in the survey area over 16 days in 2007 and 6 days in 2008. In comparison, during nonfeeding periods only about 350 whales were estimated. The closest sightings to the seismic operation were made in 2008, including 2 days, while seismic work was conducted nearby, and 4 days after the seismic operation was moved away to reduce disturbance to the feeding whales. On a broad scale, there was no apparent change in whale distribution during feeding periods with seismic versus no seismic when subtle differences in detectability were taken into account (Robertson et al., 2016a). Some feeding bowheads tolerated sound levels as high as 170–180 dB re 1 μPa (rms) but most avoided areas where received sound levels exceeded 160 dB, which was about 13–15 km from the source (Fig. 23.5; see also, Chapter 35).

Observations from circling aircraft near full-scale seismic vessels showed that few whales avoided the seismic operation when it was more than 7.5 km away, but when the vessel was approaching the whales, changes in behavior occurred at least 5–10 km away (Ljungblad et al., 1988). During summer feeding in the Canadian Beaufort Sea, subtle changes in behavior occurred up to 75 km from a seismic operation (Richardson et al., 1986). It is likely that differences in behavior also occur at long distances during autumn migration, but Ljungblad et al. (1988) were not able to record these. Overall, largely traveling whales avoided areas within about 20 km of active seismic operations with received levels of about 120–130 dB re 1 μPa (rms) (Miller et al., 1997, 1999).

The presence of whales closer to seismic operations than those existing 20 km apart as described in the study of Miller et al. (1999) has good and bad consequences. The good consequence is that whales would not be excluded from important feeding areas, such as occurred in 2007 and 2008. The bad, and potentially more significant, consequence is that whales are routinely exposed to higher sound levels than previously expected because they do not avoid the operation by as large a distance as expected, potentially resulting in deleterious effects such as temporary or permanent hearing impairment or stress (Rolland et al., 2012). In some cases, this has also displaced whales away from indigenous hunting areas.

Richardson et al. (1985) found differences in surfacing, respiration, and dive data when whales were and were not exposed to (largely seismic) human activities but could not discriminate differences by seasons or behavior states. From earlier and newer data, Robertson et al. (2013) showed that seismic operations had a greater effect on behavior of traveling than on feeding or socializing whales.

Feeding whales near drilling operation

Richardson and Malme summarized observations of bowheads around drilling and production sites, in summer in the Canadian Beaufort Sea (BCB stock), largely of feeding whales. One whale was sighted within 200 m of an operating drillship, and others within about 5 km (Richardson et al., 1990). Bowheads have often been seen within 10–20 km of drillships during behavior studies and aerial surveys (both authors, personal observation). Whale hunters tell us that feeding whales tend to be easier to approach than migrating whales during the autumn hunt. Similar tolerant behavior was evident during bowhead whale tagging operations near Pt. Barrow (J. C. George, pers. comm.).

Whales during social and sexual activities

Although there have not been experiments to examine impacts of industrial activities near bowheads engaged in sexual activities or bowheads that are socializing, the few observations suggest that whales engaged in these activities are less likely to show avoidance behavior. Robertson et al. (2013), however, found that socializing whales significantly reduced surface times when exposed to seismic sound, suggesting that their social activities were probably affected to at least some degree.

Mother–calf reactions

Bowhead mothers and calves (Fig. 23.6) were not subject to experimental observation, so scarce data exist. On September 10, 1986, a mother–calf pair swam eastward toward an ice field at an unusually rapid speed of 7.5 km/h, while one industry vessel was travelling 15 km east of them and one 25 km to the southwest. Fig. 23.7 shows the number of blows/surfacing and duration of dive for the mother during the behavior observation session. Behavior observed during the first six sequences is typical of highly disturbed whales, which is unusual at such long distances from a disturbance source. The behavior returned to "normal" during the last four sequences when both vessels were about 22 km southwest of the whales (Koski and Würsig, personal observation). Possible explanations of these strong reactions at great distance are that the whales were between two vessels that were approaching from opposite directions and/or that mothers with calves react at longer distances (and lower sound levels) to disturbance than nonmothers.

In the early 1990s, the authors observed a migrating mother and calf reacting to an approaching Iñupiaq whaling vessel in the spring lead off Pt. Barrow, Alaska. We had watched the whales for about 30 minutes as they migrated through open water along the north side of the lead at typical migrating speeds and behaviors for mother–calf pairs. They abruptly increased speed of travel with white water behind them, decreased dive durations, and decreased number of blows per surfacing, which traits are typical of high levels of disturbance. We then saw a whaling vessel approach rapidly at 25–30 km distance. To not interfere with the hunt, we terminated observations, so did not see subsequent behavior as the vessel approached more closely (Koski and Würsig, personal observation).

FIGURE 23.6 A bowhead whale mother and calf taken during August 2019 in Cumberland Sound. *Source: Photo taken by Ricky Kilabuk, Pangnirtung HTO.*

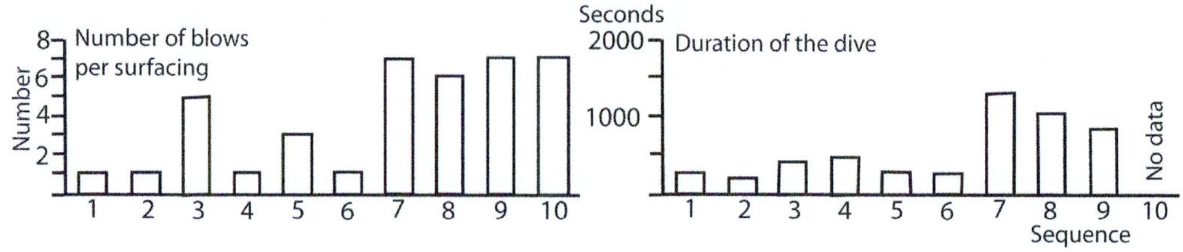

FIGURE 23.7 Respiration, surface, and dive data for a 14.9 m bowhead whale mother accompanied by a calf (see Fig. 23.4), September 10, 1986. Note the behavioral change between sequence 6 and 7; the significance is explained in the text. *Source: Redrawn from Koski, W.R., Johnson, S.R., 1987. Responses of Bowhead Whales to an Offshore Drilling Operation in the Alaskan Beaufort Sea, Autumn 1986: Behavioral Studies and Aerial Photogrammetry. Rep. by LGL Ltd., King City, ON, for Shell Western E & P Inc., Anchorage. 129 p.*

These few observations of disturbance reactions by mothers and calves suggest that they are more responsive to potential sources of disturbance and avoid them at perhaps greater distances than other whales. However, the two incidents described previously have special circumstances that may not be representative of most other cases where whales might be exposed to vessels, so they may be extremes.

Summary

Bowhead whale behavior changes when they are exposed to industrial activities in the Beaufort Sea but migrate west across the Beaufort Sea during autumn nonetheless (Davies, 1997; Koski et al., 2011; MacDonald et al., 2012; LGL Alaska Research Associates, Inc. et al., 2014; Robertson et al., 2016a) and appear not to be excluded from important feeding areas because of apparent greater tolerance of industrial activities when food is abundant (Koski et al., 2011). We do not have accurate determinations of distances when whales completely avoid different activities, because of differences in individual whales, reactions to activities, and differences when whales are engaged in variable activity states. However, most bowhead whales avoid intense industrial sounds [>160 dB re 1 μPa (rms)] by 3–15 km, depending on sound intensity and activity state, and may avoid those areas by 20 km or more when migrating (Koski and Johnson, 1987; Miller et al., 1997, 1999; Davies, 1997). Whales change behavior when exposed to intense sounds (Richardson et al., 1986, 1995c; Robertson et al., 2016b), which generally cause them to avoid the source of the disturbance by 3–15 km, and more subtle changes in calling and surfacing and dive behavior are sometimes seen at distances up to 75 km from activities (Richardson et al., 1986; Blackwell et al., 2015, 2017), although whales much closer sometimes have not altered their behavior. The changes in behavior affect our ability to detect whales, which has complicated interpretation of sighting data during aerial surveys conducted around industrial activities (Robertson et al., 2016a). These changes in behavior, termed "skittish behavior" by subsistence whale hunters, can also impact their ability to detect whales in the presence of industrial activities. This is a concern to whale hunters because it can affect their ability to successfully harvest whales when they are skittish (Ahmaogak, 1989). Recent studies (e.g., Ellison et al., 2012) have shown that reactions of marine mammals to sound are strongly affected by "context," that is, the situation or conditions (presence and location of disturbance sources, whether they are approaching or moving away, their behavior state—i.e., feeding, migrating, socializing—and whether it is a mother with calf) under which the animal is exposed. Further supporting this conclusion are the findings of Macdonald et al. (2012) at Northstar Island. This study indicated that migrating/traveling bowheads sometimes react to industrial sounds at low, and possibly below ambient, received levels.

Conclusion, syntheses, and knowledge needed

1. The dominant activity state of bowhead whales involves feeding, whether mixed with traveling or focused solely on feeding. Estimates of food ingestion by bowheads suggest that they have lower metabolic rates than other mysticetes or they could not survive in the Arctic given the density of their prey (Thomson et al., 2002a,b). Because of the low overall productivity, bowheads have slow growth rates, delayed sexual and physical maturity, store energy in their blubber that allows them to survive long periods without significant food, and delay having calves until they have sufficient energy reserves to successfully raise the calf to weaning. Given the observed time spent feeding versus other activities such as migrating between summer feeding and overwintering areas, they probably could not spend a lot more time feeding than they already do.
2. Bowhead and right whales are the only habitual skim feeders, chugging forward with the largest mouths and longest baleen plates of any whales. But, gray whales at times also skim feed when food is abundant. The balaenid and gray whales also have large testes and a general scramble competition mating system, where several males attempt to gain mating access to one female, who appears to make mating difficult and thereby is likely exhibiting a form of female choice. This system appears different from that of humpback whales, where there is also surface-active scramble competition, but apparently of a much more forceful "aggressive" nature. We do not yet know enough about the mating system of other rorquals to compare bowhead whale mating systems to their systems.
3. Environmental conditions, particularly ice cover, have more influence on the distance from shore that whales travel in a given year than do industrial activities. Despite oil and gas exploration over the past several decades, whales still use and return to the same habitat areas each year. Changing conditions due to large-scale effects of climate change are discussed in Chapter 24.
4. Data needs: While we know many behavioral aspects of bowhead whales, we do not yet know what kind of cultural information (if any in meaningful manner) may be passed on during their potentially very long lives, probably the longest living mammals on the Earth. Studies of similarity and differences in behaviors among different populations especially on vocal repertoire may help to fill in this data-gap.

More data are needed to quantify the effect of industrial activities based on a suite of variables. The surfacing and dive behavior of subadult whales is significantly different for adult whales and mothers with calves. Where data exist, distance appears to be weakly correlated to surface and dive times (Koski and Johnson, 1987) but too few data exist to quantify the relationship given daily variation, differences among seasons, variation with whale size/status, and variation related to behavior state. More data are needed on behavior of bowhead whales in the presence of industrial activities along with the size of whales, distance to activities, and received sound levels near whales to better evaluate and predict changes in behavior and impacts on whales. In particular, more data are needed on the noisier activities such as managing ice and anchor handling to better understand relationships between distances to the disturbance and sound levels received by the whales. These noisy activities have the potential to displace whales and affect behavior at greater distances than the industrial activities that have received the most study.

Acknowledgments

We thank the many Arctic and near-Arctic researchers who have helped to elucidate the behaviors, natural and relative to human actions, of bowhead whales. While much was already known from logbooks of whalers and indigenous knowledge of First Nations people, it was not until the late 1970 that systematic behavioral observations were gathered from ice platforms, aircraft, and then drones in more recent years. Information on whale behavior, distribution, and movement has also been obtained from acoustic monitoring of bowhead whales, which have provided more insights into visual observations. We especially thank W. John Richardson for his many years of guidance of research, analysis, and writing, as well as the editors of this book, J.G.M. "Hans" Thewissen and J. Craig George.

References

Accardo, C.M., Ganley, L.C., Brown, M.W., Duley, P.A., George, J.C., Reeves, R.R., et al., 2018. Sightings of a bowhead whale (*Balaena mysticetus*) in the Gulf of Maine and its interactions with other baleen whales. J. Cetacean Res. Manage. 19, 23–30.

Ahmaogak, G.N., Sr., 1989. Protecting the habitat of the bowhead whale. In: Rey, L., Alexander, V. (Eds.), Proceedings of the Sixth Conference of The Comité Arctique Int. 13–15 May 1985, E. J. Brill, pp. 593–597.

Ashjian, C.J., Braund, S.R., Campbell, R.G., George, J.C., Kruse, J., Maslowski, W., et al., 2010. Climate variability, oceanography, bowhead whale distribution, and Iñupiat subsistence whaling near Barrow, Alaska. Arctic 63, 179–194.

Blackwell, S.B., Nations, C.S., McDonald, T.L., Greene Jr., C.R., Thode, A.M., Guerra, M., et al., 2015. Effects of airgun sounds on bowhead whale calling rates in the Alaskan Beaufort Sea. Mar. Mammal. Sci. 29, E342–E365. Available from: https://doi.org/10.1111/mms.12001.

Blackwell, S.B., Nations, C.S., Thode, A.M., Kauffman, M.E., Conrad, A.S., Norman, R.G., et al., 2017. Effects of tones associated with drilling on bowhead whale calling rates. PLoS One 12 (11), e0188459. Available from: https://doi.org/10.1371/journal.pone.0188459.

Bockstoce, J.R., 1986. Whales, ice, and men: The history of whaling in the Western Arctic. Univ. of Washington Press, Seattle, p. 400.

BOWPHOT, 2014. Bowhead Whale Photograph Database Archived at the North Slope Borough. Department of Wildlife Management, Barrow, AK.

Brown, M.W., Brault, S., Hamilton, P.K., Kenney, R.D., Knowlton, A.R., Marx, M.K., et al., 2001. Sighting heterogeneity of right whales in the western North. Atlantic: 1980–1992. J. Cetacean Res. Manage. 2, 245–250. Spec. Iss.

Brownell Jr., R.L., Ralls, K., 1986. Potential for sperm competition in baleen whales. Int. Whal. Comm. 8, 97–112. Spec. Iss.

Burns, J.J., Montague, J.J., Cowles, C.J., 1993. The Bowhead Whale. Allen Press, Lawrence, KS, Spec. Publ. #2. The Soc. for Mar. Mamm.

Citta, J.J., Quakenbush, L.T., Okkonen, S.R., Druckenmiller, M.L., Maslowski, W., Clement-Kinney, J., et al., 2015. Ecological characteristics of core-use areas used by Bering–Chukchi–Beaufort (BCB) bowhead whales, 2006–2012. Prog. Ocean. 136, 201–222. Available from: http://dx.doi.org/10.1016/j.pocean.2014.08.012.

Davies, J.R., 1997. The Impact of an Offshore Drilling Platform on the Fall Migration Path of Bowhead Whales: A GIS-Based Assessment (Unpubl. M.Sc. thesis). Western Washington University, 52 p.

De Robertis, A., Schell, C., Jaffe, J.S., 2003. Acoustic observations of the swimming behavior of the euphausiid *Euphausia pacifica* Hansen. ICES J. Mar. Sci. 60, 885–898. Available from: https://doi.org/10.1016/S1054-3139(03)00070-5.

Ellison, W.E., Southall, B.L., Clark, C.W., Frankel, A.F., 2012. A new context-based approach to assess marine mammal behavioral responses to anthropogenic sounds. Conserv. Biol. 26, 21–28. Available from: https://doi.org/10.1111/j.1523-1739.2011.01803.x.

Fellner, W., Bauer, B., Harely, H., Heidi, E., 2006. Cognitive implications of synchrony in dolphins: a review. Aquat. Mammal. 32, 511–516. Available from: https://doi.10.1578/AM.32.4.2006.511.

Ferguson, S.H., Higdon, J.W., Westdal, K.H., 2012a. Prey items and predation behavior of killer whales (*Orcinus orca*) in Nunavut, Canada based on Inuit hunter interviews. Aquat. Biosyst. 8, 1–16.

Ferguson, S.H., Kingsley, M.C.S., Higdon, J.W., 2012b. Killer whale (*Orcinus orca*) predation in a multi-prey system. Popul. Ecol. 54, 31–41. Available from: https://doi.org/10.1007/s10144-011-0284-3.

Fish, F.E., Goetz, K.T., Rugh, D.J., Brattström, L.V., 2013. Hydrodynamic patterns associated with echelon formation swimming by feeding bowhead whales (*Balaena mysticetus*). Mar. Mammal. Sci. 29, E498–E507. Available from: https://doi.org/10.1111/mms.12004.

Folkow, L.P., Blix, A.S., 1992. Metabolic rates of minke whales (*Balaenoptera acutorostrata*) in cold water. Acta Physiol. Scand. 146, 141–150.

Fortune, S.M.E., Koski, W.R., Higdon, J.W., Trites, A.W., Baumgartner, M.F., Ferguson, S.H., 2017a. Evidence of molting and the function of "rock-nosing" behavior in bowhead whales in the eastern Canadian Arctic. PLoS ONE 12 (11), e0186156. 15 p.

Fortune, S.M.E., Koski, W.R., Higdon, J.W., Trites, A.W., Baumgartner, M.F., Ferguson, S.H., 2017b. Correction: evidence of molting and the function of "rock-nosing" behavior in bowhead whales in the eastern Canadian Arctic. PLoS One 13 (2), e92813. 1 p.

Funk, D.W., Rodrigues, R., Ireland, D.S., Koski, W.R., 2007. Chapter 12, Discussion, conclusions, and assessment of potential effects of seismic activities on marine mammals in the Chukchi and Beaufort Seas. In: Funk, D.W., Rodrigues, R., Ireland, D.S., Koski, W.R. (Eds.), Joint Monitoring Program in the Chukchi and Beaufort Seas, Jul–Nov 2006. 42 p. LGL Alaska Report P891-2 from LGL Alaska Research Associates, Inc., LGL Ltd., Greeneridge Sciences Inc., Bioacoustics Research Program, Cornell University, and Bio-Wave Inc. For Shell Offshore Inc., ConocoPhillips Alaska Inc., and GX Technology, and National Marine Fisheries Service, U.S. Fish and Wildlife Service. 316 p. plus Appendices.

George, J.C., 2009. Growth, Morphology and Energetics of Bowhead Whales (*Balaena mysticetus*) (unpublished Ph. D. thesis). University of Fairbanks, Fairbanks, AK, 168 p.

George, J.C., Bada, J., Zeh, J., Scott, L., Brown, S.E., O'Hara, T., et al., 1999. Age and growth estimates of bowhead whales (*Balaena mysticetus*) via aspartic acid racemization. Can. J. Zool. 77, 571–580.

George, J.C., Follmann, E., Zeh, J., Sousa, M., Tarpley, R., Suydam, R., et al., 2011. A new way to estimate the age of bowhead whales (*Balaena mysticetus*) using ovarian corpora counts. Can. J. Zool. 89, 840–852. Available from: https://doi.org/10.1139/Z11-057.

Griffiths, W.B., Thomson, D.H., 2002. Chapter 5, Species composition, biomass, and local distribution of zooplankton relative to water masses in the Eastern Alaskan Beaufort Sea. In: Richardson, W.J., Thomson, D.H. (Eds.), Bowhead Whale Feeding in the Eastern Alaskan Beaufort Sea: Update of Scientific and Traditional Information. pp. 5.1–5.72. OCS Study MMS 2002-012. LGL Rep. TA2196-7. Rep. From LGL Ltd., King City, Ontario, for US Minerals Management Service, Anchorage, Alaska and Herndon Virginia, USA. xliv + 420pp. [Available from National Technical Information Service, Springfield, Virginia, USA, Rep. No. NTIS PB2004-101568].

Griffiths, W.B., Thomson, D.H., Bradstreet, M.S.W., 2002. Chapter 6, Zooplankton and water masses at bowhead whale feeding locations in the Eastern Beaufort Sea. In: Richardson, W.J., Thomson, D.H. (Eds.), Bowhead Whale Feeding in the Eastern Alaskan Beaufort Sea: Update of Scientific and Traditional Information. pp. 6.1–6.44. OCS Study MMS 2002-012. LGL Rep. TA2196-7. Rep. From LGL Ltd., King City, Ontario, for US

Minerals Management Service, Anchorage, Alaska and Herndon Virginia, USA. xliv + 420pp. [Available from National Technical Information Service, Springfield, Virginia, USA, Rep. No. NTIS PB2004-101568].

Heide-Jørgensen, M.P., Laidre, K.L., Wiigs, Ø., Postma, L., Dueck, L., Bachmann, L., 2010. Large-scale sexual segregation of bowhead whales. Endang. Species Res. 13, 73–78.

Heide-Jørgensen, M.P., Laidre, K.L., Quakenbush, L.T., Citta, J.J., 2011. The northwest passage opens for bowhead whales. Biol. Lett. Available from: https://doi.org/10.1098/rsbl.2011.0731.

Heide-Jørgensen, M.P., Laidre, K.L., Nielsen, N.H., Hansen, R.G., Røstad, A., 2013. Winter and spring diving behavior of bowhead whales relative to prey. Anim. Biotelem. 1, 15. Available from: https://doi.org/10.1186/2050-3385-1-15.

Ignatyev, S.M., 1997. Pelagic fishes and their macroplankton prey: swimming speeds. In: Proceedings of Forage Fishes in Marine Ecosystems, Alaska Sea Grant Program, AK-SK-9-01. pp. 31–39.

Koski, W.R., Johnson, S.R., 1987. Responses of Bowhead Whales to an Offshore Drilling Operation in the Alaskan Beaufort Sea, Autumn 1986: Behavioral Studies and Aerial Photogrammetry. Rep. by LGL Ltd., King City, ON, for Shell Western E & P Inc., Anchorage, 129 p.

Koski, W.R., Davis, R.A., Miller, G.W., Withrow, D.E., 1993. Chapter 7, Reproduction. In: Burns, J.J., Montague, J. J., Cowles, C.J. (Eds.), The Bowhead Whale. Allen Press, Lawrence, KS, pp. 239–274, Spec. Publ. 2, Society for Marine Mammalogy.

Koski, W.R., George, J.C., Mactavish, B.D., Acker, R., Davis, A.R., Suydam, R., et al., 2005. Observations of bowhead whale mating in the Bering Sea. In: Presentation at the 16th Biennial Conference on the Biology of Marine Mammals. San Diego, CA, 12–16 December 2005.

Koski, W.R., Funk, D.W., Ireland, D.S., Lyons, C., Christie, K., Macrander, A.M., et al., 2009. An update on feeding by bowhead whales near an offshore seismic survey in the Central Beaufort Sea. In: Paper SC/61/BRG3 Presented to the International Whaling Commission Scientific Committee. Madeira, Portugal, 31 May–12 June 2009, 24 p.

Koski, W.R., George, J.C., Zeh, J., Brandon, J.R., 2010. Preliminary analyses on identification of yearling bowhead whales (*Balaena mysticetus*) in aerial photographs. In: Paper SC/62/BRG29 Presented to the International Whaling Commission Science Scientific Committee. Agadir, Morocco, 31 May–11 June 2010, 13 p.

Koski, W.R., Brandon, J.R., Ireland, D.S., Funk, D.W., Macrander, A.M., 2011. Aerial sighting distributions of bowhead whales in the central Beaufort Sea relative to seismic activity during 2007, 2008 and 2010. In: Presentation at the 19th Biennial Conference on the Biology of Marine Mammals. Tampa, FL, 28 November–2 December 2011.

Krutzikowsky, G.K., Mate, B.R., 2000. Dive and surfacing characteristics of bowhead whales (*Balaena mysticetus*) in the Beaufort and Chukchi seas. Can. J. Zool. 78, 1182–1198.

Lambertsen, R.H., Rasmussen, K.J., Lancaster, W.C., Hintz, R.J., 2005. Functional morphology of the mouth of the bowhead whale and its implications for conservation. J. Mammal. 86, 342–352.

LGL Alaska Research Associates, Inc., JASCO Applied Sciences, Inc., Greeneridge Sciences, Inc., 2014. Joint Monitoring Program in the Chukchi and Beaufort Seas, 2012. LGL Alaska Final Report P1272-2 for Shell Offshore, Inc. ION Geophysical, Inc., and Other Industry Contributors, for the Nat. Mar. Fish. Serv., and U.S. Fish Wildl. Serv. 320 p. plus Appendices.

Ljungblad, D.K., Leatherwood, S., Dahlheim, M.E., 1980. Sounds recorded in the presence of an adult and calf bowhead whale. Mar. Fish. Rev. 42, 86–87.

Ljungblad, D.K., Würsig, B., Swartz, S.L., Keene, J.M., 1988. Observations on the behavioral responses of bowhead whales (*Balaena mysticetus*) to active geophysical vessels in the Alaskan Beaufort Sea. Arctic 41, 183–194.

Miller, G.W., Davis, R.A., Koski, W.R., Crone, M.J., Rugh, D.J., Fraker, M.A., 1992. Calving intervals of bowhead whales—an analysis of photographic data. Rep. Int. Whal. Comm. 42, 501–506.

Miller, G.W., Elliot, R.E., Koski, W.R., Richardson, W.J., 1997. Northstar Marine Mammal Monitoring Program, 1996: Marine Mammal and Acoustical Monitoring of a Seismic Program in the Alaskan Beaufort Sea. Report from LGL Limited, King City, ON, to BP Exploration (Alaska) Inc., Anchorage, AK, pp. 5.1–5.115 (Chapter 5).

Miller, G.W., Elliot, R.E., Koski, W.R., Richardson, W.J., 1999. Chapter 5, Whales. In: Marine Mammal and Acoustical Monitoring Program of Western Geophysical's Open-Water Seismic Program in the Alaskan Beaufort Sea, 1998: 90-Day Report. Report from LGL Limited, King City, ON, to Western Geophysical, Houston TX and National Marine Fisheries Service, Anchorage, AK and Silver Spring, MD, pp. 5.1–5.59.

Mitchell, E.D., Reeves, R.R., 1982. Factors affecting abundance of bowhead whales *Balaena mysticetus* in the eastern Arctic of North America, 1915-1980. Biol. Conserv. 22, 59–78.

Mocklin, J.A., Rugh, D.J., Moore, S.E., Angliss, R.P., 2011. Using aerial photography to investigate evidence of feeding by bowhead whales. Mar. Mammal. Sci. First published: 04 September 2011. Available from: https://doi.org/10.1111/j.1748-7692.2011.00518.x.

Moore, S., George, J.C., Sheffield, G., Bacon, J., Ashjian, C., 2010. Bowhead whale distribution and feeding near Barrow, Alaska, in late summer 2005–06. Arctic 63, 195–205.

Nunavut Wildlife Management Board. 2000. Of the Inuit Bowhead Knowledge Study. Final Report. Nunavut Wildlife Management Board, PO Box 1379, Iqaluit, NU, Canada X0A 0H0.

O'Hara, T.M., George, J.C., Tarpley, R.J., Burek, K., Suydam, R.S., 2002. Sexual maturation in male bowhead whales (Balaena mysticetus). J. Cetacean Res. Manage 4 (2), 143–148.

Pack, A.A., Salden, D.R., Ferrari, M.J., Glockner-Ferrari, D.A., Herman, L.M., Stubbs, H.A., et al., 1998. Male humpback whale dies in competitive group. Mar. Mammal. Sci. 14, 861–873.

Payne, R.S., 1995. Among Whales. Bantam Doubleday Books, New York.

Reeves, R.R., Mitchell, E., Mansfield, A., McLaughlin, M., 1983. Distribution and migration of the bowhead whale, *Balaena mysticetus*, in the eastern North American Arctic. Arctic 36, 5–64.

Richardson, W.J., Fraker, M.A., Würsig, B., Wells, R.S., 1985. Behavior of bowhead whales *Balaena mysticetus* summering in the Beaufort Sea: reactions to industrial activities. Biol. Conserv. 32, 195–230.

Richardson, W.J., Würsig, B., Greene Jr., C.R., 1986. Reactions of bowhead whales, *Balaena mysticetus*, to seismic exploration in the Canadian Beaufort Sea. J. Acoust. Soc. Am. 79, 1117–1128.

Richardson, W.J., Würsig, B., Greene Jr., C.R., 1990. Reactions of bowhead whales, *Balaena mysticetus*, to drilling and dredging noise in the Canadian Beaufort Sea. Mar. Environ. Res. 29 (2), 135–160.

Richardson, W.J., Greene Jr., C.R., Hanna, J.S., Koski, W.R., Miller, G.W., Patenaude, N.J., et al., 1995a. Acoustic Effects of Oil Production Activities on Bowhead and White Whales Visible During Spring Migration Near Pt. Barrow, Alaska—1991 and 1994 Phases. OCS Study MMS 95-0051. Rep. from LGL Ltd., King City, ON, for U. S. Minerals Management Service, Herndon, VA, 539 p.

Richardson, W.J., Finley, K.J., Miller, G.W., Davis, R.A., Koski, W.R., 1995b. Feeding, social, and migration behavior of bowhead whales, *Balaena mysticetus*, in Baffin Bay vs. the Beaufort Sea—regions with different amounts of human activity. Mar. Mammal. Sci. 11, 1–45.

Richardson, W.J., Greene Jr., C.R., Malme, C.I., Thomson, D.H., 1995c. Marine Mammals and Noise. Academic Press, San Diego, CA, p. 576.

Robertson, F.C., Koski, W.R., Thomas, T.A., Richardson, W.J., Würsig, B., Trites, A.W., 2013. Seismic operations have variable effects on dive-cycle behavior of bowhead whales in the Beaufort Sea. Endang. Species Res. 21, 143–160.

Robertson, F.C., Koski, W.R., Brandon, J.R., Thomas, T.A., Trites, A.W., 2016a. Correction factors to account for the availability bias of bowhead whales exposed to seismic operations in the Beaufort Sea. J. Cetacean Res. Manage. 15, 35–44.

Robertson, F.C., Koski, W.R., Trites, A.W., 2016b. Behavioral responses affect distribution analyses of bowhead whales in the vicinity of seismic operations. Mar. Ecol. Prog. Ser. 549, 243–262.

Rolland, R.M., Parks, S.E., Hunt, K.E., Castellote, M., Corkeron, P.J., Nowacek, D.P., et al., 2012. Evidence that ship noise increases stress in right whales. Proc. R. Soc. B. Available from: https://doi.org/10.1098/rspb.2011.2429.

Rosa, C., Zeh, J., George, J.C., Botta, O., Zauscher, M., Bada, J., et al., 2013. Age estimates based on aspartic acid racemization for bowhead whales (*Balaena mysticetus*) harvested in 1998–2000 and the relationship between racemization rate and body temperature. Mar. Mammal. Sci 29, 424–445.

Saupe, S.M., Schell, D.M., Griffiths, W.B., 1989. Carbon isotope ratio gradients in western Arctic zooplankton. Mar. Biol. 103, 427–432.

Schell, D.M., Saupe, S.M., 1993. Chapter 12, Feeding and growth as indicated by stable isotopes. In: Burns, J.J., Montague, J.J., Cowles, C.J. (Eds.), The Bowhead Whale. Allen Press, Lawrence, KS, pp. 491–509, Spec. Publ. 2, Society for Marine Mammalogy.

Shpak, O.V., Paramonov, A.Y., 2018. The bowhead whale, *Balaena mysticetus* Linnaeus, 1758, in the Western Sea of Okhotsk (2009–2016): distribution pattern, behavior, and threats. Biol. Morya 44, 210–218. Available from: https://doi.org/10.1134/S1063074018030082.

Simon, M., Johnson, M., Tyack, P., Madsen, P.T., 2009. Behaviour and kinematics of continuous ram filtration in bowhead whales (*Balaena mysticetus*). Proc. R. Soc. B 276, 3819–3828.

Sumich, J.L., 1983. Swimming velocities, breathing patterns and estimated costs of locomotion in migrating gray whales *Eschrichtius robustus*. Can. J. Zool. 61, 647–652.

Thomas, T.A., Koski, W.R., Richardson, W.J., 2002a. Chapter 15, Correction factors to calculate bowhead whale numbers form aerial surveys of the Beaufort Sea. In: Richardson, W.J., Thomson, D.H. (Eds.), Bowhead Whale Feeding in the Eastern Alaskan Beaufort Sea: Update of Scientific and Traditional Information. pp. 15:1–15:28. OCS Study MMS 2002-012. LGL Rep. TA2196-7. Rep. from LGL Ltd., King City, ON, for US Minerals Management Service, Anchorage, Alaska and Herndon Virginia, USA. xliv + 420pp. [Available from National Technical Information Service, Springfield, Virginia, USA, Rep. No. NTIS PB2004-101568].

Thomas, T.A., Koski, W.R., Richardson, W.J., Würsig, B., 2002b. Chapter 14, Surfacing, respiration and dive cycles of bowhead whales in the eastern Alaskan Beaufort Sea during late summer and autumn as related to age. In: Richardson, W.J., Thomson, D.H. (Eds.), Bowhead Whale Feeding in the Eastern Alaskan Beaufort Sea: Update of Scientific and Traditional Information. pp. 14:1–14:19. OCS Study MMS 2002-012. LGL Rep. TA2196-7. Rep. from LGL Ltd., King City, Ontario, for US Minerals Management Service, Anchorage, Alaska and Herndon Virginia, USA. Vol. 1 xliv + 420pp. [Available from National Technical Information Service, Springfield, Virginia, USA, Rep. No. NTIS PB2004-101568].

Thomas, T.A., Koski, W.R., Elliott, T., 2007. Chapter 8, Beaufort Sea aerial surveys in support of seismic operations. In: Funk, D.W., Rodrigues, R., Ireland, D.S., Koski, W.R., (Eds.), Joint Monitoring Program in the Chukchi and Beaufort Seas, Jul–Nov 2006. 18 p. LGL Alaska Report P891-2 from LGL Alaska Research Associates, Inc., LGL Ltd., Greeneridge Sciences Inc., Bioacoustics Research Program, Cornell University, and Bio-Wave Inc. for Shell Offshore Inc., ConocoPhillips Alaska Inc., and GX Technology, and National Marine Fisheries Service, U.S. Fish and Wildlife Service. 316 p. plus Appendices.

Thomson, D.T., 2002. Chapter 22, Energetics of bowheads. In: Richardson, W.J., Thomson, D.H. (Eds.), Bowhead Whale Feeding in the Eastern Alaskan Beaufort Sea: Update of Scientific and Traditional Information. pp. 22:1–22:40. OCS Study MMS 2002-012. LGL Rep. TA2196-7. Rep. from LGL Ltd., King City, ON, for U.S. Minerals Manage. Serv., Anchorage, AK, and Herndon, VA. Vol. 2, 277 p. [Available from National Technical Information Service, Springfield, Virginia, USA, Rep. No. NTIS PB2004-101568].

Trivers, R.L., 1971. The evolution of reciprocal altruism. Quart. Rev. Biol. 46, 35–57.

Tyack, P., Whitehead, H., 1982. Male competition in large groups of wintering humpback whales. Behaviour 83, 132–154.

van der Hoop, J.M., Nousek-McGregor, A.E., Nowacek, D.P., Parks, S.E., Tyack, P., Madsen, P.T., 2019. Foraging rates of ram-filtering North Atlantic right whales. Funct. Ecol. Available from: https://doi.org/10.1111/1365-2435.13357.

Werth, A.J., 2004. Models of hydrodynamic flow in the bowhead whale filter feeding apparatus. J. Exper. Biol. 207, 3569–3580.

Wetzel, D.L., Reynolds III, J.E., Mercurio, P., Givens, G.H., Pulster, E.L., George, J.C., 2017. Age estimation for bowhead whales, Balaena mysticetus, using aspartic acid racemization with enhanced hydrolysis and derivatization procedures. J. Cetacean Res. Manage 17, 9–14. 2017.

Whitehead, H., 1985. Why whales breach. Sci. Am. 252, 84–93.

Würsig, B., 1989. The behavior of baleen whales. Sci. Am. 256, 102–107.

Würsig, B., 2000. Leviathan love. In: Bekoff, M. (Ed.), The Smile of the Dolphin. Discovery Books, New York.

Würsig, B., Clark, C.W., 1993. Chapter 5, Behavior of bowhead whales. In: Burns, J.J., Montague, J.J., Cowles, C.J. (Eds.), The Bowhead Whale. Allen Press, Lawrence, KS, pp. 157–199, Spec. Publ. 2, Society for Marine Mammalogy.

Würsig, B., Dorsey, E.M., Fraker, M.A., Payne, R.S., Richardson, W.J., Wells, R.S., 1984. Behavior of bowhead whales, *Balaena mysticetus*, summering in the Beaufort Sea: surfacing, respiration, and dive characteristics. Can. J. Zool. 62, 1910–1921.

Würsig, B., Dorsey, E.M., Fraker, M.A., Payne, R.S., Richardson, W.J., 1985. Behavior of bowhead whales, *Balaena mysticetus*, summering in the Beaufort Sea: a description. Fish. Bull. 83, 357–377.

Würsig, B., Guerrero, J., Silber, G., 1993. Social and sexual behavior of bowhead whales in fall in the western Arctic: a re-examination of seasonal trends. Mar. Mammal. Sci. 9, 103–110.

Würsig, B., Koski, W.R., Richardson, W.J., 1999. Whale riding behavior: assisted transport for bowhead whale calves during spring migration in the Alaskan Beaufort Sea. Mar. Mammal. Sci. 15, 204–210.

Würsig, B., Koski, W.R., Thomas, T.A., Richardson, W.J., 2002. Chapter 12, Activities and behavior of bowhead whales in the Eastern Alaskan Beaufort Sea during late summer and autumn. In: Richardson, W.J., Thomson, D.H. (Eds.), Bowhead Whale Feeding in the Eastern Alaskan Beaufort Sea: Update of Scientific and Traditional Information. pp. 12.1–12.38. OCS Study MMS 2002-012. LGL Rep. TA2196-7. Rep. from LGL Ltd., King City, Ontario, for US Minerals Management Service, Anchorage, Alaska and Herndon Virginia, USA. xliv + 420pp. [Available from National Technical Information Service, Springfield, Virginia, USA, Rep. No. NTIS PB2004-101568].

Zeh, J.E., Clark, C.W., George, J.C., Withrow, D., Carroll, G.M., Koski, W.R., 1993. Chapter 11, Current population size and dynamics. In: Burns, J.J., Montague, J.J., Cowles, C.J. (Eds.), The Bowhead Whale. Allen Press, Lawrence, KS, pp. 409–489, Spec. Publ. 2, Society for Marine Mammalogy.

CHAPTER

24

Ecological variation in the western Beaufort Sea

M.C. Ferguson[1,2], *J.T. Clarke*[1,3], *A.A. Brower*[1,3], *A.L. Willoughby*[1,3] *and S.R. Okkonen*[4]

[1]Alaska Fisheries Science Center, National Marine Fisheries Service, National Oceanic and Atmospheric Administration, Seattle, WA, United States [2]School of Aquatic and Fishery Sciences, University of Washington, Seattle, WA, United States [3]Joint Institute for the Study of the Atmosphere and Ocean, University of Washington, Seattle, WA, United States [4]Institute of Marine Science, University of Alaska Fairbanks, Fairbanks, AK, United States

Introduction

Variability in bowhead whale (*Balaena mysticetus*) feeding opportunities in the western Beaufort Sea (west of 140°W, south of 72°N) is one of the main drivers of spatiotemporal variation in their distribution and density over periods of days, weeks, months, years, and decades. Lowry (1993) recognized the influence of the physical environment on bowhead whale distribution: "Bowheads often occur in oceanographically complex areas where arctic and subarctic marine water masses interact with terrestrial runoff." Detailed information on the movement and behavior of individual bowhead whales from telemetry studies support the importance of food availability on bowhead whale distribution (Chapter 4).

The long-held model of a "year in the life" of a Bering–Chukchi–Beaufort (BCB) Seas bowhead whale included an annual journey from wintering grounds in the Bering Sea, across migration routes in the northeastern Chukchi Sea and the western Beaufort Sea during spring (Fig. 24.1), ultimately reaching foraging grounds in the eastern Beaufort Sea. There, the majority of the population resided during summer and early autumn (Moore and Reeves, 1993; Chapter 4). The return migration westward and southward included feeding along Chukotka (Bogoslovskaya et al., 1982; Miller et al., 1986; Moore and Reeves, 1993) before the whales returned to the Bering Sea (Moore and Reeves, 1993).

By the early 2000s, extensive analyses of stomach contents, in conjunction with traditional ecological knowledge and observations from aerial surveys, confirmed that bowhead whales

The Bowhead Whale
DOI: https://doi.org/10.1016/B978-0-12-818969-6.00024-8

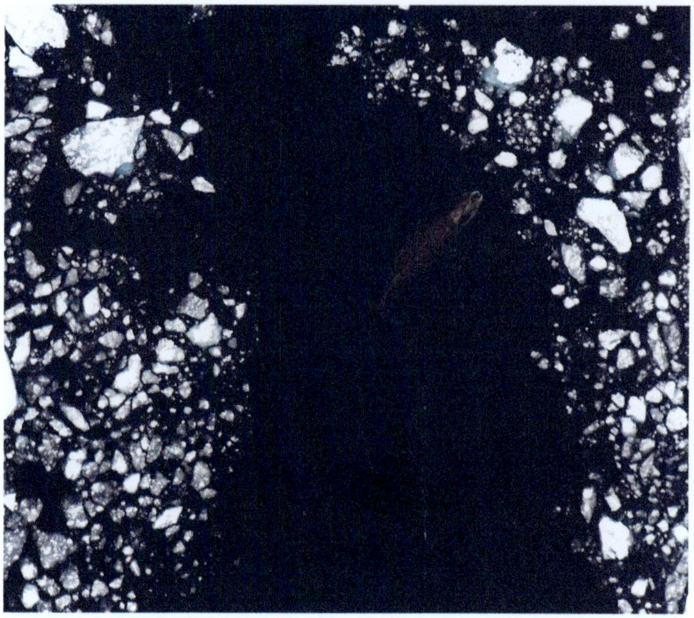

FIGURE 24.1 A bowhead whale swims through broken sea ice in the western Beaufort Sea. This photograph was taken by Amelia Brower (NMFS permit number 14245) during the Spring 2011 bowhead whale aerial photo-identification survey (Mocklin et al., 2015), funded by the North Slope Borough. The shadow of the survey plane is visible behind the whale.

regularly feed during their autumn migration across the Alaskan Beaufort Sea (Lowry et al., 2004; Chapter 4). In fact, "this entire region should be considered an integral part of the summer-autumn feeding range of bowhead whales" (Lowry et al., 2004).

We report on four decades of aerial surveys and multidisciplinary research on bowhead whales and the western Beaufort Sea ecosystem (Fig. 24.2). This has revealed considerable annual variation, in which bowhead whales are observed to feed extensively in some years and seemingly not at all in others. The locations of some feeding aggregations are consistent across years, but other locations support feeding aggregations only occasionally. While some "rules" regarding feeding behavior hold true, both local and distant forcings lead to high interannual variability. As a result, every year is somewhat different.

A trove of information about BCB bowhead whales exists, including traditional ecological knowledge of subsistence hunters (Ashjian et al., 2010; Stephen R. Braund and Associates, 2010; Chapter 34), logbooks from commercial whaling ships (Fraker and Bockstoce, 1980), and data from scientific research. Guided by the need to better understand potential impacts to this population resulting from anthropogenic activities in offshore regions in the United States and Canadian Beaufort Sea, research efforts were initiated (Montague, 1993), funded by several agencies [US Bureau of Ocean Energy Management (BOEM), US Department of Interior (DOI), US National Marine Fisheries Service] and the oil and gas industry. Bowhead whale research funded specifically by DOI (including BOEM) focused on areas of interest to offshore energy exploration, development, and production during the summer and autumn open water season. In the mid- to late 1970s, scientific studies were conducted in the eastern Beaufort Sea (Fraker and Bockstoce, 1980). In the late 1970s, offshore energy interests and, therefore, scientific study area boundaries expanded westward, including the western Beaufort Sea, northeastern Chukchi Sea, and northern Bering Sea (Moore et al., 2000).

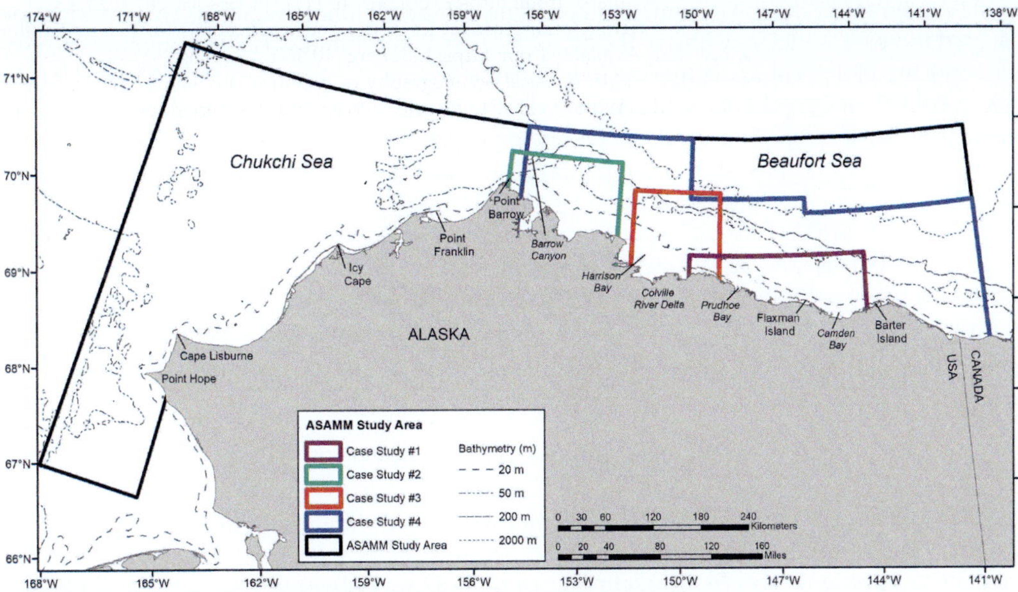

FIGURE 24.2 Aerial Surveys of Arctic Marine Mammals (ASAMM) study area, with polygons showing the locations of each of the four case studies described in the text.

The Aerial Surveys of Arctic Marine Mammals (ASAMM) project began in 1979 in the western Beaufort Sea to monitor BCB bowhead whales and continued through autumn 2019 (Clarke et al., 2019). As one of the longest uninterrupted documented time series on the spatiotemporal distribution, density, behavior, and habitat of a marine mammal population, ASAMM provides the foundation for investigating and understanding the patterns and mechanisms we discuss in this chapter.

It is not trivial to meet the energy needs of a population of over 16,000 bowhead whales (Chapter 6), with some animals reaching 100,000 kg and 19 m in body length (George et al., 1999; Chapter 7). The need to find productive foraging grounds (Lowry, 1993) and possibly refuge from predators (Chapter 29), along with maternally directed site fidelity, help explain why BCB bowhead whales undertake a 6000-km round-trip migration every year.

The BCB bowhead whale population has a diverse diet, including euphausiids, copepods, mysids, amphipods, and isopods (Lowry, 1993; Lowry et al., 2004; Moore et al., 2010; Chapter 28). Bowhead whales feed at the surface, mid-water, or benthically, using long baleen to filter prey from the water and mud (Lowry, 1993; Chapter 14). In the western Beaufort Sea, bowhead whales use all three feeding strategies (Mocklin et al., 2011; Moore et al., 2010). Bowhead whales are designed to endure one or more years when food is limited by drawing on energy reserves in their blubber (Burns, 1993; George, 2009; Lowry and Frost, 1984; Chapter 16).

Our understanding of the BCB bowhead whale has changed over the past couple decades, due to longer ecological time series, advances in technology, and changes in the arctic ecosystem (Citta et al., 2015; Chapters 4 and 27). We now have decades of observations capturing a longer seasonal period in individual years, incorporating more examples of interannual variability inherent in the ecosystem, and spanning different climate regimes. This information has helped increase understanding of the breadth of spatiotemporal variability in the arctic ecosystem and the biological and physical mechanisms forcing those patterns. ASAMM data are uniquely suited for investigating bowhead whale ecology across local (kilometers to tens of kilometers) to basin-wide (hundreds of kilometers) spatial scales and daily, seasonal, interannual, and longer temporal scales, resolving both ephemeral and general patterns (Clarke et al., 2018).

Our objectives are threefold. First, examine BCB bowhead whale relative density, distribution, and feeding, and associated spatiotemporal variability, in the western Beaufort Sea. Second, discuss biological and physical mechanisms that contribute to these phenomena. Third, compare historical (before 2000) and recent (after 2000) findings on bowhead whale ecology in the western Beaufort Sea.

Aerial Surveys field methods

ASAMM observers collected visual line-transect survey data from Grumman Goose, De Havilland Twin Otter, and Turbo Commander aircraft flying over the western Beaufort Sea from 1979 to 2019 and eastern Chukchi Sea from 1982 to 1991 and 2008 to 2019 (Fig. 24.2; Clarke et al., 2019). Although survey area boundaries and timing varied across years, the western Beaufort Sea was consistently surveyed during autumn (September to mid-October) of every year starting in 1982. Annual summer (July–August) surveys in the western Beaufort Sea began in 2012. Since 2009, ASAMM surveys in the eastern Chukchi Sea began in the first week of July and were flown until the last week of October. The exceptions to this uninterrupted coverage occurred in 2013, when a partial shutdown of the US government resulted in a gap in survey coverage from 1 to 19 October, and in August 2019, when the study area boundaries shifted eastward to focus on the Beaufort Sea and Amundsen Gulf to estimate the abundance of BCB bowhead whales. Survey methods remained largely consistent throughout the time series. Clarke et al. (2019) provide detailed survey methods.

Bowhead whale seasonal distribution in the western Beaufort Sea

Bowhead whale distribution in the western Beaufort Sea from 2000 to 2018 exhibited a shoreward and westward progression from July to October, evident in spatially explicit relative density models (Clarke et al., 2019). This seasonal variability in distribution is likely linked to spatial and temporal variability in the availability of prey. The outer continental shelf waters are influenced by the Beaufort shelfbreak jet, a northern branch of current pathways that advects nutrients and prey from the Bering Sea to the Beaufort Sea (Pickart, 2004; Chapter 25). The large copepod species *Calanus hyperboreus* and *Calanus glacialis* are found in high densities in outer shelf waters of the eastern Alaskan Beaufort Sea during some years (Griffiths and Thomson, 2002). These species undertake seasonal vertical migrations, ascending to shallower waters during spring and descending during summer or autumn (Darnis and Fortier, 2014; Griffiths and Thomson, 2002). In the extreme western Beaufort Sea, euphausiids are entrained nearshore and are often the dominant prey of bowhead whales harvested at Utqiaġvik in autumn (Ashjian et al., 2010; Chapter 26).

From 2000 to 2018, the majority of bowhead whale sightings during July were located in the eastern half of the study area (Fig. 24.3A). The highest relative densities in July were over the outer continental shelf (50–200 m depth), ~45–90 km offshore, from the eastern boundary of the study area to Barter Island (~140°W to ~142.5°W) (Fig. 24.3A). The shoreward transition in bowhead whale distribution began in August, when whales moved onto the inner continental shelf (0–50 m depth), where local winds create upwelling and establish fronts that aggregate prey (Druckenmiller et al., 2018; Pickart et al., 2013) in some years. During August, bowhead whales were closest to shore north of Barter Island (~142°W to ~144°W) (Fig. 24.3B). The highest predicted relative densities in August were found in three areas: an area centered on Barter Island, 90 km long and extending up to 30 km offshore; north of the Colville River Delta, 15–60 km offshore; and north of Dease Inlet, from the barrier islands to 30 km offshore (Fig. 24.3B). In September, bowhead whale relative density was highest, and whales were closest to shore, just outside the barrier islands from Camden Bay to Prudhoe Bay (~144°W to ~149°W) and from Smith Bay to Dease Inlet (Fig. 24.3C). In October, the highest predicted relative densities ranged from ~152.5°W to north of Dease Inlet (Fig. 24.3D). Relatively high densities were found outside the barrier islands north of Prudhoe Bay (~146°W to ~148.5°W) and nearshore northwest of Cape Halkett, extending to the mouth of Barrow Canyon. Bowhead whale distribution was farther offshore between Cape Halkett and Point Barrow in October than in September.

The location of high-density areas and the seasonality of the bowhead whale distribution in the western Beaufort Sea differ in the recent period compared to the historical period. Bowhead whale distribution was relatively closer to shore from Harrison Bay to Point Barrow during the historical period and is presently relatively closer to shore from Barter Island to Oliktok Point (Fig. 24.4A and B). Contrary to the recent period, bowhead whales in the historical period were infrequently seen on the inner continental shelf between Camden Bay and Oliktok Point in autumn (Fig. 24.5A and B; Moore and Reeves, 1993). Furthermore, although high-density areas existed in shallow continental shelf waters in the western portion of the study area during September in both the historical and recent periods, the spatial extent of the high-density area in the historical period (Fig. 24.5A) appears to be confined toward the west compared to the recent period (Fig. 24.3C).

Past interpretations of some bowhead whale high-density areas during the historical period were driven largely by the number of groups sighted, without taking survey effort or group size into consideration. For example, spatially stratified bowhead whale sighting rates (not weighted by group size) in the western Beaufort Sea for autumn 1979–89, all years pooled, were highest in the westernmost stratum located between Smith Bay and Point Barrow and second highest in the easternmost stratum between Herschel and Barter Islands (Moore and Reeves, 1993). When both effort and group size are considered, whale densities in the eastern half of the study area were considerably lower than in the western half during the historical period (Fig. 24.5A and B).

Changes in the seasonality of bowhead whale distribution in the western Beaufort Sea during recent years are manifested in two ways (Clarke et al., 2018). First, during the historical period, the transition from the continental slope and outer continental shelf habitat to inner continental shelf habitat was not observed until September. In most recent years, this shoreward shift occurred during August. Second, during the historical period, bowhead whales were not observed in the western Beaufort Sea in July (although this might partially be explained by limited survey effort), and observations in August were

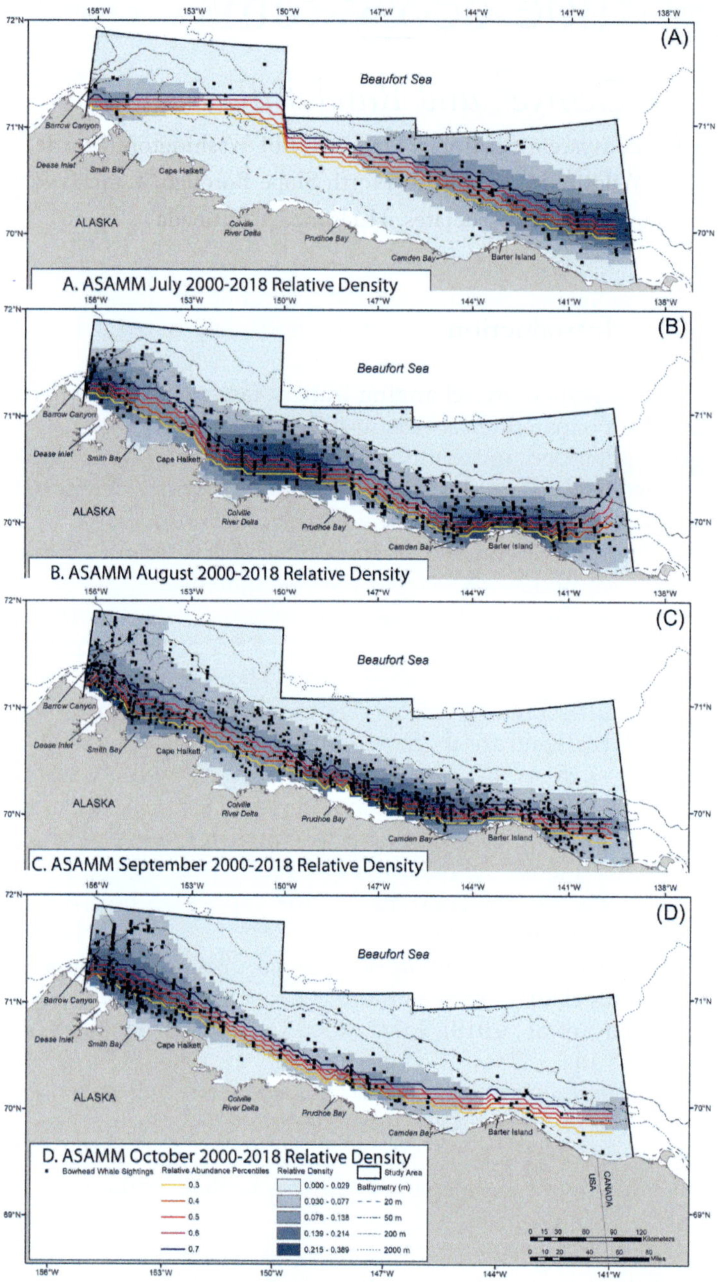

FIGURE 24.3 Bowhead whale sightings, relative density predictions, and relative abundance percentiles from spatially explicit relative density models for July (A), August (B), September (C), and October (D), 2000–2018. The relative abundance percentiles denote the offshore boundaries within which the corresponding proportions of the predicted number of bowhead whales are distributed. Note the shift of high whale densities toward the West from July to October.

limited to the eastern portion of the study area (Moore et al., 2000). During the recent period, there have been numerous bowhead whale sightings in the western Beaufort Sea during July (Fig. 24.3A), and their distribution extends across the western Beaufort Sea during August (Fig. 24.3B; Clarke et al., 2018).

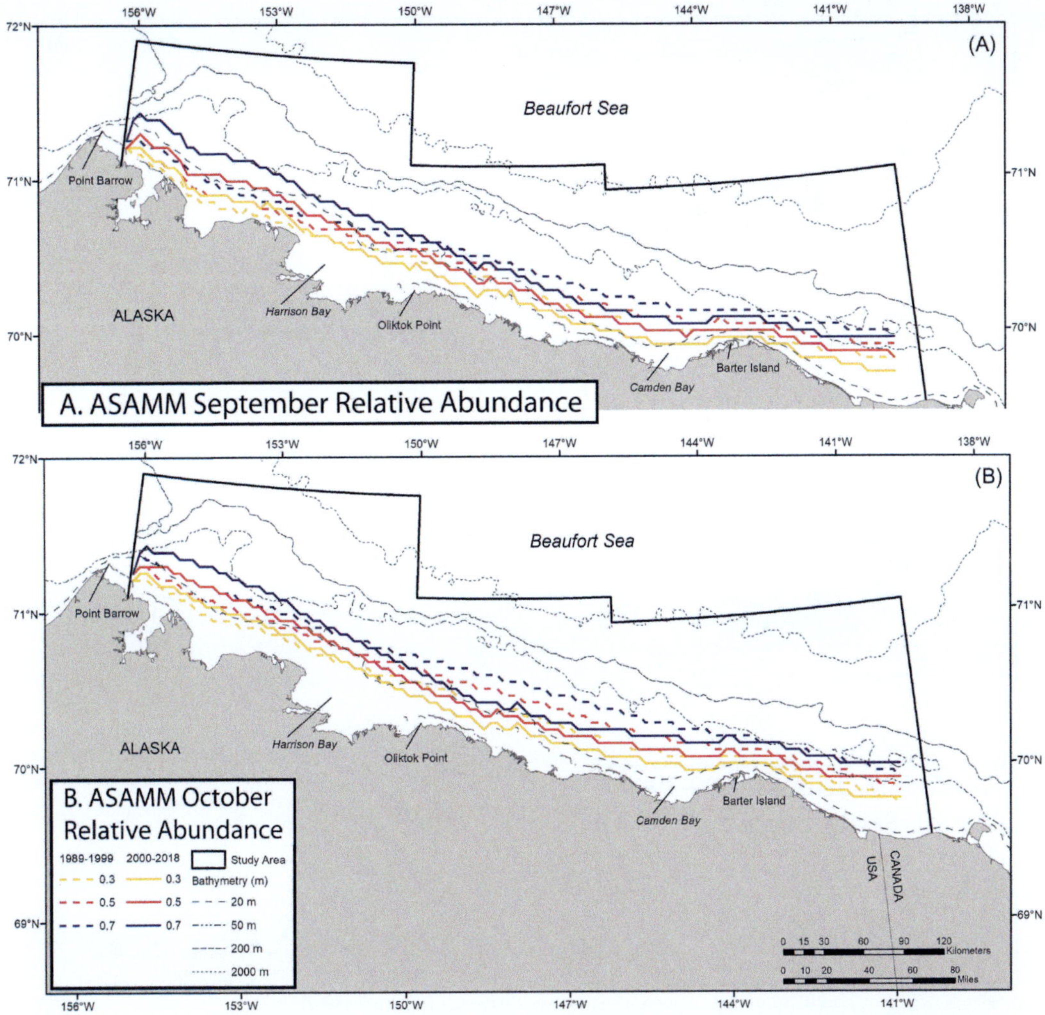

FIGURE 24.4 Bowhead whale relative abundance percentiles during the historical (1989–99) and recent (2000–18) periods for September (A) and October (B). These percentiles were derived from spatially explicit models of bowhead whale relative density.

Mechanisms driving interannual variability in whale distribution, density, and seasonality

Considerable interannual variability is evident in bowhead whale distribution and density in the western Beaufort Sea (Clarke et al., 2018; Moore, 2000; Moore and Reeves, 1993; Okkonen et al., 2018). The following four case studies exemplify this variability and the biological and physical interactions driving them. We studied the extreme changes in distribution, density, and seasonality and believe that feeding opportunities play a central role.

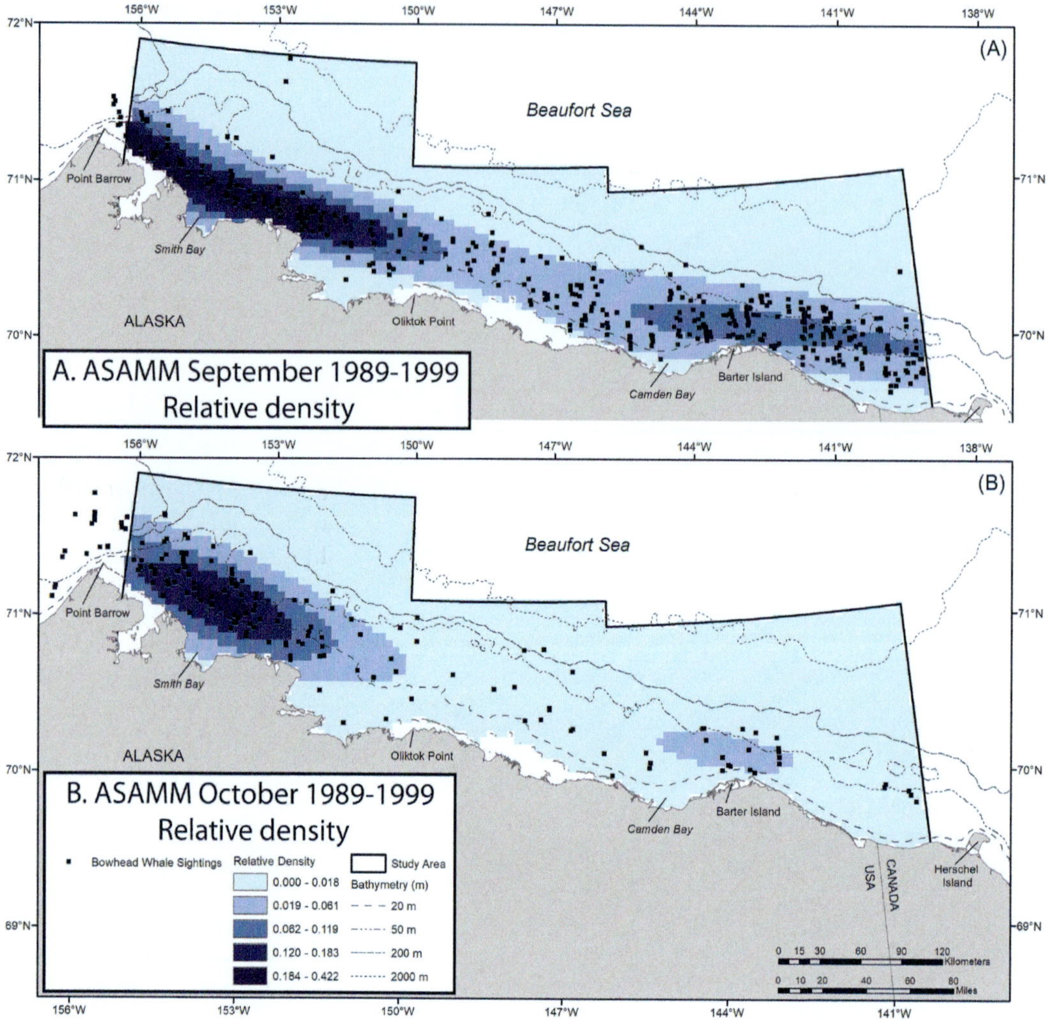

FIGURE 24.5 Bowhead whale relative density predictions in the western Beaufort Sea during September (A) and October (B), 1989–99. These predictions were derived from a spatially explicit model of bowhead whale relative density.

Case study I: Nearshore feeding aggregations

The first case study examines high densities of bowhead whales found feeding unusually nearshore (within 10 km) between Prudhoe Bay and Camden Bay (144°W–150°W; Fig. 24.2), during 1997 and 2014 (Clarke et al., 2015; Okkonen et al., 2018). These nearshore waters of the central Alaskan Beaufort Sea are not typically high-density or foraging areas for bowhead whales during summer or autumn (Clarke et al., 2015). In 1997, bowhead whales were observed feeding in this region from 3 September to 18 October; feeding/milling bowhead whales comprised 57% of all bowhead whales observed in the region (Treacy, 1998). In 2014, bowhead whales were observed feeding in this region from 17 August to 3 October; feeding/milling bowhead whales comprised 71% of all bowhead whales observed in the region (Clarke et al., 2015). These atypical feeding aggregations coincided with periods of high river discharge following upwelling events (Okkonen et al., 2018). Prior upwelling conditions are "necessary precursors" to enhance foraging opportunities for bowhead whales, as average bowhead whale group sizes are larger following upwelling events, regardless of river discharge (Okkonen et al., 2018). High river discharge establishes and maintains a nearshore front, aggregating prey and supporting the highest densities of feeding bowhead whales in these areas (Okkonen et al., 2018).

Case study II: Krill trap

The focal point of the second case study encompasses a wedge of the Beaufort Sea shelf between Barrow Canyon and the barrier islands, now known as the "krill trap" area (Ashjian et al., 2010; Chapter 26; Fig. 24.2). In some years, this area is alive with dense aggregations of feeding bowhead whales for periods that can last up to 10 days, with fewer whales in the area at other times; lower density feeding aggregations also can be intermittently present for much of autumn (Ashjian et al., 2010; Clarke et al., 2015; Ferguson et al., 2016; Moore et al., 2010; Okkonen et al., 2020). Here, the right combination of winds, currents, and bathymetry drives euphausiids (krill) and copepods from deeper waters up onto the shelf, where they then become hydrographically trapped by the Alaskan Coastal Current. These euphausiids are likely transported here from the Bering Sea (Berline et al., 2008, and references therein) and advected northward through the Chukchi Sea by northbound currents. When the krill trap is set and full of prey, bowhead whales tend to be found in large groups, primarily feeding (Ferguson et al., 2016; Okkonen et al., 2011). Once prey are advected onto the shelf, they may remain there for long periods in some years. A comparison of bowhead whale group sizes on days in 2015 when the krill trap was active with data from surveys conducted 5 days prior showed that bowhead whale group sizes were larger when the krill trap was active, implying that the bowhead whale aggregation response was rapid (Ferguson et al., 2016). ASAMM has observed bowhead whale feeding aggregations during active krill trap conditions from late July to late October.

When the krill trap is not active, fewer bowhead whales are seen in the area, and those that are sighted are typically swimming or resting (Ferguson et al., 2016). If euphausiids fail to reach the area in appreciable numbers, the trap can be set but may effectively be devoid of prey and escape the attention of bowhead whales. For example, during autumn 2018, ASAMM conducted surveys when physical conditions were conducive to krill trap formation. However, a relatively large aggregation, 30 feeding bowhead whales, was observed on only one day (2 September), and several autumn sightings were >50 km offshore (Clarke et al., 2019). The year 2018 was atypical for the recent period because sea ice cover was extensive in the western Beaufort Sea until early September (Clarke et al., 2019).

Due to the lack of real-time oceanographic sampling, it is unknown how many euphausiids were advected north into the krill trap area in 2018; however, the lack of bowhead whales in the krill trap area and the high bowhead whale sighting rate offshore suggests that relatively few euphausiids were available, limiting bowhead whale feeding opportunities in the area.

Case study III: Summer feeding aggregations in Harrison Bay

The third case study examines the effects of biophysical forcing in the eastern Beaufort Sea on the timing of the westbound bowhead whale migration and, therefore, the whales' arrival in the western Beaufort Sea.

Throughout nearly four decades of aerial surveys over the western Beaufort Sea, bowhead whales were rarely encountered in the waters of Harrison Bay relative to other parts of the study area. On August 26, 2016, ASAMM observers sighted 498 bowhead whales—a record number of bowhead whales sighted on any single day, in the entire history of the project—due primarily to a large, dense aggregation of feeding whales in the vicinity of Harrison Bay (Clarke et al., 2017; Fig. 24.2).

During July, August, and September of a typical year, easterly winds over ice-free waters promote upwelling off Cape Bathurst in the eastern Beaufort Sea, bringing copepods onto the shelf (Walkusz et al., 2012; Williams and Carmack, 2008). Copepods drift westward toward feeding bowhead whales off the Tuktoyaktuk Peninsula, a core-use area based on satellite-tagged bowhead whales (Citta et al., 2015). Copepods are available to bowhead whales on the Tuktoyaktuk shelf only until late summer or early autumn, when the copepods undertake ontogenetic migrations to deeper waters of the slope and basin. The onset of the westward autumn migration for bowhead whales typically corresponds to the timing of reduced copepod availability on the Tuktoyaktuk shelf, generally in mid- to late September (Citta et al., 2015; Fraker et al., 1978; Richardson et al., 1987).

During summer 2016, upwelling-favorable winds were lacking in the eastern Beaufort Sea (Clarke et al., 2018). This likely affected the availability of copepods to bowhead whales in the eastern Beaufort Sea and prompted an early westward migration. Simultaneously, in late August in the western Beaufort Sea, a frontal system formed in Harrison Bay, likely resulting from record high freshwater discharge from the Colville River (USGS National Water Information System, 2016), in combination with upwelling-favorable winds (NOAA National Data Buoy Center, 2016; Weather Underground, 2016; Clarke et al., 2018). These physical forcings presumably aggregated bowhead whale prey, leading to record-breaking bowhead whale sightings during ASAMM surveys from 24 to 27 August. Most of the whales observed during these surveys were found in high densities north of Harrison Bay, with additional aggregations northeast of Point Barrow (Clarke et al., 2017, 2018). The abnormally large aggregations also were observed east of Point Barrow and in Harrison Bay by local hunters in boats (F. Brower and M. Donovan, pers. comm., 2020). During this period in late August 2016, the number of whales encountered per kilometer of transect effort was twice as high as any encounter rate for August and 1.5 times greater than any encounter rate for September or October in the entire ASAMM time series (1979–2019). Of the 498 whales sighted during the 26 August survey, 87% were feeding—extraordinary for August in this region.

Case study IV: Bowhead whales and sea ice

The fourth case study describes the complicated association between sea ice and bowhead whale distribution and density in the western Beaufort Sea (Fig. 24.2). Moore (2000) analyzed ASAMM data from 1982 to 1991 and found that bowhead whales in the western Beaufort Sea were farther offshore, in slope habitat, during years with heavy sea ice cover, compared to years with light or moderate sea ice cover. However, Moore et al. (2000) noted that bowhead whale associations with sea ice are not constant and inferred, "bowhead prey availability may be influenced more by bathymetry, and its concomitant hydrography, than by ice."

Sea ice characteristics in the western Beaufort Sea have changed dramatically since the beginning of the satellite record in 1979 (Druckenmiller et al., 2018). From 1979 to 2014, the western Beaufort Sea experienced the most dramatic decrease in sea ice cover, compared to the entire range of the BCB bowhead whale. The number of days during the bowhead whale peak use period when at least 85% of the sea surface was ice free has increased by 13 days/decade near Point Barrow and 20 and 25 days/decade on the western Beaufort Sea shelf and slope, respectively (Druckenmiller et al., 2018). Furthermore, October historically has exhibited the greatest interannual variability in sea ice cover in the western Beaufort Sea, although that is changing as the number of open-water days increases (Druckenmiller et al., 2018). Similar to Moore et al. (2000), Druckenmiller et al. (2018) found that the distance from shore of bowhead whales in the western Beaufort Sea decreased as the fraction of open water increased. Druckenmiller et al. (2018) hypothesized that this association may result from "increased foraging opportunities closer to shore due to increased upwelling along the shelf break when ice cover is farther off." Bowhead whales might also remain farther offshore in years with heavy sea ice to avoid being trapped against the coastline.

The autumn 2019 migration across the western Beaufort Sea was unexpectedly far offshore, and relatively few bowhead whales were seen (Chapter 27). The median offshore distance of bowhead whale sightings was 46 km (ranging from 3 to 92 km from the coast), centered on the 46-m isobath, seemingly contrary to the findings of both Moore (2000) and Druckenmiller et al. (2018). In the western Beaufort Sea, northern Chukchi Sea, and Arctic Ocean, May to October 2019 experienced record-breaking sea surface temperature maxima, based on satellite sea surface temperature data (Thoman, 2019). Sea ice loss in the western Beaufort Sea in 2019 broke records in the early part of the year (through July), although it did not match the absolute minimum extent later in the season that was reached in 2012 [National Snow and Ice Data Center (NSIDC), 2019]. ASAMM's summer 2019 bowhead whale sighting distributions and densities largely aligned with previous years'. However, autumn 2019 was the first time during the recent period that the bowhead whale distribution in the western Beaufort Sea overlaid those from the historical heavy ice years: 1983, 1988, 1991, and 1992 (Chapter 27). Anecdotal evidence indicates that some bowhead whales departed the eastern Beaufort Sea for their westward migration as expected in 2019 but took a northerly route, outside boundaries of the ASAMM study area, arriving off the Chukotka Peninsula by early- to mid-October (E. Zdor, pers. comm., 2019). Furthermore, some whales might have delayed their departure from the eastern Beaufort Sea. During an ASAMM flight conducted on October 29, 2019, between Camden Bay and

Oliktok Point, the team sighted 30 bowhead whales, including 8 calves, on the inner shelf, which is more typical of sightings earlier in autumn in recent years.

Were the same mechanisms operating during the heavy ice years of the historical period and the anomalously warm autumn of 2019? Was the uncharacteristic 2019 autumn migration a reflection of enhanced foraging opportunities in the eastern Beaufort Sea or reduced foraging opportunities in the western Beaufort Sea? Did the anomalously warm waters over the continental shelf in the western Beaufort Sea exceed the whales' physiological tolerance (e.g., Chambault et al., 2018), pushing the whales farther north into cooler waters? The bowhead whale migration during autumn 2019 remains a mystery. Variability is a defining characteristic of the Arctic, but the parameters appear to be changing.

Conclusions

Our understanding of the mechanisms driving interannual variability in BCB bowhead whale distribution, density, and behavior in the western Beaufort Sea has increased over the past two decades due to continued monitoring and multidisciplinary collaborations. This additional understanding makes the system slightly more predictable. We highlight the principal discoveries next.

First, the western Beaufort Sea is not simply a westward migratory corridor for bowhead whales as it was often perceived to be during the historical period. It is also a feeding area, although bowhead whale feeding in this region is spatiotemporally dynamic and ephemeral (Lowry et al., 2004).

Second, forces affecting bowhead whale foraging in the western Beaufort Sea are both remote and local. Remote forces act temporally and spatially distant from foraging in the western Beaufort Sea: the speed and direction of summer winds over the Chukchi Sea influence the circulation pathways by which euphausiids are carried from the Bering Sea to the western Beaufort Sea and, therefore, the subsequent arrival times of euphausiids at Point Barrow (Berline et al., 2008). Similarly, summer wind conditions in the eastern Beaufort Sea likely influence the availability of zooplankton prey in Canadian waters and the subsequent timing of the bowhead whale westward migration to Point Barrow (Clarke et al., 2017). Local forces act temporally and spatially proximate to foraging: winds, currents, fronts, upwelling, and river discharges promote local aggregation of zooplankton for efficient foraging by bowhead whales.

Third, relative to the 1980s and the early 1990s, bowhead whale seasonality in the western Beaufort Sea has changed: the old August is the new July, as events tend to occur earlier now. Furthermore, the number and location of bowhead whale high-density areas have changed over time. The earlier shoreward shift and extensive longitudinal distribution of bowhead whales in the region during the recent period suggest that some bowhead whales may now return to the western Beaufort Sea earlier or remain in the region throughout the summer. This shift could be related to enhanced foraging opportunities due to ecosystem changes associated with the increased number of open-water days, or it could be the result of the expansion to previously marginal areas due to the larger BCB bowhead whale population. Of course, in a dynamic marine system, the ecosystem will continue to change.

Fourth, the relationship between bowhead whale distribution in the western Beaufort Sea and sea ice cover during summer and autumn is not simple. This conclusion would have been very different had the ASAMM time series ended in 2018. The most extreme examples of autumn bowhead whale migrations far offshore in the region include years in the 1980s and the early 1990s when sea ice cover was heaviest and, unexpectedly, in 2019 when sea ice cover reached record-breaking minima during spring and summer.

Anthropogenic influences, such as seismic testing and offshore drilling activity, may also influence spatiotemporal patterns of bowhead whales (Chapters 23, and 35). The effects of anthropogenic and environmental factors on bowhead whales may be confounded, especially lacking sufficient understanding of the underlying environmental variability inherent in the ecosystem.

Adaptation to environmental variability in the Arctic has shaped the life history and anatomical features of bowhead whales: extreme longevity (Keane et al., 2015; George et al., 1999), thick blubber, slow reproduction, and a limited migratory pattern whereby they winter within the sea ice in arctic waters (Chapter 7). While all arctic ecosystems are highly variable, bowhead whales have relatively predictable migration patterns that are important to the communities who depend on them for food security and cultural and spiritual enrichment. Improving our understanding and capacity to predict arctic variability is fundamentally important to sound natural resource management. Over four decades of ASAMM surveys have taught us that the environment is always changing, sometimes in unexpected ways, making a strong argument for continued monitoring.

Acknowledgments

The ASAMM project was funded by the Bureau of Ocean Energy Management (BOEM) Alaska OCS Region and its predecessors. Data collected over the four-decade time span benefitted from the valuable contributions of biologists, pilots, mechanics, and support of personnel and canines too numerous to mention but to whom we are extremely grateful. We particularly would like to thank Sue Moore for her always insightful feedback, and Craig George and Hans Thewissen for their patience and assistance.

References

Ashjian, C.J., Braund, S.R., Campbell, R.G., George, J.C., Kruse, J., Maslowski, W., et al., 2010. Climate variability, oceanography, bowhead whale distribution, and Iñupiat subsistence whaling near Barrow, Alaska. Arctic 63 (2), 179–194.

Berline, L., Spitz, Y.H., Ashjian, C.J., Campbell, R.G., Maslowski, W., Moore, S.E., 2008. Euphausiid transport in the western Arctic Ocean. Mar. Ecol. Prog. Ser. 360, 163–178.

Bogoslovskaya, L.S., Votrogov, L.M., Krupnik, I.I., 1982. The bowhead whale off Chukotka: migrations and aboriginal whaling. Rep. Int. Whal. Comm. 32, 391–399.

Burns, J.J., 1993. Epilogue. In: Burns, J.J., Montague, J.J., Cowles, C.J. (Eds.), The Bowhead Whale. Spec. Publ. 2, Society for Marine Mammalogy, pp. 745–764.

Chambault, P., Albertson, C.M., Patterson, T.A., Hansen, R.G., Tervo, O., Laidre, K.L., et al., 2018. Sea surface temperature predicts the movements of an Arctic cetacean: the bowhead whale. Sci. Rep. 8, 9658. Available from: https://doi.org/10.1038/s41598-018-27966-1.

Citta, J.J., Quakenbush, L.T., Okkonen, S.R., Drunkenmiller, M.L., Maslowski, W., Clement-Kinney, J., 2015. Ecological characteristics of core-use areas used by Bering–Chukchi–Beaufort (BCB) bowhead whales, 2006–2012. Prog. Oceanogr. 136, 201–222. Available from: https://doi.org/10.1016/j.pocean.2014.08.012.

Clarke, J.T., Brower, A.A., Ferguson, M.C., Kennedy, A.S., Willoughby, A.L., 2015. Distribution and relative abundance of marine mammals in the eastern Chukchi and western Beaufort Seas, 2014. In: Annual Report, OCS Study BOEM 2015-040. National Marine Mammal Laboratory, Alaska Fisheries Science Center, NMFS, NOAA, Seattle, WA.

Clarke, J.T., Brower, A.A., Ferguson, M.C., Willoughby, A.L., 2017. Distribution and Relative abundance of marine mammals in the eastern Chukchi and western Beaufort Seas, 2016. In: Annual Report, OCS Study BOEM 2017-078. Marine Mammal Laboratory, Alaska Fisheries Science Center, NMFS, NOAA, Seattle, WA.

Clarke, J.T., Ferguson, M.C., Willoughby, A.L., Brower, A.A., 2018. Bowhead whale and beluga distributions, sighting rates, and habitat associations in the western Beaufort Sea, summer and fall 2009–2016, with a retrospective comparison to 1982–1991. Arctic 71 (2), 115–138. Available from: https://doi.org/10.14430/arctic4713.

Clarke, J.T., Brower, A.A., Ferguson, M.C., Willoughby, A.L., 2019. Distribution and relative abundance of marine mammals in the eastern Chukchi and western Beaufort Seas, 2018. In: Annual Report, OCS Study BOEM 2019-021. Marine Mammal Laboratory, Alaska Fisheries Science Center, NMFS, NOAA, Seattle, WA.

Darnis, G., Fortier, L., 2014. Temperature, food and the seasonal vertical migration of key arctic copepods in the thermally stratified Amundsen Gulf (Beaufort Sea, Arctic Ocean). J. Plankton Res. 36 (4), 1092–1108. Available from: https://doi.org/10.1093/plankt/fbu035.

Druckenmiller, M.L., Citta, J.J., Ferguson, M.C., Clarke, J.T., George, J.C., Quakenbush, L., 2018. Trends in sea-ice cover within bowhead whale habitats in the Pacific Arctic. Deep Sea Res., II 152, 95–107. Available from: https://doi.org/10.1016/j.dsr2.2017.10.017.

Ferguson, M., Okkonon, S., Clarke, J., Willoughby, A., Brower, A., Ashjian, C., et al., 2016. Observation of bowhead whale foraging near Barrow, Alaska, in 2015 support the krill trap model. In: Poster Presented at the Alaska Marine Science Symposium. January 2016, Anchorage, AK.

Fraker, M.A., Bockstoce, J.R., 1980. Summer distribution of bowhead whales in the eastern Beaufort Sea. Mar. Fish. Rev. 42 (9–10), 57–64.

Fraker, M.A., Sergeant, D.E., Hoek, W., 1978. Bowhead and white whales in the southern Beaufort Sea. In: Beaufort Sea Project Technical Report 4. Department of Fisheries and the Environment, Sidney, BC, 114 p.

George, J.C., 2009. Growth, Morphology and Energetics of Bowhead Whales (*Balaena mysticetus*) (Ph.D. thesis). University of Alaska Fairbanks. <http://hdl.handle.net/11122/9031> (accessed 02.12.19.).

George, J.C., Bada, J., Zeh, J., Scott, L., Brown, S.E., O'Hara, T., et al., 1999. Age and growth estimates of bowhead whales (*Balaena mysticetus*) via aspartic acid racemization. Can. J. Zool. 77 (4), 571–580. Available from: https://doi.org/10.1139/z99-015.

Griffiths, W.B., Thomson, D.H., 2002. Chapter 5—Species composition, biomass, and local distribution of zooplankton relative to water masses in the eastern Alaskan Beaufort Sea. In: Richardson, W.J., Thomson, D.H (Eds.), Bowhead Whale Feeding in the Eastern Alaskan Beaufort Sea: Update of Scientific and Traditional Information, vol. 1. OCS Study MMS 2002-012; LGL Rep. TA2196-7. Report from LGL Ltd., King City, ON, for U.S. Minerals Management Service, Anchorage, AK, and Herndon, VA, pp. 5-1–5-42.

Keane, M., Semeiks, J., Webb, A.E., Bickham, J.W., Thomsen, B., de Magalhães, J.P., 2015. Insights into the evolution of longevity from the bowhead whale genome. Cell Rep 10, 112–122. Available from: https://doi.org/10.1016/j.celrep.2014.12.008.

Lowry, L.F., 1993. Foods and feeding ecology. In: Burns, J.J., Montague, J.J., Cowles, C.J. (Eds.), The Bowhead Whale. Spec. Publ. 2, Society for Marine Mammalogy, pp. 201–238.

Lowry, L.F., Frost, K.J., 1984. Foods and feeding of bowhead whales in western and northern Alaska. Sci. Rep. Whales Res. Inst. 35, 1–16.

Lowry, L.F., Sheffield, G., George, J.C., 2004. Bowhead whale feeding in the Alaskan Beaufort Sea, based on stomach contents analyses. J. Cetacean Res. Manage. 6 (3), 215–223.

Miller, R.V., Rugh, D.J., Johnson, J.H., 1986. The distribution of bowhead whales, *Balaena mysticetus*, in the Chukchi Sea. Mar. Mammal Sci. 2 (3), 214–222.

Mocklin, J.A., Rugh, D.J., Moore, S.E., Angliss, R.P., 2011. Using aerial photography to investigate evidence of feeding by bowhead whales. Mar. Mammal Sci. 23 (3), 602–619. Available from: https://doi.org/10.1111/j.1748-7692.2011.00518.x.

Mocklin, J., Vate Brattström, L., Tudor, B., George, J.C., Givens, G.H., 2015. Update on the 2011 Bowhead Whale Aerial Abundance Spring Survey (BAASS) Photoanalysis. Paper SC-66a-BRG03 presented to the International Whaling Committee Scientific Committee, July 2015 (unpublished). Available from: https://www.iwc.int.

Montague, J.J., 1993. Introduction. In: Burns, J.J., Montague, J.J., Cowles, C.J. (Eds.), The Bowhead Whale. Spec. Publ. 2, Society for Marine Mammalogy, pp. 1–21.

Moore, S.E., 2000. Variability of cetacean distribution and habitat selection in the Alaskan Arctic, autumn 1982-91. Arctic 53 (4), 448–460.

Moore, S.E., Reeves, R.R., 1993. Distribution and movement. In: Burns, J.J., Montague, J.J., Cowles, C.J. (Eds.), The Bowhead Whale. Spec. Publ. 2, Society for Marine Mammalogy, pp. 313–386.

Moore, S.E., DeMaster, D.P., Dayton, P.K., 2000. Cetacean habitat selection in the Alaskan Arctic during summer and autumn. Arctic 53 (4), 432–447.

Moore, S.E., George, J.C., Sheffield, G., Bacon, J., Ashjian, C.J., 2010. Bowhead whale distribution and feeding near Barrow, Alaska, in late summer 2005–06. Arctic 63 (2), 195–205.

National Snow and Ice Data Center (NSIDC), 2019. <http://nsidc.org/arcticseaicenews/sea-ice-comparison-tool/> (accessed 05.12.19.).

NOAA National Data Buoy Center, 2016. <https://www.ndbc.noaa.gov/> (accessed 02.12.19.).

Okkonen, S.R., Ashjian, C.J., Campbell, R.G., Clarke, J.T., Moore, S.E., Taylor, K.D., 2011. Satellite observations of circulation features associated with a bowhead whale feeding 'hotspot' near Barrow, Alaska. Remote Sens. Environ. 115 (8), 2168–2174. Available from: https://doi.org/10.1016/j.rse.2011.04.024.

Okkonen, S.R., Clarke, J.T., Potter, R.A., 2018. Relationships among high river discharges, upwelling events, and bowhead whale (*Balaena mysticetus*) occurrence in the central Alaskan Beaufort Sea. Deep Sea Res., II 152, 195–202. Available from: https://doi.org/10.1016/j.dsr2.2016.11.015.

Okkonen, S.R., Ashjian, C., Campbell, R.G., Alatalo, P., 2020. Krill diel vertical migration: a diagnostic for variability of wind forcing over the Beaufort and Chukchi Seas. Prog. Oceanogr. 181, 102265. Available from: https://doi.org/10.1016/j.pocean.2020.102265.

Pickart, R.S., 2004. Shelfbreak circulation in the Alaskan Beaufort Sea: Mean structure and variability. J. Geophys. Res. 109 (C4). Available from: https://doi.org/10.1029/2003JC001912.

Pickart, R.S., Schulze, L.M., Moore, G.W.K., Charette, M.A., Arrigo, K.R., vanDijken, G., et al., 2013. Long-term trends of upwelling and impacts on primary productivity in the Alaskan Beaufort Sea. Deep Sea Res., I 79, 106–112. Available from: https://doi.org/10.1016/j.dsr.2013.05.003.

Richardson, W.J., Davis, R.A., Evans, C.R., Ljungblad, D.K., Norton, P., 1987. Summer distribution of bowhead whales, *Balaena mysticetus*, relative to oil industry activities in the Canadian Beaufort Sea, 1980–84. Arctic 40 (2), 93–104.

Stephen R. Braund and Associates, 2010. Subsistence mapping of Nuiqsut, Kaktovik, and Barrow. In: Final Report, OCS Study MMS 2009-003. Stephen R. Braund and Associates, Anchorage, AK.

Thoman, R., 2019. <https://www.flickr.com/photos/snapandaccap/albums/72157709845937092>.

Treacy, S.D., 1998. Aerial surveys of endangered whales in the Beaufort Sea, fall 1997. OCS Study MMS 98-0059. USDOI, MMS, Alaska OCS Region, Anchorage, AK.

USGS National Water Information System, 2016. <https://www.usgs.gov/nwis-national-water-information-system> (accessed 02.12.19.).

Walkusz, W., Williams, W.J., Harwood, L.A., Moore, S.E., Stewart, B.E., Kwasniewski, S., 2012. Composition, biomass and energetic content of biota in the vicinity of feeding bowhead whales (*Balaena mysticetus*) in the Cape Bathurst upwelling region (south eastern Beaufort Sea). Deep Sea Res., I 69, 25–35.

Weather Underground, 2016. <https://www.wunderground.com/history> (accessed 02.12.19.).

Williams, W.J., Carmack, E.C., 2008. Combined effect of wind-forcing and isobath divergence on upwelling at Cape Bathurst Beaufort Sea. J. Mar. Res. 66, 645–663.

SECTION II

The bowhead ecosystem

CHAPTER

25

Physical Oceanography

T.J. Weingartner, S.R. Okkonen and S.L. Danielson

College of Fisheries and Ocean Sciences, University of Alaska Fairbanks, AK, United States

Introduction

Bowhead whales are an ice-associated species that thrive in cold water (<2°C) and sea ice–dominated ecosystems (Chambault et al., 2018; Fig. 25.1). The International Whaling Commission recognizes four bowhead whale stocks that inhabit areas with seasonally varying ice covers, which are intimately linked to either the Pacific or Atlantic Oceans (Chapter 3). The Pacific and Atlantic Oceans shape bowhead habitats through control of the annual sea ice cycle, which is critical in producing the cold water preferred by bowheads, and by providing carbon, nutrients, and planktonic organisms upon which the bowheads depend. These same connections provide migratory corridors for potential predators and competitors and are the advective routes by which contaminants, pollutants, and the oceanic consequences of climate change are propagated into bowhead habitats. This chapter provides an overview of the physical oceanography of each of these regions with an emphasis on the upper layer (100–150 m) currents, water masses, and the physical mechanisms that may aggregate prey and thus optimize bowhead foraging.

Bering–Chukchi–Beaufort Seas

The Bering–Chukchi–Beaufort stock seasonally ranges across the western Arctic Ocean (Fig. 25.2). A brief generalized summary of their migratory pattern follows. In winter (December–March) the whales inhabit the Gulf of Anadyr and St. Lawrence Island area on the northern Bering Sea shelf. In spring (April–May) they migrate northeastward across the Chukchi Sea and into the deep Canada Basin enroute to Amundsen Gulf and the Mackenzie Beaufort shelf. Here most remain through summer (May–August) before beginning their return migration westward across the Alaskan Beaufort shelf (August–November) and the northern Chukchi Sea. From there, they migrate southward and thence southeastward along the Chukotka Peninsula (September–January), before

The Bowhead Whale
DOI: https://doi.org/10.1016/B978-0-12-818969-6.00025-X

FIGURE 25.1 The bowhead life cycle is tightly linked to the annual expansion and retreat of sea ice. Ice formation is dynamic, with older ice (white, top of photo) being rimmed by newly formed ice (gray), and isolated square blocks being sutured into new floes (bottom). Photo by Vicki Beaver (NOAA/North Slope Borough, NMFS Permit No. 14245).

returning to the northern Bering shelf to overwinter (Chapter 4). The oceanographic conditions in each of these regions differ distinctly from one another, but each is ultimately linked, atmospherically and oceanographically, to the Pacific Ocean. This Pacific connection maintains an ecosystem/environmental continuum throughout the western Arctic Ocean. The atmospheric connection is primarily via the Aleutian Low, the seasonally varying position and strength and interactions of which with the polar high-pressure cell affect regional meteorological conditions. The oceanographic link originates in the North Pacific Ocean, which feeds the Bering Sea basin and shelf, and subsequently the western Arctic Ocean through Bering Strait.

Bering Sea and Shelf

North Pacific waters enter the Bering Sea via the Alaska Coastal Current on the Gulf of Alaska shelf and via the Alaskan Stream, a limb of the Pacific Ocean's Subarctic Gyre. Waters from the stream flow through the Aleutian Island passes and then northward via the Bering Slope Current, which flows adjacent to the broad Bering shelf. The slope current exchanges waters between the shelf and basin along its course and flows counterclockwise around the Bering Sea basin before returning to the North Pacific through Kamchatka

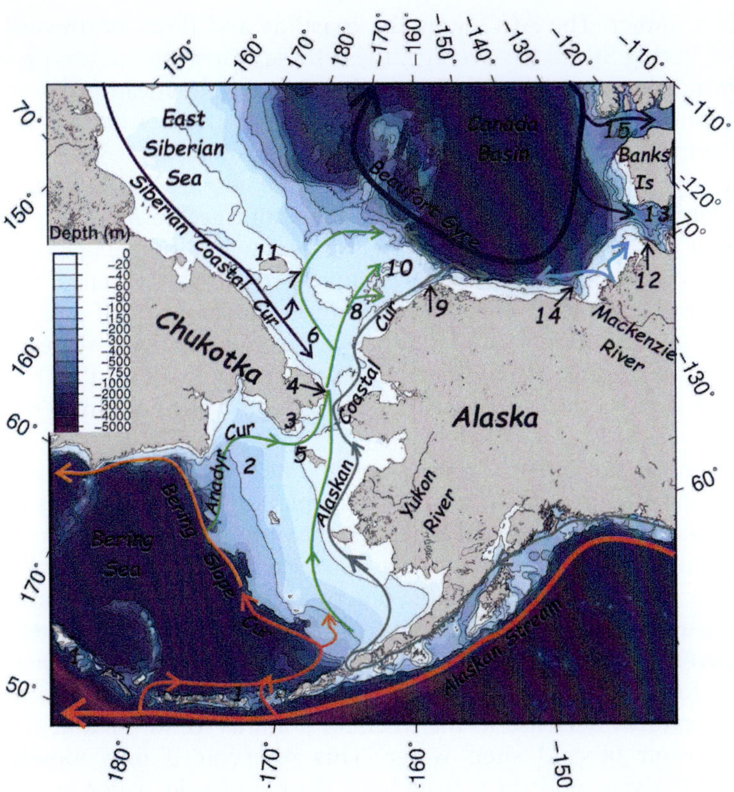

FIGURE 25.2 Bathymetric map of the Bering–Chukchi–Beaufort seas overlain with the major currents with warmest (red) and coldest (blue). Major geomorphic features are the Aleutian Island Archipelago (1), Gulf of Anadyr (2), Anadyr (3) and Bering (4) straits, St. Lawrence Island (5), Hope (6) and Herald (7) valleys, Central Channel (8), Barrow Canyon (9), Hanna Shoal (10), Wrangel Island (11), Cape Bathurst (12), Amundsen Gulf (13), Mackenzie Canyon (14), and McClure Strait (15).

Strait. Some of these waters eventually enter the Okhotsk Sea (see below) while the remainder merges with the Oyashio Current (Stabeno et al., 1999). Of particular importance to the western Arctic is that some waters from the Bering Slope Current are upwelled into the Gulf of Anadyr. The upwelled Anadyr Water is cold (<2°C), salty, nutrient-rich, and an important source of marine carbon and Pacific plankton species for both the Bering and Chukchi shelves. Most of the Anadyr Water flows northward through Anadyr Strait and then the western channel of Bering Strait, while some Anadyr Water enters the Bering shelf south of St. Lawrence Island (Danielson et al., 2014). Two other water masses also flow northward toward the Strait across the Bering shelf (Danielson et al., 2014). Bering Shelf Water is drawn from the central shelf (and is a mix of Bering Slope Current waters and other shelf waters modified by atmospheric exchanges). The nutrient load of the moderately saline Bering Shelf Water varies seasonally; its surface waters are relatively low in summer due to phytoplankton consumption but rich at depth and in winter through regeneration. Summer surface waters over the central shelf can be warm (>4°C), but the bottom waters comprising Bering Shelf Water are drawn from the Bering shelf's "cold pool", a nearly 30-m deep, bottom layer of ≤2°C water over the central Bering shelf. The Alaskan Coastal Current (ACC) flows along the west coast of Alaska and is derived from the Gulf of Alaska's ACC and the rivers of western Alaska. ACC waters are dilute, nutrient- and plankton poor, laden with refractile terrestrial carbon derived from western Alaska rivers, and warm (>6°C to 10°C) in summer. The ACC hugs the coastline and flows northward through the eastern channel of Bering Strait. On average, these three water masses are propelled northward over the Bering shelf and through Bering Strait due to the mean pressure gradient between the Pacific and Atlantic Oceans. North of the Strait, the Anadyr and Bering Shelf waters mix and form Bering Sea Water (BSW) on the Chukchi shelf.

The Bering shelf is enormously productive and supports diverse commercial and subsistence fisheries as well as large numbers of marine mammals and seabirds. Over the southern and central portions of this shelf, the trophic system is dominated by pelagic production, but benthic production increases in importance moving northward. Hence, the marine ecosystem of the northern Bering shelf consists of a complex assemblage of pelagic and benthic productivity with part of the benthic production sustained by ice algae in spring. However, an important contribution to the benthos is pelagic production that has been advected from the Gulf of Anadyr and subsequently deposited on the bottom to the south and north of Bering Strait.

Sea ice begins advancing from the north and forming over the Bering Sea shelf (but not over the continental slope) in December, coincident with the arrival of bowhead whales at the end of their annual migration. Ice typically reaches its maximum annual extent in March. Numerous leads permeate the ice cover over the northern Bering shelf and two prominent polynyas form under strong northerly winds: one over the Gulf of Anadyr and the other south of St. Lawrence Island. The permeability of the ice pack and the likely prevalence of zooplankton upwelled into the Gulf of Anadyr or overwintering on the shelf establishes a setting conducive for bowhead foraging throughout winter (Citta et al., 2015). Moreover, water column temperatures decrease to the freezing point as ice forms and the polynyas establish a vast reservoir of cold shelf water. This reservoir is only slowly drained throughout the rest of the year and thus provides a source of cold water to the Chukchi Sea via Bering Strait.

Chukchi Sea

The shallow (~50 m) Chukchi shelf extends ~800 km northward from Bering Strait to the shelfbreak on about the 200 m isobath. On average, the flow of Pacific-derived waters here is northward, opposes the prevailing northeasterly winds, and is channeled by the bathymetry along three principal pathways (Weingartner et al., 2005; Woodgate et al., 2005). A western branch flows northwestward through Hope Valley and thence northward through Herald Valley (Pickart et al., 2010). Most of this flow turns eastward at the shelfbreak, but some flows across the central shelf. A second branch flows northward through the Central Channel shelf and then splits; some water flows eastward toward the Alaskan coast while the remainder flows northeastward around Hanna Shoal. The third branch flows northeastward along the Alaskan coast into Barrow Canyon at the junction of the Chukchi and Beaufort shelves and is a continuation of the ACC from the Bering shelf. Within the canyon the ACC is joined by waters flowing eastward from the central shelf. The merged flow then accelerates rapidly down canyon and enters the Arctic Ocean. This swift flow involves large horizontal velocity gradients, which, in winter, are probably essential in forming the complex system of lead prevalent in the ice cover of Barrow Canyon.

The nutrient and carbon loads transported along these branches differ from one another (Hansell et al., 1993; Danielson et al., 2017). The Herald Valley outflow is saltier, colder, and richer in nutrients and marine-derived carbon than the waters transported in the ACC, whereas waters crossing the central shelf have intermediate properties. In winter, shelf waters decrease to the freezing point and salinities increase due to salt distilled from growing sea ice. The seasonal changes in shelf salinities have important implications for the disposition of Pacific waters into the Arctic Ocean. Fresher and warmer low-density summer waters are confined to the upper 50 m of the shelfbreak and slope, whereas near-freezing, saline winter waters descend to 75–150 m depth along the continentals slope and eventually enter the halocline of the Western Arctic Ocean. A variety of processes redistribute (and mix) Pacific Waters around the western Arctic Ocean where they retain their distinct nutrient and carbon signatures. These include the clockwise circulation of the Beaufort Gyre, wind-forced excursions of shelf water into the basin, and the formation of subsurface eddies from the currents along the continental slopes of the Chukchi and Beaufort Seas (Watanabe, 2011). Pacific waters are eventually transported into the North Atlantic through Fram Strait and into Baffin Bay though McClure Strait and Amundsen Gulf.

There are two other important aspects of the Chukchi shelf circulation. The first is the Siberian Coastal Current (SCC) that originates from the massive river discharges in the East Siberian Sea, which flows southeastward along the Siberian coast into the Chukchi Sea. The SCC carries cold, dilute, nutrient-poor ice-melt and river waters, and sea ice from the East Siberian and Laptev Seas. The current lies within ~60 km of the Chukotkan coast and is bounded on its offshore side by a meandering front (Weingartner et al., 1999). SCC waters mix with northward-flowing waters from Bering Strait, with the resultant subsequently transported into Hope Valley. The second aspect is that the upper layer waters over the outer Chukchi shelf (north of Hanna Shoal) and slope are flowing westward on average (above the eastward shelfbreak current carrying waters from Herald Canyon; Li et al., 2019) and include wind-driven shelf waters from the Beaufort Sea, outflow from Barrow Canyon, and the cold, low-salinity waters of the Arctic Ocean's polar mixed layer within the southern limb of the large clockwise Beaufort Gyre flowing around the Canada Basin.

Historically portions of the Chukchi shelf remained ice-covered for most of the year, with ice present 72°N in September when maximum open water area occurs. Late summer ice typically consisted of multiyear ice that had drifted into the region from the main pack or heavily deformed first-year ice that had grounded on Hanna Shoal, along the north coast of Wrangel Island, and in some years along the Chukotka coast. In the past decade or so the shelf has been largely ice-free through October or mid-November. Ice formation proceeds southward through fall so that the shelf is ice-covered by December. Strong winter winds can create extensive polynyas, along the west coast of Alaska under easterlies and south of Wrangel Island under northerlies. Freezing in the polynyas produces salty, cold waters, which spread across the shelf and can linger at depth through much of the following summer as they slowly drain into the Arctic Ocean.

Ice retreat begins in May in Bering Strait and progresses northward with embayments, associated with the three current pathways carrying warm Pacific waters northward, forming along the ice edge. Primary production along the retreating ice edge is enhanced due to the abundance of nutrients and the onset of water column stratification. In some years, easterly winds may lead to open water along the Alaskan coast as early as May, with melt then proceeding from east to west as well as south to north. Much of this primary production occurs in early spring, settles to the bottom, and sustains a rich and productive benthic ecosystem (Grebmeier, 1993).

Beaufort Sea Shelf

The Beaufort Sea shelf extends ~1200 km eastward from Point Barrow to Cape Bathurst at the entrance to Amundsen Gulf. Its northern boundary is the deep Canada Basin. On average, the entire shelf is subject to northeasterly winds, which force westward flow and an onshelf and westward drift of first- and multiyear sea ice from the basin. The winds promote upwelling of nutrient- and zooplankton-laden subsurface waters at the shelfbreak (Schulze and Pickart, 2012), which is essential to biological production and establishing favorable foraging areas for bowheads. Nevertheless, the Beaufort shelf consists of two dynamically different subregions: the Mackenzie shelf and the Alaskan Beaufort shelf. The Mackenzie shelf is a broad rectangular platform (width ~120 km; length ~550 km) bordered by Cape Bathurst to the east, Mackenzie Canyon to the west, and the Mackenzie River delta to the south. Its hydrography is profoundly influenced by the massive year-round runoff from the Mackenzie River. Under sufficiently strong northeasterly winds the Mackenzie plume is driven westward onto the Alaskan Beaufort shelf and offshore into the Canada Basin (Macdonald et al., 1999) with its signature detected as far west as the northern Chukchi Sea (Guay and Falkner, 1998).

The Alaskan Beaufort shelf is ~80 km wide with little bathymetric variability. Numerous, small, shallow arctic rivers, the watersheds of which are entirely underlain by permafrost punctuate the coast. The annual discharge cycle of these rivers consists of a 2-week-long spring freshet, commencing in early June, during which ~75% of the annual discharge occurs (Weingartner et al., 2017). Thereafter, discharge rapidly decreases and is negligible from October to April.

Sea ice may cover the shelf throughout the year, although in recent years most of the shelf has been ice-free from July to October. Freeze-up begins in October near the coast and progresses offshore while at the same time, wind-driven pack ice is advected onto the outer shelf from the Canada Basin. The shelf ice cover consists of two distinct components: freely drifting pack ice over the middle and outer shelf and virtually immobile landfast ice on the inner shelf. Landfast ice is anchored to the coast and extends 20–40 km offshore, typically grounding along the 20 m isobath. Along its seaward boundary, the landfast ice is heavily deformed due to collisions with the pack ice. Deformation leads to a thick, heavily ridged ice zone that can extend well offshore, which, under the prevailing winds, is generally devoid of leads until summer melt. The different ice regimes profoundly affect the underlying shelf circulation. Wind-forced currents over the mid- and outer shelf regions can be vigorous within the drifting pack ice region, whereas currents are feeble beneath the immobile landfast ice. The landfast ice also influences the spreading of under ice plumes as these generally extend farther offshore in the presence of landfast ice compared to its absence.

Polynyas rarely form on the Beaufort shelf, except within the Cape Bathurst and Banks Island regions. These polynyas form intermittently in winter and spring and are large leads created when easterly winds force pack ice westward and away from the more stable ice of Amundsen Gulf and along Banks Island. The leads are important bowhead spring foraging areas because the ice-free areas allow the penetration of sunlight into the water column and because nutrient-rich deeper waters are upwelled into the surface layer, which enhances primary (and secondary) production (Williams and Carmack, 2008; Walkusz et al., 2012).

Breakup of ice usually begins in the Mackenzie Delta in May or June as sediment-laden river waters overflow the ice surface, decreasing the albedo and accelerating melt. The heat in this strongly stratified plume accelerates ice melt along the coast, shelf, and slope compared to other areas and thus plays an important role in seasonal ice retreat. Similarly, but to a lesser extent, warm Pacific waters rounding Point Barrow in the ACC erode ice in the western Beaufort Sea in spring. Thus the eastern and western ends of the Beaufort shelf melt before the central portion of the shelf, where the landfast ice cover may remain intact until early July and after the spring freshet.

Water properties are controlled by the annual freeze-thaw cycle and inflows from the oceanic and coastal boundaries. In winter, temperatures are at the freezing point and near-freezing waters remain at depth on the shelf year-round. However, in the highly stratified river plumes, temperatures can be 5°C–10°C above the freezing point. Shelf salinities are relatively high, whereas the river plumes are extremely fresh and shallow being ~10 m thick on the Mackenzie shelf and ~2 m thick on the Alaskan shelf. Once the landfast ice detaches from the coast, plume and ambient waters can begin to mix. However, cross-shelf gradients in water properties are often large and embedded in a multiplicity of fronts that influence biological production and the aggregation of organisms.

Okhotsk Sea

The Okhotsk Sea (Fig. 25.3) is a marginal sea in the northwest Pacific Ocean and comparable to the Bering Sea in that it is seasonally ice-covered, similarly sized, and at comparable latitudes and contains a broad continental shelf subject to large tidal excursions. Each sea supports abundant fisheries and each receives the discharge from large rivers (which in the Okhotsk Sea is the Amur River on the northern shelf) and each communicates with the North Pacific Ocean through island archipelagos. In the Sea of Okhotsk the Kuril Island Archipelago trends southwestward from the Kamchatka Peninsula to Hokkaido Island that delimits the southern boundary of the Okhotsk Sea. The deep basin's bathymetry is complicated by subbasins and submarine ridges. The broad (~400 km wide) continental shelf is confined to the northern basin and occupies ~40% of the Okhotsk Sea's area. The shelf contains a number of banks and is indented by two prominent submarine valleys (Derjugin's Hollow and the TINRO Basin) that may channel slope waters onto the shelf. Shelikhov Gulf in the northeast and Sakhalin Gulf in the northwest comprise >25% of the shelf area and both gulfs are important bowhead summer foraging areas (Ivaschenko and Clapham, 2010; Shpaka and Paramonov, 2018). These authors suggest that bowheads overwinter in the basin east of Sakhalin Island and along the southwest side of the Kamchatka Peninsula near Kruzenshtern Strait. Fedoseev (1984) suggests that bowheads will overwinter in leads and along the ice edge.

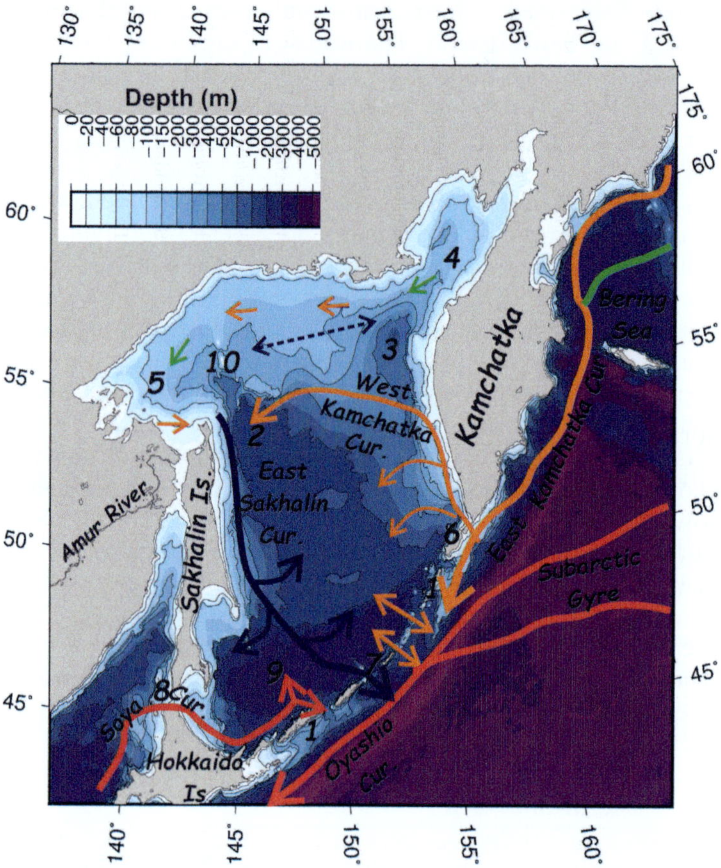

FIGURE 25.3 Bathymetric map of the Okhotsk Sea overlain with the major warm (red), cooler (orange), and cold (blue) currents. Major geomorphic features are the Kuril Island Archipelago (1), Derjugin's (2) and TINRO (3) trough and basin, Shelikhov (4) and Sakhalin (5) Gulfs, Kruzenshtern (6), Bussol (7), and Soya (8) straits, Kuril Basin (9), and Kasherov Bank (10). The dashed two-headed arrow signifies strong oscillatory tidal currents, while the small, green arrows denote a weak, net drift associated with buoyancy-forced shelf currents. The orange two-headed arrows in the Kuril chain indicate bilateral exchange between the North Pacific Ocean and Okhotsk Sea.

The syntheses of Talley and Nagata (1995) and Radchenko et al. (2010) indicate that the large-scale circulation over the Okhotsk Sea basin is counterclockwise, although details are lacking due to a dearth of data. Some of the waters flowing southward from the Bering Sea in the East Kamchatka Current enter Kruzenshtern Strait at the southern tip of the Kamchatka Peninsula. This relatively warm and nutrient-rich inflow provides a pathway by which zooplankton from the North Pacific Ocean enter the Okhotsk basin. The remainder continues southward, merges with waters from the Subarctic Gyre, and enters the Oyashio Current. Upon entering the Okhotsk Sea the inflow continues northward as the West Kamchatka Current along the west side of the Kamchatka Peninsula. Thereafter it flows westward along the northern continental slope (and possibly over the outer shelf) before turning southward to form the East Sakhalin Current along Sakhalin Island (Ohshima et al., 2002). The East Sakhalin Current veers eastward offshore of Sakhalin Island transporting water and ice into the Pacific Ocean. All of these currents are weak, populated by numerous eddies, and subject to wind-driven variability. The basin circulation varies seasonally and is stronger in fall–winter when winds are strong and northerly and weak in summer when winds are mild. In winter Okhotsk waters cool, absorb oxygen, sink, and exit the basin at a greater density (and depth) than when they first enter from the East Kamchatka Current. The outflow, which primarily leaves the Okhotsk Sea through Bussol Strait, forms the intermediate waters (between 200 and 800 m depth) that pervade the northern North Pacific Ocean. Along the north coast of Hokkaido, the Soya Current transports warm waters from the Japan Sea through Soya Strait and into the Oyashio, with some also entering the Kuril Basin.

Little is known about the northern shelf circulation. Markina and Chernyavsky (1984) depict a sluggish westward drift over the shelf, consistent with the forcing tendency provided by a river discharge. However, this flow is easily modified by the winds and the mean drift is small compared to the very large shelf tidal currents. Tidal current magnitudes increase markedly as the shelf shallows and are especially large in Shelikhov and Sakhalin Gulfs and around Kasherov Bank. (Kowalik and Polyakov, 1998). Large tidal velocity gradients often establish quasistationary fronts that intersect the bottom and are of biological importance. These fronts are likely more prominent in summer when river discharge is high and winds are weak. In winter the fronts will weaken or possibly vanish as discharge decreases and strong winds and convective cooling promote vertical mixing. The water column is typically well-mixed shoreward of the fronts. Seaward of the fronts the water column is more strongly stratified with warm (10°C–15°C) dilute waters occupying the upper 30 m (NOAA National Centers for Environmental Information, 2019a). Below this 30-m deep surface layer temperatures decrease quickly from 0°C to −1.8°C at ~100 m. (This cold layer pervades the Okhotsk Sea. The coldest waters are in the northwest basin and cold core temperatures increase to ~0°C in the southern basin and near the Kuril Island passes.) The cold temperatures are remnant from the previous winter and formed during convective cooling and sea ice formation. Winter convection promotes deep mixing that brings nutrients to the surface layer so that by spring the surface waters are primed for the spring bloom. Although phytoplankton may exhaust surface nutrients in summer, the colder waters just below the stratified layer are nutrient-rich and may be entrained into the surface layer by tidal mixing or episodic wind events.

Sea ice begins forming as the frequency of cold, northerly winds increases in November (Ohshima et al., 2006) and progresses from north to south through winter. Ice is not only formed in situ but also advected southward in the East Sakhalin Current and by the winds. Historically, the maximal extent of the ice cover was attained in March when ice covered much of the basin. Nevertheless, open water is typically maintained along western Kamchatka and the Kuril Islands due to the warm waters in the West Kamchatka Current. Interannual variability in ice extent is large and highly correlated with autumn and winter air temperatures (Ohshima et al., 2006) and by the rate at which warm Pacific waters enter the Okhotsk Sea. Fall and winter air temperatures have steadily increased over the past 30 years leading to a 20% reduction in sea ice extent. In addition, the duration of the ice cover has decreased such that the period of maximum ice cover has shifted from mid-March to February (Radchenko et al., 2010).

Ice retreats in the deep basin beginning in April as solar insolation increases, although the retreat rate is a function of March winds (Nihashi et al., 2011). Retreat is more rapid under northerly (offshore) March winds, which disperses floes and reduces ice concentrations. As the open water fraction increases in March (prior to the onset of melt), more solar radiation enters the upper ocean and enhances ocean warming and ice melt rates.

In contrast, southerly winds compress the ice pack, thus reducing the amount of solar radiation entering the water column and the rate of ice melt. This description applies to the broader Okhotsk Sea, but there are significant spatial differences. For example, in the shallow, bays on the northwestern shelf ice may persist for up to 9 months of the year and late summer surface temperatures may range up to 16°C (Shpaka and Paramonov, 2018). Nevertheless, the deeper reaches of these bays may still contain cold waters formed the previous winter.

Nordic Seas

The East Greenland–Svalbard–Barents (EGSB) Sea bowhead and the East Canada–West Greenland (ECWG) stocks inhabit waters influenced by both the Arctic and Atlantic Oceans (see Chapter 3). Although the stocks are separate, the large-scale circulation connects the two regions: some of the waters inhabited by the EGSB stock along eastern Greenland are eventually transported along western Greenland and into Baffin Bay. Consequently, we first describe the oceanographic conditions of the Nordic Seas (Fig. 25.4) encountered by the EGSB stock and then describe the oceanographic habitat of the ECWG stock.

The Nordic Seas lie between the Greenland–Iceland–Scotland Ridge and the Arctic Ocean (Fig. 25.4). They consist of the deep (>1000 m) basins comprising the Greenland–Iceland–Norwegian (GIN) Sea and the Barents Sea, a deep (~200 to 400 m) shelf sea situated between northern Norway and the Arctic Ocean. The GIN Sea basin is divided by a series of mid-ocean ridges: the Jan Mayen Ridge in the south, the Mohns Ridge in the center, and the Knipovich Ridge that runs northward toward Svalbard along the eastern side of Fram Strait. The ridge system effectively separates warm, salty, nutrient-rich, North Atlantic-derived surface waters in the east from cold, fresh, nutrient-

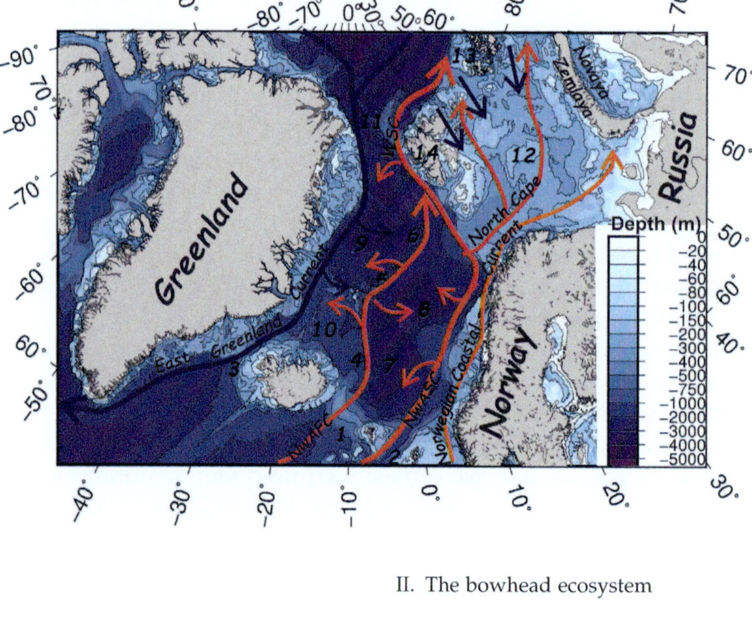

FIGURE 25.4 Bathymetric map of the Nordic Seas with major warm (red), coastal (orange), and cold (blue) currents. Major geomorphic features are the Wyville-Thomson (1), Scotland-Faroe (2) ridges, Denmark Strait (3), Jan Mayen (4), Mohns (5), and Knipovich (6) ridges, Norwegian (7), Lofoten (8), Greenland (9), and Iceland basins (10), Fram Strait (11), Barents Sea (12), Franz Josef Land (13) and the WSC, NwASC, and NwAFC. *NwAFC*, Norwegian-Atlantic Front Current; *NwASC*, Norwegian-Atlantic Slope Current; *WSC*, West Spitsbergen Current.

poor surface waters of Arctic origin in the west. The bathymetry also includes several zonally oriented ridges that form smaller basins: the Norwegian and Lofoten Basins in the east and the Greenland and Iceland basins in the west.

North Atlantic waters enter the Nordic Seas via the Norwegian-Atlantic Slope Current (NwASC), flowing across the >500 m deep Scotland–Faroe Ridge, and the Norwegian-Atlantic Front Current (NwAFC) entering over the shallower Wyville Thomson ridge between Iceland and the Faroe Islands (Björk et al., 2001; Piechura and Walczowski, 1995). The northward heat and mass transports in these currents are similar (Hansen et al., 2008), with temperatures in the upper 100 m averaging ~10°C. The NwASC proceeds along the Norwegian continental slope. The NwAFC extends along the Mohns and Knipovich ridges on the west side of the Lofoten Basin, although some of its waters leak westward into the Greenland Sea. The mean temperatures in the uppermost 100 m of both currents are ~5°C on crossing the Lofoten Basin. These currents converge at about 75°N and flow northward as the West Spitsbergen Current (WSC) along the slope and the deep (>200 m) shelf with average temperatures of <3°C. After entering Fram Strait, some of the WSC rounds the west side of Svalbard and continues eastward along the Eurasian continental slope to feed the circumpolar flow of Atlantic waters around the Arctic Ocean. The remainder turns westward, encounters the cold (<0°C), dilute waters flowing southward from the Arctic Ocean, and recirculates back into the GIN Sea (Schauer et al., 2004). The strong temperature and salinity contrasts between the Atlantic and Arctic waters in Fram Strait result in fronts and eddies as the denser, recirculating Atlantic waters sink beneath the Arctic waters.

Although more than half of the Atlantic inflow enters the WSC, a substantial fraction crosses onto the Barents Sea shelf via the North Cape Current and thus provides a critical supply of nutrients and zooplankton to the Barents Sea. As these waters cross the southern Barents shelf, they cool from ~3°C to 0°C (Loeng, 1991). They eventually sink and enter the Arctic Ocean either between Svalbard and Franz Josef Land or along the western side of Novaya Zemlya. Cold (<0°C), fresher Arctic waters encompass the northern half of the Barents shelf, with the Polar Front separating Arctic and Atlantic waters. This front's location is coupled to the bathymetry and thus independent of the position of the sea ice (Harris et al., 1998; Barton et al., 2018). Although winds can force sea ice across the front, the southernmost ice limit is set by the temperature of the Atlantic inflow. In recent years, this inflow has warmed and prevented ice from extending onto the southern shelf even in winter (Barton et al., 2018). In addition to these deep-water currents, the northward-flowing Norwegian Coastal Current transports waters from the North and Baltic along the Norwegian continental shelf and then eastward across the Barents shelf, hugging the northern coast of Norway and Russia (Mork, 1981). Upon entering the Barents Sea, these waters are warm (>2°C) and thus inhibit the formation of landfast ice along theses coasts.

While the eastern side of Fram Strait is the main gateway through which Atlantic waters enter the Arctic Ocean, the western side is the main pathway by which Arctic waters enter the GIN Sea and the North Atlantic Ocean. Arctic waters are carried southward in the East Greenland Current (EGC) along the continental slope, through the Denmark Strait, and round the southern tip of Greenland (Woodgate et al., 1999). The EGC transports >90% of the sea ice exported from the Arctic Ocean (Rudels et al., 1999) and it entrains warmer Atlantic waters that have recirculated from the NwAFC in the GIN Sea. The resultant mixture weakens the density stratification within the current, which inhibits ice formation

(Aagaard et al., 1985). The current also absorbs an enormous volume of the glacial meltwater runoff from eastern Greenland (Sutherland and Pickart, 2008). Upon rounding southern Greenland the EGC entrains warmer, saltier waters from the Irminger Sea and then enters the Labrador Sea to feed the West Greenland Current (WGC).

Sea ice begins forming in fall in the northern Greenland Sea and progresses southward through winter. The ice primarily remains in the EGC and consists of a mixture of multi-year and first-year ice exported from the Arctic Ocean, icebergs calved from Greenland's glaciers, and ice formed in situ. The ice extends southward as a long, but ever-narrowing tongue hugging the continental shelf and slope. The maximum median sea ice extent occurs in March with ice reaching the southern tip of Greenland. Ice concentrations diminish eastward across the Greenland Sea as warmer Atlantic waters are encountered resulting in the Lofoten and Norwegian basins remain ice-free throughout winter. Sea ice extent varies tremendously from year to year as a function of atmospheric conditions, heat flux in the Atlantic inflow, and the amount of low-salinity water exported from the Arctic Ocean. In some years an extensive bulge in sea ice concentration spreads eastward across the Greenland Sea between ~71°N and 72°N and can occasionally extend westward around southern Greenland (Germe et al., 2011). Superimposed on this interannual variability in ice extent is a downward trend of ~5%/year in ice extent since 1979. The decrease results from numerous factors: the loss of thick, multiyear floes, warmer air temperatures, and, in recent years, an increase in heat transport from the Atlantic Ocean (Muilwijk et al., 2018). Thus there has been open water year-round in eastern Fram Strait in recent years and a marked reduction in sea ice over the continental slope north of Svalbard, precisely where the EGSB stock resides (Wigg and Bachmann, 2007). For similar reasons the duration and extent of ice cover in the Barents Sea have also decreased over time.

Primary production in the Nordic Seas is highly seasonal and includes tremendous spatial variability. It depends upon nutrient supply, stratification, and the influence on light levels, which are largely controlled by the duration and spatial extent of sea ice. Thus primary production is low in the EGC due to heavy ice concentrations and the dearth of nutrients in its surface waters and low in the Greenland Sea because of the weak stratification. Exceptions to these generalities are the marginal ice zones in spring where abundant nutrients at the surface and sufficient stratification can lead to explosive blooms. Within the WSC region the advection of phytoplankton by the WSC appears to be far more important than local production in establishing phytoplankton biomass (Vernet et al., 2019). Recent increases in Atlantic Water transport (and thus phytoplankton) may lead to an increase in zooplankters in the bowhead foraging areas of Fram Strait, Svalbard, and Franz Josef Land. The Barents Sea shelf supports important commercial fisheries and is one of the more productive pelagic marine ecosystems in the Arctic. Some of this productivity, which also depends upon the Atlantic inflow, is exported northward into bowhead foraging areas and eventually the Arctic Ocean.

East Canada–West Greenland

The ECWG bowhead stock summer habitat includes Baffin Bay (Fig. 25.5) where depths range from 200 to 2000 m and the northern portion of the much shallower

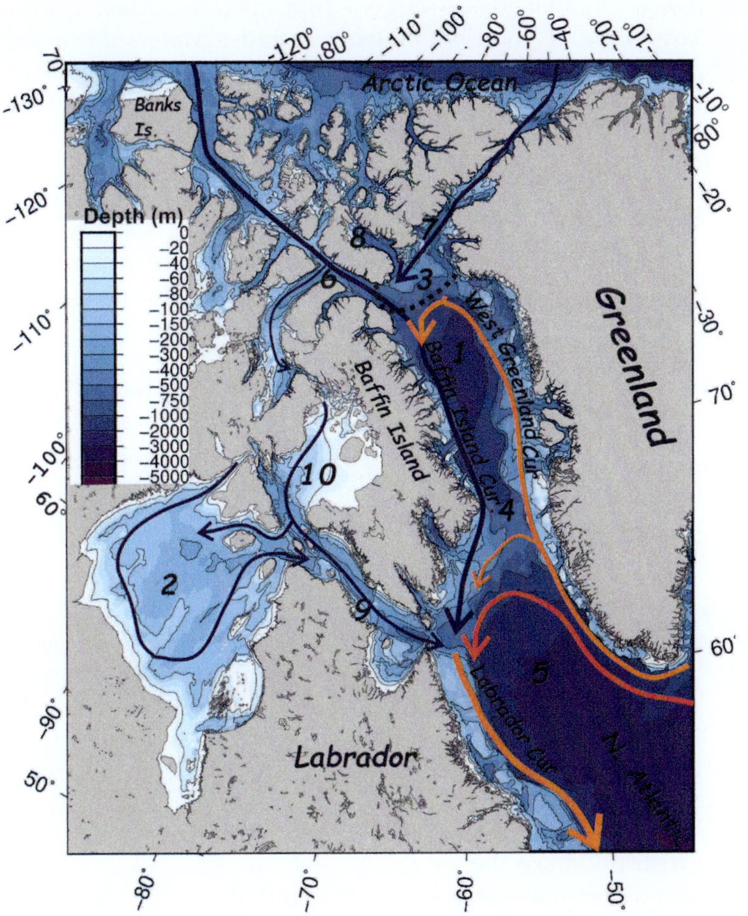

FIGURE 25.5 Bathymetric map of the Eastern Canada–Western Greenland area with major warm (red), moderately warm (orange), and cold (blue) currents. Major geomorphic features are Baffin Bay (1), Hudson Bay (2), North Water Polyna (3; outlined by dashed line), Davis Strait (4), Labrador Sea (5), Barrow (6) and Nares (7) straits, Jones Sound (8), Hudson Strait (9), and Foxe Basin (10).

(100–200 m) Hudson Bay. This stock overwinters in the North Water Polynya of northern Baffin Bay, along the west coast of Greenland, and Davis Strait (200–600 m deep), and the northern Labrador Sea. This extensive habitat is influenced by inflows from the Atlantic and Arctic Oceans.

The Atlantic connection is sustained by the WGC that flows northward along the west Greenland slope and shelf as a continuation of the EGC. The warmer, saltier portion of the WGC circulates counterclockwise around the northern Labrador Sea, while the cooler, fresher fraction (derived from the Arctic Ocean outflow through Fram Strait and glacial runoff and ice discharged from Greenland) continues northward along the east side of Davis Strait into Baffin Bay (Curry et al., 2011). The northern branch extends to ~78°N and then turns west to join the cold, fresh, southward-flowing Baffin Island Current (BIC). BIC waters are cooled and freshened by the addition of Arctic waters most of which enter Baffin Bay through Barrow (125 m sill depth) and Nares Strait (~300 m sill depth; Münchow and Melling, 2008) and Jones Sound. The Arctic waters mainly derive from the upper 150 m of the Western Arctic Ocean and are largely comprised modified Pacific Ocean waters, although the Barrow Strait contribution also includes waters from the Mackenzie shelf (Melling et al., 2008). The BIC flows southward along the western side of Davis Strait and thence into the Labrador Sea, where it mixes with warmer Atlantic waters to form the Labrador Current. The ECWG temperature distribution reflects the various water mass contributions. Summer temperatures in the upper 100 m of the WGC cool from ~4°C at the southern tip of Greenland to ~1°C in northern Baffin Bay, whereas the comparable winter values are ~2.5°C and −1.5°C, respectively (Tang et al., 2004; NOAA National Centers for Environmental Information, 2019b). Throughout the year however, the subsurface (50–100 m) temperatures in Baffin Bay remain fairly constant and <0°C due to intense cooling and deep convection in winter (Hamilton and Wu, 2013).

South of Baffin Island, the Labrador Current incorporates the outflow from Hudson Strait, which includes cold, fresh waters from Foxe Basin and Hudson Bay. Within Hudson Bay, the circulation is counterclockwise and mainly driven by the large discharge from the numerous rivers along the bay's periphery and the seasonally varying wind field (Ingram and Prinsenberg, 1998). The circulation is comparatively swift near the coast, but much weaker in the interior. The large riverine runoff and ice melt strongly stratify northern Hudson Bay. Consequently, temperatures average ~4°C at the surface in summer, and ~ − 1°C at 100 m. In winter the water column is homogeneous and near the freezing point (Hamilton and Wu, 2013).

There are three sources of ice in Baffin Bay and the Labrador Sea. The first two are associated with transport of ice by the mean circulation and involve the influx of ice through the various straits entering the region and glacial ice calved from the Greenland. Ice thicknesses from these sources are >1.5 m and in the case of glacial ice can be many meters in thickness. The third source is locally formed ice, which typically is <1.5 m thick (Hamilton and Wu, 2013). Ice begins forming in the northern end of Baffin Bay in September and growth progresses southward through fall and winter. Freezing ends in Baffin Bay in April. The spatial distribution of the ice cover is asymmetric, however, and is conditioned by the temperature differences between the BIC and WGC. Hence, Baffin Bay ice is advected southward within the BIC and the Labrador Current in a narrowing tongue, with its average maximum extent being the southeastern tip of Labrador (although in some years it can extend much farther). The eastern side of Davis Strait remains ice-free due to the warmer temperatures of the WGC (Curry et al., 2011). However, even within the domain of the WGC ice extent can be highly variable, with heavy ice concentrations reaching as far south as 62°N, or open water conditions extending to 69°N.

The ice cover in Baffin Bay consists of two distinct modes. Within the central bay the interannual variability in ice cover is low and this region remains ice-covered for ~10 months/year. In most years, northern Baffin Bay includes an extensive area (85,000 km^2) of open water in late winter and spring due the North Water Polynya. This polynya forms once an ice dam is established in Nares Strait, which effectively blocks ice movement from the Arctic Ocean into Baffin Bay (Yao and Tang, 2003). Once the dam forms, the prevailing northerly winds propel a southward flow of ice over northern Baffin Bay that leads to open water. Similarly, landfast ice dams often form in Barrow Strait and Jones Sound, resulting in expansion of the North Water Polynya into these regions. The polynya is ecologically significant insofar as the open water of the polynya allows primary production to begin earlier here in spring than elsewhere in Baffin Bay. Consequently, the polynya region supports a rich marine ecosystem and provides important habitat for numerous marine mammals, including bowhead whales (Nielsen et al., 2015).

Ice retreat and breakup commences in April coincident with the increase in solar radiation and warmer air temperatures. Ice retreat is also asymmetrical due to the prevailing currents with retreat proceeding slowly within the BIC and more rapidly within the WGC. Although this asymmetry reflects the differences in temperatures between the two current systems, ice is also continually discharged into the BIC well into June or July due to the collapse of ice dams and the breakup of landfast ice from the various sounds and straits around Baffin Bay.

Ice begins to form in northern Foxe Basin in early October and growth progresses southward through fall, with complete freeze-up attained in southeastern Hudson Bay and the mouth of Hudson Strait by early December. Although first-year ice covers the entire region, the thickness distribution is not uniform due to the effect of winds and currents on ice motion, and river runoff. For example, ice tends to be thicker along the southern and eastern coasts of Hudson Bay than in the central bay due to convergence in ice drift associated with the counterclockwise circulation and northerly winter winds. Melt begins in May and proceeds from south to north, although this general pattern is complicated by ice thickness differences established during the previous winter. Moreover, runoff over the surface of near-shore ice tends to accelerate ice melt in coastal regions. Nevertheless, Hudson Bay is typically ice-free by late June through mid-July, while northern Foxe Basin becomes ice-free in August.

Some physical processes that aggregate bowhead prey

The large-scale circulation patterns described previously provide the physical background for bowhead habitats. However, additional processes, generally operating at much smaller scales, are involved in aggregating zooplankters in sufficient concentrations to permit efficient foraging by filter-feeding bowhead whales (Goldbogen et al., 2013). For the most part, these processes occur at ocean fronts (water mass boundaries), which are often coupled to topography, major current systems, and/or with coastal or bottom boundaries. Fronts have large horizontal density (and other property) gradients, involve a complex three-dimensional circulation, and may be quasipermanent or transient. Transient fronts include those associated with tides, wind events, river discharges, and marginal ice zones and can vary over periods ranging from the tidal, to wind events, to the seasonal. Frontal zone widths range from ~1 km in estuaries to ~10–20 km in fronts associated with coastal upwelling, marginal ice zones, the shelfbreak, or buoyancy-forced currents (such as the SCC), to the 50–100 km width of the permanent Arctic Front that separates Atlantic and Arctic waters in Fram Strait. Frontal characteristics include a swift along-frontal flow field, and much weaker cross-frontal circulation cells involving upwelling and convergent flow on either side of the front. Upwelled nutrients enhance primary production, while horizontal cross-frontal convergence forms dense phytoplankton and zooplankton aggregations. Along the Arctic Front and within the marginal ice zone, the convergent fronts are shallow and far from coastal or bottom boundaries. Here bowhead feeding is confined to the surface waters.

Several studies have examined bowhead foraging habitats in some detail. For example, the front along the shallow SCC slants upward from the bottom near the coast to the surface over a distance of ~60 km. Convergent flow at the base of the front causes krill to accumulate in a narrow band because they remain at the bottom to avoid the well-lit surface layer. Consequently, bowhead foraging is concentrated along the foot of the front (Citta et al., 2017; Moore et al., 1995). Similarly, Okhotsk Sea bowheads forage along the bottom on the west coast of Academy Bay in the Gulf of Sakhalin (Rogachev et al., 2008). Here, the estuarine circulation results in an outflow of river and meltwaters at the surface and an inflow of cold, salty waters and oceanic zooplankton at depth. Due to tidal effects, a weak cross-estuarine flow is established, which is eastward at the surface and westward at depth. The deeper waters, enriched in zooplankton, converge along the bottom on the western side of the bay and produce a lucrative feeding ground.

A diversity of upwelling-related processes on the Beaufort shelf appears to create, at least episodically, favorable foraging areas. At Cape Bathurst, easterly winds promote westward flow around the Cape. The winds, in conjunction with westward divergence of the isobaths promote vigorous upwelling of nutrients onto the shelf and establish a region of high biological productivity (Williams and Carmack, 2008) favored by bowheads (Citta et al., 2015). Upwelling in conjunction with fronts along the perimeter of riverine plumes may also provide foraging opportunities (Chapter 24). Over the Alaskan Beaufort shelf, Okkonen et al. (2018) find episodic foraging events associated with upwelling pulses that carry zooplankton onto the shelf and contemporaneous anomalously large discharges from the small rivers. As upwelling winds relax, the riverine fronts intensify and presumably aggregate zooplankton in sufficient numbers suitable for efficient foraging. This same mechanism may be at play on the central Mackenzie shelf with its much larger river discharge and consequently much stronger fronts (Walkusz et al., 2010). Citta et al. (2015) have identified this region to be a major summer bowhead foraging area.

During their summer return migration, bowheads often feed over the western Beaufort shelf. These feeding opportunities appear to depend upon the duration and magnitude of upwelling-favorable winds. The winds move krill from the Beaufort slope across the shelf and trap them near the coast, while simultaneously forcing westward flow over the shelf. If the upwelling winds persist for too long, however the krill are advected into Barrow Canyon and do not accumulate (Okkonen et al., 2011). If, however, the winds relax, the krill are retained on the shelf and convergence of the westward flow over the Beaufort shelf with the northerly flow through Barrow Canyon causes them to aggregate in dense patches. It should be noted that while upwelling typically enhances biological productivity, sustained upwelling can be counterproductive to bowhead foraging. In the cases cited previously, persistent upwelling can weaken fronts and, over the western Beaufort shelf, prevent flow convergence between the shelf and Barrow Canyon. Indeed, in some years persistent upwelling favorable winds may prevent the SCC from entering the Chukchi Sea (Weingartner et al., 1999). In recent years there has been an increasing trend in the strength and duration of easterly winds over the northern Chukchi and southern Beaufort regions (Pickart et al., 2013) attributed to diminishing Arctic sea ice cover (Comiso et al., 2008). River discharge cycles may also be altered as a consequence of changes in seasonal precipitation rates and melting permafrost. As a consequence, bowhead foraging opportunities over the Beaufort shelf and along the Chukotkan coast may diminish or be altered in the future.

Acknowledgement

The authors thank Craig George for his many insights and discussions concerning polar ecosystems and to his able assistance and helpful advice in their research programs over many years.

References

Aagaard, K., Swift, J.H., Carmack, E.C., 1985. Thermohaline circulation in the arctic Mediterranean seas. J. Geophys. Res. 90, 4833–4846.

Barton, B., Lenn, Y.-D., Lique, C., 2018. Observed atlantification of the Barents Sea causes the Polar Front to limit the expansion of winter sea ice. J. Phys. Oceanogr. 48, 1849–1866. Available from: https://doi.org/10.1175/JPO-D-18-0003.1.

Björk, G., Gustafsson, B.G., Stigebrandt, A., 2001. Upper layer circulation of the Nordic seas as inferred from the spatial distribution of heat and freshwater content and potential energy. Polar Res. 20 (2), 161–168.

Chambault, P., Albertsen, C.M., Patterson, T.A., Hansen, R.G., Tervo, O., Laidre, K.L., et al., 2018. Sea surface temperature predicts the movements of an Arctic cetacean: the bowhead whale. Sci. Rep. 8, 9658. Available from: https://doi.org/10.1038/s41598-018-27966-1.

Citta, J.J., Lori, T., Quakenbush, L.T., Okkonen, S.R., Druckenmiller, M.L., Maslowski, W., et al., 2015. Ecological characteristics of core-use areas used by Bering–Chukchi–Beaufort (BCB) bowhead whales, 2006–2012. Prog. Oceangr. 136, 201–222.

Citta, J.J., Okkonen, S.R., Quakenbush, L.T., Maslowski, W., Osinski, R., George, J.C., et al., 2017. Oceanographic characteristics associated with autumn movements of bowhead whales in the Chukchi Sea. Deep Sea Res. II 152, 121–131. Available from: https://doi.org/10.1016/j.dsr2.2017.03.009.

Comiso, J.C., Parkinson, C.L., Gersten, R., Stock, L., 2008. Accelerated decline in the Arctic sea ice cover. Geophys. Res. Lett. 35, L01703. Available from: https://doi.org/10.1029/2007GL031972.

Curry, B., Lee, C.M., Petrie, B., 2011. Volume, freshwater, and heat fluxes through Davis Strait, 2004-2005. J. Phys. Oceanogr. 41, 429–436.

Danielson, S.L., Weingartner, T.W., Hedstrom, K., Aagaard, K., Woodgate, R., Curchitser, E., et al., 2014. Coupled wind-forced controls of the Bering–Chukchi shelf circulation and the Bering Strait through-flow: Ekman transport, continental shelf waves, and variations of the Pacific–Arctic sea surface height gradient. Prog. Oceanogr. Available from: https://doi.org/10.1016/j.pocean.2014.04.006.

Danielson, S.L., Eisner, L., Ladd, C., Mordy, C., Sousa, L., Weingartner, T.J., 2017. A comparison between late summer 2012 and 2013 water masses, macronutrients, and phytoplankton standing crops in the northern Bering and Chukchi Seas. Deep Sea Res. II 135, 7–26. Available from: https://doi.org/10.1016/j.dsr2.2016.05.024.

Fedoseev, G.A., 1984. Whale encounters in the ice massifs of the Sea of Okhotsk. Ekologiya 3, 81–83.

Germe, A., Houssais, M.-N., Herbaut, C., Cassou, C., 2011. Greenland Sea ice variability over 1979–2007 and its link to the surface atmosphere. J. Geophys. Res. 116, C10034. Available from: https://doi.org/10.1029/2011JC006960.

Goldbogen, J.A., Friedlaender, A.S., Calambokidis, J., McKenna, M.F., Simon, M., Nowacek, D.P., 2013. Integrative approaches to the study of baleen whale diving behavior, feeding performance, and foraging ecology. BioScience 63, 90–100. Available from: https://doi.org/10.1525/bio.2013.63.2.5.

Grebmeier, J., 1993. Studies on pelagic-benthic coupling extended onto the Russian continental shelf in the Bering and Chukchi Seas. Cont. Shelf Res. 13, 653–668.

Guay, C.K., Falkner, K.K., 1998. A survey of dissolved barium in the estuaries of major Arctic rivers and adjacent seas. Cont. Shelf Res. 8, 859–882.

Hamilton, J.M., Wu, Y., 2013. Synopsis and trends in the physical environment of Baffin Bay and Davis Strait. Can. Tech. Rep. Hydrogr. Ocean Sci. 282, 39.

Hansell, D., Whitledge, T.E., Georing, J.J., 1993. Patterns of nitrate utilization and new production over the Bering-Chukchi shelf. Cont. Shelf Res. 13, 601–627.

Hansen, B.S., Østerhus, W., Turrel, O.W., Jonsson, S., Valdimersson, H., Hatún, H., et al., 2008. The inflow of Atlantic water, heat and salt to the Nordic Seas across the Greenland-Scotland Ridge. In: Dickson, R.R., Meincke, J., Rhines, P. (Eds.), Arctic-Subarctic Ocean Fluxes: Defining the role of the Northern Seas in Climate. Springer-Verlag, Dordrecht, pp. 15–43.

Harris, C.L., Plueddemann, A.J., Gawarkiewicz, G.G., 1998. Water mass distribution and polar front structure in the western Barents Sea. J. Geophys. Res. 103, 2905–2917. Available from: https://doi.org/10.1029/97JC02790.

Ingram, R.G., Prinsenberg, S., 1998. Coastal oceanography of Hudson Bay and surrounding eastern Canadian Arctic waters coastal segment. In: Robinson, A.R., Brink, K.H. (Eds.), The Sea, Volume 11: The Global Coastal Ocean: Regional Studies and Syntheses (The Sea: Ideas and Observations on Progress in the Study of the Seas). Wiley & Sons, New York, pp. 835–861.

Ivaschenko, Y., Clapham, P., 2010. Bowhead whales *Balaena mysticetus* in the Okhotsk Sea. Mammal Rev. 40 (1). Available from: http://doi.org/10.1111/j.1365-2907.2009.00152.x.

Kowalik, Z., Polyakov, I., 1998. Tides in the Sea of Okhotsk. J. Phys. Oceanogr. 28, 1389–1409.

Li, M., Pickart, R.S., Spall, M.A., Weingartner, T.J., Lin, P., Moore, G.W.K., et al., 2019. Circulation of the Chukchi Sea shelfbreak and slope from moored timeseries. Prog. Oceanogr. 172, 14–33. Available from: http://doi.org/10.1016/j.pocean.2019.01.002.

Loeng, H., 1991. Features of the physical oceanographic conditions of the Barents Sea. In: Sakshaug, E., Hopkins, C.C.E., Britsland, N.A. (Eds.), Proceedings of the Pro Mare Symposium on Polar Marine Ecology. Polar Research, Trondheim, 10/1, S18, 12–16 May 1990.

Macdonald, R.W., Carmack, E.C., McLaughlin, F.A., Falkner, K.K., Swift, J.H., 1999. Connections among ice, run-off, and atmospheric forcing in the Beaufort Gyre. Geophys. Res. Lett. 26, 2223–2226.

Markina, N.P., Chernyavsky, V.I., 1984. Quantitative distribution of plankton and benthos in the Sea of Okhotsk. Izv. TINRO 109, 109–119 (in Russian).

Melling, H., Agnew, T.A., Falkner, K.K., Greenberg, D.A., Lee, C.M., Münchow, A., et al., 2008. Fresh-water fluxes via Pacific and Arctic Outflows cross the Canadian polar shelf. In: Dickson, R.R., Meincke, J., Rhines, P. (Eds.), Arctic-Subarctic Ocean Fluxes: Defining the role of the Northern Seas in Climate. Springer Verlag, Dordrecht, pp. 193–247.

Moore, S.E., George, J.C., Coyle, K.O., Weingartner, T., 1995. Bowhead whales along the Chukotka coast in autumn. Arctic 48, 155–160.

Mork, M., 1981. Circulation phenomena and frontal dynamics of the Norwegian Coastal Current. In: Swallow, J.C. (Ed.), Circulation and Fronts in Continental Shelf Seas. Royal Society, London, pp. 635–647.

Muilwijk, M., Smedsrud, L.H., Ilicak, M., Drange, H., 2018. Atlantic Water heat transport variability in the 20th century Arctic Ocean from a global ocean model and observations. J. Geophys. Res: Ocean 123, 8159–8179. Available from: https://doi.org/10.1029/2018JC014327.

Münchow, A., Melling, H., 2008. Ocean current observations from Nares Strait to the west of Greenland: Interannual to tidal variability and forcing. J. Mar. Res. 66, 801–833.

Nielsen, N.H., Laidre, Larsen, R.S., Heide-Jørgensen, M.P., 2015. Identification of potential foraging areas for bowhead whales in Baffin Bay and adjacent waters. Arctic 68 (2), 169–179. Available from: https://doi.org/10.14430/arctic4488.

Nihashi, S., Ohshima, K.I., Nakasato, H., 2011. Sea-ice retreat in the Sea of Okhotsk and the ice-ocean albedo feedback effect on it. J. Oceanogr. 67. Available from: https://doi.org/10.1007/s10872-011-0056-x.

NOAA National Centers for Environmental Information, 2019a. Hydrochemical Atlas of the Sea of Okhotsk, International Ocean Atlas Series, Volume 3, <https://www.nodc.noaa.gov/OC5/okhotsk/start_ok.html>.

NOAA National Centers for Environmental Information, 2019b. <https://www.nodc.noaa.gov/OC5/regional_climate/arctic/>.

Ohshima, K.I., Wakatsuchi, M., Fukamachi, Y., Mizuta, G., 2002. Near-surface circulation and tidal currents of the Okhotsk. Sea observed with satellite-tracked drifters. J. Geophys. Res. 107 (C11), 3195. Available from: https://doi.org/10.1029/2001JC001005.

Ohshima, K.I., Nihashi, S., Hashiya, E., Watanabe, T., 2006. Interannual variability of sea ice area in the Sea of Okhotsk: Importance of surface heat flux in fall. J. Meteorol. Soc. Jpn. 84, 907–919.

Okkonen, S.R., Ashjian, C., Campbell, R.G., Clarke, J.T., Moore, S.E., Taylor, K.D., 2011. Satellite observations of circulation features associated with a bowhead whale feeding 'hotspot' near Barrow, Alaska. Remote Sens. Environ. 115, 2168–2174.

Okkonen, S.R., Clarke, J.T., Potter, R.A., 2018. Relationships among high river discharges, upwelling events, and bowhead whale (*Balaena mysticetus*) occurrence in the central Alaskan Beaufort Sea. Deep Sea Res. II 152, 195–202. Available from: https://doi.org/10.1016/j.dsr2.2016.11.015.

Pickart, R.S., Pratt, L.J., Torres, D.J., Whitledge, T.E., Proshutinsky, A.Y., Aagaard, K., et al., 2010. Evolution and dynamics of the flow through Herald Canyon in the western Chukchi Sea. Deep Sea Res. II 57, 5–26.

Pickart, R.S., Schulze, L.M., Moore, G.W.K., Charette, M.A., Arrigo, K.R., van Dijken, G., et al., 2013. Long-term trends of upwelling and impacts on primary productivity in the Alaskan Beaufort Sea. Deep. Sea Res. I 79, 106–121. Available from: https://doi.org/10.1016/j.dsr.2013.05.003.

Piechura, J., Walczowski, W., 1995. The Arctic Front: structure and dynamics. Oceanologia 37, 47–73.

Radchenko, V.I., Dulepova, E.P., Figurkin, A.L., Katugin, O.N., Ohshima, K., Nishioka, J., et al., 2010. Status and trends of the Sea of Okhotsk region, 2003-2008. In: McKinnell, S.M., Dagg, M.J. (Eds.), Marine Ecosystems of the North Pacific Ocean, 2003-2008. North Pacific Marine Science Organization, PICES Special Publication 4, pp. 268–299.

Rogachev, K.A., Carmack, E.C., Foreman, M.G.G., 2008. Bowhead whales feed on plankton concentrated by estuarine and tidal currents in Academy Bay, Sea of Okhotsk. Cont. Shelf Res. 28, 1811–1826.

Rudels, B., Friedrich, H.J., Quadfasel, D., 1999. The arctic circumpolar boundary current. Deep Sea Res. II 46, 1023–1062.

Schauer, U., Fahrbach, E., Osterhus, S., Rohardt, G., 2004. Arctic warming through the Fram Strait: Oceanic heat transport from 3 years of measurements. J. Geophys. Res. 109, C06026. Available from: https://doi.org/10.1029/2003JC001823.

Schulze, L.M., Pickart, R.S., 2012. Seasonal variation of upwelling in the Alaskan Beaufort Sea: Impact of sea ice cover. J. Geophys. Res. 117, C06022. Available from: https://doi.org/10.1029/2012JC007985.

Shpaka, O.V., Paramonov, A.Y., 2018. The bowhead whale, *Balaena mysticetus Linnaeus, 1758*, in the Western Sea of Okhotsk (2009–2016): distribution pattern, behavior, and threats. Russ. J. Mar. Biol. 44 (3), 210–218.

Stabeno, P.J., Schumacher, J.D., Ohtani, K., 1999. The physical oceanography of the Bering Sea: A summary of physical, chemical, and biological characteristics, and a synopsis of research on the Bering Sea. In: Loughlin, T.R., Ohtani, K. (Eds.), Dynamics of the Bering Sea: A Summary of Physical, Chemical, and Biological Characteristics, and a Synopsis of Research on the Bering Sea. North Pacific Marine Science Organization (PICES), University of Alaska Sea Grant, AK-SG-99-03, pp. 1–28.

Sutherland, D.A., Pickart, R.S., 2008. The East Greenland Coastal Current: structure, variability, and forcing. Prog. Oceanogr. 78, 58–77.

Talley, L.D., Nagata, Y., 1995. The Okhotsk Sea and Oyashio Region (Report of Working Group I) PICES Scientific Report #2. North Pacific Marine Sciences Organization (PICES), 227 p.

Tang, C.L., Ross, C.K., Yao, T., Petrie, B., Detracey, B.B., Dunlap, E., 2004. The circulation, water masses and sea-ice of Baffin Bay. Prog. Oceanogr. 63, 183–228.

Vernet, M., Ellingsen, I.H., Seuthe, L., Slagstad, D., Cape, M.R., Matrai, P.A., 2019. Influence of phytoplankton advection on the productivity along the Atlantic water inflow to the Arctic Ocean. Front. Mar. Sci. Available from: https://doi.org/10.3389/fmars.2019.00583.

Walkusz, W., Paulic, J.E., Kwasniewski, S., Williams, W.J., Wong, S., Papst, M.H., 2010. Distribution, diversity and biomass of summer zooplankton from the coastal Canadian Beaufort Sea. Polar Biol. 33, 321–335.

Walkusz, W., Williams, W.J., Harwood, L.A., Moore, S.E., Stewart, B.E., Kwasniewski, S., 2012. Composition, biomass and energetic content of biota in the vicinity of feeding bowhead whales (*Balaena mysticetus*) in the Cape Bathurst upwelling region (south eastern Beaufort Sea). Deep Sea Res. I 69, 25–35.

Watanabe, E., 2011. Beaufort shelf break eddies and shelf-basin exchange of Pacific summer water in the western Arctic Ocean detected by satellite and modeling analyses. J. Geophys. Res. 116, C08034. Available from: https://doi.org/10.1029/2010JC006259.

Weingartner, T.J., Danielson, S., Sasaki, Y., Pavlov, V., Kulakov, M., 1999. The Siberian Coastal Current: a wind- and buoyancy-forced Arctic coast current. J. Geophys. Res. 104, 26697–29713.

Weingartner, T.J., Aagaard, K., Woodgate, R.A., Danielson, S., Sasaki, Y., Cavalieri, D., 2005. Circulation on the north central Chukchi Sea shelf. Deep Sea Res. II. Available from: https://doi.org/10.1016/j.dsr2.2005.10.015.

Weingartner, T., Danielson, S.L., Potter, R.A., Trefry, J.H., Mahoney, A., Savoie, M., et al., 2017. Circulation and water properties in the landfast ice zone of the Alaskan Beaufort Sea. Cont. Shelf Res. 148, 185–198. Available from: https://doi.org/10.1016/j.csr.2017.09.001.

Wigg, O., Bachmann, L., 2007. Spitsbergen bowhead whales revisited. Mar. Mammal Sci. 23 (3), 688–693. Available from: https://doi.org/10.1111/j.1748-7692.2007.02373.x.

Williams, W.J., Carmack, E.C., 2008. Combined effect of wind-forcing and isobath divergence on upwelling at Cape Bathurst Beaufort Sea. J. Mar. Res. 66, 645–663.

Woodgate, R.A., Fahrbach, E., Rohardt, G., 1999. Structure and transport of the East Greenland Current at 75°N from moored current meters. J. Geophys. Res. 104, 18059–18072.

Woodgate, R.A., Aagaard, K., Weingartner, T., 2005. A Year in the Physical Oceanography of the Chukchi Sea: moored measurements from autumn 1990-91. Deep Sea Res. II 52, 3116–3149.

Yao, T., Tang, C.L., 2003. The formation and maintenance of the North Water Polynya. Atmos. Ocean 41 (3), 187–201. Available from: https://doi.org/10.3137/ao.410301.

CHAPTER

26

Biological environment

C.J. Ashjian[1], R.G. Campbell[2] and S.R. Okkonen[3]

[1]Department of Biology, Woods Hole Oceanographic Institution, Woods Hole, MA, United States [2]Graduate School of Oceanography, University of Rhode Island, Kingston, RI, United States [3]College of Fisheries and Ocean Sciences, University of Alaska Fairbanks, Fairbanks, AK, United States

Introduction

The Arctic and sub-Arctic seas that form the core habitats for the bowhead whale provide a rich and abundant supply of planktonic prey. Both physical and biological characteristics drive the availability of the large copepods and euphausiids (krill) that are common in bowhead whale diets. Currents carry plankton and the nutrients that sustain primary productivity and ultimately zooplankton production into bowhead whale habitats (Chapter 25). Smaller scale physical processes such as shelf break upwelling of plankton onto the shallow shelf and convergent fronts between water types generate patches of aggregated plankton that permit bowhead whales to feed efficiently. Zooplankton behavior such as diel vertical migration (DVM) or swimming to layers of phytoplankton food also increases their availability to the whales.

Here, the biological characteristics important to bowhead whales of the four primary bowhead habitat regions are considered: (1) The Bering–Chukchi–Beaufort Seas (BCB stock, see Chapter 3), (2) the East Canada-West Greenland (ECWG stock), (3) the Sea of Okhotsk (OKH stock), and (4) East Greenland–Svalbard–Barents Seas (EGSB stock). These regions are classified either as Arctic or sub-Arctic and share common characteristics, the most notable being seasonal sea ice and the presence of large, lipid-rich zooplankton that are key bowhead whale prey. Seasonal sea ice drives the phenology of the Arctic, with many species having life histories timed to coincide with the annual retreat of sea ice. The life histories and ecology of the dominant bowhead whale prey species in conjunction with patterns of primary production and with the physical environment determine the regional abundance of the prey as well as their responses to climatic environmental change (Fig. 26.1).

The Bowhead Whale
DOI: https://doi.org/10.1016/B978-0-12-818969-6.00026-1

FIGURE 26.1 Bowhead whales of the Bering–Chukchi–Beaufort stock live in an ecosystem with other marine mammals, including walrus, bearded seal, spotted seal, ringed seal, and beluga (seen here). All feed on invertebrate and vertebrate prey of their Arctic ecosystem. *Source: Photo by Vicki Beaver (NOAA/North Slope Borough, NMFS Permit No. 14245).*

General characteristics

Primary production in all regions is driven by ice algae, primarily diatoms growing on the underside of or interstitially in sea ice, and by phytoplankton, including diatoms, dinophytes, chrysophytes, and prymnesiophytes. The annual cycle in sea ice and snow cover drives the timing of algal blooms when sunlight reaches the underside of the sea ice (ice algae) and upper water column (phytoplankton). Ice-associated enhancement of primary production occurs at ice edges in marginal ice zones, in polynyas, and under melt ponds and leads. Primary production also is increased through oceanic physical drivers, including input of nutrients through advection or vertical mixing and shelf break upwelling (Sakshaug, 2004; Pickart et al., 2013; Vernet et al., 2019). Pelagic secondary producers include microzooplankton (ciliates, flagellates), mesozooplankton (e.g., copepods, euphausiids or krill), and macrozooplankton (amphipods, gelatinous forms). Large-bodied copepods, and euphausiids when present, usually comprise the greatest portion of zooplankton biomass. Primary production, especially large accumulations of ice algae, not consumed in the water column by the zooplankton falls to the sea floor to be consumed by benthic organisms (Klages et al., 2004; Grebmeier, 2012). Both water column and benthic secondary producers sustain the upper trophic levels, including marine mammals, seabirds, and fish.

Although a wide range of zooplankton is consumed by bowhead whales, most of their diet is composed of a few large zooplankton species (Chapter 28). These species are found primarily in the deeper shelves, slope, and basin with their presence near-shore resulting from physical translocations. Large-bodied copepod species of the genus *Calanus* that are key bowhead whale prey dominate zooplankton biomass in all regions. Two *Calanus* species (*Calanus hyperboreus* and *Calanus glacialis*) are considered Arctic endemics, although *C. glacialis* also is endemic in some sub-Arctic regions. *C. hyperboreus* is found primarily along deeper slopes and in basins while *C. glacialis* is associated with shelves and slopes (Falk-Petersen et al., 2009). Two additional sub-Arctic *Calanus* spp. also are found in bowhead whale habitats; *Calanus finmarchicus* in eastern regions influenced by water from the sub-Arctic Atlantic and *Calanus marshallae* in western regions influenced by the sub-Arctic Pacific (Wassmann et al., 2015). In the Bering and Okhotsk Seas, three *Neocalanus* copepod species (*Neocalanus cristatus*, *Neocalanus flemingeri*, and *Neocalanus plumchrus*) also can be important in deeper water off of the shelf. The southern Okhotsk Sea also contains the temperate species *Calanus pacificus*. All of these species are relatively large (2.5 mm or larger adult female prosome length) and follow multiple year life cycles (2–3 years from egg to adult through 14 life stages) in the cold sub-Arctic and Arctic seas (Conover, 1988).

These copepod species have an obligate overwintering or resting stage termed diapause in which they reduce their metabolism dramatically and subsist off of stored lipid reserves accumulated during the summer-fall growing period (Falk-Petersen et al., 2009). Critical to successful diapause is accumulation of sufficient lipid to last through winter until the spring productive season. This accumulation of lipid also makes these species a high energy food source for their predators. Diapausing copepods also are vulnerable to predation because their low metabolism prevents effective escape. Overwintering occurs at depth, well below the euphotic zone, and is not thought to be successful on shallow (e.g., 50 m) shelves. Diapause ceases during spring, just prior to the reproductive season. *C. hyperboreus* reproduce at their overwintering depths (usually deeper than ~200 m) and use stored lipid for egg production. The lipid-rich eggs float from depth to the surface to hatch in time to exploit the spring algal blooms. *C. glacialis* starts reproduction later, when sea ice diminishes, and the species relies on ice algae and early phytoplankton blooms and potentially stored lipid (Daase et al., 2013). The *Neocalanus* spp. reproduce at depth (400–600 m) only in their native sub-Arctic habitat, whereas *N. plumchrus* and *N. cristatus* reproduce in late summer through winter and *N. flemingeri* reproduces in early winter (Miller et al., 1984; Miller and Clemons, 1988). Although *N. flemingeri* is able to reproduce in Arctic waters, recruitment is limited due to low hatching success of their eggs (Matsuno et al., 2015). In their native habitats, *C. finmarchicus* and *C. marshallae* reproduce in the upper water column, following diapause and in association with the spring bloom. These species may sustain reproduction throughout the productive season, producing multiple generations per year. However, they appear unable to successfully reproduce when advected into Arctic water (e.g., Hirche and Kosobokova, 2007).

Euphausiids, or krill, are another prominent prey item for bowhead whales. They do not appear to be endemic to the Arctic proper but rather are advected there from sub-Arctic source regions (Dalpadado and Skjoldal, 1991; Berline et al., 2008). Four species of *Thysanoessa* (*Thysanoessa inermis*, *Thysanoessa raschii*, *Thysanoessa longicaudata*, and *Thysanoessa longipes*) and, in the eastern Arctic, the Atlantic species *Meganyctiphanes norvegica* and *Nematosceles megalops* are found in bowhead whale habitats. Together with *Calanus* spp., they are critical prey for the younger life stages of commercially important pelagic fishes. Euphausiids follow a 2- to 3-year life cycle, transitioning through 14–15 life stages, and store modest amounts of lipid for overwintering but they do not enter a diapause stage as do the Calanidae (Siegel, 2000). Similar to *Calanus* spp., the euphausiids reproduce in spring coincident with the retreat of seasonal sea ice and the ice algal and phytoplankton blooms. *T. inermis* may utilize stored reserves to produce eggs, permitting early reproduction, but *T. raschii* requires food to support reproduction and thus initiates reproduction later in the spring. Euphausiids, because of their larger body size, are better swimmers than copepods. This, coupled with better vision because of their compound eyes (copepods have light sensing organs but no eyes), make them more elusive to plankton nets so that their abundance may be underestimated due to net avoidance during sampling.

All of these copepods and euphausiids are omnivores, consuming ice algae, phytoplankton, and microzooplankton, and, for euphausiids, detritus. Their preference for herbivory varies between species and with the relative availability of the different food types. For instance, *C. glacialis* preferentially consumes microzooplankton over algae, particularly during summer months when the smaller phytoplankton species are abundant (Campbell et al., 2009). By contrast, *C. hyperboreus* appears to consume the different food types indiscriminately or perhaps even with a preference for phytoplankton. Euphausiids consume not only algae and microzooplankton but also the smaller life stages of copepods and they practice detritivory. Both copepods and euphausiids exploit the under-ice environment, utilizing ice algae that sloughs off of the sea ice or on the algae while it is attached and with euphausiids feeding on epontic algal films on the underside of the sea ice.

The vertical distributions of the bowhead's planktonic prey generally follow three different patterns: Association with physical and biological characteristics of the water column, DVM, and ontogenetic vertical migration. Both copepods and euphausiids may be associated with specific types of water mass and follow the vertical distribution of that water. Both may be found in the euphotic zone where their algal and microzooplankton prey are most abundant. Both frequently are found in conjunction with layers or patches of their phytoplankton and microzooplankton prey that form at density interfaces, often at the base of the euphotic zone.

All of these major prey taxa engage in light-associated DVM, spending periods of daylight at depth and migrating up to the euphotic zone during darkness. DVM is believed to be a strategy to avoid predation by visual predators in the euphotic zone. Because of DVM, a substantial proportion of the zooplankton biomass/whale prey relocates between the near surface and depth on a diel basis. There have been reports of sustained daily/ DVMs during the darkness of Arctic winter, however, the amplitude and involved biomass are reduced.

Ontogenetic vertical migration refers to the seasonal descent to depth for overwintering well below the euphotic zone, often at depths of several hundred meters. Studies of dense aggregations of overwintering *C. finmarchicus* in deep holes on the Nova Scotia shelf suggest that there is a critical depth to which the animals are obligated to migrate for successful overwintering (Sameoto and Herman, 1990). Such dense aggregations can form when overwintering animals are constrained by the sea floor from migrating deeper. Similar processes likely drive overwintering aggregations of *Calanus* spp. and *Neocalanus* spp. in bowhead whale habitats.

Biogeography of bowhead habitats

The Bering, Chukchi, and Beaufort Seas

The Bering–Chukchi–Beaufort Seas encompass three distinct ocean regions coupled by strong advection (Chapter 25). The biogeography of the Bering Sea is sub-Arctic while the Chukchi and Beaufort both are classified as Arctic. Both the Bering and Beaufort Seas contain shelf, slope, and basin regions while the Chukchi Sea is a shelf-sea, with water depths of 30–50 m. While the Beaufort and Bering Seas contain a range of hydrographic features, the Chukchi Sea is dominated by northward flow along three pathways from Bering Strait to the northern edges of the Chukchi Shelf. Both the Chukchi and Bering Seas have high primary production (Bering: $>230\,gC/m^2/year$; Chukchi: up to $>400\,gC/m^2/year$; Sakshaug, 2004; Grebmeier, 2012; Mathis et al., 2014). By contrast, primary production in the Beaufort Sea, including over the Canada Basin, is much lower (Beaufort Shelf: $30-70\,gC/m^2/year$; Sakshaug, 2004) and the biomass of the secondary producers is reduced relative to the Chukchi and Bering Seas. Primary production in the Bering and Beaufort Seas is sustained through a combination of regenerated, resuspended nutrients and upwelling along the shelf and slope while Chukchi Sea production also is sustained by nutrients carried in the dominant currents from the Bering Sea (Chapter 25).

The prevailing currents of the Chukchi Sea carry Bering Sea zooplankton (*C. glacialis*, *C. marshallae*, *Neocalanus* spp., and *Thysanoessa* spp.) northward (e.g., Nelson et al., 2014; Wassmann et al., 2015). These same currents, particularly the central and western pathways, also carry euphausiids believed to be transported into the Chukchi Sea. Little is known about abundances of euphausiids in the western Bering Sea/Gulf of Anadyr, however, high abundances appear to be advected into the Chukchi Sea through the western side of Bering Strait, based on bowhead lingering locations along the Chukotka Coast and in the northern central Chukchi Sea (Chapter 4) and limited descriptions of zooplankton communities (e.g., Ershova et al., 2015). The northward currents in the Chukchi thus link the three ecosystems and deliver significant bowhead prey biomass to the northern regions. Overwintering of copepods and euphausiids occurs in the deep basins of the Bering and Beaufort Seas as well as on the outer Bering Sea shelf. Upwelling of zooplankton, including Bering sea euphausiids, occurs along the Chukchi and Beaufort shelf breaks, with their occurrence sometimes extending far onto the shelf (Ashjian et al., 2010; Smoot and Hopcroft, 2017), providing feeding opportunities for bowheads along its length. Upwelling also occurs to the south and west of St. Lawrence Island, in the Bering Sea. Zooplankton from both the Canada Basin and the Chukchi slope are regularly transported into the Chukchi Sea in the prevailing southward currents along the western side of Barrow Canyon or in episodic southward flow across the entire canyon.

Sea of Okhotsk

The Sea of Okhotsk is essentially a microcosm of a larger, sub-Arctic–Arctic ocean system, with regions proscribed by mesoscale current features and exhibiting sub-Arctic and Arctic flora and fauna (Pinchuk and Paul, 2000; Dulepova and Radchenko, 2004; Shuntov et al., 2019). It is highly productive (100–200 gC/m^2/year; Sakshaug, 2004), particularly in the north, supporting substantial pelagic and benthic fisheries. The northern region is characterized by Arctic fauna (*C. glacialis*) while the central and southern regions contains sub-Arctic (*N. plumchrus* and *N. cristatus*) and even temperate species (*C. pacificus*). The euphausiid *T. raschii* also has been reported for the northern and central regions. Zooplankton in the southern Okhotsk is influenced by inflow of water, and intrinsic plankton, from the adjacent North Pacific. Zooplankton biomass is correlated with water temperature, with greater biomass in colder years (Dulepova, 2008).

East Greenland–Svalbard–Barents Seas

The biogeography of the EGSB is defined to a great extent by the Norwegian-Atlantic Slope Current (NwASC) and the East Greenland Current (EGC). The northward flowing NwASC carries Atlantic water and biota into the southern Barents Sea, north along western Svalbard, and along the Barents slope to the east north of Svalbard. This same current recirculates to the west across the Greenland Sea, joining the EGC that flows southward out of the Arctic Ocean along the eastern coast of Greenland. The NwASC carries nutrients, elevated phytoplankton biomass, and North Atlantic species, including *C. finmarchicus* while the EGC carries few nutrients, very little phytoplankton biomass, and the Arctic

C. hyperboreus and *C. glacialis* (Wassmann et al., 2015; Vernet et al., 2019). As a consequence of the NwASC inflow, the Barents Sea and region near Svalbard are highly productive (up to 200 gC/m^2/year; Sakshaug, 2004). North Atlantic water also extends into the southern Barents Sea so that the Atlantic species *C. finmarchicus* is abundant while Arctic species dominate in the northern Barents Sea (Tande, 1991) with the two regions separated hydrographically by the Polar Front. *C. finmarchicus* is found north of Svalbard as non-native, especially along the slope north and east of Svalbard (Kosobokova and Hirche, 2009), although establishment in those regions is not successful (Hirche and Kosobokova, 2007; Wassmann et al., 2015). Overwintering of the large *Calanus* spp. occurs in the deep Greenland and Norwegian Seas, off of the shelves. Euphausiids (*T. inermis, T. raschii, T. longicaudata, M. norvegica,* and *N. megalops*) are common in the southern Barents Sea but are found in lower numbers in the northern Barents Sea and around Svalbard (Dalpadado and Skjoldal, 1991). The prominent fjords in the western Svalbard archipelago contain the Arctic *C. glacialis* as well as the Atlantic *C. finmarchicus* and both boreal (*T. inermis, T. raschii*) and Atlantic (*T. longicaudata, M. norvegica,* and *N. megalops*) euphausiids (Gluchowska et al., 2016; Hirche et al., 2016; Lischka and Hagen, 2016). Although historically believed to be nonreproductive in the Svalbard fjords, recent evidence suggests euphausiids may now be able to successfully reproduce there (Buchholz et al., 2012).

The EGC flows along the eastern coast of Greenland, advecting both the Arctic *Calanus* species from the Arctic and Greenland Sea and the Atlantic *C. finmarchicus* from the retroflected NwASC southwards through the Greenland–Iceland–Norwegian Seas to Fram Strait and the waters around Iceland (Michel et al., 2015). Primary production varies along the EGC, with northern regions being of low productivity because of ice cover and enhanced productivity at marginal ice zones and along the coast of Greenland (Michel et al., 2015). Along the East Greenland Shelf, Atlantic Intermediate Water is upwelled onto the shelf through troughs and canyons, bringing *C. finmarchicus* and *C. hyperboreus* onto the shelf (Ashjian et al., 1997; Michel et al., 2015).

East Canada–West Greenland

The East Canada–West Greenland (ECWG) region extends from the deep oceanic Baffin Bay west to the shallower Hudson Bay, connected by the channels of the Canadian Archipelago. Similarly to the EGSB region, the ECWG is characterized by a confluence of a major, northward Atlantic influenced current and a southward Arctic outflow (Michel et al., 2015). The West Greenland Current (WGC) flows north along the western coast of Greenland to northern Baffin Bay where it turns south, joining with water outflowing from the Canadian Archipelago. Water and plankton also exit the Arctic through Nares Strait at the northern end of Baffin Bay. The zooplankton of eastern Baffin Bay, along the West Greenland Coast, contains both Arctic and Atlantic *Calanus* spp., with *C. finmarchicus* advected there in the WGC, and euphausiids. *C. finmarchicus* can be the biomass dominant species along the western Greenland shelf and in the fjord systems. The importance of *C. finmarchicus* decreases moving to the north and along western Baffin Bay; the species is absent from the Canadian Archipelago and Hudson Bay. The Canadian Archipelago is one of two major outflow shelves of the Arctic Ocean (Michel et al., 2015) and as such the zooplankton community there is distinctly Arctic in nature. Both Atlantic and Arctic *Calanus* spp. are important in Hudson Strait and Fox Basin, where Atlantic Water from the Labrador Sea joins Arctic water from the Archipelago (Estrada et al., 2012). The euphausiid *T. raschii* also is present, although as a low proportion of the zooplankton biomass. Hudson Bay itself is dominated by smaller copepods species, with few *Calanus* spp. found.

Formation of feeding hotspots

The "continuous filter feeding" strategy of bowhead whales requires high concentrations or patches of their plankton prey in order to feed efficiently (Chapter 14). Zooplankton is generally too dispersed to provide suitable feeding opportunities unless concentrated by physical mechanisms, biological characteristics such as zooplankton behavior, distributions, and life histories or combinations of the two. Each of the bowhead whale habitats exhibits locations where biological distributions and/or behavior intersect with physical processes to recurrently produce favorable feeding environments or "hotspots" for the bowhead whales. Wind-driven upwelling along shelf breaks can transport high abundances of large, lipid-rich zooplankton (*Calanus* spp.) or krill onto shelves, concentrating them and generating a bowhead whale feeding hotspot such as occurs near Cape Bathurst (e.g., Walcusz et al., 2012). Upwelling of nutrients also enhances primary production (Pickart et al., 2013) that in turn attracts zooplankton to feed, creating a layer of elevated bowhead whale prey. Convergent frontal regions between different water masses or resulting from tidal and estuarine currents also concentrate zooplankton. In the northern Sea of Okhotsk (Academy Bay), estuarine and tidal circulation coupled with DVMs concentrates copepods and support bowhead whale feeding (Rogachev et al., 2008). Bowhead whales congregate at fronts between river discharge and shelf water along the Beaufort Shelf following shelf break upwelling, presumably to feed on zooplankton patches that collect at those convergences (Okkonen et al., 2018). Zooplankton behavior also can generate patches of bowhead prey. In Disko Bay, Greenland, dense aggregations at depth of presumably diapausing *C. finmarchicus* are targeted by diving bowhead whales (Laidre et al., 2007). Dense aggregations of overwintering copepods and of euphausiids also have been observed near bottom in Kongsfjorden, Svalbard (Hirche et al., 2016). On larger scales, polynyas, formed either convectively (upwelling of warmer water) or mechanically (winds or current forced divergence of ice) (Williams et al., 2007), are known sites of elevated primary and secondary production and provide enhanced prey as well as reduced sea ice cover for bowhead whale feeding. For bowhead whales, the North Water Polynya in northern Baffin Bay and Northeast Water Polynya on the NE corner of Greenland all contain both the Arctic *Calanus* spp. and the Atlantic *C. finmarchicus* and provide feeding opportunities for a wide range of upper trophic level species (Boertmann et al., 2015; Michel et al., 2015). Smaller recurrent polynyas important to bowhead whales include the St. Lawrence Island polynya on the Bering Shelf, Whalers Bay to the north of Svalbard, and the Cape Bathurst Polynya in the Canadian Beaufort (Karnovsky et al., 2007; Falk-Petersen et al., 2014).

The region near Point Barrow, Alaska, is known as a core use area for bowhead whales during their fall southward migration from the Canadian Arctic to the Bering Sea (Citta et al., 2015; Chapters 4 and 24). Over the last 15 years, studies have identified the physical mechanisms that produce a feeding hotspot there and persistently attract bowhead whales. Local winds and currents interact to concentrate euphausiids (krill) into dense patches on the shallow shelf near Point Barrow, Alaska (Ashjian et al., 2010), a sequence of events now commonly called the "krill trap." Krill are carried in the prevailing currents north through the Chukchi Sea and along the Beaufort Sea slope, however, they are quite dispersed. Winds from the east initiate upwelling along the Beaufort shelf, bringing water and krill onto the shelf especially through shelf break troughs, and push the northeastward flowing Alaskan Coastal Current into the center of Barrow Canyon, away from the eastern wall (Fig. 26.2A). Bowhead whales may be in the region but are not observed in large numbers (Okkonen et al., 2011). Continued easterly winds move krill across the shelf inshore, with DVM of water column scatters typical of krill occurring when the krill trap is active (Fig. 26.3). If easterly winds persist, Beaufort shelf waters with newly upwelled krill are carried westward off-shelf past Point Barrow into Barrow Canyon where the krill are entrained in a southwestward flow into the Chukchi Sea (Fig. 26.2B). Bowhead whales may recognize the impending formation of the feeding hotspot. If the upwelling winds are followed by south/southwesterly winds or gentle winds from any direction, the Alaskan Coastal Current returns to lie along the eastern canyon wall, preventing water and krill from escaping around Point Barrow (Fig. 26.2C). Because of momentum, the shelf currents continue to the west, creating a convergence zone on the shelf to the north and east of Point Barrow where krill are concentrated into dense patches (Fig. 26.2C). The krill trap is considered to be "on" for days when upwelling winds have been followed by calm or southerly winds. Krill abundances are substantially greater on the shelf when the krill trap is on than when it is off (Fig. 26.4) and bowhead whales are found in high abundances in that region (Okkonen et al., 2011).

The formation of the krill trap hotspot in fall relies on the occurrence of the specific wind sequence that aggregates the krill (easterly winds followed by calm or southerly winds) and possibly equally important, a sufficient supply of krill that have been advected to the Beaufort Sea from the Bering Sea. This upstream source of krill in turn depends on

FIGURE 26.2 Interaction of winds and currents that drive the Utqiaġvik krill trap. The perspective in this figure is looking at the northern coast of Alaska and Point Barrow from the northeast, over the Canada Basin. Bottom topography is from IBCAO3 (Jakobsson et al., 2012). The occurrence of easterly winds is shown by the *stylized cloud figure*. Krill are shown as the *pink shading*. The Alaskan Coastal Current is shown as the *blue arrows*.

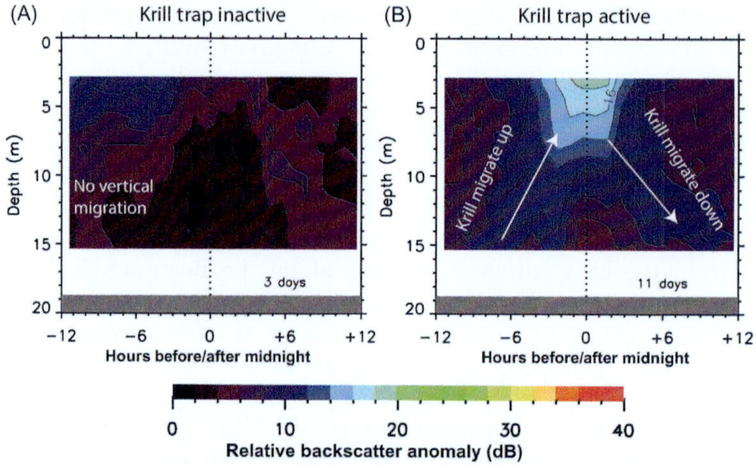

FIGURE 26.3 Daily vertical distribution of water column backscatter when the krill trap is "off" (A) compared to when the krill trap is "on" (B). "On" refers to days when upwelling winds have been followed by calm or southerly winds. A distinct DVM signal is observed when the krill trap is on. Backscatter, the strength of the reflection of sound on objects in the water (such as krill) is proportional to the accumulated mass of those objects. Backscatter was measured using a 300 kHz RDI acoustic Doppler current profiler moored for 2 weeks in August to early September 2009 at 19 m bottom depth on the Beaufort Shelf at ~71.35°N, 155.23°W (see Okkonen, 2013, for more information).

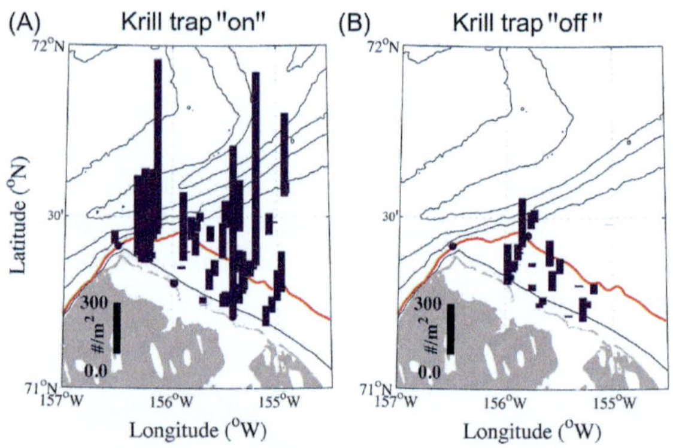

FIGURE 26.4 Abundances of krill on the shelf north and east of Point Barrow when the krill trap is "on" (A) and when the krill trap is "off" (B). Krill abundances collected using a ¼ m² Tucker Trawl in August–September 2009–15 (see Ashjian et al., 2013, for more information).

successful recruitment in the Bering Sea and survival during the transit through the Chukchi Sea. Bowhead whales predictably feed near Point Barrow during their fall southward migration, presumably utilizing the krill trap feeding hotspot, however, conditions in the Western Arctic in recent years suggest that efficacy of the krill trap hotspot may be becoming less reliable. The required periodic oscillation between upwelling and relaxed wind regimes may be transitioning to a regime characterized by more persistent, less variable winds. Sea ice extent and seasonal coverage are undergoing remarkable reductions, so much so that the Bering Sea cold pool has not formed in recent winters (Stabeno and Bell, 2019; Duffy-Anderson et al., 2019). This sea ice reduction is not conducive to successful krill recruitment, since recruitment success of large-bodied crustaceans, including krill in the Bering Sea is higher during years with cold, icy winters than those that are warmer with little ice (Eisner et al., 2014; Kimmel et al., 2018).

Climatically driven environmental changes

Ongoing climatically driven environmental change will have impacts on the biological environment of the Arctic bowhead whale habitats (see Chapter 27). Decreasing sea ice thickness and seasonal sea ice extent and changes in the timing of seasonal sea ice formation and retreat can significantly modify the timing of primary production, with important consequences to the zooplankton whose life histories are tightly tied to sea ice phenology (Wassmann and Reigstad, 2011). Increasing advection of sub-Arctic waters and elevated water temperatures may facilitate the northward movement and successful recruitment of sub-Arctic species (Wassmann et al., 2011). The unusual absence of bowheads in the Point Barrow region in fall 2019 may be indicative of the kinds of changes that may occur in the coming decades (Chapter 24). Ultimately, sustained environmental changes could result in shifts in zooplankton community composition that may or may not be favorable as prey for bowhead whales.

Acknowledgements

We gratefully acknowledge the many people and organizations that supported and encouraged the fieldwork at Utqiaġvik. Funding for that work was provided by the U.S. National Science Foundation, the U.S. National Oceanic and Atmospheric Administration, the U.S. Bureau of Ocean Energy Management, the U.S. National Oceanographic Partnership Program, the UAF Coastal Marine Institute, and the WHOI James M. and Ruth P. Clark Arctic Research Initiative Fund.

References

Ashjian, C., Campbell, R.G., Okkonen, S., Alatalo, P., 2013. Broad-scale oceanography. In: Shelden, K.E.W., Mocklin, J.A. (Eds.), Bowhead Whale Feeding Ecology Study (BOWFEST) in the Western Beaufort Sea. Final Report, OCS Study BOEM 2013-0114. National Marine Mammal Laboratory, Alaska Fisheries Science Center, NMFS, NOAA, Seattle, WA, pp. 153–194.

Ashjian, C., Smith, S., Bignami, F., Hopkins, T., Lane, P., 1997. Distribution of zooplankton in the Northeast Water Polynya during summer 1992. J. Mar. Syst. 10, 279–298.

Ashjian, C.J., Braund, S.R., Campbell, R.G., George, J.C., Kruse, J., Maslowski, W., et al., 2010. Climate variability, oceanography, bowhead whale distribution, and Iñupiat subsistence whaling near Barrow, AK. Arctic 63, 179–194.

Berline, L., Spitz, Y.H., Ashjian, C.J., Campbell, R.G., Maslowski, W., Moore, S.E., 2008. Euphausiid transport in the Western Arctic Ocean. Mar. Ecol. Prog. Ser. 360, 163–178.

Boertmann, D., Kyhn, L., Witting, L., Heide-Jorgensen, M.P., 2015. A hidden getaway for bowhead whales in the Greenland Sea. Polar Biol. 38, 1315–1319. Available from: https://doi.org/10.1007/s00300-015-1695-y.

Buchholz, F., Werner, T., Buchholz, C., 2012. First observation of krill spawning in the high Arctic Kongsfjorden, west Spitsbergen. Polar Biol. 35, 1273–1279. Available from: https://doi.org/10.1007/s00300-012-1186-3.

Campbell, R.G., Sherr, E.B., Ashjian, C.J., Plourde, S., Sherr, B.F., Hill, V., et al., 2009. Mesozooplankton prey preference and grazing impact in the Western Arctic Ocean. Deep Sea Res. II 56, 1274–1289.

Citta, J.J., Quakenbush, L.T., Okkonen, S.R., Druckenmiller, M.L., Maslowski, W., Clement-Kinney, J., et al., 2015. Ecological characteristics of core areas used by western Arctic bowhead whales, 2006–2012. Prog. Oceanogr. 136, 201–222. Available from: https://doi.org/10.1016/j.pocean.2014.08.012.

Conover, R.J., 1988. Comparative life histories in the genera *Calanus* and *Neocalanus* in high latitudes of the northern hemisphere. Hydrobiologia 167/168, 127–142.

Daase, M., Falk-Petersen, S., Varpe, Ø., Darnis, G., Søreide, J.E., Wold, A., et al., 2013. Timing of reproductive events in the marine copepod *Calanus glacialis*: a pan-Arctic perspective. Can. J. Fish. Aquat. Sci. 70, 871–884. Available from: https://dx.doi.org/10.1139/cjfas-2012-0401.

Dalpadado, P., Skjoldal, H.R., 1991. Distribution and life history of krill from the Barents Sea. In: Sakshaug, E. Hopkins, C.C.E., Øritsland, N.A. (Eds.), Proceedings of the Pro Mare Symposium on Polar Marine Ecology, Trondheim, Norway, May 12–16, 1990. Polar Res. 10, 443–460.

Duffy-Anderson, J.T., Stabeno, P., Andrews III, A.G., Ciecel, K., Deary, A., Farley, E., et al., 2019. Responses of the northern Bering Sea and southeastern Bering Sea pelagic ecosystems following record-breaking low winter sea ice. Geophys. Res. Lett. 46, 9833–9842. Available from: https://doi.org/10.1029/2019GL083396.

Dulepova, E.P., 2008. Dynamics of structural and functional characteristics of zooplankton in the Northern Sea of Okhotsk in the "Warm" and "Cold" years. Oceanology 48, 378–382. Available from: https://doi.org/10.1134/S0001437008030090.

Dulepova, E.P., Radchenko, V., 2004. Okhotsk Sea. In: McKinnell, S.M. (Ed.), Marine Ecosystems of the North Pacific. PICES Special Publication, pp. 95–111.

Eisner, L.B., Napp, J.M., Mier, K.L., Pinchuk, A.I., Andrews III, A.G., 2014. Climate-mediated changes in zooplankton community structure for the eastern Bering Sea. Deep Sea Res. II 109, 157–171. https://doi.org/10.1016/j.dsr2.2014.03.004.

Ershova, E.A., Hopcroft, R.R., Kosobokova, K.N., Matsuno, K., Nelson, R.J., Yamaguchi, A., et al., 2015. Long-term changes in summer zooplankton communities of the western Chukchi Sea, 1945–2012. Oceanography 28, 100–115. Available from: http://dx.doi.org/10.5670/oceanog.2015.60.

Estrada, R., Harvey, M., Gosselin, M., Starr, M., Galbraith, P.S., Straneo, F., 2012. Late-summer zooplankton community structure, abundance, and distribution in the Hudson Bay system (Canada) and their relationships with environmental conditions, 2003–2006. Prog. Oceanogr. 101, 121–145. Available from: https://doi.org/10.1016/j.pocean.2012.02.003.

Falk-Petersen, S., Pavlov, V., Berge, J., Cottier, F., Kovacs, K.M., Lydersen, C., 2014. At the rainbow's end: high productivity fueled by winter upwelling along an Arctic shelf. Polar Biol. 38, 5–11. Available from: https://doi.org/10.1007/s00300-014-1482-1.

Falk-Petersen, S., Mayzaud, P., Kattner, G., Sargent, J.R., 2009. Lipids and life strategy of Arctic *Calanus*. Mar. Biol. Res. 5, 18–39.

Gluchowska, M., Kwasniewski, S., Prominska, A., Olszewska, A., Goszczko, I., Falk-Peterson, S., et al., 2016. Zooplankton in Svalbard fjords on the Atlantic–Arctic boundary. Polar Biol 10, 1785–1802.

Grebmeier, J.M., 2012. Shifting patterns of life in the Pacific Arctic and Sub-Arctic Seas. Annu. Rev. Mar. Sci. 4, 63–78. Available from: https://doi.org/10.1146/annurev-marine-120710-100926.

Hirche, H.-J., Kosobokova, K., 2007. Distribution of *Calanus finmarchicus* in the northern North Atlantic and Arctic Ocean—expatriation and potential colonization. Deep Sea Res. II 54, 2729–2747. Available from: https://doi.org/10.1016/j.dsr2.2007.08.006.

Hirche, H.J., Laudien, J., Bucholz, F., 2016. Near-bottom zooplankton aggregations in Kongsfjorden: implications for pelago-benthic coupling. Polar Biol. 39, 1897–1912. Available from: https://doi.org/10.1007/s00300-015-1799-4.

Jakobsson, M., Mayer, L.A., Coakley, B., Dowdeswell, J.A., Forbes, S., Fridman, B., et al., 2012. The international Bathymetric Chart of the Arctic Ocean (IBCAO) Version 3.0. Geophys. Res. Lett. 39, L12609. Available from: https://doi.org/10.1029/2012GL052219.

Karnovsky, N., Ainley, D.G., Lee, P., 2007. The impact and importance of production in polynyas to top-trophic predators: three case histories. In: Smith, W.O.Jr, Barber, D.G. (Eds.), Polynyas—Windows to the World, 74. Elsevier Oceanography Series, pp. 391–410. Available from: https://doi.org/10.1016/S0422-984(06)74002-8..

Kimmel, D.G., Eisner, L., Wilson, M.T., Duffy-Anderson, J.T., 2018. Copepod dynamics across warm and cold periods in the eastern Bering Sea: implications for walleye pollock (*Gadus ghalcogrammus*) and the Oscillating Control Hypothesis. Fish. Oceanogr. 27 (2), 143–158. Available from: https://doi.org/10.1111/fog.12241.

Klages, M., Boetius, A., Christensen, J.P., Deubel, H., Piepenburg, D., Schwew, I., et al., 2004. The benthos of Arctic seas and its role for the organic carbon cycle at the seafloor. In: Stein, R., Macdonald, R.W. (Eds.), The Organic Carbon Cycle in the Arctic Ocean. Springer-Verlag, Berlin, Heidelberg, pp. 139–167.

Kosobokova, K., Hirche, H.-J., 2009. Biomass of zooplankton in the eastern arctic Ocean—a base line study. Prog. Oceanogr. 82, 265–280. Available from: https://doi.org/10.1016/j.pocean.2009.07.006.

Laidre, K.L., Heide-Jørgensen, M.P., Nielsen, T.G., 2007. Role of the bowhead whale as a predator in West Greenland. Mar. Ecol. Prog. Ser. 346, 285–297. Available from: https://doi.org/10.3354/meps06995.

Lischka, S., Hagen, W., 2016. Seasonal dynamics of mesozooplankton in the Arctic Kongsfjord (Svalbard) during year-round observations from August 1998 to July 1999. Polar Biol. 39, 1859–1878. Available from: https://doi.org/10.1007/s00300-016-2005-z.

Mathis, J.T., Grebmeier, J.M., Hansell, D.A., Hopcroft, R.R., Kirchman, D.L., Lee, S.H., et al., 2014. Carbon biogeochemistry of the Western Arctic: primary production, carbon export, and the controls on ocean acidification. In: Grebmeier, J.M., Maslowski, W. (Eds.), The Pacific Arctic Region: Ecosystem Status and Trends in a Rapidly Changing Environment. Springer Science + Business Media Dordrecht. Available from: http://dx.doi.org/10.1007/978-94-017-8863-2-9.

Matsuno, K., Yamaguchi, A., Hirawake, T., Nshino, S., Inoue, J., Kikuchi, T., 2015. Reproductive success of Pacific copepods in the Arctic Ocean and the possibility of changes in the Arctic ecosystem. Polar Biol. 38, 1075–1079.

Michel, C., Hamilton, J., Hansen, E., Barber, D., Reigstad, M., Iacozza, J., et al., 2015. Arctic Ocean outflow shelves in the changing Arctic: a review and perspectives. Prog. Oceanogr. 139, 66–88.

Miller, C.B., Clemons, M.J., 1988. Revised life history analysis for large grazing copepods in the subarctic Pacific Ocean. Prog. Oceanogr. 20, 293–313.

Miller, C.B., Frost, B.W., Batchelder, H.P., Clemons, M.J., Conway, R.E., 1984. Life histories of large, grazing copepods in a subarctic ocean gyre: *Neocalanus plumchrus*, *Neocalanus cristatus* and *Eucalanus bungii* in the Northeast Pacific. Prog. Oceanogr. 13, 201–243.

Nelson, R.J., Ashjian, C.J., Bluhm, B.A., Conlan, K.E., Gradinger, R.R., Grebmeier, J.K., et al., 2014. Biodiversity and biogeography of the lower trophic taxa of the Pacific Arctic Region: sensitivities to climate change. In: Grebmeier, J.M., Maslowski, W. (Eds.), The Pacific Arctic Region: Ecosystem Status and Trends in a Rapidly Changing Environment. Springer Science + Business Media Dordrecht, pp. 269–336. Available from: http://dx.doi.org/10.1007/978-94-017-8863-2-10.

Okkonen, S., 2013. Moorings. In: Shelden, K.E.W., Mocklin, J.A. (Eds.), Bowhead Whale Feeding Ecology Study (BOWFEST) in the Western Beaufort Sea. Final Report, OCS Study BOEM 2013-0114. National Marine Mammal Laboratory, Alaska Fisheries Science Center, NMFS, NOAA, Seattle, WA, pp. 143–152.

Okkonen, S.R., Ashjian, C.J., Campbell, R.G., Clarke, J., Moore, S.E., Taylor, K.D., 2011. Satellite observations of circulation features associated with the Barrow area bowhead whale feeding hotspot. Remote Sens. Environ. 115, 2168–2174.

Okkonen, S.R., Clarke, J.T., Potter, R.A., 2018. Relationships among high river discharges, upwelling events, and bowhead whale (*Balaena mysticetus*) occurrence in the central Alaskan Beaufort Sea. Deep Sea Res. II 152, 195–202. Available from: https://doi.org/10.1016/j.dsr2.2016.11.015.

Pickart, R.S., Schulze, L.M., Moore, G.W.K., Charette, M.A., Arrigo, K., van Dijken, G., et al., 2013. Long-term trends of upwelling and impacts on primary productivity in the Alaskan Beaufort Sea. Deep Sea Res. I 79, 106–121.

Pinchuk, A.I., Paul, A.J., 2000. Zooplankton of the Okhotsk Sea: A Review of Russian Studies. University of Alaska Sea Grant College Program, Fairbanks, AK, p. 62.

Rogachev, K.A., Carmack, E.C., Foreman, M.G.G., 2008. Bowhead whales feed on plankton concentrated by estuarine and tidal currents in Academy Bay, Sea of Okhotsk. Continental Shelf Res. 28, 1811–1826.

Sakshaug, E., 2004. Primary and secondary production in the Arctic Seas. In: Stein, R., Macdonald, R.W. (Eds.), The Organic Carbon Cycle in the Arctic Ocean. Springer-Verlag, Berlin, Heidelberg, pp. 57–81.

Sameoto, D.D., Herman, A.W., 1990. Life cycle and distribution of *Calanus finmarchicus* in deep basins on the Nova Scotia shelf and seasonal changes in *Calanus* spp. Mar. Ecol. Prog. Ser. 66, 225–237.

Shuntov, V.P., Ivanov, O.A., Dulepova, E.P., 2019. Biological resources in the Sea of Okhotsk large marine ecosystem: their status and commercial use. Deep Sea Res. II 163, 33–45. Available from: https://doi.org/10.1016/j.dsr2.2019.01.006.

Siegel, V., 2000. Krill (Euphausiacea) life history and aspects of population dynamics. Can. J. Fish. Aquat. Sci. 57 (Suppl. 3), 130–150.

Smoot, C.A., Hopcroft, R.R., 2017. Cross-shelf gradients of epipelagic zooplankton communities of the Beaufort Sea and the influence of localized hydrographic features. J. Plankton Res. 36, 65–78. Available from: https://doi.org/10.1093/plankt/fbw080.

Stabeno, P.J., Bell, S.W., 2019. Extreme conditions in the Bering Sea (2017–2018): record-breaking low sea-ice extent. Geophys. Res. Lett. 46, 8952–8959. Available from: https://doi.org/10.1029/2019GL083816.

Tande, K.S., 1991. Calanus in North Norwegian fjords and in the Barents Sea. In: Sakshaug, E. Hopkins, C.C.E., Øritsland, N.A. (Eds.), Proceedings of the Pro Mare Symposium on Polar Marine Ecology, Trondheim, Norway, May 12–16, 1990. Polar Res. 10, 389–407.

Vernet, M., Ellingsen, I.H., Seuthe, L., Slagstad, D., Cape, M., Matrai, P.A., 2019. Influence of phytoplankton advection on the productivity along the Atlantic water inflow to the Arctic Ocean. Front. Mar. Sci. Available from: https://doi.org/10.3389/fmars.2019.00583.

Walcusz, W., Williams, W.J., Harwood, L.A., Moore, S.E., Steward, B.E., Kwasniewski, S., 2012. Composition, biomass and energetic content of biota in the vicinity of feeding bowhead whales (*Balaena mysticetus*) in the Cape Bathurst upwelling region (south eastern Beaufort Sea). Deep Sea Res. I 69, 25–35. Available from: https://doi.org/10.1016/j.dsr.2012.05.016.

Wassmann, P., Kosobokova, K.N., Slagstad, D., Drinkwater, D.F., Hopcroft, R.R., Moore, S.E., et al., 2015. The contiguous domains of Arctic Ocean advection: trails of life and death. Prog. Oceanogr. 139, 42–65.

Wassmann, P., Duarte, C.M., Agustí, S., Sejr, M.K., 2011. Footprints of climate change in the Arctic marine ecosystem. Glob. Change Biol. 17, 1235–1249. Available from: https://doi.org/10.1111/j.1365-2486.2010.02311.x.

Wassmann, P., Reigstad, M., 2011. Future Arctic Ocean seasonal ice zones and implications for pelagic-benthic coupling. Oceanography 24 (3), 220–231. Available from: https://doi.org/10.5670/oceanog.2011.74.

Williams, W.J., Carmack, E.C., Ingram, R.G., 2007. Physical oceanography of polynyas. In: Smith, W.O.Jr, Barber, D.G. (Eds.), Polynyas—Windows to the World, vol. 74. Elsevier Oceanography Series, pp. 55–85., doi: 10.1016/S0422-984(06)74002-8.

CHAPTER

27

Bowhead whale ecology in changing high-latitude ecosystems

Sue E. Moore[1], *J.C. George*[2] *and Randall R. Reeves*[3]

[1]Department of Biology, Center for Ecosystem Sentinels, University of Washington, Seattle, WA, United States [2]Department of Wildlife Management, North Slope Borough, Utqiaġvik, AK, United States [3]Okapi Wildlife Associates, Hudson, QC, Canada

Introduction

Arctic and subarctic marine ecosystems are changing much faster than predicted (Overland et al., 2019). Since the advent of satellite records in 1979, sea-ice surface cover has diminished by about 50% at the September minimum, with a roughly 75% year-round reduction in thickness of multiyear ice. This reduction has not been linear; rather, there were dramatic step-changes of sea-ice loss in late summer 2007 and 2012, and in winter 2018 and 2019 (https://nsidc.org/arcticseaicenews/). Meanwhile, seawater temperatures have been rising at an alarming rate, being pushed upward both by solar insolation that is no longer reflected by sea ice and by transport of warm ocean water from the south into subarctic and arctic regions (e.g., Stabeno and Bell, 2019; Haug et al., 2017). The combination of sea-ice loss and warmer seawater has reset the clock on ecological processes in subarctic and arctic seas (e.g., Duffy-Anderson et al., 2019; Eriksen et al., 2017). Overall, primary productivity is initiated earlier in spring, with enormous blooms sometimes encountered under thin sea ice or in newly open water. Although changes in primary productivity vary on local or regional scales, the biophysical impacts of reduced sea ice, increased ocean temperatures, and shifting atmospheric and ocean dynamics can drive swift and fundamental changes near the base of marine food webs where bowhead whales feed on zooplankton.

Four bowhead whale stocks are recognized, each of which occupies a distinct subarctic or arctic region (Chapters 3–5). Climate change has altered the physical and biological attributes of all of these regions (Moore et al., 2019; Chapters 25 and 26), in turn leading to changes in prey and predator dynamics (Duffy-Anderson et al., 2019; Chapters 28 and 29) and underwater soundscapes (Halliday et al., 2018). Like all marine mammals, bowhead

The Bowhead Whale
DOI: https://doi.org/10.1016/B978-0-12-818969-6.00027-3

whales respond to habitat alterations through shifts in their behavioral ecology (e.g., phenology, seasonal distribution, feeding hotspots), which can result in changes in body condition and overall health (Moore, 2018). Here, we summarize recent changes to the regional ecosystem for each stock, starting with the largest [Bering–Chukchi–Beaufort (BCB)] and ending with the smallest, lowest-latitude, and most isolated [Okhotsk Sea (OKH)]. Our goal is to compare and contrast the ecologies of the four stocks. This brief review of a very broad topic is bolstered by frequent reference to other chapters in this volume, where the interested reader can find more in-depth coverage of specific topics. We close with an appraisal of bowhead population status using a simple framework (Moore and Reeves, 2018) as a means of gauging the species' resilience to rapid ecosystem alteration.

Bowhead whale ecology in regional ecosystems

The bowhead whale is the only mysticete species that is endemic to subarctic and arctic seas and therefore adapted to cold-water habitat dominated by sea ice for substantial portions of the year, including giving birth there (Fig. 27.1). Nevertheless, bowheads have no difficulty migrating and feeding in open water, where the nearest sea ice can be hundreds of kilometers away. The role that sea ice plays in the species' ecology is complex, both because the structure of arctic ecosystems is broadly driven by how the extent, timing, and thickness of ice governs the flow of primary production (e.g., Arrigo and van Dijken, 2015) and because the presence of ice can act to deter predators (Chapter 29) while also ameliorating underwater acoustic soundscapes (Chapter 22).

Loss of sea ice has become symbolic of climate warming and all bowhead stocks now regularly inhabit open-water regions from at least late summer through autumn (Fig. 27.2). When the extreme increases in ocean heat now evident in all high-latitude seas are added to sea-ice loss (Chapter 25), it becomes clear that bowhead whales are living in habitats that are undergoing rapid alteration. Below, we provide a brief overview of the region inhabited and ecosystem perturbations experienced by each of the four stocks. Except for the OKH stock, much of this information is drawn from Moore et al. (2019) and the references therein.

Bering–Chukchi–Beaufort stock

The BCB stock numbers roughly 17,000 animals (Chapter 6), with a geographical range that extends from the Bering Sea north across the Chukchi Sea, west into the East Siberian Sea, and east into the Beaufort Sea (Chapter 4). The northern Bering, Chukchi, and East Siberian seas have broad, shallow (~50 m) continental shelves, while the Beaufort Sea has a narrow shelf and steep slope culminating in the deep (~3000 m) Canadian Basin. The region as a whole is an inflow-shelf system, with the narrow and shallow (85 km wide-×50 m deep) Bering Strait being the only gateway for Pacific water to enter the Arctic (Chapter 25). Pacific inflow is primarily northward, with increased transport of warm water into the Chukchi and East Siberian seas evident since the late 1990s (Woodgate, 2018). Sea ice covers this region for 5–7 months of the year, typically reaching maximum and minimum areal extent in March and September, respectively (Chapter 25). Until 2018,

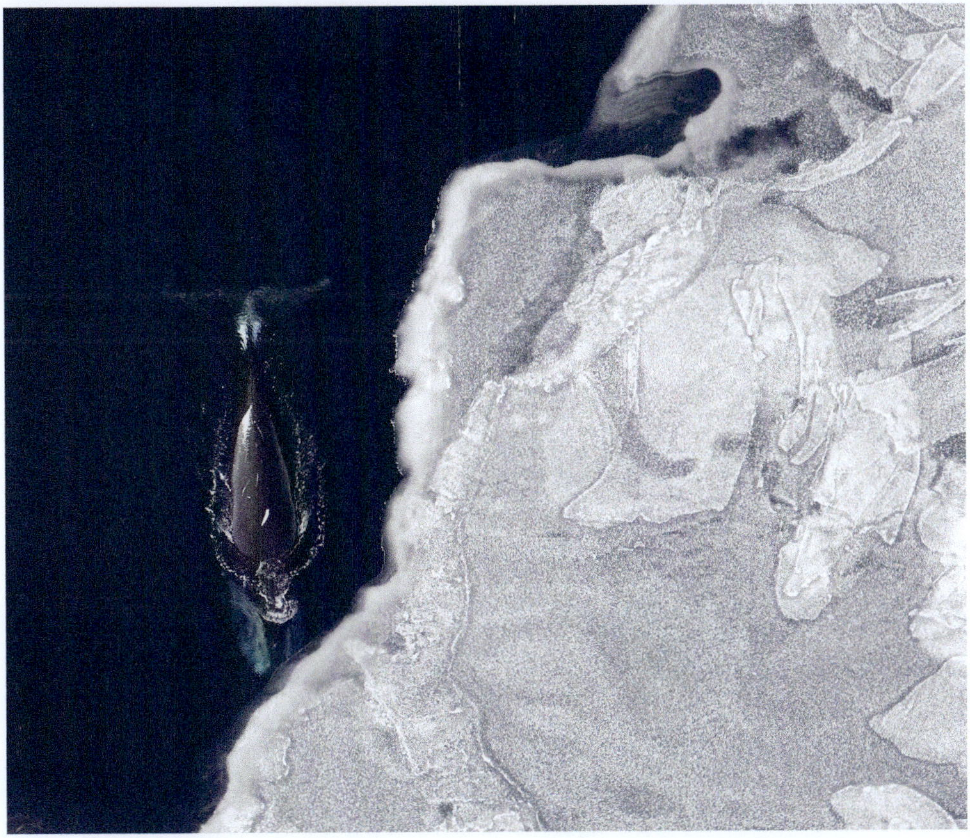

FIGURE 27.1 With the loss of arctic sea ice and warming seas, bowhead whale habitats are changing rapidly, raising questions regarding the resilience and adaptability of the species. *Source: Photo by Brenda Rone (NOAA/ North Slope Borough, NMFS Permit No. 14245).*

dramatic reductions in the persistence of sea ice were driven more by late ice formation than by early ice retreat, but exceptionally low winter sea-ice cover in the Bering Sea since then may have disrupted this pattern (Stabeno and Bell, 2019).

The BCB stock undertakes a predictable, age-structured seasonal migration (Chapter 4). The whales leave the northern Bering Sea in March–April, calve during migration, feed primarily in the Canadian Beaufort Sea in May–July, with feeding continuing during the August–October autumn migration across the Alaskan Beaufort and Chukchi seas, and into the northern Bering Sea from November to February. Acoustic sampling provides evidence that some whales deviate from this pattern, with bowhead calls recorded near the Chukchi Plateau from March through August (Moore et al., 2012). BCB bowheads commonly feed on copepods and krill during summer and autumn in the Beaufort Sea, with winter feeding in the northern Bering Sea and Anadyr Gulf inferred from dive patterns (Chapter 4) and confirmed by postmortem examinations (Chapter 28). Unexpectedly, through 2015, reduction in sea ice was significantly correlated with improved bowhead

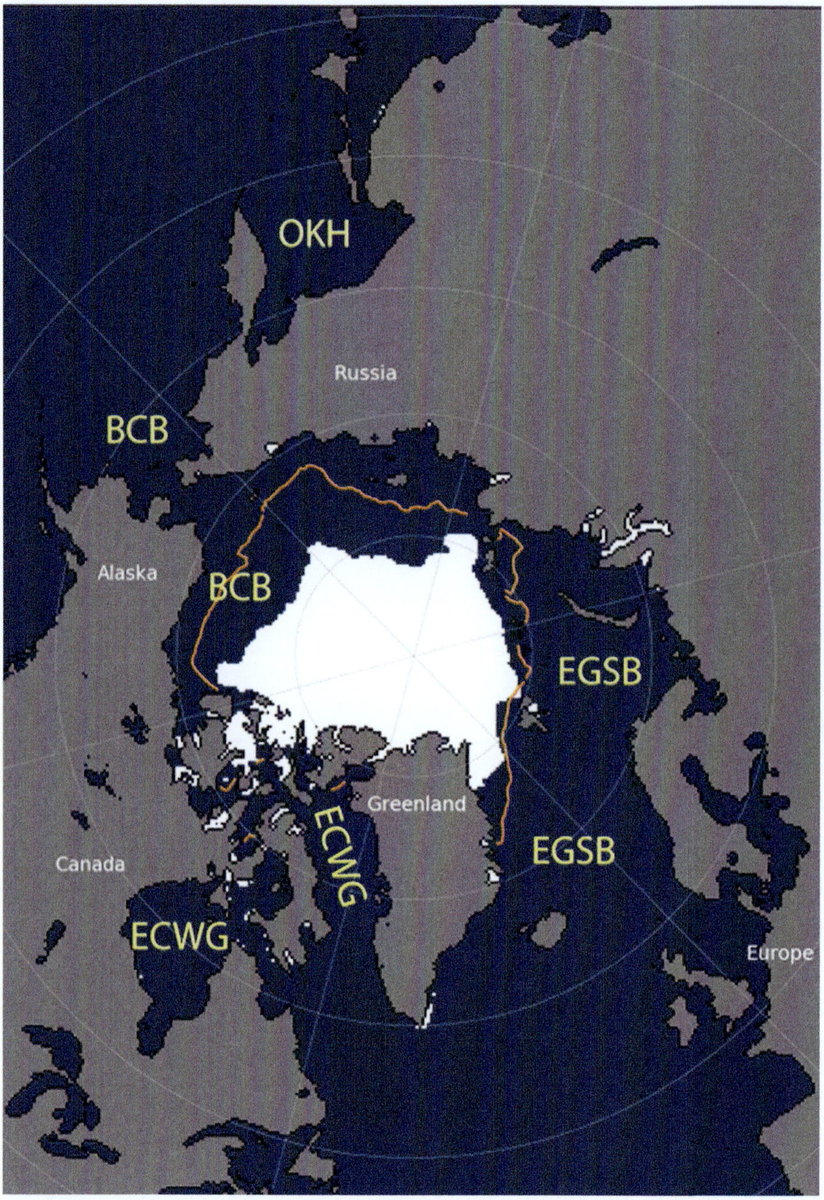

FIGURE 27.2 Example sea-ice minimum from September 2019, depicting the large open-water areas now commonplace in the regions inhabited by each of the four stocks of bowhead whales. Magenta line indicates median September ice edge, 1981–2010. *Source: Sea-ice map from National Snow and Ice Data Center.*

whale body condition (especially in juvenile whales), increased upwelling, and increased late-summer feeding in the western Beaufort Sea, most notably in years when wind conditions trapped krill there (Chapter 26). Satellite-telemetry studies have identified six core-use habitats and contrasting patterns of whale distribution during the autumn migration across the Chukchi Sea that correspond to the presence (or absence) of Bering Sea water, which often contains krill (Chapter 4). Further, with sea-ice formation now often delayed, tagged bowheads sometimes linger in the Chukchi Sea through December, with at least two whales known to have overwintered there in 2017–18 (Chapter 4).

East Canada–West Greenland stock

The East Canada–West Greenland (ECWG) stock numbers roughly 6500 animals, with a geographical range that extends from the Labrador Sea (maximum depth >3500 m) into Davis Strait (sill depth c. 800 m) and north to Baffin Bay (maximum depth >2000 m) (Chapter 5). This stock also occupies roughly the eastern half of the Canadian Arctic Archipelago, including Lancaster Sound, Gulf of Boothia, Foxe Basin, northern Hudson Bay, and Hudson Strait (Chapter 5). The region is primarily an outflow-shelf system, with cold saline waters exiting Nares Strait, Lancaster Sound and Jones Sound to flow southward along Baffin Island as the Baffin Island Current (Chapter 25). The Labrador Current brings cold arctic water across the wide (>150 km) Labrador and Newfoundland shelves to the Grand Banks, the southernmost penetration of polar waters in the northern hemisphere.

The ECWG stock overwinters in areas of unconsolidated sea ice in Hudson Strait, along the east coast of Baffin Island, and in Disko Bay (Matthews and Ferguson, 2015). Spring and summer migrations do not conform to a strict schedule but rather follow a complex, sex- and age-structured movement pattern, influenced by the recession of sea ice and the circulation of cold arctic water (Chapter 5). ECWG whales appear to feed year-round, primarily on benthic and epibenthic calanoid copepods and mysids (Chapter 28; Pomerleau et al., 2018), with isotope and fatty acid analyses suggesting that they also feed on pelagic prey including krill (Matthews and Ferguson, 2015). In autumn the whales swim toward the wintering grounds ahead of sea-ice formation, with some movement across Davis Strait from Canada to Greenland throughout the winter (Chapter 5). The fluidity of ECWG bowhead movements coupled with the complex topography and oceanography of this region makes it difficult to generalize about associations with shifts in ecosystem dynamics.

East Greenland–Svalbard–Barents Sea stock

The East Greenland–Svalbard–Barents Sea (EGSB) stock numbers roughly 350 animals, with a geographical range that extends north from Iceland and east from Greenland, past Svalbard and Franz Josef Land to Severnaya Zemlya (Chapter 5). The narrow shelves along East Greenland and around Iceland drop off steeply to a deep (> 5000 m) basin. The region includes both outflow and inflow currents (near East Greenland and western Svalbard, respectively), resulting in highly variable ocean conditions (Chapter 25). Atlantic water that flows north turns eastward after passing Svalbard and enters the Barents Sea (mean depth ~200 m) through the broad and deep (450 km × 2500 m) Fram Strait, supporting a highly productive ecosystem where Atlantic and arctic waters meet (Haug et al., 2017).

Details on seasonal movements are not well known for this stock. The results of acoustic sampling suggest that Fram Strait contains important habitat, with bowhead calls detected there year-round (Moore et al., 2012) and complex songs recorded nearly every hour from October to April (Chapter 22). This and the seasonal movements of a single satellite-tagged female bowhead that returned to Fram Strait in December after swimming south in April to summer in steep-slope waters off East Greenland are consistent with migration patterns described by commercial whalers in the 16th and 17th centuries (Hacquebord, 2001). As for the BCB stock, sea-ice extent and upwelling are likely important features that provide EGSB bowheads with good foraging opportunities. The recent reduction in sea ice, in combination with mid-winter upwelling events, may be creating good foraging conditions similar to those that existed during the peak period of whaling near Svalbard (approximately 1680–1790), when the summer ice edge retreated to north of 79°N (Falk-Petersen et al., 2014). A number of effects of ecosystem shifts on zooplankton and fish have been documented in the Barents Sea (e.g., Eriksen et al., 2017), but the extremely limited ecological data on EGSB bowheads precludes making inferences about the effects of such shifts on the whales.

Okhotsk Sea stock

The OKH stock may number 200–300 animals, all of which reside year-round in the semienclosed OKH, bounded by mainland Russia to the north and west, the Kamchatka peninsula to the east, and Japan and the Kuril Islands to the south (Chapter 5). The sea is connected to the Pacific Ocean by roughly 50 straits and passes in the Kuriles and encompasses the deep (3900 m) Kuril Basin in the south, shoaling to shallow continental shelves in Shelikhov Gulf in the north and the Shantar Archipelago in the southwest. Sea-ice production has diminished substantially in recent years, resulting in a dramatic reduction of winter ice extent (Chapter 25). The interaction of the northward-flowing West Kamchatka current in the east and the southward-flowing East Sakhalin current in the west generates a generally cyclonic movement of water.

The OKH bowhead stock is the least known in terms of ecology. Most recent observations are opportunistic and limited to waters around the Shantar Archipelago (Shpak and Paramonov, 2018) and in Shelikhov Gulf, with little known about whale movements to and from these areas (Chapter 5). The whales seem to prefer shallow-coastal habitat in both regions, perhaps due to feeding opportunities there or as a way of avoiding or responding to predation by killer whales. With regard to feeding, Rogachev et al. (2008) reported that a combination of estuarine and tidal currents concentrates zooplankton in Academy Bay, in the Shantar Archipelago. Specifically, near-bottom, cold-water intrusions from the northern OKH associated with the estuarine circulation transport herbivorous arctic calanoid copepods and other zooplankton into the region where they are concentrated by tidal currents, facilitating bowhead feeding. It is not known whether recent ecosystem perturbations are altering these types of ocean dynamics there or in other parts of the OKH.

Bowhead whale status and resilience

How resilient are bowhead whales to the rapid changes ongoing in all subarctic and arctic regions? Although eventual outcomes are hard to predict, we used a simple framework (Moore and Reeves, 2018) as a means to assess the status and resilience of each of the four bowhead stocks (Table 27.1). The BCB stock had the best rank (1.0), based on its large population size and spatial range, demonstrated flexibility in behavior and diet, and good health record (Chapters 4 and 30). The ECWG stock had the next-best rank (1.5), given its smaller population size and range (scores 2 and 2, respectively), with diet flexibility and health given the same scores as the BCB stock, although based on examinations of comparatively few animals. The EGSB and OKH ranks were based solely on their small population sizes (scores 3 and 4, respectively) and relatively large-open range (EGSB: score 2) and small-restricted range (OKH: score 5). While this framework is obviously a very simple one, we feel the results represent a reasonable first cut at assessing the relative status and resilience of the four stocks. The framework also highlights the paucity of information for the two smallest bowhead whale stocks, a situation that could be remedied with increased research on their ecology and health.

The ultimate question is: what does the future hold for bowhead whales? Our simple ranking approach does not capture the variety and complexity of possibilities that arise when ecosystems shift into new states (e.g., Overland et al., 2019; Duffy-Anderson et al., 2019; Eriksen et al., 2017). Will there be sufficient food and can bowheads modify their behavior to find novel prey aggregations and adapt to new prey species? What about the possibilities of competition for habitat and prey from other baleen whales (as well as marine fish and seabirds), and the potential for increased predation pressure from killer whales?

Biophysical differences in regional marine ecosystems will influence how each population fares in the current era of rapid change, and surprises are likely. For example, through 2016, it seemed as though BCB bowhead whales and subarctic mysticete species (e.g., gray, humpback, fin, and minke whales) were all faring well in the Pacific Arctic system (Moore, 2016). Increases in within-region primary productivity together with the transport

TABLE 27.1 Vulnerability rankings for the four bowhead whale stocks. Each of the four metrics is ranked from 1 to 5, with a rank of 1 representing a large, migratory population exhibiting behavioral plasticity (movements and diet) and resistance to disease and stress, and a rank of 5 representing the opposite.

Metric	BCB	ECWG	EGSB	OKH
Population size	1	2	3	4
Range	1	2	2	5
Behavior	1	1	Unknown	Unknown
Health	1	1	Unknown	Unknown
Status and resilience	**1**	**1.5**	**2.5**	**4.5**

Status and resilience = metric total/number of metrics used (Moore and Reeves, 2018). *BCB*, Bering–Chukchi–Beaufort; *ECWG*, East Canada–West Greenland; *EGSB*, East Greenland–Svalbard–Barents Sea; *OKH*, Okhotsk Sea.

of nutrients and prey through Bering Strait seemed to be serving the needs of both the arctic endemic bowhead and the seasonally migrant subarctic mysticetes. Then, in the nearly ice-free autumn of 2019, bowhead whales were not seen in the western Beaufort Sea where they are usually common, that is, the few whales seen did not occupy continental-shelf habitat but remained offshore over the slope (Fig. 27.3; Chapter 24). The abrupt shift offshore and lack of late-summer feeding were most dramatic in the BCB core-use habitat near Point Barrow, while whale distribution and feeding in western Canadian (two core habitats) and eastern Russian (one core habitat) waters were similar to the previous four decades (Chapter 24). We can only speculate that the available prey base was insufficient to draw the whales onto the Alaskan-Beaufort shelf in 2019, but other causal factors such as elevated ocean temperatures or the presence of killer whales (Chapter 29) cannot be ruled out. Also, other aspects of the 2019 bowhead spring migration and summer foraging patterns were similar to past years, and a partial deviation in habitat use in any one season may be inconsequential to the overall health of the BCB stock. Of note, the 2019 summertime sighting rates were high for subarctic mysticetes in the Chukchi Sea, indicating advection of prey at least that far north in the Pacific Arctic system. Although foraging competition between bowheads and subarctic (especially humpback and fin whales, which often target krill) may increase over time in all regions, dissimilar migratory cycles seem to be limiting such competition, at least for the moment (Moore, 2016; Chapter 5). Even so, other health-related factors, such as the recent and unexpected discovery of kidney worms in harvested BCB

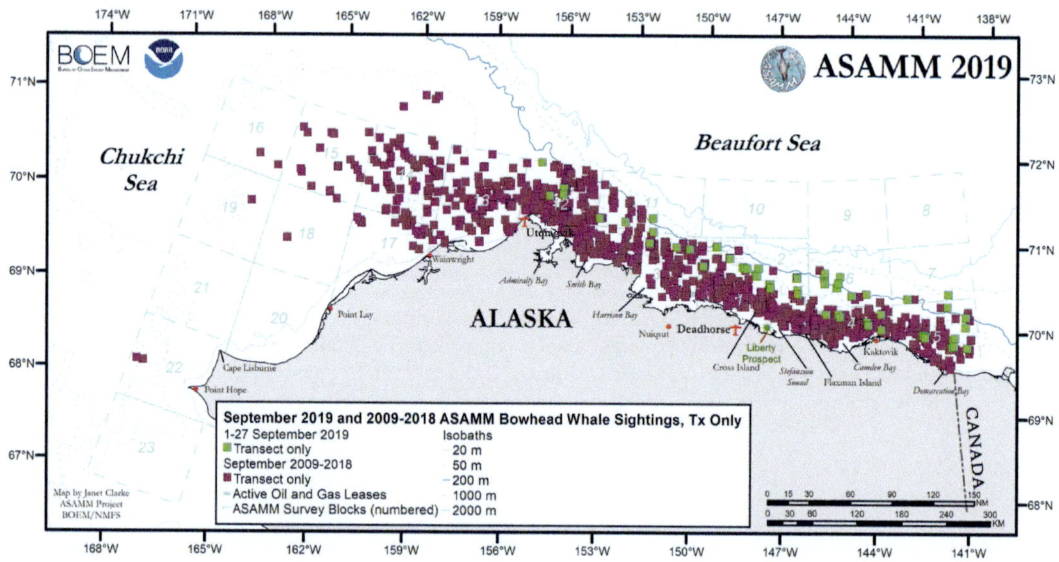

FIGURE 27.3 Bowhead whale distribution in September 2019 (green) compared to September 2009–18 (purple) during the autumn migration across the Alaskan Beaufort Sea. The offshore distribution along the continental slope in 2019 is reminiscent of the distribution during heavy-ice years in the 1980s and early 1990s, but reasons for this in an "ice-free" autumn are unknown. Tx Only, data only from aerial transects. *Source: Courtesy Aerial Surveys of Arctic Marine Mammals (ASAMM, NOAA/AFSC/MML, and BOEM).*

bowheads (Chapter 30), may be related to increased exposure to temperate-region species as the arctic marine ecosystem warms (e.g., Duffy-Anderson et al., 2019).

The bowhead whale appears to be a resilient species, as exemplified by their 4–5 million years' history (Chapters 2 and 39), the recovery of the BCB stock and the evidently strong increase in numbers of ECWG whales after both populations were greatly depleted by commercial whaling (Chapter 33). It has been suggested that the near-extirpation of the EGSB stock in the 17th–19th centuries resulted in a regime shift in that ecosystem, due to the huge quantities of zooplankton that became available to other marine mammals, birds, and fish (Hacquebord, 2001). While Hacquebord's hypothesis is difficult to test, a similar suggestion for the Southern Ocean, where intensive 20th century commercial whaling created a "krill surplus," has been challenged (Suma et al., 2014). The modeling results of Suma et al. (2014) indicate that changes in ocean conditions from 1975 to 1995 likely surpassed in importance the ecological effects of baleen whale depletion, although those authors noted that a decrease in the quantity of feces released into the water column by Antarctic whales might have led to reduced primary productivity in some areas due to decreased iron bioavailability.

Looking farther back in time, the survival of bowhead whale lineages during the rapid climate changes of the Pleistocene–Holocene transition (~11,000 years ago) suggests that the species can adapt to significant ecosystem alterations by shifting its range to track suitable habitat (Foote et al., 2013). The survival of bowheads depends on the existence of such habitat, the specific attributes of which remain rather poorly understood. For example, suitable bowhead habitat is usually defined as ocean regions with sea ice, even though the whales can thrive during periods with expansive areas of open water in summer and autumn. While recent modeling results suggest that late-summer ice-free conditions could prevail throughout the Arctic as soon as mid-century (Overland et al., 2019), the formation of sea ice in winter will likely continue and therefore some high-latitude areas may remain more suited to bowhead whales than to other mysticetes. In addition, year-round feeding and low metabolism may act to buffer the species against periods of prey scarcity. Indeed, physiological evidence that bowheads can extend fasting is supported by bio-energetic analyses (Chapter 16). Furthermore, Burns (1993) speculated that extremely thick blubber buffered bowheads against years, or even multiple successive years, of low prey production, allowing them to inhabit regions that other species simply cannot. However, there are costs associated with low metabolism, including slow body growth, extremely delayed age at sexual maturity and extended longevity (Chapter 7), which is a viable life history strategy only when competition and predation-risk are low. Therefore a strategy involving lowered metabolism, slow reproductive rates, and large lipid reserves and thick blubber may not be particularly advantageous in a warming or ice-free Arctic. What is far more difficult to anticipate is a radical change in food webs that can result from shifts in the range and abundance of other marine taxa such as fish and seabirds, which can respond to climate signals quickly and deplete invertebrate prey resources (e.g., Haug et al., 2017; Eriksen et al., 2017; Duffy-Anderson et al., 2019). Increased predation on bowheads, concomitant with killer whale range extensions, may also abruptly alter the suitability of regional habitats (Chapter 29). Modeling studies might help predict some aspects of food-web tipping points, but observing the integrated responses of marine fish, birds, and mammals should provide direct insights to what is actually happening in high-latitude marine ecosystems (e.g., Moore et al., 2014).

Finally, this chapter has not touched upon the negative effects on bowhead whales of human activities, which are rapidly increasing in the far north as sea-ice wanes and becomes a much less formidable impediment to navigation (e.g., Reeves et al., 2014). How the underwater noise associated with human activities affects the whales has been a focus of much research (Chapter 35), with ship strikes and entanglement in fishing gear (Chapter 36) and exposure to diseases and contaminants (Chapters 30 and 37) potentially becoming significant threats. Direct removals, such as the indigenous subsistence harvest, are carefully managed and modest in relation to bowhead numbers and not likely to affect the two hunted stocks, BCB and ECWG (Chapter 6).

The long-term future of the bowhead whale, as a species, will depend on the ability of each of the four stocks to persist in the face of similar pressures (i.e., ecological shifts and anthropogenic stressors) that vary in intensity by region. Given our simple ranking framework (Table 27.1), the small and isolated OKH stock seems to be facing the greatest challenges, while the more numerous, northerly distributed, and broader-ranging BCB and ECWG stocks may be less vulnerable in the near term. Foote et al. (2013) projected a northward expansion of bowhead whales in the North Atlantic during the Holocene, when suitable core habitat, as defined by sea-ice concentration, became available. Those authors predicted that suitable bowhead habitat will shrink by roughly 50% by 2100, which will in turn influence the species' population dynamics. Importantly, Foote et al. (2013) also noted that responses to climate warming are likely species-specific. Moreover, ecosystem variability at a regional scale will undoubtedly confound general predictions about climate-change impacts on bowhead whales, now and into the future.

References

Arrigo, K.R., van Dijken, G.L., 2015. Continued increase in Arctic Ocean primary production. Prog. Oceanogr. 136, 60.

Burns, J.J., 1993. Preface, p xxiii-xxxvi. In: Burns, J.J., Montague, J.J., Cowles, C.J. (Eds.), The Bowhead Whale. The Society for Marine Mammalogy, Special Publication Number 2.

Duffy-Anderson, J.T., Stabeno, P., Andrews III, A.G., Kieciel, K., et al., 2019. Responses of the northern Bering Sea and the southeast Bering Sea pelagic ecosystems following record-breaking low winter sea ice. Geophys. Res. Lett. 46. Available from: https://doi.org/10.1029/2019GL083396.

Eriksen, E., Skjoldal, H.R., Gjosaeter, J., Primicerio, R., 2017. Spatial and temporal changes in the Barents Sea pelagic compartment during the recent warming. Prog. Oceanogr. 151, 206–226.

Falk-Petersen, S.V., Pavlov, J., Berge, P., Cottier, K.M., Kovacs, C., Lydersen, 2014. At the rainbow's end: high productivity fueled by winter upwelling along an Arctic shelf. Polar Biol. 38, 5–11.

Foote, A.D., Kaschner, K., Schultze, S.E., Garileo, C., et al., 2013. Ancient DNA reveals that bowhead whale lineages survived late Pleistocene climate change and habitat shifts. Nat. Commun. 4, 1677.

Hacquebord, L., 2001. Three centuries of whaling and walrus hunting in Svalbard and its impact on the Arctic ecosystem. Environ. History 7 (2), 169–185.

Halliday, W.D., Insley, S.J., de Jong, T., Mouy, X., 2018. Seasonal patterns in acoustic detections of marine mammals near Sachs Harbour, Northwest Territories, Arctic Science, vol. 4. NRC Research Press, pp. 259–278.

Haug, T., Bogstad, B., Chierici, M., Gjosaeter, E.H., Hallfredsson, et al., 2017. Future harvest of living resources in the Arctic Ocean north of the Nordic and Barents Seas: a review of possibilities and constraints. Fish. Res. 188, 38–57.

Matthews, C.J., Ferguson, S.H., 2015. Seasonal foraging behaviour of Eastern Canada-West Greenland bowhead whales: an assessment of isotopic cycles along baleen. Mar. Ecol. Prog. Ser. 522, 269–286.

Moore, S.E., 2016. Is it 'boom times' for baleen whales in the Pacific Arctic region? Biol. Lett. 12. Available from: http://dx.doi.org/10.1098/rsbl.2016.0251.

Moore, S.E., 2018. Climate change. In: Würsig, B., Thewissen, J.G.M., Kovacs, K. (Eds.), Encyclopedia of Marine Mammals, third ed. Elsevier, pp. 194–197.

Moore, S.E., Reeves, R.R., 2018. Tracking arctic marine mammal resilience in an era of rapid ecosystem alteration. PLoS Biol. 16. Available from: https://doi.org/10.1371/journal.pbio.2006708.

Moore, S.E., Logerwell, E., Eisner, L., Varlety Jr., E.V., et al., 2014. Marine fishes, birds and mammals as sentinels of ecosystem variability and reorganization in the Pacific Arctic region. In: Grebmeier, J.M., Maslowski, W. (Eds.), The Pacific Arctic Region: Ecosystem Status and Trends in a Rapidly Changing Environment. Springer, pp. 337–392.

Moore, S.E., Stafford, K.M., Melling, H., Berchok, C., et al., 2012. Comparing marine mammal acoustic habitats in Atlantic and Pacific sectors of the High Arctic: year-long records from Fram Strait and the Chukchi Plateau. Polar Biol. 35, 475–480.

Moore, S.E., Haug, T., Vikingsson, G.A., Stenson, G.B., 2019. Baleen whale ecology in arctic and subarctic seas in an era of rapid habitat alteration. Prog. Oceanogr. 176, 102–118. Available from: https://doi.org/10.1016/j.pocean.2019.05.010.

Overland, J.E., Dunlea, J.E., Box, R., Corell, M., Forsius, V., Kattsov, M.S., et al., 2019. The urgency of Arctic change. Polar Sci. 21, 6–13.

Pomerleau, C., Matthews, C.J.D., Gobell, C., Stern, G.A., Ferguson, S.H., McDonald, R., 2018. Mercury and stable isotope cycles in baleen plates are consistent with year-round feeding in two bowhead whale (*Balaena mysticetus*) populations. Polar Biol. 41, 1881–1893.

Reeves, R.R., Ewens, P.J., Agbayani, S., Heide-Jorgensen, M.P., Kovacs, K.M., et al., 2014. Distribution of endemic cetaceans in relation to hydrocarbon development and commercial shipping in a warming Arctic. Mar. Pol. 44, 375–389.

Rogachev, K.A., Carmack, E.C., Foreman, M.G.G., 2008. Bowhead whales feed on plankton concentrated by estuarine and tidal currents in Academy Bay, Sea of Okhotsk. Cont. Shelf Res. 28, 1811–1820.

Shpak, O.V., Yu, P., 2018. The bowhead whale, *Balaena mysticetus*, Linnaeus, 1758, in the western Sea of Okhotsk (2009–2016): distribution pattern, behavior and threats. Russ. J. Mar. Biol. 44 (3), 210–218.

Stabeno, P.J., Bell, S., 2019. Extreme conditions in the Bering Sea (2017–2018): record-breaking low sea-ice extent. Geophys. Lett. 46. Available from: https://doi.org/10.1029/2019GL083816.

Suma, S., Pakhomov, E.A., Pitcher, T.J., 2014. Effects of whaling on the structure of the Southern Ocean food web: insights on the "krill surplus" from ecosystem modeling. PLoS One 9 (12), e114978. Available from: https://doi.org/10.1371/journal.pone.0114978.

Woodgate, R.A., 2018. Increases in the Pacific Inflow to the Arctic from 1999 to 2015 and insights to seasonal trends and driving mechanisms from year-round Bering Strait mooring data. Prog. Oceanogr. 160, 124–154.

CHAPTER

28

Diet and prey

Gay Sheffield[1] and J.C. George[2]

[1]Alaska Sea Grant, College of Fisheries and Ocean Sciences, University of Alaska Fairbanks, Nome, AK, United States [2]Department of Wildlife Management, North Slope Borough, Utqiaġvik, AK, United States

Introduction

Like other baleen whales, bowheads must find relatively high-density prey patches to feed efficiently and meet their nutritional needs (Lowry, 1993). The abundance and locations of zooplankton concentrations has a strong influence on bowhead feeding locations. The bowhead whale's long baleen plates with fine "fringe hairs" are highly specialized for feeding on small zooplankton such as copepods, euphausiids, mysids, and amphipods (Chapter 14). Bowheads are often described as water column and surface feeders, yet they will feed near the seafloor in relatively shallow nearshore waters (Lowry, 1993; Lowry et al., 2004), presumably accessing copepods in diapause (Laidre et al., 2007) or other appropriate prey. Evidence of feeding near the seafloor includes, epibenthic prey collected from the stomachs of bowheads (Lowry, 1993; Lowry et al., 2004; Sheffield and George, 2013, NSB-DWM data), dive data from telemetry studies (Heide-Jørgensen et al., 2013; Citta, 2015a), and direct observations (Mocklin et al., 2011).

Behaviors associated with feeding include whales diving in the same general area (Ljungblad et al., 1987), synchronous diving/surfacing (Moore et al., 2010), mud on the body (Mocklin et al., 2011), a fecal plume (Tomilin, 1957), echelon swimming (Moore et al., 2010; Fish et al., 2013), head lunging (Moore et al., 2010; Fig. 12.1), and/or open mouth surface swimming (Fig. 28.1) (Scammon, 1874; Würsig et al., 1989).

Across the Holarctic range of the bowhead, several methods have been employed to document bowhead feeding. This evidence includes behavioral observations made from manned aircraft (Ljungblad et al., 1987; Würsig et al., 1989; Landino et al., 1994; Smultea et al., 2012; Rugh et al., 2013; Clarke et al., 2015; Chapter 24), unmanned aircraft (Ferguson et al., 2018; Fortune, 2018), ships (Bodfish, 1936; Moore et al., 1995; Melnikov and Fedorets, 2016), shore-based observations (Finley, 1990; Melnikov et al., 2004; Melnikov and Zdor, 2018), sea ice (Carroll et al., 1987), analyses of aerial photographs (Mocklin et al., 2011),

The Bowhead Whale
DOI: https://doi.org/10.1016/B978-0-12-818969-6.00028-5

FIGURE 28.1 Two bowhead whales surface-feeding, approximately 30 km east of Utqiaġvik, Alaska, Beaufort Sea. Note their sideways orientation at the surface to maximize ingestion of prey. The whales were in very shallow water and their flukes strokes were churning up bottom sediment, as evidenced by circles of muddy water behind the whales. *Source: Photo by Suzie Hanlan (NOAA/North Slope Borough), August 27, 2017.*

embedded satellite-linked tags (Heide-Jørgensen et al., 2013; Citta et al., 2015b, 2018; Chapters 4 and 5), and short-term suction tags (Baumgartner, 2013). Additionally, quantitative documentation of the composition of zooplankton species near feeding bowheads has also been used to indirectly determine the possible diet of bowhead whales (Baumgartner, 2013; Melnikov and Fedorets, 2016). Regardless, bowheads apparently "do a better job" of finding and catching zooplankton than researchers (Thomson, 2002; Moore et al., 2010).

The most direct evidence of feeding and the types of prey consumed comes from examining the stomach of freshly harvested whales. Coastal bowhead whaling communities offer an important source of information on bowhead feeding and diet. Experienced whalers can provide expert observations in situations unavailable to western researchers, long-term knowledge, and make their harvested whales accessible for examination. The ability to access freshly harvested bowhead whales has led to a more detailed understanding of the prey consumed in several locations and during different seasons.

This chapter has three objectives. First, to provide an overview of diet information for each bowhead stock. Second, to provide historic and current knowledge of diet and feeding obtained from the stomach contents, feces, and other tissues sampled from harvested bowhead whales. Third, to provide insights on emerging issues likely to affect bowhead diet and prey in response to the ongoing and significant reduction in sea ice extent, quality, and duration—especially the Bering–Chukchi–Beaufort Seas (BCB) population (Chapter 27).

Diet research methods

The first written records of bowhead diet occurred as early as CE 1250, from an unknown Norwegian author, "This fish is very clean in its choice of food; for people say that is subsists wholly on mist and rain—and whatever falls into the sea from the air above" (Walløe, 1999). As Dryden (1913) noted in Annus Mirabilis (CE 1666) the whales "give no chase, but swallow in the fry, which through their gaping jaws mistake the way."

Commercial whalers provided some of the first written documentation of bowhead diet and foraging ecology (Scoresby, 1820; Scammon, 1874; Bodfish, 1936). Unfortunately, little detailed information on bowhead diet was documented during the intensive commercial bowhead whaling era in the 17th and early 20th centuries (Scoresby, 1820; Bockstoce, 1986; Sheldon and Rugh, 1995; Higdon, 2010; Ivashchenko and Clapham, 2010; Chapter 33).

Detailed information regarding bowhead diet and relatively nearshore feeding areas have been known by indigenous coastal peoples that have engaged the bowhead for subsistence purposes for millennia (Tomilin, 1957; Stoker and Krupnik, 1993; Savelle and McCartney, 1999; Galginaitis and Koski, 2002; Noongwook, 2007; George et al., 2008; Higdon, 2010). Understanding the diet and/or feeding behavior of a harvested bowhead whale comes from several sources that include: observations of the whale behavior and the ecological conditions throughout the hunt, as well as a physical examination of the whale including its stomach, organs, and overall body condition.

Several indigenous communities allow access to their harvested bowheads which provide extensive biological specimens, body measurements, and behavioral data for research (Chapter 32). Flensing and butchering a bowhead requires extensive knowledge and intensive labor as the whaling community processes skin, blubber, meat, organs, and appendages (Chapter 31). Depending on the size of the landed whale, comprehensive data/specimen collection, including access to the gastrointestinal tract, may take hours to days.

Several research methodologies are typically used to determine diet and/or feeding behavior from a harvested whale and include the following.

Body measurements (morphometrics)

Following the methods of Rice and Wolman (1971) on gray whales, body measurements from harvested bowhead whales demonstrate seasonal girth differences associated with seasonal feeding (George et al., 2015; Chapter 7). Bowhead whales fatten seasonally, as evidenced by a higher girth-at-length in fall specimens than in spring specimens. George et al. (2015) demonstrated differences in body condition associated with ice density, upwelling-favorable winds, and duration of open water in the summer feeding grounds.

Fatty acid analyses

Fatty acid signature analysis is used to describe the feeding habits of marine mammals based on the assumption that chemical components of their prey are incorporated largely unaltered into an animal's soft tissues (Iverson et al., 2002). Fatty acid analyses of bowhead whale blubber (Budge et al., 2008; Pomerleau et al., 2014), muscle (Budge et al., 2008), and skin (Budge et al., 2008) have added new information to the long-term understanding of bowhead whale diet and physiology. Horstmann-Dehn and George (2013) concluded, however, that changes in the fatty acid composition of prey during their journey through the gastrointestinal tract make fatty acid analysis ineffective for bowhead whales (Chapter 12).

Stable isotope analyses

Stable isotope analyses are based on the assumption that isotope ratios of $\delta^{13}C$ and $\delta^{15}N$ in the soft tissues of an animal reflect the isotopic ratios in the prey consumed (Hobson and Schell, 1998). Soft tissues used in bowhead analyses included baleen (Schell et al., 1989a; Schell and Saupe, 1993; Hobson and Schell, 1998; Lee et al., 2005; Matthews and Ferguson, 2015; Sensor et al., 2018), muscle (Hoekstra et al., 2002; Lee et al., 2005; Horstmann-Dehn et al., 2012), skin (Horstmann-Dehn et al., 2012; Pomerleau et al., 2012), and digestive contents (Horstmann-Dehn and George, 2013). Due to the effects of seasonal fasting and feeding in baleen whales, variation in isotopic baselines need to be better understood for diet-related analyses to reflect the actual diet (Aguilar et al., 2014). While earlier studies suggested late fall and winter feeding from the Bering Sea occurs, the relative importance of the Chukchi and Beaufort seas as feeding areas varied (Schell and Saupe 1993, Hoekstra et al., 2002; Lee et al., 2005). Possible explanations of the disparities included: (1) prey enriched in the Bering Sea were advected to the Chukchi Sea to the western Alaskan Beaufort Sea region and (2) researchers assumed different tissue turnover rates which affect interpretation of the relative importance of regional feeding areas (Hoekstra et al., 2002; Lee et al., 2005). Though not yet used for bowheads, stable isotope analyses of feces have potential to determine prey taxa and short-term diet (Arregui et al., 2017).

Blood

Blood serum chemistry analyses are able to separate feeding from nonfeeding whales (Chapter 11). At the field level, samples of lipemic blood are commonly noted from harvested whales that were recently feeding and have a large volume of prey in their stomach (G. Sheffield, unpublished data).

Feces

Gross examination of the bowhead small intestines and/or colon can provide evidence of short-term feeding status and limited prey identification (Lowry, 1993; Sheffield and George, 2013). Bowhead whale fecal samples have been used to detect algal biotoxins (e.g., saxitoxin and domoic acid) (Lefebvre et al., 2016) as well as determine digestive efficiency (Horstmann-Dehn and George, 2013; Chapter 12).

Stomach contents

The ram filter-feeding behavior used by bowheads is energetically expensive, thus prey in the stomach represents an investment in time and energy by the whale (George, 2009;

Chapter 16). An examination of the stomach contents of a recently harvested bowhead whale provides direct evidence as to: (1) whether or not the whale had fed prior to death and/or (2) what its last meal consisted of. Though flensing and butchering practices may differ between whaling communities, the exposure of the stomach occurs after the skin, blubber, lateral musculature, and an entire side of ribs has been removed (Fig. 28.2A).

The stomach is not often utilized for subsistence purposes and typically must be removed from consumable whale products before examination of its contents. The contents are accessed by cutting into the forestomach (when possible), they are photo-documented (Fig. 28.2C), a quantitative estimate of total volume is made, qualitative condition of the contents is noted, and an aliquot is collected into a portable container (Fig. 28.2B). It is frozen

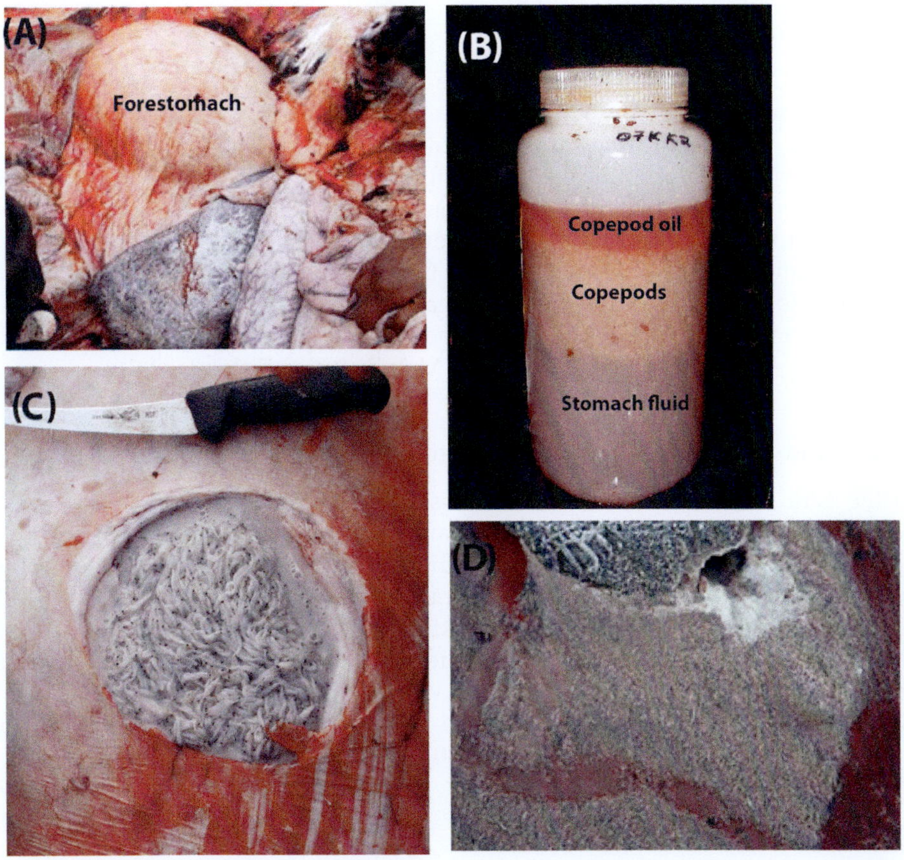

FIGURE 28.2 (A) The stomach of a subadult (9.2 m length) bowhead whale partially exposed during the butchering processing at Kaktovik, Alaska, fall 2001 (NSB-DWM 2000KK1). (B) Orange-colored copepod oil, copepods, and digestive fluid in 1 L container from the forestomach of a bowhead whale near Kaktovik during 2007 (NSB-DWM 2007KK2). (C) A stomach full of undigested euphausiids being sampled near Utqiaġvik, Alaska during fall 2009 (NSB-DWM 2009B8). (D) Krill spilled onto the beach from the stomach of an adult (16.1 m) female bowhead whale (NSB-DWM 1751) harvested during early January 2017 that had been feeding near Savoonga, Alaska. *Source: (A, D) Photos by G. Sheffield. (B, C) Photo by J. C. George.*

for later rinsing, sorting, and examination in a laboratory setting. Stomach and intestinal samples provide direct evidence of feeding but several issues can make interpretation difficult. Stomach contents can be compromised during the hunt by regurgitation, weapon damage, and/or ingestion of blood and debris. During butchering, inadvertent knife damage, spillage, or damage of prey items can occur. Additional problems with the interpretation of diet samples include not accounting for the state of digestion in the stomach (Sheffield et al., 2001), lack of data on total volume during field sampling, as well as relatively small sample sizes of harvested whales in smaller whaling communities.

Diet and feeding in four bowhead stocks

Okhotsk Sea stock

Okhotsk Sea (OKH) bowheads, unlike those of other stocks, live in an ice-free environment during the summer months. Although the Sea of Okhotsk was briefly a focus of intensive commercial whaling during the 19th and early 20th centuries (Ivashchenko and Clapham, 2010), we found no published diet information on OKH bowhead stomach examinations. The waters of the Shantar islands region of the Okhotsk Sea are a major summering and feeding area for OKH bowheads (Rogachev et al., 2008; Shpak and Paramonov, 2015, 2018; Melnikov and Fedorets, 2016; Chapter 5). Quantitative documentation of the composition of zooplankton species near feeding bowheads and in areas known to be frequented by OKH bowheads indicate that aggregates of copepods (and other zooplankton) are likely an important food resource (Rogachev et al., 2008; Melnikov and Fedorets, 2016).

East Greenland–Svalbard–Barents Sea stock

From the 17th to the 20th centuries, the East Greenland–Svalbard-Barents Sea (EGSB) stock was almost eliminated by commercial harvest (Ross 1993; Chapter 33) and little historical information exists on EGSB bowhead diet. Christensen et al. (1992) suggested that older reports indicated euphausiids were an important food item. Based on seasonal bowhead sightings and documentation of zooplankton in Fram Strait, Wiig et al. (2007) suggest bowheads are feeding on copepods along the ice edge during spring. Satellite data from a tag embedded in one female EGSB bowhead whale indicated southward summer movements in response to the enormous southward outflow of Arctic water in Fram Strait that contained zooplankton prey (Lydersen et al., 2012).

East Canada–West Greenland stock

The East Canada–West Greenland (ECWG) stock remains decimated by an extensive commercial harvest from the 17th to early 20th centuries (Higdon, 2010). Scoresby (1820) provides the earliest written account of ECWG bowhead prey from the whale itself "the very few instances in which I have been enabled to open their stomachs, squillae or shrimps were the only substances discovered." Stomach content data exist for seven bowheads harvested for subsistence purposes (Pomerleau et al., 2011; Heide-Jørgensen et al., 2012).

During spring 2009–2010, stomachs examined from four adult bowheads (three females, one male) harvested in Disko Bay, West Greenland were estimated to be 50%–100% full, primarily with large (> 3 mm) copepods, likely *Calanus hyperboreus* (Heide-Jørgensen et al., 2012). These findings agree with diving behaviors documented by Laidre et al. (2007). During fall of 1994, 2008, and 2009 stomachs were examined from three whales harvested in separate locations in eastern Canada. Two whales, an adult female and an immature male, contained mostly mysids and one immature female provided evidence of feeding on or near the seafloor (Pomerleau et al., 2011).

The isotopic composition of ECWG bowhead skin reveals limited foraging on ice-associated zooplankton (Pomerleau et al., 2012). Isotopic analysis of baleen plates suggested ECWG whales forage throughout their range, primarily on their summering grounds with indications of foraging at a reduced rate during winter (Matthews and Ferguson, 2015). Fatty acid profiles of ECWG whales were similar between sexes and between age classes suggesting mixing during summer or selective feeding were occurring, and indicated calanoid copepods were an important prey item (Pomerleau et al., 2014). Using aerial video and telemetry studies, Fortune (2018) indicated that ECWG bowhead whales feed throughout the year, but with peak feeding occurring during July–September with Cumberland Sound, Nunavut an important area to ECWG bowhead whales due to the presence of large cold-water copepods.

Bering–Chukchi–Beaufort Seas stock

Due to the active and continuous subsistence whale harvest in the Bering, Chukchi, and Beaufort seas and the collaborative working relationships between whaling crews and regionally based biologists, the diet of the BCB stock is the best documented. Routine sampling of harvested whales began in the early 1970s by the federal government and moved to the North Slope Borough Department of Wildlife Management during 1981 (Albert, 2001). Examinations of harvested bowhead whales has primarily been conducted in the Beaufort and Chukchi seas with information on the feeding status and/or diet data gathered from hundreds of whales (Lowry, 1993; Lowry et al., 2004; Moore et al., 2010; Sheffield and George, 2013, NSB-DWM data).

Since 1969, a broad array of taxa have been identified from the gastrointestinal tract of BCB bowhead whales throughout their range (Table 28.1). Typically, the primary diet of BCB bowheads has consisted of small zooplankton such as copepods, euphausiids, mysids, and amphipods (Lowry, 1993; Lowry et al., 2004, Sheffield and George, 2013, NSB-DWM data). The whales close proximity to shore during the harvest and/or seasonal location of prey may explain the relatively common occurrence of epibenthic prey and sediments found in stomach samples.

Milk in the stomach has been documented in nine calves from the Beaufort Sea since 1991 (four females, four males, one sex undetermined) (Lowry and Sheffield, 2002; Moore et al., 2010, Sheffield and George, 2013, NSB-DWM data). These small whales were 5.3–7.2 m in length and with the exception of two calves examined during May; all were examined during the fall. The largest volume of milk estimated in a stomach was about 12 L. One young male (NSB-DWM 2008KK1) not only contained milk in his stomach but also had copepod

TABLE 28.1 Prey and other items consumed by BCB bowhead whales harvested during 1966–2012. Locations and seasons: CBF, Canadian Beaufort Fall, CF, Cross Island, Fall; CKS, Chukchi Sea, Spring; KF, Kaktovik, Fall; SHK, Shaktoolik; SLIS, Saint Lawrence Island, Spring; SLIF, Saint Lawrence Island, Fall; US, Utqiaġvik, Spring; UF, Utqiaġvik, Fall. Data from Johnson et al., 1966 (0); Lowry, 1993 (1), Lowry et al., 2004 (2); NSB-DWM unpublished data 2001-2005 (3); Pomerleau et al., 2011 (4); Sheffield and George, 2013 (5). The table is read vertically, left column first then the right column.

Prey type and location	References	Prey type and location	References
CNIDARIA		**Amphipods** (continued)	
"Jellyfish"^{UF}		*Harpinia* sp.$^{\text{US, CKS}}$	[1,2]
Scyphozoa$^{\text{US, UF}}$	[2]	*Hippomedon denticulatus*^{SLIS}	[1]
Hydroida^{UF}	[5]	Hyperiid amphipods$^{\text{US, UF, KF}}$	[1–,3,5]
ANNELIDA		*Hyperia* sp.$^{\text{US, KF}}$	[1,2]
Polychaete worms$^{\text{UF, US, SLIF, CKS}}$	[0,2,3,5]	*Hyperia galba*$^{\text{US, KF}}$	[1,2]
Capitellidae^{KF}	[5]	*Hyperia medusarum*$^{\text{US, UF, KF}}$	[1,2,5]
Maldanidae^{KF}	[5]	*Hyperoche medusarum*$^{\text{US, UF}}$	[1,2,5]
Pectinariidae^{KF}	[5]	*Isaeidae*^{UF}	[5]
Repantia^{CKS}		Lysianassidae$^{\text{UF, KF}}$	[2,3,5]
MOLLUSCA		*Maera* sp.^{UF}	[3]
Snails$^{\text{KF, UF, SLIS, CKS}}$	[0,2,3,5]	*Melita* sp.$^{\text{UF, SLIS}}$	[2,3,5]
Colus sp.^{UF}	[3]	*Melita quadrispinosa*^{UF}	[2]
Crytonatica sp.^{UF}	[3]	*Monoculodes* sp.$^{\text{UF, KF}}$	[1–3,5]
Diplodanta sp.^{UF}	[3	*Monoculodes zervoni*^{UF}	[1,2]
Limacina helicina$^{\text{US, KF}}$	[1–3]	*Monoculopsis* sp.^{UF}	[3]
Margarites sp.^{US}	[2]	*Munnopsis* sp.^{KF}	[1]
Natica sp.^{US}	[2]	Oedicerotidae^{UF}	[2,3,5]
Natica clausa$^{\text{US, CKS}}$	[1,2]	*Onisimus* sp.$^{\text{UF, KF, US, SLIS}}$	[2,5]
Neptunea sp.^{US}	[2]	*Onisimus glacialis*^{KF}	[2]
Oenopota sp.^{UF}	[3]	*Onisimus litoralis*^{KF}	[2]
Trochidae^{KF}	[3,5]	*Onisimus nanseni*^{KF}	[2]

(Continued on next page, left column)

(Continued on next page, right column)

TABLE 28.1 (Continued)

Prey type and location	References
Clams[UF, SLIS]	[2,3,5]
Astarte sp.[US, KF]	[2,5]
Cyclocardia crebricostata[US]	[3]
Ennucula tenuis[UF, SLIS]	[2,3]
Liocyma fluctuosa[US]	[2]
Pelecypoda[SLIS]	[0]
Nuculana sp.[US, SLIS]	[1,2,5]
Tellinidae[US]	[2]
Yoldia sp.[US, UF]	[2]
PRIAPULA	
Priapulidae[KF, UF]	[3,5]
ECHIURA[UF]	[3]
CHELICERATA	
Pycnogonidae[KF]	[2]
CRUSTACEA[SLIF, CKS]	[0,3]
Ostracods[US, UF]	[1,2,5]
Conchoecia sp.[UF]	[3]
Copepods[US, UF, KF, CF, SLIS, SLIF]	[1–3,5]
Aetideidae[UF, US]	[2,3,5]
Calanus sp.[US, UF, KF, SLIF, CBF]	[1,2,4,5]
Calanus cristatus[US]	[1,2]
Calanus finmarchicus[KF]	[2]
Calanus glacialis[US, UF, KF, CF, SLIF, SLIS, CKS]	[1–3,5]
Calanus hyperboreus[US, UF, KF, CF]	[1–3,5]
Orchonome sp.[SLIS, UF]	[1,3]
Parathemisto sp.[UF, US, KF]	[2,3,5]
Parathemisto abyssorum[US, KF, CF]	[1–3]
Parathemisto libellula[UF, KF, CF, SLIF]	[1–3,5]
Paroediceros sp.[UF]	[5]
Pleustes sp.[UF]	[5]
Podoceridae[UF]	[2,5]
Pontoporeia sp.[UF]	[5]
Pontoporeia femorata[UF, SLIS]	[1–3,5]
Protomedeia sp.[UF]	[2,3]
Rhacotropis sp.[US, UF, KF]	[1,2]
Rozinante sp.[UF]	[1–3]
Rozinante fragilis[UF, KF]	[1,2]
Stegocephalidae[UF]	[3]
Stenothoidae[UF]	[5]
Themisto spp.[CBF]	[4]
Westwoodilla sp.[UF]	[3]
Weyprechtia sp.[UF, US, KF, CF]	[2,5]
Weyprechtia heuglini[UF, KF]	[2]
Weyprechtia pinguis[UF, KF]	[2]
Synopiidae[SLIS]	[1]
Euphausiids[US, UF, KF, CF, SLIF]	[1–3,5]
Thysanoessa sp.[US, UF, KF]	[2,3,5]
Thysanoessa inermis[US, UF, KF]	[1–3,5]

(Continued on next page, left column)

(Continued on next page, right column)

TABLE 28.1 (*Continued*)

Prey type and location	References	Prey type and location	References
Chiridius obtusifrons$^{\text{US, KF}}$	[1,2]	*Thysanoessa raschii*$^{\text{US, UF, KF, CF, SLIF}}$	[1–3,5]
Chiridius polaris ^{UF}	[5]	**Decapods**$^{\text{UF, KF, CF, SLIS}}$	[2,3,5]
Derjuginia tolli^{KF}	[1,2]	*Argis* sp.$^{\text{US, UF}}$	[2,5]
Euchaetidae^{KF}	[2]	*Argis lar* ^{UF}	[3,5]
Euchaeta sp.$^{\text{US, KF, CKS}}$	[1,2]	*Chionoecetes opilio*^{UF}	[3]
Euchaeta glacialis$^{\text{US, UF, KF, CF}}$	[1,2]	Crangonidae$^{\text{UF, KF, SLIS, SLIF}}$	[2,3,5]
Harpacticoid^{UF}	[3]	*Crangon dalli*^{UF}	[3]
Harpacticus sp.^{UF}	[3]	*Eualus* sp.$^{\text{US, UF}}$	[2,5]
Heterorhabdus sp.^{KF}	[1,2]	*Eualus fabrici* $^{\text{UF, SLIS}}$	[1,5]
Limnocalanus grimaldii$^{\text{KF, UF}}$	[1–3,5]	*Eualus gaimardi*$^{\text{UF, KF}}$	[2,3,5]
Limnocalanus macrurus^{CBF}	[4]	*Heptacarpus* sp.^{UF}	[2]
Metridia sp.$^{\text{US, KF, CKS}}$	[1,2]	Hippolytidae$^{\text{UF, KF, SLIS, SLIF}}$	[2,3,5]
Metridia lucens^{KF}	[1,2]	*Hyas* sp. (Megalopa)^{UF}	[3]
Metridia longa$^{\text{US, UF, KF}}$	[1–3,5]	Hermit crab^{UF}	[5]
Pareuchaeta sp.$^{\text{UF, KF}}$	[5]	Paguridae^{UF}	[5]
Pareuchaeta glacialis$^{\text{UF, US, KF}}$	[3,5]	Paguridae (Megalopa)^{UF}	[3]
Pseudocalanus sp.$^{\text{US, KF}}$	[1–3]	*Pandalidae*$^{\text{UF, SLIF}}$	[3,5]
Mysids$^{\text{US, UF, KF}}$	[2,3,5]	*Pandalus* sp.^{UF}	[3,5]
Boreomysis arctica^{US}	[3]	*Pandalus goniurus*^{KF}	[2]
Mysis sp.$^{\text{UF, KF}}$	[2,3,5]	*Sabinea* sp.^{UF}	[2]
Mysis litoralis$^{\text{US, UF, KF, SLIS}}$	[1,2,5]	*Sabinea septemcarinata*$^{\text{UF, KF}}$	[2,5]
Mysis oculata$^{\text{UF, US, KF, CF, SLIS, CBF}}$	[2–5]	*Sclerocrangon* sp.^{UF}	[2]
Neomysis sp.$^{\text{US, UF}}$	[2,5]	*Sclerocrangon boreas*^{UF}	[2,3]
Neomysis rayii$^{\text{US, UF}}$	[1–3,5]	"Shrimp" parts$^{\text{UF, KF, SLIS, SHK}}$	[1,3,5]

(*Continued on next page, left column*)

(*Continued on next page, right column*)

TABLE 28.1 (*Continued*)

Prey type and location	References
Cumaceans$^{\text{KF, CF, UF}}$	[2,3,5]
Brachydiastylis resima^{US}	[2]
Diastylis sp.$^{\text{UF, US, CF, KF, CKS}}$	[2,3,5]
Diastylis bidentata$^{\text{UF, SLIS}}$	[1,5]
Diastylis dalli^{KF}	[2]
Diastylis galbra^{KF}	[2]
Diastylis sulcata^{UF}	[3,5]
Eudorellopsis sp.^{UF}	[3,5]
Leuconidae^{UF}	[1,2]
Leucon sp.^{UF}	[1–3]
Leucon nasica^{US}	[2]
Isopods$^{\text{KF, UF}}$	[2,3]
Munnopsis sp.^{KF}	[2]
Saduria sp.$^{\text{UF, KF}}$	[2]
Saduria entomon$^{\text{UF, US, KF}}$	[2,3,5]
Amphipods$^{\text{US, KF, CF, UF, SLIF}}$	[2,3,5]
Gammarid amphipods$^{\text{US, UF, KF, CF, SLIS}}$	[1,2]
Acanthostepheia sp.$^{\text{US, UF, KF}}$	[1,2,5]
Acanthostepheia behringiensis$^{\text{UF, US, KF}}$	[1–3,5]
Acanthostepheia incarinata^{KF}	[2]
Acanthostephia malmgreni^{UF}	[2]
Aceroides sp.^{UF}	[3]
Aceroides latipes^{KF}	[2]
Anonyx sp.$^{\text{US, UF, KF, CF}}$	[2,3,5]

(*Continued on next page, left column*)

Prey type and location	References
Decapod zoea$^{\text{UF, KF, CF}}$	[1–3]
Crangonidae zoea^{UF}	[3]
Hippolytidae zoea$^{\text{UF, KF}}$	[3]
Paguridae zoea$^{\text{US, UF, CF, KF}}$	[3]
Pandalidae zoea^{UF}	[3]
Pisinae zoea^{UF}	[3]
"Crabs"^{SLIS}	[1]
Barnacles	
Barnacle^{UF}	[5]
ECHINODERMATA	
Echinoidea^{CKS}	[0]
Ophiuroidea^{US}	[2]
Ophiura sp.$^{\text{UF, US, KF}}$	[3,5]
VERTEBRATA$^{\text{BF, K}}$	[2]
Fishes$^{\text{UF, US, SLIS}}$	[1,3,5]
Agonidae^{UF}	[2,3]
Ammodytes sp.^{UF}	[2]
Ammodytes hexapterus^{UF}	[3,5]
Anisarchus medius^{UF}	[5]
Boreogadus sp.^{UF}	[2]
Boreogadus saida$^{\text{US, UF, KF}}$	[2,3,5]
Cottidae$^{\text{UF, SLIS}}$	[1–3,5]
Eleginus gracilis$^{\text{UF, US}}$	[5]
Gadidae$^{\text{CF, UF, KF}}$	[2,5]

(*Continued on next page, right column*)

TABLE 28.1 *(Continued)*

Prey type and location	References
Anonyx compactus[SLIS]	[1]
Anonyx nugax[US]	[1,2]
Ampeliscidae[UF]	[5]
Ampelisca sp.[UF]	[5]
Ampelisca macrocephala[UF, CKS]	[1,2]
Apherusa sp.[UF]	[3,5]
Apherusa glacialis[KF]	[2]
Arrhis sp.[UF]	[2]
Atylus sp.[US, UF]	[3,5]
Atylus atlassovi[SLIS]	[1]
Atylus carinatus[UF, KF]	[1,2]
Atylus collingi[UF]	[3]
Bathymedon sp.[SLIS]	[1]
Boeckosimus affinis[UF]	[2]
Boeckosimus krassini[UF]	[2]
Boeckosimus litoralis[KF]	[2]
Byblis sp.[US, UF]	[2]
Erichthonius sp.[US]	[2]
Eusirus sp.[UF, US]	[3,5]
Eusirus cuspidatus[US, UF]	[3]
Gammarid	[3,5]
Gammarus sp.[US, UF, KF]	[2,3,5]
Gammarus zaddachi[UF]	[1,2]
Gammaracanthus sp.[UF]	[2]
Gammaracanthus loricatus[UF, KF]	[2]
Gymnacanthus tricuspis[UF, KF]	[3,5]
Hippoglossoides robustus[UF]	[3]
Icelinus sp.[UF]	[2]
Icelus sp.[UF]	[5]
Lepidopsetta bilinicata[UF]	[2]
Limanda proboscidea[UF]	[3]
Lumpenus fabricii[UF]	[3]
Lycodes sp.[UF]	[2]
Myoxocephalus sp.[UF]	[2,5]
Myoxocephalus quadricornis[KF]	[2]
Myoxocephalus scorpius[KF]	[5]
Pleuronectidae[UF]	[2]
Pungitius pungitius[KF]	[2]
Stichaeidae[UF]	[2]
Ulcina olrikii[UF]	[5]
Zoarcidae[UF]	[2]
PHAEOPHYCEAE	
plant material[UF]	[2,3]
OTHER (not comprehensive)	
baleen hairs[US, UF, KF, CF]	[2]
bird feathers[UF, KF]	[1–3]
plastic[UF, KF]	[1,2]
wood[UF, KF]	[2]
sediments[US, UF, KF, SLIS]	[0,1,2]
polychaete "tube"[US]	[3]

(Continued on first page, right column)

debris in his small intestine suggesting this 7.2 m whale was alternating nursing with foraging (NSB-DWM data). Another calf, examined at Utqiaġvik (NSB-DWM 2013B13), contained a mix of milk and invertebrates and suggests calves begin to consume zooplankton at 5–6 months of age (Chapter 13).

The documentation of ingested plastics, though not an objective in previous diet studies, has been noted in several stomachs. First documented in a subadult male (NSB-DWM 1986KK3) harvested near Kaktovik during late September 1986, the stomach contained a "12 × 12 cm piece of plastic sheet" (Lowry, 1993), and then in the stomach of a second subadult male (NSB-DWM 1989B8) harvested near Utqiaġvik during early October 1989 which contained "a piece of plastic approx. 5 × 8 cm about the thickness of a cigarette wrapper" (Lowry et al., 2004). More recently, a third subadult male (NSB-DWM 2017B13) harvested near Utqiaġvik during early October 2017 contained a piece of "plastic sheet 3 × 3 cm" (NSB-DWM, unpublished data).

Description of BCB stock seasonal feeding by region

The annual migration cycle of the bowhead seasonal transverses three separate arctic seas—from the northern Bering Sea to the Eastern Beaufort Sea—and back again (Chapter 4).

Northern Bering Sea—fall

Indigenous whaling communities know bowheads feed when they return to the northern Bering Sea during fall (Hazard and Lowry, 1984; Noongwook, 2007). The first details of prey types consumed came from intestinal samples of three adult female bowheads (NSB-DWM 2005S5-7) harvested near Saint Lawrence Island during November 2005 in which euphausiids were the dominant prey (Sheffield, 2008). Analysis of three additional fecal samples during late fall (Nov. 2010, Dec. 2012) identified copepods (*Calanus* sp.) and euphausiids (*Thysanoessa raschii*) as important prey (Sheffield and George, 2013).

Northern Bering Sea—winter

Many marine mammal species actively feed in the northern Bering Sea during winter—including bowhead whales. Saint Lawrence Island whalers were aware that bowhead whales commonly fed in the northern Bering Sea (Hazard and Lowry, 1984) during late fall and winter (Noongwook, 2007, NSB-DWM data); however, it was not commonly known to the research community even though reported (e.g., during February, five bowhead whales were observed feeding nearshore for 3 days near Gambell; F. Kangingok 1985 to NSB-DWM). Since 1990, northern Bering Sea whalers have been extending their harvest season into late fall and winter months (Noongwook, 2007) in response to the evolving conditions associated with the ongoing reduction in sea ice quality, extent, and duration. This situation has allowed the first opportunity to sample winter prey. During January 2017, Saint Lawrence Island whalers landed an adult female bowhead (NSB-DWM 2017S1) in remarkably sea ice-free conditions. The whalers noted this whale was feeding, "swimming slowly in a circle near the surface" before the hunt initiated. Once landed, a copious amount of semisolid crimson brown feces was released from the anus. During subsequent butchering onshore, >16 L of freshly consumed krill (mysids/euphausiids) were expelled from the throat (Sheffield, 2017; Fig. 28.2).

Further evidence of winter feeding has been inferred from satellite-linked tags applied to bowheads which indicate the whales spend a high proportion of time at or near the seafloor near the Gulf of Anadyr during winter, strongly suggesting they are feeding (Citta et al., 2015a).

Northern Bering Sea—spring

Spring in the northern Bering Sea for BCB whales is marked by mating activities, calving (Tomilin, 1957; Koski, 1993), as well as beginning the northward migration (Noongwook, 2007) through the Chukchi Sea and into the Beaufort Sea (Chapter 4). Whalers are familiar with bowheads feeding during spring in the shallow waters east and west of Saint Lawrence Island (Noongwook, 2007), including the northern Gulf of Anadyr (Melnikov et al., 2004). During 1978–94, the stomach (n = 16) and intestines (n = 2) of 18 bowheads were opportunistically examined by State of Alaska, North Slope Borough, and National Marine Fisheries Service biologists during spring (April–May). The first summary diet information was of "a few euphausiid-like creatures" from a subadult male (1978S1) harvested near Saint Lawrence Island during April 1978 (Lowry and Frost, 1984). Hazard and Lowry (1984) provided the first identification of prey sampled from stomach contents of a bowhead near Saint Lawrence Island (NSB-DWM 1982G2) that had been feeding on or near the benthos. Additional examinations of gastrointestinal tracts revealed "a full colon" from an adult female bowhead whale (NSB-DWM 1980G1) near Gambell during May 1980 (Lowry and Frost, 1984), shrimp fragments in feces of a subadult male bowhead whale (NSB-DWM 1980SH1) harvested near Shaktoolik in eastern Norton Sound during May 1980 (Frost and Lowry, 1981), and the removal of a large (15 cm) piece of wood from the intestine of a subadult male (86G1; NSB-DWM19). More recently, a subadult female (NSB-DWM 1912W1) examined near Wales during April 2012 had an empty stomach (NSB-DWM data).

During April 2007–12, analysis of one stomach sample and 15 fecal samples indicated copepods (most commonly, *Calanus glacialis*) were a frequent prey (Sheffield and George, 2013). Euphausiids were not identified in any of the bowheads sampled during spring and were the only prey taxa showing a seasonal difference (Sheffield and George, 2013).

In the Northern Bering Sea, bowhead whales feed during spring (April–May) before their northward migration and continue upon returning in the fall (November–December). No difference in the proportion of whales feeding was detected between spring and fall migrations (Sheffield, 2008; Sheffield and George, 2013). Although caution should be taken when interpreting the small sample number of examined whales ($n = 16$), the data indicate that a significantly larger proportion of bowhead whales are feeding more often in the northern Bering Sea (73%) before their migration begins than when they are actively transiting past Utqiaġvik (10%) to their summer range in the eastern Beaufort Sea (Sheffield and George, 2013).

The implications and importance of the samples from relatively few whales from the northern Bering Sea should not be underestimated as they document what bowhead whales eat and confirm local indigenous knowledge of feeding behavior and diet (Chapter 34). These results also support previous isotope studies, *all* of which have produced consistent results supporting the importance of the Bering and Chukchi seas as important feeding areas (Schell and Saupe, 1993; Hoekstra et al., 2002; Lee et al., 2005). Location and dive data from satellite-linked tags demonstrated individuals spent much of their time at or near the seafloor for up to six months in the Bering Sea prior to their northbound migration in spring (Citta et al., 2015b).

Chukchi Sea—spring

Examinations of bowheads harvested along the eastern Chukchi Sea coast near Kivalina, Point Hope, Wainwright, and Utqiaġvik during the spring migration suggest only occasional feeding occurs (NSB-DWM data).

During 1992–1994, two female bowheads were harvested during spring (April–May) near Kivalina and their stomachs examined. One whale (NSB-DWM 92K1) had been feeding and one (NSB-DWM 1994K1) had not (NSB-DWM data). At Point Hope, Johnson et al. (1966) provided the first records of spring diet based on stomach examinations of three bowheads (M39, M1571, M1527) harvested during 1960 and 1961. They described two empty stomachs and one that contained fragmentary remains of epibenthic prey such as "... polychaeta, repantia, gastropods, crustacea, echinoidea, and sand and gravel." During 1974–2015, additional stomachs and/or intestines from 17 bowhead whales harvested near Point Hope during spring (April–May) were examined. Field notes indicated that only 6% had been feeding prior to death, primarily on "shrimp" (NSB-DWM data). Lowry (1993) described the stomach contents of two subadult males sampled at Point Hope during spring 1978 and 1979. One whale (NSB-DWM 1978H1) contained one ampeliscid amphipod (NSB-DWM 1978H1) and the other whale (NSB-DWM 79H3) contained a single bivalve. Another subadult male examined during spring 2015 (NSB-DWM 2015H2) had not been feeding (NSB-DWM data).

In the northeastern Chukchi Sea, whaling captains from Wainwright observe whales feeding during the spring migration (Huntington and Quakenbush, 2009a). During 1980–2019, stomachs and/or intestines from 25 bowheads harvested near Wainwright were examined during spring (April–May). One subadult female (NSB-DWM 1988WW1) had been feeding primarily on copepods during April 1988 (Lowry, 1993). Unpublished NSB-DWM field notes for the remainder indicated that only 32% had been feeding, primarily on "krill."

Beaufort Sea—spring

Bowheads arrive to the Beaufort Sea during spring and continue their migration east to their summer range. Whaling captains report bowheads usually migrate rapidly past Utqiaġvik during spring, though some occasionally stop and feed (Huntington and Quakenbush, 2009b). Based on examinations of over 150 bowhead stomachs, the NE Chukchi and extreme western Beaufort sea are not heavily utilized as feeding areas during spring with more than 70% of the examined whales having not fed (Lowry et al., 2004; Sheffield and George, 2013, NSB-DWM data), although significant feeding occasionally occurs (Carroll et al., 1987). Copepods occur more often in the diet during spring than fall near Utqiaġvik (Lowry et al., 2004; Sheffield and George, 2013, NSB-DWM data). Due to the broad shorefast ice shelf east of Utqiaġvik during spring, two whaling communities (Nuiqsut and Kaktovik) hunt only during fall and do not have access to migrating whales during the spring.

Beaufort Sea—summer

During the summer months, bowheads typically occupy the Eastern Beaufort Sea (Harwood et al., 2017; Citta et al., 2015a) and feed. This knowledge is based on large-scale research efforts (i.e., aerial surveys, satellite-linked tagging, etc.) associated with offshore oil and gas development and which typically occur throughout the summer months (Chapters 4 and 24, this volume). Bowheads are sometimes seen near Utqiaġvik, Nuiqsut, and Kaktovik by coastal residents during summer (Galginaitis and Koski, 2002; Huntington and Quakenbush, 2009a,b; Shelden and Mocklin, 2013). Typically, no bowhead whales are harvested during summer. However, two whales (NSB-DWM 86B8 and 2013B2) were harvested near Utqiaġvik during July 1986 and July 2013, but the stomachs were not examined (NSB-DWM data; Suydam et al., 2014).

In the Canadian Beaufort Sea, one stomach was examined from a subadult male bowhead harvested near Shingle Point during July 1996. This whale was feeding primarily on copepods (primarily *Limnocalanus macrurus*) with smaller amounts of amphipods and mysids (Pomerleau et al., 2011). Results of this singular whale agreed with fatty acid analyses (using skin and blubber) indicating bowheads feed extensively during their summer in the Beaufort Sea, and that all age classes consume significant amounts of copepods (Budge et al., 2008).

Isotopic results have produced inconsistent results regarding the importance of the Beaufort Sea as a feeding area for BCB bowheads—especially during their summer occupation of the eastern Beaufort Sea. Isotopic analyses of baleen suggest the eastern Beaufort Sea is not the primary feeding area for BCB whales (Schell and Saupe, 1993). Isotopic analyses of muscle and/or visceral fat suggest whales feed at the same magnitude during summer and fall (Schell et al., 1989b) with no seasonal changes. Similar analyses by Hoekstra et al. (2002) suggested both the Bering and Beaufort seas are important feeding areas. Additional isotopic analysis of muscle indicated that most food for bowheads (especially adults) was not derived from the Beaufort Sea (Lee et al., 2005). Richardson and Thomson (2002) however, concluded the unequivocal importance of feeding in the Eastern Beaufort Sea based on direct observations of summer feeding, analyses of stomach contents, and evidence of summer fattening based on a comparison of spring versus fall photogrammetric images.

Beaufort Sea—fall

Considerable evidence indicates that the BCB population feeds while migrating westward during fall across the Beaufort Sea (Lowry et al., 2004; Chapters 4, 24, and 26). Observations by whalers from three communities (Utqiaġvik, Nuiqsut, and Kaktovik) in the Beaufort Sea indicate whales feed on occasional large swarms of zooplankton with specific feeding areas frequented by many whales (Galginaitis and Koski, 2002; Huntington and Quakenbush, 2009a,b; Chapter 24).

The importance of the Eastern Alaskan Beaufort Sea as a feeding area has been well documented due to Kaktovik whalers providing detailed information and access to their landed whales for over four decades. Over 75% of bowhead whales harvested near Kaktovik contained food in their stomachs (Lowry et al., 2004). There were no sex or age-related differences in the occurrence of prey eaten (Lowry et al., 2004). Copepods are the dominant prey of whales harvested near Kaktovik (Lowry et al., 2004; Sheffield and George, 2013) (Fig. 28.2). Lowry et al. (2004) identified the most commonly consumed copepods as *C. hyperboreus* and *C. glacialis* whereas Sheffield and George (2013) identified copepods *C. hyperboreus* and *Pareuchaeta glacialis* as the most commonly consumed. During fall, euphausiids and hyperiid amphipods occurred significantly more often in the stomachs of whales harvested near Utqiaġvik than in whales harvested near Kaktovik (Lowry et al., 2004). During 2007–2012, only 54% of whales near Kaktovik were feeding during fall (Sheffield and George, 2013), which is less than expected based on results in Lowry et al. (2004).

During 2007–2012, copepods dominated the prey by volume for 86% of the samples ($n = 13$). Euphausiids (*T. raschii*) dominated, by volume, a single stomach sample from a young adult male (NSB-DWM 09KK2). This whale was harvested during the latter part of September 2009 and had eaten isopods and >15 fish which included: Arctic cod (*Boreogadus saida*), Arctic staghorn sculpin (*Gymnocanthus tricuspis*), and Shorthorn sculpin (*Myoxocephalus scorpius*). Euphausiids, fish, and isopods were not identified in any other 2009 stomach samples from Kaktovik (Sheffield and George, 2013). Copepods dominated (60%) the prey by volume for three of the five bowheads examined near Kaktovik during the fall from 2013 to 2016 (NSB-DWM data).

West of Kaktovik, in the central Alaskan Beaufort Sea, whalers from Nuiqsut travel offshore during late August and September to Cross Island, to harvest bowhead whales. Like bowheads examined near Kaktovik, bowheads primarily consume copepods near Cross Island. During 1987–2017, stomachs from 15 bowheads harvested near Nuiqsut during fall (September–October) were examined and/or sampled. Results indicated nine whales had been feeding on copepods, four stomachs were empty, and two were determined "uncertain" as one stomach sample contained only a trace amount of copepods, and field notes for another stomach described an undetermined amount of "red-colored invertebrates" (NSB-DWM data). These results may underestimate the feeding status and/or prey as three of the samples mentioned above had been compromised during shipping (NSB-DWM data). Regardless, these results confirm observations and reports by Nuiqsut whalers' that whales harvested near Cross Island typically have full stomachs (Huntington, 2013). Their observations of whale behavior, prey conditions, and their ecological assessment of Camden Bay and the waters near Cross Island suggest these are important feeding areas for bowhead whales during fall: "In Camden Bay, bowheads stay to feed rather than migrating straight through. The plankton can be plentiful and it is a prime feeding ground, with excellent feeding. Feeding whales leave an oily sheen behind them, from the plankton on which they are grazing. This 'whale trail' can be seen in the water, marking the path of the whale" (Huntington, 2013).

Typically, the fall migration continues near Utqiaġvik before the whales enter and continue across the Chukchi Sea westward to the northern waters of the Chukotka peninsula. Wind-driven upwelling during fall significantly contributes to concentrating bowhead whale prey nearshore, resulting in the waters near Point Barrow being an important feeding area for bowheads (Ashjian et al., 2010; Chapter 26) and thus an important area for whale hunters as well. Based on stomach examinations, bowhead whales commonly feed near

Utqiaġvik during the fall (Lowry et al., 2004; Moore et al., 2010; Sheffield and George, 2013, NSB-DWM data). Bowhead whales feed significantly more often near Utqiaġvik on their return to the western Beaufort Sea during the fall than in spring when migrating eastward (Lowry et al., 2004; Sheffield and George, 2013, NSB-DWM data). Regarding the breadth of the Alaskan Beaufort Sea as a feeding area, there was no difference in the proportion of whales feeding during fall near Utqiaġvik in the western Beaufort Sea than in the eastern Beaufort Sea near Kaktovik (Lowry et al., 2004). Euphausiids (predominantly *T. raschii*) occurred in almost every whale sampled from Utqiaġvik and were dominate by volume in >85% of the stomachs, whereas copepods were dominant by volume in only 5% of stomachs (Fig. 28.2, Lowry et al., 2004). Results from Moore et al. (2010) indicated only about half the whales examined had full stomachs, as well as a marked difference between 2005 results, with euphausiids the dominant prey by volume in only four of 11 bowheads, and 2006 results with euphausiids dominant in 10 of 13 stomachs. Epibenthic taxa occurred in less than 10% of the whales and included clam, snail, priapulid worm, echiurid worm, polychaete worm, ostracod, and small fish (Moore et al., 2010).

During fall 2007–12, the proportion of feeding whales near Utqiaġvik was higher than expected, based on findings in Lowry et al. (2004), and 100% whales had fed recently when harvested (Sheffield and George, 2013). Copepods (primarily *C. glacialis*), amphipods, and euphausiids (primarily *T. raschii*) each occurred in more than half of the samples. However, the percent-by-volume diet results near Utqiaġvik indicate a switch from a relatively consistent annual dominance of either euphausiids or copepods during 2007–10, to an eclectic mix of prey that included mysids, isopods, amphipods, and fish in quantities indicating these taxa were not incidentally ingested. The stomach contents of two subadult male bowheads (NSB-DWM 2011B12, 2011B13) harvested during late October 2011 were dominated by fish (Sheffield and George, 2013). A 1 L stomach sample of 2011B12 included the remains of over 45 fish and included: Arctic cod (*B. saida*), stout eelblenny (*Anisarchus medius*), Arctic alligator fish (*Ulcina olrikii*), and unidentified sculpins (*Icelus* sp.; Cottidae). The stomach of 2011B13 contained the remains of several hundred fish that included Arctic cod, Saffron cod (*Eleginus gracilis*), Arctic staghorn sculpin (*G. tricuspis*) as well as sculpin (*Myoxocephalus* sp.) and undetermined cod (Gadidae, juveniles). Sheffield and George (2013) noted the number and proportion of fish in this sample had not been observed in any previous postmortem examinations. Near Utqiaġvik, the occurrence of fish as prey has been reported as minor components of samples otherwise dominated by euphausiids or copepods—but during 2011–2012 this was not always the case.

During fall 2013–16, krill, amphipods, and copepods occurred most frequently (84%) in harvested bowheads (NSB-DWM data). Interestingly, small fish were present in 37% of the whales and may indicate increased prey diversity near Utqiaġvik during fall (NSB-DWM data). Fall 2016 was unusual in that all seven stomachs with percent-prey-by volume data were dominated by copepods (NSB-DWM data).

During spring–fall 2019, the western and northern Alaska marine ecosystem experienced a significant reduction in seasonal sea ice which resulted in unprecedented and sustained warming ocean temperatures in the Bering, Chukchi, and western Beaufort seas. During fall 2019, the BCB bowhead whales demonstrated an unprecedented shift in their timing and migration route by heading offshore before they reached the western Beaufort Sea and thus avoided many important feeding areas—including near Utqiaġvik (Clarke et al., 2020; Chapter 27). This novel event resulted in a drastic change in harvest timing, increased safety risk, financial costs, and raised food security concerns in the Utqiaġvik whaling community as the season progressed (Hertz, 2019). The one subadult male (NSB-DWM 2019B10) harvested on 16 November marked the latest fall harvest ever documented near Utqiaġvik and the examined stomach was "very full with euphausiids" (NSB-DWM data). Whether the unprecedented offshore location of the fall sea ice, reduction of advected zooplankton from the south, or the presence of killer whales (as observed by whalers) caused this unusual situation remains unknown.

Chukchi Sea—fall

During autumn, the BCB population enters the Chukchi Sea and continues westward offshore toward the northern Chukotka peninsula (Tomilin, 1957; Moore et al., 1995; Melnikov et al., 2004; Quakenbush et al., 2012). The first documented fall harvest in the northeastern Chukchi Sea near Wainwright occurred during October 2010 (Suydam et al., 2011), with a subadult female bowhead (NSB-DWM 2010WW3) that, based on examination of the stomach, had been feeding heavily on copepods (NSB-DWM data).

Unfortunately, there is a lack of available diet data from Chukotkan whaling communities during the fall with which to describe the actual diet of bowhead whales in this region. Commercial whalers of the 19th century described the southern Chukchi Sea during fall as the "cow yard"—a place where the largest and most oil-laden female bowheads aggregated (Bockstoce, 1986, p. 94). During fall, a large adult female would very likely be pregnant or lactating; that mature females gathered in this region strongly suggests feeding occurs. Tomilin (1957) reported fall time was the best time to harvest bowheads along the Chukotka peninsula, as the whales were most numerous, and would linger in the Chukchi Sea until "forced" south through the Bering Strait by the arrival of sea ice from the north. Considerable behavioral evidence suggests that bowhead whales feed along the Chukotkan coast of the Chukchi Sea during fall (Tomilin, 1957; Moore et al., 1995; Melnikov et al., 2004; Quakenbush et al., 2010; Melnikov and Zdor, 2018; Citta et al., 2018; Chapter 4). Melnikov and Zdor (2018) suggested the bowhead whales southward migration to the northern Bering Sea begins with the formation of the sea ice and ends with the complete freezing of the waters off the Chukotka Peninsula. They suggested the bowheads extended occupation of the area was related to feeding, supported by shore-based observations of feeding whales. These multiple lines of evidence support isotope studies which consistently suggested the importance of the Bering and Chukchi seas as feeding areas (Schell and Saupe, 1993; Hoekstra et al., 2002; Richardson and Thomson, 2002; Lee et al., 2005).

Future concerns

The continuing reduction in sea ice quality, extent, and duration continues to have profound environmental, ecological, and industrial impacts on the northern marine ecosystems throughout the range of all bowhead whales (Stevenson and Lauth, 2019; Post et al., 2019; Huntington et al., 2020; Chapters 27, 35 and 36). As entire marine ecosystems transition, we expect changes in the timing and routes of bowhead migrations, feeding behaviors, and/or diet, as the bowheads respond to ecological and environmental pressures or novel opportunities throughout their range. The dependence of the bowhead whale on arctic species of zooplankton may make them vulnerable to potential shifts in prey composition (Pomerleau et al., 2012). Bowhead whales carry a considerable mass of blubber that serves not only as insulation from the arctic marine environment but is an energy reserve during periods when prey is lacking (Burns, 1993). Horstmann-Dehn and George (2013) estimated that a subadult bowhead in poor feeding conditions could survive a year or more—with adults likely able to endure longer periods (George, 2009). Additionally, Chambault et al. (2018) described potentially strict environmental and physiological parameters which may constrain bowhead feeding as Arctic sea surface temperatures increase above 2°C (in areas devoid of sea ice).

Bering–Chukchi–Beaufort bowhead specific concerns

Currently, loss of seasonal ice coverage in the northern Bering Sea (NBS) (Thoman et al., 2020) has severely weakened the thermal barrier between the northern and southern Bering Sea ecosystems causing massive shifts in marine fish and invertebrate populations (Stevenson and Lauth, 2019). Additional ecological pressure on BCB bowheads may be via increased interspecific competition as numerous zooplankton feeders (i.e., jellyfish, Alaska pollock) migrate northwards (Stevenson and Lauth, 2019) as well as intraspecific competition if zooplankton productivity is diminished with the retreat of the sea ice. George et al. (2015) showed sea ice loss and winds favoring upwelling of zooplankton in the Beaufort Sea were positively correlated with BCB bowhead body condition during 1990–2012 and possibly an acceleration in their rate of population increase (Givens et al., 2016) though they did not speculate on future trends.

Currently, changes in the availability, timing, or types of zooplankton in the NBS are unknown. The increased open water season, however, has already advanced the timing of the BCB bowhead spring migration by one month (Noongwook, 2007), consistent with abundance surveys at Point Barrow (George et al., 2012). For Saint Lawrence Island whalers, harvest opportunities end earlier due to changing weather patterns (Noongwook, 2007).

Increasing summer water temperatures in the NBS will likely increase the frequency and scale of toxic algal bloom events (Natsuike et al., 2017) and by default, contaminate zooplankton with biotoxins (e.g., saxitoxin and domoic acid)—a potential threat to BCB bowhead whales (Lefebvre et al., 2016; Chapter 30). Of 13 species of Alaskan marine mammals tested, Lefebvre et al. (2016) demonstrated BCB bowheads had the highest prevalence of biotoxins to date.

Lastly, the reduction of NBS sea ice coverage coincides with increased opportunities for industrial large vessel traffic throughout the BCB bowhead range (Reeves et al., 2014). Potential threats to BCB bowhead prey, physical ability to feed, and/or accessibility to feeding areas include: accidental discharges, increased risk of oil-fouling in Arctic waters (Stimmelmayr et al., 2018), deflections due to loud underwater noises (Reeves et al., 2012), increased injuries from large line entanglements and/or other marine debris (George et al., 2017; Rolland et al., 2019; Chapters 35 and 36).

Conclusions

Based on examination of the review of bowhead diet and prey types, etc., we conclude:

- Bowhead whales feed primarily on small zooplankton such as copepods, euphausiids, mysids, and amphipods. It is not uncommon, however, for bowhead whales to forage near the seafloor and ingest epibenthic prey in certain regions and seasons.
- Many research methods are used to infer diet and feeding solely from whale behaviors. A thorough examination of a harvested whale provides direct evidence of diet and feeding behavior.
- Indigenous whaling communities provide observations in situations unavailable to western researchers, extensive local knowledge, and access to freshly harvested bowheads resulting in direct observations of bowhead health, diet, and feeding behavior.
- Gastrointestinal samples provide information on whether whales were recently feeding and what their last meal consisted of. Interpretation of diet from biological specimens can be complicated by the stage of digestion, variation in stomach sizes, few estimates of total stomach volume, and small sample sizes in some communities and regions.
- The waters of the Shantar islands region of the Okhotsk Sea are a major summering and feeding area for OKH bowheads and copepods are likely an important food. Documentation of zooplankton near EGSB bowheads suggests copepods and other zooplankton are important prey. Based on limited ECWG bowhead stomach content data, copepods (primarily *C. hyperboreus*) are important to adults during spring whereas fall diet data indicate mysids and epibenthic prey are important.
- Based on examinations of harvested whales, BCB bowhead whales feed throughout the year—though with a reduction in feeding effort during the northward spring migration from the Bering Sea to the Beaufort Sea. That is, the northern Bering, Beaufort, and Chukchi seas are integral to bowhead feeding/diet as they annually travel their range. Seasonal and geographic differences in bowhead prey consumed exist in Alaskan waters. There is no difference in the diet of males and females. Recent diet data from the western Beaufort Sea indicate a potential increase in the diversity of prey types consumed.
- The continuing reduction in sea ice quality, extent, and duration continues to have profound environmental, ecological, and industrial impacts on the northern marine ecosystems throughout the range of all bowhead whales. During 2019, the western and northern Alaska marine ecosystem experienced unprecedented and sustained warming ocean temperatures in the Bering, Chukchi, and western Beaufort seas. For BCB bowheads, shifting maritime ecological, environmental, and industrial conditions may change the whales (and indigenous whalers) accessibility to feeding areas, the types of prey available, and potentially cause food security and health issues.

Acknowledgments

The extensive documentation of BCB bowhead diet has been possible because of the countless collaborative contributions of expert knowledge and shared enthusiasm by Alaskan bowhead whaling communities, biologists, and other researchers. We thank the Alaska Eskimo Whaling Commission and the Whaling Captains Associations members of Gambell, Savoonga, Diomede, Wales, Kivalina, Point Hope, Point Lay, Wainwright, Utqiaġvik, Nuiqsut, and Kaktovik. Whether on the sea ice or beach, it is the camaraderie at each landed whale that helps guide us to a better understanding of important bowhead feeding areas and their diet. We thank Chris Stark and William Walker who identified many of the prey items. Work was conducted under NMFS Permits 814-1899, 481-1464, 17350, and 21386.

References

Aguilar, A., Gimenez, J., Gomez-Campos, E., Cardona, L., Borrell, A., 2014. δ^{15}N value does not reflect fasting in mysticetes. PLoS One 9 (3), e92288. Available from: https://doi.org/10.1371/journal.pone.0092288.

Albert, T.F., 2001. The influence of Harry Brower, Sr., an Inupiaq Eskimo hunter, on the bowhead whale research program conducted at the UIC-NARL faculty by the North Slope Borough. In: Norton, D. (Ed.), Fifty More Years Below Zero: Tributes and Mediations for the Naval Arctic Research Laboratory's First Half Century at Barrow. University of Alaska Fairbanks, Fairbanks, AK (xvi. 576pp., ill., maps, 20 cm). ISBN 0-919034098-5.

Ashjian, C.J., et al., 2010. Climate variability, oceanography, bowhead whale distribution, and Inupiat subsistence whaling near Barrow, Alaska. Arctic 63, 179–194.

Arregui, M., Borrell, A., Vikingsson, G., Olafsdottir, D., Aguilar, A., 2017. Stable isotope analysis of fecal material provides insight into the dies of baleen whales. Paper SC/67A/EM/05 Presented to the International Whaling Commission Scientific Committee, 10pp.

Baumgartner, B., 2013. Tagging and fine-scale oceanography. In: Shelden, K.E.W., Mocklin, J.A. (Eds.), Bowhead Whale Feeding Ecology Study (BOWFEST) in the Western Beaufort Sea. Final Report, OCS Study BOEM 2013. National Marine Mammal Laboratory, Alaska Fisheries Science Center, NMFS, NOAA, Seattle, WA, pp. 195–222. <http://www.afsc.noaa.gov/nmml/cetacean/bwasp/bowfest.php>.

Bockstoce, J.R., 1986. Whales, Ice, and Men: The History of Whaling in the Western Arctic. University of Washington Press, 400 pp.

Bodfish, H.H., 1936. Chasing the Bowhead. Harvard University Press, Cambridge, MA, 281 pp.

Budge, S.M., Springer, A.M., Iverson, S.J., Sheffield, G., Rosa, C., 2008. Blubber fatty acid composition of bowhead whales, *Balaena mysticetus*: Implications for diet assessment and ecosystem monitoring. J. Exp. Mar. Biol. Ecol. 359, 40–46.

Burns, J.J., 1993. Epilogue. In: Burns, J.J., Montague, J.J., Cowles, C.J. (Eds.), The Bowhead Whale. Society for Marine Mammalogy, Lawrence, KS, pp. 745–764. Special Publication No. 2, 787 pp.

Carroll, G.M., George, J.C., Lowry, L.F., Coyle, K.O., 1987. Bowhead whale (*Balaena mysticetus*) feeding near Point Barrow, Alaska, during the 1985 spring migration. Arctic 40, 105–110.

Chambault, P., Albertsen, C.M., Patterson, T.A., Hansen, R.G., Tervo, O., Laidre, K.L., et al., 2018. Sea surface temperature predicts the movements of an Arctic cetacean: the bowhead whale. Sci. Rep. 8, 9658. Available from: https://doi.org/10.1038/s41598-018-27966-1.

Christensen, I., Haug, T., Øien, N., 1992. A review of feeding and reproduction in large baleen whales (Mysticeti) and sperm whales *Physeter microcephalus* in Norwegian and adjacent waters. Fauna Norv. Ser. A 13, 39–48.

Citta, J.J., Quakenbush, L.T., George, J.C., Small, R.J., Heide-Jørgensen, M.P., Brower, H., et al., 2015a. Winter movements of bowhead whales (*Balaena mysticetus*) in the Bering Sea. Arctic 65 (1), 13–34.

Citta, J.J., Quakenbush, L.T., Okkonen, S., Druckenmiller, M.L., Maslowski, W., Clement-Kinney, J., et al., 2015b. Ecological characteristics of core-use areas used by Bering-Chukchi-Beaufort (BCB) bowhead whales, 2006–2012. Prog. Oceanogr. 136, 201–222.

Citta, J.J., Okkonen, S.R., Quakenbush, L.T., Maslowski, W., Osinki, R., George, J.C., et al., 2018. Oceanographic characteristics associated with autumn movements of bowhead whales in the Chukchi Sea. Deep Sea Res. II: Top. Stud. Oceanogr. 152, 121–131.

Clarke, J.T., A.A. Brower, M.C. Ferguson, A.S. Kennedy, and A.L. Willoughby. 2015. Distribution and relative abundance of marine mammals in the Eastern Chukchi and Western Beaufort seas, 2014. Annual Report, OCS Study BOEM 2015-2040. <http://www.afsc.noaa.gov/nmml/PDF/2014_ASAMM_FinalAnnualReport_19october15_clean.pdf>.

Clarke, J.T., Brower, A.A., Ferguson, M.C., Willoughby, A.L., Rotrock, A.D., 2020. Distribution and relative abundance of marine mammals in the Eastern Chukchi Sea, Eastern and Western Beaufort Sea, and Amundsen Gulf, 2019. Annual Report, OCS Study BOEM 2020. Marine Mammal Laboratory, Alaska Fisheries Science Center, NMFS, NOAA, Seattle, WA.

Dryden, J., 1913. Annus Mirabilis—the year of wonders, 1666. In: Sargeaunt, J. (Ed.), The Poems of John Dryden. Oxford University Press, London, NY.

Ferguson, M.C., Angliss, R.P., Kennedy, A., Lynch, B., Willoughby, A., Helker, V., et al., 2018. Performance of manned and unmanned aerial surveys to collect visual data and imagery for estimating arctic cetacean density and associated uncertainty. J. Unmanned Veh. Syst. 6 (3), 128–154. Available from: https://doi.org/10.1139/juvs-2018-0002.

Fish, F., Goetz, K.T., Rugh, D.J., Brattstrom, L.V., 2013. Hydrodynamic patterns associated with echelon formation swimming by feeding bowhead whales (*Balaena mysticetus*). Mar. Mamm. Sci. 29 (4), 498–507. Available from: https://doi.org/10.1111/mms.12004.

Finley, K.J., 1990. Isabella Bay, Baffin Island: an important historical and present-day concentration area for the endangered bowhead whale (*Balaena mysticetus*) of the Eastern Canadian Arctic. Arctic 43 (2), 137–352.

Fortune, S.M.E., 2018. Diet, Feeding Behaviour and Habitat-Use Patterns of Bowhead Whales in the Eastern Canadian Arctic (Electronic theses and dissertations (ETDs) 2008 +). University of British Columbia. Available from: https://doi.org/10.14288/1.0375660.

Frost, K.J., Lowry, L.F., 1981. Feeding and trophic relationships of bowhead whales and other vertebrate consumers in the Beaufort Sea. Alaska Dep. Fish & Game, Fairbanks, AK, Contract No. 80–ABC–00160 with Natl. Mar. Fish. Serv., National Marine Mammal Lab., Seattle, WA. Final Rep. dated April 1, 106 pp. + 3 appendices.

Galginaitis, M.S., Koski, W.R., 2002. Kaktovikmiut whaling: historical harvest and local knowledge of whale feeding behavior. In: Richardson, W.J., Thomson, D.H. (Eds.), Bowhead Whale Feeding in the Eastern Alaskan Beaufort Sea: Update of Scientific and Traditional Information, vol. 1. OCS Study MMS 2002-012; LGL Rep. TA2196-7. Rep. from LGL Ltd, King City, Ont., for U.S. Minerals Management Service, Anchorage, AK, and Herndon, VA, pp. 2-1–2-30.

George, J.C. 2009. Growth, Morphology, and Energetics of Bowhead Whales (*Balaena mysticetus*) (Ph.D. dissertation). University of Alaska-Fairbanks, Fairbanks, Alaska, 168 pp.

George, J.C., Druckenmiller, M.L., Laidre, K.L., Suydam, R., Person, B., 2015. Bowhead whale body condition and links to summer sea ice and upwelling in the Beaufort Sea. Prog. Oceanogr. 136, 250–262.

George, J.C., Sheffield, G., Reed, D.J., Tudor, B., Stimmelmayr, R., Person, B.T., et al., 2017. Frequency of injuries from line entanglements, killer whales, and ship strikes on Bering-Chukchi-Beaufort seas bowhead whales. Arctic 70 (1), 37–46. Available from: https://doi.org/10.14430/arctic4631.

George, J.C., Wetzel, D., O'Hara, T.M., Robertson, K., Dehn, L., Leduc, R., et al., 2008. An analysis of ancient bowhead whale Mangtak from Gambell Alaska: What can it tell us? Paper SC/60/E2 Presented to the International Whaling Commission Scientific Committee, 11 pp.

George, J.C., Herreman, J., Givens, G.H., Suydam, R., Mocklin, J., Clark, C., et al., 2012. Brief review of the 2010 and 2011 bowhead whale abundance surveys near point Barrow, Alaska. Report to the IWC SC SC/64/AWMP7.

Givens, G.H., Edmondson, S.L., George, J.C., Suydam, R., Charif, R.A., Rahaman, A., et al., 2016. Horvitz-Thompson whale abundance estimation adjusting for uncertain recapture, temporal availability variation and intermittent effort. Environmetrics 27 (3), 134–146.

Harwood, L.A., Quakenbush, L.T., Small, R.J., George, J.C., Pokiak, J., Pokiak, C., et al., 2017. Movements and inferred foraging by bowhead whales in the Canadian Beaufort Sea during August and September, 2006–2012. Arctic 70 (2), 161–176.

Hazard, K.W., Lowry, L.F., 1984. Benthic prey in a bowhead whale from the northern Bering Sea. Arctic 37, 166–168.

Heide-Jørgensen, M.P., Laidre, K.L., Nielsen, N.H., Hansen, R.G., Rostad, A., 2013. Winter and spring diving behavior of bowhead whales relative to prey. Anim. Biotelemetry 1, 15.

Heide-Jørgensen, M.P., Garde, K.E., Nielsen, N.H., Andersen, O.N., Hansen, S.H., 2012. A note on biological data from the hunt of bowhead whales in west Greenland 2009–2011. 1:15. J. Cetacean Res. Manage. 12 (3), 329–333.

Hertz, N., 2019. 25 Oct Whalers in Utqiaġvik Can't Remember Hunting This Late Without Landing a Bowhead. Alaska Public Radio, <https://www.alaskapublic.org/2019/10/25/a-month-into-utqiagviks-whaling-season-none-have-been-landed/>.

Higdon, J.W., 2010. Commercial and subsistence harvests of bowhead whales (*Balaena mysticetus*) in eastern Canada and West Greenland. J. Cetacean Res. Manage. 11 (2), 185–216.

Hobson, K.A., Schell, D.M., 1998. Stable carbon and nitrogen isotope patterns in baleen from eastern Arctic bowhead whales (*Balaena mysticetus*). Can. J. Fish. Aquat. Sci. 55 (12), 2601–2607. Available from: https://doi.org/10.1139/f98-142.

Hoekstra, P.F., Dehn, L.A., George, J.C., Solomon, K.R., Muir, D.C.G., O'Hara, T.M., 2002. Trophic ecology of bowhead whales (*Balaena mysticetus*) compared with that of other arctic marine biota as interpreted from carbon-, nitrogen-, and sulfur-isotope signatures. Can. J. Zool. 80, 223–231.

Horstmann-Dehn, L., George, J.C., 2013. Bowhead whale digestive efficiency. In: Shelden, K.E.W., Mocklin, J.A. (Eds.), Bowhead Whale Feeding Ecology Study (BOWFEST) in the western Beaufort Sea. Final Report, OCS Study BOEM 2013. National Marine Mammal Laboratory, Alaska Fisheries Science Center, NMFS, NOAA, Seattle, WA, pp. 279–317. <http://www.afsc.noaa.gov/nmml/cetacean/bwasp/bowfest.php>.

Horstmann-Dehn, L., Follmann, E.H., Rosa, C., Zalensky, G., George, J.C., 2012. Stable carbon and nitrogen isotope ratios in muscle and epidermis of arctic whales. Mar. Mamm. Sci. 28 (2), 173–190. Available from: https://doi.org/10.1111/j.1748-7692.2011.00503.x.

Huntington, H.P., 2013. Traditional knowledge regarding bowhead whales and Camden Bay, Beaufort Sea, Alaska. Report to the North Slope Borough Department of Wildlife Management, Barrow, AK.

Huntington, H.P., Quakenbush, L.T., 2009a. Traditional knowledge of bowhead migratory patterns near Wainwright, Alaska. Report to the Alaska Eskimo Whaling Commission and the Wainwright Whaling Captains Association, Eagle River, AK.

Huntington, H.P., Quakenbush, L.T., 2009b. Traditional knowledge of bowhead migratory patterns near Kaktovik and Barrow, Alaska. Report to the Barrow and Kaktovik Whaling Captains Associations and the Alaska Eskimo Whaling Commission, Eagle River, AK.

Huntington, H.P., Danielson, S.L., Wiese, F.K., Baker, M., Boveng, P., Citta, J.J., et al., 2020. Evidence suggests potential transformation of the Pacific Arctic Ecosystem is underway. Nat. Clim. Change. Available from: https://doi.org/10.1038/s41558-020-0695-2.

Ivashchenko, Y., Clapham, P., 2010. Bowhead whales *Balaena mysticetus* in the Okhotsk Sea. Mammal Rev. 40 (1), 65–89.

Iverson, S.J., Lowry, L.F., Sheffield, G., 2002. Fatty acids in bowhead whales and potential prey from the Alaskan Beaufort Sea. In: Richardson, W.J., Thomson, D.H. (Eds.), Bowhead Whale Feeding in the Eastern Alaskan Beaufort Sea: Update of Scientific and Traditional Information. OCS Study MMS 2002-012; LGL Rep. TA2196-7. Rep. from LGL Ltd., King City, Ont., for U.S. Minerals Manage. Serv., Anchorage, AK, and Herndon, VA, P. 19:1–23. vol. 1, xliv + 420 pp; vol. 2, 277 pp.

Johnson, M.L., Fiscus, C.H., Ostenson, B.T., Barbour, M.L., 1966. Marine mammals. In: Wilimovsky, N.J., Wolfe, J.N. (Eds.), Environment of the Cape Thompson Region, Alaska. U.S. Atomic Energy Commission, Oak Ridge, TN, pp. 877–924.

Koski, W.R., 1993. Reproduction. In: Burns, J.J., Montague, J.J., Cowles, C.J. (Eds.), The Bowhead Whale. Society for Marine Mammalogy, Lawrence, KS, pp. 239–274. Special Publication No. 2, 787 pp.

Laidre, K.L., Heide-Jørgensen, M.P., Neilsen, T.G., 2007. Role of the bowhead whale as a predator in West Greenland. Mar. Ecol. Prog. Ser. 346, 285–297.

Landino, S.W., Treacy, S.D., Zerwick, S.A., Dunlap, J.B., 1994. A large aggregation of bowhead whales (*Balaena mysticetus*) feeding near Point Barrow, Alaska in late October 1992. Arctic 47 (3), 232–235.

Lee, S.H., Schell, D.M., McDonald, T.L., Richardson, W.J., 2005. Regional and seasonal feeding by bowhead whales *Balaena mysticetus* as indicated by stable isotope ratios. Mar. Ecol. Prog. Ser. 285, 271–287.

Lefebvre, K., Quakenbush, L., Frame, E., Huntington, K., Sheffield, G., Stimmelmayr, R., et al., 2016. Prevalence of algal toxins in Alaskan marine mammals foraging in a changing arctic and subarctic environment. Harmful Algae 55, 13–24. Available from: https://doi.org/10.1016/j.hal.2016.01.007.

Ljungblad, D.K., Moore, S.E., Clarke, J.T., Bennett, J.C., 1987. Distribution, abundance, behavior, and bioacoustics of endangered whales in the Alaskan Beaufort and Eastern Chukchi Seas, 1979–86. Final Technical Report 1177. Accession Number ADA183934, Naval Ocean Systems Center, San Diego, CA, 391 pp. <https://apps.dtic.mil/dtic/tr/fulltext/u2/a183934.pdf>

Lowry, L.F., 1993. Foods and feeding ecology. In: Burns, J.J., Montague, J.J., Cowles, C.J. (Eds.), The Bowhead Whale. Society for Marine Mammalogy, Lawrence, KS, pp. 201–238. Special Publication No. 2, 787 pp.

Lowry, L.F., Sheffield, G., George, J.C., 2004. Bowhead whale feeding in the Alaskan Beaufort Sea, based on stomach content analysis. J. Cetacean Res. Manage. 6 (3), 215–223.

Lowry, L.F., Frost, K.J., 1984. Foods and feeding of bowhead whales in western and northern Alaska. Sci. Rep. Whales Res. Inst. Tokyo 35, 1–16.

Lowry, L., Sheffield, G., 2002. Stomach contents of bowhead whales harvested in the Alaskan Beaufort Sea. In: Richardson, W.J., Thomson, D.H. (Eds.), Bowhead Whale Feeding in the Eastern Alaskan Beaufort Sea:

Update of Scientific and Traditional Information. OCS Study MMS 2002-012; LGL Rep. TA2196-7. Rep. from LGL Ltd., King City, Ont., for U.S. Minerals Manage. Serv, Anchorage, AK and Herndon, VA (Chapter 18; vol. 1, xliv + 420 pp.; vol. 2, 277 pp.).

Lydersen, C.C., Freitas, O., Wiig, L., Bachmann, M.P., Heide-Jorgensen, R., Swift, et al., 2012. Lost highway not forgotten: Satellite tracking of a bowhead whale (*Balaena mysticetus*) from the critically Endangered Spitsbergen stock. Arctic 62 (1), 76–86.

Matthews, C.J.D., Ferguson, S.H., 2015. Seasonal foraging behaviour of Eastern Canada-West Greenland bowhead whales: An assessment of isotopic cycles along baleen. Mar. Ecol. Prog. Ser. 522, 269–286. Available from: https://doi.org/10.3354/meps11145.

Melnikov, V.V., Litovka, D.I., Zagrebin, I.A., Zelensky, G.M., Ainana, L.I., 2004. Shore-based counts of bowhead whales along the Chukotka Peninsula in May and June 1999–2001. Arctic 57 (3), 290–298.

Melnikov, V.V., Fedorets, Y.V., 2016. The distribution of zooplankton and bowhead whales, *Balaena mysticetus* Linnaeus, 1758, in Akademiya Bay, Sea of Okhotsk. Russ. J. Mar. Biol. 42 (3), 216–221.

Melnikov, V., Zdor, E., 2018. Observations of bowhead whales along the Northern Chukotka peninsula 2010–2012 with comparisons to 1994–1996 and 2002–2005. J. Cetacean Res. Manage. 18, 81–91.

Mocklin, J.D.J., Rugh, S.E., Moore, Angliss, R.P., 2011. Using aerial photography to investigate evidence of feeding by bowhead whales. Mar. Mamm. Sci. 28 (3), 602–619. Available from: https://doi.org/10.1111/j.1748-7692.2011.00518.x.

Moore, S.E., George, J.C., Coyle, K.O., Weingartner, T.J., 1995. Bowhead whales along the Chukotka Coast in autumn. Arctic 48 (2), 155–160.

Moore, S.E., George, J.C., Sheffield, G., Bacon, J., Ashjian, C.J., 2010. Bowhead whale distribution and feeding near Barrow, Alaska, in late summer 2005–06. Arctic 63 (2), 195–205.

Natsuike, M., Matsuno, K., Hirawake, T., Yamaguchi, A., Nishino, S., Imai, I., 2017. Possible spreading of toxic *Alexandrium tamarense* blooms on the Chukchi Sea shelf with the inflow of Pacific summer water due to climatic warming. Harmful Algae 61, 80–86.

Noongwook, G., the Native Village of Gambell, the Native Village of Savoonga, Huntington, H.P., George, J.C., 2007. Traditional knowledge of the bowhead whale (*Balaena mysticetus*) around St. Lawrence Island, Alaska. Arctic 60, 47–54.

Pomerleau, C., Lesage, V., Ferguson, S.H., Winkler, G., Petersen, S.D., Higson, J.W., 2012. Prey assemblage isotopic variability as a tool for assessing diet and the spatial distribution of bowhead whale *Balaena mysticetus* foraging in the Canadian eastern Arctic. Mar. Ecol. Prog. Ser. 469, 161–174. Available from: https://doi.org/10.3354/meps10004.

Pomerleau, C., Lesage, V., Winkler, G., Rosenberg, B., Ferguson, S.H., 2014. Contemporary diet of bowhead whales (*Balaena mysticetus*) from the Eastern Canadian Arctic inferred from fatty acid biomarkers. Arctic 67 (1), 84–92.

Pomerleau, C., Ferguson, S., Walkusz, W., 2011. Stomach contents of bowhead whales (*Balaena mysticetus*) from four locations in the Canadian Arctic. Polar Biol. 17 (2), 310–330.

Post, E., Alley, R.B., Christensen, T.R., Macias-Fauria, M., Forbes, B.C., Gooseff, M.N., et al., 2019. The polar regions in a 2°C warmer world. Sci. Adv. 5 (12), eaaw9883. Available from: https://doi.org/10.1126/sciadv.aaw9883. <http://advances.sciencemag.org/>.

Quakenbush L., Citta, J.J., George, J.C., 2012. Seasonal movements of the Bering-Chukchi-Beaufort stock of bowhead whales: 2006–2011 satellite telemetry results. Report to the International Whaling Commission Scientific Committee, Report SC/64/BRG1.

Quakenbush, L.T., Citta, J.J., George, J.C., Small, R.J., Heide-Jørgensen, M.P., 2010. Fall and winter movements of bowhead whales (*Balaena mysticetus*) in the Chukchi Sea and within a potential petroleum development area. Arctic 63 (3), 289–307.

Reeves, R., Rosa, C., George, J.C., Sheffield, G., Moore, M., 2012. Implications of Arctic industrial growth and strategies to mitigate future vessel and fishing gear impacts on bowhead whales. Mar. Policy 36, 454–462. Available from: https://doi.org/10.1016/j.marpol.2011.08.005.

Reeves, R.R., Ewins, P.J., Agbayani, S., Heide-Jørgensen, M.P., Kovacs, K.M., Lydersen, C., et al., 2014. Distribution of endemic cetaceans in relation to hydrocarbon development and commercial shipping in a warming Arctic. Mar. Policy 44, 375–389.

Rice, D.W., Wolman, A.A., 1971. The Life History and Ecology of the Gray Whale (*Eschrichtius robustus*). The American Society of Mammalogists, Lawrence, KS, Special Publication No. 3, 142 pp.

Richardson, W.J., Thomson, D.H. (Eds.), 2002. Bowhead whale feeding in the Eastern Alaskan Beaufort Sea: update of scientific and traditional information. OCS Study MMS 2002-012; LGL Rep. TA2196-7. Rep. from LGL Ltd., King City, Ont., for U.S. Minerals Manage. Serv., Anchorage, AK, and Herndon, VA (vol. 1, xliv + 420 pp.; vol. 2, 277 pp).

Rogachev, K.A., Carmack, E.C., Foreman, M.G.G., 2008. Bowhead whales feed on plankton concentrated by estuarine and tidal currents in Academy Bay, Sea of Okhotsk. Continental Shelf Res. 28, 1811–1826.

Rolland, R.M., Graham, K.M., Stimmelmayr, R., Suydam, R., George, J.C., 2019. Chronic stress from fishing gear entanglement is recorded in baleen from a bowhead whale (*Balaena mysticetus*). Mar. Mamm. Sci. 35 (4), 1625–1642. Available from: https://doi.org/10.1111/mms.12596.

Ross, W.G. 1992. Commercial whaling in the North Atlantic sector, In: Burns J.J., Montague, J.J., Cowles, C.J., (Eds.), The Bowhead Whale, Society for Marine Mammalogy, Special Publication 2, pp. 511–561.

Rugh, D.J., Goetz, K.T., Mocklin, J.A., Brattstrom, L.V., Sheldon, K.E.W., 2013. Aerial surveys. In: Shelden, K.E.W., Mocklin, J.A. (Eds.), Bowhead Whale Feeding Ecology Study (BOWFEST) in the Western Beaufort Sea. Final Report, OCS Study BOEM. National Marine Mammal Laboratory, Alaska Fisheries Science Center, NMFS, NOAA, Seattle, WA, pp. 17–74. <http://www.afsc.noaa.gov/nmml/cetacean/bwasp/bowfest.php>.

Savelle, J., McCartney, A.P., 1999. Thule Eskimos subsistence and bowhead whale procurement. In: Stiner, M.C. (Ed.), Human Predators and Prey Mortality. Westview Press, Boulder, CO, pp. 201–216.

Scammon, C.M., 1874. The Marine Mammals of the Northwestern Coast of North America. Reprinted in 1968 by Dover Publications, Inc., New York.

Schell, D.M., Saupe, S.M., Haubenstock, N., 1989a. Bowhead whale (*Balaena mysticetus*) growth and feeding as estimated by $\delta^{13}C$ techniques. Mar. Biol. 103 (4), 433–443.

Schell, D.M., Saupe, S.M., Haubenstock, N., 1989b. Natural isotope abundances in bowhead whale (*Balaena mysticetus*) baleen: markers of aging and habitat usage. In: Rundel, P.W., Ehleringer, J.R., Nagy, K.A. (Eds.), Stable Isotopes in Ecological Research. Ecological Studies (Analysis and Synthesis), vol. 68. Springer, New York.

Schell, D.M., Saupe, S.M., 1993. Feeding and growth as indicated by stable isotopes. In: Burns, J.J., Montague, J.J., Cowles, C.J. (Eds.), The Bowhead Whale. Society for Marine Mammalogy, Lawrence, KS, pp. 491–509. Special Publication No. 2, 787 pp.

Scoresby, W., 1820. An Account of the Arctic Regions with a History and Description of the Northern Whale-Fishery. Archibald Constable, Edinburgh (vol. I, 464 pp.; vol. II viii + 574 pp. + 22 pls).

Sensor, J.D., George, J.C., Clementz, M.T., Lovano, D., Waug, D.A., Givens, G.H., et al., 2018. Age estimation in bowhead whales using tympanic bulla histology and baleen isotopes. Mar. Mamm. Sci. 34 (2), 347–364. Available from: https://doi.org/10.1111/mms.12476.

Sheffield, G., 2008. Bowhead whale diet investigation: St. Lawrence Island, Bering Sea. Final Federal Aid Report. Grant T1, Project No. 3, 12 pp.

Sheffield, G., 2017. Bowhead whale (17S1) harvested near Savoonga, January 2017. Report to Savoonga Whaling Captains Association, University of Alaska Fairbanks, Alaska Sea Grant, Marine Advisory Program, Nome, AK, 7 pp.

Sheffield, G., George, J.C., 2013. Diet studies. In: Shelden, K.E.W., Mocklin, J.A. (Eds.), Bowhead Whale Feeding Ecology Study (BOWFEST) in the Western Beaufort Sea. Final Report, OCS Study BOEM 2013. National Marine Mammal Laboratory, Alaska Fisheries Science Center, NMFS, NOAA, Seattle, WA, pp. 245–269. <http://www.afsc.noaa.gov/nmml/cetacean/bwasp/bowfest.php>.

Sheffield, G., Fay, F.H., Feder, H., Kelly, B.P., 2001. Laboratory digestion of prey and interpretation of walrus stomach contents. Mar. Mamm. Sci. 17 (2), 310–330.

Shelden, K.E.W., Mocklin, J.A. (Eds.), 2013. Bowhead Whale Feeding Ecology Study (BOWFEST) in the Western Beaufort Sea. Final Report, OCS Study BOEM 2013-0114. National Marine Mammal Laboratory, Alaska Fisheries Science Center, NMFS, NOAA, Seattle, WA, <http://www.afsc.noaa.gov/nmml/cetacean/bwasp/bowfest.php>.

Sheldon, K.E.W., Rugh, D.J., 1995. The bowhead whale, *Balaena mysticetus*: its historic and current status. Mar. Fish. Rev. 57 (3-4), 1–20.

Shpak, O.V., Paramonov, A.Y., 2015. Observations of bowhead whales (*Balaena mysticetus*) in Shantar region of the Okhotsk Sea; potential threats for population recovery. Mar. Mamm. Holarct. 2, 334–342.

Shpak, O.V., Paramonov, A.Y., 2018. The bowhead whale, *Balaena mysticetus* Linnaeus, 1758, in the Western Sea of Okhotsk (2009–2016): distribution pattern, behavior, and threats. Russ. J. Mar. Biol. 44 (3), 210–218. Available from: https://doi.org/10.1134/S1063074018030082.

Smultea, M.A., Fertl, D., Rugh, D.J., Bacon, C.E., 2012. Summary of systematic bowhead surveys conducted in the U.S. Beaufort and Chukchi seas, 1975–2009. NOAA Technical Memorandum NMFS-AFSC-237, US Department of Commerce, NOAA-NMFS Alaska Fisheries Science Center, Seattle, WA, 49 pp.

Stevenson, D.E., Lauth, R.R., 2019. Bottom trawl surveys in the northern Bering Sea indicate recent shifts in the distribution of marine species. Polar Biol. 42, 407–421. Available from: https://doi.org/10.1007/s00300-018-2431-1.

Stimmelmayr, R., Ylitalo, G.M., Sheffield, G., Beckmen, K., Burek-Huntington, K., Metcalf, V., et al., 2018. Oil fouling in three subsistence-harvested ringed (*Phoca hispida*) and spotted (*Phoca largha*) seals from the Bering Strait region, Alaska: polycyclic aromatic hydrocarbon bile and tissue levels and pathological findings. Mar. Pollut. Bull. 130 (2018), 311–323. Available from: https://doi.org/10.1016/j.marpolbul.2018.02.040.

Stoker, S.W., Krupnik, I.I., 1993. Subsistence whaling. In: Burns, J.J., Montague, J.J., Cowles, C.J. (Eds.), The Bowhead Whale, first ed. Society for Marine Mammalogy, Lawrence, KS, pp. 579–629. Special Publication No. 2, 787 pp.

Suydam, R.S., George, J.C., Rosa, C., Person, B., Hanns, C., Sheffield, G., 2011. Subsistence harvest of bowhead whales (*Balaena mysticetus*) by Alaskan Eskimos during 2010. Annual Report to the Scientific Committee of the International Whaling Commission. Paper SC/63/BRG2, Department of Wildlife Management, North Slope Borough, Barrow, Alaska, 7 pp.

Suydam, R.S., George, J.C., Person, B., Hanns, C., Stimmelmayr, R., Pierce, L., et al., 2014. Subsistence harvest of bowhead whales (*Balaena mysticetus*) by Alaskan Eskimos during 2013. Annual Report to the Scientific Committee of the International Whaling Commission. Paper SC/65b/BRG08, Department of Wildlife Management, North Slope Borough, Barrow, Alaska, 10 pp.

Thoman, R., Bhatt, U., Bieniek, P., Brettschneider, B., Brubaker, M., Danielson, S., et al., 2020. The record low Bering Sea ice extent in 2018: content, impacts, and an assessment of the role of anthropogenic climate change. Chapter 10, Special Report—Explaining Extreme Events from a Climate Perspective. Bulletin of the American Meteorological Society, Boston, MA, pp. S18–S19, <http://ametsoc.net/eee/2018/10_Thoman0175.pdf>.

Thomson, D.H., 2002. Energetics of bowhead whales. In: Richardson, W.J., Thomson, D.H. (Eds.), Bowhead Whale Feeding in the Eastern Alaska Beaufort Sea: Update of Scientific and Traditional Information. For OCS Study MMS 2002-012; LGL Rep. TA 2196-7. Rep. From LGL Ltd., King City. Ontario Department of Interior, Mineral Management Service, Anchorage, AK and Herndon, VA, pp. 22-1–22-40. (vol. 1, xliv + 420 pp.; vol. 2; 277pp.).

Tomilin, A.,G., 1957. Mammals of the USSR and adjacent countries. In: Heptner, V.G. (Ed.), Volume IX Cetacea. Reprinted in 1967 by Israel Program for Scientific Translations, Jerusalem 1967, Moscow, 715 pp.

Walløe, L., 1999. Translation to English with annotations of the description of whales given in Konungs skuggsjá (The King's mirror). Norway, about 1250. Early descriptions of whales. J. Cetacean Res. Manage. 1 (2), iii–iv. and 2(1), iii–iv.

Wiig, Ø., Bachmann, L., Janik, V.M., Kovacs, K.M., Lydersen, C., 2007. Spitsbergen bowhead whales revisited. Mar. Mamm. Sci. 23 (3), 688–693.

Würsig, B., Dorsey, E.M., Richardson, W.J., Wells, R.S., 1989. Feeding, aerial and play behaviour of the bowhead whale, *Balaena mysticetus*, summering in the Beaufort Sea. Aquat. Mamm. 15 (1), 27–37.

CHAPTER

29

Predators and impacts of predation

Greg A. Breed

Institute of Arctic Biology, University of Alaska Fairbanks, Fairbanks, AK, United States

Introduction

Baleen whales have few natural predators, the two most significant are killer whales and sharks. In the Arctic, polar bears also occasionally harass ice-entrapped cetaceans, including bowhead whales. Although there are many records of sharks predating small odontocete cetaceans, shark predation on baleen whales is rare, and the few records available involve young, debilitated, or entangled animals (Stroud and Roffe, 1979; Long and Jones, 1996; Mazzuca et al., 1998; Taylor et al., 2013). Cookie-cutter sharks (*Isistius brasiliensis*) also feed on cetaceans of all sizes, including baleen whales (Mikhalev, 1997), but the feeding behavior of this small, warm-water shark is more parasitic than predatory. No cookie-cutter shark scars have been found on bowhead whales.

The large shark species believed responsible for the few confirmed attacks on baleen whales (white, mako, and tiger), are restricted to warmer waters south of bowhead distributions. Sleeper and Greenland sharks are cold-water and Arctic species found within bowhead ranges large enough to potentially predate them. These slow swimmers have different jaw morphology from the aforementioned southerly shark species, and although they may be effective ambush predators of seals (Lucas and Natanson, 2010; Leclerc et al., 2012), they do not appear to attack cetaceans. No bowhead whales have been reported predated by sharks or carry shark bite scars. Polar bear claw scars do occasionally occur on Bering–Chukchi–Beaufort Seas (BCB) bowheads (Fig. 29.1E and F). Polar bears are known to be significant predators of other Arctic cetaceans (e.g., Lowry et al., 1987), and polar bears might occasionally kill entrapped bowhead neonates. However, given the evidence in hand, the likelihood that this predation is significant is small.

Killer whales, by contrast, are major predators of baleen whales (Jefferson et al., 1991), though there is remarkable lack of consensus about the degree killer whales impact baleen

The Bowhead Whale
DOI: https://doi.org/10.1016/B978-0-12-818969-6.00029-7

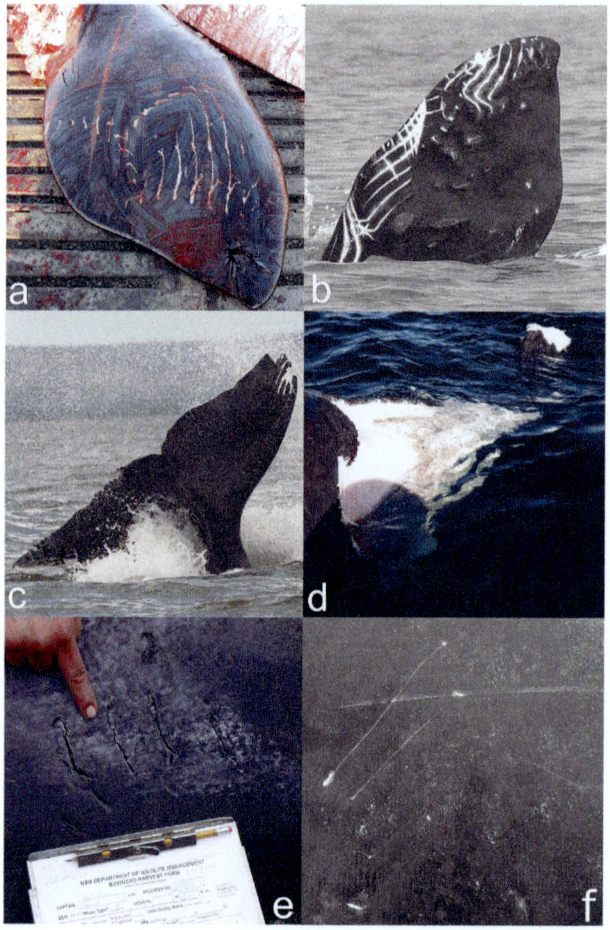

FIGURE 29.1 Scarring caused by killer whales. (A) Bowhead pectoral fin from a whale taken in the Barrow (Utqiaġvik) harvest demonstrating classic killer whale rake marks on a Bering–Chukchi–Beaufort individual. (B) Killer whale rake marks on an OKS bowhead pectoral fin. (C) Killer whale scarred flukes on an OKS bowhead. (D) Yearling bowhead whale from the ECWG stock killed by killer whales with partial removal of the blubber layer in Cumberland Sound, Baffin Island. Blubber on the right flank and peduncle is missing along with part of the fluke. This animal was also missing a pectoral fin, which had been severed perimortem. (E) Polar bear injury from carcass feeding and (F) healed polar bear scar on bowhead whales taken in the Barrow harvest for comparison. Polar bear claw scars are more irregular and more closely spaced than killer whale rake marks, and the resulting white streaks are much narrower, if they occur at all. *Source: (A) Photo by L. Pierce, NSB-DWM. (B and C) Photo by Olga Shpak. (D) From Young, B.G., Fortune, S.M.E., Koski, W.R., Raverty, S.A., Kilabuk, R., Ferguson, S.H., 2020. Evidence of killer whale predation on a yearling bowhead whale in Cumberland Sound, Nunavut, Arct. Sci., 6, 53–61. Available from: https://doi.org/10.1139/as-2019-0014. (E and F) Photos by Craig George.*

whale populations or other ecosystem members (e.g., Mitchell and Reeves, 1982; Jefferson et al., 1991; Estes et al., 1998; Springer et al., 2003; DeMaster et al., 2006; Trites et al., 2007). Similarly, although there is strong evidence killer whales effectively predate bowhead

FIGURE 29.2 Killer whales patrol Eclipse Sound on the north tip of Baffin Island, August 2018. *Source: Photo by Maha Ghazal.*

whales across the Arctic, estimating the importance of these effects is fraught with uncertainty and it remains unclear how killer whale predation impacts bowhead population sizes, distributions, or habitat selection (Fig. 29.2).

Uncertainty in estimating effects of killer whale predation is driven by poorly constrained estimates of killer whale population size and distribution in the Arctic. This is further confounded by the tendency for killer whale family units and populations to diverge into ecotypes that become specialized on different prey through cultural segregation (Foote et al., 2016). Even among lineages that target marine mammals, it is unclear how often they take baleen whales relative to other prey. However, targeted research documenting the behavior and demography of Arctic killer whales and their predatory impact on Arctic endemic marine mammals, including bowhead whales, has intensified in the past 10 years. Here, I summarize this recent body of work and the evidence killer whale predation is impacting bowhead stocks.

Evidence of predation

The Inuit have long recognized bowhead whale avoidance of killer whales, including retreat into ice cracks and shallow water, who call the behavior "aarlirijuk" (fear of killer whales; Finley, 1990; Hay et al., 2000). Such indigenous knowledge (IK) provides historical context and suggests predation of bowhead whales by killer whales is not a new phenomenon (at least in some areas used by the East Canada-West Greenland (ECWG) Stock). It is difficult to evaluate the historical intensity or ecological importance of killer whale predation on bowhead whales from the available IK, but it was observed frequently enough that specific antipredator behaviors were recognized by Inuit hunters (Chapter 34).

Modern evidence of killer whale predation and its effects has generally come from three sources: direct observations of killer whale attacks (and secondhand accounts) (e.g., Mitchell and Reeves, 1982; George and Suydam, 1998; Reeves et al., 2006; Higdon et al., 2012; Shpak and Paramonov, 2018), bowhead carcasses bearing lethal injuries consistent with killer whales (e.g., George and Suydam, 1998; Shpak and Paramonov, 2018; Willoughby et al., 2018; Young et al., 2020), and sightings of live individuals bearing healed scars consistent with killer whales (e.g., Reinhart et al., 2013; George et al., 1994, 2017). More recently, satellite telemetry has been used to infer nonlethal and risk effects over larger spatiotemporal scales (Breed et al., 2017; Matthews et al., 2020).

Attack accounts and carcass data

Direct observations offer impressive, vivid accounts of killer whale attacks, and the natural history observations made during these chance encounters have been invaluable in understanding the predatory tactics of killer whales and antipredator behaviors of bowheads (Mitchell and Reeves, 1982; Higdon et al., 2012; Shpak and Paramonov, 2018). Such direct observations have most often been reported by Inuit hunters in areas used by the ECWG stock (Finley, 1990; Mitchell and Reeves, 1982; George and Suydam, 1998; Higdon et al., 2012). In the Eastern Canadian Arctic (ECA), these accounts are mostly second hand (Mitchell and Reeves, 1982; George and Suydam, 1998; Higdon et al., 2012), and the details somewhat difficult to confirm. However, several well-documented firsthand observations of killer whale attacks on both beluga and narwhal made by scientists (Laidre et al., 2006; Westdal et al., 2017), and telemetry data reported in Matthews et al. (2020) align with these accounts.

By contrast, fewer directly observed killer whale attacks have been reported in BCB bowheads (George and Suydam, 1998). Since 2012, however, one to three bowhead carcasses bearing killer whale injuries have been found nearly every year during the ASAMM (Aerial Surveys of Arctic Marine Mammals—an extensive aerial survey of the Alaskan Beaufort and Chukchi Seas conducted most years; Chapter 24). Prior to 2012, bowhead carcasses with killer whale injuries had never been reported in the ASAMM (Willoughby et al., 2018).

Recent, excellent firsthand observations have been made of killer whale attacks on bowheads from the Okhotsk (OKS) stock (Chapter 5; Shpak and Paramonov, 2018). Relative to other areas, the frequency of directly observed attacks, the number of bowhead carcasses discovered with fatal killer whale injuries, and the generally high number mammal-eating killer whales observed in the area suggest killer whale predation may be markedly higher on this population than others (Chapter 5). This may be due to longer ice-free periods compared to regions used by other stocks along with steady sea ice degradation over the past decade, which provide killer whales better, more prolonged access.

Scars

Scars borne by living bowheads provide the best available index of predation intensity and change through time. Scars applied by killer whales are permanent, unique, and conspicuous, forming bright white rake marks generally distinct from ship strike or entanglement scars (George et al., 1994; Reinhart et al., 2013). Scars are easily detected on necropsy but also via traditional boat-based photo id (Finley, 1990; Reinhart et al., 2013) and rarely in aerial photographs (George et al., 1994, 2017). Scars are most common on flukes but also regularly appear on pectoral fins and other areas.

Finley (1990) reported that 31% of bowheads bore killer whale rake marks during 4 years of observation and photography in the mid-1980s from a shore-based camp on Isabella Bay, Baffin Island. During that period of observations, Finley also witnessed two killer whale attacks. Analyzing a more extensive set of photo id data from the ECWG stock, Reinhart et al. (2013) found a lower average scarring rate of 10.2% between 1986 and 2012. However, only 2% of individuals were found to be scarred in 1986, 9% in 2008, and 15% were scarred in 2012. The discrepancy between Finley (1990) and Reinhart et al. (2013) may be due to the sample of whales observable from shore, which may have been biased toward older individuals in Finley's (1990) study.

In the BCB stock, 8% of individuals landed in the traditional harvest between 1990 and 2012 bore killer whale scars (George et al., 2017). However, scarring rates were sharply higher in the 2002–12 period as compared to 1990–2002, and the rate of scarring on large whales (>17 m) was 40% or higher (Fig. 29.3). Reports from Reinhart et al. (2013) and George et al. (2017) indicate similar scarring rates in the ECWG and BCB stocks. This is inconsistent with directly observed predation events, which are much rarer in the BCB stock compared to the ECWG population. This difference may indicate differing seasonality in predation. In the BCB stock, winter predation along the Bering Sea ice edge may be important but essentially unobservable (Citta et al., 2012). Summer predation might also be occurring farther from shore in the Chukchi and western Beaufort, where predation

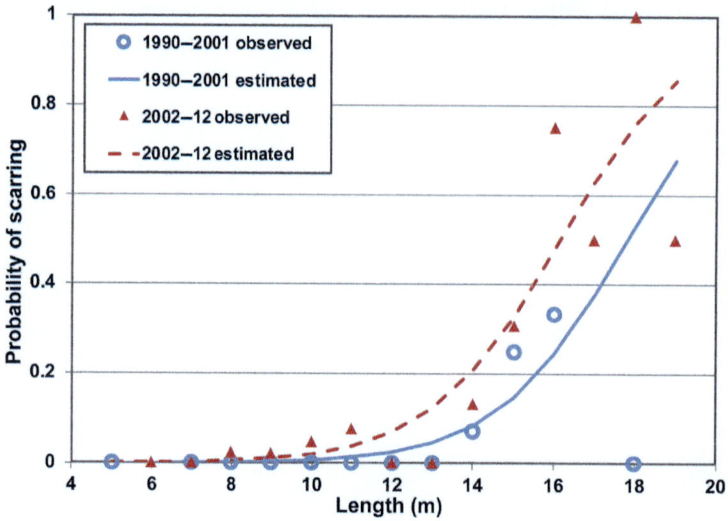

FIGURE 29.3 Killer whale scarring rates by age on bowheads landed in Barrow (Utqiagvik), Alaska, between 1990 and 2012. *Source: Modified from George, J.C., Sheffield, G., Reed, D.J., Tudor, B., Stimmelmayr, R., Person, B.T., et al., 2017. Frequency of injuries from line entanglements, killer whales, and ship strikes on Bering-Chukchi-Beaufort seas bowhead whales. Arctic 70, 37–46.*

events are less observable than in the deep fjords and narrow passages of the ECA, that force whales into close proximity of human settlements where predation events are more likely to be observed.

In both BCB and ECWG stocks, larger and older individuals are much more likely to bear scars (Fig. 29.3, Reinhart et al., 2013; George et al., 2017). In the ECA, females tend to be scarred more than males, which may be a consequence of calf defense (Reinhart et al., 2013), while males were more likely to be scarred in the BCB stock. Across all regions, very few young animals carry scars. The pattern could imply that attacks are infrequent, and individuals take many years to accumulate scars. However, most authors suggest this pattern is due to differential survival and/or targeting of young animals by killer whales. These very young animals do not survive predation events, while older, larger individuals do, and thus survive to carry observable evidence of the attack (Reinhart et al., 2013; George et al., 2017). This hypothesis is supported by a recent spate of carcasses found with lethal killer whale injuries, a disproportionate number of which have been yearling or young of year individuals (Willoughby et al., 2018; Young et al., 2020).

Although scars are a useful proxy for killer whale predation, there are some important caveats. As noted above, differential survival across demographic groups introduces classic survival bias, making it difficult to assess scar acquisition rate as animals age—a parameter that would be needed to estimate absolute attack rate from scar data. Additionally, scars may be acquired during capture practice by killer whales; mature adult killer whales are known to train juveniles in prey capture techniques without intention of taking prey (Guinet and Bouvier, 1995). Rake mark scars might also result from killer whales probing the vulnerability of individual bowheads, and although individuals identified as weak may escalate into a true predation attempt, many other probes may end if a potential target responds vigorously, but leave them scarred nonetheless.

Effects of predation

Direct mortality

Given the data in hand, the direct mortality inflicted by killer whales on bowheads (also called "density-mediated" or "consumptive" effects) is hard to assess in absolute terms. Ferguson et al. (2012) estimated that killer whales may take as many as 50 bowheads in and around Hudson Bay in a typical summer, which would be a large number given the population size (see Chapter 6). Similarly, Lefort et al. (2020) estimated that killer whales in Lancaster Sound require the caloric equivalent of nearly 1000 narwhal during the period they inhabit the region in the summer, and some fraction of this energetic demand is likely met by predating bowheads. However, both of these model-derived estimates require a number of assumptions about killer whale prey preference and behavior. The estimates are highly uncertain and the true number may be as few as 5 or 10 bowheads. Except for the ECA, no attempts to estimate absolute predation rates have been made, and are likely not possible given available data.

The similar, relatively high scarring rates on ECWG and BCB stocks, suggest predation high enough to impact calf and juvenile survival in these stocks, which could slow population growth in both populations. However, given the great age to which bowheads live, likely small impact on adult survival (given preference for juveniles and calves), and limited seasonal exposure to the peak ice-free period, these density-mediated effects are probably not a threat to population viability of the ECWG and BCB stocks.

Although there are no direct observations of killer whale attacks on East Greenland-Svalbard-Barents Sea (EGSB) stock bowheads, mammal-eating killer whales frequent these areas and it likely experiences killer whale predation of similar intensity to the ECWG stock (Lydersen et al., 2012). Density-mediated impacts on the OKS stock appears more substantial (see Chapter 5), perhaps higher than all other stocks owing to the relatively lower sea ice and higher exposure to killer whales. This higher level of predation pressure may be limiting population growth.

Risk effects

The effects of predators on prey populations are multifaceted and assuming direct mortality is the only substantial effect of killer whale predation is likely misleading. Trait-mediated effects, also known as "fear" or "risk" effects (Hairston et al., 1960; Lima, 1998; Lima and Bednekoff, 1999; Werner and Peacor, 2003), may be consequential. Traits altered via risk effects include reproductive rates, stress levels, morphology, and coloration, but the most easily observed changes are behavioral. Under predation threat, prey often use different, less risky but less resource-rich habitats, forage less, forage on different resources, and use refugia habitats to greater extents. In most cases, these changes incur additional energy expenditure and/or reduced resource acquisition. Risk effects can collectively alter the overall distribution of prey at the individual and population level (Menge and Sutherland, 1987; Werner and Peacor, 2003). In other systems, risk effects reduce energy intake, increase stress, and decrease reproductive success. These costs can be the dominant aspect of predator–prey interactions, as they are felt by every individual in the population and not just those killed in predation events (Preisser et al., 2005; Creel and Christianson, 2008). Under the right conditions, the population-level impact can be equal to, or greater than, the consumptive effects (Werner and Peacor, 2003; Preisser et al., 2005).

Risk effects of killer whales on other marine mammals have received little attention. Directly observed killer whale attacks on a variety of marine mammals, including bowheads, have provided key details of antipredator behavior. Similarly, recent killer whale vocalization playback experiments show that vocalizations from mammal-eating killer whales elicit intense behavioral responses in most marine mammal prey species (e.g., Cummings and Thompson, 1971; Fish and Vania, 1971; Curé et al., 2013, 2015; Isojunno et al., 2016). Under predation threat, Arctic cetaceans, including bowhead whales, flee into shallow water near coasts or into denser sea ice (Westdal et al., 2017; Breed et al., 2017; Matthews et al., 2020; Chapter 7). As an antipredator behavior, fleeing into sea ice appears highly effective due to Arctic killer whales being strongly averse to ice (Matthews et al., 2011). Fleeing into shallow, near-shore waters likely also offers some protection. In very shallow water, bowheads are likely less vulnerable from attacks coming from directly beneath them, which may be particularly dangerous. Sound also attenuates more rapidly in shallow water, and waves crashing on beaches may provide some acoustic masking, making prey less detectable.

In both directly observed attacks and playback experiments, the temporal scope of observation is very short, a few minutes to a few hours, and at most a day. It is easy to assume from these short observations that the interactions themselves are also short, and that prey return to relatively normal behavior after predation events/attacks. Recent work using synchronous satellite telemetry deployments on killer whales and narwhal, beluga, and bowheads has found that Arctic marine mammals, including bowhead whales, intensely change their behavior and distribution in response to predation threat over much longer time scales (Westdal et al., 2017; Breed et al., 2017; Matthews et al., 2020). Matthews et al. (2020) demonstrated that, like other arctic marine mammals, bowhead whales changed their movement behavior and habitat selection patterns to move out of open water and flee into sea ice and shallow, near-shore habitats (Fig. 29.4), and also found that, in the absence of killer whales, bowheads in the ECA from the ECWG stock had no preference for sea ice condition—they were equally likely to select open water as they were to select areas of moderate or dense sea ice. This suggests that, under predator-free conditions, bowheads select habitat based on the distribution of other resources (i.e., food), which during late summer in the Canadian Arctic, did not seem correlated with sea ice. After the arrival of satellite-tagged killer whales, tagged bowheads immediately moved into, and strongly selected for, dense sea ice deep in estuaries, fjords, and bays. These effects were especially notable for two other reasons. First, killer whales appeared to be able to elicit behavioral changes consistent with risk effects from long distances (10s or even a 100 km), and second, they were protracted, affecting the behavior of bowhead whales for weeks. This pattern suggests bowhead whales move into refugia habitats to avoid predation as soon as they receive information about predator threat, which may not have been via a direct encounter with killer whales, and remained in these refugia habitats for weeks when killer whales are present in the Arctic.

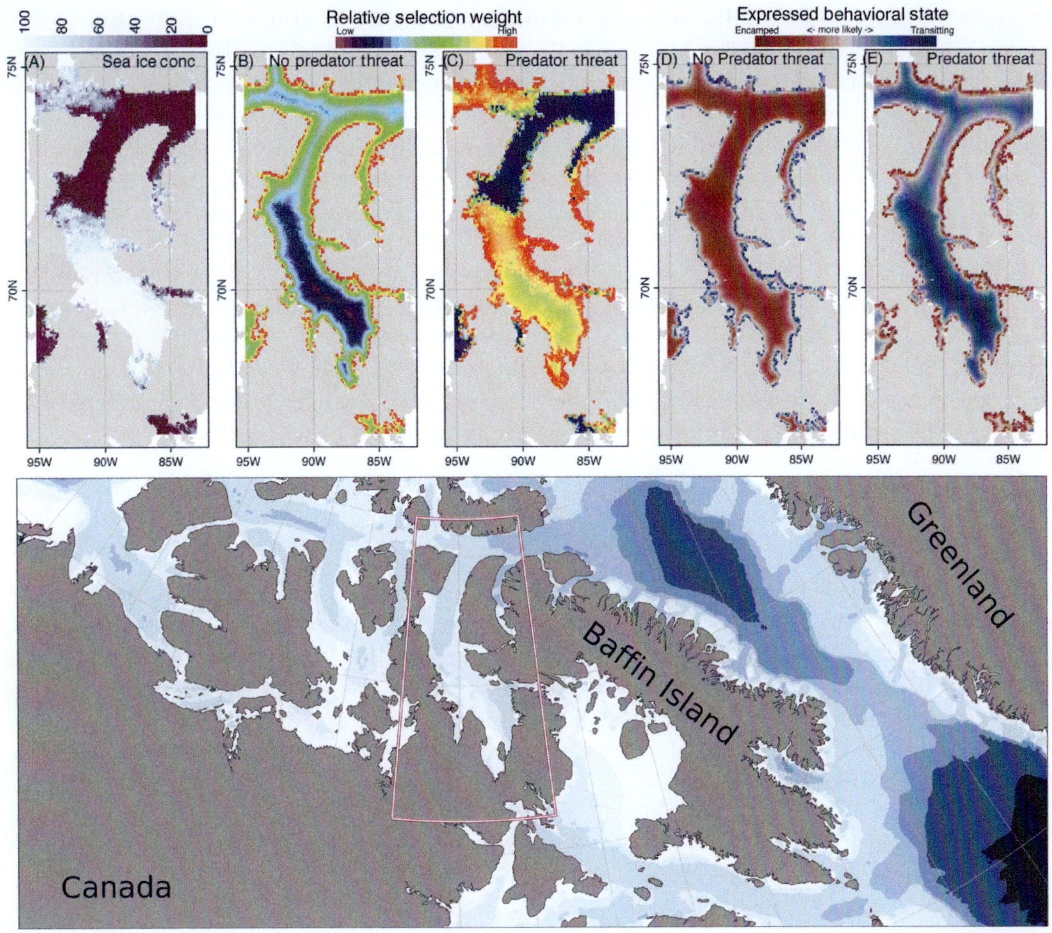

FIGURE 29.4 Analyzing killer and bowhead whale contemporaneously tracked by satellite telemetry, Matthews et al. (2020) assessed changes in habitat selection and behavior of bowhead whales in the Gulf of Boothia under predation threat and in predator-free conditions. (A) Sea ice conditions used to make predictions—from July 13, 2013. Predicted relative probability of habitat selection under those ice conditions with no predation threat (B) and under predation threat (C). Predicted behavioral state under given ice conditions in (A) with no predation threat (D) and under threat (E). *Source: Modified from Matthews, C.J., Breed, G.A., LeBlanc, B., Ferguson, S. H., 2020. Killer whale presence drives bowhead whale selection for sea ice in Arctic seascapes of fear. Proc. Natl. Acad. Sci. 117(12), pp.6590–6598. authors retain copyright.*

Given that bowhead whales forage extensively during the open water season (Lowry, 1993; Lowry et al., 2004; Moore and Huntington, 2008; Pomerleau et al., 2012), these risk effects likely result in lost foraging opportunities and may represent a significant energetic loss. Although bowhead whales have immense blubber stores that buffer periodic disruptions in foraging (Lindstedt and Boyce, 1985), the brief, intense pulse of productivity during the Arctic open water season may contribute disproportionately to their annual energetic requirements (Matthews and Ferguson, 2015). Risk effects during this period may therefore incur considerable energetic cost, particularly for calves and juveniles, which have higher mass-specific energetic requirements than adults, and for lactating females, whose gross energy requirements are more than double those of other adults (Fortune et al., 2013). The population-level impacts of lost foraging opportunities and reduced calf or juvenile growth could therefore be considerable. These effects may be elicited by a very small number of killer whales across a large number of bowheads, given the distances the effects were felt (Breed et al., 2017; Matthews et al., 2020).

However, unlike most other baleen whales which fast through the winter, bowheads often engage in extensive winter feeding in both the Eastern Canadian-Western Greenland (Nielsen et al., 2015) and BCB stocks (Lowry, 1993; Lowry et al., 2004; Citta et al., 2012; Chapter 28). Winter sea ice is extensive affording protection from predation during foraging and/or easy escape habitat near winter foraging areas. Such winter foraging may (at least partially) mitigate disruption of summer foraging by killer whales.

Predation and Arctic warming

Killer whales had a regular summer presence in some areas of the Arctic prior to the recent warming and loss of sea ice, including both east and west coasts of Greenland, many areas of the Northeast Atlantic, and some parts of the ECA. However, regular was probably not annual. In most years, sea ice likely blocked killer whales from entering many areas, including most areas west of Baffin Island, all of Hudson Bay and adjacent basins and inlets. In the Pacific, historical summer presence of killer whales north of the Bering Strait was likely rare.

This situation is rapidly changing. Degraded summer sea ice is permitting temperate marine mammal species with limited or no historical Arctic presence regular summer access (Moore and Huntington, 2008; Wang and Overland, 2009, Chapter 27), including killer whales. In the ECA, remarkable increases in killer whales have been well documented. Once restricted to areas along the northeast coast of Baffin Island, where they were seen at most a few times a decade, killer whales now appear every year and reside for 1–2 months (Higdon and Ferguson, 2009; Ferguson et al., 2012). They are now present annually throughout Hudson Bay and adjacent basins. Rough estimates from genetic mark-recapture and photo ID suggest there are around 180 individuals (Lefort et al., 2020). In addition to the well-documented increase in ECA killer whale presence, passive acoustic data show commensurate increasing occurrence of killer and other temperate whales in the Chukchi and Western Beaufort Seas (Moore and Huntington, 2008; George et al., 2004; Stafford, 2019; Moore et al., 2019), coincident with increases in killer whale scars on bowheads reported by George et al. (2017) and carcasses found with killer whale injuries reported by (Willoughby et al., 2018). In addition to sea ice loss permitting more extensive Arctic access to ice-intolerant species, these apparent numerical increases may also be a reflection of recovering killer whale stocks following industrial whaling, that decimated them particularly in the Atlantic.

Increases in Arctic killer whale presence, however, have been uneven across the Arctic. Summer killer whale presence has markedly increased in the Chukchi Sea and ECA.

However, similar increases have not been observed in the Canadian Beaufort Sea—key summer habitat for BCB bowheads (Moore and Laidre, 2006; Higdon et al., 2012; Moore et al., 2019). Point Barrow seems to be an important biogeographic barrier.

Dense summer sea ice near the North Pole can still occasionally be blown by summer storms southward across the open Arctic Ocean to contact the coast at Point Barrow, blocking an open water return route from the Canadian Beaufort to the North Pacific for ice-intolerant species. These sporadic events pose a serious summer ice entrapment hazard for any species that move past Point Barrow not adapted to sea ice, including killer whales (Matthews et al., 2011, 2019). Consequently, killer whales are likely reticent to pass Point Barrow; these long-lived species carry cultural memory that may inform them of the entrapment risk if they proceed past this biogeographic boundary into the Eastern Beaufort Sea. However, such events are becoming increasingly rare, and the entrapment risk is diminishing. It seems only a matter of time before the large population of mammal-eating killer whales of the North Pacific begin migrating into the Canadian Beaufort during summer given the large beluga and bowhead prey populations summering there.

Given climate predictions and the patterns observed in the ECA, an increased summer presence of killer whales in the Eastern Beaufort Sea is all but certain in the near future. This change could strongly redistribute the summer range of BCB bowheads and potentially come with large effects on population demographics and growth potential. There is also risk that traditional hunters will lose access, as bowheads move their migration offshore to stay within remaining sea ice as a refuge from killer whale threat.

Most models estimating the effects of sea ice loss on bowhead whales and other ice-associated marine mammals assess changes to bottom-up influences, including thermal stress, changes in the phenology, community composition, and distribution of primary production and zooplankton (Tynan and DeMaster, 1997; Laidre et al., 2008; George et al., 2015; Citta et al., 2015, 2018; Chambault et al., 2018). Studies have accordingly assessed when and where bowheads feed (e.g., Pomerleau et al., 2012; Matthews and Ferguson, 2015) or resource partitioning with other baleen whales (Laidre and Heide-Jørgensen, 2012) to understand the impacts of altered prey communities or competitive interactions as more temperate species arrive earlier and expand northward into Arctic waters. Reliance on sea ice as a predator refuge, however, has generally not been considered important in these predictions (but see Corkeron and Connor, 1999, regarding the potential influence of predation threat on the evolution of bowhead migration timing and routes). For bowhead whales, these effects could be equal to, or more important than, bottom-up forces driven by sea ice loss, particularly as Arctic killer whale presence increases and summer sea ice refuge habitats diminish. Despite their capacity to buffer foraging disruptions, protracted predator disturbance over spatial and temporal scales to which bowheads are not adapted could eventually impact reproductive success and even adult survival (e.g., Boonstra et al., 1998; Fortune et al., 2013).

Acknowledgments

This chapter was greatly improved by conversations, ideas, feedback, and assistance of John Citta, Craig George, and Steve Ferguson.

References

Boonstra, R., Hik, D., Singleton, G.R., Tinnikov, A., 1998. The impact of predator-induced stress on the snowshoe hare cycle. Ecol. Monogr. 68, 371–394.

Breed, G.A., Matthews, C.J., Marcoux, M., Higdon, J.W., LeBlanc, B., Petersen, S.D., et al., 2017. Sustained disruption of narwhal habitat use and behavior in the presence of Arctic killer whales. Proc. Natl. Acad. Sci. U. S. A. 114, 2628–2633.

Chambault, P., Albertsen, C.M., Patterson, T.A., Hansen, R.G., Tervo, O., Laidre, K.L., et al., 2018. Sea surface temperature predicts the movements of an arctic cetacean: the bowhead whale. Sci. Rep. 8, 9658.

Citta, J.J., Quakenbush, L.T., George, J.C., Small, R.J., Heide-Jørgensen, M.P., Brower, H., et al., 2012. Winter movements of bowhead whales (*Balaena mysticetus*) in the Bering Sea. Arctic 65, 13–34.

Citta, J.J., Quakenbush, L.T., Okkonen, S.R., Druckenmiller, M.L., Maslowski, W., Clement-Kinney, J., et al., 2015. Ecological characteristics of core-use areas used by Bering-Chukchi-Beaufort (BCB) bowhead whales, 2006–2012. Prog. Oceanogr. 136, 201–222.

Citta, J.J., Okkonen, S.R., Quakenbush, L.T., Maslowski, W., Osinski, R., George, J.C., et al., 2018. Oceanographic characteristics associated with autumn movements of bowhead whales in the Chukchi Sea. Deep. Sea Res. II Top. Stud. Oceanogr. 152, 121–131.

Corkeron, P.J., Connor, R.C., 1999. Why do baleen whales migrate? Mar. Mamm. Sci. 15, 1228–1245.

Creel, S., Christianson, D., 2008. Relationships between direct predation and risk effects. Trends Ecol. Evol. 23, 194–201.

Cummings, W.C., Thompson, O., 1971. Gray whales, *Eschrichtius robustus*, avoid the under water sounds of killer whales, *Orcinus orca*. Fish. Bull. 69, 525–530.

Curé, C., Antunes, R., Alves, A.C., Visser, F., Kvadsheim, P.H., Miller, P.J., 2013. Responses of male sperm whales (*Physeter macrocephalus*) to killer whale sounds: implications for anti-predator strategies. Sci. Rep. 3, 1579.

Curé, C., Doksaeter Sivle, L., Visser, F., Wensveen, P.J., Isojunno, S., Harris, C.M., et al., 2015. Predator sound playbacks reveal strong avoidance responses in a fight strategist baleen whale. Mar. Ecol. Prog. Ser. 526, 267–282.

DeMaster, D.P., Trites, A.W., Clapham, P., Mizroch, S., Wade, P., Small, R.J., et al., 2006. The sequential megafaunal collapse hypothesis: testing with existing data. Prog. Oceanogr. 68, 329–342.

Estes, J.A., Tinker, M.T., Williams, T.M., Doak, D.F., 1998. Killer whale predation on sea otters linking oceanic and nearshore ecosystems. Science 282, 473–476.

Ferguson, S.H., Higdon, J.W., Westdal, K.H., 2012. Prey items and predation behavior of killer whales (*Orcinus orca*) in Nunavut, Canada based on Inuit hunter interviews. Aquat. Biosyst. 8, 1.

Finley, K.J., 1990. Isabella Bay, Baffin Island: an important historical and present-day concentration area for the endangered bowhead whale (*Balaena mysticetus*) of the eastern Canadian Arctic. Arctic 43, 137–152.

Fish, J.F., Vania, J.S., 1971. Killer whale, *Orcinus orca*, sounds repel white whales, *Delphinapterus leucas*. Fish. Bull. Natl. Ocean. Atmos. Adm. 69, 531–535.

Foote, A.D., Vijay, N., Ávila-Arcos, M.C., Baird, R.W., Durban, J.W., Fumagalli, M., et al., 2016. Genome-culture co-evolution promotes rapid divergence of killer whale ecotypes. Nat. Commun. 7, 11693.

Fortune, S.M., Trites, A.W., Mayo, C.A., Rosen, D.A., Hamilton, P.K., 2013. Energetic requirements of north Atlantic right whales and the implications for species recovery. Mar. Ecol. Prog. Ser. 478, 253–272.

George, J., Zeh, J., Suydam, R., Clark, C., et al., 2004. Abundance and population trend (1978–2001) of western Arctic bowhead whales surveyed near Barrow, Alaska. Mar. Mamm. Sci. 20, 755–773.

George, J.C., Suydam, R., 1998. Observations of killer whale (*Orcinus orca*) predation in the northeastern Chukchi and western Beaufort Seas. Mar. Mamm. Sci. 14, 330–332.

George, J.C., Philo, L.M., Hazard, K., Withrow, D., Carroll, G.M., Suydam, R., 1994. Frequency of killer whale (*Orcinus orca*) attacks and ship collisions based on scarring on bowhead whales (*Balaena mysticetus*) of the Bering-Chukchi-Beaufort Seas stock. Arctic 47, 247–255.

George, J.C., Druckenmiller, M.L., Laidre, K.L., Suydam, R., Person, B., 2015. Bowhead whale body condition and links to summer sea ice and upwelling in the Beaufort Sea. Prog. Oceanogr. 136, 250–262.

George, J.C., Sheffield, G., Reed, D.J., Tudor, B., Stimmelmayr, R., Person, B.T., et al., 2017. Frequency of injuries from line entanglements, killer whales, and ship strikes on Bering-Chukchi-Beaufort seas bowhead whales. Arctic 70, 37–46.

Guinet, C., Bouvier, J., 1995. Development of intentional stranding hunting techniques in killer whale (*Orcinus orca*) calves at Crozet Archipelago. Can. J. Zool. 73, 27–33.

Hairston, N.G., Smith, F.E., Slobodkin, L.B., 1960. Community structure, population control, and competition. Am. Nat. 94, 421–425.

Hay, K., Aglukark, D., Igutsaq, D., Ikkidluak, J., Mike, M., 2000. Final Report of the Inuit Bowhead Knowledge Study. Nunavut Wildlife Management Board.

Higdon, J.W., Ferguson, S.H., 2009. Loss of Arctic sea ice causing punctuated change in sightings of killer whales (*Orcinus orca*) over the past century. Ecol. Appl. 19, 1365–1375.

Higdon, J.W., Hauser, D.D.W., Ferguson, S.H., 2012. Killer whales (*Orcinus orca*) in the Canadian Arctic: distribution, prey items, group sizes, and seasonality. Mar. Mamm. Sci. 28, E93–E109.

Isojunno, S., Curé, C., Kvadsheim, P.H., Lam, F.-P.A., Tyack, P.L., Wensveen, P.J., et al., 2016. Sperm whales reduce foraging effort during exposure to 1–2 kHz sonar and killer whale sounds. Ecol. Appl. 26, 77–93.

Jefferson, T.A., Stacey, P.J., Baird, R.W., 1991. A review of killer whale interactions with other marine mammals: predation to co-existence. Mammal Rev. 21, 151–180.

Laidre, K.L., Heide-Jørgensen, M.P., 2012. Spring partitioning of Disko Bay, West Greenland, by Arctic and Subarctic baleen whales. ICES J. Mar. Sci. 69, 1226–1233.

Laidre, K.L., Heide-Jørgensen, M.P., Orr, J.R., 2006. Reactions of narwhals, *Monodon monoceros*, to killer whale, *Orcinus orca*, attacks in the eastern Canadian Arctic. Can. Field Nat. 120, 457–465.

Laidre, K.L., Stirling, I., Lowry, L.F., Wiig, Ø., Heide-Jørgensen, M.P., Ferguson, S.H., 2008. Quantifying the sensitivity of Arctic marine mammals to climate-induced habitat change. Ecol. Appl. 18, S97–S125.

Leclerc, L.-M.E., Lydersen, C., Haug, T., Bachmann, L., Fisk, A.T., Kovacs, K.M., 2012. A missing piece in the arctic food web puzzle? Stomach contents of Greenland sharks sampled in Svalbard, Norway. Polar Biol. 35, 1197–1208.

Lefort, K.J., Garroway, C.J., Ferguson, S.H., 2020. Killer whale abundance and predicted narwhal consumption in the Canadian Arctic. Global Change Biology 26, 4276–4283.

Lima, S.L., 1998. Nonlethal effects in the ecology of predator–prey interactions. BioScience 48, 25–34.

Lima, S.L., Bednekoff, P.A., 1999. Temporal variation in danger drives antipredator behavior: the predation risk allocation hypothesis. Am. Nat. 153, 649–659.

Lindstedt, S.L., Boyce, M.S., 1985. Seasonality, fasting endurance, and body size in mammals. Am. Nat. 125, 873–878.

Long, D.J., Jones, R.E., 1996. White shark predation and scavenging on cetaceans in the eastern North Pacific Ocean. Great White Sharks: The Biology of *Carcharodon carcharias*. Academic Press, San Diego, CA, pp. 293–307.

Lowry, L., Burns, J., Nelson, R., 1987. Polar bear, *Ursus maritimus*, predation on belugas, *Delphinapterus leucas*, in the Bering and Chukchi Seas. Can. Field Nat. 101, 141–146.

Lowry, L.F., 1993. Foods and Feeding Ecology. The Bowhead Whale, vol. 2. Society for Marine Mammalogy, pp. 201–238, Special Publication.

Lowry, L.F., Sheffield, G., George, J.C., 2004. Bowhead whale feeding in the Alaskan Beaufort Sea, based on stomach contents analyses. J. Cetacean Res. Manage. 6, 215–223.

Lucas, Z., Natanson, L., 2010. Two shark species involved in predation on seals at Sable Island, Nova Scotia, Canada. Proc. Nova Scotian Inst. Sci. 45, 64–88.

Lydersen, C., Freitas, C., Wiig, Ø., Bachmann, L., Heide-Jørgensen, M.P., Swift, R., et al., 2012. Lost highway not forgotten: satellite tracking of a bowhead whale (*Balaena mysticetus*) from the critically endangered Spitsbergen stock. Arctic 65, 7686.

Matthews, C.J., Ferguson, S.H., 2015. Seasonal foraging behaviour of eastern Canada-West Greenland bowhead whales: an assessment of isotopic cycles along baleen. Mar. Ecol. Prog. Ser. 522, 269–286.

Matthews, C.J., Raverty, S.A., Noren, D.P., Arragutainaq, L., Ferguson, S.H., 2019. Ice entrapment mortality may slow expanding presence of arctic killer whales. Polar Biol. 42, 639–644.

Matthews, C.J.D., Breed, G.A., LeBlanc, B., Ferguson, S.H., 2020. Killer whale presence drives bowhead whale selection for sea ice in Arctic seascapes of fear. Proc. Natl. Acad. Sci. 117 (12), 6590–6598.

Matthews, C.J.D., Luque, S.P., Petersen, S.D., Andrews, R.D., Ferguson, S.H., 2011. Satellite tracking of a killer whale (*Orcinus orca*) in the eastern Canadian Arctic documents ice avoidance and rapid, long-distance movement into the North Atlantic. Polar Biol. 34, 1091–1096.

Mazzuca, L., Atkinson, S., Nitta, E., 1998. Deaths and entanglements of humpback whales, *Megaptera novaeangliae*, in the main Hawaiian Islands, 1972–1996. Pac. Sci. 52, 1–13.

Menge, B.A., Sutherland, J.P., 1987. Community regulation: variation in disturbance, competition, and predation in relation to environmental stress and recruitment. Am. Nat. 130, 730–757.

Mikhalev, Y.A., 1997. Humpback whales *Megaptera novaeangliae* in the Arabian Sea. Mar. Ecol. Prog. Ser. 149, 13–21.

Mitchell, E.D., Reeves, R.R., 1982. Factors affecting abundance of bowhead whales *Balaena mysticetus* in the eastern Arctic of North America, 1915–1980. Biol. Conserv. 22, 59–78.

Moore, S.E., Huntington, H.P., 2008. Arctic marine mammals and climate change: impacts and resilience. Ecol. Appl. 18, S157–S165.

Moore, S.E., Laidre, K.L., 2006. Trends in sea ice cover within habitats used by bowhead whales in the western Arctic. Ecol. Appl. 16, 932–944.

Moore, S.E., Haug, T., Vikingsson, G.A., Stenson, G.B., 2019. Baleen whale ecology in arctic and subarctic seas in an era of rapid habitat alteration. Prog. Oceanogr. 17, 102118.

Nielsen, N.H., Laidre, K., Larsen, R.S., Heide-Jørgensen, M.P., 2015. Identification of potential foraging areas for bowhead whales in Baffin Bay and adjacent waters. Arctic 68, 169–179.

Pomerleau, C., Lesage, V., Ferguson, S., Winkler, G., Petersen, S., Higdon, J., 2012. Prey assemblage isotopic variability as a tool for assessing diet and the spatial distribution of bowhead whale *Balaena mysticetus* foraging in the Canadian Eastern Arctic. Mar. Ecol. Prog. Ser. 469, 161–174.

Preisser, E.L., Bolnick, D.I., Benard, M.F., 2005. Scared to death? The effects of intimidation and consumption in predator–prey interactions. Ecology 86, 501–509.

Reeves, R.R., Berger, J., Clapham, P.J., 2006. Killer whales as predators of large baleen whales and sperm whales. Whales, Whaling and Ocean Ecosystems. University of California Press, Berkeley, CA, pp. 174–187.

Reinhart, N., Ferguson, S., Koski, W., Higdon, J., LeBlanc, B., Tervo, O., et al., 2013. Occurrence of killer whale *Orcinus orca* rake marks on Eastern Canada-West Greenland bowhead whales *Balaena mysticetus*. Polar Biol. 36, 1133–1146.

Shpak, O., Paramonov, A.Y., 2018. The bowhead whale, *Balaena mysticetus* Linnaeus, 1758, in the western Sea of Okhotsk (2009–2016): distribution pattern, behavior, and threats. Russ. J. Mar. Biol. 44, 210–218.

Springer, A.M., Estes, J., Van Vliet, G., Williams, T., Doak, D., Danner, E., et al., 2003. Sequential megafaunal collapse in the North Pacific Ocean: an ongoing legacy of industrial whaling? Proc. Natl. Acad. Sci. U. S. A. 100, 12223–12228.

Stafford, K.M., 2019. Increasing detections of killer whales (*Orcinus orca*), in the Pacific Arctic. Mar. Mamm. Sci. 35, 696–706.

Stroud, R.K., Roffe, T.J., 1979. Causes of death in marine mammals stranded along the Oregon coast. J. Wildl. Dis. 15, 91–97.

Taylor, J.K., Mandelman, J.W., McLellan, W.A., Moore, M.J., Skomal, G.B., Rotstein, D.S., et al., 2013. Shark predation on North Atlantic right whales (*Eubalaena glacialis*) in the southeastern United States calving ground. Mar. Mamm. Sci. 29, 204–212.

Trites, A.W., Miller, A.J., Maschner, H.D., Alexander, M.A., Bograd, S.J., Calder, J.A., et al., 2007. Bottom-up forcing and the decline of Steller sea lions (*Eumetopias jubatus*) in Alaska: assessing the ocean climate hypothesis. Fish. Oceanogr. 16, 46–67.

Tynan, C.T., DeMaster, D.P., 1997. Arctic climatic change: potential effects on marine mammals. Arctic 50, 308–322.

Wang, M., Overland, J.E., 2009. A sea ice free summer Arctic within 30 years? Geophys. Res. Lett. 36, L07502.

Werner, E.E., Peacor, S.D., 2003. A review of trait-mediated indirect interactions in ecological communities. Ecology 84, 1083–1100.

Westdal, K.H., Davies, J., McPherson, A., Orr, J., Ferguson, S.H., 2017. Behavioural changes in belugas (*Delphinapterus leucas*) during a killer whale (*Orcinus orca*) attack in southwest Hudson Bay. Can. Field Nat. 130, 315–319.

Willoughby, A.L., Clarke, J.T., Ferguson, M.C., Stimmelmayr, R., Browe, A.B., 2018. Bowhead Whale Carcasses in the eastern Chukchi and Western Beaufort Seas, 2009–2017. SC/67B/AWMP/02, International Whaling Commission.

Young, B.G., Fortune, S.M.E., Koski, W.R., Raverty, S.A., Kilabuk, R., Ferguson, S.H., 2020. Evidence of killer whale predation on a yearling bowhead whale in Cumberland Sound, Nunavut. Arct. Sci. 6, 53–61. Available from: https://doi.org/10.1139/as-2019-0014.

CHAPTER

30

Diseases and parasites

R. Stimmelmayr[1,2], D. Rotstein[3], Gay Sheffield[4], H.K. Brower, Jr[5] and J.C. George[1]

[1]Department of Wildlife Management, North Slope Borough, Utqiaġvik, AK, United States
[2]Institute of Arctic Biology, University of Alaska Fairbanks, Fairbanks, AK, United States
[3]Marine Mammal Pathology Services, Olney, MD, United States [4]Alaska Sea Grant, College of Fisheries and Ocean Sciences, University of Alaska Fairbanks, Nome, AK, United States
[5]Alaska Eskimo Whaling Commission, Utqiaġvik, AK, United States

Introduction

Inupiaq, Siberian Yupik, and Inuit communities of the Arctic have a tradition in hunting bowhead whales that reaches back several thousands of years (Fig. 30.1). Central to Inuit culture, hunted bowhead whales have undergone rigorous health evaluation by hunters and communities through various iterative processes guided by traditional customary practices. These traditional customary practices are similar to western meat hygiene designed to provide a thorough health assessment of the catch and by default, a public health judgment on food safety. With the establishment of the bowhead whale harvest monitoring program in Utqiaġvik in 1972, led by the North Slope Borough leadership, Alaska Eskimo Whaling Commission (AEWC), and the North Slope Borough Department of Wildlife Management (NSB-DWM), collaborative research on the health of bowhead whales between the AEWC, Village Whaling Captains Association, Whaling Captains Wives Association, veterinarians and research scientists have continued to build on the in-depth Inuit knowledge of the bowhead whale. This collaborative effort between indigenous hunters, scientists, and regulatory agencies has provided a unique opportunity to study natural causes of morbidity and mortality of this extraordinary Arctic baleen whale. This chapter synthesizes historic and current knowledge about disease conditions and parasites of the bowhead whale. The information provided is a coproduction of knowledge reflecting perspectives of Inuit indigenous and local knowledge, veterinary medicine, and biology.

The Bowhead Whale
DOI: https://doi.org/10.1016/B978-0-12-818969-6.00030-3

FIGURE 30.1 A bowhead whale has been caught by Utqiaġvik Iñupiat, and the butchering has started on the ice edge. Much of what is known about the health of bowhead whales comes from the study of animals caught in this way. *Source: Photo by North Slope Borough.*

Infectious diseases

Viruses and bacteria

Few infectious agents are present that could impact bowhead health or pose a public health risk. Among the viral agents previously identified by serology, vesivirus (family Caliciviridae) figures prominently. In summary, neutralizing antibodies in serum (Smith et al., 1987; O'Hara et al., 1998) and colostrum (Harms, 1993) have been detected against San Miguel sea lion virus (SMSV strains 1, 4, 5, 8, 9, 10, 12), vesicular exanthema of swine virus (VESV strains B, C 1934B, F55, J56, K), and mink calicivirus. Other viral agents tested but failed to be detected included SMSV serovars (2, 6, 7, 11, 13), VESV serovars (A48, B51, C52, D53, E54, G55, H54, I55, K54) Tillamook calicivirus, dolphin morbillivirus, orthomyxoviruses (influenza A and B viruses), and paramyxovirus (Newcastle disease). Based on findings by O'Hara et al. (1998), prevalence for VESV strains (F55; J56; 1934B) was 47% (17/36), 8% (3/36), and 14% (5/36). For SMSV (8, 12), prevalence was 6% (2/36) and 3% (1/36). Although titers indicate prior exposure and/or infection, no disease was recognized in these bowhead whales. Type-specific antibodies to various caliciviruses have been detected in other baleen whales from the North Pacific (Bossart and Duignan, 2018). In a recent retrospective study where select tissues (liver, kidney, spleen, and lung) from bowhead whales ($n = 59$; 2011–15) were tested by polymerase chain reaction (PCR) for a suite of viral pathogens (morbillivirus, adenovirus, paramyxovirus, influenza A virus, Pan-corona virus; herpes virus); adenovirus (10%; 6/59) was the only viral agent detected. Adenoviruses were previously isolated from bowhead and sei whales during the early 1980s (Smith et al., 1987). Sequencing of current viral isolates and comparison with available Genbank material revealed diverse genotypes, with three of the adenoviruses detected being novel (Sanchez and Stimmelmayr, unpublished data). For the other three specimens, a comparison search of a 292 bp sequence in NCBI/BLAST/Blastn database revealed the best match (89% query coverage; 72% similarity) with vespertilionid adenovirus 2 (Genbank accession KM043097.1). Comparison of a 310 bp sequence in NCBI/BLAST/Blastn database revealed the best match (66% Query coverage; 75% similarity) with bat adenovirus; Genbank accession JX065124.1. Comparison search of a 313 bp sequence in NCBI/BLAST/Blastn database revealed the best match (99% query coverage: 75% similarity) with bottlenose dolphin adenovirus (Genbank accession KR024710.1). No histopathological lesions were observed in adenovirus-positive bowhead whales. There is only limited evidence that adenovirus infection causes disease (e.g., self-limiting gastroenteritis) in cetaceans (Rubio-Guerri et al., 2015). A cetacean pox virus (CePV-2) specific to the bowhead whale was previously isolated from skin lesions of one bowhead whale (4.5%; 1/22 examined) (Bracht et al., 2006). In a recent retrospective study, 2.2% (3/139) of examined skin lesions were suspected for viral etiology. Combined data suggests that most skin lesions in bowhead whales are not caused by poxvirus infection.

Bacterial agents that have epizootic and/or zoonotic potential (e.g., *Brucella* sp., *Leptospira* sp.) have not been found in the BCB stock of bowheads. This is based on early serological studies (Smith et al., 1987) and recent studies using PCR (Sanchez and Stimmelmayr, unpublished data). Leptospirosis is an important infectious disease of pinnipeds; however, evidence for interspecies transmission to cetaceans is lacking. A novel halophytic *Leptospira* has recently been isolated from a dead Southern right whale calf (*Eubalaena australis*, Grune Loffler et al., 2015). Marine *Brucella ceti*, on the other hand, appears to be widespread, having been isolated from various cetacean hosts and ocean regions (Guzmán-Verri et al., 2012). Three *Brucella* isolates (*B. ceti* porpoise type, *B. ceti* dolphin type, and *Brucella pinnipedialis*) have been reported in baleen whales, suggesting equal susceptibility to either of the marine strains. Brucellosis-associated lesions in cetaceans can involve the central nervous system (meningoencephalomyelitis), cardiovascular system (endocarditis), reproductive system (placentitis; abortion, epididymitis, and orchitis), integumentary system (abscessation), immune system (hepatomegaly, splenomegaly, and lymph node enlargement), and skeletal system (osteomyelitis; arthrosis). Recent serological and gross pathological evidence indicates active *Brucella* infections (granulomatous testis) in several baleen whales (common minke whale, Bryde's whale, and sei whale) in the western North Pacific (Ohishi et al., 2016). As the Arctic ecosystem continues to undergo a system-wide transformation, regularly updated surveys of bowhead whales on the prevalence of microbial agents known to globally impact cetaceans (Bossart and Duignan, 2018) are crucial to provide current baseline health data.

Parasites and commensals

Similar to other baleen whales, the bowhead whale carries parasitic and commensal organisms (Table 30.1; Fig. 30.2). Parasitism of the bowhead whale is characterized by low

TABLE 30.1 Host–parasite and commensal list reported for bowhead whales.

Metazoan parasite	Organ infected	Distribution/region
Protozoa		
Entamoeba sp.	Colon, small intestines	Alaska, United States
Flagellate form	Colon	Alaska, United States
Giardia spp.*	Colon	Alaska, United States
Cryptosporidium	Skeletal muscle	Alaska, United States
Sarcocystis sp.*		Alaska, United States
Cestoda		
Phyllobothrium delphi	Blubber	Europe
Phyllobothrium physeteris (possibly identical with delphini)	Blubber, skin	Europe
Trematoda		
Ogmogaster plicatus	Intestines	Alaska, United States
Brachycladium goliath	Liver	Europe
Acanthocephala		
Bolbosoma balaenae	Intestines	Europe
Nematoda		
Anisakis type larvae	Forestomach	Alaska, United States
Anisakid		
Contracaecum or Anisakis	Forestomach	Alaska, United States
Crassicauda sp.*	Kidney, renal arteries	Alaska, United States
Amphipoda		
Cyamus ceti	Skin, body orifices, scars, wounds	Alaska, United States
Diatoms	Skin, baleen	
Cocconeis		Alaska, United States
Stauroneis		Alaska, United States
Navicula		Alaska, United States
Gomphonema		Alaska, United States

Asterisk indicates unpublished data from authors.

*Data from Delyamure, S.L., 1955. Helminthofauna of marine mammals (Ecology and Phylogeny). Skrjabin, K.I. (Ed.) [Gel'mintofauna morskikh mlekopitayuschchikh v svete ikh ekologii i filogenii] Translated from Russian by the Israel Program for Scientific Translations, Jerusalem, 1968; Migaki, G., Heckmann, R.A., Albert, T.F., 1982. Gastric nodules caused by "Anisakis type" larvae in the bowhead whale (*Balaena mysticetus*). J. Wildl. Dis. 18, 353–357; Heckmann, R., Jensen, L., Warnock, R., Coleman, B., 1987. Parasites of the bowhead whale, *Balaena mysticetus*. Gt. Basin Nat. 47, 355–372; Henk, W.G., Mullan, D.L., 1996. Common epidermal lesions of the bowhead whale, *Balaena mysticetus*. Scanning Microsc. 10, 905–916; Hughes-Hanks, J.M., Rickard, L.G., Panuska, C., Saucier, J.R., O'Hara, T.M., Dehn, L., et al., 2005. Prevalence of *Cryptosporidium* spp. and *Giardia* spp. in five marine mammal species. J. Parasitol. 91, 1225–1228. Erratum 91, 1357.*

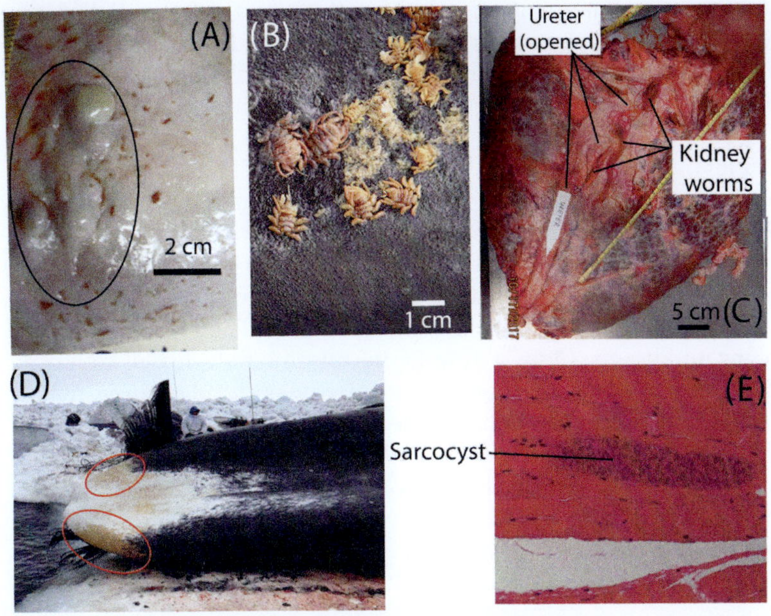

FIGURE 30.2 Parasites and commensals on bowhead whale. (A) Gastric nodules in the forestomach (NSB-DWM 2016B4). (B) Cyamids (whale lice) on skin of whale (NSB-DWM 2004B3). (C) Kidney and opened ureter with kidney worms (NSB-DWM 2017B21). (D) Yellow patches on chin caused by diatoms. (E) Sarcocyst in histological slide of muscle tissue. *Source: Photos by R. Stimmelmayr except E by D. Rotstein and B by C. George.*

species diversity with none of the described species being unique to the bowhead whale (Delyamure, 1955; Felix, 2013; Flores-Cascante et al., 2019). Most information about the bowhead whale parasite ecology comes from long-term assessment of the Bering–Chukchi–Beaufort Sea (BCB) stock (Chapter 3 and comparative analyses with other stocks are not yet possible. Parasites, like other infectious agents, are considered climate-sensitive, and thus their distribution, prevalence, and host range in the Arctic and subarctic are expected to change (Chapter 27).

Protozoa

Cryptosporidium and *Giardia* are parasitic single-celled protozoans implicated in infectious diarrheal episodes in a variety of mammals, including humans. Negative health effects due to persistent protozoal infection in cetaceans are not known. Both agents have been reported in the North Atlantic right whale, bowhead whale, and common minke whale (Hughes-Hanks et al., 2005; Reboredo-Fernández et al., 2015). Based on findings from a recent retrospective parasite screening study in BCB bowhead whales ($n = 159$; 2002–15), prevalence for both protozoa in bowhead whales has decreased over time: in *Giardia* from 33% (1998–2001) to 22% (2002–15), and in *Cryptosporidium* from 5.1% (1998–2001) to 0% (2002–15) (Stimmelmayr et al., 2018). Most current *Giardia* spp. prevalence estimates in bowhead whales are comparable to *Giardia* prevalence observed in free-swimming fin and sei whales (17.6% combined) from the Archipelago of the Azores (Hermosilla et al., 2016). Of 47 positive *Giardia* cases, 5 yielded amplification products and shared sequence similarity with Assemblage A and B (Lappin and Stimmelmayr, unpublished data). Marine contamination with human feces is recognized as a potential pathogen pathway for the introduction of *Giardia* assemblages A and B (both known to infect humans) to marine mammal populations (Heyworth, 2016; Reboredo-Fernández et al., 2015). Two unknown types of Protozoa, one amoeba form, *Entamoeba* sp. and one flagellate form have been described in bowhead whales (Heckmann et al., 1987). *Entamoeba* has also been described for free-ranging blue whale, fin whale, sei whale with a prevalence of ~65% (Hermosilla et al., 2016).

Sarcocysts are rare in skeletal muscle tissue of bowhead whales (Fig. 30.2E). No gross lesions were associated with their tissue presence. Cysts were irregular round to elongate, with a length of 100–200 μm, a wall of 50–100 μm thick, and numerous dacryoid bradyzoites. No inflammatory response was observed, though there were occasional myofibers that were disrupted and infiltrated by inflammatory cells with no cysts present in the degenerate myofibers. Cysts have not been genetically characterized. The marine life cycle of *Sarcocystis* species remains unknown. Only a few of its intermediate hosts including ringed seal (Migaki and Albert, 1980), bearded seal, polar bear, and beluga whale (Stimmelmayr and Rotstein, unpublished data) have been identified in northern Alaska. Rare occurrence of intramuscular sarcocysts in bowhead whales suggests that they are accidental intermediate hosts for this parasite. Exposure of bowhead whales through environmental marine contamination with sporulated sporocysts excreted by a definitive host is a likely source. Intramuscular sarcocysts have been previously described in the sei whale (Akao, 1970) and small cetaceans (De Guise et al., 1993; Ewing et al., 2002).

Helminths: cestodes

The cestode fauna of the bowhead whale is limited to one species. Phyllobothriid cysts of the *Phylobothrium delphi* type, that preferentially parasitizes the perigenital blubber region in cetaceans have been reported from bowhead whales from Europe (Delyamure, 1955) but not in the BCB stock. It is possible that cyst detection during postmortem examination of Alaskan bowhead whales has been biased as the perigenital blubber skin region is removed in toto and discarded as it is not consumed by the hunters. Adult worms have been found in sharks suggesting predation and scavenging of infected cetaceans play a role in the life cycle of this parasite.

Helminths: trematodes

Of the 10 digenea genera known to occur in cetaceans, only two species, *Brachycladium goliath* (van Beneden, 1858, homotypic synonym: *Lecithodesmus goliath*) and *Ogmogaster plicatus* have been identified in bowhead whales from Europe (Delyamure, 1955). Pathological lesions associated with liver fluke infection (*Lecithodesmus goliath*) in small cetaceans include chronic cholangitis and chronic hepatitis (Dailey and Stroud, 1978). In gray whales, rare fluke infection was associated with chronic cholangitis (Rice and Wolman, 1971). *Ogmogaster plicatus* specimens have been isolated from ingesta of several BCB bowhead whales (Shults, 1979; Heckmann et al., 1987), but neither eggs nor adult specimens were detected during a recent retrospective fecal parasite survey ($n = 159$; 2002–15; Ballweber and Stimmelmayr, unpublished data). Trematodes, in low numbers, associated with mild gastroenterocolitis were observed on histopathology in an individual bowhead whale (NSB-DWM 2010WW4; unpublished data). Transmission of these parasites, as with all other cetacean trematodes, has not been studied in detail. Fish may be a second intermediate host. Bowhead whales ingest fish only occasionally and probably accidentally (Chapter 28).

Acanthocephala

The acanthocephalan genus *Bolbosoma* has only been reported in bowhead whales from Europe (Delyamure, 1955). Pathological lesions associated with acanthocephalan infection, including intestinal mucosal ulceration and localized mucosal abscessation, have been described for small cetaceans (Dailey and Stroud, 1978) and gray whales (Rice and Wolman, 1971; Dailey et al., 2000). In recent years, intestinal lesions in humans due to rare *Bolbosoma* infection (Acanthocephaliasis) have been reported from Japan and indicate a rare zoonotic potential (Kaito et al., 2019), probably linked to the ingestion of undercooked and raw fish and squid.

Nematodes

Only two nematode species, *Anisakis* and *Crassicauda* have been identified in bowhead whales of the BCB stock. *Anisakis* roundworm infection consists of the larvae and/or adults in bowhead whale stomach contents and occurs at 17% of individuals (Sheffield et al., 2016). Adult *Anisakis* worms are found in the lumen of the forestomach and do not seem to affect the health of the whales, but infection with *Anisakis* larvae has been associated with gastric lesions (Migaki et al., 1982). Single to multiple granulomas caused by anisakid type larvae have been observed in immature bowhead whales (NSB-DWM 2016B3 and 2016B14; Sheffield et al., 2016). In gray whales with mature and larval *Anisakis*, no gastrointestinal lesions have been reported (Dailey et al., 2000). Gastrointestinal anisakidosis (human infection with *Anisakis* larvae) is linked to the consumption of raw/undercooked fish and squid. Bowhead whales most likely acquire infection through consumption of marine krill and copepods which are intermediate hosts, and less likely through fish and squid, which are paratenic hosts.

A variety of *Crassicauda* species have been documented in various baleen whales, including the southern right whale and the bowhead whale (Delyamure, 1955; Skrjabin, 1969). The previously ascribed presence of *Crassicauda crassicauda* in bowhead whales has been contested by Baylis (1916) and others due to marked inconsistencies between subsequent authors as to the species of whale (i.e., blue whale, fin whale, minke whale, bowhead whale, and sei whale) from which the original specimen was obtained. Since 2014, kidney worm infection has been documented in BCB bowhead whales (Stimmelmayr, 2015; George et al., 2017; Stimmelmayr et al., 2018). Species identification was based on morphological criteria and phylogenetic analysis performed on 18S sequences of two specimens, which allowed us to assign them to *Crassicauda* spp. (Stimmelmayr, Verocai, and Baird, unpublished data). Morphologically, the two specimens had cephalic papillae arrangement similar to that of *Crassicauda boopis*, *C. crassicauda*, *Crassicauda costata*, and *Crassicauda tortilis* (Baylis, 1916; Skrjabin, 1966, 1969; Lambertsen, 1985). The egg size range (length between 52.5 and 57.5 μm, and diameter between 32.5 and 37.5 μm) partially overlaps with that of the mentioned species. Spicules were absent in the caudal extremity of the one male specimen, similar to *C. boopis* and *C. tortilis*. Vestigial spicules are present in *C. costata*. Given the limitation of the 18S sequence for definitive molecular species identification (see Marcer et al., 2019), additional nematode molecular markers have been added to improve species discrimination, and morphological features are being analyzed. The life cycle of *Crassicauda* is not well understood. For its indirect life cycle, proposed host infection occurs by larval ingestion through various intermediate hosts (prey items) and for the direct transmission, calves are infected by larval ingestion through urine-contaminated milk (Lambertsen, 1985, 1986). Presence of adult nematodes in the kidneys of weaning fin whale calves has provided support for the whale to calf transmission. Recent isolation of larval *C. boopis* from the intestinal lumen of a fin whale calf (Marcer et al., 2019) provides support for the direct life cycle. Gross and histopathological renal examination of a mature pregnant female bowhead whale (NSB-DWM 2019B9) with severe *Crassicauda* infection and her full-term fetus (NSB-DWM 2019B9F) did not provide evidence for in utero transmission (Stimmelmayr and Rotstein, unpublished data). Congenital transmission of nematodes with a tissue migratory phase has been shown for humans and dogs (Costa-Macedo and Rey, 1990) and a possible transplacental route of infection has been considered for *Crassicauda* in baleen whales (Lambertsen, 1992). Given the emerging nature of this host–parasite relationship and the incomplete understanding of the life cycle of *Crassicauda*, much research is needed to further characterize the ecology of this unique parasite. Arctic climate change (Chapter 27) may be setting the stage for an evolving host–parasite relationship in the BCB bowhead whale stock.

Renal lesions associated with mild to severe *Crassicauda* infection have been confirmed in 23 bowhead whales landed in Utqiaġvik (2013–19 spring), predominantly in immature whales (76%). In fin whales, a high prevalence was also seen in calves (Lambertsen, 1986). It is unclear whether the age distribution reflects sampling bias (~70% immature; Chapter 32) or indicates age-related host–parasite effect. A sex-biased parasitism, a common phenomenon among many host–parasite relationships (Klein, 2004) cannot be ruled out as of yet. Predominant gross renal lesions in bowhead whales with mild to severe crassicauda infection include renal arteritis with and without partial lumen occluding thrombi, single to multiple renal granulomas and renal cysts (Stimmelmayr and Rotstein, unpublished data). Such lesions are similar to what has been described for Cuvier beaked whales (Díaz-Delgado et al., 2016) and less similar to what has been reported for fin, humpback, and blue whales (Lambertsen, 1986). Results from analysis of limited serum chemistry for two bowhead whales (NSB-DWM 2016B16 and 2016B22) suggest that renal lesions were unlikely to have impacted renal function. Creatinine values (1.9 mg/dL; 4.0 mg/dL) were within reported ranges for bowhead whales, while BUN (blood urea nitrogen) for both whales (110 mg/dL; 97 mg/dL) exceeded reported ranges (Chapter 11). Serum electrolytes of NSB-DWM 16B22 (potassium 9.0 meq/L; chloride 122 meq/L) were within the previously reported ranges. Redundancy of kidney tissues (i.e., size and weight) and mild to moderate *Crassicauda* infection may explain the apparent lack of an effect of these lesions on kidney function markers in the two bowhead whales. Renal failure due to severe *Crassicauda* infection has been described for fin whales (Lambertsen, 1992).

Amphipoda

Cyamids, commonly known as whale lice, are benign skin feeding cetacean ectoparasites (Fig. 30.2B). They range in size from 3 to 30 mm and spend their entire lives on a whale (Rowntree, 1996). Based on morphological identification, bowhead whales carry

Cyamus ceti (synonym of *Cyamus mysticeti* Liitken, 1870; Dall, 1872), as do gray whales and right whales (Felix, 2013). Cyamids are associated with natural body crevices (blowholes, genital slit, palpebral fissure, mouth gape, gum line, flipper insertion), scars, and wound cavities. Since cyamids have no free-swimming stage, infection depends on direct physical contact between whales. Cyamid prevalence and body burden of BCB bowhead whales is around 20% with an average body burden of 1–5 whale lice (Von Duyke et al., 2016). Heavy cyamid infestations (~50–100 cyamids) are uncommon, but have been observed in bowhead whales entangled in fishing gear, bowhead whales with old entanglement injuries, and previously harpooned bowhead whales. In right whales, blowhole-associated cyamid burden is correlated with long-term entanglement events and other injuries (Rolland et al., 2007a). Cyamid burden has been visually assessed on free-ranging Okhotsk Sea (OKS) stock of bowhead whales and appears to be greater than what is known for BCB bowhead whales (Shpak and Stimmelmayr, 2017).

Diatoms

Diatoms, single-celled benthic microalgae, are present in the aquatic environment and known to colonize whale skin. Diatoms are considered epizoic organisms of cetaceans. Eroded skin areas tend to harbor more diatoms (Heckmann et al., 1987; Henk and Mullan, 1996). Heavy colonization of whale skin with diatoms results in yellow to ochre skin discoloration, mostly notable on any white skin and is occasionally observed in BCB and OKS bowhead whales (Fig. 30.2D). During commercial whaling, large balaenopterid whales with heavy loads of diatoms turning them yellow were called "algae whales" (Japan), and "sulfur bottoms" (Alaska; blue whale). Epizoic diatoms species identified from bowhead whale skin include the genera *Cocconeis*, *Stauroneis*, *Gomphonema*, and *Navicula* (Heckmann et al., 1987; Henk. and Mullan, 1996). Diatom species composition on large whales is reflective of their feeding grounds (Nemoto, 1956).

Seasonal skin molting as reported for three bowhead whale stocks (OKS, Chernova et al., 2016; Eastern Canada-West Greenland, Fortune et al., 2017; BCB, Rehorek et al., 2019) could provide a regulatory mechanism for shedding cyamids, epizoic diatom, as well as photodamaged skin (Martinez-Levasseur et al., 2013; Shpak and Stimmelmayr, 2017).

Noninfectious diseases

This section describes noninfectious lesions in the bowhead whale, except for those that are anthropogenic (bycatch; shipstrike; blast injury, see Chapter 36), predatory (Chapter 29), and toxin related (Chapter 37). Lesions are described by organ system with the exception of neoplastic lesions which are discussed separately. Reported findings are based on gross examinations and study of standard and abnormal tissue samples (Table 30.2) collected for histopathological examination from landed bowhead whales harvested between 1979 and 2019 in Utqiaġvik, Kaktovik, Wainwright, Saint Lawrence Island (Gambell, Savoonga), Cross island, and Point Hope.

TABLE 30.2 Summary of range of lesions observed in tissue specimens collected from hunted bowhead whales ($n = 382$; 1979–2019).

	Lesions observed	% Affected
Cardiovascular system *($n = 200$)*		
Contraction band necrosis	8	4.00%
Granulomatous myocarditis	1	0.50%
Interstitial myocarditis	3	1.50%
Attenuation	2	1.00%
Hemorrhage	2	1.00%
Respiratory system *($n = 262$)*		
Congestion	8	3.1%
Hemorrhage (harvest related)	63	24.0%
Alveolar histiocytosis	4	1.5%
Interstitial fibrosis	6	2.3%
Pulmonary edema	1	0.4%
Iatrogenic intravascular skeletal muscle and bone (harvest related)	1	0.4%
Bronchopneumonia	1	0.4%
Thrombosis	1	0.4%
Interstitial pneumonia	1	0.4%
Pleural tags	1	0.4%
Angiomatosis	1	0.4%
Pulmonary vernix caseosa	5	1.9%
Pleural fibrosis	1	0.4%
Fetal atelectasis	2	0.8%
Nervous system *($n = 6$)*		
Subdural hematoma (harvest related)	2	33.3%
Endocrine system *($n = 39$)*		
Adrenalitis	1	2.6%
Adrenal cortical hyperplasia	2	5.1%
Adrenal hemorrhage	1	2.6%
Adrenal gland cortical attenuation	1	2.6%
Thyroid gland follicular cysts	1	2.6%
Thyroid gland adenoma	1	2.6%
Thyroid gland atrophy	1	2.6%

(Continued)

TABLE 30.2 (Continued)

	Lesions observed	% Affected
Hemato/lymphoreticular system *(n = 268)*		
Sinus histiocytosis	27	10.1%
Hemosiderosis	1	0.4%
Congestion	11	4.1%
Eosinophilic lymphadenitis	1	0.4%
Lymphoid hyperplasia	20	7.5%
Fibrosis, lymph node	5	1.9%
Extramedullary hematopoiesis	3	1.1%
Reactive lymph node	1	0.4%
Lymphoid depletion	1	0.4%
Preneoplastic	1	0.4%
Lymphadenitis	2	0.7%
Lymph node edema	2	0.7%
Splenic periarterial hemorrhage	3	1.1%
Hyalinosis spleen	2	0.7%
Thymic cysts	1	0.4%
Digestive system *(n = 57)*		
Pancreatitis	2	3.5%
Pancreatic atrophy	1	1.8%
Pancreatic nodular hyperplasia	1	1.8%
Granulomatous and eosinophilic gastritis/gastroenteritis	2	3.5%
Gastritis	6	10.5%
Gastric fibrosis and mucosal epithelial hyperplasia	2	3.5%
Gastric submucosal abscessation	1	1.8%
Gastroenteric trematodiasis (Ogmogaster)	1	1.8%
Alimentary lymphosarcoma	1	1.8%
Mucosal-associated lymphoid hyperplasia	1	1.8%
Enteritis	10	17.5%
Small intestine congestion	1	1.8%
Colitis	5	8.8%
Colonic submucosal fibrosis	1	1.8%

(Continued)

TABLE 30.2 (Continued)

	Lesions observed	% Affected
Epidermal inclusion cyst (glossal)	1	1.8%
Palate compression and erosion	1	1.8%
Rope injury: glossal compression, fibrosis, glossitis	3	5.3%
Glossal epithelial hydropic swelling	1	1.8%
Intraepithelial vesicular stomatitis (fetus)	1	1.8%
Glossal hyperplasia and fibroplasia	2	3.5%
Musculoskeletal system *(n = 303)*		
Hemorrhage with or without myofiber necrosis (harvest related)	14	4.6%
Myofiber degeneration (harvest related)	43	14.2%
Sarcocystis	3	1.0%
Compression atrophy (adipose)	1	0.3%
Urinary system *(n = 288)*		
Intraductal mineral	23	8.0%
Granulomatous nephritis	1	0.3%
Congestion	9	3.1%
Interstitial fibrosis	14	4.9%
Hemorrhage (harvest-related)	3	1.0%
Chronic infarct	1	0.3%
Interstitial nephritis	3	1.0%
Perirenal hemorrhage	1	0.3%
Periglomerular fibrosis	1	0.3%
Crassicaudiasis	12	4.2%
Renal fibrosis	2	0.7%
Cystitis	1	0.3%
Hydronephrosis	1	0.3%
Renal nephrolithiasis	1	0.3%
Renal cyst with atrophy and hydronephrosis	1	0.3%
Umbilical congestion	1	0.3%
Reproductive system *(n = 200)*		
Testicular atrophy	8	4.0%
Testicular fibrosis	1	0.5%

(Continued)

TABLE 30.2 (Continued)

	Lesions observed	% Affected
Epididymal fibrosis	1	0.5%
Ovarian congestion/hemorrhage	1	0.5%
Mesovarial angioleiomyoma	1	0.5%
Mesovarial abscess	1	0.5%
Mastitis	1	0.5%
Placenta congestion and hemorrhage	1	0.5%
Placenta edema	2	1.0%
Endometrial hyperplasia	1	0.5%
Penile epithelial hyperplasia	1	0.5%
Hepatobiliary system *(n = 293)*		
Congestion	63	21.5%
Hemosiderosis	13	4.4%
Hemosiderosis subset: Kupffer cell hemosiderosis	13	4.4%
Periportal hepatitis	2	0.7%
Extramedullary hematopoiesis	4	1.4%
Hepatitis	4	1.4%
Portal fibrosis	4	1.4%
Fibrosis	2	0.7%
Central vein fibrosis	1	0.3%
Bile duct hyperplasia	2	0.7%
Hepatic lipidosis	6	2.0%
Lipomatosis w/ or w/o heterotrophic bone formation	14	4.8%
Extracapsular fatty adhesion	1	0.3%
Chronic passive congestion (hepatocellular atrophy)	1	0.3%
Sinusoidal distension	1	0.3%
Capsular fibrosis	1	0.3%
Integumentary system *(n = 139)*		
Erosive dermatitis	7	5.0%
Erosion (noninflammatory)	4	2.9%
Blubber congestion and hemorrhage	3	2.2%
Fat necrosis and saponification	2	1.4%

(Continued)

TABLE 30.2 (Continued)

	Lesions observed	% Affected
Steatitis (infarcted adipose)	2	1.4%
Sclerosing (infarcted adipose)	1	0.7%
Dermatitis	18	12.9%
Bacterial dermatitis (subset of dermatitis)	2	1.4%
Mixed (protozoal/bacterial/algae)	3	2.2%
Pox virus	3	2.2%
Perivascular dermatitis	1	0.7%
Suppurative and ulcerative dermatitis	2	1.4%
Dermal fibrosis (eschar)	9	6.5%
Epidermal laceration/trauma	2	1.4%
Hyperkeratosis	10	7.2%
Epithelial depression (trauma)/superficial	1	0.7%
Epithelial degeneration	1	0.7%
Edema (intercellular)	1	0.7%
Epithelial hyperplasia	10	7.2%
Epithelial necrosis	4	2.9%
Hydropic degeneration	2	1.4%
Fibroma	1	0.7%

Not all organs were collected for every individual sampled, hence totals vary. Sample includes 73% (262/359) immature (<40′; 12.2 m), 27% (97/359) mature (>40′ >12.2 m), 52% (190/363) females, and 48% (173/363) males. These statistics exclude fetuses, calves, and specimens with incomplete data.

Neoplasia

Similar to other baleen whales, observation of neoplastic lesions in bowhead whales are rare occurrences. Affected organs include the liver, the female genital tract, the small intestines, the kidney, and the thyroid gland. With the exception of liver-associated benign neoplastic lesions, respective tumors represent single cases.

Benign fatty masses (lipomas; myelolipomas; Fig. 30.3A) of the liver are common in bowhead whales (Stimmelmayr et al., 2017). Observed lesions, single or multiple white to yellow pink white nodules, are well demarcated ranging in size 0.5–3 cm and are commonly found on the diaphragmatic aspect of the liver. The masses extend into the liver tissue. Observed hepatic lesions are not associated with extensive atrophy and/or destruction of surrounding hepatic parenchyma. Furthermore, lesions are not associated with other significant disease in examined bowhead whales. The pathogenesis and exact

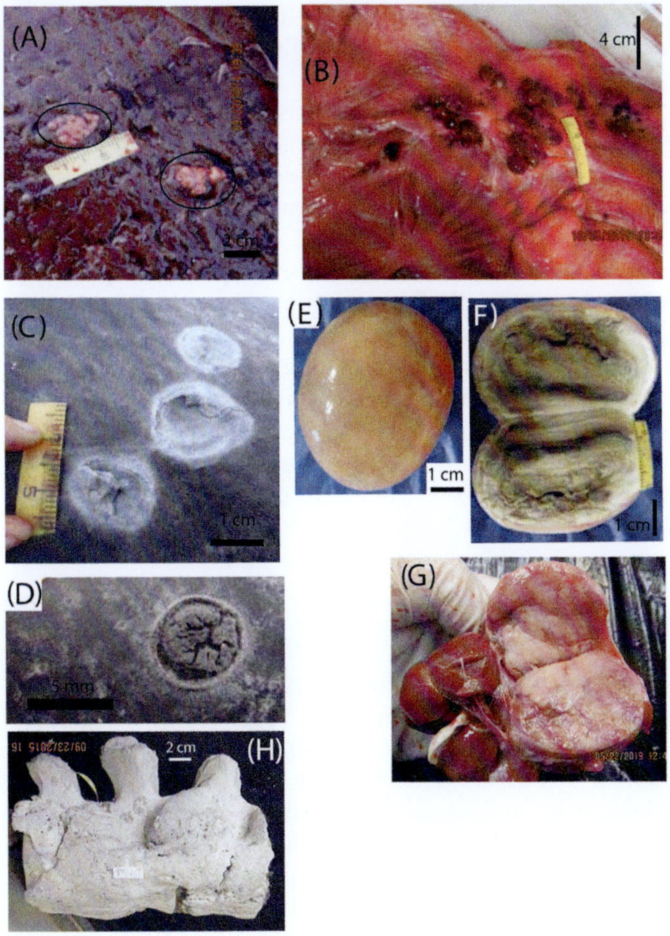

FIGURE 30.3 Neoplasia, integumentary, and skeletal lesions of bowhead whale. (A) Hepatic lipoma (NSB-DWM 2014B8). (B) Angioleiomyoma of the mesovarium (NSB-DWM 2015B24). (C) Ulcerative circular depressions on skin. (D) Raised skin plug. (E and F) Encapsulated fat mass, with internal necrosis (F). (G) Renal carcinoma (opened) with two normal reniculi (NSB-DWM 19B6). (H) Ankylosed thoracic vertebrae. *Source: Photos by R. Stimmelmayr.*

cell origin of these benign fatty tumors in bowhead whales remain undetermined. Lipomas in various anatomical locations (brain, stomach, intestines, muscle) have been reported in other baleen whales (Newman and Smith, 2006).

A case of an angioleiomyoma, a rare vascular variant of leiomyoma, was observed in a mature pregnant female (NSB-DWM 2015B24; Fig. 30.3B; Stimmelmayr and Rotstein, unpublished data). The benign tumor presented as multiple (100) discrete sessile burgundy red round masses ranging in size from 0.5 to 1 cm arising from the mesovarium. Uterine leiomyoma, but not angioleiomyoma are among the most commonly reported benign tumors of the female genital tract in cetaceans (Newman and Smith, 2006; St. Leger et al., 2018). A single stalked and pedunculated cystic white vaginal fibroma (4×2.5 cm^2) was observed in a mature female bowhead whale (NSB-DWM 2004B11). Fibromas of the vagina are uncommonly reported in cetaceans (Newman and Smith, 2006). A discrete thyroid gland adenoma was observed on histopathology in a bowhead whale (NSB-DWM

2014B7; Stimmelmayr, 2015). Thyroid adenomas (adenomatous hyperplasia) have been previously reported in aged toothed whales (Newman and Smith, 2006; St. Leger et al., 2018) but not baleen whales.

A renal carcinoma (4 × 3.5 cm) involving a single renicule was observed in a mature pregnant female (NSB-DWM 2019B6; Fig. 30.3G, Stimmelmayr and Rotstein, unpublished data). The neoplastic lesion was concurrent with bilateral moderate to severe kidney worm infection. Chronic inflammation caused by urogenital parasitism has been associated with urinary tract malignancy in humans (Kuehn et al., 2016). Neoplastic lesions associated with crassicauda infection in other whales have not been reported. Rare renal adenomas but not renal carcinoma have been described in dolphins (Gonzales-Viera et al., 2015). A suspect alimentary lymphosarcoma with no evidence for metastasis to other organs was diagnosed on histopathology in a large mature lactating female bowhead whale (NSB-DWM 1996B4; Stimmelmayr and Rotstein, unpublished data). Immunohistochemical examination for lymphoid markers failed, most likely reflecting the length of tissue storage in formalin (~24 years). Lymphosarcoma has been rarely reported in small cetaceans involving the lymph nodes, uterus, and central nervous system (Bossart et al., 1997; Arbelo et al., 2014; Díaz-Delgado et al., 2015a).

Integumentary system

A variety of epidermal lesions (shallow lacerations, circular depressions, and epidermal sloughing; Fig. 30.3) of mostly unknown etiology commonly occur in bowhead whales (Henk and Mullan, 1996; Haldiman et al., 1985; Philo et al., 1993). Specific microbiome alterations were associated with lesioned versus nonlesioned skin (Shotts et al., 1990). For example, *Corynebacterium* spp., *Acinetobacter* sp. and *Moraxella* sp. isolates dominated lesional skin, while yeast, *Candida* spp. were found on both normal and lesional skin. Most of these epidermal lesions range in color from gray to the typical black of bowhead whale skin. Untypical yellow discoloration from heavy diatom colonization involving the white chin patch area is occasionally observed (see "Diatoms" section). Anomalous white coloration (albinism) of bowhead whales, one in a full-term fetus and in two free-ranging bowhead whale, have been reported by Iñupiat hunters (McVay, 1973; Nerini et al., 1984; Eugene Brower, Charlie Hopson, Harry Brower Jr., personal communication). Pigmentation disorders including albinism have been observed in other baleen whales (blue, gray; humpback, fin) and in a broad range of toothed whales (Fertl et al., 2004; Methion and Díaz López, 2019).

Encapsulated fat necrosis in bowhead whales are a rare lesions associated with external (subcutis) and internal (body cavities) fat depots (Stimmelmayr et al., submitted for publication). These firm ellipsoid to round nodules present as whitish to pink in color, smoothly surfaced, and ranging in diameter from 6 to 15 cm (Fig. 30.3E and F, NSB-DWM 2014B15 and 2015KK1). Histopathologically, the nodules are composed of variably degenerated and necrotic adipocytes covered by a dense fibrous connective tissue capsule. The exact mechanisms of development of these encapsulated peritoneal and subcutaneous bodies in bowhead whales remains to be determined. Similar type lesions were described in other baleen whales during the time of commercial whaling in the Antarctic and southern Africa (Cockrill, 1960; Uys and Best, 1966).

Musculoskeletal system

Two cases of presumed congenital scoliosis (deformed vertebral column) without noticeable impact on swimming behavior have been observed in free-swimming bowhead whales (Mocklin et al., 2012; Vicky Beaver, personal communication). Rare cases of spinal ankylosis affecting multiple vertebrae have been observed in a Danish museum specimen (St. Leger, personal communication) and in two BCB bowhead whales (Nader, personal communication; Fig. 30.3H). Ankylosis of the axial skeleton has been reported in a southern right whale (La Sala et al., 2012). Since whole skeletons were not available, impact on locomotion and underlying etiology (degenerative; congenital, inflammatory/infectious, and trauma) of the observed vertebral abnormalities could not be assessed. Skeletal and body wall defects (rostral deformities; various abnormal deviations of the vertebral column; spina bifida, rib fusion, hernias) have been reported in baleen whales (for review see Lockyer, 1984; Kompanje, 1999; Groch et al., 2012) and small cetaceans (Berghan and Visser, 2000). The prevalence and etiology of congenital defects in cetaceans have not been determined, but similar to other species, multiple factors (genetic, infectious, and environmental) probably play a role. Other lesions associated with skeletal muscle were limited to myofiber degeneration (harvest related) and sarcocystosis.

Cardiovascular and respiratory system

Few myocardial lesions have been observed in harvested bowhead whales. With the exception of one unusual case of granulomatous myocarditis (NSB-DWM 2002B3; Stimmelmayr, 2015) with apparent gross lesions in a mature female bowhead whale, the remainder of myocardial lesions including myocardial contraction band necrosis (CBN) and interstitial myocarditis were strictly histological findings. CBN is part of a suite of pathological changes (myocardial renal lesions) brought on by massive catecholamine release (alarm reaction) in stranded cetaceans (Cowan and Curry, 2008; Groch et al., 2018). Respiratory lesions recorded in bowhead whales include isolated observations of lesions likely indicative of previous inflammatory processes of viral, parasitic or bacterial origin (i.e., alveolar histiocytosis; interstitial fibrosis; bronchopneumonia; interstitial pneumonia; pleural fibrosis; angiomatosis). Amniotic pearl aspiration was present in fetal lung tissue. In a mature whale that had been previously struck, a chronic reactive process in the thoracic cavity which involved the lungs was associated with an unexploded bomb that had penetrated the thoracic cavity (NSB-DWM 1998B23; unpublished data).

Digestive system

Upper digestive lesions involving the baleen feeding apparatus, tongue, and oral cavity in bowhead whales are rarely observed (Fig. 30.4A and B). Fractures of baleen plates causing lip and tongue abrasions were observed as a sequela to the unilateral fracture of the mandible in an immature female bowhead whale (Philo et al., 1990). Two other cases of ulcerated skin with secondary infection, located at the caudal aspect of the mouth were similarly attributed to physical trauma from baleen tips (Migaki, 1979 cited in Philo et al., 1993; Fig. 30.4B). Hypertrophy of the gray gum line seems to be mostly cosmetic without any obvious effect (NSB-DWM 2014B9; Stimmelmayr, 2015). Oral abscessation with

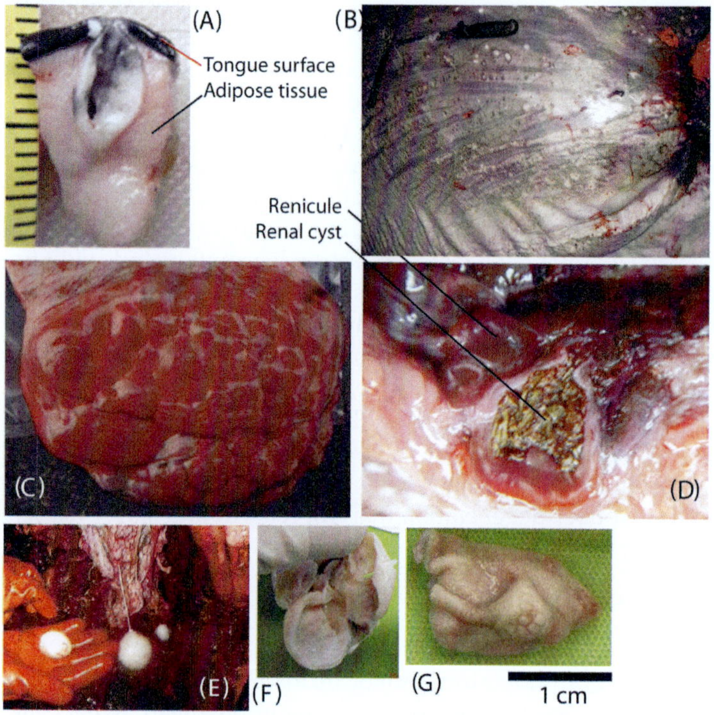

FIGURE 30.4 Lesions of the digestive and urogenital system of the bowhead whale. (A) Inclusion cyst of the tongue (NSB-DWM 2017KK2). (B) Ulcers on the surface of the tongue (NSB-DWM 2011B11). (C) Cobblestone liver (NSB-DWM 2016B19). (D) Renal cyst with worm debris. (E) Cavitary encapsulated fat necrosis (NSB-DWM 2015KK1). (F) Renal cyst (NSB-DWM 2003B5). (G) Lymphoid hyperplasia in intestine (with scale bar). *Source: Photos by R. Stimmelmayr except E by G. Sheffield.*

Clostridium perfringens and an unspecified *Fusobacterium*, involving the gumline and baleen plates of the right rostrum in a mature female has been reported (Philo et al., 1993). Baleen plate shedding of unknown etiology has been observed in sei, blue, and fin whales (Tomilin and Smyshlyayev, 1968). A single case of multiple epidermal inclusion cysts of the tongue has been observed in an immature bowhead whale (NSB-DWM 2017KK2; Fig. 30.4A). It is likely that these glossal cysts were acquired (traumatic implantation of epidermal tissue); however, the clustered presence of multiple cysts could suggest a genetic background. Epidermal inclusion cysts of the neck have been reported in two killer whales (Kamiya et al., 1979). Extensive acute and chronic circular depressed and raised ulcerative lesions of unknown etiology involving the oral cavity, anterior dorsal body of the tongue, and skin along the jawline were observed in an immature male bowhead (NSB-DWM 2011B11; Stimmelmayr, 2015).

With the exception of a suspect intestinal volvulus with marked tissue necrosis and peritonitis in an immature female bowhead whale (Heidel and Albert, 1994), inflammatory disease conditions of the gastrointestinal tract (i.e., gastritis, enteritis, and colitis) are mostly uncommon (Fig. 30.4E), generally mild and appear self-limiting. Isolated lymphoid hyperplasia of gut-associated lymphoid tissue alone or in combination with ulcerative enteritis is sometimes observed by whaling captains' wives during traditional food preparation in the small intestines and is most likely caused by parasite larvae tissue migration (NSB-DWM 2019B1and 2015B13). Gastric nodules associated with anisakid larvae infection described elsewhere (see "Parasites and Commensals" section).

Nonfood item ingestion by bowhead whales has been documented, including wood, mud, stones, and plastics (Lowry et al., 2004; Chapter 28). Plastics found in the intestines from four immature bowhead whales included clear plastic sheeting and black plastic from industrial type garbage bags (NSB-DWM 2013B6; NSB-DWM 2019N1 Thomas Napageak, personal communication). Obstructive potential of ingested plastic was low given their small size (less than 20 cm in greatest length). Bottom feeding of young bowhead whales (Chapter 28) may increase the likelihood of ingesting marine debris.

Liver lesions recorded in bowhead whales are probably indicative of previous inflammatory processes of parasitic, viral, or bacterial origin (i.e., hepatic lipidosis, hepatitis, bile duct hyperplasia, and fibrosis). Excessive stored intracytoplasmic iron (hemosiderosis) and extramedullary hematopoiesis were infrequently observed. In the absence of other indicators of systemic and hepatic disease, it is likely that they are physiological in nature and reflective of diving physiology associated processes of iron metabolism and red blood cell demands. A single case of a broad adhesion of liver to abdomen was observed in an immature whale. There was focally limited distorted liver architecture with cobblestone appearance and cysts contained yellowish fluid (Fig. 30.4C; NSB-DWM 2016B19). Histopathology was without significant findings. The pancreas in bowhead whales, a relatively small organ has been infrequently sampled. A few pancreatic lesions have been observed, including pancreatitis, pancreas atrophy, and pancreatic nodular hyperplasia.

Urinary system

Tissue samples from kidneys are regularly collected from landed bowhead whales and a variety of nonspecific findings associated with various etiologies have been observed in bowhead whales over the years including renal intraductal mineralization, renal/interstitial fibrosis, interstitial nephritis, congestion and hemorrhage, hydronephrosis, and chronic renal infarct. Given the enormous size and weight of kidneys (~70–120 cm length; 21–48 kg, 42–108 lb), limited tissue sampling without complete external examination and dissection probably underestimates prevalence of subtle renal lesions. The emergence of kidney worm infections has led to a more thorough examination of complete kidneys of harvested individuals (see "Nematodes" section).

Renal fibrosis is commonly encountered lesion in older bowhead whales (Rosa et al., 2008). The potential contributory role of renal cadmium levels in the development of renal fibrosis (interstitial fibrosis) remains unresolved (see Chapter 37). Renal intraductal mineralization is not uncommon and has been observed in immature and mature bowhead whales. Similar type lesions have been observed in dolphins (Mackey et al., 2003). In dogs, excess phosphorus in the diet has been linked to renal mineralization. It is possible that this is the case for the bowhead whale too, as it primarily feeds on euphausiids which are characterized by a high phosphorus and calcium levels (seeChapter 28). The source of the interrenicular nephritis in two immature female and one immature male immature whales was not determined, but parasitic infection was likely based on the presence of eosinophils. A single case of localized renal infarction (interstitial fibrosis with tubular loss and periglomerular fibrosis) was observed in a mature male bowhead whale (NSB-DWM 2011B16). Renal infarction has occasionally been reported in other cetaceans (Díaz-Delgado et al., 2018;

Jepson et al., 2005). A simple renal cyst with atrophy and hydronephrosis was observed in a large mature female with concurrent multiple myocardial abscesses (NSB-DWM 2002B3). Renal cysts (Fig. 30.4F) have been mostly observed in bowhead whales with concurrent kidney worm infection, but *Crassicauda* infection was absent in this animal. Cystic renal disease is a common condition in small cetaceans (Gonzales-Viera et al., 2015).

Cases of renal nephrolithiasis (kidney stones) have been observed in bowhead whales (e.g., NSB-DWM 2018B15; Stimmelmayr, 2015). Renal calculi (~20) were composed of 100% calcium oxalate and varied in size (<1–4.1 mm in diameter). Calcium oxalate crystals have been previously described for bowhead urine (Chapter 11). No parasites were found in the kidneys of whale NSB-DWM 2018B15. Nephrolithiasis is common in captive bottlenose dolphins with a 100% ammonium acid urate stone composition (Venn-Watson et al., 2010). Bladder-associated lesions are rare with only two cases of cystitis observed one grossly in an immature female (NSB-DWM 2013B7) and one on histopathology in a large (~57′10″; 17.62 m) mature female bowhead whale (NSB-DWM 2017G2). Enlarged bladders, probably of neurogenic etiology with 20–50 L of urine content were observed in two bowhead whales (NSB-DWM 2015B2; 2018KK2).

Reproductive system

Reproductive disorders are uncommon in both sexes. Two cases of developmental sexual disorders (DSD) due to impaired androgen synthesis or action have been reported for bowhead whales (Tarpley et al., 1995). External genitalia were unambiguously female (short genital slit; presence of mammary slits with teats; see Chapter 7) with presence of hypoplastic gonads (testes) and absence of prostate gland, epididymides, and Muellerian duct derivatives. A single case of Muellerian agenesis/hypoplasia complex has recently been observed in a full-term female bowhead whale fetus (NSB-DWM 2019B9F). This abnormality was characterized by a noncommunicating uterine horn with abnormal uterine folding pattern (cobblestone vs normal longitudinal folds; Fig. 30.5C) and ipsilateral renal hypoplasia. DSDs are rare in cetaceans (Stimmelmayr et al., 2019).

The presence of vestigial nipples in male bowhead whales has not been systematically assessed, but mammary slits with or without rudimentary nipples do occur (see Chapter 10). In blue, fin and gray whales, male nipples appear to be consistently present (Clarke, 2005).

Signs of testicular atrophy in a mature bowhead were likely associated with senescence (14.91 m total body length), but were also observed in a few immature, most likely pubertal bowhead whales (total body length 12.4–13.4 m; O'Hara et al., 2002). Given that no other comorbidities/reproductive anomalies were present in these pubertal whales, underlying etiology of testicular atrophy remains undetermined. Single cases of testicular fibrosis and focal epididymis fibrosis of unknown etiology in an immature whale (NSB-DWM 2012B11) have been observed. With the exception of testicular lesions associated with *Brucella* infection testicular lesions (e.g., orchitis, epididymal necrosis) are rare in baleen whales (Tomilin and Smyshlyayev, 1968; Uys and Best, 1966). Penile disorders include shortened penis (~30 cm length) with "contorted penile glans" and marked atrophy of the testes in a mature bowhead whale (NSB-DWM 1990B2; Philo et al., 1993) and a case of penile epithelial hyperplasia concurrent with pigmentation disorder of unknown etiology in an immature bowhead whale (NSB-DWM 2011B11).

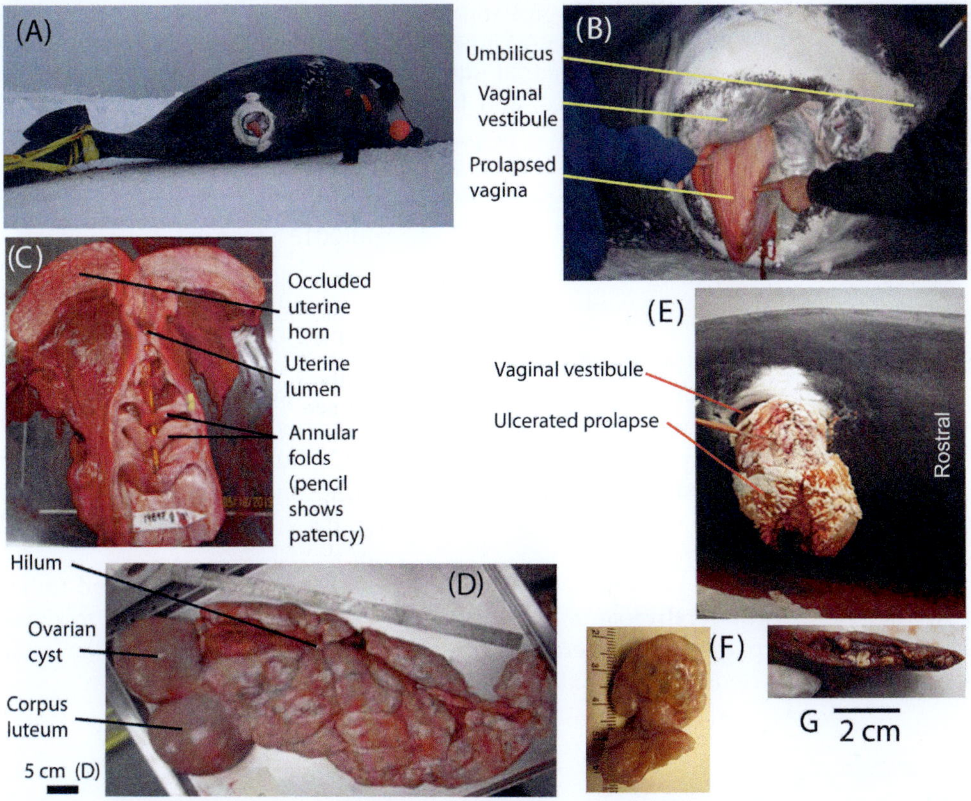

FIGURE 30.5 Disorders of the reproductive and immune systems. (A and B) Prolapsed vagina (NSB-DWM 2007B9) in overview (A) and close-up (B). (C) Noncommunicating uterine horn in fetus (NSB-DWM 2019B9F). (D) Ovary with cyst and corpus luteum (NSB-DWM 2013B1). (E) Ulcerated vaginal prolapse. (F) Cystic vaginal fibroma (NSB-DWM 2004B11, scale in mm), (F) Focal fibrosis of epididymis (NSB-DWM 2012B11). *Source: Photos by R. Stimmelmayr except A and B by J. C. George and E by Herbert Kinneeveauk.*

Single ovarian fluid-filled and solid cysts (possibly follicular and luteinized cysts) have been observed in two large females with early pregnancy (NSB-DWM 2013B1 and 2012S7; Stimmelmayr, 2015) and one nonpregnant mature female (Lara Horstmann, personal communication). In the pregnant females cyst diameter was comparable to the active corpus luteum ($\sim 14.5 \times 13.5$ cm) on the same ovary. Ovarian cysts are rare in baleen whales (fin whale Tomilin and Smyshlyayev, 1968).

Dystocia is suspected in one whale with a full-term fetus (NSB-DWM 2007B9) which showed vaginal prolapse (first pseudocervix; Fig. 30.5A and B). In a similar case, a large female presented with an ulcerated necrotic prolapsed vagina, however, pregnancy was not confirmed (2018, Point Hope, Herbert Kinneeveauk, personal communication). Parturition-associated complications (dystocia, stillbirth, abortions, perinatal death) have been documented in various baleen whales stocks (Slijper, 1949; Cockrill, 1960; Stephen et al., 1978, McAloose et al., 2016). An unusual endometrial type lesion was observed in a mature postpartum female (NSB-DWM 2002B2; Stimmelmayr, 2015). Multiple pedunculated masses (polyps) ranging in size 1–3 cm and smaller sessile masses were present in the left uterine horn. The polyps on the endometrial surface were fibrovascular with interspersed erythrocytes, collagen, hemosiderophages, and basophilic mineral. Hemorrhage had taken place in the endometrium with associated hemosiderosis being consistent with a postpartum/involuting uterus. We speculate that the endometrial fibrovascular nodules are likely regressed sites of placental/maternal connection in a diffuse epitheliochorial placenta (see Chapter 13).

Immune and endocrine system

Mesenteric lymph nodes (MLNs) are regularly collected from landed bowhead whales and a variety of nonspecific findings associated with various etiologies have been observed in individual bowhead whales including sinus histiocytosis (sinus hyperplasia), congestion, lymphoid hyperplasia (Fig. 30.4G), lymphnode fibrosis, lymph node edema, lymphoid depletion, activated lymph node, and hyaline deposits. Extramedullary hematopoiesis (EMH) within MLN was observed in two immature whales (NSB-DWM 2006B14 and 2006B16).

Few endocrine lesions are present in bowhead whales including thyroid gland follicular cysts, thyroid gland atrophy, thyroid gland adenoma (see "Neoplasia" section), adrenalitis, adrenal cortical hyperplasia, and adrenal cortical atrophy (Table 30.2). A similar suite of lesions have been reported in beluga whales (Lair et al., 2016).

Special senses

Lens yellowing (Fig. 30.6B and C) occurs with aging in baleen whales (blue and fin whale Nishiwaki, 1950; bowhead whale George et al., 1999; gray whale Stimmelmayr, 2019). Although brunescent lenses still transmit light, short wavelength light transmission is reduced (Artigas et al., 2012; Najjar et al., 2014). Cetaceans have a blue light shifted

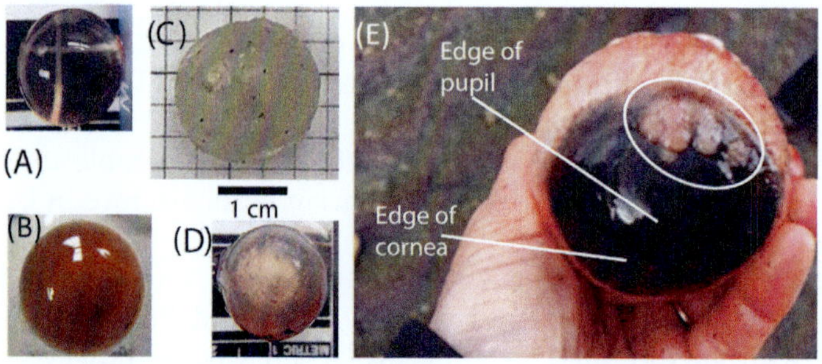

FIGURE 30.6 Disorders of the visual system of the bowhead whale. (A) Clear lens of a healthy individual. (B) Brunescent lens (NSB-DWM 2001B10). (C) Yellow lens of a gray whale (NSB-DWM 2018GW0815FD). (D) Lens with cataract (NSB-DWM 2017B6). (E) Eyeball with nodular granulomatous episcleritis (NSB-DWM 2019KK2). *Source: Photos by R. Stimmelmayr except E by G. Scheffield.*

photopigment rhodopsin which improves visual acuity in dim light environments (Chapter 18). Similar to other baleen whales (Nishiwaki, 1950; Harms et al., 2008; Panfilov, 1975, Stimmelmayr, 2019) cataracts (NSB-DWM 2017B6 and 2019KK2; Rolland et al., 2019, Fig. 30.6D) and corneal clouding (NSB-DWM 1996B14; Qian, 1997) are rare ocular disease conditions in the bowhead whale (Fig. 30.6C). An unusual case of bilateral nodular granulomatous episcleritis was observed in an immature female bowhead whale (Fig. 30.6E; NSB-DWM 2019KK2). Rare lesions of the spiral ganglion of the ear in bowhead whales have been documented (Sensor, 2017) and are consistent with hearing damage.

Conclusions

Over the last 40 years the basic knowledge of diseases in bowhead whales has continued to expand, and it provides valuable baseline data to assess future trends. In the face of the dramatic ecosystem transformation that the subarctic and Arctic are undergoing, continued health assessment of the bowhead whales is essential. This will provide timely data for the Inuit communities who depend on this whale for nutritional, cultural, and spiritual well-being; and for the scientists and agencies engaged in management and conservation of the bowhead whale.

Acknowledgments

This research on the health of bowhead whales could not have been done without the long-standing support and guidance by the whaling captains, the whaling communities, and the Alaska Eskimo Whaling Commission (AEWC) and the many NSB-DWM staff and visiting scientists who have assisted with tissue and data collection over the years. We especially thank Harry Brower Sr. and Dr. Tom Albert, the AEWC, and the NSB leadership for having the vision to establish the bowhead whale harvest monitoring program. Special thanks also to Taqulik Hepa and all previous department directors for their essential departmental support over many years.

References

Akao, S., 1970. A new species of *Sarcocystis* parasitic in the whale *Balaenoptera borealis*. J. Protozool. 17, 290–294.

Arbelo, M., De Los Monteros, A.E., Herraez, P., Suarez-Bonnet, A., Andrada, M., Rivero, M., et al., 2014. Primary central nervous system T-cell lymphoma in a common dolphin (*Delphinus delphis*). J. Comp. Pathol. 150, 336–340.

Artigas, J.M., Felipe, A., Navea, A., Fandiño, A., Artigas, C., 2012. Spectral transmission of the human crystalline lens in adult and elderly persons: color and total transmission of visible light. Invest. Ophthalmol. Vis. Sci. 53, 4076–4084.

Baylis, H.A., 1916. On *Crassicauda crassicauda* (Crepl.) [Nematoda] and its hosts. Ann. Mag. Nat. Hist. 17, 144–148.

Berghan, J., Visser, I.N., 2000. Vertebral column malformations in New Zealand delphinids with a review of cases worldwide. Aquat. Mamm. 26, 17–25.

Bossart, G.D., Duignan, P.J., 2018. Emerging viruses in marine mammals. CAB Rev. 13, 052. <http://www.cabi.org/cabreviews>.

Bossart, G.D., Ewing, R., Herron, A.J., Cray, C., Mase, B., Decker, S.J., et al., 1997. Immunoblastic malignant lymphoma in dolphins: histologic, ultrastructural, and immunohistochemical features. J. Vet. Diagn. Invest. 9, 454–458.

Bracht, A.J., Brudek, R.L., Ewing, R.Y., Manire, C.A., Burek, K.A., Rosa, C., et al., 2006. Genetic identification of novel poxviruses of cetaceans and pinnipeds. Arch. Virol. 151, 423–438.

Chernova, O.F., Shpak, O.V., Kiladze, A.B., Azarova, V.S., Rozhnov, V.V., 2016. Summer molting of bowhead whales *Balaena mysticetus* Linnaeus, 1758, of the Okhotsk Sea population. Doklady Biol. Sci. 471, 261–265.

Clarke, R., 2005. Male nipples in blue and fin whales and their absence in sperm whales. Aquat. Mamm. 31, 124–132.

Cockrill, W.R., 1960. Pathology of Cetacea. A veterinary study on whales. Part I and II. Br. Vet. J. 116, 133–189.

Costa-Macedo, L., Rey, L., 1990. *Ascaris lumbricoides* in neonate: evidence of congenital transmission of intestinal nematodes. Rev. Inst. Med. Trop. São Paulo 32, 351–354. Available from: https://doi.org/10.1590/S0036-46651990000500007.

Cowan, D.F., Curry, B.E., 2008. Histopathology of the alarm reaction in small odontocetes. J. Comp. Pathol. 139, 24–33.

Dailey, M., Stroud, R., 1978. Parasites and associated pathology observed in cetaceans stranded along the Oregon coast. J. Wildl. Dis. 14, 503–511.

Dailey, M.D., Gulland, F.M.D., Lowenstine, L.J., Silvagni, P., Howard, D., 2000. Prey, parasites and pathology associated with the mortality of a juvenile gray whale (*Eschrichtius robustus*) stranded along the northern California coast. Dis. Aquat. Org. 42, 111–117.

De Guise, S., Lagacé, A., Girard, C., Béland, P., 1993. Intramuscular *Sarcocystis* in two beluga whales and an Atlantic white-sided dolphin from the St. Lawrence estuary, Quebec, Canada. J. Vet. Diagn. Invest. 5, 296–300.

Delyamure, S.L., 1955. Helminthofauna of Marine Mammals (Ecology and Phylogeny) Skrjabin, K.I. (Ed.). [Gel'mintofauna morskikh mlekopitayuschchikh v svete ikh ekologii i filogenii] Translated from Russian by the Israel Program for Scientific Translations, Jerusalem, 1968.

Díaz-Delgado, J., Fernández, A., Xuriach, A., Sierra, E., Bernaldo de Quirós, Y., Mompeo, B., et al., 2016. Verminous arteritis due to Crassicauda sp. in Cuvier's Beaked Whales (*Ziphius cavirostris*). Vet. Pathol. 53, 1233–1240.

Díaz-Delgado, J., Fernández, A., Sierra, E., Sacchini, S., Andrada, M., Vela, A.I., et al., 2018. Pathologic findings and causes of death of stranded cetaceans in the Canary Islands (2006–2012). PLoS One 13, e0204444.

Díaz-Delgado, J., Sierra, E., Arbelo, M., Suárez-Bonnet, A., Suárez-Santana, C., Grau-Bassas, E., et al., 2015a. Primary uterine T-cell lymphoma with metastasis in an Atlantic spotted dolphin (*Stenella frontalis*), Canary Islands, Spain. J. Wildl. Dis. 51, 538–541.

Ewing, R., Zaias, J., Stamper, M.A., Bossart, G.D., Dubey, J.P., 2002. Prevalence of *Sarcocystis* sp. in stranded Atlantic white-sided dolphins (*Lagenorhynchus acutus*). J. Wildl. Dis. 38, 291–296.

Felix, J.R., 2013. Reported Incidences of Parasitic Infections in Marine Mammals from 1892 to 1978, 20. Zea E-Books, <http://digitalcommons.unl.edu/zeabook/20>.

Fertl, D., Barros, N.B., Rowlett, R.A., Estes, S., Richlen, M., 2004. An update on anomalously white cetaceans, including the first account for the pantropical spotted dolphin (*Stenella attenuata graff-mani*). Lat. Am. J. Aquat. Mamm. 3, 163–166.

Flores-Cascante, L., Gómez-Gutiérrez, J., Gómez, M.D.C., et al., 2019. Helminth load in feces of free-ranging blue and fin whales from the Gulf of California. Act. Parasitol. 64, 625–637.

Fortune, S.M.E., Kosk, W.R., Higdon, J.W., Trites, A.W., Baumgartner, M.F., Ferguson, S.H., 2017. Evidence of molting and the function of "rock-nosing" behavior in bowhead whales in the eastern Canadian Arctic. PLoS One 12, e0186156.

George, J.C., Bada, J., Zeh, J., Scott, L., Brown, S., O'Hara, T., et al., 1999. Age and growth estimates of bowhead whales (*Balaena mysticetus*) via aspartic acid racemization. Can. J. Zool. 77, 571–580.

George, R., Stimmelmayr, R., Brower, A., Clarke, J., Ferguson, M., Von Duyke, A., et al., 2017. 2016 Health report for the BCB bowhead whales—preliminary findings. Submitted as Paper SC67A/AWMP to the Scientific Committee of the International Whaling Commission, 2017.

Gonzales-Viera, O., Ruoppolo, V., Marigo, J., Carvalho, V.L., Groch, K.R., Bertozzi, C.P., et al., 2015. Renal lesions in cetaceans from Brazil. J. Comp. Pathol. 152, 345–354.

Groch, K.R., Díaz-Delgado, J., Marcondes, M.C.C., et al., 2018. Pathology and causes of death in stranded humpback whales (*Megaptera novaeangliae*) from Brazil. PLoS One 13, e0194872.

Groch, K.R., Marcondes, M.C.C., Colosio, A.C., Catão-Dias JL, J.L., 2012. Skeletal abnormalities in humpback whales *Megaptera novaeangliae* stranded in the Brazilian breeding ground. Dis. Aquat. Org. 101, 145–158.

Grune Loffler, S., Rago, V., Martínez, M., Uhart, M., Florin-Christensen, M., Romero, G., et al., 2015. Isolation of a seawater tolerant *Leptospira* spp. from a Southern Right Whale (*Eubalaena australis*). PLoS One 10, e0144974.

Guzmán-Verri, C., González-Barrientos, R., Hernández-Mora, G., Morales, J.A., Baquero-Calvo, E., Chaves-Olarte, E., et al., 2012. *Brucella ceti* and brucellosis in cetaceans. Front. Cell. Infect. Microbiol. 6 (2), 3.

Haldiman, J.T., Henk, W.G., Henry, R.W., Albert, T.F., Abdelbaki, Y.Z., Duffield, D.W., 1985. Epidermal and papillary dermal characteristics of the bowhead whale (*Balaena mysticetus*). Anat. Rec. 211, 391–402.

Harms, C.A., 1993. Composition of prepartum mammary secretions of two bowhead whales (*Balaena mysticetus* L.). J. Wildl. Dis. 29, 94–97.

Harms, C.A., Lovewell, G.N., Rotstein, D.S., 2008. Presumed hyperglycemic cataracts in a live stranded minke whale (*Balaenoptera acutorostrata*) calf. Mar. Mamm. Sci. 24, 388–397.

Heckmann, R., Jensen, L., Warnock, R., Coleman, B., 1987. Parasites of the bowhead whale, *Balaena mysticetus*. Gt. Basin Nat. 47, 355–372.

Heidel, J.R., Albert, T.F., 1994. Intestinal volvulus in a bowhead whale, *Balaena mysticetus*. J. Wildl. Dis. 30, 126–128.

Henk, W.G., Mullan, D.L., 1996. Common epidermal lesions of the bowhead whale, *Balaena mysticetus*. Scanning Microsc. 10, 905–916.

Hermosilla, C., Liliana, M.R., Silva, R., Kleinertz, S., Priet, R., Silva, A.M., et al., 2016. Endoparasite survey of free-swimming baleen whales (*Balaenoptera musculus*, *B. physalus*, *B. borealis*) and sperm whales (*Physeter macrocephalus*) using non/minimally invasive methods. Parasitol. Res. 115, 889.

Heyworth, M.F., 2016. *Giardia duodenalis* genetic assemblages and hosts. Parasite. 23, 13.

Hughes-Hanks, J.M., Rickard, L.G., Panuska, C., Saucier, J.R., O'Hara, T.M., Dehn, L., et al., 2005. Prevalence of *Cryptosporidium* spp. and *Giardia* spp. in five marine mammal species. J. Parasitol. 91, 1225–1228.

Jepson, P.D., Deaville, R., Patterson, I.A., Pocknell, A.M., Ross, H.M., Baker, J.R., et al., 2005. Acute and chronic gas bubble lesions in cetaceans stranded in the United Kingdom. Vet. Pathol. 42, 291–305.

Kaito, S., Sasaki, M., Goto, K., Matsusue, R., Koyama, H., Nakao, M., et al., 2019. A case of small bowel obstruction due to infection with *Bolbosoma sp.* (Acanthocephala: Polymorphidae). Parasitol. Int. 68, 14–16.

Kamiya, T., Tobayama, T., Nishiwaki, M., 1979. Epidermal cyst in the neck of a killer whale. Sci. Rep. Whales Res. Inst. 31, 93–94.

Klein, S.L., 2004. Hormonal and immunological mechanisms mediating sex differences in parasite infection. Paras. Immunol. 26, 247–264.

Kompanje, E.J.O., 1999. Considerations on the comparative pathology of the vertebrae in Mysticeti and Odontoceti: evidence for the occurrence of discarthrosis, zygarthrosis, infectious spondylitis and spondyloarthritis. Zool. Meded. Leiden 73, 99–130.

Kuehn, J., Lombardo, L., Janda, W.M., Hollowell, C.M., 2016. Giant kidney worms in a patient with renal cell carcinoma. BMJ Case Rep. bcr2015212118.

La Sala, L.F., Pozzi, L.M., McAloose, D., Kaplan, F.S., Shore, E.M., Kompanje, E.J., et al., 2012. Severe soft tissue ossification in a southern right whale *Eubalaena australis*. Dis. Aquat. Org. 102, 149–156.

Lair, S., Measures, L.N., Martineau, D., 2016. Pathologic findings and trends in mortality in the beluga (*Delphinapterus leucas*) population of the St Lawrence Estuary, Quebec, Canada, from 1983 to 2012. Vet. Pathol. 53, 22–36.

Lambertsen, R.H., 1985. Taxonomy and distribution of a *Crassicauda* species (Nematoda: Spirurida) infecting the kidney of the common fin whale (*Balaenoptera physalus* Linne, 1758). J. Parasitol. 71, 485–488.

Lambertsen, R.H., 1986. Disease of the common fin whale (*Balaenoptera physalus*): crassicaudiosis of the urinary system. J. Mammal. 67, 353–366.

Lambertsen, R.H., 1992. Crassicaudosis: a parasitic disease threatening the health and population recovery of large baleen whales. Rev. Sci. Tech. 11, 1131–1141.

Lockyer, C.H., 1984. Review of baleen whale (Mysticeti) reproduction and implication for management. Rep. Int. Whal. Comm. 6 (Spec Issue), 27–50.

Lowry, L.F., Sheffield, G., George, J.C., 2004. Bowhead whale feeding in the Alaskan Beaufort Sea, based on stomach contents analyses. J. Cetacean Res. Manage. 6, 215–223.

Mackey, E., Oflaz, R., Epstein, M.B., Buehler, B.J., Porter, T., Rowles, S.A., et al., 2003. Elemental Composition of liver and kidney tissues of rough-toothed dolphins (*Steno bredanensis*). Arch. Environ. Contam. Toxicol. 44, 0523.

Marcer, F., Negrisolo, E., Franzo, G., Tessarin, C., Pietrobelli, M., Marchiori, E., 2019. Morphological and molecular characterization of adults and larvae of *Crassicauda spp.* (Nematoda: Spirurida) from Mediterranean fin whales *Balaenoptera physalus* (Linnaeus, 1758). Int. J. Parasitol. Parasites Wildl. 9, 258–265.

Martinez-Levasseur, L.M., Birch-Machin, M.A., Bowman, A., Gendron, D., Weatherhead, E., Knell, R.J., et al., 2013. Whales use distinct strategies to counteract solar ultraviolet radiation. Sci. Rep. 3, 2386.

McAloose, D., Rago, M.V., Di Martino, M., Chirife, A., Olson, S.H., Beltramino, L., et al., 2016. Post-mortem findings in southern right whales *Eubalaena australis* at Península Valdés, Argentina, 2003–2012. Dis. Aquat. Org. 12 (119), 17–36.

McVay, S., 1973. Stalking the Arctic whale. Am. Sci. 61, 24–37.

Methion, S., Díaz López, B., 2019. First record of atypical pigmentation pattern in fin whale *Balaenoptera physalus* in the Atlantic Ocean. Dis. Aquat. Org. 135, 121–125.

Migaki, G., Albert, T.F., 1980. Sarcosporidiosis in the ringed seal. J. Am. Vet. Med. Assoc. 177, 917–918.

Migaki, G., Heckmann, R.A., Albert, T.F., 1982. Gastric nodules caused by "Anisakis type" larvae in the bowhead whale (*Balaena mysticetus*). J. Wildl. Dis. 18, 353–357.

Mocklin, J., J.C. George, M. Ferguson, L. Vate Brattstroem, V. Beaver, B. Rone, et al. 2012. Aerial photography of bowhead whales near Barrow, Alaska, during the 2011 spring migration. Scientific Committee of the International Whaling Commission, Paper SC/64/BRG3.

Najjar, R.P., Chiquet, C., Teikari, P., Cornut, P.-L., Claustrat, B., Denis, P., et al., 2014. Aging of non-visual spectral sensitivity to light in humans: compensatory mechanisms? PLoS One 9, e85837. Available from: https://doi.org/10.1371/journal.pone.0085837.

Nemoto, T., 1956. On the diatoms of the skin film of whales in the northern Pacific. Sci. Rep. Whales Res. Inst. 2, 99–132.

Nerini, M.K., Braham, H.W., Marquette, W.M., Rugh, D.J., 1984. Life history of the bowhead whale, *Balaena mysticetus* (Mammalia: Cetacea). J. Zool. 204, 443–468.

Newman, S.J., Smith, S.A., 2006. Marine mammal neoplasia: a review. Vet. Pathol. 43, 865–880.

Nishiwaki, M., 1950. Determination of the age of Antarctic blue and fin whales by the colour changes in crystalline lens. Sci. Rep. Whales Res. Inst. 4, 115–161.

O'Hara, T.M., George, J.C., Tarpley, R.J., Burek, K., Suydam, R.S., 2002. Sexual maturation in male bowhead whales (*Balaena mysticetus*) of the Bering Sea stock. J. Cetacean Res. Manage. 4, 143–148.

O'Hara, T.M., House, C., House, J.A., Suydam, R.S., George, J.C., 1998. Viral serological survey of bowhead whales in Alaska. J. Wildl. Dis. 34, 39–46.

Ohishi, K., Bando, T., Abe, E., Kawai, Y., Fujise, Y., Maruyama, T., 2016. Long-term and large-scale epidemiology of *Brucella* infection in baleen whales and sperm whales in the western North Pacific and Antarctic Oceans. J. Vet. Med. Sci. 78, 1457–1464.

Panfilov, B.G., 1975. A fin whale, *Balaenoptera physalus*, with complete cataracts of both eyes. Zool. Zhurn. 54, 1739–1740.

Philo, L.M., Hanns, C., George, J.C., 1990. Fractured mandible and associated oral lesions in a subsistence-harvested bowhead whale (*Balaena mysticetus*). J. Wildl. Dis. 26, 125–128.

Philo, L.M., Schotts Jr, E.B., George, J.C., 1993. Morbidity and mortality. In: Burns, J.J., Montague, J.J., Cowles, C.J. (Eds.), The Bowhead Whale. Allen Press, Lawrence, KS, pp. 275–307.

Qian, Z.C., 1997. Firsts record of an eye disease in the bowhead whale (*Balaena* mysticetus). J. Ocean. Limnol. 15, 192.

Reboredo-Fernández, A., Ares-Mazás, E., Martínez-Cedeira, J.A., Romero-Suances, R., Cacciò, S.M., Gómez-Couso, H., 2015. *Giardia* and *Cryptosporidium* in cetaceans on the European Atlantic coast. Parasitol. Res. 114, 693–698.

Rehorek, S.J., Stimmelmayr, R., George, J.C., Suydam, R., McBurney, D.M., Thewissen, J.G.M., 2019. Structure of the external auditory meatus of the Bowhead whale (*Balaena mysticetus*) and its relation to their seasonal migration. J. Anat. 234, 201–215.

Rice, D.W., Wolman, A.A., 1971. The life history and ecology of the gray whale (*Eschrichtius robustus*). Am. Soc. Mammal. Spec. Publ. 3, 1–42.

Rolland, R.M., Graham, K.M., Stimmelmayr, R., Suydam, R.S., George, J.C., 2019. Chronic stress from fishing gear entanglement is recorded in baleen from a bowhead whale (*Balaena mysticetus*). Mar. Mamm. Sci. 35, 1625–1642.

Rolland, R.M., Hamilton, P.K., Marx, M.K., Pettis, H.M., Angell, C.M., Moore, M.J., 2007a. External perspectives on right whale health. In: Kraus, S.D., Rolland, R.M. (Eds.), The Urban Whale: North Atlantic Right Whales at the Crossroads. Harvard University Press, Cambridge, MA, pp. 273–309.

Rosa, C., Blake, J.E., Bratton, G.R., Dehn, L.A., Gray, M.J., O'Hara, T.M., 2008. Heavy metal and mineral concentrations and their relationship to histopathological findings in the bowhead whale (*Balaena mysticetus*). Sci. Total Environ. 25, 165–178.

Rowntree, V.J., 1996. Feeding, distribution, and reproductive behavior of cyamids (Crustacea: Amphipoda) living on humpback and right whales. Can. J. Zool. 74, 103–109.

Rubio-Guerri, C., García-Párraga, D., Nieto-Pelegrín, E., Melero, M., Álvaro, T., Valls, M., et al., 2015. Novel adenovirus detected in captive bottlenose dolphins (*Tursiops truncatus*) suffering from self-limiting gastroenteritis. BMC Vet. Res. 7 (11), 53.

Sensor, J.D., 2017. Hearing and Age Estimation in Two Species of Arctic Whale (Ph.D. thesis). Kent State University, Kent, OH, 144 pp.

Sheffield, G., Stimmelmayr, R., Rausch, R., George, J.C., 2016. *Anisakis* spp. in legally harvested BCB bowhead whales (*Balaena mysticetus*): preliminary findings from stomach content analysis and post mortem field examinations. Scientific Committee of the International Whaling Commission, Paper SC/66b/E/06.

Shotts Jr, E.B., Albert, T.F., Wooley, R.E., Brown, J., 1990. Microflora associated with the skin of the bowhead whale (*Balaena mysticetus*). J. Wildl. Dis. 26, 351–359.

Shpak, O.V., Stimmelmayr, R., 2017. Preliminary image analysis of acute and chronic injuries, parasites, and skin conditions in the Okhotsk bowhead whale (*Balaena mysticetus*) stock in the western Okhotsk Sea. Scientific Committee of the International Whaling Commission, Paper SC67a/E/01.

Shults, L., 1979. *Ogmogaster antarcticus* Johnston, 1931 (Trematoda: Notocotylidae) from the bowhead whale, *Balaena mysticetus* L., at Barrow, Alaska. Can. J. Zool. 57, 1347–1348.

Skrjabin, A.S., 1966. A new *Crassicauda* (*Crassicauda delamureana* n. sp.) a parasite of the sei whale. Probl. Parazit.i, Kraevaya Parazitologiya i Prirodnaya Ochagovost Transmissivnikh Boleznei. Akad. Nauk Ukransk. SSSR, 100–107 (in Russian).

Skrjabin, A.S., 1969. Novaya nematoda *Crassicauda costata* n. sp.—parazit yshnovo kita. Parazitologiya 3, 258–265 (in Russian).

Slijper, E.J., 1949. On some phenomena concerning pregnancy and parturition of the Cetacea. Bijdr. Dierk. 28, 415–448.

Smith, A.V., Skilling, D.E., Benirschke, K., Albert, T.F., Barlough, J.E., 1987. Serology and virology of the bowhead whale (*Balaena mysticetus*). J. Wildl. Dis. 23, 92–98.

St. Leger, J., Raverty, S., Mena, A., 2018. Cetacea. In: Terio, K.A., McAloose, D., St. Leger, J. (Eds.), Pathology of Wildlife and Zoo Animals. Academic Press, pp. 533–568.

Stephen, B., White, H., Griese, J., 1978. Notes on lengths, weights, and mortality of gray whale calves. J. Mammal. 59, 440–441.

Stimmelmayr, R., 2015. Health assessment of subsistence harvested Bering-Chukchi-Beaufort Seas bowhead whales (*Balaena mysticetus*): an overview. Scientific Committee of the International Whaling Commission, Paper SC/66a/E8.

Stimmelmayr, R., 2019. Incidental findings of lens yellowing and cataracts in beach cast gray whales (*Eschrichtius robustus*), Alaska. In: 2020 Alaska Marine Science Symposium, Anchorage (abstract).

Stimmelmayr, R., Rotstein, D., Seguel, M., Gottdenker, N., 2017. Hepatic lipomas and myelolipomas in subsistence-harvested bowhead whales, *Balaena mysticetus*, Alaska (USA): a case review 1980-2016. Dis. Aquat. Org. 127, 71–74.

Stimmelmayr, R., George, J.C., Willoughby, A., Brower, A., Clarke, J., Ferguson, M., et al., 2018. 2017 Health report for the Bering-Chukchi-Beaufort Seas. Scientific Committee of the International Whaling Commission, Paper SC/66B/AWMP/08.

Stimmelmayr, R., Ferrer, T., Rotstein, D.S., 2019. Persistent Müllerian duct syndrome in a beluga whale *Delphinapterus leucas*. Dis. Aquat. Org. 136, 273–278.

Stimmelmayr, R., Rotstein, D., Sheffield, G., Brower, S., George, J.C., submitted for publication. Subcutaneous encapsulated fat necrosis and loose cavitary fat bodies in bowhead whales (*Balaena mysticetus*) from Alaska. Dis. Aquat. Org.

Tarpley, R.J., Jarrell, G.H., George, J.C., Cubbage, J., Stott, G.G., 1995. Male pseudohermaphroditism in the bowhead whale, *Balaena mysticetus*. J. Mammal. 76, 1267–1275.

Tomilin, A.G., Smyshlyayev, M.I., 1968. Some Factors Affecting Whale Mortality (Diseases of Cetaceans). Byull. Mosk. O-va. Ispyt. Prir., Otd. Biol. 3.

Uys, C.J., Best, P.B., 1966. Pathology of lesions observed in whales flensed at Saldanha Bay, South Africa. J. Comp. Pathol 76, 407–412.

Venn-Watson, S., Smith, C.R., Johnson, S., Daniels, R., Townsend, F., 2010. Clinical relevance of urate nephrolithiasis in bottlenose dolphins, *Tursiops truncatus*. Dis. Aquat. Org. 89, 167–177.

Von Duyke, A., Stimmelmayr, R., Sheffield, G., Sformo, T., Suydam, R., Givens, G.H., et al., 2016. Prevalence and abundance of cyamid "Whale Lice" (*Cyamus ceti*) on subsistence harvested bowhead whales (*Balaena mysticetus*). Arctic 69, 331–340.

SECTION III

Interactions with humans

CHAPTER

31

Whale hunting in Indigenous Arctic cultures

H.P. Huntington[1], *C. Sakakibara*[2], *G. Noongwook*[3], *N. Kanayurak*[4], *V. Skhauge*[5], *E. Zdor*[6], *S. Inutiq*[7] *and B. Lyberth*[8]

[1]Ocean Conservancy, Eagle River, AK, United States [2]Oberlin College, Oberlin, OH, United States [3]Savoonga Whaling Captains Association, Savoonga, AK, United States [4]Department of Wildlife Management, North Slope Borough, Utqiaġvik, AK, United States [5]Sireniki, Chukotka, Russia [6]University of Alaska Fairbanks, Fairbanks, AK, United States [7]Iqaluit, NU, Canada [8]Kalallit Nunaanni Aalisartut Piniartullu Kattuffiat (KNAPK; Association of Fishers and Hunters of Greenland), Nuuk, Greenland

Introduction

This chapter explores what whaling means to whalers and their communities, from a brief review of the literature and a series of first-person accounts by Indigenous authors. The idea of Indigenous peoples speaking for themselves is, of course, not a new one, but its practice has been sporadic at best in the scholarly world. In contrast to ethnographic and anthropological studies, first-person accounts typically reflect the view of only the individual writing or speaking. Their emotional power draws on deeply personal experiences and beliefs, which can be a useful complement to generalized accounts of an entire community or people. By including such essays at its center, this chapter attempts to reflect and promote a wider trend, with contributions not just about, but also by, those whose lives intersect with bowhead whales (Fig. 31.1).

Bowhead whaling in the scholarly literature

Bowhead whaling in Native cultures has been explored and documented through myriad lenses in interdisciplinary and international contexts. In Alaska, Iñupiat and

The Bowhead Whale
DOI: https://doi.org/10.1016/B978-0-12-818969-6.00031-5

FIGURE 31.1 The village of Point Hope (Alaska) has a long history of harvesting bowhead whales. Lower jaws of whales are hauled to the field where the nalukataq is celebrated, a feast of thanks for the successful harvest. *Source: Photo by Lori Quakenbush.*

St. Lawrence Island Yupik have been sharing the bowhead whale amongst themselves for over a thousand years (VanStone, 1958; Bockstoce, 1977; Burch, 1981; Chance, 1990; Turner, 1990; Worl, 1980; Jolles, 2002; Bodenhorn, 2003; Kishigami, 2013). In this way, Indigenous social and environmental ethics are deeply rooted in the physical and spiritual relationships with and admiration for the bowhead whale. This massive creature holds a central place in the subsistence cycle, and is the focus of an annual set of cultural and social events and practices that rhythmically organize community life (Lantis, 1938; Lowenstein, 1981; Nelson, 1981; Hess, 1999; Fienup-Riordan, 2000; Ikuta, 2011; Sakakibara, 2020). Local traditions assert that the whale gives itself to the whaling captain who respects the animal (Lowenstein, 1993; Zumwalt, 1988; Brewster 2004). In return, the whaling captain honors and shares the whale within and beyond the village (Foote, 1992; Turner, 1993; Brewster, 2004; Foote, 2009). The people get together to help one another, to serve the entire community, and to share the whale (Brower, 1942; Bodfish, 1991; Edwardson, 2004). Fulfillment and happiness are in the air each time the whale is involved. Through the hunting itself, the associated ceremonies and other events, and the generous distribution of meat and other body parts of the whale, the bowhead remains central to life and worldview in the whaling communities of Alaska (Blackman, 1989; Sakakibara, 2017).

While bowhead whaling in other Arctic regions such as Chukotka, the Northwest Territories, Nunavut, and Greenland is irregular compared with Alaska (Freeman et al., 1992; Stevenson, 1997; Freeman et al., 1998; Heide-Jørgensen and Laidre, 2006; Ikeya, 2013; Heyes and Helgen, 2014; Bogoslovskaya et al., 2016), the significance of the animal is revealed by contemporary subsistence hunters-artists such as Tim Pitsiulak from Cape Dorset as well as Leetia Alivaktuk and Elisapee Ishulutaq from Panniqtuuq (Pangnirtung), Nunavut, to name but a few (Boyd, 2018), and by sections of this chapter.

Studying bowheads requires researchers to cross the academic boundaries between the natural sciences, social sciences, and humanities to fully grasp what the whale means to human survival and flourishing in the Arctic (Brewster 2004; George et al., 2004, 2015). The whale also enables and requires a collaboration between subsistence hunters and scientists (Noongwook et al., 2007; Quakenbush et al., 2010). The 1977 formation of the Alaska Eskimo Whaling Commission (AEWC) enabled the whalers to negotiate with the federal government and the International Whaling Commission (IWC) about management plans and the bowhead whale catch limits or whale quota system (Chapters 32 and 38). The politicization of Native whaling identity played a surprising role in spurring collaboration between Western science and Indigenous knowledge, which resulted in the successful defense of subsistence whaling rights (Huntington, 1992; Albert, 2001; Brewster, 2004).

Indigenous whaling in the Arctic has regained momentum since the 1970s, rejuvenating the connection between people and whale. In the meantime, global climate change coupled with the Arctic gas and oil development and other industrial activity poses a new threat to their multispecies relationship (Wohlforth, 2004; Druckenmiller et al, 2010; Kishigami, 2010; Thompson et al, 2012; Gearheard et al, 2013; Sakakibara, 2020; Chapter 38). Warming temperatures and disappearing sea ice suggest that the future of the bowhead is uncertain (Chapter 27), but a variety of studies shed light on the resilience of Arctic marine mammals and Indigenous cultures. This is shown by climate change and resilience of the Arctic marine mammals (Moore and Huntington 2008; George et al., 2015); the interplay of Iñupiat subsistence, climate variability, and bowhead whale distribution (Ashjian et al., 2010); recent distribution, sighting rates, and habitat association of the bowhead whale and gray whale in the Eastern Chukchi Sea (Clarke et al., 2016); and transformative human–whale relations in Indigenous cultures (Sakakibara 2020).

Exploring the significance of human–whale relations through the voices of Arctic Indigenous whalers is crucial as they inform us of how to shape our relationships with nonhuman animals.

Bowhead whaling as lived experience

Here, six contemporary Indigenous authors describe experiences with bowhead whales. Each account reflects what bowhead whaling means to those who live where bowheads are found, spanning a range of interaction from the annual hunt in the Alaska communities of Savoonga and Utqiaġvik, to more sporadic hunts in Sireniki, Chukotka and Clyde River, Nunavut, to a largely dormant tradition in Greenland. These are the stories of individuals, surrounded by family and friends, as their engagement with the bowhead whale creates meaning and community.

Contributors

George Noongwook is a whaling captain from Savoonga, Alaska. He served for many years as Savoonga's delegate to the Alaska Eskimo Whaling Commission, holding the positions of chairman and vice-chairman at various times. He has also been involved in many research projects about the St. Lawrence Island Yupik and their culture and practices.

Nicole Kannik Kanayurak is Iñupiaq from Utqiaġvik, Alaska. Her parents are Lloyd and Abby Kanayurak. Nicole works on wildlife management at the North Slope Borough as the Deputy Director. Nicole enjoys engaging in Indigenous comanagement of natural resources in the Arctic. She has a master's degree in Marine Affairs from the University of Washington and earned her bachelor's degree from Dartmouth College. Nicole is a current Fulbright Arctic Scholar as well.

Valerii Skhauge is a hereditary hunter from Sireniki. He has a wife Nataliia, two sons, Yurii and Vadim, and daughter Veronika. Eduard Zdor is a Chukotkan Indigenous scholar and activist. He is a former Executive Secretary of the Association of Traditional Marine Mammal Hunters of Chukotka (ChAZTO). He has been involved in joint Alaska-Chukotka research projects as a partner and principal investigator for over 15 years.

Sandra Inutiq is currently the Chief Negotiator for the Qikiqtani Inuit Association's Tallurutiup Imanga Inuit Impact and Benefit Agreement. Inutiq received her law degree from Akitsiraq Law School in 2005, and in 2006 she became the first Inuk woman in Nunavut to pass the bar exam. In the past, she has worked as legal counsel for the Government of Nunavut, as the Director of Policy for the Office of the Languages Commissioner, and served as the Official Languages Commissioner for Nunavut. Inutiq has lived in various communities in Nunavut. She spent her childhood in Kangiqtualuk outpost camp and Clyde River, in her youth she moved to Iqaluit, where she now lives.

Bjarne Ababsi Lyberth is a biologist and senior adviser on environment with Kalallit Nunaanni Aalisartut Piniartullu Kattuffiat, the Association of Fishers and Hunters in Greenland. He lives in Nuuk, Greenland.

Aghveq angyiiquq: the gift of the whale on St. Lawrence Island, Alaska

By George Noongwook

Our relationship to bowhead whales is passed on from generations before, how we relate to them and other marine mammals. Our ancestors realized how important whales are because they provide sustenance to everyone. All people are just as important for survival because everybody gets involved by cooperating and sharing of this "gift from the Maker." In that sense it creates a bond for all people, the bowhead whale and thus elation and celebrations to follow once the season for gathering is completed. Gathering sustenance requires observations of the whole ecosystem including the heavens and stars that denote that a season will soon begin.

Prior to the beginning of this very important event, the whalers prepare their equipment and most importantly the *angyaq*, the skinboat, which literally means "for the pursuit of a gift." When you are handing a gift to someone, you say *Ang*, or "Here's your gift." The suffix *-yaq* means to pursue. Two hunting stories show this aspect of our relationship to bowhead whales in practice.

My first experience of these beliefs and stories happened when I went on a whale hunt as a boy. In 1961, my mother said, "We need food," so she had prepared hunting gear, warm clothing, and a white covering for my parka. Because my father was in basic training in Ft. Ord, California, which meant 6 months away from home, she said her cousin Harold Koonooka was coming on a dogsled to take me to Gambell to participate in whaling in hopes I would earn a share to bring home.

At Camp Tapghuq, where we stopped to have tea and refreshments, Harold told me when we got to Gambell he would drop me off at Homer Apatiki's. Homer was the youngest brother of my grandpa Noongwook. Homer and his wife had prepared food in a sealskin bag to take with me. He said tomorrow Tapisak, the eldest of the Kaningoks, and the crew that was going to take me whaling would come to get me. They are part of my clan, the Pugughileghmiit. Apparently, the clan met and decided which crew needed the most help. Tapisak came in the early morning and I was ready to go and took the bag and went with him. It was still dark but they kept an eye on the Big Dipper they call *Tungtut* (caribou) and they had markers on Gambell Mountain for when and where the sun, moon, and Tungtut would be and they knew when to get ready, to be aware that it would be daylight soon.

Atleghuq, the son of Tapisak, was our captain, Egmighun the engineer. Vernon Sloowko, Junior (everyone called him Junior) and myself manned the three sealskin floats of different sizes: small, medium, and large, with long rawhide ropes coiled around below the flippers. Allaggaq was the striker and he wore cold weather military gear (he had just returned from war, they told me later).

We shoved off and it was already daylight by that time and pretty soon we were approaching the ice pack and our captain saw a bay in the pack ice and wanted us to park. Not more than 12 feet away from the bay, a big bowhead whale popped out right beside our boat out of nowhere and stayed there, his neck right by the harpooner. We were waiting for Allaggaq to strike it, ready to throw the floats as instructed. It seemed like the whale was waiting for the strike, too. And after a while, I watched Allaggaq put his harpoon back in the boat and as soon as he did that the whale splashed him with water and got him all wet but none of us was splashed. The captain asked him, "Why didn't you harpoon the whale, it was a gift?"

Not long after someone else got a whale so I was able to get a share to take home with me. I think our striker had flashbacks about killing during the war because that was not the only time he had not struck a whale.

In the early 1970s, Nathan Noongwook restarted whaling from Pugughileq, on the south side of the island. I was in the Army at the time, but I heard about it from my family. Everyone wanted to be in his crew, since he was the first to get a whale there. I remember a hunt there from the mid-1980s. There were skin boats lined up east to west along the edge of the shorefast ice near Pugughileq. They all had their masts up and sails attached ready to shove off if a bowhead whale spouted close to your boat. The easternmost boat had the best vantage position because the bowheads were migrating east to west.

I was a crewman for Chester Noongwook (Tapghaghmii) our captain and Dwight Noongwook his younger brother was manning the outboard motor. My father Joseph Noongwook (Qukaghmii) was a middle crewman in charge of floats, and Dwight was also in charge of two floats. I was the lookout directly in front of the mast, and Elvin Noongwook was our striker.

As we waited, we noticed there was activity at the easternmost boat. They had shoved off and headed west with full sails and we looked for their quarry. When we saw the bowhead, it had already moved west away from them so they went back to wait for the next opportunity. The next boat shoved off under full sail toward the area where the whale was last seen. In the meantime, we began timing each spout to anticipate how far the whale was moving. The second boat missed their opportunity and went back. The next boat shoved off and went to where they last saw the bowhead. Realizing the whale had advanced further west, they went back. Since we had timed each spout and how far the whale had advanced, Chester gave us the order to shove off and head west with full sails.

Then from the rear, he said, "Angyiigaqiinkut sama!" "Look down, the whale is under our boat." As I was at the best vantage point and it was sunny, I could see the whale, maybe 20 feet north of us, suspended and watching us. Then I saw it begin to move toward our boat and climb toward us, so I told Elvin to get his harpoon ready to throw. But instead, it went under our boat and spouted on the left side, hugging our boat directly in front of Elvin, but he was warned never to harpoon if your quarry is on the left side because our float lines are on the right side of the boat. The whale knew the strike would not happen. It went down beneath our boat, then went back to the north side and suspended itself while looking at us and began to swim toward us. Elvin was ready but it went under our boat and spouted again hugging our boat.

The whale seemed to come up in front of every crew member. Finally at the last sequence I saw it coming toward our boat but it didn't spout so I directed us directly downwind and sure enough it spouted just ahead of us. We came next to it and Elvin threw his darting gun into the whale. His bomb didn't detonate, but the harpoon stayed in. Because of the cooperative nature of the hunt, other boats came to finish the hunt and bring home the gift of a whale.

An Aġviq foundation

By Nicole Kanayurak

Can you smell fall coming? It smells like the crisp ocean mist infiltrating the community that awakes excitement for early mornings making coffee and breakfast well before the sun rises for a boat load of whalers. Can you see whale snow? It looks like large airy cotton ball snowflakes falling that you hope stick to the ground and triggers alertness throughout the community that the bowhead whales are here without a word spoken.

The meaning of the bowhead whale, *aġviq*, to Inupiaq comes full circle in our lives. *Aġviq* is an Inupiaq child's first taste when parting from a mother's milk as the baby teethes on the *akikaq*, whale flipper, to soothe the gums. Whale bones arch around the cross of a whaling captain's grave.

The meaning of the *aġviq* is spiritual. It gives itself to us by the grace of God. A whaling family must abide by the Inupiaq values and have good spirit throughout the year in order to be successful at harvesting a bowhead whale. These Inupiaq values include humility, respect for nature, respect for your elders and one another, compassion, sharing, cooperation, and avoidance of conflict just to name a few. These values guide our lives and help shape our principles for food security.

As a young girl when serving the bowhead whale at our Thanksgiving and Christmas feasts, I was taught to serve the elders with the upmost respect and listen to the guidance they provide. These feasts are where our entire communities gather to celebrate as one (Fig. 31.2). As a server of the whale I was taught to share among everyone, give whale servings freely with a smile on my face, and work together with others as a team in the distribution. The whale nourishes our bodies, minds, and souls.

Whaling binds families together. Rather than family crests, Inupiaq have family whaling crew flags. These flags represent an Inupiaq family and their ancestors. The symbols marked on whaling gear identify what family it belongs to. My father's flag is designed and modeled after his Grandfather, Ralph Aveoganna's flag. I am able to identify what family a

FIGURE 31.2 Inupiaq mask. The nose and mouth of this mask resemble blowholes and mouth opening of a bowhead whale when seen from above, see Figs. 14.1 and 21.2 (this volume) for comparison. Provenance: Point Hope. Collection of Sheldon Jackson Museum, Sitka, Alaska (SJ-II-K-86). *Source: Photograph by Jacqueline Fernandez-Hamberg.*

person belongs to by the flag on their whaling crew jacket. And when a crew successfully harvests a whale, a portion of the whale, *niŋiq*, is given to each whaling crew member.

Whaling is not just about the harvest of a whale. It is about going out on the land to harvest caribou in order to make sinew with the caribou tendons, braid *ivalu* thread, and sew the fermented bearded seal skins that makeup the boat to go whaling. Whaling is about making sure the whalers have the white snow shirt parkas to blend in with the ice, so the whale does not see them.There are many intricacies to whaling as a livelihood and its meaning to our communities. In the Spring, landing the skin boat ashore, and apuġauti, the family/community feast that marks this occasion, represents the successful completion of the hunt by the whaling crew. It is a place and time of celebration for the safety of the whalers on land and the natural bounty given to the community. Mikigaq, a fermented whale delicacy, coupled with bread and nigliq and qaugaq (goose and duck) soup are served. A few weeks later is Nalukataq, the community blanket toss festival, where community members and iglaaqs (guests from neighboring communities) come to celebrate and share in the feast of the bowhead whale. Nalukataq is a place where I enjoy spending time with my grandmother and family in sharing a meal and then dancing just as our ancestors did.

This past year I learned how to braid the *ivalu* thread. Braiding of the ivalu is both an art form and a creation that literally keeps a boat afloat. The process, from the pulling out of the caribou tendons from the legs, cleaning and drying, to making the housing tip of the thread and adding thread by thread to make a strand length that goes from the tip of your nose stretched out to the end of your fingertips, takes about 4 months but is only a small morsel of life in a whaling community. Whaling is timeless and priceless to Inupiaq. There are countless hours, roles of individuals, and bellies fed. To name a few roles, there are the:

> caribou hunters, the seal hunters, the tool makers, the equipment providers, the bakers, the candy baggers, the skin boat sewers, the seal skin scrappers, the parka sewers, the servers, the harpooner, the aquti or boat steerer, the captain and his wife, the search and rescue radio communication liaison, the biologist, the advocate, and the drummers and dancers.

Each with their own knowledge and expertise as ornate as the ivalu (braider).

A person I look up to in my whaling family is my grandmother, Isabel Kanayurak, because she is one who wears many hats in our whaling community and serves selflessly and humbly. She makes the housing for the thread, she spends countless hours sewing boats late into the night, she is a search and rescue communication liaison, she has taught my family how to cut up and store the seal, whale, and other animals we harvest, she is a dancer, and now she is the elder the prays for the safety of the whalers as they go out whaling just as her mother did.

I was always shy as a kid but going to a whaling captain's home who harvested a whale to receive a serving makes you feel as welcome as going into your grandmother's house for your favorite dinner and dessert. Going to receive a serving is an opportunity to see your great aunts, uncles, and cousins at one time from both parent's families at one place without having to arrange a gathering. Opening up your home to share with the community, without turning anyone away, while very hard work, is pure joy.

My ancestor's housing foundations were literally made with whale bones and whale oil heated their homes to bring them warmth. Whaling continues to support the foundation of our homes and brings warmth to the community and all our relations.

The first bowhead whale after a 20 year ban

Told by Valerii Skhauge and recorded by Eduard Zdor

For many years our traditional hunt was greatly interfered with. We were not allowed to hunt the bowhead whale at all and we could only watch as a government ship captured gray whales. ... We were placed into a more confined community that did not permit us to practice our traditional ways. The pressures associated with these cultural changes have allowed the rise of poor behaviors (such as alcohol abuse) and a great erosion of our sense of cultural identity (especially among our young people). ... By returning to a more traditional way of life, we can restore our cultural identity and increase our sense of worth as individuals. We can hunt to our families. By again living with Nature, we will work together, as in times past. We will perform the hunting, sharing, festivals and obtain the foods that nourish our body and our spirit.

Statement by Vladimir Etylin, Chukotkan Indigenous activist and member of the Chukotkan and Russian legislatures, at the International Whaling Commission, October 16, 1997.

In Soviet times, we were forbidden to hunt bowhead whales. The authorities justified the ban because of the danger to the whalers, though no one asked our opinion. After the collapse of the USSR, the *Zvezdny* whaling ship no longer brought gray whales to our village of Sireniki. On the other hand, the prohibition on traditional whaling was lifted and we could again hunt whales ourselves.

We got our first gray whale in the middle of the 1990s. We were used to the taste of gray whale meat because the whaling ship brought gray whales to the village for many years. However, the elders always dreamed of a bowhead whale. One winter day we were discussing the upcoming summer hunting season. Pyotr Typykhkak came to our hunting base and proposed hunting a bowhead whale. In the spring, we prepared our equipment. The bowhead whale is much larger than the gray whale. To harpoon it, we use a larger toggled-head harpoon, a heavier shaft, and a stronger and longer line. Luckily, our Alaska relatives sent us darting guns as a gift.

I remembered the date of the first bowhead whale hunt because it was Border Guard Day. On May 28, it was sunny, the sea was calm like a mirror. We went to sea early in the morning. Our elders told us how to hunt and helped us prepare. Most importantly, Pyotr, an experienced whaler and a respected elder, went hunting with us.

We took five boats. Vladimir Typykhkak was the captain and helmsman of the first Whaleboat. His team was Valery Kanikhin—mechanic; Dmitry Typykhkak—first shooter; and Anatoly Kuilgin—second shooter. I was near the helmsman, helping navigate and find whales. In the second Whaleboat there were: Alexander Inmugie—captain and helmsman; Vitaliy Talpugye—the mechanic; Andrey Talpugye—the first shooter; and Vasily Palkintin—the second shooter. In the third Whaleboat were: Victor Mienkov—captain and helmsman; Sergey Gorbunov—mechanic; Konstantin Vashurin—first shooter; and Vladimir Talpuki—second shooter. In the *Lund*: Nikolai Panaugye—helmsman; Andrei Einelkut—shooter; and Pyotr Typykhkak as an elder. In the Progress: Oleg Isakov—helmsman; and Afanasy Makovnev—shooter.

We were lucky. We quickly found a bowhead whale about 5 km from the village. He appeared on the surface just next to our whaleboat. Dmitry was ready and immediately threw a "pushkan" (darting gun) into him. One buoy was tied to a line at 10 m, and the second at 50 m. The whale dived and, as often happens in such cases, did not appear on the surface for a long time. All boats went in different directions to be ready to finish off the whale. Finally, after 35–40 minutes, we found him. He was already dead. A projectile had hit the kidney, we later determined.

Everyone shouted for joy. Pyotr spoke words of gratitude in our language, which we always spoke on the hunt. He said it was a great day because our community finally got a bowhead whale again, for the first time since the 1970s. He reminded us that my father got the last bowhead whale before the ban. Pyotr said, finally, we again have the right to live as our ancestors and do what is our life.

After we found the whale, our whaleboat towed the whale, since we had harpooned it. The rest of the boats attached to the towline in front of us. The bowhead was towed like a gray whale, by the tail. Our parents used to tow headfirst. Because the current is strong, we towed the whale for around 4 hours.

Finally, in the early evening we landed the whale on shore. The excited villagers were already waiting there, having seen us with binoculars. The whale was huge and we tied ropes to the head and tail and tractors rolled the whale up the beach. Pyotr and several old women cut a small piece of whale, spoke ceremonial words of gratitude, and returned the piece to the sea. I was excited and busy pulling the whale and I only heard fragments of the thanksgiving.

Then we started to butcher the whale. Everyone who was there took part in the butchering. We cut off layers of mantak from the ridge to the abdomen with a width of about 50 cm and a length of about 4 m. We removed a few ribs and pulled out the stomach and intestines using a tractor. The organs are delicious food. We finished butchering around 4 or 5 in the morning. Each family took whatever they wanted from any part of the whale. There were no customary rules used to distribute mantak and meat.

We went home tired but happy. My mother was ecstatic, she laughed and seemed to cry. She said, "Finally, I can't believe that the village got a bowhead whale again." From childhood, she was very fond of the kidneys of a bowhead whale. Although I am accustomed to the taste of gray whale meat and mantak, I prefer bowhead. My youngest daughter tasted the mantak and meat and said, "Yummy."

The feeling of the whole community working together and sharing was unforgettable. This is how our ancestors lived, this is how we live, and I hope our children will live like that.

Arvangniarniq: bowhead whale hunt in Clyde River, Nunavut, August 2014

By Sandra Inutiq

As soon as I heard my home community of Clyde River was interested in a bowhead whale hunt, I knew I wanted to be there, to be a part of my community in such a significant moment. In the months leading up to it, I was very excited; a week prior to the hunt, I was starting to lose sleep with excitement.

I knew it would be a monumental moment, a reclamation of part of ourselves. Bowhead whaling is a practice almost forgotten, but still alive in stories and oral history. The traditional hunt is unlike any other because it tests the cohesiveness of a community. It takes a large team to pull it off: a definitive characteristic of Inuit. Teamwork is how we had not only survived but thrived in our environment for thousands of years.

Only one member of the team, Niore Iqalukjuak, had taken part in such a hunt before. But I believed in our abilities as a hunting society, that the hunter instinct would kick in. The skill, astuteness, intuitiveness, alertness, even if everyone was not known to be *angunatturiktuq* (an able hunter), would work—especially if everyone is to play a role: *atuutiqaqtuinnauniaqtugu*.

At a discussion about the strategy of the hunt, I was overwhelmed by the terminology I had not been exposed to. The hunters spoke of the bowhead whale hunt and the tools and methods that were to be used: *anguvigaq, niutaq, aaqsiiq*: a whole sector of language related to this hunt that had been sleeping was awakened in the community. It was a moment to see the interconnectedness of culture, language, and identity, and the significance of our oral history.

The morning the hunt began, we were told to meet at the breakwater in the morning. There was a buzz in the air, and excitement for what was to possibly come. The sun was out smiling down at the hunters and the community. When it was time to leave, people were quick to get in the boats, antsy with anticipation.

We stopped down the inlet from the community in Akulaanga. Some climbed the hill Pamialluat to see the ice conditions ahead. People ate, talked, teased, and laughed. There was energy like no other, the collaborative spirit was in the air. We then headed for Umiujaq, a small island in front of Cape Christian in the area where the hunt was planned, and landed the boats. We sent a climber to the top of the hill to scope out the scene for *arviit* (bowhead whales, plural of *arvik*). Larry Kautuq caught a seal. We ate, and then set up camp. After the tents were up, it was clear we would not be leaving to go hunting that evening. Everyone settled in, played cards, or mingled outside.

The next day, the wind and ice conditions were still not ideal, another test of our patience. We waited until that evening, when the hunting team decided that Captain David Iqaqrialu would drive his four-wheeler to an area further north of the cape. There was less ice there, there had also been sightings of *arviit*. First he spotted some large ones, excitement growing among the hunters as he described the whales to us over the VHF radio. He finally described a third one which seemed smaller, and there was the complete silence of anticipation. The captain would direct the hunt from the hill, which seemed selfless as he would not directly partake. He would have the advantage of perspective from which to direct the parties, while Sandy Kautuq the assistant captain directed the hunt on the water.

The captain announced that Sandy's canoe was headed toward a bowhead. The chase was on and the adrenaline kicked in. In our boat, in the moments when a bit of panic would set in, Niore would chuckle at the chaos. Sometimes he would give calm advice over the radio. Jerry Natanine was so excited he could not restrain himself from being comical and silly. He made us laugh throughout. Noah Kautuq kept singing *akuqtujuuk anngutivuuk*, a song of hope, even as in true Inuk fashion he improvised a way to put together the pieces of the whale gun that had expanded overnight in the humidity of the boat. Our youngest, Anthony Iqaqrialu, cheered on and encouraged each harpooner.

Then Sandy's boat was ready, and in no time Roger Etuangat harpooned it. Things seemed chaotic for the first while as the hunters and the crews worked to catch their senses. Overwhelming excitement had overtaken them. The hunt of this colossal mammal was clearly more of a challenge than we all had anticipated. After a while, the hunters seemed to get into a rhythm with the whale. Self-doubt seemed to dissipate and the hunters emerged. That energy seemed to overtake everyone involved. At last, Esa Piungittuq set off the flare gun to signal the *arvik* was dead.

How beautiful the whale was. Feeling emotional, I thanked it for giving itself, told its spirit we deeply appreciate you. The cheers could be heard on the radio. It looked like the whole community was strewn across the hill with their arms up in the air. What an image to have imprinted in the mind.

We could not resist tasting the *maktak*, sharing the elation and gratitude. Bringing the whale home, I had to remind myself to be patient and understanding that this was a moment and experience which serves a purpose for every individual as it does for me. Our hunting and land skills have been largely compromised, against a backdrop of the idea of abandoning our ways having faced strong assimilation efforts. Clearly for some of us, our hunting skills having been more comprised than others. This is a reflection of our society.

Taking part and witnessing the hunt was truly an amazing experience. It had a profound effect on me, that can only be described by saying a sleeping part of my spirit was awakened. I have a whole new respect for our existence as a hunting society, and for those during the hunt who managed to shine and show their adeptness and quick judgment to make this a successful hunt.

Bowhead whaling in Greenland

By Bjarne Ababsi Lyberth

The earliest known evidence of whaling in Greenland is a carving in a piece of bone found near present-day Moriusaq in the Avanersuaq area (Thule area). The piece of bone, about 5 cm long, has a carving that shows a boat that seems to be framed, presumably an umiaq (Fig. 31.3). In the front is a person with a harpoon and paddlers are behind him. There is plenty of concrete evidence of whaling done in Greenland by the Thule culture, the latest Inuit immigrants who arrived around CE 1200, like special harpoon heads and the iconic drysuit made of skin. After attaching a float to the whale with a harpoon, the hunter would jump on the whale to kill it with a lance. The drysuit would keep him afloat should he fall off or the whale dive.

Despite the archaeologic evidence, there is little mention of whaling nor bowhead whaling in the oral traditions. In the national romantic trilogy "Qooqa," "Tulluartoq I and II," by Ole Brandt, first published in 1971, describing life in a hunting community in West Greenland in the early 18th century, the heroic family sometimes catches minke whales from the beach. As minke whales are smaller and abundant in southwestern Greenland, they may have been preferred over bowheads that are much bigger and only abundant in the Disko Bay area. Likewise the abundance of beluga whales and narwhal as well as

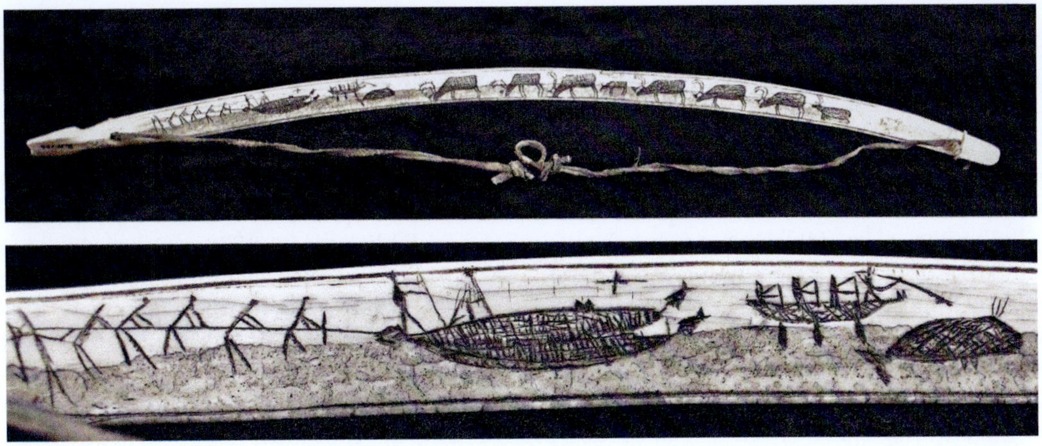

FIGURE 31.3 Iñupiat mask representing a bowhead whale. Note the resemblance of the nose and mouth of this mask to that of the blowhole and mouth of a bowhead whale (see Fig. 18.1, this volume, for comparison). Collection of the Sheldon Jackson Museum (Sitka, Alaska).

occasional occurrences of harbor porpoise, dolphins, pilot whales, and other smaller whales may have played a role in the preference for more conveniently sized whales.

Whaling may have occurred on occasions when suitable size whales were spotted at a convenient distance from the community. The arrival of seasonal commercial European whalers in the 17th century meant a decrease in whale populations but also easy access to meat and mattak of whales caught by commercial whalers, whose main objectives were oil from the blubber, baleen, and other special products for the European and North American markets.

By the beginning of the 20th century, Greenlanders' access to meat, blubber, and mattak from larger whales were dependent on supplies from harpoon vessels. The use of umiaqs (longer skin boats that were traditionally used for whaling) was limited to a few families, and access to whales was thus probably limited partly due to overharvest by European whalers and partly because knowledge about traditional whaling was more or less lost. But commercial whaling was losing its significance: products from whales were being replaced by other products, and decreasing whale populations probably played a role, too. But the Greenlanders still needed meat and mattak from whales, so on September 5, 1919, hunter Jeremias Geisler from Skansen on Disko Island proposed acquiring a steam engine-powered harpoon boat to be placed in Godhavn (now Qeqertarsuaq). The proposal was forwarded to the North Greenland Council, and forwarded again to the colonial Agency of Greenland, which treated the proposition seriously, but nothing really happened until 1924. In the 1920s, a transition from hunting to fisheries was taking place in southwestern Greenland and meat supplies became scarce, but blubber was still important for oil production, so the Agency bought the Norwegian whaler *Sonja*, which had been operating in the Antarctic. The supplies of whale meat, blubber, and mattak to the communities was a success. Best known from this era is the song "Sunnia Kalippoq" ("*Sonja* Is Towing") by Peter Olsen, which later became the tune of the Kalaallit Nunaata Radioa (national radio of Greenland).

In 1938, three brothers from Aasiaat bought the fishing vessel *Auveq* from Denmark. It was used for shrimp fishing and also for hunting walrus, beluga, and narwhals. In 1948 it was supplied with a harpoon cannon and thus became the first Greenlandic-owned harpoon vessel.

Following the establishment of the International Whaling Commission and requirements that large whales be killed by grenade harpoons, and also the increased importance of inshore fisheries, vessels with harpoon cannons have been acquired by many Greenlanders.

Following a moratorium on bowhead whales established in 1938, Greenland was granted a quota of four bowhead whales in 2009. This was celebrated together with the introduction of the Self Rule Government the same year. Two bowhead whales were harvested in Disko Bay and the meat and mattak were distributed to all of Greenland as part of the celebration.

Whaler Otto "Aqqalu" Mathiesen owns a harpoon vessel and has been whaling for minke, fin, and humpback whales for decades. He recounts his first hunt for a bowhead whale in 2009:

In 2009, the bowhead hunt reopened for the first time since 1938. I was of course thrilled to be asked to participate in the hunt. Two whales were to be distributed among all Greenlanders (as part of the celebration of the Self Rule Government). We caught them west of Qeqertarsuaq (Disko Island). After a great butchering we brought meat to Ilulissat that was distributed to Greenlandic towns with the help of hunting officers and municipality workers. I think all Greenlandic towns got a share of the first bowhead back then.

I participated in five successful hunts in total, two boats each hunt, sometimes three. Butchering is the challenge, though the hunt isn't that difficult. Our vessel back then had a strong winch so we could remove the flippers by lifting them before bringing the whale on the beach. Once on the beach, the flippers are so hard to take off.

Although people also didn't know the bowhead well, they did enjoy the meat and mattak on national day probably all over Greenland, but the last couple of years it's not so appealing. Though we see many of them every year, there's nowhere to sell and butcher. The baleen is easy to sell though: one whale has about 500 of them, and it's 1000 kroner each [about 150 US dollars as of September 2019].

We don't know the bowhead as well as minke and fin whales, we just picked one that was surfacing continuously (pussinnartoq) and put the harpoon in it when we were close enough. We were two vessels. They're much easier compared with fin whales that really put up a fight, but sometimes you only need to give them one harpoon. I have no count of how many fin whales I've caught.

We were picked as whalers to catch the bowhead together with whalers from Aasiaat and Qeqertarsuaq (Jess Johansen).

The hunt took place around first of May. The municipal workers had removed the ice foot on the beach and, using wires and hoists, we got them on the beach. I don't know where we got the strength. It took three days to butcher, our crew together with municipal workers from Qeqertarsuaq. We also took off the jaws to be raised somewhere.

The Greenlandic name for bowhead whale is "arfivik" which means "real whale," but as Aqqalu mentions challenges such as butchering, transport, freezing capacity as well as the Greenlandic people's preferences for other whale species has limited our interest for arfivik. None has been caught in Greenland since 2015.

Discussion

The significance of bowhead whaling for Arctic coastal communities from Chukotka to Greenland is beyond question. The whaling tradition has survived government-imposed interruptions in Chukotka and Nunavut and regulatory threat in Alaska. The practice of whaling continues to evolve, with new boats and tools, shifts in timing and location brought about by a changing climate, and even changing food preferences, as in Greenland. At the same time, the bowhead whale hunt continues to create deeper meaning for those who practice it, from the teamwork of a boat crew and their interactions on the water with the whales, to the preparations and sharing and hard work that take place onshore. These themes are apparent in the writings of both Indigenous authors and visiting scholars.

Throughout the range of the bowhead whale, whaling communities have taken back control over their practices, defining whaling in their own terms for their own purposes, rather than attempting to suit the expectations or views of outsiders with regard to what constitutes "tradition" or "respect" regarding hunting. This effort remains a work in progress. The growing role of Indigenous authors and scholars and whalers in research and writing about bowhead whales and bowhead whaling is but one indication of the importance that the human–whale relationship continues to have in sustaining, enriching, and bringing together the communities of Arctic coasts.

Acknowledgments

We are grateful to all the whalers and their communities in the Arctic who have supported, participated in, and contributed to scholarly work on whaling cultures. Without their generosity, kindness, openness, and patience, little research of this kind would have been possible.

References

Albert, T.F., 2001. The influence of Harry Brower, Sr., an Iñupiaq Eskimo hunter, on the bowhead whale research program conducted at the UIC-NARL facility by the North Slope Borough. In: Norton, D.W. (Ed.), Fifty More Years Below Zero. University of Alaska, Fairbanks, pp. 265–278.

Ashjian, C.J., Braund, S.R., Campbell, R.G., George, J.C., Kruse, J., Maslowski, W., et al., 2010. Climate variability, oceanography, bowhead whale distribution, and Iñupiat subsistence whaling near Barrow, AK. Arctic 63 (2), 179–194.

Blackman, M.B., 1989. Sadie Brower Neakok: An Iñupiaq Woman. University of Washington, Seattle, WA.

Bockstoce, J.R., 1977. Eskimos of Northwest Alaska in the Early Nineteenth Century. University of Oxford Pitt Rivers Museum, Oxford, UK.

Bodenhorn, B., 2003. Fall whaling in Barrow, Alaska: a consideration of strategic decision-making. In: McCartney, A.P. (Ed.), Indigenous Ways to the Present: Native Whaling in the Western Arctic. University of Utah, Salt Lake City, UT, pp. 277–306.

Bodfish Sr., W., 1991. Kusiq: An Eskimo Life History from the Arctic Coast of Alaska. University of Alaska, Fairbanks, AK.

Bogoslovskaya, L., Slugin, I., Zagrebin, I., Krupnik, I., 2016. Maritime Hunting Culture of Chukotka: Traditions and Modern Practices. National Park Service Shared Beringian Heritage Program, Anchorage, AK.

Boyd, L., 2018. Tim Pitsiulak: Drawings and Prints From Cape Dorset. Pomegranate, Portland, OR.

Brewster, K., 2004. The Whales, They Give Themselves: Conversations With Harry Brower, Sr. University of Alaska, Fairbanks, AK.

Brower, C.D., 1942. Fifty Years Below Zero: A Lifetime of Adventure in the Far North. Dodd, Mead, New York.

Burch Jr., E.S., 1981. The Traditional Eskimo Hunters of Point Hope, Alaska: 1800–1875. North Slope Borough, Barrow, AK.

Chance, N.A., 1990. The Iñupiat and Arctic Alaska: An Ethnography of Development. Holt, Rinehart and Winston, New York.

Clarke, J.T., Kennedy, A.S., Ferguson, M.C., 2016. Bowhead and gray whale distributions, sighting rates, and habitat associations in the eastern Chukchi Sea, summer and fall 2009–15, with a Retrospective Comparison to 1982–91. Arctic 69 (4), 359–377.

Druckenmiller, M.L., Eicken, H., George, J.C., Brower, L., 2010. Assessing the shorefast ice: Iñupiat whaling trails off Barrow, Alaska. In: Krupnik, I., Aporta, C., Gearheard, S., Laidler, G.L., Holm, L.K. (Eds.), SIKU: Knowing Our Ice: Documenting Inuit Sea-Ice Knowledge and Use. Springer, New York, pp. 203–228.

Edwardson, D.D., 2004. Uqsruaġnaq: Whale Snow. Charlesbridge, Watertown, MA.

Fienup-Riordan, A., 2000. Where the Echo Begin: and Other Oral Traditions From Southwestern Alaska Recorded by Hans Himmelheber. University of Alaska, Fairbanks, AK.

Foote, B.A., 1992. The Tigara Eskimos and Their Environment. North Slope Borough, Barrow, AK.

Foote, B.A., 2009. Point Hope Alaska: Life on Frozen Water: Photographs 1959–1962. University of Alaska, Fairbanks, AK.

Freeman, M.M.R., Wein, E.E., Keith, D.E., 1992. Recovering Rights: Bowhead Whales and Inuvialuit Subsistence in the Western Canadian Arctic. Studies on Whaling Number 2. Canadian Circumpolar Institute and Fisheries Joint Management Committee, Edmonton.

Freeman, M.M.R., Bogoslovskaya, L., Caulfield, R.A., Egede, I., Krupnik, I.I., Stevenson, M.G., 1998. Inuit, Whaling, and Sustainability. Altamira Press, London.

Gearheard, S., Holm, L.K., Huntington, H., Leavitt, J.M., Mahoney, A.R., Opie, M., et al., 2013. The Meaning of Ice: People and Sea Ice in Three Arctic Communities. International Polar Institute, Hanover, NH.

George, J.C., Huntington, H.P., Brewster, K., Eicken, H., Norton, D.W., Glenn, R., 2004. Observation on shorefast ice dynamics in Arctic Alaska and the responses of the Iñupiat hunting community. Arctic 57 (4), 363–374.

George, J.C., Druckenmiller, M.L., Laidre, K.L., Suydam, R., Person, B., 2015. Bowhead whale body condition and links to summer sea ice and upwelling in the Beaufort Sea. Prog. Oceanogr. 136, 250–262.

Heide-Jørgensen, M.P., Laidre, K., 2006. Greenland's Winter Whales: The Beluga, the Narwhal and Bowhead Whale. Ilinniusiorfik Undervisningsmiddelforlag, Nuuk, Greenland.

Hess, B., 1999. Gift of the Whale: The Iñupiat Bowhead Hunt, a Sacred Tradition. Sasquatch Books, Seattle, WA.

Heyes, S.A., Helgen, K.M. (Eds.), 2014. Mammals of Ungava & Labrador: The 1882–1884 Fieldnotes of Lucien M. Turner Together With Inuit and Innu Knowledge. Smithsonian, Washington, DC.

Huntington, H.P., 1992. Wildlife Management and Subsistence Hunting in Alaska. University of Washington, Seattle, WA.

Ikeya, K., 2013. Whale hunting and use among the Chukchi in northeastern Siberia. In: Kishigami, N., Hamaguchi, H., Savelle, J.M. (Eds.), Anthropological Studies of Whaling. Senri Ethnograph. Stud., 84. pp. 121–136.

Ikuta, H., 2011. Embodied knowledge, relations with the environment, and political negotiation: St. Lawrence Island Yupik and Iñupiaq dance in Alaska. Arctic Anthropol. 48 (1), 54–65.

Jolles, C.Z., 2002. Faith, Food, and Family in a Yupik Whaling Community. University of Washington, Seattle, WA.

Kishigami, N., 2010. Climate change, oil and gas development, and Inupiat whaling in northwest Alaska. Études Inuit Stud. 34 (1), 91–107.

Kishigami, N., 2013. Sharing and Distribution of Whale Meat and Other Edible Whale Parts by the Inupiat Whalers in Barrow, Alaska, USA, National Museum of Ethnography, Osaka, Japan.

Lantis, M., 1938. The Alaskan whale cult and its affinities. Am. Anthropol. 40 (3), 438–464.

Lowenstein, T., 1981. Some Aspects of Sea Ice Subsistence Hunting in Point Hope, Alaska: A Report for the North Slope Borough's Coastal Zone Management Plan. North Slope Borough, Barrow, AK.

Lowenstein, T., 1993. Ancient Land: Sacred Whale: The Inuit Hunt and Its Rituals. Farrar, Straus and Giroux, New York.

Moore, S.E., Huntington, H.P., 2008. Arctic marine mammals and climate change: impacts and resilience. Ecol. Appl. 18 (2), s157–s165.

Nelson, R.K., 1981. Harvest of the Sea: Coastal Subsistence in Modern Wainwright: A Report for the North Slope Borough's Coastal Management Program. North Slope Borough, Barrow, AK.

Noongwook, G., The Native Village of Savoonga, the Native Village of Gambell, Huntington, H.P., George, J.C., 2007. Traditional knowledge of the bowhead whale (*Balaena mysticetus*) around St. Lawrence Island, Alaska. Arctic 60 (1), 47–54.

Quakenbush, L.T., Citta, J.J., George, J.C., Small, R.J., Heide-Jørgensen, M.P., 2010. Fall and winter movements of bowhead whales (*Balaena mysticetus*) in the Chukchi Sea and within a potential petroleum development area. Arctic 63 (3), 289–307.

Sakakibara, C., 2017. People of the whales: climate change and cultural resilience among Iñupiat of Arctic Alaska. Geograph. Rev. 107 (1), 159–184.

Sakakibara, C., 2020. Whale Snow: Iñupiat, Climate Change, and Multispecies Resilience in Arctic Alaska. University of Arizona, Tucson, AZ.

Stevenson, M.G., 1997. Inuit, Whalers, and Cultural Persistence: Structure in Cumberland Sound and Central Inuit Social Organization. Oxford University, Oxford.

Thompson, R., Ahtuangaruak, R., Cannon, C., Kingik, E., 2012. We will fight to protect the Arctic Ocean and our way of life. In: Banerjee, S. (Ed.), Arctic Voices: Resistance at the Tipping Point. Seven Stories, New York, pp. 301–334.

Turner, E., 1990. The whale decides: Eskimos' and ethnographer's shared consciousness in the ice. Études Inuit Stud. 14, 39–42.

Turner, E., 1993. American Eskimo celebrate the whale: structural dichotomies and spirit identities among the Iñupiat of Alaska. Drama Rev. 37 (1), 98–114.

VanStone, J.W., 1958. Commercial whaling in Arctic Ocean. Pac. Northwest Q. 49 (1), 1–10.

Wohlforth, C., 2004. The Whale and the Supercomputer: On the Northern Front of Climate Change. North Point, New York.

Worl, R., 1980. The North Slope Iñupiat whaling complex. In: Kotani, Y., Workman, W. (Eds.), Alaska Native Culture and History. Senri Ethnological Studies 4. National Museum of Ethnology, Osaka, pp. 305–320.

Zumwalt, R.L., 1988. The return of the whale: Nalukataq, the Point Hope whale festival. In: Falassi, A. (Ed.), Time out of Time: Essays on the Festival. University of New Mexico, Albuquerque, NM, pp. 261–276.

CHAPTER

32

Current indigenous whaling

Robert Suydam and J.C. George

Department of Wildlife Management, North Slope Borough, Utqiaġvik, AK, United States

Introduction

Bowhead whales have been hunted by Inuit for at least several millennia in the Arctic and Subarctic. Although exact estimates vary, whale hunting became central to these cultures at least 1000 years ago (Whitridge, 1999) (Fig. 32.1). Stoker and Krupnik (1993) provided a thorough overview of the origins and development of precontact (i.e., prior to interactions with Europeans and Americans) whaling, including methods and technology. Commercial hunting of bowheads initiated in about 1540 near southern Labrador, Canada and then moved to Svalbard, Norway, when the whales near Labrador were overhunted (Ross, 1993). Operations expanded to Davis Strait and Hudson Bay by the late 1600s and 1860, respectively (Ross, 1993). Eventually, bowheads were discovered by commercial whalers in the North Pacific in the 1840s in both the Okhotsk and the Bering Seas (Bockstoce and Burns, 1993). The commercial harvest of bowheads resulted in the substantial decline of all stocks that in turn decreased the availability of bowheads and caused major disruption to many Inuit communities. Commercial bowhead whaling finally ceased in the early 1900s due to the discovery of petroleum, decimation of the stocks, and the discovery of alternative materials to replace baleen (Bockstoce, 1986; Woodby and Botkin, 1993). Chapter 33 provides additional details about commercial whaling on bowheads on all four stocks. The stocks are described in detail in Chapter 3.

Exposure of the Inuit people in the Pacific to Yankee whalers and missionaries decimated Inuit populations and significantly altered cultures in many intentional and unintentional ways. Native communities suffered tremendously because of exposure to disease, introduction of alcohol, and resource overutilization, including on bowheads (Stoker and Krupnik, 1993; Bockstoce, 1986; Chapter 33). Stoker and Krupnik (1993) summarized the many alterations that European and American commercial whalers had on Inuit hunters and whaling methods. Even with those challenges, subsistence hunting for bowheads continued, especially in Alaska and Chukotka, and communities survived in part by relying on other species and resources.

The Bowhead Whale
DOI: https://doi.org/10.1016/B978-0-12-818969-6.00032-7

FIGURE 32.1 Successful crew and their family celebrate the harvest of *aġviq* (bowhead whale) at Utqiaġvik. Inuit peoples of the Arctic have depended on the bowhead harvest for food, fuel, and building materials for millennia.These societies also developed a deep cultural and spiritual connection with the whale. *Source: Photo, North Slope Borough, Utqiaġvik.*

Despite the negative impacts of commercial whaling, new technologies and weapons were also introduced to Arctic communities that helped make subsistence hunting more successful and efficient (Ross, 1993; Bockstoce and Burns, 1993). With the cessation of commercial whaling, two bowhead stocks have made a strong recovery while two have not. Bowhead populations adjacent to West Greenland, in parts of Arctic and Subarctic Canada, western and northern Alaska, United States, and Chukotka, Russian Federation, are large enough to sustain harvests. Those harvests are managed by the International Whaling Commission (IWC: for Greenland, Alaska, and Chukotka) to conserve whale stocks and to meet the nutritional and cultural needs of Inuit. The hunt in Canada is managed domestically (see Chapter 39).

Chapter 31 provides details about Inuit attitudes about subsistence hunting with some mention of methods. In this chapter, we present information about the level of current subsistence whaling of bowhead whales, over the past 40–50 years. We also provide information about the location and dates of the hunts and the size and sex of whales landed.

Data

Data for Chukotka, Greenland, and Canada were obtained from the IWC. For Alaska, biologists from the US National Marine Fisheries Service collected harvest data from 1973to 1981.

During those years, scientists routinely collected data and samples at Utqiaġvik (formerly called Barrow) and Point Hope but only intermittently at the other villages (Braham, 1995). Hunting captains or the Alaska Eskimo Whaling Commission (AEWC) provided data for the other villages (Braham, 1995). The North Slope Borough began collecting harvest data and samples in 1982 in collaboration with the AEWC and continues through the present (Albert, 1988). Until ~1984, biologists were stationed in most villages to measure and collect biological samples from harvested whales. After that time, scientists primarily measured and sampled whales harvested at Utqiaġvik. Beginning regularly in the late 1990s, whales were closely examined at Kaktovik. By 2007 whales were also regularly measured and sampled at Gambell and Savoonga on Saint Lawrence Island with the assistance from the Alaska Department of Fish and Game, the University of Alaska Marine Advisory Program, and Alaska Sea Grant. The AEWC and individual captains provided data, typically, including sex, standard length (i.e., straight-line measurement from the tip of the rostrum to the fork in the tail), date landed, and fate of struck and lost whales, from the other villages. Whales examined by biologists often included considerably more data such as additional measurements and biological samples. Because so many more whales have been landed in Alaska, we emphasize data from those hunts in this chapter.

East Greenland, Svalbard, Barents Seas and Okhotsk Sea stocks

No subsistence hunting occurs or occurred on whales from the East Greenland, Svalbard, Barents Seas stock. In the Okhotsk Sea, bowheads ceased being hunted for subsistence probably in the 1800s (Stoker and Krupnik, 1993) because of the decimation of the stock from commercial whaling. The populations in both locations continue to be small (see Chapter 6), and no hunting is currently being considered, to the best of our knowledge.

East Canada–West Greenland stock

Greenland

To help adjust for the considerable environmental variability, the whale harvest in West Greenland is a multispecies hunt, including fin, humpback, minke, and bowhead whales. However, bowheads are taken in low numbers by hunters in western Greenland. A quota was established for the Greenlandic portion of that hunt by the IWC beginning in 2007 (https://iwc.int/greenland). Hunting is managed domestically by the Ministry of Fisheries, Hunting and Agriculture and supervised locally by the Fisheries Licence and Control Authority, but with the involvement of the Organization of Fishermen and Hunters, the municipalities, the Greenland Institute of Natural Resources, and the Ministry of Nature and Environment (Chapter 39). A total of nine whales, one male, seven females, and one with sex undetermined have been taken off West Greenland, primarily near Disko Bay, since 1973, but only one of those between 1973 and 2007. The hunts in Greenland occur from small whaling boats (~11 m long) using deck-mounted harpoon cannons and penthrite projectiles (see Greenland's description of hunt: https://iwc.int/greenland). Because of substantial tides, small remote communities, and the large size of bowheads, it is difficult to butcher and process those whales. Edible parts from whales are primarily sold locally in Inuit communities in Greenland (https://iwc.int/greenland).

Canada

In Canada the hunt is managed by the Department of Fisheries and Oceans (DFO). Few whales were harvested during the 1950s to the 1990s, one to seven whales per decade, until the stock had grown to a size where takes would be sustainable. Since 1970, 45 bowheads have been landed, including 14 males, 18 females, and 13 with sex undetermined. The first recent and officially sanctioned hunt occurred in 1996. Kishigami (2015) provides some details for the 24 whales hunted and landed with Canadian government permission between 1996 and 2014. DFO works with indigenous hunters to evaluate conservation risk, approve the hunts, and manage the harvests (Chapter 39). The hunt in Canada occurs from small open skiffs, 5–7 m long, with handheld weapons (i.e., shoulder guns, harpoons, and rifles; Kishigami, 2015). The products from the hunt are shared locally and with neighboring communities.

Bering–Chukchi–Beaufort Seas stock

Whales from this stock are taken primarily in Alaska and Chukotka. The quota for the subsistence harvest (other than in Canada) has been set by the IWC since 1978 (Table 32.1). The quota was initially very low but has increased over time as the IWC better understood the need for bowhead muktuk (i.e., skin and a thin layer of outer blubber), meat and other parts of the whales, and the new data showing a growing and sizable whale population (see Chapter 39). The landed quota has remained stable since 1995 (Table 32.1). Fig. 32.2 is a map showing the locations of villages that hunt bowheads from the BCB seas stock.

Canada

The most recent hunts in northwest Canada occurred during the summer of 1991 and 1996 by residents of Aklavik, NWT, Canada, after a hiatus in hunting bowheads of about 50 years (Kishigami, 2015). They harvested a male whale, ~11 m long near Shingle Point in the western Mackenzie River Delta in each of those years. The butchering proved to be very difficult because of challenges getting the whale near to shore and transporting the whale parts back to town, ~110 km away. Primarily because of those challenges, no other bowheads have been harvested in northwest Canada since that time.

Chukotka

In Russia, Inuit hunters in Chukotka are the only people allowed to hunt whales in that country. Seven bowheads were landed between 1972 and 1975. From 1998 to 2018,

TABLE 32.1 International Whaling Commission quota for Bering–Chukchi–Beaufort Seas bowheads.

Year	Landed whales	Strikes	Carryover strikes (i.e., previously unused strikes)	Reference
1978	14	20	–	International Whaling Commission (IWC, 1979)
1979	18	27	–	International Whaling Commission (IWC, 1979)
1980	18	26	–	International Whaling Commission (IWC, 1980)
1981–83	45 total	65 total	–	International Whaling Commission (IWC, 1981)
1984–85	–	43 total	–	International Whaling Commission (IWC,1984)
1985–87[a]	–	26/year	≤6/year	International Whaling Commission (IWC, 1986)
1987–88	–	32, 35, respectively		International Whaling Commission (IWC, 1988)
1989–91	41/year	44/year	≤3/year	International Whaling Commission (IWC, 1989)
1992–94	41/year	54/year	Any unused strikes from the previous block up to 10% of the total allowed in that block	International Whaling Commission (IWC, 1992)
1995–98	204 total	68/67/66/65[b]	10/year	International Whaling Commission (IWC, 1995)
1998–2002	280	67/year	15/year	International Whaling Commission (IWC, 1998)
2003–07	280 total	67/year	15/year	International Whaling Commission (IWC, 2003a,b)
2008–12	280 total	67/year	15/year	International Whaling Commission (IWC, 2007)
2013–18	336 total	67/year	15/year	International Whaling Commission (IWC, 2012)
2019–25	392 total	67/year	33/year (i.e., 50% of annual strike limit)	International Whaling Commission (IWC, 2018)

[a]1985 quota revised in 1986.
[b]68 strikes in 1995, 67 in 1996, 66 in 1997, and 65 in 1998; in 1997 a new quota was established for 1998–2002. Thus the strike limit for 1998 was 67/year.

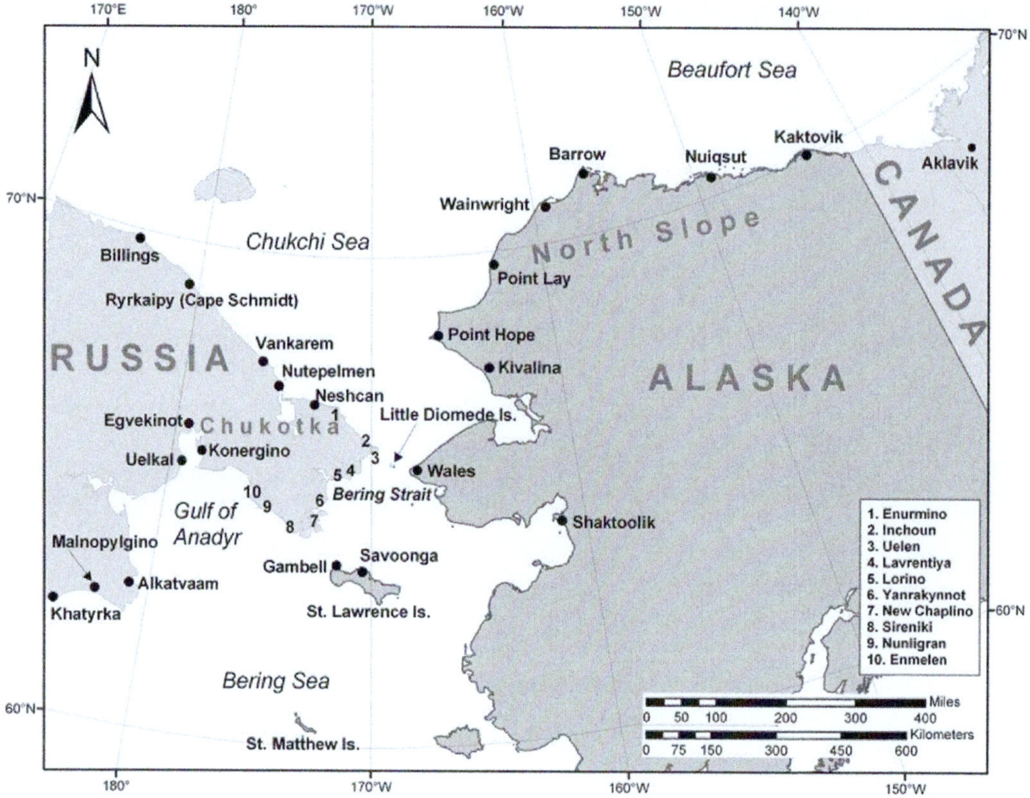

FIGURE 32.2 Map of bowhead hunting communities adjacent to the Bering, Chukchi, and Beaufort Seas (DeMaster and Rauch, 2013).

24 more were landed. Gray whales are taken in much larger numbers, which provides resources for Chukotkan villages to meet their nutritional and cultural needs. Both species are taken from small motorized skiffs or skin boats (~3 to 5 m long) during summer months (https://iwc.int/russian-federation).

Alaska

Some harvest data from Alaska have previously been published (e.g., Marquette et al., 1982), while others are submitted annually to the IWC (e.g., Suydam et al., 2016). Here we present data from 1974 to 2018.

The subsistence hunts typically take place in spring and autumn as whales migrate between the Bering and Beaufort Seas. Most of the hunts are subjected to considerable environmental interference from weather (e.g., strong winds, fog, and extreme temperatures), sea ice thickness, large seas, and sea ice concentration. The success of the hunt is greatly affected by these factors and shows considerable variation by year and location

FIGURE 32.3 A darting gun, with attached harpoon and connected by a rope to a float, used for hunting bowheads. *Source: Photo, North Slope Borough, Utqiaġvik.*

(George et al., 2003). Locations of the hunt are usually within 50 km of the village; however, some villages stage their hunts at more distant locations. During the spring, Gambell hunters travel ~75 km to the southwest corner of Saint Lawrence Island. During the autumn, Nuiqsut hunters travel 130 km to Cross Island. Both of those locations are closer to the migratory path of bowheads. The hunters use either walrus or seal skin covered boats or small motorized skiffs (~5 to 8 m long), and darting and shoulder guns are the primary weapons used (Fig. 32.3; Fig. 34.1).

Numbers and timing of whales harvested

In total, 1493 whales were landed at 12 Alaskan Native villages (Fig. 32.2) between 1974 and 2018 (Table 32.2). The village harvests ranged from 748 bowheads landed at Utqiaġvik to two landed at Little Diomede and one landed at Shaktoolik.

At present, there are 11 AEWC villages. The villages that hunt during the spring migration include (from southwest to northeast) Gambell, Savoonga, Little Diomede, Wales, Kivalina, Point Hope, Point Lay, Wainwright, and Utqiaġvik. Point Lay became a member of the AEWC in 2008 and landed its first whale in more than 70 years in 2009. Those villages typically hunt from the edge of shorefast ice. The spring hunt occurs as whales migrate northeast through the spring lead system in the sea ice along the northwestern coast of Alaska, typically from early April to early June. Shaktoolik is not a member of the AEWC but landed a bowhead whale in May 1980. In that year, there were very unusual ice conditions that blocked the Bering Strait until mid-May, which altered the timing of

TABLE 32.2 Summary of the number of bowhead whales landed in Alaska by village, and in Chukotka, Russia from 1974 to 2018.

Year	Barrow	Gambell	Kaktovik	Kivalina	Little Diomede	Nuiqsut	Point Hope	Point Lay[a]	Savoonga	Shaktoolik[b]	Wainwright	Wales	Chukotka[c]	Landed total
1974	9	2	2	0	0	0	7		0	0	1	0		21
1975	10	1	0	0	0	0	4		0	0	0	0		15
1976	23	1	2	0	0	0	12		7	0	3	0		48
1977	20	2	2	1	0	0	2		0	0	2	0		29
1978	4	1	2	0	0	0	2		1	0	2	0		12
1979	3	0	5	0	0	0	3		0	0	1	0		12
1980	9	1	1	0	0	0	0		2	1	1	1		16
1981	4	1	3	0	0	0	4		2	0	3	0		17
1982	0	2	1	0	0	1	1		1	0	2	0		8
1983	2	1	1	0	0	0	1		1	0	2	1		9
1984	4	0	1	1	0	0	2		2	0	2	0		12
1985	5	1	0	0	0	0	1		1	0	2	1		11
1986	8	3	3	0	0	1	2		0	0	3	0		20
1987	7	2	0	1	0	1	5		1	0	4	1		22
1988	11	2	1	0	0	0	5		0	0	4	0		23
1989	10	0	3	0	0	2	0		1	0	2	0		18
1990	11	4	2	0	0	0	3		5	0	5	0		30
1991	12	1	2	1	0	1	6		0	0	4	1		28
1992	22	4	3	1	0	2	2		4	0	0	0		38
1993	23	4	3	0	0	3	2		1	0	5	0		41
1994	16	1	3	2	0	0	5		2	0	4	1		34
1995	19	4	4	1	0	4	1		4	0	5	1		43
1996	24	3	1	0	0	2	3		2	0	3	0		38

1997	30	3	4	0	0	3	4		1	0	3	0		48
1998	25	0	3	0	0	4	3		3	0	3	0	1	42
1999	24	1	3	0	1	3	2		3	0	5	0	1	43
2000	18	0	3	0	0	4	3		1	0	5	1	1	36
2001	27	2	4	0	0	3	4		3	0	6	0	1	50
2002	22	2	3	0	0	4	0		5	0	1	0	3	40
2003	16	1	3	0	0	4	4		2	0	5	0	3	38
2004	21	3	3	0	0	3	3		0	0	4	0	1	38
2005	29	2	3	0	1	1	7		7	0	4	1	2	57
2006	22	0	3	0	0	4	0		0	0	2	0	3	34
2007	20	4	3	0	0	3	3		4	0	4	0	0	41
2008	21	2	3	0	0	4	2		4	0	2	0	2	40
2009	19	1	3	0	0	2	1	1	3	0	1	0	0	31
2010	22	6	3	0	0	4	2	0	5	0	3	0	2	47
2011	18	4	3	0	0	3	3	1	2	0	4	0	0	38
2012	24	5	3	0	0	4	5	1	8	0	4	1	0	55
2013	22	2	3	0	0	4	6	0	6	0	3	0	1	47
2014	18		3	0	0	5	6	0	3	0	3	0	0	38
2015	24	1	4	0	0	3	3	0	1	0	3	0	0	39
2016	22	1	3	0	0	4	7	1	2	0	7	0	1	48
2017	21	2	3	0	0	4	10	1	2	0	7	0	0	50
2018	27	3	3	0	0	3	7	0	1	0	3	0	0	47
Total	748	88	114	8	2	93	158	5	103	1	142	10	23	1493

[a]*Point Lay became a member of the AEWC in 2008 and landed its first whale in more than 70 years in 2009.*
[b]*Shaktoolik is not a member of the AEWC but landed a whale in 1980.*
[c]*Chukotka, Russian Federation, began sharing the IWC bowhead quota with Alaska, United States in 1998.*
AEWC, Alaska Eskimo Whaling Commission; *IWC*, International Whaling Commission.
Data source: Alaska Eskimo Whaling Commission, the North Slope Borough, and the National Marine Fisheries Service.

migration and temporarily altered the distribution of whales (Johnson et al., 1981). They have not struck a bowhead since then.

Utqiaġvik also hunts during the autumn migration, as do Nuiqsut and Kaktovik (Fig. 32.1). The autumn hunt occurs in "open-water and/or broken-ice conditions" as whales migrate west along the Beaufort Sea or southwest along northeastern Chukchi Sea coasts of northern Alaska. The autumn hunt usually occurs from August to October. Because of difficult environmental conditions during the spring, especially due to deteriorating shorefast sea ice from a warming climate, Wainwright, Point Lay, and Point Hope have begun hunting opportunistically in the autumn. While crews in all three villages have attempted to hunt, only Wainwright landed whales. Their first was landed in autumn 2010 with a few whales landed periodically since then (Suydam et al., 2011, 2012, 2013, 2016).

Since about 2000, Savoonga and Gambell, villages on Saint Lawrence Island, have been hunting more frequently during the late autumn and early winter (Fig. 32.4A and B). Hunters on Saint Lawrence Island indicate that there are more whales near Saint Lawrence Island now than in the past. There may also be more of an opportunity now and the future to hunt in late fall and early winter because of greater ice retreat in summer contributing to later ice formation in the autumn (Noongwook et al., 2007). In addition, spring ice is breaking up more rapidly than in the past, which can reduce opportunities for hunters to pursue whales or walruses. For these reasons, it is likely that Saint Lawrence Island will continue and perhaps increase their hunting effort during the fall and winter months as weather and ice conditions allow.

The bowhead quota was implemented by the IWC in 1978. Immediately following the quota, the number of whales struck and landed dropped dramatically (Table 32.2, Fig. 32.4C). The quota increased (Table 32.1) in response to more accurate and increasing abundance estimates of the bowhead population (Givens et al., 2016; George et al., 2004). In addition, the IWC's increased understanding of the importance bowheads played in meeting cultural and nutritional needs of Inuit people also contributed to the increased quota (Braund, 1992; Stoker and Krupnik, 1993). The number of landed and struck and lost bowheads stabilized in the late 1990s (Fig. 32.4C and D). Two of the more notable changes at the village level included (1) a decrease in the number of whales landed by Point Hope after the quota was implemented and (2) a substantial increase in the number of whales landed at Utqiaġvik and Nuiqsut after about 1990.

Efficiency of the hunt

Because of the nature of the hunt, some whales are struck but then lost and not harvested. For example, struck whales may escape from hunters under shorefast ice or into the floating pack ice. Also, the harpoon and float might pull free of a struck whale, and hunters are unable to locate the animal again, or occasionally, struck and killed whales may sink and be lost. In response to concerns expressed by the IWC, the efficiency (number of whales landed/number of whales struck) of the harvest has increased since 1978 and has stabilized at about 80% since about the late 1990s (Fig. 32.4D). The efficiency increased for several reasons: (1) enhanced training conducted by senior captains of the AEWC on where to strike a whale; (2) improved communication for alerting other crews that a whale had been struck. Therefore, other crews can search for the struck whales or

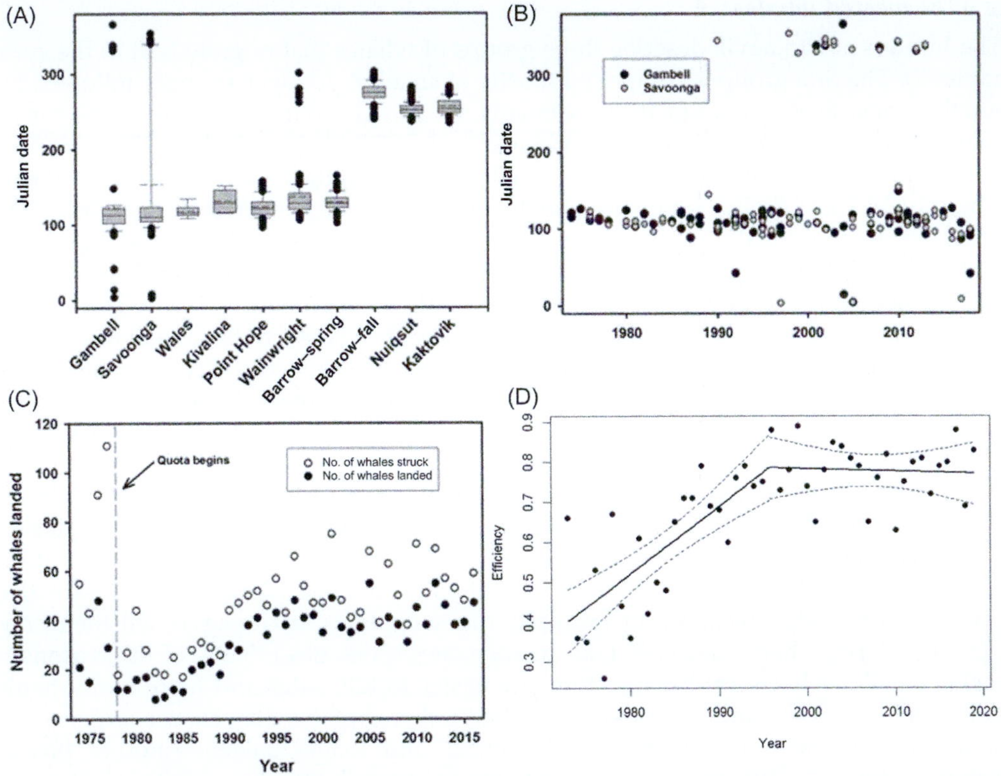

FIGURE 32.4 Statistics for whales harvested in Alaska from 1974 to 2018. (A) Box plots of bowhead whale harvest dates (Julian date; where days of the year are numbered sequentially, starting with January 1) by village. Most villages hunt in the spring but four villages [Gambell, Savoonga, Barrow (Utqiaġvik), and Wainwright] have both a spring and autumn/winter hunt. Because of the number of whales landed at Barrow, the spring and autumn hunts are plotted separately. Two villages (Kaktovik and Nuiqsut) hunt only in autumn. Shaktoolik only landed one whale in 1980, while Little Diomede and Point Lay have only landed a few. Because of small sample sizes, those three villages are not included in this figure. (B) Julian dates that bowhead whales have been landed at Saint Lawrence Island by the villages of Gambell and Savoonga, 1974–2018. (C) Number of bowhead whales landed by Alaskan Eskimos, 1974–2016. The quota was implemented in 1978. (D) Efficiency (no. of whales landed/no. of whales struck) of the bowhead whale harvest by Alaskan Eskimos, 1974–2019.

can come and help tow the whale; (3) efforts by some captains to only strike smaller whales, which tend to be easier to harvest and tow to the ice edge or shore; (4) enhanced efforts to locate and retrieve struck whales using (a) aircraft to locate struck whales and (b) dive teams to retrieve whales that sank; and (5) a program to improve the weaponry. The efficiency has now stabilized and may not increase much more primarily because of the nature of the hunt. Traditional methods are still being used to hunt whales in very challenging environmental conditions. While traditional methods have been modified and new methods have been implemented, there will always be some whales that are lost for a myriad of reasons, but mainly due to environmental effects (e.g., weather and sea ice).

Size of harvested whales

The hunters at Utqiaġvik describe three groups of whales that migrate past in the spring (Chapter 7). The first group is comprised mostly of smaller, younger animals, followed by a group that consists of animals of all sizes. Finally, the last group to pass in the spring consists primarily of large females that are often pregnant or have recently given birth. Hunters avoid mothers with calves because that is a traditional practice, and the IWC does not allow the taking of calves. Periodically, pregnant females and calves near independence from their mothers are unintentionally landed. Data from the spring harvests at Utqiaġvik (Fig. 32.5A), Point Hope, and Wainwright (Fig. 32.5B) show this same pattern. The average lengths of harvested whales increased over the season. Mostly small whales are caught early in the season and mostly large whales are caught late in the season. The change in the length of harvested whales over the course of the season is mainly driven by the availability of whales (Koski et al., 2006). The autumn hunt shows the reverse pattern. In Utqiaġvik the lengths of whales decreased over the autumn season (Fig. 32.5A) similar to observations at Kaktovik (Koski et al., 2004).

The average lengths of harvested whales differed among villages (Fig. 32.5C). Lengths of whales harvested at Utqiaġvik and Point Hope were generally smaller than whales harvested at Gambell, Savoonga, and Wainwright (Fig. 32.5C). The reasons for these differences are not clear. One explanation is that hunters from some villages may preferentially select larger or smaller whales. For example, many captains at Utqiaġvik tell their crews not to strike large whales and only take smaller ones (pers. obs.). Some of the reasons for choosing smaller whales are because they are easier to kill, safer and more easily hauled onto the ice (or ashore), and the muktuk is softer. Other captains may choose larger whales because they provide more maktak and muscle, and longer baleen. Another possible explanation for size difference among villages is accessibility. For example, numerous larger whales may be more common near Saint Lawrence Island, and thus the hunters from Gambell and Savoonga may take those that are most available. Braham (1995) also observed that Wainwright hunted larger whales than the other villages.

Sex ratios

Based on the harvest, there does not appear to be sexual segregation during the spring or autumn migration, with the exception that large, often pregnant females pass late in the spring migration. Both males and females are taken throughout April and May (Fig. 32.5D) and August to October (Fig. 32.5E). Overall, there was an equal sex ratio ($\chi^2 = 3.67$, $P = 0.06$), although more females ($n = 745$) were recorded than males ($n = 673$). This was especially the case for Gambell, Savoonga, and Wainwright, the villages that tend to harvest larger whales (Fig. 32.5B), although Wainwright also landed more females than males. Braham (1995) found an overall equal sex ratio between 1973 and 1993; however, when broken down by sex, size [<13 m (presumed immature) and >13 m (presumed mature)], and season, he concluded that during spring, more large females and fewer large males were taken than due to chance alone.

Suydam and George (2012) found no differences in the proportion of males and females in the fall harvest ($\chi^2 = 2.07$, $P = .15$), but there was a difference in the spring. More large females and fewer large males were taken than by chance ($\chi^2 = 6.95$, $P = .008$) (Fig. 32.5D

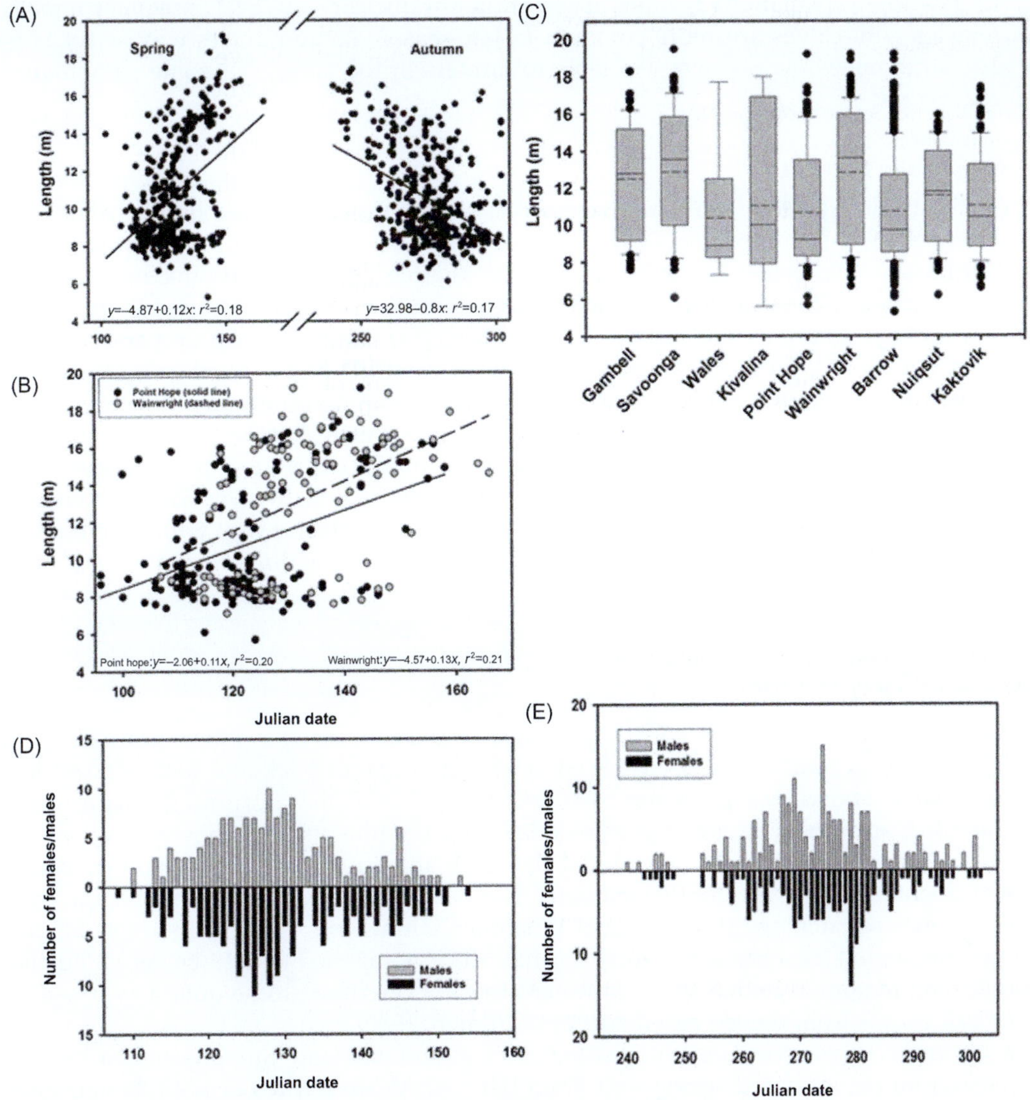

FIGURE 32.5 (A) Julian dates and lengths of bowhead whales landed at Barrow (Utqiaġvik) during the spring and autumn hunts 1974–2018. Data from both hunts were fit with linear regression to show the increase/decrease in the size of whales landed as the season progressed. (B) Julian dates and lengths of bowhead whales landed at Wainwright and Point Hope during the spring, 1974–2018. Data from both villages were fit with linear regression to show the increase/decrease in the size of whales landed as the season progressed. (C) Box plots of the lengths of bowhead whales taken by Alaskan Eskimos in each village 1974–2018. Data from Little Diomede, Point Lay, and Shaktoolik were not plotted because few whales were landed at these villages. (D–E) Number of male and female bowhead whales taken by Alaskan Eskimos at Utqiaġvik, Alaska during the spring (D) and autumn (E) hunt.

and E). Possible explanations for this observation are unclear; however, pregnant females and females with calves are more common in late-season, some hunters may select large females, large males may be less available to hunters in the spring, or some combination of these factors.

Sharing of the harvest

Whales are thought to give themselves to deserving crews, which the Inupiat refer to as "the gift of the whale" (Brewster 2004). In Alaska the successful captain and crew supervise the distribution of the whale at the butchering site, sharing with the crews that help tow and butcher the whale. The whale is then distributed to the entire community primarily through a series of festivals and other means. The level of sharing is unprecedented in many societies. Academic research confirms the importance of whale hunting and "mixed subsistence-cash economies" across centuries; and sharing and hunting cooperation continues "to act as central features of the local culture" (BurnSilver et al., 2016; Kofinas et al., 2016).

After distribution to the community, the whale is further distributed across many communities. In some cases, shares from a single whale were sent to over 60 other villages and cities across Alaska (Kofinas et al., 2016). The whale is given away with no monetary compensation. Therefore the whaling captains sustain people within their own and many other communities.

Sustainability of the hunt

The IWC Scientific Committee (SC) had considerable concern about the sustainability of the BCB bowhead hunt in the 1970s based on the increasing struck and lost rate and low but unreliable abundance estimates that led to the 1977 moratorium (Braham, 1995; Chapter 6). A small harvest quota was negotiated for the nine villages; however, even several years later in 1981, the SC considered the stock at great risk. A population dynamics paper presented at that meeting concluded, "the Bering Sea population will decline even in the absence of catching" (International Whaling Commission, 1981; p. 134). The SC further recommended the only safe course for this stock is for the "take to be zero." Further complicating matters was that the recruitment rate was assumed to be quite low based on visual calf counts from the ice-based surveys (Tillman, 1980).

In the early 1980s more accurate estimates of abundance and calf production became available from the ice-based surveys off Point Barrow, AK, in particular with the inclusion of acoustic surveillance (Zeh et al., 1993; Chapter 6). The quota was adjusted upward as a result (Chapter 6). By the early 1990s the first statistical evidence of an increasing abundance trend was published that was a "game changer" in terms of scientific confidence that the BCB hunt was sustainable (Zeh et al., 1991).

A strike limit algorithm (SLA) for assessing safe harvest levels, was adopted in 2002 (International Whaling Commission, 2003a,b; pp. 20–21) and is discussed in greater detail in Chapter 39. The SLA was thoroughly tested and the SC has repeatedly recommended it as the "best tool" for giving management advice (Chapter 39).

While the SLA is quite complex, the following simple example is helpful for illustrating the harvest risk. The 2011 abundance estimate accepted by the IWC SC for the BCB stock is about 16,820 whale (CV = 5%) with a rate of increase of 3.7% (Givens et al 2016, Chapter 6). Considering that the harvest quota has been stable at 67/year since 1995, the removal rate is less than 0.5%/year, even in seasons where the maximum number of strikes was used.

Summary and conclusions

The current indigenous hunts for bowhead whales are a blending of traditional and modern methods (Chapters 31 and 35). Skin covered boats have likely been used for millennia, and many whaling implements were introduced in Alaska by Yankee whalers and improved over time. Modern communication and global positioning systems (GPS) have contributed to more efficient and safe hunts. Hunting techniques have modernized (Stoker and Krupnik, 1993), but the nutritional and cultural needs have remained similar.

Bowhead whales continue to provide for these needs of Inuit in the Arctic and Subarctic, as they have for millennia. While bowheads are harvested in western Greenland, northern Canada, and Chukotka, they are not the main resource utilized by Inuit in those places. Other whale species, terrestrial resources, and aquatic resources contribute substantially to the needs of those communities. In Alaska, bowheads play more of a central role in the culture of Inuit, including Inupiat, Yupik, and Siberian Yupik.

While there may be selectivity in hunts in some villages, the harvest seems to generally reflect the characteristics of the bowhead population. During spring, small whales tend to be harvested early in the migration and larger animals later. During the autumn, larger animals are more available early in the migration. Small whales and large whales are both harvested, and males and females are harvested equally.

In 1978 the IWC implemented a quota system that resulted in many fewer bowheads being struck and landed. The quota increased over time as the nutritional and cultural needs of Inuit communities were realized and as understanding of the bowhead population size and increasing trend improved. The current quota on landed whales has been stable since 1998 (Table 32.1).

The population of bowheads in West Greenland and Canada has increased and exceeds 6,000 animals (Chapter 6). The hunts in those areas are small and represent only a small fraction (<0.01%) of the bowhead population.

The BCB population has grown steadily for at least 50 years and may have been increasing since the cessation of the commercial harvests. A landed quota of 56 whales per year and a strike quota of 67 per year in Alaska and Chukokta represents <0.5% of the population.

Both the BCB and East Canada–West Greenland stocks are recovering or have recovered even with the presence of a subsistence hunt. The current indigenous hunts of bowhead whales are sustainable.

Acknowledgments

We thank the Alaska Eskimo Whaling Commission and local hunters for providing data on the subsistence bowhead hunt in Alaska. We especially thank the captains' associations and hunters from Utqiaġvik, Point Hope,

Nuiqsut, Gambell, Savoonga, and Kaktovik for their support and providing us access to their whales for examinations and sampling. There are many people who have helped collect measurements and samples of harvested bowheads. We are deeply thankful for their efforts. Harvest data for Greenland, Chukotka, and Canada were provided by Cherry Allison from the IWC Secretariat. The North Slope Borough (NSB) and the AEWC (through grants from US National Oceanic and Atmospheric Administration) provided financial support. We are indebted to past Mayors and Directors/Deputy Directors of Wildlife Management of the NSB for their dedicated support. The current NSB Mayor, Harry K. Brower, Jr., and Director Raynita Taqulik Hepa, Deputy Director, Nicole Kanayurak, and Jessica Lefevre, legal counsel for the AEWC, have given encouragement and support for decades. Finally, we would like to thank our wives, Leslie Pierce and Cyd Hanns, for their patience and support through the process of writing these chapters and also for their involvement and enthusiasm in collecting data and samples from harvested bowheads.

References

Albert, T.F., 1988. The role of the North Slope Borough in arctic environmental research. Arct. Res. U.S. (2), 17–23.

Bockstoce, J.R., 1986. Whales, Ice, and Men: The History of Whaling in the Western Arctic. University of Washington Press, Seattle, WA.

Bockstoce, J.R., Burns, J.J., 1993. Commercial whaling in the north pacific sector. In: Burns, J.J., Montague, J.J., Cowles, C.J. (Eds.), The Bowhead Whale. Allen Press, Lawrence, KS, Special Publication Number 2, Society of Marine Mammalogy.

Braham, H.W., 1995. Sex and size composition of bowhead whales landed by Alaskan Eskimo whalers. In: McCartney, A.P. (Ed.), Hunting the Largest Animals, Native Whaling in the Western Arctic and Sub-Arctic. The Canadian Circumpolar Institute, University of Alberta, Edmonton, Canada, pp. 281–313, Studies in Whaling No. 3. Occasional Publication No. 36.

Braund, S.R., 1992. Traditional Alaska Eskimo whaling and the bowhead quota. Arct. Res. 6 (Fall), 37–42.

Brewster, K., 2004. [Ed.]) The Whales They Give Themselves: Conversations With Harry Brower, Sr. University of Alaska Press, Fairbanks, AK, 232 pp.

BurnSilver, S., Magdanz, J., Stotts, R., Berman, M., Kofinas, G., 2016. Are mixed economies persistent or transitional? Evidence using social networks from Arctic Alaska. Am. Anthropol. 118 (1), 121–129. Available from: https://doi.org/10.1111/aman.12447.

DeMaster, D.P., Rauch III, S.D., 2013. Final environmental impact statement for issuing annual quotas to the Alaska Eskimo Whaling Commission for a subsistence hunt on bowhead whales for the years 2013 through 2018. Prepared by U.S. Department of Commerce National Oceanic and Atmospheric Administration, National Marine Fisheries Service.

George, J.C., Braund, S., Brower Jr., H., Nicolson, C., O'Hara, T.M., 2003. Some observations on the influence of environmental conditions on the success of hunting bowhead whales off Barrow, Alaska. In: McCartney, A.P. (Ed.), Indigenous Ways to the Present: Native Whaling in the Western Arctic. The University of Utah Press, Salt Lake City, UT, pp. 255–276, Studies in Whaling No. 6: Occasional Publication. Canadian Circumpolar Institute Press: No. 54.

George, J.C., Zeh, J., Suydam, R., Clark, C., 2004. Abundance and population trend (1978–2001) of western Arctic bowhead whales surveyed near Barrow, Alaska. Mar. Mammal Sci. 20, 755–773.

Givens, G.H., Edmondson, S.L., George, J.C., Suydam, R., Charif, R.A., Rahaman, A., et al., 2016. Horvitz-Thompson whale abundance estimation adjusting for uncertain recapture, temporal availability variation and intermittent effort. Environmetrics 26, 1–16.

International Whaling Commission, 1979. 29th Report of the IWC 29:34.

International Whaling Commission, 1980. 30th Report of the IWC 30:39.

International Whaling Commission, 1981. 31st Report of the IWC 31:18.

International Whaling Commission, 1984. 34th Report of the IWC 34:32.

International Whaling Commission, 1986. 36th Report of the IWC 36:26.

International Whaling Commission, 1988. 38th Report of the IWC 38:31.

International Whaling Commission, 1989. 39th Report of the IWC 39:32.

International Whaling Commission, 1992. 42nd Report of the IWC 42:49.

International Whaling Commission, 1995. 45th Report of the IWC 45:52.
International Whaling Commission, 1998. 48th Report of the IWC 48:51.
International Whaling Commission, 2003a. Annual Report of the IWC, 2003:171.
International Whaling Commission, 2003b. Report of the scientific committee. J. Cetacean Res. Manage. 5 (Suppl.), 1–92.
International Whaling Commission, 2007. Annual Report of the IWC, 2007:155.
International Whaling Commission, 2012. Annual Report of the IWC, 2012:178.
International Whaling Commission, 2018. Annex P. In: Amendments to the Schedule Adopted at the 67th Meeting of the IWC. pp. 1–2. Available from: <iwc.int>.
Johnson, J.H., Braham, H.W., Krogman, B.D., Marquette, W.M., Sonntag, R.M., Rugh, D.J., 1981. Bowhead whale research: June 1979 to June 1980. Rep. Int. Whal. Comm 31, 461–475.
Kishigami, N., 2015. Revival of Inuit bowhead hunts in Arctic Canada. Jpn. Rev. Cult. Anthropol. 16, 43–58.
Kofinas, G., BurnSilver, S.B., Magdanz, J., Stotts, R., Okada, M., 2016. Subsistence sharing networks and cooperation: Kaktovik, Wainwright, and Venetie, Alaska. In: BOEM Report 2015-023 DOI; AFES Report MP 2015-02. School of Natural Resources and Extension, University of Alaska Fairbanks.
Koski, W.R., George, J.C., Sheffield, G., Glaginaitis, M.S., 2004. Subsistence harvest of bowhead whales at Kaktovik, Alaska. In: Paper SC/56/BRG23 Submitted to the Scientific Committee of the International Whaling Commission.
Koski, W.R., Rugh, D.J., Punt, A.E., Zeh, J., 2006. An approach to minimize bias in estimation of the length frequency distribution of bowhead whales (*Balaena mysticetus*) from aerial photogrammetric data. J. Cetacean Res. Manage. 8 (1), 45–54.
Marquette, W.M., Braham, H.W., Nerini, M.K., Miller, R.V., 1982. Bowhead whale studies, autumn 1980-spring 1981: harvest biology and distribution. Rep. Int. Whal. Comm 32, 357–370.
Noongwook, G., the Native Village of Savoonga, the Native Village of Gambell, Huntington, H.P., George, J.C., 2007. Traditional knowledge of the bowhead whale (*Balaena mysticetus*) around St. Lawrence Island, Alaska. Arctic 60, 47–54.
Ross, W.G., 1993. Commercial whaling in the North Atlantic sector. In: Burns, J.J., Montague, J.J., Cowles, C.J. (Eds.), The Bowhead Whale. Allen Press, Lawrence, KS, Special Publication Number 2, Society of Marine Mammalogy.
Stoker, S.W., Krupnik, I.I., 1993. Subsistence whaling. In: Burns, J.J., Montague, J.J., Cowles, C.J. (Eds.), The Bowhead Whale. Allen Press, Lawrence, KS, Special Publication Number 2, Society of Marine Mammalogy.
Suydam, R.S., George J.C., 2012. Preliminary analysis of subsistence harvest data concerning bowhead whales (*Balaena mysticetus*) taken by Alaskan Natives, 1974 to 2011. In: Paper SC/64/AWMP8 Presented to the Scientific Committee of the International Whaling Commission. p. 13.
Suydam, R.S., George, J.C., Person, B., Hanns, C., Sheffield, G., 2011. Subsistence harvest of bowhead whales (*Balaena mysticetus*) by Alaskan Eskimos during 2010. In: Paper SC/63/BRG2 Presented to the Scientific Committee of the International Whaling Commission.
Suydam, R.S., George, J.C., Person, B., Hanns, C., Stimmelmayr, R., Pierce L., et al., 2012. Subsistence harvest of bowhead whales (*Balaena mysticetus*) by Alaskan Eskimos during 2011. In: Paper SC/64/BRG2 Presented to the Scientific Committee of the International Whaling Commission.
Suydam, R.S., George, J.C., Person, B., Hanns, C., Stimmelmayr, R., Pierce, L., et al., 2013. Subsistence harvest of bowhead whales (*Balaena mysticetus*) by Alaskan Eskimos during 2012. In: Paper SC/65a/BRG19 Presented to the Scientific Committee of the International Whaling Commission. p. 7.
Suydam, R.S., George, J.C., Person, B., Ramey, D., Stimmelmayr, R., Sformo, T., et al., 2016. Subsistence harvest of bowhead whales (*Balaena mysticetus*) by Alaskan Eskimos during 2015. In: Paper SC/66b/BRG3 Presented to the Scientific Committee of the International Whaling Commission. p. 10.
Tillman, M.F., 1980. Introduction: a scientific perspective of the bowhead whale problem. Mar. Fish. Rev. 42 (9–10).
Whitridge, P., 1999. The prehistory of Inuit and Yupik whale use. Revista de arqueología americana, pp. 99–154.
Woodby, D.A., Botkin, D.B., 1993. Stock sizes prior to commercial whaling. In: Burns, J.J., Montague, J.J., Cowles, C.J. (Eds.), The Bowhead Whale. Allen Press, Lawrence, KS, pp. 387–407.
Zeh, J.E., Clark, C.W., George, J.C., Withrow, D., Carroll, G.M., Koski, W.R., 1993. Current population size and dynamics. In: Burns, J.J., Montague, J.J., Cowles, C.J. (Eds.), The Bowhead Whale. Allen Press, Lawrence, KA.
Zeh, J.E., George, J.C., Raftery, A.E., Carroll, G.M., 1991. Rate of increase, 1978–1988, of bowhead whales, *Balaena mysticetus*, estimated from ice-based census data. Mar. Mammal Sci. 7, 105–122.

CHAPTER

33

Commercial whaling

J.G.M. Thewissen[1] *and J.C. George*[2]

[1]Department of Anatomy and Neurobiology, Northeast Ohio Medical University, Rootstown, OH, United States [2]Department of Wildlife Management, North Slope Borough, Utqiaġvik, AK, United States

Introduction

To understand the current condition of the four extant bowhead stocks, one must consider the history of commercial whaling on these populations by the nations of Europe and the United States (Fig. 33.1). The history of bowhead whaling has been extensively researched and published and also has been approached quantitatively by studying annual catches, cruise tracks, kill locations, and oil yield per whale (e.g., Townsend, 1935; Bockstoce and Botkin, 1983; de Jong, 1983; Bockstoce, 1986; Ross, 1993; Woodby and Botkin, 1993). The purpose of this chapter is not to provide a detailed historical record but instead to summarize the 400-year exploitation and near-extirpation of the circumpolar bowhead stocks. We also discuss the current status of the species and the effect commercial whaling has had on indigenous societies that depend on it. A summary of catches of bowheads over time is provided in Fig. 33.2 and is based on data from Ross (1993), *Bockstoce* and Burns (1993) (see also Bockstoce and Botkin, 1983), Woodby and Botkin (1993), Mitchell (1977), Ivashchenko et al. (2011), and Aguilar (1986). These authors are explicit about strengths, shortcomings, and assumptions in the catch data, and we will not repeat them here. Regardless, these records can be used to describe the broad patterns that we are interested in.

Commercial whaling on bowhead whales (also called Greenland right whales) was mainly executed by commercial whalers of the nations of Europe and the United States, whereas catches by Russian commercial whalers were minor (Ivashchenko, Y., pers. comm.). Commercial exploitation started in 1540 when bowhead whales were hunted by Basques near southern Labrador and whaling ended in the beginning of the 20th century. Our review ends at the start of World War I in 1914, when all four populations were heavily depleted (see Chapter 3). A number of factors affected the hunt over this 400-year period, but the overall pattern was consistent: as one regional population was driven to near extinction, the whalers moved on to another, until the hunt ceased to be commercially viable (Ross, 1993; Bockstoce and Burns, 1993).

The Bowhead Whale
DOI: https://doi.org/10.1016/B978-0-12-818969-6.00033-9

FIGURE 33.1 Dutch bowhead whaling in Svalbard in the 17th century. A bowhead whale is attacked in the center, spewing blood, another one is towed back to the ship (to the left of center ship), and yet another butchered in the water (to the left of ship on right). Source: *Painting by Abraham Storck (1654–1708), 'Whaling Grounds in the Polar Sea.' The original is at the Rijksmuseum, Amsterdam, reprinted with permission.*

Whaling prior to modern industrial whaling was inherently dangerous, especially before 1900, as whalers operated far from homeports in remote and desolate areas. The basic pattern was similar over the four centuries of commercial hunting. The whalers pursued bowheads in small, double-ended whaling boats that were human powered. When a whale was within striking range, they fastened to it with a hand-thrown harpoon and line attached to a float or to the boat itself. When the animal tired, they killed it by repeatedly thrusting a long, handheld lance through the vital organs (Fig. 33.1). The use of an explosive charge fired into the whale began in the 1860s. Bowheads are relatively docile and tend not to attack whaling skiffs, as many other species do after being harpooned, but some did. Bockstoce (1986) noted that once harpooned "a whale could turn ugly and try to demolish a boat with its tail." Accidents occurred frequently as men fell overboard, harpoon lines pulled men and boats underwater, boats were damaged against ice, lost in fog, and whaling guns exploded or misfired (Bockstoce, 1986). The ships were frequently beset by ice, and if the whaling ships did not plan to overwinter the likely consequence was death by starvation during winter. However, as long as whaling was profitable, it continued (Bockstoce, 1986; Ross, 1993).

As whaling technology advanced, the hunt changed. Reinforced hulls allowed ships to penetrate heavy sea ice to pursue whales. In the 1500s and 1600s, whales had to be killed near the coast for the blubber to be rendered on shore or transported to homeport. Around 1750 tryworks were invented: structures and equipment that allowed sailors to render blubber aboard ship. This facilitated work far from shore or homeports. Further

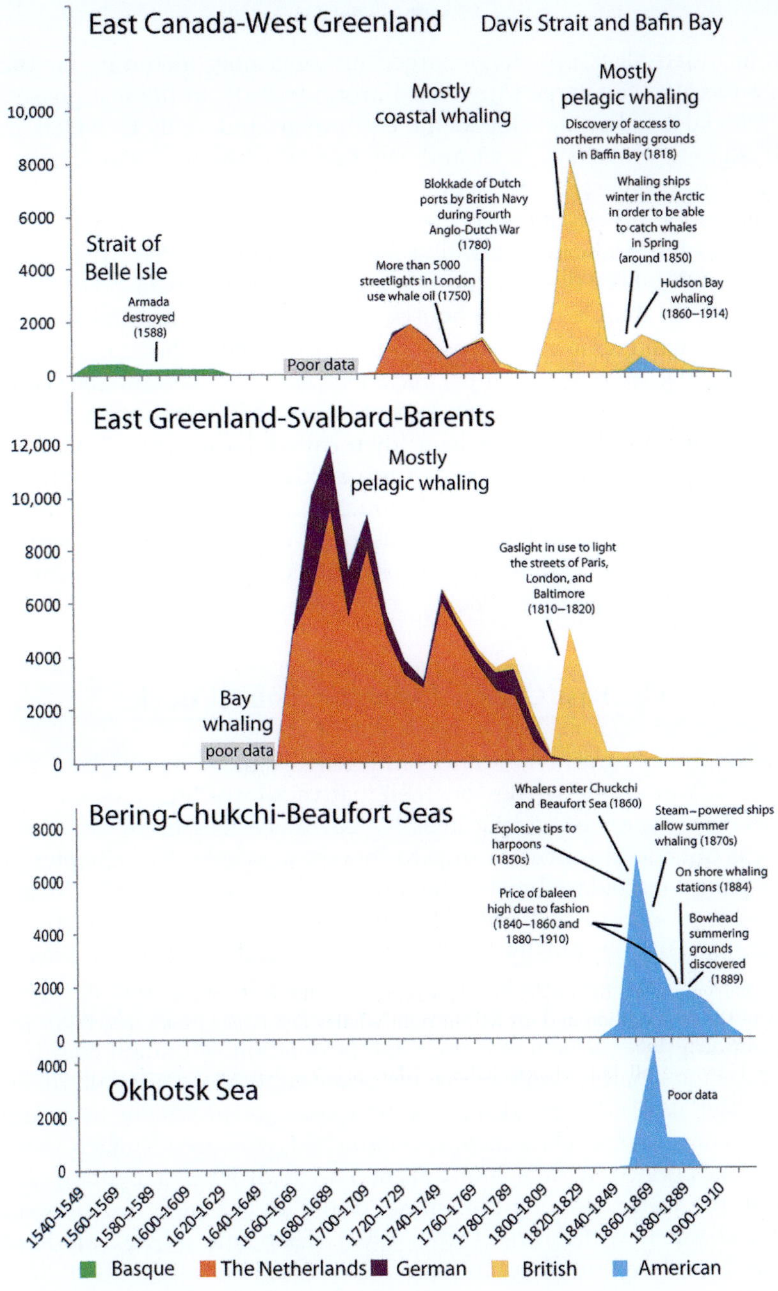

FIGURE 33.2 Estimated numbers of bowhead whales killed and harvested between 1540 and 1914 for each of the four stocks. Totals are summed by decade and the contributions of the various nations stacked. Source: *Data for East Canada–West Greenland and East Greenland–Svalbard–Barents Sea stock from Ross, W.G., 1993. Commercial whaling in the North Atlantic sector. In: Burns, J.J., Montague, J.J., Cowles, C.J. (Eds.), The Bowhead Whale. Spec. Publ. 2, Society for Marine Mammalogy, pp. 511–561, and most data for Bering–Chukchi–Beaufort Seas stock from Woodby, D.A., Botkin, D.B., 1993. Stock sizes prior to commercial whaling. In: Burns, J.J., Montague, J.J., Cowles, C.J. (Eds.), The Bowhead Whale. Spec. Publ. 2, Society for Marine Mammalogy, pp. 387–407. and Okhotsk Sea stock from Mitchell, E.D., 1977. Initial population size of bowhead whale* (Balaena mysticetus) *stocks: cumulative catch estimates. In: Paper SC/29/33 Presented at the IWC Scientific Committee. 113 p. (unpublished) (see also review by Ivashchenko et al., 2011). Data for 1530–1640 catches near Strait of Belle Isle and Labrador from Aguilar, A., 1986. A review of old Basque Whaling and its effects on the right whales (Eubalaena glacialis) of the North Atlantic. Rep. Int. Whal. Comm. Spec. 10, 191–199.*

developments included black powder–charged hunting tools, including the darting gun, a harpoon with an explosive charge introduced around 1860s. In the last phase of bowhead whaling, commercial whalers developed the equipment and skills to winter in the Arctic, allowing their ship to winter in bays and lagoons (Bockstoce, 1986). These innovations served to hasten the commercial extirpation of the stocks.

Commercially important products of the bowhead whale were oil and baleen. Oil was produced by trying blubber: heating it until it liquefied and separated from the tissue matrix. Bowheads have more blubber than any other species, up to approximately 45%–50% of their body weight. Whale oil illuminated the houses and cities of Europe and the eastern United States and was their main source of light until the early 1800s. The importance of whale oil diminished when gaslight (coal gas lamps) and then petroleum became the choice for illumination because they, were increasingly less expensive and had fewer odors.

The other important product was baleen (then called "whalebone," or simply "bone"). Typically, there are more than 600 plates within a bowhead whale's mouth, and the longest of these can be more than 4 m in length, much longer than the baleen of other species (Chapter 14). Baleen, flexible and tough, was used for corset stays, horsewhips, measuring tapes, umbrella rods, and a variety of purposes for which plastic and spring steels are now used (Brower, 1942; Bockstoce, 1986).

The East Canada–West Greenland stock

Basque whalers, from southern France and northwestern Spain, had exhausted the stocks of North Atlantic right whales on their native shores and crossed the Atlantic to hunt in the Strait of Belle Isle and Gulf of St. Lawrence, beginning around 1540 (Laist, 2017). This area is outside the current range for bowhead whales (see Chapter 5), but it was not during this period of colder climates, the little Ice Age (Mann, 2002). Occasionally now, bowhead whales are found in such southern latitudes (Accardo et al., 2018). In the past, it was thought that the Basques mostly took right whales in these southern areas, but genetic analyses have now shown that bowhead whales were their most common quarry (Rastogi et al., 2004; McLeod et al., 2008). The whales from the Strait of Belle Isle were probably part of the East Canada–West Greenland (ECWG) population, although genetic differences between stocks are too small to confirm this beyond doubt (McLeod et al., 2012).

Tuck and Grenier (1981) described that the Basques spotted whales from high points on land or spotting boats (called chalupas) and launched whaleboats when they were seen. The whale was harpooned with a line attached to a drogue, a floating device, which dragged through the water, slowing the animal down and tiring it out, allowing the whalers to follow and kill it with lances (Ross, 1993). Then, the whale was towed to land, butchered, and oil rendered from its blubber.

Aguilar (1986) estimated that between 300 and 500 whales were harvested annually by the Basques, mostly based on the number of ships involved and the quantities of oil and baleen on return voyages. Over time, the hunt diminished in importance, and after 1580 ships often returned to Europe half-full (Aguilar, 1986; Barkham, 1984). The hunt was abandoned around 1630. Causes for the decline in the hunt include the reduction of the stock, leading the Basque whalers to seek opportunities elsewhere (e.g., as whalers in

Svalbard). Geopolitical factors played a role as equipment, expertise, and skilled man power was lost when whaling ships and whalers were part of the doomed Spanish Armada in 1588, and Spain became less able to defend its maritime exploits (Proulx, 1986).

A second phase of whaling on the ECWG stock occurred in areas north of Labrador, concentrating on Davis Strait and Baffin Bay. Danes and Dutch pursued this stock near the end of the 17th century on the west coast of Greenland after the East Greenland–Svalbard–Barents (EGSB) Sea stock had been depleted. Similar to whaling near Svalbard (discussed later), they chased and killed bowheads using small boats and butchered them after pulling them alongside their ship. The blubber was not rendered on the ship but transported back to homeport in barrels. From the early days of the Davis Strait–Baffin Bay whaling until around 1820, whalers focused on whales migrating north on the eastern side of Davis Strait in spring, which has less ice than the western side (see Chapter 5). This resulted in mostly large males being caught, as females and young animals used migratory routes further west. With Greenland being a Danish territory, Danes and local Inuit whaled from land bases, whereas whalers of other nations (mainly Dutch, German, British, and American) were ship based. Whalers of all nations interacted with local Inuit along the west coast of Greenland, trading goods and knowledge. The Dutch effort, by far the largest in this region in the 18th century, faltered due to exhaustion of the stock, competition with other countries, tariffs of goods transported to foreign markets, and wars at home with neighboring countries (de Jong, 1978).

As Britain became the most powerful force on the oceans, its whaling efforts increased. Around 1820 a route to the northern summering areas of bowheads became known. Whalers could now pursue the bowhead whales as they migrated throughout most of their range, more or less in a counterclockwise direction. Around 1850 ships started to winter in the Arctic. This permitted earlier access to whales in spring, when the route to those areas from the south was still blocked by ice. Land bases were also established. The Scottish whaling station Kekerten, in Cumberland Sound, was established around 1860 and functioned for more than half a century (Stevenson, 1984). It was occupied year-round, and most inhabitants were Inuit families who provided labor for catching whales and processing blubber.

While the 18th century whaling took mainly mature males migrating along eastern routes, the effort in the 19th century shifted west and caught mostly females and immature individuals. Late in the whaling season, when the ice was gone from the Canadian coast, whalers hunted close to land, catching "rocknosers" young bowheads that migrated very close to rocky shorelines.

The Hudson Bay bowhead fishery constitutes a final phase of the exploitation of the ECWG stock and commenced when ships were able to winter there in 1860. Ships engaged in this hunt were mostly American and British. They reached Hudson Bay in July and were able to whale for 2 or 3 months before the sea ice formed. With ships still locked in ice the next spring, whaling could start in May, using whaling boats and sleds to bring blubber and baleen back to the ship. Whalers in Hudson Bay worked closely with the Inuit, who supplied the crews with fresh meat, skin clothing, and other necessities. Whaling crews often included Inuit hunters. Overall, the Hudson Bay commercial whaling effort was small, with fewer than 1000 whales taken and it ended around 1914.

The East Greenland–Svalbard–Barents Sea stock

In the early 1600s the naval powerhouses of Britain and the Netherlands explored the northern part of the North Atlantic and the Arctic Ocean in order to find a new way to the Orient. With the geographical insights acquired, and the help of Basque knowledge and man power, whaling by these nations started in this region in the second decade of the 17th century, with other nations joining in (Germany, France, and Denmark). Basque whaling in the Strait of Belle Isle had declined, but oil was needed for lighting and to make soap, as, in Europe, hygiene had improved and white collars were fashionable. Whale oil filled the demand. Competition between British and Dutch whalers, between incorporated whalers and unincorporated ones, and even between whaling towns within a single nation continued until eventually warships created a semblance of order although maybe not justice (Hacquebord and de Bok, 1981). Superior naval power enabled whalers of one nation to capture whaling tools and whale products from other whalers.

Whaling by British and Dutch initially involved launching whaling boats and trying blubber in cauldrons on the coast of the islands of the Svalbard archipelago, including Spitsbergen. Dead whales were towed to the whaling ship and butchered, and blubber floated back to shore for trying, a process called "bay whaling." Because of the difficulties of transport, only whales close to land were caught. No year-round settlements were established, whalers would return to homeport with barrels of oil, and the ships would be used for other purposes during winter. By the mid-1620s the Dutch dominated whaling, initially tolerating Danish, but not English whalers. The Dutch called their main settlement Smeerenburg ("smeren" means to grease or smear). Some 20 whaling ships might spend the summer here, with hundreds of men ashore rendering blubber, and many others involved in the hunt of whales. By 1640 the heyday of bay whaling had passed, most likely as the result of overhunting. Whales appeared to have changed their behavior, avoiding areas where whalers worked (de Jong, 1983). The end of the bay whaling coincided with changes in distribution and amounts of ice in Arctic waters, which concentrated whales in areas that might not be accessible to the whalers, although Ross (1993), points out that climate data are inconclusive in explaining the reduction of the fishery.

As bay whaling declined, pelagic whaling increased in importance, surpassing it around 1650. It was developed by Dutch whalers who did not have access to shore-based processing facilities. Ships involved in so-called ice-fishing were built specifically for whaling operations, with strengthened hulls to withstand impact from the ice. Now, far from land, whales were butchered along the side of the ship, and blubber transported back to European homeports for rendering at the end of the season. Whalers sailed between East Greenland and Svalbard, working along the ever-changing edge of the pack ice (see map of whaling grounds in Chapter 5). They followed the whales as they migrated in a roughly counterclockwise path, going north in March and April far from shore, arriving at northern feeding grounds in May, and returning south along the coast of East Greenland in late June, July, and August. Then, the whalers would hunt at the bowhead wintering grounds before returning to homeport. The whalers knew that different cohorts of whales (older single males, females with young, immature animals) migrated using different routes and migrated at different times (de Jong, 1983). Dutch whaling declined toward the end of the 17th century, and English and German whalers replaced it while stocks lasted.

The Bering–Chukchi–Beaufort stock

The bowhead fishery in the Atlantic Ocean originated when North Atlantic right whales became scarce, and a similar story developed in the Pacific. The spermaceti oil of sperm whales could be used to make candles that burned bright and were odorless, and whalers entered the Pacific to hunt sperm whales after Atlantic stocks were depleted. Then, around 1840, hoop skirts became fashionable in London and Paris, and the prices of baleen rose as it was used to make hoops. Sperm whalers responded by pursuing right whales in the Pacific. Around 1845 they became aware that the bowheads, known from high latitudes of the Atlantic, also inhabited high latitudes of the Pacific. The first whaling ship, the *Superior*, crossed alone into the Bering Sea in 1848 and found bowheads in abundance. A year later, 50 ships cruised the Bering Sea for bowheads; and by 1852, over 200 ships were engaged (Bockstoce and Burns, 1993).

The vast majority of hunting effort was by American ships, initially from New England and Long Island. These Yankee whalers would leave homeport in autumn, sail around Cape Horn, and impact the Pacific. In the early years, bowheads might be encountered as the whalers crossed the Aleutian Island chain in the southern Bering Sea, during spring of the following year. In the first phase of the fishery, approximately one-third of the total harvest of the nearly 19,000 whale killed in all years, was taken in summer in the northern Bering Sea and Bering Strait region, and the ships returned south in late summer. Catches crashed in 1855, whalers believed that whales responded to the hunting pressure by fleeing hunting grounds where great numbers were killed in earlier years. The whalers soon realized that many whales were migrating north in summer through the Bering Strait. In the 1860s whalers entered the Chukchi Sea in July as the sea ice retreated, and whales were caught during migration, in summer and autumn. In 1888 summering grounds were discovered in the eastern Canadian Beaufort Sea (Bockstoce, 1986) and hunting shifted there. Whaling ships, often using steam power, spent three or more years away from their homeport and wintered in the arctic. San Francisco became a whaling port for these whalers and crews included men of many nations. Shore-based stations became more common and were able to hunt whales long before ships could reach them. Shore stations were ethnically diverse. The shore whaling station near Point Hope was named "Jabbertown," as many languages could be heard there. Bockstoce (1986) noted, "two hundred Eskimos from many communities, as well as about two dozen whalemen, including American whites and blacks, Cape Verdeans, Portuguese, Japanese, Germans, and Irish." These stations later developed into trading posts employing local Iñupiat, and settlements developed around them (Brower, 1942).

Ice congestion in the Beaufort and Chukchi Seas allowed hunted whales to escape by diving under ice floes and to the pack ice where whalers could not pursue them. Fast killing became important. In the 1860s the harpoon initially used to fasten the boat to the whale was fitted with an explosive charge, called the "darting gun." The darting gun fired a small bomb into the whale as the harpoon tip entered the body, sometimes killing the whale instantly. Whalers also used "shoulder guns," heavy, short-barreled weapons that fired a similar bomb into the whale, generally from a distance of less than 20 m. Harvested whales were towed back to the ship, their baleen racks removed and their blubber stripped and rendered to oil on the ship. However, catches dropped and the stock was depleted during the 1870s. Ships were lost in storms and crushed by sea ice. As the US Civil War broke out, it became dangerous for Union whalers to pass by southern ports. In addition, the Confederate clipper *Shenandoah* destroyed over 30 Union ships in the Bering Sea, including some after the war had ended in 1865 because its captain was unaware of this fact (Bockstoce, 1986).

The 1880s saw a number of changes. While the price of whale oil dropped, partly due to competition with petroleum, the price of baleen rose. Narrow waists became popular in women's fashion, and baleen was used to enable corsets to achieve this shape. But the Bering–Chukchi–Beaufort (BCB) Sea stock was so depleted that the industry was doomed, and when baleen prices dropped in 1908, the large-scale bowhead hunt soon ended.

The Okhotsk Sea stock

Whaling ships first entered the Sea of Okhotsk in 1845, as this area was within the range of the Pacific right whale and a known isolated stock of bowheads existed there. The size of the population of bowheads here and estimates of the numbers caught during the next four decades vary by more than a factor of five (see review by Ivashchenko et al., 2011). Mitchell (1977) provided what may be the best supported estimate. These whalers did not distinguish right whales and bowheads in their records, and they might hunt the Okhotsk (OKS) and Bering-Chukchi-Beaufort (BCB) stocks in the same season. In the Okhotsk Sea, whales were concentrated in shallow bays in summer, and ice may have been less of a hindrance to the whalers. As a result, the Okhotsk stock was quickly depleted, especially between 1855 and 1857, a period that followed a few years of poor yields in the Bering Sea. With catches severely reduced, whaling ended, and whaleships resumed hunting the BCB stock. At the beginning of World War I, several hundred whales may have survived, and estimates of current abundance remain low (Cooke et al., 2018; Chapter 6). Soviet whalers continued hunting this stock until the 1960s (Ivashchenko et al., 2011).

Effect on indigenous people

The impact of commercial bowhead whaling on indigenous people across the Arctic was considerable and overall detrimental to their culture, society and populations. Whaling of the EGSB stock did not involve indigenous people, but that of the East Canada-West Greenland (ECWG) and BCB stocks did to significant degrees. European whalers exchanged information and knowledge with indigenous people, and many were employed by the European whalers as laborers. For the WGEC stock, indigenous people moved to be near settlements of Europeans, and this exposed them to diseases that appeared with whaling ships (Ross, 1973). At times, it also led to starvation, as Europeans exhausted hunted species using superior weapons. Bockstoce (1986) describes the precipitous decline and changes to indigenous societies in the Bering Strait region from foreign diseases, alcohol, and major reductions in the bowhead and walrus populations by American commercial whalers. Starvation was widespread throughout the BCB region in the 1860s to the 1880s particularly on Saint Lawrence, Diomede, and King Islands, where few terrestrial resources are available. On Saint Lawrence Island, 40%–50% of the indigenous population died in the winter of 1887–88.

To supplement reduced catches of bowheads, starting in about 1860, the whalers shifted their attention to walrus, landing over 150,000 animals (Bockstoce and Botkin, 1982) while killing perhaps twice as many. With bowhead numbers already reduced in the Bering Sea, the walrus-dependent indigenous peoples of Alaska and Chukotka suffered greatly. This led to the abandonment of settlements and widespread starvation. American whalers were aware of the hardships caused by the reductions of bowheads and walrus. In 1870 Captain Frederick A. Barker was shipwrecked on the Chukotka coast where indigenous people took his crew into their homes through the winter. As cited by Bockstoce (1986), he wrote, "Should I ever come to the Arctic Ocean to cruise again, I will never catch another walrus, for these poor people along the coast have nothing else to live upon.... I felt like a guilty culprit while eating their food with them, that I have been taking bread out of their mouths, yet although they knew that the whaleships are doing this, they still were ready to share all they had with us."

Changes to native societies accelerated across the United States and eastern Canadian Arctic starting in the 1880s, in part by the establishment of permanent shore-based whaling stations (Bockstoce, 1986; Brower, 1942). Permanent shore-based stations meant a continuous interaction with the outside world (Brower, 1942), generally with a negative effect on the livelihood of the indigenous peoples who depended on bowhead whales and irrevocably altering their culture.

Discussion

Economics was the engine that drove 400 years of commercial whaling and the near-extermination of the circumpolar bowhead whale stocks. A review of the catch records for all four stocks indicates that more than 136,000 bowheads were killed during the commercial whaling period. Supplies of whale products grew and shrank as stocks were discovered and exhausted, and demands grew and shrank as other, cheaper or better, alternatives for these products were found, or as once-popular products became less so (Bockstoce and Burns, 1993).

Secondary to economics, politics played a role. Basque whaling, and thereafter Dutch whaling, was reduced in importance as the naval powers of Spain and the Netherlands, respectively, diminished. The US purchase of Alaska from Russia was driven in part by commercial whaling interests (Bockstoce, 1986). The discovery of new areas with whales was of obvious importance, such as those near Baffin Island, in Hudson Bay, and in the Beaufort Sea. Finally, technology and innovation also played a role. Pelagic whaling of the WGSB stock and the ice-bound whaling on Canada's east coast and the Beaufort Sea relied on ships that could withstand impacts from ice. Trying blubber on board, explosive charges to kill whales, and steam power benefited the hunt. Finally, climate change may have played a role, as bowhead ranges were altered as the seas warmed after the Little Ice Age (Mann, 2002).

Those factors: economics, politics, discovery, technology, and innovation, exhausted the stocks of bowheads. The damage done has not been repaired: the WGSB stock is in precarious shape even a century and a half after significant whaling of it ceased. Nations and whalers of historical times lacked a long-term view of the impact of their actions.

They were not concerned with sustainability, there was no oversight and regulation, and only short-term gain was important.

As a result, all stocks of bowheads were greatly reduced, and some were nearly lost. While the outlook for the BCB and ECWG stocks is optimistic, the EGSB and OKS stocks remain precariously depleted (Cooke et al., 2018; Chapters 5 and 6).

Commercial whaling introduced disease, foreign foods, firearms, commercialism, alcohol, and unfamiliar religious practices into indigenous cultures, changing them permanently. The impact of bowhead commercial whaling was largely devastating to the culture and livelihood of the indigenous peoples. Bowhead subsistence whaling by indigenous people of Chukotka, Alaska, Canada, and Greenland does continue, using a combination of traditional and modern tools (Chapter 32).

Biological implications

The dramatic declines of bowheads across the arctic impacted the ecosystem as well. Bowheads are a major predator of invertebrates in the Arctic and were part of the cycling of nutrients in their ecosystem. The release of marine mammal feces into the upper water column benefits the entire ecosystem (Roman and McCarthy, 2010). It is possible that the decimation of bowheads in the high Arctic forced killer whales to feed more intensively on other smaller prey leading to a trophic cascade as discussed by Springer et al. (2003).

Whaling also affected the population structure of the stocks. The largest whales were taken in the earliest years of the American commercial whaling (Bockstoce and Burns, 1993), which skewed sex the ratios (since females are larger than males). In historical times the range of bowheads extended further south than current (see Chapter 4). Bockstoce and Burns (1993) assumed these areas were within the normal feeding range of BCB bowheads and speculated that several subpopulations with separate summer feeding areas existed. These may have been lost as a result of over exploitation, although the changing climate or disturbances by hunting may have altered distribution patterns too (Mann, 2002; Bockstoce et al., 2005).

References

Accardo, C.M., Ganley, L.C., Brown, M.W., Duley, P.A., George, J.C., Reeves, R.R., et al., 2018. Sightings of a bowhead whale (*Balaena mysticetus*) in the Gulf of Maine and its interactions with other baleen whales. J. Cetacean Res. Manage 19, 23–30.

Aguilar, A., 1986. A review of old Basque Whaling and its effects on the right whales (*Eubalaena glacialis*) of the North Atlantic. Rep. Int. Whal. Comm. Spec. (10), 191–199.

Barkham, S., 1984. The Basque whaling establishments in Labrador 1536-1632 – a summary. Arctic 37, 515–519.

Bockstoce, J.R., 1986. Whales, Ice and Men. Univ. Washington Press, p. 400.

Bockstoce, J.R., Botkin, D.B., 1982. The harvest of Pacific walruses by the pelagic whaling industry, 1848, 1914. Arct. Alp. Res 14, 183–188.

Bockstoce, J.R., Botkin, D.B., 1983. The historical status and reduction of the western Arctic bowhead whale (*Balaena mysticetus*) population by the pelagic whaling industry, 1848-1914. Rep. Int. Whal. Comm. Spec (5), 107–141.

Bockstoce, J.R., Burns, J.J., 1993. Commercial whaling in the North Pacific Sector. In: Burns, J.J., Montague, J.J., Cowles, C.J. (Eds.), The Bowhead Whale. Spec. Publ. 2, Society for Marine Mammalogy, pp. 563–577.

Bockstoce, J.R., Botkin, D.B., Philp, A, Collins, B.W., George, J.C., 2005. The geographic distribution of bowhead whales, *Balaena mysticetus*, in the Bering, Chukchi, and Beaufort seas: evidence from whaleship records. Mar. Fisher. Rev. 67, 31849–31914.

Brower, C.D., 1942. Fifty Years Below Zero: A Lifetime of Adventure in the Far North. Dodd, Mead and Co, Inc, New York.

Cooke, J.G., Brownell Jr., R.L., Shpak, O.V., 2018. *Balaena mysticetus* Okhotsk Sea subpopulation. In: The IUCN Red List of Threatened Species 2018: e.T2469A50345920. Available from: https://doi.org/10.2305/IUCN.UK.2018-1.RLTS.T2469A50345920.en.

de Jong, C., 1978. Geschiedenis van de oude Nederlandse walvisvaart, Deel 2, Bloei en achteruitgang. Bloei en achteruitgang 1642–1872. Johannesburg.

de Jong, C., 1983. The hunt of the Greenland Whale: a short history and statistical sources. Repts. Intern. Whal. Comm. Spec. Iss 5, 83–106.

Hacquebord, L., de Bok, R. 1981. Spitsbergen 79° N. B., een Nederlandse expeditie in het spoor van Willem Barentsz. Elsevier, Amsterdam, 160

Ivashchenko, Y.V., Clapham, P.J., Brownell Jr., R.L., 2011. Soviet illegal whaling: the devil and the details. Mar. Fish. Rev 73, 1–19.

Laist, D.W., 2017. North Atlantic Right Whales, From Hunted Leviathan to Conservation Icon. Johns Hopkins Univ. Pr, Baltimore, MD, p. 432.

Mann, M.E., 2002. Little Ice Age. In: MacCracken, M.C., Perry, J.S. (Eds.), Volume 1, The Earth System: Physical and Chemical Dimensions of Global Environmental Change, Vol 1 of the Encyclopedia of Global Environmental (Ted Munn, ed.). Wiley, Chichester, pp. 504–509.

McLeod, B.A., Brown, M.W., Moore, M.J., Stevens, W., Barkham, S.H., Barkham, M., et al., 2008. Bowhead whales, not right whales, were the primary target of 16th to 17th-century Basque whalers in the western North Atlantic. Arctic 61, 61–75.

McLeod, B.A., Frasier, T.R., Dyke, A.S., Savelle, J.M., White, B.N., 2012. Examination of ten thousand years of mitochondrial DNA diversity and population demographics in bowhead whales (*Balaena mysticetus*) of the central Canadian Arctic. Mar. Mammal Sci 28, E426–E443.

Mitchell, E.D., 1977. Initial population size of bowhead whale (*Balaena mysticetus*) stocks: cumulative catch estimates. In: Paper SC/29/33 Presented at the IWC Scientific Committee. 113 p. (unpublished).

Proulx, J.-P., 1986. Whaling in the North Atlantic From Earliest Times to the Mid-19th Century. Natl. Hist. Parks Sites Branch, Parks Canada, Environment Canada, p. 117.

Rastogi, T., Brown, M.W., McLeod, B.A., Frasier, T.R., Grenier, R., Cumbaa, S.L., et al., 2004. Genetic analysis of 16th century whale bones prompts a revision, of the impact of Basque whaling on right and bowhead whales in the western North Atlantic. Can. J. Zool 82, 1647–1654.

Roman, J., McCarthy, J.J., 2010. The whale pump: marine mammals enhance primary productivity in a coastal basin. PLoS One 5. Available from: https://doi.org/10.1371/journal.pone.0013255.

Ross, W.G., 1973. Whaling and the decline of native populations. Arctic Anthropol. 14, 1–8.

Ross, W.G., 1993. Commercial whaling in the North Atlantic sector. In: Burns, J.J., Montague, J.J., Cowles, C.J. (Eds.), The Bowhead Whale. Spec. Publ. 2, Society for Marine Mammalogy, pp. 511–561.

Springer, A.M., Estes, J.A., Van Vliet, G.B., Williams, T.M., Doak, D.F., Danner, E.M., et al., 2003. Sequential megafaunal collapse in the North Pacific Ocean: an ongoing legacy of industrial whaling? Proc. Natl. Acad. Sci. USA 100, 12223–12228.

Stevenson, M., 1984. Inuit, Whalers, and Cultural Persistence: Structure in Cumberland Sound and Central Inuit Social Organization. Oxford Univ. Pr, p. 400.

Townsend, C.H., 1935. The distribution of certain whales as shown by logbook records of American whaleships. Zoologica 19, 1–5.

Tuck, J.A., Grenier, R., 1981. A 16th century Basque whaling station in Labrador. Sci. Am 245, 180–190.

Woodby, D.A., Botkin, D.B., 1993. Stock sizes prior to commercial whaling. In: Burns, J.J., Montague, J.J., Cowles, C.J. (Eds.), The Bowhead Whale. Spec. Publ. 2, Society for Marine Mammalogy, pp. 387–407.

CHAPTER

34

Indigenous knowledge in research and management

H.P. Huntington[1], S.H. Ferguson[2], J.C. George[3], G. Noongwook[4], L. Quakenbush[5] and J.G.M. Thewissen[6]

[1]Ocean Conservancy, Eagle River, AK, United States [2]Fisheries and Oceans Canada, Central and Arctic Region, Winnipeg, MB, Canada [3]Department of Wildlife Management, North Slope Borough, Utqiaġvik, AK, United States [4]Savoonga Whaling Captains Association, Savoonga, AK, United States [5]Alaska Department of Fish and Game, Fairbanks, AK, United States [6]Department of Anatomy and Neurobiology, Northeast Ohio Medical University, Rootstown, OH, United States

Introduction

Experience and insight have been a source of knowledge throughout human history. Only in the past few centuries have scientific methods been developed and applied to questions of practice and theory. Even today, experiential knowledge is far broader than the relatively limited scope of rigorous scientific research. Such knowledge is known by many names, including Indigenous Knowledge (IK, which will be used herein), traditional knowledge, local knowledge, and other variants (Huntington and Fox, 2005). For Indigenous whale hunters in the Arctic, the knowledge gained by experience is paramount for safety and success (e.g., Nageak, 1991). Although visitors and newcomers to the Arctic have long relied on Indigenous assistance (e.g., Amundsen, 1908), only in recent decades has the knowledge of whalers and other practitioners elsewhere in the world attracted formal academic attention (e.g., Johannes, 1981; Mymrin, 1999; Hay et al., 2000). Indeed, the collaborations between Iñupiat and Yupik whalers and scientists in northern Alaska are among the best examples of sharing knowledge and developing new knowledge beyond the capabilities of either group alone. This chapter reviews the ways that scientists and managers have engaged with the knowledge of Arctic whalers, what has been learned thereby, and how the resulting knowledge has contributed to understanding and management of bowhead whales (Fig. 34.1).

The Bowhead Whale
DOI: https://doi.org/10.1016/B978-0-12-818969-6.00034-0

IK has often been defined as being based on experience and shared among members of a community (e.g., Berkes, 1999). It is thus more than what an individual has learned over a lifetime, and typically has been accumulated through many generations living in an area and continuing similar practices. The sharing of IK can occur in many ways, from observation and practicing what one observes to stories and adages that encapsulate key points (e.g., Noongwook et al., 2007). In addition to direct experience, IK may also draw on insights from dreams or the spiritual journeys of shamans (e.g., Helander, 2004). IK may thus include practical information about harvesting animals or staying safe on the land, ice, or water, as well as cultural and ethical values about respecting animals and sharing the harvest (e.g., Huntington et al., 2017). Practitioners of IK rarely make such distinctions themselves, instead of viewing their knowledge as an integral and indivisible aspect of their place in the world (e.g., Pierotti and Wildcat, 2000).

The empirical basis for much IK helps connect IK with scientific knowledge (SK) (e.g., Carmack and Macdonald, 2008), as examples in this chapter will demonstrate. The spiritual and ethical aspects of IK are harder to reconcile with scientific approaches, which can complicate attempts to link the two ways of understanding (e.g., Kawagley, 1995; Noongwook et al., 2007). The fact that IK has allowed communities to thrive in the Arctic for thousands of years is a strong reason not to discount observations and explanations that appear unlikely from a scientific point of view. Practitioners of IK literally stake their lives on its accuracy and relevance. For activities that are undertaken time and again,

FIGURE 34.1 Iñupiat hunters stand at the edge of the lead, watching for bowhead whales. Their craft, an umiaq, and gear lie ready. Upturned ice blocks are used to camouflage the hunters, who believe that whales can see them. *Source: Photo was taken around 1980 by Bill Hess, copyright, published in "The Gift of the Whale."*

a high degree of confidence may not be sufficient for long-term safety. Thus, many stories based on IK may emphasize rare or anomalous events or conditions, rather than the merely typical, so that practitioners will be prepared for the unusual (e.g., Eicken, 2010). In scientific terms, this emphasis may give disproportional weight to outliers, but returning home safely 95% or even 99% of the time is not adequate for a whaler (George et al., 2004). Being ready for the 100-year event may thus be more vital in the long-term than knowing the "normal" conditions in which even the unskilled may be comfortable.

At the same time, IK is not a static body of knowledge but an ever-developing one, based on past experiences as well as one's own observations and insights (e.g., Gadamus et al., 2015). There is every reason to incorporate new ideas and tools into one's repertoire (Sakakibara, 2020), to increase the chances of success and the likelihood of safe return. Combined with a deep curiosity about the world around them, Indigenous whalers are thus often eager to learn from scientists and others and to try new methods, without discarding the tried and true. What whalers seek, however, is not advice but information, which they can use, interpret or ignore as they see fit. The two-way sharing of information is a critical part of the successful collaborations that have characterized much bowhead research.

There are many ways to consider and characterize scientific and management engagement with IK (e.g., Ford and Martinez, 2000). Here we consider four modes of interaction. First, the informal use of IK in research and practice, as outsiders who are newcomers learn from local experience. Second, the formal documentation of IK using methods from the social sciences to record what Indigenous people say about bowhead whales. Third, the intentional application of IK to biological and ecological studies to expand scientific understanding. Fourth, the intentional application of IK to environmental management policies and actions to achieve better outcomes. Such distinctions are artificial in that many examples involve more than one of these modes, but nonetheless useful in considering the motivations and practicalities of developing collaborations of these kinds.

In this chapter, we start with a review of each of these four modes and continue with a closer look at selected cases that illustrate key aspects of Indigenous-scientific collaboration and trends over time. The chapter concludes with a discussion of these matters and some ideas for the future.

Modes of engaging Indigenous knowledge concerning bowhead whales

Informal use

IK about bowhead whales and whaling developed over many generations, encompassing information not only about the whales and their behavior but also the whaling environment, including sea ice, currents, weather, prey and predators, and other species in the ecosystem (e.g., Hay et al., 2000). When European explorers and traders arrived in the Arctic, they took to varying degrees the opportunity to learn from the peoples they encountered, about navigation, clothing, and more (e.g., Amundsen, 1908). In the case of whales and whaling, the Yankee whalers who set up shore-based stations in northern Alaska either employed Iñupiat whalers or used their expertise to hunt from the edge of the shore fast ice (Brower, 1942). While the newcomers may have been able to figure out how to succeed on their own, it is likely they would have faced higher risks and lower success rates as they learned to recognize the dangers of the ice and the habits of the whales.

Early interactions between local peoples and scientists also included informal use of IK. In the early 1850s, for example, John Simpson documented an Utqiaġvik resident's knowledge of the Beaufort Sea coastline as part of his ethnographic studies, along with observations about hunting patterns and practices, including whaling (Appendix 7 in Bockstoce, 1988). Starting in the late 1940s, the Naval Arctic Research Laboratory north of Utqiaġvik employed local Iñupiat as laborers and assistants, leading to conversations and exchanges among locals and scientists, some of which flourished into collaborations (Norton, 2001) and the use and acknowledgment of IK (Bee and Hall, 1956). The familiarity that each side gained with the other grew into respect, which laid the groundwork for more intentional use of IK in bowhead whale studies in the 1970s and 1980s (Albert, 2001). The early interactions, however, did not involve systematic methods and often omitted much recognition of the contributions of IK and specific individuals, though there are exceptions (e.g., Irving and Paneak, 1954; Irving et al., 1967).

Formal documentation

The documentation of IK, typically through social science methods, developed with the growth of scientific and scholarly recognition of the merits of IK. Perhaps because close collaboration between whalers and scientists in Alaska was already producing such strong results, dedicated documentation efforts concerning IK of bowhead whales were first carried out in Chukotka (Mymrin, 1999) and Nunavut (Hay et al., 2000) in the 1990s. Using interview methods, Huntington (1998) developed a similar project documenting IK about beluga whales in Alaska and Chukotka (Huntington et al., 1999; Mymrin et al., 1999). The Chukotka study recorded Siberian Yupik knowledge about bowhead migratory patterns, behavior, reproduction, interactions with other species and the physical environment, influence of human activities, hunting and sharing practices, and appropriate actions of hunters to respect the whales. The study reflected the resumption of bowhead whale hunting following the demise of the Soviet Union. In Nunavut, a similar interest in resuming bowhead whaling combined with a divergence of Inuit observations and scientific estimates for size of the Baffin Bay bowhead stock to spur interest in a large-scale, multi-community documentation effort, emphasizing population ecology and the cultural importance of bowhead whales to Inuit. In the 2000s, documentation of IK about bowhead whales in Alaska was done to complement the information to be gained from placing satellite transmitters on a necessarily small number of animals (Huntington and Quakenbush, 2009a,b). In Nunavut, Canada, formal documentation of IK about killer whale predation on marine mammals included observations of bowhead whale avoidance behavior such as retreating to shallow waters or cows holding calves above the water (Ferguson et al., 2012; Higdon et al., 2014). Two other documentation efforts were carried out to address management questions, as described below.

Intentional application to research

Scientists have long used the observations and ideas of IK holders as the basis for research projects (e.g., Irving and Paneak, 1954). An extensive history of interactions among scientists and Iñupiat had already taken place at the Naval Arctic Research Laboratory north of Utqiaġvik (Norton, 2001). Mutual familiarity and respect developed further when bowhead whale research increased rapidly following the International Whaling Commission's (IWC) ban on subsistence whaling in 1977 (IWC, 1978). As described in several examples in the next section, scientists were willing to listen carefully to Iñupiat whalers as they designed the bowhead whale census and other studies. The many collaborations have led to major advances in scientific understanding of bowhead whale physiology and life history. The North Slope Borough Department of Wildlife Management has been the leader in both promoting and carrying out such studies. A professional staff, resident in Utqiaġvik and including Indigenous and non-Indigenous personnel, has fostered long-term relationships based on mutual respect and shared commitment among whalers, scientists, managers, and others. The quality of the research and the participation of Iñupiat and Yupik whalers in conferences and meetings have also enhanced awareness of the whalers' expertise in wider scientific circles and at the IWC. Such efforts have not been limited to northern Alaska.

Intentional application to management

In 2004, the question of stock identity in the Bering Sea was raised at the IWC. The concern was based on the possibility that Bering–Chukchi–Beaufort Seas (BCB) stock of bowheads were not a single stock but comprised two or more distinct populations (IWC, 2005). One result was a multipronged effort to determine whether there were two stocks of bowhead whales sharing the northern Bering Sea in winter and spring. One prong was the documentation of the knowledge of Yupik whalers on St. Lawrence Island concerning the migratory routes of bowhead whales in the region (see Chapter 4) and the possibility that there were behaviorally distinct groups of whales (Noongwook et al., 2007). The research was carried out in 2006 and the results published in 2007 in time for that year's meeting of the IWC Scientific Committee (see George et al., 2007). Notably, this project was led by George Noongwook, a whaler and commissioner of the Alaska Eskimo Whaling Commission (see below). The study indicated that bowheads used two principal migratory routes past Saint Lawrence Island, but did not suggest whether these were separate stocks. Subsequently, additional projects have been conducted to document IK about bowhead whales in relation to migratory patterns and behavior, as will be discussed in more detail in the next section, and with regard to the use of Camden Bay in the Beaufort Sea along Alaska's north coast (Huntington, 2013), a site of interest for offshore oil exploration and thus a potential conflict with whales and whaling.

Examples of using Indigenous knowledge and scientific knowledge together

The bowhead whale census at Utqiaġvik, Alaska

When the IWC imposed its moratorium on bowhead whaling in 1977, Iñupiat and St. Lawrence Island Yupik whalers insisted that the ban was based on inaccurate information

(Huntington, 1992; Albert, 2001). They also recognized that such assertions would carry little weight without scientific corroboration. Fortunately, the relationships established in large part through the Naval Arctic Research Laboratory gave the Iñupiat an understanding of how to work with scientists and the scientists a reason to listen carefully to what the Iñupiat were saying (Norton, 2001; Albert, 2001). One outcome was a set of revisions to the methods used for counting bowhead whales during the spring migration past Point Barrow. The whalers said that bowheads continue migrating even when sea ice cover is essentially 100%, conditions in which it is impossible to see and thus visually count the animals. The whalers also said that some bowheads migrate farther offshore than can be seen from the counting locations on the edge of the shore fast ice (~4 km), also leading to undercounting of the whales. The use of hydrophones to detect whales when visibility was poor and aerial transects to assess the breadth of the migratory path both substantiated the whalers' observations that the population estimates were biased low and led to improved population estimates for the BCB stock of bowhead whales (Albert, 2001). Ten years of the ice-based census provided sufficient data to confirm the whalers' observations of increasing numbers of whales, and of course also meant that the quota for the whale harvest could be higher (see Chapter 6).

Estimating bowhead abundance in the eastern Canadian Arctic

The North Atlantic bowhead whale population was depleted during the late 1880s and early 1900s due to commercial whaling (Lotze and Milewski, 2004). Following the cessation of commercial whaling, bowheads were considered too rare to warrant dedicated field studies, and little was known about them to non-Indigenous peoples (Finley, 2001). It was thought that Inuit hunting, killer whale predation, and habitat instability were responsible for their lack of population recovery (Mitchell and Reeves, 1982). With the initiation of oil and gas industry exploratory research, aerial surveys were funded in the mid-1970s (Davis and Koski, 1980). These surveys in conjunction with IK led to the rediscovery of fall concentrations of whales in Isabella Bay (Davis and Koski, 1980) and IK provided the background understanding of the Foxe Basin bowhead nursery (Cosens and Blouw, 2003). Scientific estimates of bowhead abundance were considered too conservative by IK (Mitchell and Reeves, 1982; DIAND, 1993; Cosens and Innes, 2000) as Inuit believed that the number of bowheads had increased significantly since the 1960s (Hay et al., 2000). A decision was made by Fisheries & Oceans Canada (DFO) to ratify and resume Inuit hunting of bowhead in the mid-1990s as a result of IK beliefs originating from the development of the Nunavut Land Claims Agreement (DIAND, 1993). This coming together of IK and science was considered a turning point in government attitudes toward IK (Finley, 2001) and IK gained legitimacy once the hunters' observations were corroborated (Koski et al., 2007; DFO, 2015).

Sense of smell

In spring, bowhead whales in the BCB stock migrate north and east along the coast of the Chukchi Sea, which is still ice covered. An open water channel, or lead, is often present and runs parallel to the coast. The whales swim through this channel, while Iñupiat hunters stay in camps on the channel's edge and launch their umiaqs (skin boats) to pursue them. Local Iñupiat knowledge indicated that the whales have a sense of smell, and in order to not alert the whales of the whalers' presence, the whalers minimized odors on the ice and even in nearby villages: no garbage was burned and latrines were constructed north of camp (i.e., so the whales would be past the camp before encountering the smell). This was expressed in the Inupiaq concept of "puvlak," which can mean "to be frightened away by smell," but can also mean "to bubble" and "to spout (for a whale)" (Starosin, 1998–present).

In contrast, scientists have stated that baleen whales have a "small, but functional nose ... resembling that of a human" (Oelschläger and Oelschläger, 2002) or that their olfactory bulb, a critical part of the olfactory hardware, "was only found in fetuses" (Dehnhardt, 2002). The inconsistency between traditional knowledge and science was resolved by the discovery of all parts of the peripheral olfactory system (including the olfactory bulbs) in bowheads, which, combined with genomic evidence, indicates that bowhead whales have a sense of smell likely to be much better than that of humans (Thewissen et al., 2011; Chapter 18).

Buoyancy

The vast changes in body size and shape that bowhead whales undergo in the first decade of life have been noted by Iñupiaq hunters. Iñupiaq terms have been adopted by scientists, including *ingutuq* for short, fat whales with a small head and short baleen, and *qairaliq* thinner whales, with larger heads and longer baleen (MacLean 2012). There are many variants of these two broad phenotypic types. *Ingutuqs* are mostly yearling whales that were weaned recently and relied on milk for their growth. *Qairaliqs* are older whales that have used body resources (such as fat stored during suckling) to build the baleen rack needed to become efficient feeders (Chapter 7). Differences in morphology were so stark that *ingutuqs* were sometimes considered separate species (Braham et al., 1980).

The changes in the amounts of fat during the first decade would be expected to affect buoyancy, and it was thought that bone (denser than water) might be used to counterbalance fat (less dense than water). As such, the *ingutuq* would be expected to have denser, or more, bone. IK provided clues to understanding this system, when hunters explained that, historically, Iñupiat used the ribs of *ingutuqs* as net weights in fishing. Indeed, George et al. (2016) found that bone volume of ribs changes over the first decade of life in a pattern that is the inverse of that of fat stores. Consistent with this finding and with the observations of the whalers, Lubetkin et al. (2008) found that bowhead whales stop growing longer from about age 2–5 years (Chapter 7), during which the transition from *ingutuq* to *qairaliq*body type occurs.

Life span

In the absence of scientific data, biologists estimated the bowhead whale life span to be about 50–70 years (Nerini et al., 1984), consistent with that of the closely related right whale. Some Indigenous whalers, however, have stated that bowheads live "two human lifetimes" or otherwise remarked on the extended longevity of bowheads (A. Brower Jr., pers. comm.). Hunters on St. Lawrence Island and in Point Hope reported seeing the same individuals, such as a distinctive mostly white whale, over several decades or even generations of whalers. The discovery of stone harpoon heads in whales landed in the late 20th century provided additional evidence of a long life span, as metal implements had replaced stone by the end of the 19th century (George et al., 1999). St. Lawrence Island whalers also found ivory whaling implements in whales taken in recent decades, drawing the same conclusion about the long lives of bowhead whales. Subsequent estimates of bowhead age from other methods provided additional evidence that bowhead whales live well over a century and perhaps twice that long (George et al., 1999). In this case, IK and scientific studies provided consistent evidence of the remarkable longevity of the bowhead whale, giving greater confidence to findings that might otherwise have been treated with great caution or even dismissed as the result of some error of analysis or interpretation.

Molting

In 2014, biologists observed bowhead whales in Cumberland Sound, Nunavut, that appeared to be molting in shallow waters. In 2016, the biologists were able to capture aerial video recordings from the same area, showing bowhead whales rubbing on rocks to help remove old skin (Fortune et al., 2017). At the time, relatively little was known to science about bowhead molting and associated behavior. A further look into historical observations made by commercial whalers in the 1800s and by Inuit and biologists more recently revealed that rock-rubbing behavior has long been known in the region (e.g., Hay et al., 2000). The use of aerial photography and biopsy sampling allowed biologists to confirm that nearly all of the whales were molting and were using rock rubbing to assist removing old skin over much of their body. The use of IK and historical observations by commercial whalers confirmed that the rock-rubbing associated with molting was neither new nor uncommon, but instead a recognized behavior of bowhead whales that assisted with a seasonal molt on the eastern coast of Baffin Island.

Satellite telemetry and complementary understanding

When a satellite telemetry study was being planned for BCB bowhead whales in the early 2000s, the Alaska Eskimo Whaling Commission was justifiably concerned that the few tagged whales would not represent all that bowheads do near shore where the whalers observe and hunt them, and suggested interviews with whalers in addition to the telemetry work (Chapter 4). Putting the satellite telemetry and IK together provides a larger scale view of bowhead behavior and adds information from the past as well as the present. Meeting with whalers for interviews also allowed the biologists to share results of the tagging project with the whalers and their communities and facilitated discussions about the movements of tagged whales. The comparison illustrates four types of complementarity:

- *The behavior of tagged whales can be compared with past patterns of migration.* No tagged whales entered Kotzebue Sound in spring, but whalers in Kivalina said that when the ice was thick offshore, bowheads used to follow the nearshore leads past Kivalina (Huntington et al., 2016).

- *Behaviors such as surface feeding and size segregation during migration cannot be detected from satellite transmitters.* Whalers in Wainwright and Utqiaġvik report that the same-sized whales generally travel together. Mating, calving, and feeding at the surface and under the ice have been seen near Wainwright and Utqiaġvik (Huntington and Quakenbush, 2009a,b).
- *Satellite telemetry can confirm distribution patterns seen by whalers.* Bowhead whales are occasionally seen north of Utqiaġvik in summer. Some tagged whales leave Canadian waters in summer, swim west to the Utqiaġvik area, and then return to Canada before the actual fall migration. Few whales are seen near Wainwright in fall (Huntington and Quakenbush, 2009a), and satellite telemetry combined with oceanographic data show that bowheads tend to avoid the Alaska Coastal Current, which is usually close to shore near Wainwright in fall (Citta et al., 2018).
- *Insights from IK can help in telemetry study design.* The size segregation that is consistently seen during spring migration and during fall migration at Utqiaġvik has allowed biologists to better understand which segment of the population the tags may represent.

Indigenous scholarship

The contributions of IK to scientific understanding of bowhead whales have, as a general trend, been recognized to a greater degree and more formally over time. The same is true for IK generally. Early on, IK holders may not have even been acknowledged in publications, or not by name. Formal documentation of IK led gradually to acknowledgment by name, and gradually to collective (i.e., a community or tribal government or organization) and individual coauthorship (see Huntington, 2006), as can be seen in some of the works cited in this chapter. The next step has been Indigenous leadership of research projects, reflected as first authorship of academic publications (e.g., Noongwook et al., 2007). The formal demands of academic research and publication are a high, but not insurmountable, barrier to those without advanced education and scientific experience, qualifications that Arctic peoples have achieved at much lower rates than the general population (Hirschberg and Petrov, 2014). Not surprisingly, collaborations remain the norm in IK research in the Arctic, though Indigenous initiative and leadership may be starting to play a larger role (e.g., Pfeifer, 2018).

The Noongwook et al. (2007) paper is notable for including an account of the methodology of acquiring and transmitting IK among members of a community, a topic largely overlooked in previous IK scholarship. The lead author's deep knowledge of IK and St. Lawrence Island Yupik culture made the description of how individuals learn IK from experience and from others possible. This was a substantial contribution to the IK literature that would likely have continued to be overlooked otherwise.

The paper also included a description of *angyii* (misspelled *angyi* in the paper), a behavior in which bowhead whales swim alongside a boat as it considers whether to give itself to the whalers, a topic for which there is at present no basis in scientific understanding. That a whale might intentionally give itself to whalers and would have the cognitive ability to deliberate this decision is more in the realm of the spiritual than the physiological or even psychological. Such behaviors and the Indigenous understanding of them are often described in the anthropological literature (e.g., Cruikshank, 1998). Again, the lead author's role in understanding this phenomenon and its significance was essential to resisting criticisms from peer reviewers that the topic should be removed from the paper. A greater role for Indigenous scholars in bowhead whale research, whether focusing on IK or not, may further expand our appreciation for the breadth of IK, even as it challenges scientists with ideas such as *angyii*.

Discussion

The combination of IK and SK employed in the study of bowhead whales in Alaska is a clear-cut example of drawing on the strengths of each to produce better common understanding and ultimately better management. Because the scientists listen to the whalers, the whalers have greater confidence in the results of the whale census and associated studies of behavior, anatomy, and more. One result is that the whalers are more likely to respect the scientists and managers as well. To date, the BCB bowhead whale population estimate has shown a robust increase, so there has been no need to propose reductions in hunting and thereby test the resilience of the whaler–manager relationship. Nonetheless, offshore oil and gas activity and commercial shipping create a need for active management efforts, in which whalers and scientists have worked together to develop appropriate regulations. It is all too easy to imagine a different scenario, in which relationships were characterized by distrust rather than respect, cooperation was elusive, and effective management of threats to whales and whaling was that much harder.

Despite or perhaps because of this strong record of practical collaboration, formal documentation of IK about bowhead whales in Alaska lagged behind such work in Chukotka and Nunavut, and behind similar studies on other marine mammals (e.g., Thomsen, 1993; Byers and Roberts, 1995; Huntington et al., 1999, 2017; Mymrin et al., 1999; Laidre et al., 2018) and environmental features such as sea ice (e.g., Krupnik et al., 2010) and weather (e.g., Gearheard et al., 2010) in Alaska and elsewhere. Indeed, the success story of the bowhead whale census, resistance to the IWC ban on bowhead whaling, and the tremendous advances in bowhead biology have often been cited in passing but have rarely been laid out in a single narrative account, and to date not in prominent publications. Indeed, while the North Slope Borough Department of Wildlife Management's website (http://www.north-slope.org/departments/wildlife-management/studies-and-research-projects/bowhead-whales/traditional-ecological-knowledge-of-bowhead-whales) describes several examples of using IK and SK, this informal history is not reflected in the "official" publications particularly about bowhead abundance (e.g., Givens et al., 2016) and anatomy (e.g., Thewissen et al., 2011) in academic journals, few of which mention the role that IK played in the study in question. Nonetheless, the design of the bowhead whale census using IK continues to be held up as a prime example of why attention to IK is important in scientific studies, even though the tale is nearly four decades old. While the Department of Wildlife Management strongly relied on and encouraged attention to IK, there was apparently a separation between daily interactions and most academic articles in the peer-reviewed literature.

Other studies of IK on different topics have helped demonstrate the level of detail that hunters, fishers, and gatherers can provide, in the Arctic and around the world (e.g., Souza and Begossi, 2007; Acebes, 2009; Liu et al., 2016). Such studies can reinforce the findings of scientific studies, complement those findings with details unavailable from other sources or methods, or point toward new areas of inquiry (Huntington et al., 2004). Climate change and its effects have raised the prominence of IK, as IK provides a record of change over a longer period of time than that available through contemporary observations, as well as details that might otherwise be overlooked (Fox et al., in review). This attention to IK, however, has not been uniformly praised (e.g., Pfeifer, 2018), as some scholars see a repetition of colonial habits in which academically trained "experts" become the voices of the field, displacing those who actually hold and acquire IK and are recognized by their peers and communities as reliable sources of information. The role of Indigenous scholars and the recognition of Indigenous expertise can be improved.

Similarly, the application of IK to environmental management is not without its critics. For example, Nadasdy (2003) found that attention to IK was at times a means of coopting hunters into accepting the dominant paradigm of professional wildlife managers. His attention to the importance of power dynamics in hunter–biologist–manager relationships is an important caution. Gadamus et al. (2015), among others, propose an alternative approach giving greater weight to Indigenous practices, knowledge, and authority. In the case of bowhead whaling in Alaska, the census can be seen as a tacit acceptance of the IWC's authority and paradigm. As long as the BCB population remains robust and the IWC issues a suitable quota, tensions between worldviews can be overlooked. If those conditions change, then such differences may resurface, and the tidy story of IK contributing to better science which then benefits the whalers may unravel. Similarly, the management of offshore oil and gas activity in northern Alaska has the stated goal of including IK in decision-making (Kendall et al., 2017), but such efforts are typically done in an agency-centered paradigm with few, if any, IK holders among those actually making the decisions.

With these cautions in mind, we can also note the limitations of IK (and recognize that SK, too, has weaknesses and shortcomings). Based on observations and practices of a particular place, IK typically has a limited spatial coverage (Gilchrist et al., 2005), though hunters from different communities may share their ideas with one another. IK, again typically but not exclusively, focuses on practical knowledge and the conditions of the place where whaling and other activities take place. It is of little use to know what whales do when they are far from one's home, but of great value to know precisely what they are likely to do in front of one's whaling camp. IK is also limited by and large to the capabilities of human senses. What whales do deep under water or over the horizon is not accessible to humans on the shore or the ice edge. Except in extreme cases, the concentrations or even the existence of chemical contaminants in whale tissues cannot be measured by taste, smell, or appearance. IK by itself has sustained whalers and whaling communities since time immemorial, but it does not encompass everything there is to know about bowhead whales. Similarly, not all whalers are equally observant or knowledgeable, and it is important to identify and recognize those who have earned the respect of their peers rather than those who simply speak most readily.

At the same time, based on the observations of the whalers, there is much yet for scientific studies to learn. In hunting camps, the color red is avoided so as not to alert the whales to the presence of the whalers. The visual acuity and color sensitivity of bowhead whales is not yet known to science, but this is an important topic when considering the potential for disturbance from ships or from industrial structures such as offshore oil platforms or coastal infrastructure. The Arctic Ocean continues to change rapidly and monitoring that change and its implications throughout the food web and ecosystem will only become more important (Chapter 27). Hundreds of whalers are on the ice and the water in many communities along the coast, providing greater consistency in skilled observation than is possible in a scientific study. Finding ways to continue to learn from the whalers is important, not just through the occasional effort to document IK, but in ways that create a more continuous and regular record of what they see and how they understand those observations.

Finally, the recognition of IK has increased, but still has a long way to go. This is true in scientific and management settings, and also in Indigenous communities themselves. One effect of colonization is the devaluation of Indigenous ways, a habit that can be absorbed by Indigenous peoples themselves as they are taught or compelled to give greater weight to the dominant culture's language, worldviews, and systems (Smith, 2012). A particularly gratifying outcome of formal attention to IK is a growing appreciation within Indigenous communities that their knowledge is indeed valuable, worth perpetuating, and worth some care as to when and how it is shared with others. Collaboration with scientists has helped Alaska whalers retain their ability to hunt bowhead whales legally, which is a notable success. There is room for further progress in recognizing IK holders through coauthorship and other manifestations of partnerships. There is also room for improving the inclusion of IK in the formal education system, in addition to the necessary ways of learning by doing and listening that have allowed IK to persist through the generations and sustain those who live by it.

Acknowledgments

We are grateful to all the whalers and their communities in the Arctic who have supported, participated in, and contributed to scholarly work on Indigenous knowledge about bowhead whales. Without their generosity, kindness, openness, and patience, little research of this kind would have been possible.

References

Acebes, J.M.V., 2009. Historical whaling in the Philippines: origins of 'indigenous subsistence whaling', mapping whaling grounds and comparison with current known distribution: a HMAP Asia Project Paper. Working Paper No. 161. Murdoch University, Perth.

Albert, T.F., 2001. The influence of Harry Brower Sr., an Iñupiaq Eskimo hunter, on the bowhead whale research program conducted at the UIC-NARL facility by the North Slope Borough. In: Norton, D.W. (Ed.), Fifty More Years Below Zero. University of Alaska Press, Fairbanks, AK, pp. 265–278.

Amundsen, R., 1908. The North West Passage. Archibald Constable, London.

Bee, J.W., Hall, E.R., 1956. Mammals of Northern Alaska on the Arctic Slope. University of Kansas, Museum of Natural History, Lawrence, KS.

Berkes, F., 1999. Sacred Ecology: Traditional Ecological Knowledge and Resource Management. Taylor & Francis, Philadelphia, PA.

Bockstoce, J. (Ed.), 1988. The Journal of Rochfort Maguire, 1852–1854. Hakluyt Society, London.

Braham, H.W., Durham, F.E., Jarrell, G.H., Leatherwood, S., 1980. Ingutuk, a morphological variant of the bowhead whale. Balaena mysticetus. Mar. Fish. Rev. 42 (9–10), 70–73.

Brower, C.D., 1942. Fifty Years Below Zero. Dodd, Mead, New York.

Byers, T., Roberts, L.W., 1995. Harpoons and Ulus: Collective Wisdom and Traditions of Inuvialuit Regarding the Beluga ("qilalugaq") in the Mackenzie River Estuary. Byers Environmental Studies, Winnipeg, MB.

Carmack, E., Macdonald, R., 2008. Water and ice-related phenomena in the coastal region of the Beaufort Sea: some parallels between Native experience and western science. Arctic 61 (3), 265–280.

Citta, J.J., Okkonen, S.R., Quakenbush, L.T., Maslowski, W., Osinski, R., George, J.C., et al., 2018. Oceanographic characteristics associated with autumn movements of bowhead whales in the Chukchi Sea. Deep Sea Res. II 152, 121–131. Available from: https://doi.org/10.1016/j.dsr2.2017.03.009.

Cosens, S.E., Blouw, A., 2003. Size-and age-class segregation of bowhead whales summering in northern Foxe Basin: a photogrammetric analysis. Mar. Mamm. Sci. 19 (2), 284–296.

Cosens, S.E., Innes, S., 2000. Distribution and numbers of bowhead whales (*Balaena mysticetus*) in Northwestern Hudson Bay in August 1995. Arctic 53, 36–41.

Cruikshank, J., 1998. The Social Life of Stories. University of Nebraska Press, Lincoln.

Davis, R.A., Koski, W.R., 1980. Recent observations of the bowhead whale in the eastern Canadian High Arctic. Rep. Int. Whal. Commn. 30, 439–444.

Dehnhardt, G., 2002. Sensory systems. In: Hoelzel, A.R. (Ed.), Marine Mammal Biology: An Evolutionary Approach. Blackwell, Oxford, UK, pp. 116–141.

DFO, 2015. Updated abundance estimate and harvest advice for the Eastern Canada-West Greenland bowhead whale population. Fisheries and Oceans Canada, Canadian Scientific Advisory Secretariat, Science Advisory Report 2015/052.

DIAND, 1993. Agreement Between the Inuit of the Nunavut Settlement Area and Her Majesty the Queen in Right of Canada. Department of Indian Affairs and Northern Development and Tungavik, Ottawa, p. 282.

Eicken, H., 2010. Indigenous knowledge and sea ice science: what can we learn from Indigenous ice users? In: Krupnik, I., Aporta, C., Gearheard, S., Laidler, G.J., Holm, L.K. (Eds.), SIKU: Knowing Our Ice. Springer, Dordrecht, pp. 357–376.

Ferguson, S.H., Higdon, J.W., Westdal, K.H., 2012. Prey items and predation behavior of killer whales (*Orcinus orca*) in Nunavut, Canada based on Inuit hunter interviews. Aquat. Biosyst. 8, 3. Available from: https://doi.org/10.1186/2046-9063-8-3.

Finley, K.J., 2001. Natural history and conservation of the Greenland whale, or bowhead, in the northwest Atlantic. Arctic 54, 55–76.

Ford, J., Martinez, D. (Eds.), 2000. Traditional ecological knowledge, ecosystem science, and environmental management. Ecol. Appl. 10 (5), 1249–1340.

Fortune, S.M.E., Koski, W.R., Higdon, J.W., Trites, A.W., Baumgartner, M.F., Ferguson, S.H., 2017. Evidence of molting and the function of "rock-nosing" behavior in bowhead whales in the eastern Canadian Arctic. PLoS One 12 (11), e0186156. Available from: https://doi.org/10.1371/journal.pone.0186156.

Fox, S., Qillaq, E., Angutikjuak, I., Tigullaraq, D.J., Kautuk, R., Huntington, H., et al., in review. Connecting understandings of weather and climate: steps towards co-production of knowledge and collaborative environmental management in Inuit Nunangat. Arctic Sci.

Gadamus, L., Raymond-Yakoubian, J., Ashenfelter, R., Ahmasuk, A., Metcalf, V., Noongwook, G., 2015. Building an indigenous evidence-base for tribally-led habitat conservation policies. J. Mar. Policy 62, 116–124. Available from: https://doi.org/10.1016/j.marpol.2015.09.008.

Gearheard, S., Pocernich, M., Stewart, R., Sanguya, J., Huntington, H.P., 2010. Linking Inuit knowledge and meteorological station observations to understand changing wind patterns at Clyde River, Nunavut. Clim. Change 100 (2), 267–294.

George, J.C., Huntington, H.P., Brewster, K., Eicken, H., Norton, D.W., Glenn, R., 2004. Observations on shorefast ice failures in Arctic Alaska and the responses of the Inupiat hunting community. Arctic 57 (4), 363–374.

George, J.C., Bada, J., Zeh, J., Scott, L., Brown, S.E., O'Hara, T., et al., 1999. Age and growth estimates of bowhead whales (*Balaena mysticetus*) via aspartic acid racemization. Can. J. Zool. 77, 1–10.

George, J.C., Stimmelmayr, R., Suydam, R., Usip, S., Givens, G., Sformo, T., et al., 2016. Severe bone loss as part of the life history strategy of bowhead whales. PLoS One 11 (6), e0156753. Available from: https://doi.org/10.1371/journal.pone.0156753.

George, J.C., Moore, S.E., Suydam, R., 2007. Summary of stock structure research on the Bering-Chukchi-Beaufort Seas stock of bowhead whales 2003–2007. Report to the Scientific Committee of the International Whaling Commission, SC/59/BRG3.

Gilchrist, G., Mallory, M., Merkel, F., 2005. Can local ecological knowledge contribute to wildlife management? Case studies of migratory birds. Ecol. Soc. 10 (1), 20. <http://www.ecologyandsociety.org/vol10/iss1/art20/>.

Givens, G.H., Edmondson, S.L., George, J.C., Suydam, R., Charif, R.A., Rahaman, A., et al., 2016. Horvitz-Thompson whale abundance estimation adjusting for uncertain recapture, temporal availability variation and intermittent effort. Environmetrics 27, 134–146.

Hay, K., Aglukark, D., Igutsaq, D., Ikkudluaq, J., Mike, M., 2000. Final Report of the Inuit Bowhead Knowledge Study, Nunavut, Canada. Nunavut Wildlife Management Board, Iqaluit.

Helander, E., 2004. Myths, shamans and epistemologies from an Indigenous vantage point. In: Helander, E., Mustonen, T. (Eds.), Snowscapes, Dreamscapes. Tampere Polytechnic Publications, Tampere.

Higdon, J.W., Westdal, K.H., Ferguson, S.H., 2014. Distribution and abundance of killer whales (*Orcinus orca*) in Nunavut, Canada—an Inuit knowledge survey. J. Mar. Biol. Assoc. U. K. 94 (6), 1293–1304.

Hirschberg, D., Petrov, A.N., 2014. Education and human capital. In: Larsen, J.N., Fondahl, G. (Eds.), Arctic Human Development Report: Regional Processes and Global Linkages. Nordic Council of Ministers, Copenhagen, pp. 349–399.

Huntington, H.P., 1992. Wildlife Management and Subsistence Hunting in Alaska. Belhaven Press, London.

Huntington, H.P., 1998. Observations on the utility of the semi-directive interview for documenting traditional ecological knowledge. Arctic 51 (3), 237–242.

Huntington, H.P., 2006. Who are the "authors" when traditional knowledge is documented? Commentary. Arctic 59 (3), iii–iv.

Huntington, H.P., 2013. Traditional knowledge regarding bowhead whales and Camden Bay, Beaufort Sea, Alaska. Unpublished report to the North Slope Borough Department of Wildlife Management.

Huntington, H.P., Quakenbush, L.T., 2009a. Traditional knowledge of bowhead whale migratory patterns near Wainwright, Alaska. Final report to the Wainwright Whaling Captains Associations, the Alaska Eskimo Whaling Commission, ConocoPhillips, Minerals Management Service, and the Coastal Marine Institute. Alaska Department of Fish and Game, Fairbanks, AK.

Huntington, H.P., Quakenbush, L.T., 2009b. Traditional knowledge of bowhead whale migratory patterns near Kaktovik and Barrow, Alaska. Final report to the Barrow and Kaktovik Whaling Captains Associations, the Alaska Eskimo Whaling Commission, ConocoPhillips, and the Minerals Management Service. Alaska Department of Fish and Game, Fairbanks, AK.

Huntington, H.P., Fox, S., 2005. The changing Arctic: indigenous perspectives. ACIA: Arctic Climate Impact Assessment. Cambridge University Press, Cambridge, pp. 61–98.

Huntington, H.P., the Communities of Buckland, Elim, Koyuk, Point Lay, and Shaktoolik, 1999. Traditional knowledge of the ecology of beluga whales (*Delphinapterus leucas*) in the eastern Chukchi and northern Bering seas, Alaska. Arctic 52 (1), 49–61.

Huntington, H.P., Quakenbush, L.T., Nelson, M., 2017. Evaluating the effects of climate change on Indigenous marine mammal hunting in northern and western Alaska using traditional knowledge. Front. Mar. Sci. 4, 319. Available from: https://doi.org/10.3389/fmars.2017.00319.

Huntington, H.P., Nelson, M., Quakenbush, L.T., 2016. Traditional knowledge regarding ringed seals, bearded seals, walrus, and bowhead whales near Kivalina, Alaska. Final report to the Eskimo Walrus Commission, the Ice Seal Committee, and the Bureau of Ocean Energy Management. Alaska Department of Fish and Game, Fairbanks, AK.

Huntington, H.P., Callaghan, T., Fox, S., Krupnik, I., 2004. Matching traditional and scientific observations to detect environmental change: a discussion on Arctic terrestrial ecosystems. Ambio Spec. Rep. 13, 18–23.

Irving, L., Paneak, S., 1954. Biological reconnaissance along the Ahlasuruk River west of Howard Pass, Brooks Range, Alaska, with notes on the avifauna. J. Wash. Acad. Sci. 44 (7), 201–211.

Irving, L., West, G.C., Peyton, L.J., Paneak, S., 1967. Migration of willow ptarmigan (*Lagopus lagopus*) in Arctic Alaska. Arctic 20 (2), 77–85.

IWC, 1978. Chairman's Report of the Twenty-Ninth Meeting. International Whaling Commission 28:22.
IWC, 2005. Report of the Scientific Committee. International Whaling Commission.
Johannes, R.E., 1981. Words from the Lagoon: Fishing and Marine Lore in the Palau District of Micronesia. University of California Press, Berkeley, CA.
Kawagley, A.O., 1995. A Yupiaq Worldview: A Pathway to Ecology and Spirit. Waveland Press, Prospect Heights, IL.
Kendall, J.J., Brooks, J.J., Campbell, C., Wedemeyer, K.L., Coon, C.C., Warren, S.E., et al., 2017. Use of traditional knowledge by the United States Bureau of Ocean Energy Management to support resource management. Czech Polar Rep. 7 (2), 151–163.
Koski, W.R., Heide-Jorgensen, M.P., Laidre, K.L., 2007. Winter abundance of bowhead whales, *Balaena mysticetus*, in the Hudson Strait, March 1981. J. Cetacean Res. Manage. 8 (2), 139–144.
Krupnik, I., Aporta, C., Gearheard, S., Laidler, G.J., Holm, L.K. (Eds.), 2010. SIKU: Knowing Our Ice. Springer, Dordrecht.
Laidre, K.L., Northey, A.D., Ugarte, F., 2018. Traditional knowledge about polar bears (*Ursus maritimus*) in East Greenland: changes in the catch and climate over two decades. Front. Mar. Sci. 5, 135. Available from: https://doi.org/10.3389/fmars.2018.00135.
Liu, M., Lin, M., Turvey, S., Li, S., 2016. Fishers' knowledge as an information source to investigate bycatch of marine mammals in the South China Sea. Anim. Conserv. 20 (2). Available from: https://doi.org/10.1111/acv.12304.
Lotze, H.K., Milewski, I., 2004. Two centuries of multiple human impacts and successive changes in a North Atlantic food web. Ecol. Appl. 14 (5), 1428–1447.
Lubetkin, S.C., Zeh, J.E., Rosa, C., George, J.C., 2008. Age estimation for young bowhead whales (*Balaena mysticetus*) using annual baleen growth increments. Can. J. Zool. 86, 525–538.
MacLean, E.A., 2012. Iñupiatun Uqaluit Taniktun Sivunniuġutiŋit, North Slope Iñupiaq to English Dictionary. University of Alaska Fairbanks, Alaska Native Languages Archives, Fairbanks, AK.
Mitchell, E.D., Reeves, R.R., 1982. Factors affecting abundance of bowhead whales *Balaena mysticetus* in the eastern Arctic of North America, 1915–1980. Biol. Conserv. 22 (1), 59–78.
Mymrin, N.I., 1999. Traditional Ecological Knowledge of Siberian Eskimo: Bowhead Whale, North of Bering Sea, Chukchi Peninsula. B.M. Zhitkov All-Russian Research Institute of Hunting and Fur Farming, Kirov, Russia.
Mymrin, N.I., the Communities of Novoe Chaplino, Sireniki, Uelen, and Yanrakinnot, Huntington, H.P., 1999. Traditional knowledge of the ecology of beluga whales (*Delphinapterus leucas*) in the northern Bering Sea, Chukotka, Russia. Arctic 52 (1), 62–70.
Nadasdy, P., 2003. Hunters and Bureaucrats: Power, Knowledge, and Aboriginal-State Relations in the Southwest Yukon. UBC Press, Vancouver.
Nageak, J.M., 1991. An unusual whaling season: an interview with Waldo Bodfish, Sr. Bodfish, Waldo Sr. Kusiq. University of Alaska, Fairbanks, AK, pp. 237–246.
Nerini, M.K., Braham, H.W., Marquette, W.M., Rugh, D.J., 1984. Life history of the bowhead whale, *Balaena mysticetus* (Mammalia: cetacea). J. Zool. 204, 443–468.
Noongwook, G., the Native Village of Savoonga, the Native Village of Gambell, Huntington, H.P., George, J.C., 2007. Traditional knowledge of the bowhead whale (*Balaena mysticetus*) around St. Lawrence Island, Alaska. Arctic 60 (1), 47–54.
Norton, D.W. (Ed.), 2001. Fifty More Years Below Zero. University of Alaska Press, Fairbanks, AK.
Oelschläger, H.H.A., Oelschläger, J.S., 2002. Brain. In: Perrin, W.F., Würsig, B., Thewissen, J.G.M. (Eds.), Encyclopedia of Marine Mammals. Academic Press, San Diego, CA, pp. 133–158.
Pfeifer, P., 2018. From the credibility gap to capacity building: an Inuit critique of Canadian Arctic research. North. Public Aff. 6 (1), 29–34.
Pierotti, R., Wildcat, D., 2000. Traditional ecological knowledge: the third alternative (commentary). Ecol. Appl. 10 (5), 1333–1340.
Sakakibara, C., 2020. Whale Snow: Iñupiat, Climate Change, and Multispecies Resilience in Arctic Alaska. University of Arizona Press, Tucson.
Smith, L.T., 2012. Decolonizing Methodologies: Research and Indigenous Peoples, second ed. Zed Books, New York.
Souza, S.P., Begossi, A., 2007. Whales, dolphins, or fishes? The ethnotaxonomy of cetaceans in São Sebastião, Brazil. J. Ethnobiol. Ethnomed. 3, 9. Available from: https://doi.org/10.1186/1746-4269-3-9.
Starosin, S., 1998–present. The Tower of Babel: An Etymological Database Project. <http://starling.rinet.ru>.

Thewissen, J.G.M., George, J., Rosa, C., Kishida, T., 2011. Olfaction and brain size in the bowhead whale (*Balaena mysticetus*). Mar. Mamm. Sci. 27, 282–294.

Thomsen, M.L., 1993. Local knowledge of the distribution, biology, and hunting of beluga and narwhal: a survey among Inuit hunters in West and North Greenland. Greenland Hunters' and Fishermen's Association, Greenland Home Rule Authorities, and Inuit Circumpolar Conference, Nuuk, Greenland.

CHAPTER

35

Effects of noise

Susanna B. Blackwell[1,2] and Aaron M. Thode[3]

[1]Greeneridge Sciences, Inc., Santa Barbara, CA, United States [2]University of California, Santa Cruz, CA, United States [3]Marine Physical Laboratory, Scripps Institution of Oceanography, University of California, San Diego, CA, United States

Introduction

As a result of global changes in climate, the Arctic waters that are home to bowhead whales are increasingly ice-free during summer months (Fig. 35.1). A warming planet has direct impacts on bowheads' environment, such as shifts in prey species (see Chapter 27), but indirect effects are also taking place, through modification of the way humans utilize the Arctic. For centuries the pack ice protected the Arctic against most ship-based anthropogenic activities. With ever-lengthening open-water seasons in the summer (Stroeve et al., 2014), routine commercial vessel traffic through the Arctic will become a reality in the next few decades (Smith and Stephenson, 2013; Chapter 27). Diminishing sea ice is also allowing increased exploration for oil and gas, which will likely result in the construction and operation of industrial infrastructures, such as oil-production islands or oil platforms. All these activities have and will continue to produce noise in bowhead whales' underwater habitat. Thus knowledge about the whales' reactions to anthropogenic noise becomes vital for understanding their long-term prospects.

An excellent introduction to concepts of physical acoustics is provided by Richardson and Malme (1993), as well as a summary of industrial noise sources and short-term disturbance reactions of bowhead whales to various man-made sounds. The latter topic is also included in Chapter 23 and summarized more generally in Richardson et al.'s (1995) book. This material will not be repeated here. Instead, this chapter focuses on recent studies of the effects of noise on bowhead acoustic communication, which is thought to be an important sensory modality for bowhead whales year-round (see Chapter 22). Due to the paucity of studies elsewhere, our discussion is restricted to the Bering–Chukchi–Beaufort Seas (BCB) stock in summer and fall, when whales are migrating in mostly ice-free water, and thus most sources of underwater noise arise from wind-driven air–sea interactions or

The Bowhead Whale
DOI: https://doi.org/10.1016/B978-0-12-818969-6.00035-2

FIGURE 35.1 A bowhead whale among ice that is just beginning to form (Iñupiat: *sikuliaq*). With the warming of the Arctic and reduction of sea ice, human activities such as shipping, fishing, and oil exploration are increasing. Noise produced by these activities has been shown to affect the whales' behavior including acoustic behavior. *Source: Photo by Brenda Rone (NOAA/North Slope Borough, NMFS Permit No. 14245).*

human causes. Despite these caveats, this chapter will reveal that a great deal has been learned about effects of various types of noise on bowhead whale acoustic activity.

Sources of noise in bowhead whale habitats

When listening to sound using an acoustic sensing device, or *hydrophone*, many noise sources are present. Some noise arises from the measurement process itself, such as electrical interference from the electronic recording circuitry. This *self-noise* is not discussed further here, although it is worth noting that at certain times in mid-winter, the Arctic Ocean is so quiet that self-noise can dominate the recordings. *Ambient noise* arises from actual acoustic sources in the ocean and can originate from natural sources, such as the grinding and shifting of ice floes in the winter, wind agitation of the ocean surface, or even animals themselves. Noise can also be produced by human industrial or commercial activities, referred to here as *anthropogenic noise*.

Potential sources of anthropogenic noise in bowhead whale habitats are numerous, ranging from pile-driving and generators near the coasts, to sounds associated with shipping and oil and gas production offshore. This chapter limits our discussion to three main noise sources, one natural and two anthropogenic. First, we focus on ambient wind-driven noise during ice-free conditions, because it has always been part of bowhead whales' environment and therefore serves in some sense as both a control and a reference. Second, we focus more generally on continuous sources of industrial sound. These include an omnipresent sound source in most of the world's oceans: vessel noise, which is predicted to increase in the Arctic. Finally, we discuss impulsive air gun sounds, because they are one of the strongest man-made sound sources in the bowhead whale hearing range and are also increasingly used in the Arctic. Each section below reviews how bowheads adjust their call production and source level in response to these sounds.

Ambient wind-driven noise

Since the mid-20th century, a number of studies have characterized the relationship between wind speed and ambient noise levels (Knudsen et al., 1948; Wenz, 1962; Ross, 1976; see Richardson et al., 1995 for a review). Wind is generally the most important nonbiological determinant of baseline ambient sound levels at frequencies between about 100 Hz and 10 kHz. In most world oceans the lower frequency spectrum (10–100 Hz) becomes dominated by distant commercial shipping noise (Hildebrand, 2009). In the ice-free Arctic, however, ocean shipping is still negligible, though this is changing, as will be discussed next. Furthermore, ocean swell is relatively small in the Arctic basin, resulting in low levels of surf noise along the coasts. The rarity of shipping and surf noise ensures that wind remains the dominant natural noise mechanism encountered by bowhead whales over their typical acoustic bandwidth (see Chapter 22). The close-knit relationship between wind speed and low-frequency sound in Arctic waters is illustrated in Fig. 35.2 for a 25-day period in summer 2012 in the Beaufort Sea. In this simple example, changes in wind speed explain about 74% of the changes in ambient sound levels.

If we omit the large spike on the 24th (caused by a vessel), the received levels (SPL) roughly span between 85 and 115 decibels, measured relative to 1 microPascal (abbreviated as dB re 1 μPa). The decibel is a logarithmic measure, so when received levels increase by nearly 30 dB, that is, the range of broadband level variation in Fig. 35.2, the environment's acoustic intensity increases by almost a factor of 1000. These shifts in sound level are predominantly driven by fluctuations in wind-driven noise. How do bowhead whales adapt their vocal activity over such large natural variations?

Thode et al. (2020) used a 7-year acoustic dataset of about 500,000 bowhead calls to investigate how such variations in ambient sound levels affected two components of whale communication: call densities (calls per unit time per unit area) and the source level (SL) of calls, with the latter component quantifying the "loudness" of the call. Data were collected using DASARs (Directional Autonomous Seafloor Acoustic Recorders, Greene et al., 2004), which combine an omnidirectional sensor with two orthogonal particle velocity sensors (1 kHz sampling rate on all channels). DASARs are deployed as equilateral triangular grids, and their ability to estimate a sound's direction allows for localization of whale calls

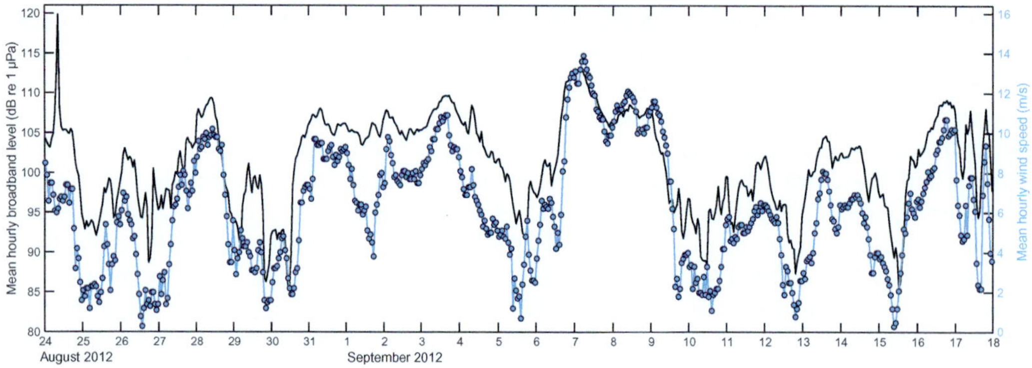

FIGURE 35.2 Relationship between wind speed and underwater sound levels. Mean hourly broadband levels (10–450 Hz, *black line*) and wind speed (*blue dots*) are shown as a function of date for 25 days in August–September 2012 in the Beaufort Sea. The sound data were collected by a seafloor recorder at location 70.54°N, 146.64°W (38 m depth), while the wind data were collected by the Camden Bay Met Buoy station, located 7.8 km south-southeast of the recorder.

through triangulation, provided that a call is detected concurrently by two or more recorders. This localization ability has been key in elucidating effects of anthropogenic sounds on calling behavior, and several of the studies presented in this chapter are based on DASAR data.

The results of the Thode et al. (2020) study showed that call density rose with increasing wind-driven noise levels, as measured in the same bands of frequencies as the calls (Fig. 35.3A). A 30 dB increase in ambient noise roughly doubled call density. In addition, when background levels were low, increases in noise were matched 1:1 by increases in the SL of calls (Fig. 35.3E). The whales thereby maintained their *communication space*, which is defined (Clark et al., 2009) as "space over which an individual animal can be heard by other conspecifics, or a listening animal can hear sounds from other conspecifics." This phenomenon, wherein a species' SL increases with background noise level, has also been observed in humans and other animal taxa, and is known as the "Lombard effect" (Brumm and Zollinger, 2011). As noise levels increased further, however, the mean SL no longer kept pace (Fig. 35.3E), possibly due to physiological limits in the sound production mechanism. Bowhead whales are thus highly responsive to changes in their acoustic environment, even under natural noise conditions.

Continuous industrial sounds

"Continuous industrial sounds" may seem like a hodgepodge category, but they have a few noteworthy common characteristics. First, the fact that they are continuous is important and distinguishes them from impulsive sounds, such as air gun pulses or explosions, which are thought to have a different effect on the hearing and behavior of marine mammals (Southall et al., 2007). The second common thread is that continuous industrial sounds generally include tones at frequencies related to power generation (60 Hz) or

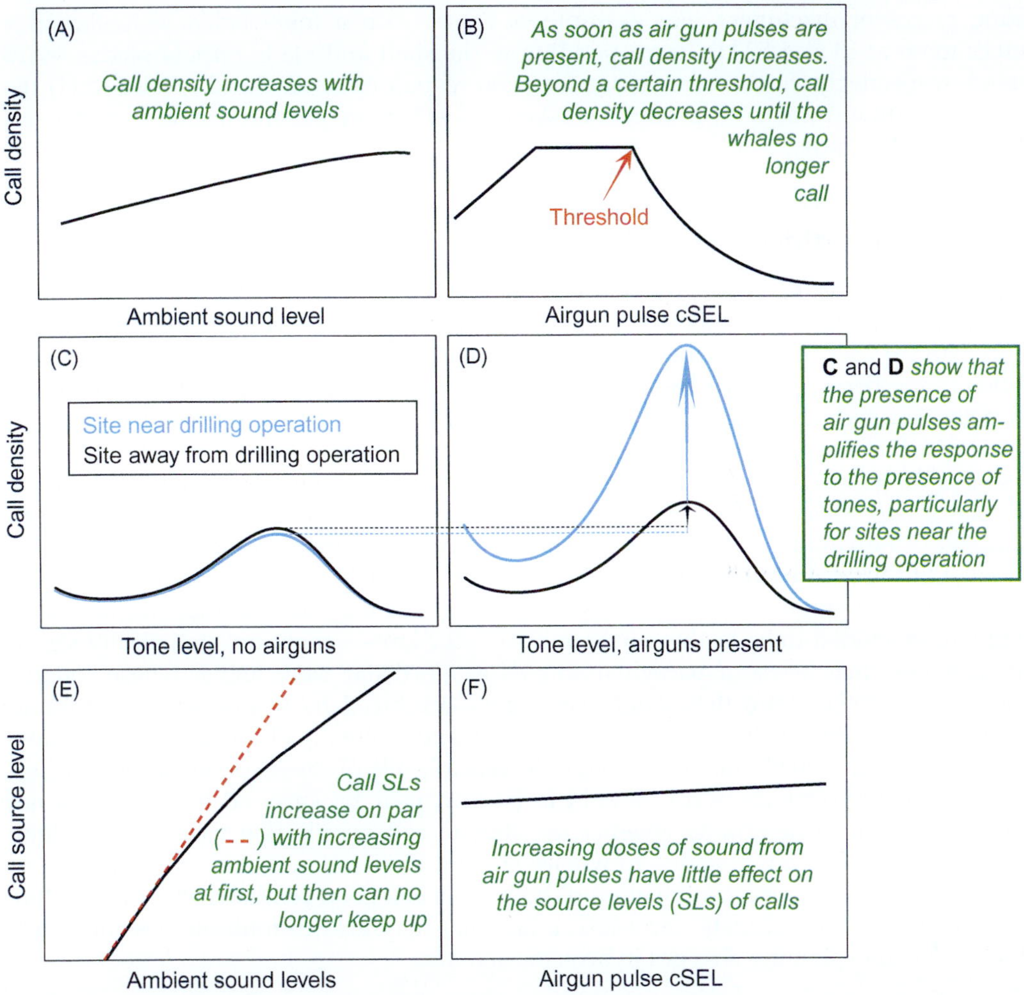

FIGURE 35.3 Conceptual summary of key bowhead whale behavioral responses to various types of noise. (A) Effect of ambient wind-driven noise on call density (calls per unit time per unit area); (B) effect of air gun pulses, from both nearby and distant seismic exploration, on call density; (C) effect of the presence of tones from machinery on call density, for sites near (*blue line*) and away (*black line*) from a drilling operation; (D) same as (C) but with air gun pulses present in the background; (E) effect of ambient wind-driven noise on call SL; (F) effect of air gun pulses on call SL. *cSEL*, cumulative sound exposure level; *SL*, source level. *Source: From (A, E, and F) Thode, A.M., Blackwell, S.B., Conrad, A.S., Kim, K.H., Marques, T., Thomas, L., et al., 2020. Roaring and repetition: how bowhead whales adjust their call density and source level (Lombard effect) in the presence of natural and seismic airgun survey noise. J. Acoust. Soc. Am. 147, 2061; (B) Blackwell, S.B., Nations, C.S., McDonald, T.L., Thode, A.M., Mathias, D., Kim, K.H., et al., 2015. Effects of airgun sounds on bowhead whale calling rates: evidence for two behavioral thresholds. PLoS One 10 (6), e0125720; (C and D) Blackwell, S.B., Nations, C.S., Thode, A.M., Kauffman, M.E., Conrad, A.S., Norman, R.G., et al., 2017. Effects of tones associated with drilling activities on bowhead whale calling rates. PLoS One 12 (11), e0188459.*

rotating pieces of machinery. For example, a vessel with a four-bladed propeller may produce tones at 13 and 52 Hz, corresponding to the shaft and blade rates (4 blades × 13 turns/s), respectively. In this section, we focus on responses of bowhead whales to (1) the sounds emanating from an oil-production island and (2) more general responses to vessels and their tonal sounds.

Sounds from an artificial oil-production island

Northstar is an artificial oil-production island built in shallow water (12 m) northwest of Prudhoe Bay, Alaska. It is located at the southern edge of the whales' migration corridor in August–September, when bowheads travel westward toward their wintering grounds in the Bering Sea. Northstar produces fluctuating levels of sound that depend on daily activities, such as the use of generators and other motors, drilling, vessel traffic, and construction. Starting in 2000, an array of DASARs was deployed northeast of Northstar to investigate the effects of island sounds on the distribution of bowhead calls. Richardson et al. (2012) and McDonald et al. (2012) reported results for 2001–04. They did this by relating the 5th quantile of the offshore distances of calls, i.e., the locations at the southernmost edge of the migration where only 5% of whale calls were closer to shore, to various measures of anthropogenic sound near Northstar. Results showed that the 5th quantile of calling whales tended to be farther offshore by 0.7–2.2 km—a slight but statistically significant change—when levels of particular sounds at Northstar were higher. These sounds, as measured ~450 m from the island, were tones produced by machinery (in 2003 and 2004), transient noise due to vessel traffic (in 2002), and generally elevated levels of industrial noise in the 28–90 Hz range (in 2001). Received levels of these sounds at the whales, several km farther offshore, were thought to be low, near background levels. It was not possible, however, to determine whether the shift in call distribution was due to calling whales physically moving offshore, to whales ceasing their calling behavior but maintaining their original route, or a combination of both. This is a good reminder that only calling whales are detected by acoustic recorders, a fact that may lead to multiple possible interpretations when explaining changes in acoustic activity.

Vessels and other tonal sources

At low frequencies (5–500 Hz), vessels are a major contributor to noise in world oceans (Hildebrand, 2009). Several studies (Ross, 1993; Andrew et al., 2002; McDonald et al., 2006) have reported substantial increases in the levels of certain low-frequency bands (e.g., 30–50 Hz) over spans of decades and have attributed it to increases in the number and gross tonnage of commercial vessels. In the Arctic, vessel noise is still relatively modest, yet it is currently the most common source of anthropogenic continuous noise in the ice-free environment of bowhead whales. Vessel traffic is projected to increase over the next few decades. The Northern Sea Route (NSR), along the north coast of Eurasia, already supports economically viable ship transits, and the Northwest Passage (NWP) will likely be used routinely by mid-century (Smith and Stephenson, 2013; Chapter 27). As summarized in a chapter by Richardson and Malme (1993) and Chapter 23, there is ample evidence that bowhead whales react to approaching vessels of all sizes, with avoidance behaviors, changes in heading and speed, and changes in surfacing and breathing rates. Effects on calling are less well known but Blackwell et al. (2017) quantified changes in calling rate as a function of the presence and "dose" of tones from a large drilling operation in the Beaufort Sea in 2012. In this study, tones were produced by vessel engines and their propellers, as well as other (unidentified) machinery. Using a "tone index" that quantified the tonal sound power every 10 minutes, Blackwell et al. (2017) showed that with increasing tone levels, bowhead calling rates initially increased, peaked, and then decreased (Fig. 35.3C).

Vessels are clearly a source of disturbance that needs further quantification, a task that was undertaken by Hauser et al. (2018) in a modeling study. They found that feeding areas and migration paths for the BCB and ECWG (East Canada–West Greenland) stocks of bowhead whales overlap current or predicted shipping lanes in the NWP and NSR. This fact, combined with the whales' responsiveness to vessel disturbance and acoustic effects, and their potential sensitivity to ship strikes (Reeves et al., 2012), made them relatively vulnerable to increases in Arctic vessel traffic. The East Greenland–Svalbard–Barents Sea (EGSB) and Okhotsk Sea (OKH) stocks were not assessed due to the focus on the NWP and NSR, but this by no means implies that those populations are not subjected to vessel traffic (see, e.g., Adams and Silber, 2017). Nevertheless, more information is needed about the acoustic environment experienced by these smaller stocks.

Sounds from air guns

Air guns are devices that release high-pressure air underwater, thereby creating a high-intensity impulsive sound every 8–40 seconds—defined here as an "air gun pulse." The sound produced by an air gun travels to the seafloor and the underlying geological layers, where the sound gets reflected and refracted according to the physical characteristics of the layers. Echoes returning to the sea surface are then detected by hydrophones and analyzed to reveal geological information, such as the location of oil-bearing strata. Air guns are often used in arrays of multiple "guns," which increases the downward directivity of the generated sound; nevertheless, sound will also propagate horizontally from the sides of an array (see Gordon et al., 2003). Air gun SLs are high enough that their use generally requires the user to monitor for nearby marine mammals to mitigate for potential injuries. In addition, the sound produced is mostly (but not only) at low frequencies (10–120 Hz, but see Madsen et al., 2006) and for reasons having to do with physical acoustics, these sounds travel well underwater and can in some cases be heard hundreds or even thousands of km away (e.g., Nieukirk et al., 2012; Thode et al., 2012).

Another aspect of the sound produced by air guns, which is rarely discussed, is the reverberation that occurs between successive pulses—an effect analogous to the rolling thunder one sometimes hears after the initial thunderclap during a thunderstorm. Using DASAR data from the Beaufort Sea, Guerra et al. (2011) quantified the rise in ambient noise levels due to reverberation and found increases of 30–45 dB within 1 km of the activity, 10–25 dB within 15 km of the activity, and a few dB at 128 km range. These values naturally depend on the size and arrangement of the air gun array, as well as other factors such as the water depth, sound-speed profile, and ocean bottom composition.

Air gun sounds and their effects on bowhead whales are of particular interest here, since areas of the Arctic have long been known or suspected to include oil deposits, and several waves of seismic exploration have taken place off the Alaskan North Slope since the 1970s (Gilders and Cronin, 2000). Logistical constraints (e.g., pack ice, short open-water season) formerly made exploration and exploitation of these deposits too expensive, but a warming Arctic is changing this situation.

Early attempts to correlate the presence of air gun pulses with bowhead call densities on seafloor recorders were confusing, as calling sometimes increased and other times decreased (Richardson et al., 1986; Greene et al., 1996, 1999; Blackwell et al., 2013). Further research determined that documenting the simple presence/absence of air gun sounds was not enough; rather, the received dose of sound detected at the whale—requiring localized calls—was a crucial parameter affecting behavior. As shown in Fig. 35.4, while low doses of received sound from air gun pulses led to increased calling, higher doses had the opposite effect. In this study (Blackwell et al., 2015) the air gun pulse "dose" was defined as a cumulative sound exposure level (cSEL) over 10 minutes. The leftmost column in

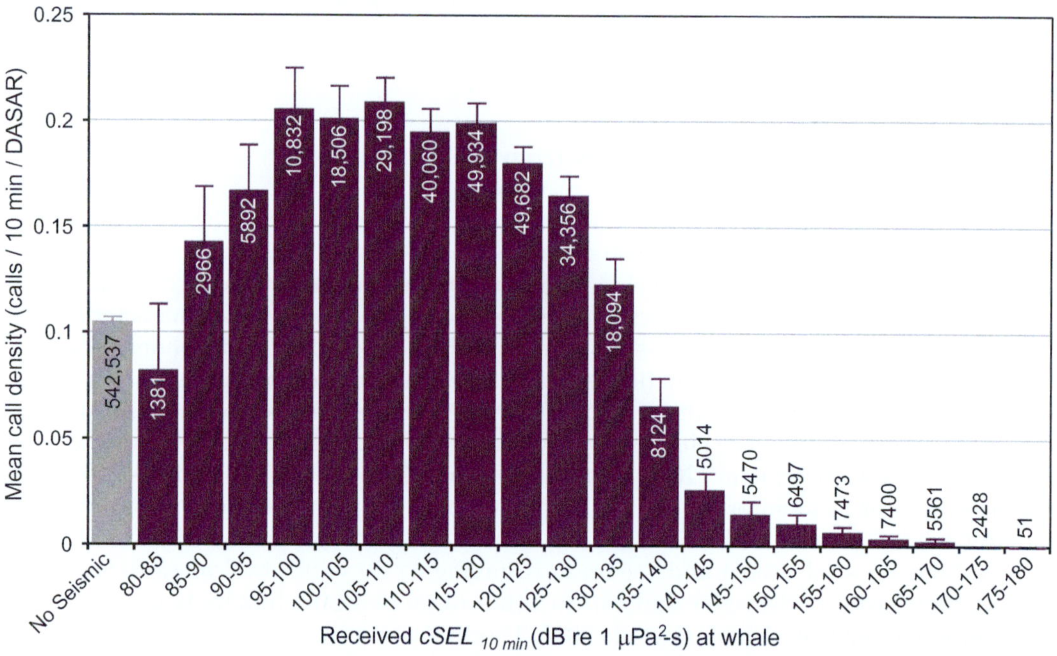

FIGURE 35.4 Call densities [in calls/10 min/DASAR area (12.6 km^2)] as a function of received cumulative sound exposure level (cSEL) from air gun pulses. The gray column shows data collected in the absence of air gun sounds. Sample sizes and upper 95% confidence limits are shown for each bar. *cSEL*, cumulative sound exposure level; *DASAR*, Directional Autonomous Seafloor Acoustic Recorders. *Source: Reproduced from Blackwell, S.B., Nations, C.S., McDonald, T.L., Thode, A.M., Mathias, D., Kim, K.H., et al., 2015. Effects of airgun sounds on bowhead whale calling rates: evidence for two behavioral thresholds. PLoS One 10 (6), e0125720.*

Fig. 35.4 (*gray bar*) shows call densities when air gun pulses were absent. Calling increased as soon as air gun pulses were detectable in the background (*purple bars*). With increasing cSEL, call density plateaued at about twice the "normal" (nonseismic) rate. At higher received levels, above cSEL of ~127 dB re 1 μPa^2 • s, call density dropped rapidly, and by cSEL of ~160 dB, the whales had stopped calling altogether (Fig. 35.3B).

Similar results were obtained in the study of Thode et al. (2020), mentioned previously. Ambient sound levels were the focus of that study but anthropogenic sounds, including air gun pulses, were present some of the time. Thode et al. (2020) showed that the appearance of weak air gun pulse sounds initially caused an increase in call density equivalent to that prompted by a 10–15 dB change in natural noise levels. With further increases in cSEL from received air gun pulses, call density dropped substantially, a very similar situation to that shown in Fig. 35.3B. Thode et al. (2020) also investigated effects of air gun pulses on call SLs and found that an increase of >40 dB cSEL in seismic air gun dosage matched SL increases of just a few dB (Fig. 35.3F). This result was not surprising, in that whales frequently call during the intervals between air gun shots, and thus would not need to increase their SLs much to remain detectable, provided that the reverberation effects previously mentioned are small.

Results from these two studies (Blackwell et al., 2015; Thode et al., 2020) were perplexing, because the received levels of air gun sound at which calling rates started to decrease were surprisingly low. For example, for the characteristics (i.e., size and firing interval) of the array used by the seismic ship in the Blackwell et al. (2015) study, this threshold corresponded to a single pulse received SEL of about 110 dB re 1 μPa^2 • s. Such a level was obtained 50–100 km from the seismic ship, far enough that classic masking (Moore, 1982) of calls by air gun pulses (or their reverberation) was likely not a good explanation for the results. In addition, as explained previously, air gun pulses are an intermittent sound source, with time for a few calls between successive shots.

Instead, a better explanation may be obtained from the field of information theory (Shannon and Weaver, 1949), which predicts that when transmitting a signal through a medium, moderate increases in noise levels can be counteracted by the repetition of the signal of interest, and also that when noise levels become large enough and signals are completely masked, then communication ceases, since no amount of repetition can increase the probability of detection. Incidentally, one of the strategies used across mammal species when communication is hampered by noise is to repeat the signal (e.g., call) more frequently. Among marine mammals, this has been shown in blue whales, who increase their calling rate in response to seismic surveying (Di Iorio and Clark, 2010), and in humpback whales, who increase the rate of feeding calls in the presence of vessel noise (Doyle et al., 2008). Bowhead responses to increases in ambient sound levels (Fig. 35.3A), air gun pulses and their reverberation (Fig. 35.3B), and tones (Fig. 35.3C and D) all fit this interpretation, with an initial increase in call density followed by a decrease at higher received levels. The whales may still be able to identify the presence of a call, but they may not be able to decode the information content of the call sufficiently for the calling behavior to be worth it. This speculative hypothesis could be tested once we know more about the information content of bowhead calls.

During the drilling operation in the Beaufort Sea that was the focus of the Blackwell et al. (2017) study mentioned earlier, distant air gun pulses were also present.

Interestingly, the presence of air gun sounds amplified the effects of tones on the call densities, the more so for locations close to the source of the sound, in this case the drilling operation (Fig. 35.3D).

Summary of short-term acoustic responses to fluctuations in noise

The studies detailed previously, conceptually summarized in Fig. 35.3, used a variety of assumptions, sound measurement units, and statistical methodology, yet show remarkable consistency in bowheads' responses to a variety of noise sources. With the onset of rising noise levels of any type (natural or anthropogenic), whales fought a potential decreased detectability of their calls—and thereby a shrinking communication space (Clark et al., 2009)—by calling more often and calling louder. At higher levels of noise, call densities decreased, in some cases until no more calls were detected. This binary response—of call densities either increasing or decreasing as a function of the dose of received sound—was clearly shown for air gun pulses and tones (Fig. 35.3B–D) and could also help one to explain the Northstar results (see section Sounds from an artificial oil production island, above). These showed a northward shift in the location of the 5th quantile of calling whales in the presence of certain sounds from the island; in other words, an apparent *decrease* in calling near Northstar. Nevertheless, such a shift could also be caused by an *increase* in calling farther offshore, as seen for low levels of sound from tones and air gun pulses. Because of differences in methods, we cannot currently tease apart these various potential interpretations of the Northstar study.

Potential long-term impacts and conclusions

Two main effects of fluctuating levels of anthropogenic noise on calling behavior are presented in this chapter: (1) increases in call density and call SL and (2) reductions in call density down to a cessation of long-range communication, along with tapering off of SLs. We can assume effect (1) will have an energetic cost, which could be negligible in some situations and important in others, while effect (2) could have a more dramatic effect, since long-distance communication may be key to every aspect of a bowhead's life, from the search for food and mates to pathways taken while migrating, including avoiding dangers in a pack-ice environment (Chapter 22). Until we know more about the exact role of communication in bowhead whales and the information content of calls, it will be difficult to gauge these effects.

Stress in mammals can be assessed by measuring concentrations of fecal glucocorticoids (fGCs) (see Chapter 12). As far as we know, measurements of stress levels have not been made in live bowheads, but they have in North Atlantic right whales (*Eubalaena glacialis*), the bowheads' closest living relative. Whereas bowhead whales live in a relatively pristine environment, the habitat of right whales has been heavily modified by human activity, mainly through shipping. An opportunistic study by Rolland et al. (2012) gives us insight into the potential effects of living in an environment that is so much noisier than the one these whales evolved in. The events of

September 11, 2001 led to a large reduction in commercial ship traffic off the Atlantic Seaboard for several days. In the Bay of Fundy, Canada, underwater noise decreased by 6 dB (i.e., a 75% decrease in acoustic intensity), particularly below 150 Hz, and concurrent sampling of fGCs in right whales showed a marked decrease in the baseline levels of this stress-related hormone metabolite. Chronic stress due to noise may therefore be one of several potential threats to all whales in areas of heavy ship traffic and should be considered in view of the unavoidable increase in shipping in bowheads' Arctic habitats in coming decades.

References

Adams, J., Silber, G.K., 2017. 2015 Vessel Activity in the Arctic, NOAA Technical Memorandum NMFS-OPR, vol. 57. National Oceanic and Atmospheric Administration Fisheries, Silver Spring, MD, 171p.

Andrew, R.K., Howe, B.M., Mercer, J.A., Dzieciuch, M.A., 2002. Ocean ambient sound: comparing the 1960s with the 1990s for a receiver off the California coast. ARLO 3 (2), 65–70.

Blackwell, S.B., Nations, C.S., McDonald, T.L., Greene Jr., C.R., Thode, A.M., Guerra, M., et al., 2013. Effects of airgun sounds on bowhead whale calling rates in the Alaskan Beaufort Sea. Mar. Mammal Sci. 29, E342–E365.

Blackwell, S.B., Nations, C.S., McDonald, T.L., Thode, A.M., Mathias, D., Kim, K.H., et al., 2015. Effects of airgun sounds on bowhead whale calling rates: evidence for two behavioral thresholds. PLoS One 10 (6), e0125720.

Blackwell, S.B., Nations, C.S., Thode, A.M., Kauffman, M.E., Conrad, A.S., Norman, R.G., et al., 2017. Effects of tones associated with drilling activities on bowhead whale calling rates. PLoS One 12 (11), e0188459.

Brumm, H., Zollinger, S.A., 2011. The evolution of the Lombard effect: 100 years of psychoacoustic research. Behaviour 148, 1173–1198.

Clark, C.W., Ellison, W.T., Southall, B.L., Hatch, L., Van Parijs, S.M., Frankel, A., et al., 2009. Acoustic masking in marine ecosystems: intuitions, analysis, and implication. Mar. Ecol. Prog. Ser. 395, 201–222.

Di Iorio, L., Clark, C.W., 2010. Exposure to seismic survey alters blue whale acoustic communication. Biol. Lett. 6, 51–54.

Doyle, L.R., McCowan, B., Hanser, S.F., Chyba, C., Bucci, T., Blue, J.E., 2008. Applicability of information theory to the quantification of responses to anthropogenic noise by southeast Alaskan humpback whales. Entropy 10, 33–46.

Gilders, M.A., Cronin, M.A., 2000. North Slope oil field development. In: Truett, J.C., Johnson, S.R. (Eds.), The Natural History of an Arctic Oil Field – Development and the Biota. Academic Press, San Diego, CA, pp. 15–33., 422p.

Gordon, J.C.D., Gillespie, D., Potter, J., Frantzis, A., Simmonds, M.P., Swift, R., et al., 2003. A review of the effects of seismic surveys on marine mammals. Mar. Technol. Soc. J. 37 (4), 14–32.

Greene Jr., C.R., Altman, N.S., Richardson, W.J., 1999. The influence of seismic survey sounds on bowhead whale calling rates. J. Acoust. Soc. Am. 106, 2280.

Greene Jr., C.R., McLennan, M.W., Norman, R.G., McDonald, T.L., Jakubczak, R.S., Richardson, W.J., 2004. Directional frequency and recording (DIFAR) sensors in seafloor recorders to locate calling bowhead whales during their fall migration. J. Acoust. Soc. Am. 116, 799–813.

Greene Jr., C.R., Richardson, W.J., Altman, N.S., 1996. Bowhead whale call detection rates versus distance from airguns operating in the Alaskan Beaufort Sea during fall migration. J. Acoust. Soc. Am. 104, 1826.

Guerra, M., Thode, A.M., Blackwell, S.B., Macrander, A.M., 2011. Quantifying seismic survey reverberation off the Alaskan North Slope. J. Acoust. Soc. Am. 130, 3046–3058.

Hauser, D.D.W., Laidre, K.L., Stern, H.L., 2018. Vulnerability of Arctic marine mammals to vessel traffic in the increasingly ice-free Northwest Passage and Northern Sea Route. Proc. Natl. Acad. Sci. U.S.A. 115, 7617–7622.

Hildebrand, J.A., 2009. Anthropogenic and natural sources of ambient noise in the ocean. Mar. Ecol. Prog. Ser. 395, 5–20.

Knudsen, V.O., Alford, R.S., Emling, J.W., 1948. Underwater ambient noise. J. Mar. Res. 7 (3), 410–429.

Madsen, P.T., Johnson, M., Miller, P.J.O., Aguilar Soto, N., Lynch, J., Tyack, P., 2006. Quantitative measures of air-gun pulses recorded on sperm whales (*Physeter macrocephalus*) using acoustic tags during controlled exposure experiments. J. Acoust. Soc. Am. 120, 2366–2379.

McDonald, M.A., Hildebrand, J.A., Wiggins, S.M., 2006. Increases in deep ocean ambient noise in the Northeast Pacific west of San Nicholas Island, California. J. Acoust. Soc. Am. 120, 711–718.

McDonald, T.L., Richardson, W.J., Greene Jr., C.R., Blackwell, S.B., Nations, C.S., Nielson, R.M., et al., 2012. Detecting changes in the distribution of calling bowhead whales exposed to fluctuating anthropogenic sounds. J. Cetacacean Res. Manage. 12, 91–106.

Moore, B.C.J., 1982. An Introduction to the Psychology of Hearing. Academic Press, London.

Nieukirk, S.L., Mellinger, D.K., Moore, S.E., Klinck, K., Dziak, R.P., Goslin, J., 2012. Sounds from airguns and fin whales recorded in the mid-Atlantic Ocean, 1999–2009. J. Acoust. Soc. Am. 131, 1102–1112.

Reeves, R., Rosa, C., George, J.C., Sheffield, G., Moore, M., 2012. Implications of Arctic industrial growth and strategies to mitigate future vessel and fishing gear impacts on bowhead whales. Mar. Policy 36, 454–462.

Richardson, W.J., Greene Jr., C.R., Malme, C.I., Thomson, D.H., 1995. Marine Mammals and Noise. Academic Press, San Diego, CA, 576p.

Richardson, W.J., Malme, C.I., 1993. In: Burns, J.J., Montague, J.J., Cowles, C.J. (Eds.), The Bowhead Whale. The Society for Marine Mammalogy, Special Publication Number 2, pp. 631–700.

Richardson, W.J., McDonald, T.L., Greene Jr., C.R., Blackwell, S.B., Streever, B., 2012. Distribution of bowhead whale calls near an oil production island with fluctuating underwater sound [Extended abstract]. In: Popper, A.N., Hawkins, A. (Eds.), The Effects of Noise on Aquatic Life. Springer, New York, pp. 303–306, 695p.

Richardson, W.J., Würsig, B., Greene Jr., C.R., 1986. Reactions of bowhead whales, *Balaena mysticetus*, to seismic exploration in the Canadian Beaufort Sea. J. Acoust. Soc. Am. 79, 1117–1128.

Rolland, R.M., Parks, S.E., Hunt, K.E., Castellote, M., Corkeron, P.J., Nowacek, D.P., et al., 2012. Evidence that ship noise increases stress in right whales. Proc. R. Soc. Lond. B Biol. Sci. 279, 2363–2368.

Ross, D., 1976. Mechanics of Underwater Noise. Pergamon, New York, 375p. Reprinted 1987, Peninsula Publ, Los Altos, CA.

Ross, D.G., 1993. On ocean underwater ambient noise. Acoust. Bull. 18, 5–8.

Shannon, C.E., Weaver, W., 1949. The Mathematical Theory of Communication. University of Illinois Press, Urbana, IL.

Smith, L.C., Stephenson, S.R., 2013. New trans-Arctic shipping routes navigable by mid-century. Proc. Natl. Acad. Sci. U.S.A. 110, E1191–E1195.

Southall, B.L., Bowles, A.E., Ellison, W.T., Finneran, J.J., Gentry, R.L., Greene Jr., C.R., et al., 2007. Marine mammal noise exposure criteria: initial scientific recommendations. Aquat. Mamm. 33, 411–522.

Stroeve, J.C., Markus, T., Boisvert, L., Miller, J., Barrett, A., 2014. Changes in Arctic melt season and implications for sea ice loss. Geophys. Res. Lett. 41, 1216–1225.

Thode, A.M., Blackwell, S.B., Conrad, A.S., Kim, K.H., Marques, T., Thomas, L., et al., 2020. Roaring and repetition: how bowhead whales adjust their call density and source level (Lombard effect) in the presence of natural and seismic airgun survey noise. J. Acoust. Soc. Am. 147, 2061.

Thode, A.M., Kim, K.H., Blackwell, S.B., Greene Jr., C.R., Nations, C.S., McDonald, T.L., et al., 2012. Automated detection and localization of bowhead whale sounds in the presence of seismic airgun surveys. J. Acoust. Soc. Am. 131, 3726–3747.

Wenz, G.M., 1962. Acoustic ambient noise in the ocean: spectra and sources. J. Acoust. Soc. Am. 34, 1936–1956.

CHAPTER

36

Fishing gear entanglement and vessel collisions

J.C. George[1], *Gay Sheffield*[2], *Barbara J. Tudor*[1], *R. Stimmelmayr*[1] *and M. Moore*[3]

[1]Department of Wildlife Management, North Slope Borough, Utqiaġvik, AK, United States [2]Alaska Sea Grant, College of Fisheries and Ocean Sciences, University of Alaska Fairbanks, Nome, AK, United States [3]Biology Department, Woods Hole Oceanographic Institution, Woods Hole, MA, United States

Introduction

The International Whaling Commission (IWC) has identified entanglement in commercial fisheries gear (termed bycatch) and vessel collisions as two of the greatest threats facing whales worldwide (Cates et al., 2017; IWC, 2017). Read et al. (2006) estimated 308,000 cetaceans die every year as a result of fisheries entanglement. A recent assessment of available data showed fisheries bycatch of air-breathing vertebrates to be global, with taxa- and gear-specific hot spots (Lewison et al., 2014). Despite their remote arctic distribution, three bowhead stocks [Okhotsk Sea (OKH), East Canada–West Greenland (ECWG), and Bering–Chukchi–Beaufort Seas (BCB)] currently overlap to some degree with large-scale commercial fisheries. Fishing operations are expected to move northward under climate change scenarios, increasing the threat of bowhead entanglement possibly to all four stocks. As the bowhead's habitat warms, a reduction of sea ice and a more extended open-water season will also increase vessel traffic (research, ecotourism, and industrial shipping) and the potential for bowhead injuries from ship strikes (Huntington et al., 2020; Reeves et al., 2012; Fig. 36.1, Fig. 36.2F).

Bowheads can become entangled in commercial nets and lines that are in use or in abandoned, lost, or discarded fishing gear, commonly known as "ghost gear." Aerial photos provide evidence of entangled bowheads dragging gear wrapped around the peduncle as well as streaming from both sides of the mouth. An entanglement may result in drowning, starvation, reduced energy, infection, and serious tissue injuries leading to

The Bowhead Whale
DOI: https://doi.org/10.1016/B978-0-12-818969-6.00036-4

FIGURE 36.1 A bowhead whale surfaces near a ship in the Okhotsk Sea. It is likely that ship strikes will become more common as shipping in the Arctic increases. *Source: Photo by Olga Shpak, copyrighted.*

death (Cassoff et al., 2011; Moore and van der Hoop, 2012). Rolland et al. (2019) found elevated glucocorticoid hormone levels in the baleen of a bowhead that had been severely entangled by fishing gear before being harvested by Alaska Native hunters, indicating that bowheads suffer physiological stress from chronic entanglement. Sublethal entanglement trauma has also been shown to negatively impact fecundity (van der Hoop et al., 2017) as well as body condition (Miller et al., 2012).

Like other balaenids, injuries form white scars that contrast with the black skin of the bowhead and appear to last for the lifetime of the animal (Philo et al., 1992, 1993; Rugh et al., 1992; George et al., 1994, 2017, 2019). In one of the first published studies of bowhead entanglement, Philo et al. (1992) described injuries thought to be from fishing gear as a series of white, linear, "spiral wrap marks," and "crisscrossed" scars on the dorsal and ventral peduncle (Fig. 36.2B,C). These authors also reported two bowheads entangled in heavy line associated with Bering Sea commercial pot fisheries—a dead-stranded bowhead found near Gambell, Alaska, during 1989 and an actively entangled whale harvested by subsistence hunters in 1990 near Utqiaġvik, Alaska. Based on observations by subsistence hunters that bowheads roll to rid themselves of harpoon lines (small diameter line), Philo et al. (1992) speculated that bowheads may try shedding a commercial large diameter line by rolling, ultimately wrapping their peduncle, mouth, or appendages. Nonlethal entanglement scarring on bowheads is typically located on the peduncle and leading edge of the flukes, but it has been observed on the pectoral flippers and mouth region (Philo et al., 1993; George et al., 2017).

The endangered North Atlantic right whale (*Eubalaena glacialis*, NARW), a close relative to the bowhead that is similar in size and body type, offers insights into the threats bowheads may face in the future, if fishing and shipping increase within their range. The NARW population continues to decline, largely due to lethal entanglement in commercial fisheries gear and poor fecundity (Knowlton et al., 2012; Kraus et al., 2005; Pace et al., 2017). Entanglement-associated lesions to NARW included deep lacerations around the body, flippers, flukes, head, and mouth. In addition, baleen plates can be mutilated, and there can be extensive chronic bone lesions where the line wraps around and constricts extremities. Chronically entangled NARWs typically endure severe weight loss and an increased and widespread cyamid burden over a number of years (Pettis et al., 2004), although an increased cyamid burden was not consistently observed in entangled, harvested bowheads (Von Duyke et al., 2016). In a recent summation of 70 NARW mortalities between 2003 and 2019 (Sharp et al., 2019), the cause of death was determined in 43 cases, including 38 (88.4%) due to anthropogenic trauma of which 22 (57.9%) were from entanglement and 16 (42.1%) from vessel strikes. Entanglement mortality in NARW has increased from 21% in the period from 1970 to 2002 (Moore et al., 2004) to 88% from 2003 to 2019 (Sharp et al., 2019).

Vessel collisions may cause blunt-impact injuries to large whales, including fractures of the skull and vertebrae as well as muscle and blubber contusions. Ships may also inflict propeller-induced wounds, including severe blubber, muscle, and bone injuries (Sharp et al., 2019) (Fig. 36.2F). To date, vessel-inflicted mortality and injuries to BCB bowheads appear to be relatively uncommon ($\sim$2%) (George et al., 1994, 2017), and little information is available for other stocks. While large vessel strikes can kill whales relatively quickly, line entanglement usually kills very slowly. A chronically entangled NARW can take an average of 6 months to die from starvation, drowning, and/or increased predation risk due to limited mobility from line constriction, severe pain, and/or autoamputation of flukes (Moore et al., 2006). Each of these scenarios represents a significant animal welfare concern (Moore and van der Hoop, 2012). Moore (2014) further notes that all seafood consumers should be concerned about whales entangled by commercial fishing gear.

Copious evidence suggests that large whales are particularly vulnerable to gear entanglement and vessel strike injuries worldwide (Cates et al., 2017). Considering the reductions in sea ice and corresponding increases in commercial vessel traffic and fishing in the Arctic, bowheads will likely be at increasing risk of entanglement and vessel strikes. The purpose of this chapter is to report current knowledge of entanglement and vessel strike injuries covering the four circumpolar bowhead stocks.

Review of fishing gear entanglement by stock

Okhotsk Sea stock

We found little published information on entanglement of the OKS stock bowheads (Table 36.1). Brownell (1999) reported a fatal entanglement associated with the Okhotsk Sea crab fishery, and there have been unverified reports of additional bowhead entanglements in the Okhotsk Sea (Ivashchenko and Clapham, 2010). Shpak et al. (2014) reported entanglement in another gear type in Udskaya Bay where in 2012 fishermen freed an adult bowhead that became entangled in a net after following her dead calf near a salmon

TABLE 36.1 Partial listing of bowhead whales entangled in fishing gear that were harvested, dead-stranded or observed swimming during the period 1969–2017, and also included are comments on gear type and other observations.

Location	Year	ID no.	Status	Comments and gear type
Osaka Bay, Japan	1969	NA	Dead	Trapped in fishing net and died (Nishiwaki and Kasuya, 1970).
Upernavik, Baffin Bay, Canada	1980	NA	Dead	Caught in beluga net and subsequently died (Kapel, 1985).
Wales, Bering Strait, AK, United States	1987	NA	Dead, beached	Two lines attached to flukes (Philo et al., 1993). Line type and thickness were not reported.
Gambell, Bering Strait, AK, United States	1989	NA	Harvested	Line wrapped around head and in mouth (Philo et al., 1992). Line 3-braid yellow polypropylene, 19 mm diameter.
Nuiqsut, Beaufort Sea, AK, United States	1990	NA	Harvested	Harvest report of a bowhead whale landed with line in mouth (NSB-DWM unpublished data). Line type and thickness were not reported.
Utqiaġvik, Chuckchi Sea, AK, United States	1990	90B6	Harvested	Two lines: one exiting mouth and one recovered in water (Philo et al., 1992). White nylon, ~19 mm diameter, appeared to be crab gear.
Utqiaġvik, Chukchi Sea, AK, United States	1993	NA	Dead	Commercial pot line wrapped around flukes (NSB-DWM unpublished data). White nylon, ~19 mm diameter.
Chukchi Sea, AK, United States	1994	94001/ 94002	Live, swimming	Swimming with line attached. Assume it was of substantial thickness to be seen during aerial survey (NMML aerial survey, May 20, 1994). Line type and thickness were not reported.
Red Dog Mine, Chukchi Sea, AK, United States	1998	NA	Dead, beached	Beachcast bowhead noted with "line on whale" (NMFS unpublished data).
Okhotsk Sea, Russia	1995	NA	Dead	Bowhead found dead entangled in cable of crab trap when crew retrieved gear (Brownell, 1999).
Utqiaġvik, AK, United States	1999	99B14	Harvested	Commercial pot line entangled in mouth, flipper and tail (NSB-DWM unpublished data). Gray nylon, ~19 mm. Over 50 m of line on whale.
Utqiaġvik, Chukchi Sea, AK, United States	2001	NA	Live, swimming	Whale entangled with a green line of substantial thickness trailing from the flukes (Brower, 2006)
Cinder River, Bering Strait, AK, United States	2003	NA	Dead, beached	Several lines (~3/4″ diameter) of different colors wrapped at the flukes (NSB-DWM unpublished data).
Pt. Barrow, Beaufort Sea, AK, United States	2003	033410	Live, swimming	Aerial photo of whale with a line of substantial thickness attached to flukes (NOAA unpublished data).
Nunavut/West Greenland	2003–06	NA	Live, swimming	Four whales reported entangled in fishing net (COSEWIC, 2009).

(*Continued*)

TABLE 36.1 (Continued)

Location	Year	ID no.	Status	Comments and gear type
Cape Dorset, Canada	2005	NA	Live, swimming	Entangled in beluga net (DFO unpublished data).
Pond Inlet, Canada	2006	NA	Live, swimming	Entangled in green rope with red buoy (COSEWIC, 2009).
Kotzebue, Chukchi Sea, AK, United States	2010	10KTZ-FD1	Dead, beached	Confirmed commercial pot gear (>30 m of line and portion of pot) recovered from mouth and peduncle (Sheffield, 2010). Flukes possibly cut off by line constriction.
Pt. Barrow, Beaufort Sea, AK, United States	2011	01161591	Live, swimming	Aerial photo of whale with a yellow line of "substantial thickness" attached to flukes (Vate Brattstrom et al., 2016).
Udskaya Bay, Okhotsk Sea, Russia	2012	NA	Live, swimming	Female became entangled in salmon net after "following" her dead calf which became beach-cast near a fishing operation. Fishermen said they cut off net and released the bowhead (Shpak et al., 2014).
Ulbansky Bay, Okhotsk Sea, Russia	2013	NA	Live, swimming	Bowhead caught in salmon fishing net and swam away with gear attached around flukes (Shpak et al., 2014).
St. Lawrence Island, Bering Sea, AK, United States	2015	2015-FD2	Dead, at sea	Found floating with confirmed commercial pot line, two vinyl floats, and a limited entry pot permit tag. Line wrapped around peduncle [Sheffield and Savoonga Whaling Captains Association (SWCA), 2015].
Baffin Bay, Qeqertarsuaq, Greenland	2017	NA	Unknown	Bycatch of one bowhead in crab pot gear [IWC/67/Rep01(2018), Annex J].
Utqiaġvik, Beaufort Sea, AK, United States	2017	17B6	Harvested	Severe entanglement: peduncle; confirmed commercial nylon line typically used. Line 19 mm thick, anchored to peduncle, pectoral fin, and through the baleen/mouth.
Utqiaġvik, Beaufort Sea, AK, United States	2017	17B8	Harvested	Severely entangled: mouth, flipper, neck wrap, confirmed commercial nylon line (19 mm) typically used in crab fishery (Rolland et al., 2019).

fishing operation. In 2013, a bowhead was observed in Ulbansky Bay swimming away from a salmon fishery with gear around its tail (Shpak et al., 2014). Shpak and Stimmelmayr (2017) provided an analysis of injuries and skin conditions of OKS bowheads, noting "wrapping linear scars at the base of the peduncle consistent with line entanglement were present on several whales." They further noted "scars were delicate and precise," suggesting that the cause of the injury was longline gear or light gillnets rather than heavy lines associated with gear used in the Bering Sea commercial pot fisheries (Fig. 36.2E).

East Greenland, Svalbard, Barents Sea stock

The East Greenland, Svalbard, Barents (EGSB) stock was nearly extirpated by European commercial whalers, and its current range is now restricted to only the high latitudes of the North Atlantic (Chapters 5 and 33). Nonetheless, Reeves et al. (2014) noted that commercial fishing now extends north to the ice edge where it might intercept EGSB bowhead whales. They also noted that the catch levels and fishing area of the Barents Sea fisheries have expanded with the reduction of sea ice. Stafford et al. (2018) noted that bowheads spend the winter in Fram Strait where they would presumably not be vulnerable to active fisheries. A year-round acoustic recorder in Western Fram Strait recorded bowheads almost daily from October to April but not in summer (Ahonen et al., 2017); thus, with no overlap with summer fisheries, they do not likely pose a threat, although entanglement caused by ghost gear cannot be discounted. We are unaware of any published records of bowheads entangled in fishing gear for this stock.

East Canada–West Greenland stock

Significant commercial fisheries are located in the waters of West Greenland and are expanding into Baffin Bay during the open-water season (Reeves et al., 2014; Table 36.1). The North Atlantic Marine Mammal Commission (NAMMCO, 2017) reported that bowheads have been entangled in snow crab pot lines between April and December off West Greenland "inshore from Upernavik and southwards" and estimate moderate risk for bowhead entanglement in crab pot gear. Overlap between crab pot fisheries and bowhead habitat is understood due to the fact that entanglements have occurred (NAMMCO, 2017). Philo et al. (1993) included a 1980 record (Kapel, 1985) of a lethal take when a small bowhead became entangled in a net used to subsistence harvest beluga whales in NW Greenland. The NAMMCO (2017) lists five bowheads fatally entangled between 1998 and 2016. The 2018 IWC report of the Subcommittee on the Non-Deliberate Human-Induced Mortality of Cetaceans includes one bowhead entangled in crab pot gear in 2017 in Baffin Bay near Qeqertarsuaq, Greenland [International Whaling Commission (IWC), Report of the Scientific Committee, 2018]. Little other entanglement data appear to be available for the ECWG stock.

Bering–Chukchi–Beaufort Seas stock

The frequency and rate of entanglement injuries on Bering-Chukchi-Beaufort Sea stock (BCB) bowheads have been reasonably well documented (George et al., 1994, 2017, 2019; Table 36.1) compared with the other bowhead stocks, although it is unclear where and how gear entanglements occur. Satellite tag data suggest that there is limited temporal and spatial overlap between the location of whales and where/when pots are set in US waters (Citta et al., 2013). The greatest overlap in US waters is with the St. Matthew Island blue king crab (*Paralithodes platypus*) fishery. In past years, this fishery ended before tagged whales arrived to the area in winter; however, if pack ice overruns active commercial fishing areas, resulting in the loss of gear, entanglement may occur. Lost pots are common and were especially common before the pot fishery was rationalized (Citta et al., 2013). It is also likely that the distribution of bowhead whales is larger than suggested by the satellite tag data, and some whales venturing south of the ice edge may encounter active pot gear (Citta et al., 2012). The exact origin of the gear recovered from bowheads is rarely known. However, during 2015, a dead entangled bowhead was recovered in the Bering Sea near Saint Lawrence Island carrying commercial pot gear that included the limited entry pot permit tag issued by the Alaska Department of Fish and Game [Sheffield and Savoonga Whaling Captains Association (SWCA), 2015]. The permit was issued for the Bering Sea commercial blue king crab fishery near St. Matthew Island during the winter season (October–February 2012/2013). It remains unknown whether this was an active set or ghost gear. Significant pot fisheries also occur in the Russian waters of the western Bering Sea, although specific details regarding the distribution and timing of the Russian pot fisheries were unavailable. The main Bering Sea bowhead wintering areas are in Russian waters, and the southernmost of these areas has been largely ice free in recent years (see Chapter 4). If some whales frequent these waters when they are ice free, they may encounter active pot gear. Changes in the location and timing of pot fisheries with declines in sea ice, both in the United States and Russian waters, need to be revisited.

While harvested whales probably do not represent a random sample of the population, about 12.2% of whales harvested between 1990 and 2012 from the BCB stock showed entanglement scars (George et al., 2017) (Fig. 36.2C). This estimate closely agrees with the percentage of entanglement scars observed in a photographic survey of migrating BCB whales conducted during spring 2011 (George et al., 2019). That survey indicated that 12.4% of the whales photographed in 2011 had entanglement scars. These authors expanded the study by analyzing photo-recapture data spanning 1985 to 2011 to estimate *the rate* at which BCB bowheads acquire entanglement scars. Using two statistical approaches, the estimated probability of a bowhead acquiring an entanglement injury was 2.2% per year (95% CI: 1.1%–3.3%). This estimate applies to larger subadult and adult whales, since young bowheads less than 10 m body length rarely carry scars suitable for photo-recapture studies. The acquisition rate (2.2% per year) was somewhat higher than expected and suggests more of the BCB population should carry scars. However, it is consistent with the frequency of scars on large and older bowheads where about 50% of whales over 17 m (both sexes) carried entanglement scars, with males having a slightly higher, but significant, rate of scarring. The authors suggest this is due to larger older whales having greater exposure time to fisheries gear and that the smaller size classes do not survive the entanglement (George et al., 2019).

It is important to note the difference between noncommercial and commercial pot gear. Noncommercial crab pots in the Bering Sea weigh ~20 kg and are typically set in relatively shallow water near shore; however, commercial pots associated with the Bering Sea pot fisheries can weigh 350 kg when empty and are typically set in deep (up to 183 m) offshore waters. The tethering of a bowhead to commercial Bering Sea crab gear will cause immediate stress and/or injury as its movements face resistance (i.e., drag). George et al. (2017) reported the line found on actively entangled bowheads harvested by Alaska Native hunters to be 19 mm in diameter and consistent with Bering Sea commercial crab gear (Fig. 36.2A,D). For NARWs, Knowlton et al. (2016) showed that entanglement severity

FIGURE 36.2 (A) Heavy line, presumably from commercial pot gear, entangled in the mouth and around the flipper of whale (NSB-DWM 2017B8) harvested near Utqiaġvik, Alaska (AK). (Photo: A. Morris, NSB DWM). (B) Scarring associated with a healed line entanglement injury located on the peduncle of an adult bowhead whale harvested near Savoonga, AK. (Photo: G. Sheffield). (C) Entanglement injuries are visible in high-resolution aerial photographs, particularly those with a clear view of the peduncle (inset). These types of images can be used to estimate the proportion of whales with scars and the scar acquisition rate (Photo: NSB-NMFS PhotoID survey). (D) Heavy line and buoys from commercial crab pot gear removed off an adult female bowhead whale floating dead near Saint Lawrence Island, July 2015. A commercial permit tag remained on the line and indicated this gear was deployed in the Central Bering Sea during winter 2012 by a US Bering Sea commercial vessel fishing for king crab. White knife and sheath are ~30 cm long and is shown for scale (Photo: G. Sheffield). (E) Bowhead in the Okhotsk Sea showing thin linear scars on the peduncle possibly associated with entanglement in salmon nets, long-line gear, or light pot gear (Photo: Olga Shpak). (F) Healed laceration to the fluke of bowhead harvested at Wainwright AK in Spring 1987. The injury was most likely caused by a ship propeller (photo: NSB-DWM).

has recently increased in correlation with increased line strength. Rope strength and equipment size need to be further addressed in bowhead entanglement, especially since we think they most likely cause lethal injuries in immature whales.

Vessel strike injuries

Vessel strike injuries are currently uncommon on BCB bowhead whales based on examinations of harvested whales. George et al. (1994) reported that the injury rate from large vessels was low at ~1% for the period 1976–92. In an updated analysis, George et al. (2017) evaluated records of nearly 500 harvested whales for ship-related injuries of which only 10 whales (~2%) showed scars associated with injuries from ship propellers. Blunt trauma was not recorded; however, Philo et al. (1993) reported a whale harvested in 1987 at Wainwright, AK, had a mid-length fracture of the right mandible. The periosteum was necrotic, and the fracture had not healed (Philo et al., 1993). Considering the great strength of the balaenid mandible, this fracture may have been caused by a ship strike (Tsukrov et al., 2009). With regard to propeller inflicted injuries to bowheads, George et al. (1994, 2017) described a whale with three (30 cm) crescent-shaped scars on the lateral surface of the head, a different individual with a large 1 m long laceration to the flukes (Fig. 36.2F), and several other whales with smaller (20 cm) propeller-type scars.

Large industrial vessel traffic in the North Atlantic is substantial and increasing in the Svalbard archipelago, for instance, which is ice free much of the year (Reeves et al., 2014). Regardless, we do not know of any records of bowheads being struck by vessels in the North Atlantic. Also, North Atlantic bowheads appear to stay associated with the ice during winter (Stafford et al., 2018) reducing the likelihood of collisions with nonicebreaking ships. Nonetheless, we note that a bowhead whale far south of its normal range has recently been feeding with right whales in the Gulf of Maine where it is exposed to high risk of a vessel strike (Accardo et al., 2018).

Reeves et al. (2014) noted the expansion of the shipping season and increases in vessel traffic in the Arctic, including regions frequented by bowheads. Shipping in the US Arctic has increased dramatically since 2010 and is expected to increase in the coming decades associated with sea ice retreat. In 2018 the US Committee on the Marine Transportation System (USCMTS, 2019) used the automatic identification system to estimate that 255 (S.D. = 26) unique vessels transited through the US Arctic and surrounding region from 2015 to 2017. Half of the vessels were barge-and-tug rigs and cargo vessels. The committee suggests the number of vessels operating in the US Arctic to more than triple by 2030. Their report, based on the US Coast Guard (USCG) data, indicated the length of the shipping season is increasing by 7–10 days per year. When extrapolated to 2030, the report suggested the current navigation period through the Bering Strait may be extended by 2.5 months (USCMTS, 2019).

Discussion and conclusions

Dramatic changes are occurring in the marine ecosystem of the Arctic due to sea ice reduction and the subsequent expansion of industrial marine traffic volume and vessel size (Reeves et al., 2014). For the ECWG stock the opening of northern shipping routes may increase the threat of ship strikes (COSEWIC, 2009), and the northward expansion of commercial fisheries may also increase the risk of entanglement. In light of increasing human activity in the Okhotsk Sea, Shpak et al. (2014) urge continued monitoring of the OKH stock. With regard to the EGSB stock, Boertmann et al. (2015) are cautiously optimistic that the stock is increasing based on observations during an aerial survey over the Northeast Water Polynya; however, anticipated industrial activity, including mineral exploration and ensuing ship traffic, warrants continued research to monitor the status and trend of this stock.

Fisheries entanglement of BCB bowheads has not significantly affected the recovery of this stock; however, the mortality rate from bycatch specifically is unknown. Similarly, relatively little information is available on entanglement mortality for the other three bowhead stocks. The lack of sea ice has, among other things, significantly diminished the annually replenished cold (typically $-2°C$) saline water that served as a thermal barrier that separated the two distinct ecosystems of the southern and northern Bering Sea (Stevenson and Lauth, 2019; Thoman et al., 2020; Huntington et al., 2020) (see Chapter 25). As a result, commercially viable fish populations moved into the northern Bering Sea, and longline fishing for Pacific Cod has occurred in the northernmost legal waters of the United States during the summer–late fall of 2019 (NOAA, 2019). It is unknown if this trend will become a permanent change (Huntington et al., 2020). The current commercial fishing moratorium established by multinational consent entitled the agreement to prevent unregulated high seas fisheries in the Central Arctic Ocean pertains only to the waters north of each signatory Exclusive Economic Zone (Schatz et al., 2019). Reeves et al. (2012) recognized that entanglement in commercial fisheries gear in the northern Bering Sea and Bering Strait region is not solely an Alaskan issue, but rather a transboundary responsibility for the United States and the Russian Federation, in which frequent and relevant communication and planning are essential for this shared marine resource.

The humane aspects of gear entanglement to large whales are often overlooked in the literature that is generally limited to statistics on entanglement rates (e.g., George et al., 2017). Clearly, whales suffer greatly when entangled in heavy line as indicated by markedly increased cortisol levels (Moore, 2014; Rolland et al., 2019; see Chapter 19). The Alaska Eskimo Whaling Commission (AEWC) is concerned about the bowhead entanglement issue not only from a food security perspective (a potential increase in lethal entanglements could negatively affect the International Whaling Commission harvest quota) but also from a humane perspective (Rolland et al., 2019). The traditional values of the Iñupiat community include the belief that an animal should never be allowed to suffer (*nagliksaaq*). Reducing gear entanglement is consistent with habitat protection under the AEWC/NOAA agreement, including protection of subsistence and cultural traditions of the Alaskan Native community. In addition, the AEWC and the Bering Sea Crabbers Association members also have begun a dialogue on entanglement issues at AEWC meetings in order to mitigate future impacts (George et al., 2019). The AEWC also passed a resolution recommending photographic surveys be conducted with drones to further document entanglement injuries and fishing gear attached to bowheads [Alaska Eskimo Whaling Commission (AEWC), 2019].

While Reeves et al. (2014) found that entanglement was not a significant "population-level threat" for the BCB stock, they acknowledged this could change dramatically if the moratorium on fishing in the Arctic Ocean and areas north of the Bering Strait ended. George et al. (2019) noted that the current abundance and population rate of increase (3.7% per year) suggest that gear entanglement is not "significantly interfering with the recovery of BCB bowheads." However, George et al. (2019) also made clear that the current statistics coupled with ecosystem changes indicate gear entanglement is now "a non-ignorable concern for BCB bowheads."

How the wintering patterns (i.e., timing and movements) of the BCB bowhead population will change with ice reduction is unclear. It is also unclear how the northward shift of commercial fishing and other maritime industries, along with a transitioning ecosystem, including a reduction in the extent, quality, and duration of their sea ice habitat will affect bowheads across the Arctic.

Considering the rapidly changing arctic habitat and in light of the severe population-level effects that gear entanglement has on NARW, immediate action should be taken to mitigate the effects of line entanglement and large vessel strikes on bowhead in the coming decades (Moore, 2019). Entanglement mitigation measures include spatial and temporal closures, adoption of fixed gear fishing practices that regulate maximum line strength, or moving toward "ropeless fishing" that uses acoustic triggers rather than end lines for bottom gear recovery. With regard to industrial shipping in the US Arctic, the USCG, International Maritime Organization, and Native Communities are collaborating to devise shipping schedules and routes to mitigate impacts to whales and risks to subsistence hunters (USCMTS, 2019).

As recommended in Reeves et al. (2012), strengthening/expanding existing bowhead whale health assessment programs as well as regional communication networks throughout western and northern Alaska coastal communities regarding bowhead strandings and/or the cause(s) of bowhead whale deaths would allow comparisons in the rate of entanglement injuries to bowhead whales. During this period of rapid ecological (and industrial) reorganization, future studies are warranted on the winter movements and potential for gear entanglement and ship strikes on all stocks of bowheads.

Acknowledgements

We thank the whaling captains and crews of the Alaska Eskimo Whaling Commission communities for allowing us to examine their bowhead whales. The data summarized here is the result of the hard work and dedication of many people that examined harvested and stranded whales over the past 40 years. We thank Randall Reeves, Scott Kraus, Frances Gulland, Rosalind Rolland, Alexander Costidis and several others for reviewing photographs and the assistance with the manuscript. Finally we thank the North Slope Borough Department of Wildlife Management and the Alaska Eskimo Whaling Commission for their long-term support of this work.

References

Accardo, C.M., Ganley, L.C., Brown, M.W., Duley, P.A., George, J.C., Reeves, R.R., et al., 2018. Sightings of a bowhead whale (*Balaena mysticetus*) in the Gulf of Maine and its interactions with other baleen whales. J. Cetacean Res. Manage. 19, 23–30.

Ahonen, H., Stafford, K.M., de Steur, L., Lydersen, C., Wiig, Ø., Kovacs, K.M., 2017. The underwater soundscape in western Fram Strait: Breeding ground of Spitsbergen's endangered bowhead whales. Mar. Pollut. Bull. 123 (1–2), 97–112. Available from: https://doi.org/10.1016/j.marpolbul.2017.09.019.

Alaska Eskimo Whaling Commission (AEWC), 2019. A Resolution to Utilize Drones to Investigate Impacts of Fishing/Crab Gear and Satellite Tags of Bowheads. Resolution 2019-05. Alaska Eskimo Whaling Commission, Utqiaġvik, AK.

Boertmann, D., Kyhn, L., Witting, L., Heide-Jørgensen, M.P., 2015. A hidden getaway for bowhead whales in the Greenland Sea. Polar Biol. 38. Available from: https://doi.org/10.1007/s00300-015-1695-y.

Brower, G., 2006. Personal Communication. Utqiaġvik, AK.

Brownell, R.L. Jr., 1999. Mortality of a bowhead whale in fishing gear in the Okhotsk Sea. In: Paper SC/51/AS27. The International Whaling Commission, Cambridge.

Cassoff, R.M., Moore, K.M., McLellan, W.A., Barco, S.G., Rotstein, D.S., Moore, M.J., 2011. Lethal entanglement in baleen whales. Dis. Aquat. Organ. 96 (3), 175–185.

Cates, K., DeMaster, D.P., Brownell Jr., R.L., Silber, G., Gende, S., Leaper, R., et al., 2017. IWC Strategic Plan to Mitigate the Impacts of Ship Strikes on Cetacean Populations: 2017–2020.

Citta, J.J., Burns, J.J., Quakenbush, L.T., Vanek, V., George, J.C., Small, R.J., et al., 2013. Potential for bowhead whale entanglement in cod and crab pot gear in the Bering Sea. Mar. Mammal Sci. 30 (2), 445–459. Available from: https://doi.org/10.1111/mms.12047.

Citta, J.J., Quakenbush, L.T., George, J.C., Small, R.J., Heide-Jorgensen, M.P., Brower, H., et al., 2012. Winter movements of bowhead whales (*Balaena mysticetus*) in the Bering Sea. Arctic 65, 1.

COSEWIC, 2009. COSEWIC Assessment and Update Status Report on the Bowhead Whale *Balaena mysticetus*, Bering–Chukchi–Beaufort Population and Eastern Canada–West Greenland Population in Canada. Committee on the Status of Endangered Wildlife, Ottawa. vii + 49 p.

George, J.C., Philo, L.M., Hazard, K., Withrow, D., Carroll, G.M., Suydam, R., 1994. Frequency of killer whale (*Orcinus orca*) attacks and ship collisions based on scarring on bowhead whales (*Balaena mysticetus*) of the Bering-Chukchi-Beaufort Seas stock. Arctic 47 (3), 247–255.

George, J.C., Sheffield, G., Reed, D.J., Tudor, B., Suydam, R., 2017. Frequency of injuries from line entanglements, killer whales, and ship strikes on Bering-Chukchi-Beaufort Seas bowhead whales. Arctic 70 (1), 37–46.

George, J.C., Tudor, B., Givens, G.H., Mocklin, J., Vate Brattström, L., 2019. Entanglement-scar acquisition rates and scar frequency for Bering-Chukchi-Beaufort Seas bowhead whales using aerial photography. Mar. Mammal Sci. 35 (4), 1304–1321Published on line. Available from: https://doi.org/10.1111/mms.12597.

Huntington, H.P., Danielson, S.L., Wiese, F.K., Baker, M., Boveng, P., Citta, J.J., et al., 2020. Evidence suggests potential transformation of the Pacific Arctic Ecosystem is underway. Nat. Clim. Change 10, 342–348. Available from: https://doi.org/10.1038/s41558-020-0695-2.

International Whaling Commission (IWC), Report of the Scientific Committee, 2017. IWC/67a/Rep01(2017)rev1. Bled, p. 63. Available from: <www.iwc.int>.

International Whaling Commission (IWC), Report of the Scientific Committee, 2018. IWC/67b/Rep01(2018). Annex J. Bled. Available from: <www.iwc.int>.

Ivashchenko, Y., Clapham, P., 2010. Bowhead whales *Balaena mysticetus* in the Okhotsk Sea. Mammal Rev. 40 (1), 65–89.

Kapel, F.O., 1985. A note on the net-entanglement of a bowhead whale (*Balaena mysticetus*) in northwest Greenland, November 1980. Rep. Int. Whal. Comm. 35, 377–378.

Knowlton, A.R., Hamilton, P.K., Marx, M.K., Pettis, H.M., Kraus, S.D., 2012. Monitoring North Atlantic right whale *Eubalaena glacialis* entanglement rates: a 30 yr retrospective. Mar. Ecol. Prog. Ser. 466, 293–302.

Knowlton, A.R., Robbins, J., Landry, S., McKenna, H.A., Kraus, S.D., Werner, T.B., 2016. Effects of fishing rope strength on the severity of large whale entanglements. Conserv. Biol. 30 (2), 318–328.

Kraus, S.D., Brown, M.W., Caswell, H., Clark, C.W., Fujiwara, M., Hamilton, P.K., et al., 2005. North Atlantic right whales in crisis. Science 309, 561–562.

Lewison, R.L., Crowder, L.B., Wallace, B.P., Moore, J.E., Cox, T., Zydelis, R., et al., 2014. Global patterns of marine mammal, seabird, and sea turtle bycatch reveal taxa-specific and cumulative megafauna hotspots. Proc. Natl. Acad. Sci. USA 111 (14), 5271–5276.

Miller, C., Best, P., Perryman, W., Baumgartner, M., Moore, M., 2012. Body shape changes associated with reproductive status, nutritive condition and growth in right whales *Eubalaena glacialis* and *Eubalaena australis*. Mar. Ecol. Prog. Ser. 459, 135–156.

Moore, M.J., 2014. How we all kill whales. ICES J. Mar. Sci. 71, 760–763. Available from: https://doi.org/10.1093/icesjms/fsu008.

Moore, M.J., 2019. How we can all stop killing whales: a proposal to avoid whale entanglement in fishing gear. ICES J. Mar. Sci. 76 (4), 781–786. Available from: https://doi.org/10.1093/icesjms/fsy194.

Moore, M.J., van der Hoop, J., 2012. The painful side of trap and fixed net fisheries: chronic entanglement of large whales. J. Mar. Biol. 2012, 1–4. Available from: https://doi.org/10.1155/2012/230653:.

Moore, M.J., Bogomolni, A., Bowman, R., Hamilton, P., Harry, C., Knowlton, A., et al., Fatally entangled right whales can die extremely slowly. In: Oceans'06 MTS/IEEE-Boston, Mass. September 18–21, 2006. ISBN: 1-4244-0115-1.

Moore, M.J., Knowlton, A.R., Kraus, S.D., McLellan, W.A., Bonde, R.K., 2004. Morphometry, gross morphology and available histopathology in North Atlantic right whale (*Eubalaena glacialis*) mortalities (1970–2002). J. Cetacean Res. Manage. 6 (3), 199–214.

NAMMCO, 2017. Review of existing knowledge on marine mammal by-catch in Greenland the government of Greenland ministry of fisheries & hunting MAY 2017. In: Paper SC/24/BYC/14/.

Nishiwaki, M., Kasuya, T., 1970. A Greenland right whale caught at Osaka Bay. Sci. Rep. Whales Res. Inst. 22, 45–62.

NOAA, 2019. BSAI Inseason Management Report. December 2019. Prepared by Josh Keaton, November 2019. NOAA Fisheries Alaska Region. <https://www.fisheries.noaa.gov/alaska/commercial-fishing/fisheries-catch-and-landings-reports>.

Pace, R.M., Corkeron, P.J., Kraus, S.D., 2017. State–space mark–recapture estimates reveal a recent decline in abundance of North Atlantic right whales. Ecol. Evol. 7, 8730–8741. Available from: https://doi.org/10.1002/ece3.3406.

Pettis, H.M., Rolland, R.M., Hamilton, P.K., Brault, S., Knowlton, A.R., Kraus, S.D., 2004. Visual health assessment of North Atlantic right whales (*Eubalaena glacialis*) using photographs. Can. J. Zool. 82 (1), 8–19. Available from: http://dx.doi.org/10.1139/z03-207.

Philo, L.M., Shotts Jr., E.B., George, J.C., 1993. Morbidity and mortality. In: Burns, J.J., Montague, J.J., Cowles, C.J. (Eds.), The Bowhead Whale. Society for Marine Mammalogy, pp. 275–312.

Philo, L.M., George, J.C., Albert, T.F., 1992. Rope entanglement of bowhead whales (*Balaena mysticetus*). Mar. Mammal Sci. 8 (3), 306–311.

Read, A.J., Drinker, P., Northridge, S., 2006. Bycatch of marine mammals in U.S. and global fisheries. Conserv. Biol. 20 (1), 163–169.

Reeves, R., Rosa, C., George, J.C., Sheffield, G., Moore, M., 2012. Implications of Arctic industrial growth and strategies to mitigate future vessel and fishing gear impacts on bowhead whales. Mar. Policy 36, 454–462. Available from: https://doi.org/10.1016/j.marpol.2011.08.005.

Reeves, R.R., Ewins, P.J., Agbayani, S., Heide-Jørgensen, M.P., Kovacs, K.M., Lydersen, C., et al., 2014. Distribution of endemic cetaceans in relation to hydrocarbon development and commercial shipping in a warming Arctic. Mar. Policy 44, 375–389. Available from: http://dx.doi.org/10.1016/j.marpol.2013.10.005.

Rolland, R.M., Katherine, M.G., Stimmelmayr, R., Suydam, R.S., George, J.C., 2019. Chronic stress from fishing gear entanglement is recorded in baleen from a bowhead whale (*Balaena mysticetus*). Mar. Mammal Sci. 35 (4), 1625–1642.

Rugh, D.J., Braham, H.W., Miller, G.W., 1992. Methods for photographic identification of bowhead whales, *Balaena mysticetus*. Can. J. Zool. 70, 617–624.

Schatz, V., Liu, N., Proelss, A., 2019. The 2018 agreement to prevent unregulated high seas fisheries in the central Arctic Ocean: a critical analysis. Int. J. Mar. Coastal Law 34, 195–244.

Sharp, S.M., McLellan, W.A., Rotstein, D.S., Costidis, A.M., Barco, S.G., Durham, K., et al., 2019. Gross and histopathologic diagnoses from North Atlantic right whale *Eubalaena glacialis* mortalities between 2003 and 2018. Dis. Aquat. Organ. 135 (1), 1–31.

Sheffield, G., 2010. A Bowhead Whale Entangled in Bering Sea Commercial Pot Gear, Chukchi Sea. Unpublished Report to the North Slope Borough Department of Wildlife Management, Barrow, AK, 9p.

Sheffield, G., Savoonga Whaling Captains Association (SWCA). 2015. Bowhead Whale Entangled in Commercial Crab Pot Gear Recovered Near Saint Lawrence Island, Bering Strait. Unpublished Report to the North Slope Borough Department of Wildlife Management, Barrow, AK, 8p.

Shpak, O., Meschersky, I., Chichkina, A.N., Kuznetsova, D., Paramonov, A.Y., Rozhnov, V., 2014. New Data on the Okhotsk Sea Bowhead Whales. Rept Sci. Comm. Int. Whaling Comm. SC/65b/BRG17, Bled, Slovenia, May 12–14, pp. 1–5.

Shpak, O.V., Stimmelmayr, R., 2017. Preliminary image analysis of acute and chronic injuries, parasites, and skin conditions in the Okhotsk Bowhead whale (*Balaena mysticetus*) stock in the western Okhotsk Sea. In: Presented to the 67th International Whaling Commission. SC/67a/E01.

Stafford, K.M., Lydersen, C., Wiig, Ø., Kovacs, K.M., 2018. Extreme diversity in the songs of Spitsbergen's bowhead whales. Biol. Lett. 14 (4), 20180056. Available from: http://doi.org/10.1111/j.1748-7692.2003.tb01321.x.

Stevenson, D.E., Lauth, R.R., 2019. Bottom trawl surveys in the northern Bering Sea indicate recent shifts in the distribution of marine species. Polar Biol. 42, 407–421. Available from: https://doi.org/10.1007/s00300-018-2431-1.

Thoman, R.L., Bhatt, U.S., Bieniek, P.A., Brettschneider, B.R., Brubaker, M., Danielson, S.L., et al., 2020. The record low bering sea ice extent in 2018: context, impacts, and an assessment of the role of anthropogenic climate change. B Am Meteorol Soc 101 (1), S53–S58.

Tsukrov, I., DeCew, J., Baldwin, K., Campbell-Malone, R., Moore, M., 2009. Mechanics of the right whale mandible: full scale testing and finite element analysis. J. Exp. Mar. Biol. Ecol. 374, 93–103.

U.S. Committee on the Marine Transportation System, 2019. A Ten-Year Projection of Maritime Activity in the U.S. Arctic Region, 2020–2030. Washington, DC, 118 p.

van der Hoop, J., Corkeron, P., Moore, M., 2017. Entanglement is a costly life-history stage in large whales. Ecol. Evol. 7 (1), 92–106.

Vate Brattstrom, L., Mocklin, J., Tudor, B, George, J.C., Givens, G.H., et al., 2016. Update on the bowhead whale aerial photoID program and 2011 aerial spring survey. In: Paper SC/66b/BRG04. The IWC Scientific Committee.

Von Duyke, A.L., Stimmelmayr, R., Sheffield, G., Sformo, T., Suydam, R., Givens, G.H., et al., 2016. Prevalence and abundance of cyamid "whale lice" (*Cyamus ceti*) on subsistence harvested bowhead whales (*Balaena mysticetus*). Arctic 69 (4), 331–340.

CHAPTER

37

Contaminants

I.R. Schultz[1], *J.L. Bolton*[1], *R. Stimmelmayr*[2] *and G.M. Ylitalo*[1]

[1]Environmental and Fisheries Sciences Division, Northwest Fisheries Science Center, National Marine Fisheries Service, National Oceanic and Atmospheric Administration, Seattle, WA, United States [2]Department of Wildlife Management, North Slope Borough, Utqiaġvik, AK, United States

Introduction

Concern over adverse health effects of contaminants in whales and subsistence consumers in Alaska and across the Arctic (Dietz et al., 2019) has stimulated monitoring efforts in recent decades (O'Hara et al., 1999; Hoekstra et al., 2005). Based on these concerns, major studies on contaminant levels in subsistence foods, including bowhead whales, began in the early 1980s by the North Slope Borough, Department of Wildlife Management (Bratton et al., 1993). Collaborations with subsistence bowhead whale hunters in Alaska (Chapters 32 and 34) have offered a rare opportunity to examine tissues from freshly harvested animals and measure contaminant concentrations across the major organ systems (Fig. 37.1).

The major classes of contaminants monitored in Bering–Chukchi–Beaufort Seas (BCB) bowhead whales continue to be petroleum-related, persistent organics, and metals. The Arctic Monitoring and Assessment Program (AMAP, 2016a,b) has also estimated the potential for emerging contaminants such as polyfluoroalkyl substances, pharmaceuticals, and personal care products to enter the Arctic environment, but little monitoring data are available for whales; therefore our focus in this chapter will be on the major classes of contaminants.

Exposure to contaminants occurs primarily through ingestion of contaminated prey or particulate matter, although for some contaminants, such as petroleum, direct contact and inhalation of volatile components is important (O'Hara and O'Shea, 2001; Helm et al., 2015; Godard-Codding and Collier, 2018). Bowhead whales are extremely long-lived animals, with a life span that may exceed 200 years and both sexes mature sexually in their

The Bowhead Whale
DOI: https://doi.org/10.1016/B978-0-12-818969-6.00037-6

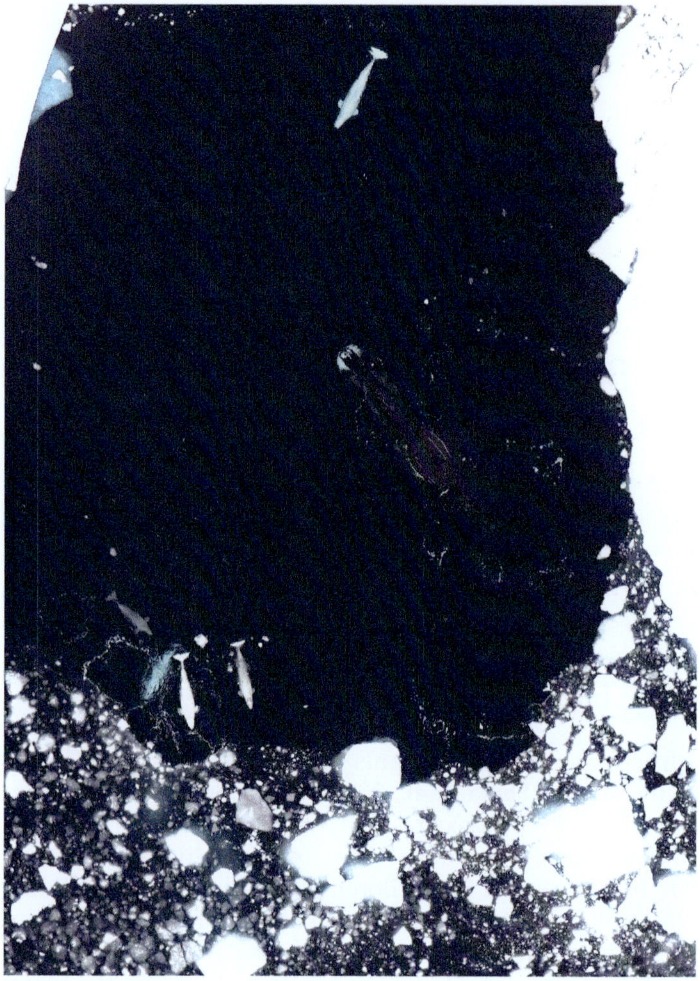

FIGURE 37.1 The Arctic marine ecosystem is very different from the other temperate and tropical ecosystems, but it is also deeply sensitive to natural and human-induced perturbations brought to it from temperate latitudes. Arctic species, such as belugas and bowheads, shown here, will be affected by anthropogenic global change. *Source: Photo by Corey Accardo (NOAA/North Slope Borough, NMFS Permit No. 14245).*

twenties (George et al., 1999; Chapter 7). Reproductive senescence is also significantly delayed, or nonexistent, compared to other mammals, with two pregnant females estimated to be 83 years and 125 years, and some possibly older (Lubetkin et al., 2012; Rosa et al., 2013; Chapter 21). This combination of extreme longevity and protracted reproductive time span can complicate interpretation of contaminant monitoring data, particularly persistent organics (Rowe, 2008). Age is a well-established factor in tissue levels of bioaccumulative substances, as a longer life span permits greater exposure. This becomes apparent when examining data from male whales, where a close relationship between age and tissue levels of contaminants is often seen (Aguilar et al., 1999). In females a significant portion of the body burden of persistent contaminants such as polychlorinated biphenyls (PCBs) can be transferred to their offspring during gestation and especially during lactation, causing tissue levels to decrease (Cockcroft et al., 1989). Another factor relevant for long-lived species such as bowhead whales is the timing of their birth cohort. Individuals born before the 1930s and prior to widespread use of many persistent contaminants may have lower than expected tissue levels compared to individuals born after significant environmental contamination had occurred (Binnington and Wania, 2014). The explanation for this phenomenon is the reduced or lack of embryonic and lactational exposure in older individuals compared to younger individuals who received substantial exposure resulting from maternal transfer.

Petroleum-related contaminants

Unintentional release of petrogenic compounds from increased marine industrial activities (e.g., oil exploration and production, shipping) occurring in the Arctic is a major concern (Geraci and St. Aubin, 1980; Dalsøren et al., 2007). Crude oil contains thousands of chemicals that persist in the marine environment with polycyclic aromatic hydrocarbons (PAHs) considered the most toxic (Krahn and Stein, 1998). In addition to petroleum products, other sources of PAHs include atmospheric releases from incomplete combustion of fossil fuels or biomass and volcanic eruptions (Balmer et al., 2019).

Cetaceans, like other vertebrates, rapidly absorb PAHs from the environment and metabolize them into more polar compounds that are concentrated in bile and excreted (Krahn et al., 1992; Varanasi et al., 1993; Beyer et al., 2010). This makes assessment of PAH exposure challenging because muscle, blubber, and other organs usually contain low levels of PAHs (Varanasi et al., 1993; Meador et al., 1995). Thus, determining recent PAH exposure in cetaceans and other vertebrates is assessed by measuring polar PAH metabolites in bile, urine, or feces (Beyer et al., 2010; Godard-Codding and Collier, 2018).

Despite their importance, only limited data on petroleum-related chemicals have been reported for Arctic marine mammals. Petroleum-related hydrocarbons, including PAHs, studied in eight different bowhead whale matrices collected from whales harvested from Utqiaġvik in 2002 were below detection in all samples analyzed (Wetzel and Reynolds, 2003). Additional studies have assessed PAHs in tissues and fluids of belugas, ice seals, polar bears, and other marine mammals in the US Arctic and have shown nondetectable to low tissue levels of PAHs (<150 ng/g, wet weight; Wetzel, 2007; Stimmelmayr et al., 2018; Ylitalo et al., in prep). Recently, PAH metabolites have been measured in fluids of marine mammals harvested in the US Arctic, with elevated biliary levels being associated with visible oiling or recent feeding behavior (Stimmelmayr et al., 2018; Ylitalo et al., in prep).

As a continuation of earlier studies on baseline tissue PAH levels for bowhead whales, we assessed concentrations of summed PAHs ($\sum$PAH) in muscle and blubber of males harvested annually from 2006 to 2015 by hunters from Utqiaġvik, Alaska. Concentrations of $\sum$PAHs in muscle and blubber were below the lower limit of quantitation (<LOQ) or, when detected, were low (<140 ng/g, wet weight; Fig. 37.2). Regardless of sampling year, we found that the median muscle $\sum$PAHs were consistently lower than the median blubber $\sum$PAHs sampled in the same year, likely due to the lower percent lipid content of muscle (0.95 ± 0.81) compared to blubber (81 ± 4.3). Concentrations of lipid-normalized $\sum$PAH levels that we determined in blubber of bowheads sampled from 2006 to 2015

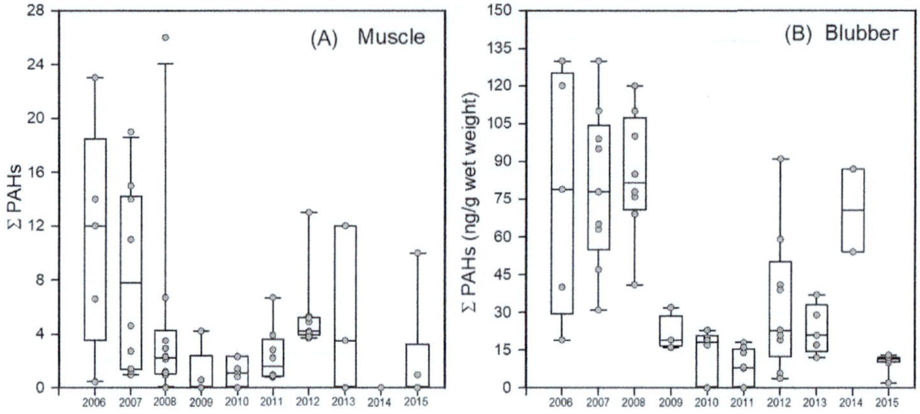

FIGURE 37.2 Concentrations of summed polycyclic aromatic hydrocarbons (∑PAHs) determined in bowhead muscle (A) and blubber samples (B) collected annually by subsistence hunters 2006–15 in Utqiaġvik, Alaska, United States. Observed data (*gray circles*), means, ranges, and one standard deviation are shown on each graph.

ranged from <LOQ to 180 ng/g, lipid weight. These values are comparable to, or lower than, the ∑PAHs reported in blubber of belugas sampled in the Canadian Arctic (Binnington et al., 2017), bottlenose dolphins from the Atlantic Ocean (Fair et al., 2010; García-Alvarez et al., 2014), and other cetacean species sampled in Asia, North America, and Europe (Fossi et al., 2014; Leung et al., 2005; Moon et al., 2012; Zhan et al., 2019). Low-molecular-weight PAHs (those containing two or three fused aromatic rings) were the only contributors to the ∑PAH values for all bowhead samples analyzed with exception of a muscle sample collected in 2006. In this sample, fluoranthene, a high-molecular-weight PAH containing four fused rings, was measured at approximately 1 ng/g wet weight. Although bowhead tissues collected from 2006 to 2007 had elevated median ∑PAH values compared to whales sampled in subsequent years, no clear temporal trends in median PAH values were noted due to PAH concentrations at or near the LOQ and variable number of tissue samples analyzed each year. Elevated concentrations of ∑PAHs were measured in blubber and muscle of bowhead whales sampled in 2006–07 compared to the same tissue levels determined in whales sampled in other years of this study. During this time frame, no major oil spills were reported in the US feeding grounds but smaller, undocumented petroleum spills may have occurred (e.g., Russia) and contributed to the PAHs measured.

Petroleum exposure in marine mammals is well known to cause acute and chronic toxicity such as epidermal inflammation and oil fouling of eyes and baleen (Bratton et al., 1993). More recent studies conducted after the 2010 Deepwater Horizon oil spill in the northern Gulf of Mexico (NGOM) have identified other adverse effects on cetaceans. A health assessment study with common bottlenose dolphins of the NGOM found that dolphins from a heavily oiled area showed evidence of adrenal toxicity and were five times more likely to have moderate-to-severe lung disease compared to dolphins from a non-oiled area (Schwacke et al., 2013). Other cetacean studies have found relationships of oil exposure and various biological effects. These include suppression of immune responses and increased susceptibility to diseases, increased prevalence of organ lesions, impaired response to stress, and reduced reproductive success (Matkin et al., 2008; Venn-Watson et al., 2015; Colegrove et al., 2016; De Guise et al., 2017; Kellar et al., 2017; Smith et al., 2017). Although a major spill has not occurred in ecosystems with bowhead populations, these whales would likely be subjected to the same exposure routes and biological effects noted for dolphins and other cetaceans. Thus continued monitoring of PAH and PAH metabolite levels along with biological measurements for bowhead whales is important to detect any changes from baseline levels.

Essential and nonessential elements

There have been several studies of elemental content of bowhead whale tissue dating back to an initial study by Byrne et al. (1985) who reported on trace metal content in different tissues collected from two individuals. A subsequent review of contaminants in bowhead whales in 2004 summarized important aspects of metal and other element concentrations in various tissues (O'Hara et al., 2004). They concluded generally that concentrations of essential elements were comparable to those measured in other marine mammal species (O'Hara et al., 2004). Most nonessential elements were also comparable, whereas mercury and lead levels were lower than in other cetacean species (O'Hara et al., 2004). Tissue levels of cadmium were also similar to other marine mammals; however, it was noted that levels are higher than those reported in terrestrial animals (O'Hara et al., 2004). A later study compared element concentrations in various bowhead tissues and reported nonessential elements such as cadmium and mercury were much higher in liver and kidney compared to muscle, heart, blubber, and other tissues (O'Hara et al., 2006). Differences in tissue concentrations for essential elements were less pronounced although blubber tended to have the lowest concentration of most elements (O'Hara et al., 2006).

As a follow-up to these earlier studies, we determined elemental content of bowhead liver and kidney tissues using inductively coupled plasma mass spectrometry as described in Miller et al. (2016). Measurements were done on tissue samples collected from over 50 individuals of varying body length, which we assume correlates with age. These results are summarized in Table 37.1. Essential elements such as potassium, sodium, sulfur, and iron are among the most abundant elements measured and do not appear to vary significantly with age. Less abundant essential elements such as calcium, copper, zinc, and magnesium also do not appear to vary with age or tissue although for many elements, concentrations were highly variable among individuals (Table 37.1). In contrast, the potentially toxic elements cadmium and mercury do show a trend of increasing concentration with age. This observation has been documented by Krone et al. (1999), who reported a mercury to selenium ratio of 1:40, similar to our observations (Table 37.1). In Fig. 37.3A, we show the relationship between liver cadmium concentrations and body length. Included on this graph are the values reported in Krone et al. (1999) and Byrne et al. (1985), which appear consistent with our results and suggest that liver cadmium content in bowhead whales, when adjusted for body length, has not increased since monitoring efforts began. Additional associations between cadmium and mercury and mercury and selenium are shown in Fig. 37.3B and C. There is a strong correlation between tissue levels

TABLE 37.1 Summary of total metal concentrations in liver and kidney from bowhead whales in the Bering–Chukchi–Beaufort Seas stock.

	Liver	Kidney
Metal	Geometric mean and range	Geometric mean and range
Silver	0.165 (0.011–11.7)	
Arsenic	6.12 (4.77–11.3)	0.804 (0.230–2.80)
Calcium	192.6 (101–782)	333 (76.4–1,490)
Cadmium	6.99 (0.078–159)	11.8 (0.184–171)
Cobalt	0.104 (0.051–0.219)	0.54 (0.021–0.811)
Copper	18.5 (9.10–233)	5.76 (1.64–19.3)
Iron	1370 (261–13,400)	243 (58.0–3350)
Mercury	0.116 (0.010–0.698)	0.055 (0.005–0.287)
Potassium	7610 (5240–10,100)	4230 (1100–8610)
Magnesium	399 (266–666)	266 (64.9–676)
Manganese	3.98 (1.34–8.11)	1.27 (0.338–16.4)
Molybdenum	1.33 (0.366–3.22)	
Sodium	5080 (3100–7440)	6500 (1660–12,300)
Sulphur	7240 (5210–8590)	4490 (1210–8550)
Selenium	3.50 (1.84–6.85)	3.26 (0.579–7.19)
Silicon	2.31 (1.26–10.1)	3.95 (1.66–423)
Strontium	0.471 (0.194–3.79)	0.934 (0.188–36.1)
Titanium	1.00 (0.494–38.9)	5.44 (0.486–158)
Vanadium	1.34 (0.111–6.84)	0.243 (0.051–1.86)
Zinc	102 (52.5–288)	60.8 (10.6–225)

Values are ng/g, dry weight of tissue. Moisture content of liver was 73.7% ± 1.9% (mean, SD) and 69.0% ± 10.9% for the kidney. Only those metals detected in ≥60% of the samples are shown.

of cadmium and mercury (Fig. 37.3B). Both elements are known to induce and become sequestered by the metal-binding protein metallothionein (MTH; Roesijadi, 1996). Given the much higher tissue levels of cadmium and the relationship shown in Fig. 37.3B, it is likely that the gradual increase in tissue cadmium levels during the lifetime of the whale causes MTH induction to sequester the metal along with mercury. The association between mercury and selenium is also well established and is dependent in part on the specific mercury species (elemental, inorganic or methyl/ethyl mercury) that is accumulated (Spiller, 2018). In recent years, it has become increasingly recognized that mercury can interfere with selenium-containing enzymes, while selenium can be viewed as

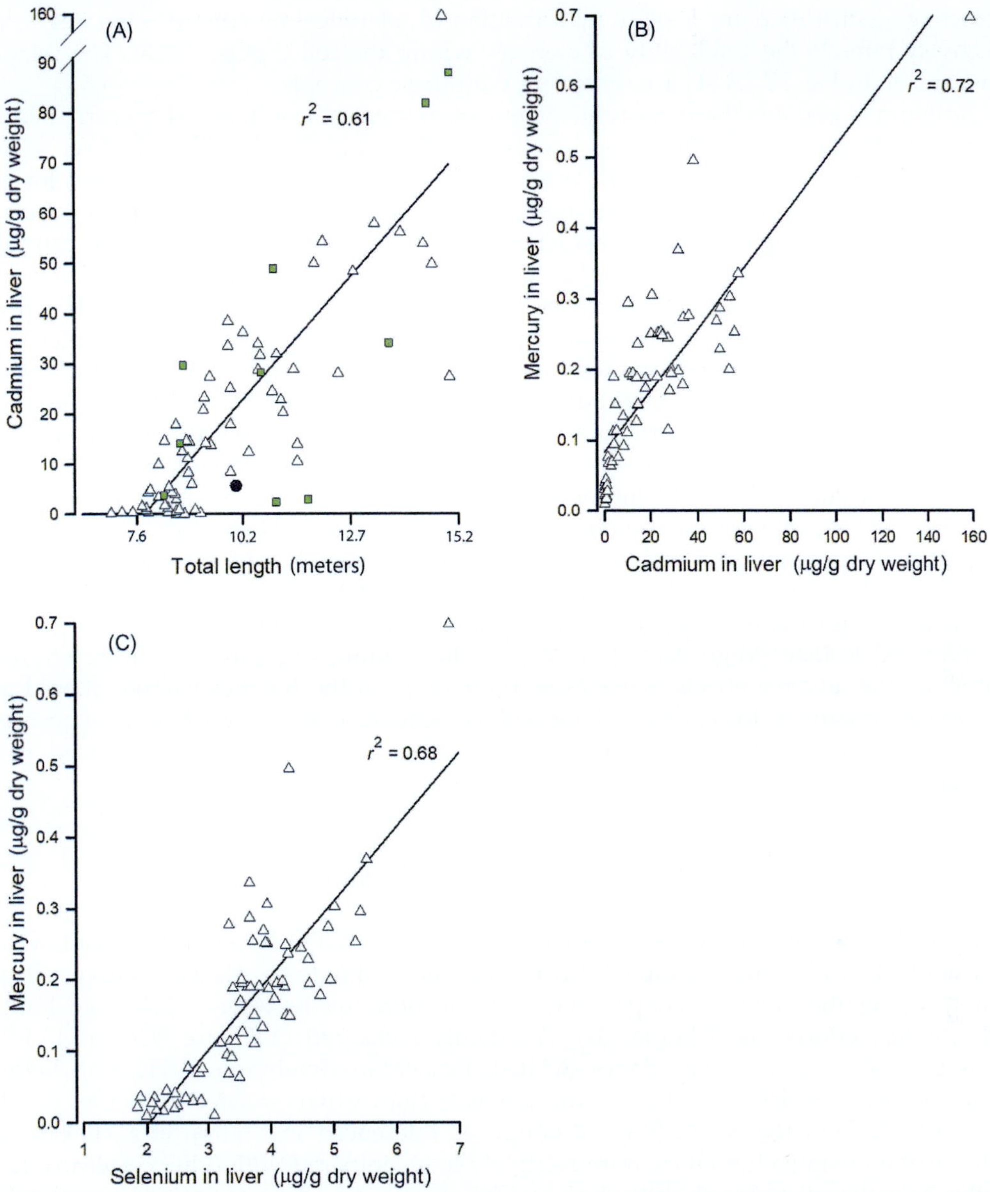

FIGURE 37.3 (A) The relationship between hepatic cadmium concentrations and body length of bowhead whale, including data from Byrne et al. (1985) (*blue circle*) and Krone et al. (1999) (*green squares*). (B) The relationship between hepatic mercury and cadmium concentrations. (C) The relationship between mercury and selenium concentrations in the liver.

protecting against mercury toxicity by formation of selenide–mercuric precipitates, which effectively remove the availability of mercury within the cell (Spiller, 2018). The relationships shown in Fig. 37.3A–C are consistent with these concepts.

Cadmium is perhaps the nonessential element of greatest toxicological concern, because of its well-known toxic effects on the mammalian kidney, bone, and other tissues (reviewed in Satarug et al., 2010). Because of the higher concentrations observed for cadmium in the bowhead kidney and liver, an emphasis has been placed on assessing whether histopathological changes associated with these tissues were linked to cadmium accumulation. It was noted that higher tissue levels of cadmium were associated with increased evidence of renal fibrosis and lung fibromuscular hyperplasia in bowheads (Rosa et al., 2008). However, these histopathological findings are also associated with senescence. Because older animals tend to have higher tissue concentrations of cadmium, it is difficult to separate potential adverse effects of cadmium from normal age-related changes in kidney and liver structure (Rosa et al., 2008). The protective role of MTH in sequestering cadmium in the bowhead kidney is still unclear although it has been suggested that whales may be better protected than other vertebrates because of a more aggressive MTH induction in response to metal exposures (O'Hara et al., 2006). There is the potential for synergistic interactions between toxic metals and organic contaminants that are also bioaccumulated by whales. For example, it has recently been shown in rats that dichlorodiphenyldichloroethylene can alter the normal expression of MTH genes in the liver and kidney (Migliaccio et al., 2019). This finding suggests that future efforts at evaluating the adverse effects of metal accumulation on the bowhead whale should also quantify expression of MTH genes along with consideration of organochlorine exposure as potential explanatory variables that can modulate progression of metal-induced tissue damage.

Persistent organic pollutants

All published data on persistent organic pollutants (POPs) levels in bowhead whales are from BCB stock whales sampled during subsistence hunts by Alaskan Native villages, demonstrating the value of cooperation and collaboration between subsistence hunters and scientific efforts (See Chapter 31). Frequently measured POPs are PCBs and chlorinated pesticides such as chlordanes (CHLs), hexachlorocyclohexanes (HCHs), dieldrin, hexachlorobenzene (HCB), and dichlorodiphenyltrichloroethane-related pesticides (DDTs). Early reports from the North Slope Borough as referenced in Bratton et al. (1993) suggested that contamination levels were relatively low compared with other cetaceans. Later studies have described biomagnification and the effects of feeding ecology and metabolism on the bioaccumulation of POPs in whales and other Arctic animals (Hoekstra et al., 2002a,b,c, 2003a,b,c, 2005; Welfinger-Smith et al., 2011). Data in those studies were from animals collected between 1997–2000 and 2005–09. More recently, Bolton et al. (2020) assessed temporal trends of POPs in blubber and muscle samples collected from males between 2006 and 2015 and is summarized in Table 37.2. Concentrations of POPs in blubber were ordered ΣHCHs > ΣCHLs ~ HCB > ΣPCBs > ΣDDTs ~ dieldrin, which is different from the order reported for bowhead blubber samples collected from 1997 to 2000

TABLE 37.2 Selected persistent organic pollutant levels in bowhead whale tissues harvested from the Bering–Chukchi–Beaufort Seas stock between 1992 and 2015.

Sample period	*n*	Sex	Tissue	% Lipid	HCH	HCB	PCB[a]	DDT	Dieldrin	CHL[b]
1992–93[c]	12	F	Blubber	75		90.3	*376.7*	131.3	110.1	109.2
	14	M	Blubber	75.6		87.3	*327.4*	129	116.7	95.6
	9	F	Liver	4.5		7.9	*20.8*	3.9	3.2	3.6
	11	M	Liver	4.9		8.2	*39.5*	6.5	4	4.4
1997–99[d]	5		Blubber	83.7	297	100	354	377		255
	5		Muscle	2.39	2.74	1.6	1.87	1.71		2.32
1997–2000[e]	71		Blubber	75.8	203	100	410	331	84	152
	23		Liver	6.6	9.5	3.1	9.1	3.7	3	5.4
2005–09[f]	3		Blubber			24	318			
	7		Mungtak (blubber)			12	143			
	3		Rendered oil			17	354			
	4		"Meat"			0.58	27			
2006–15[g]	71	M	Blubber	82	89	64	81/*111*	55	60	77
	66	M	Muscle	0.94	1.4	0.84	4.8/*5.3*	0.68	0.66	0.83

[a]*PCBs are reported as the sum of all congeners monitored (40–124; varies by reference) or the 17 most common congeners × 2 (italicized values).*
[b]*CHL: all values include* cis-chlordane, trans-*nonachlordane and oxychlordane. Some data sources include less frequently detected CHLs (heptachlor epoxide,* trans-chlordane, cis-*nonachlor and nonarchlor III), see cited sources.*
[c]*O'Hara et al. (1999).*
[d]*Hoekstra et al. (2005).*
[e]*Hoekstra et al. (2002a,b,c).*
[f]*Welfinger-Smith et al. (2011).*
[g]*Bolton et al. (2020).*
All values are mean net weight concentrations (ng/g). Sex is listed when reported in the data source. *HCB*, hexachlorobenzene; *HCH*, hexachlorocyclohexane; *PCB*, polychlorinated biphenyl; *CHL*, Chlordane.

(Hoekstra et al., 2002a, 2003b, 2005; O'Hara et al., 1999), which was ΣPCBs > ΣDDTs ≥ ΣHCHs ≥ ΣCHLs. Concentrations of polybrominated diphenylether (PBDE) flame retardants, reported for the first time for the 2006–15 samples, are very low (<2 ng/g wet weight in most samples) with frequent nondetects, with BDE 99 being the predominant congener when detected.

Temporal POP trends in bowhead blubber (Bolton et al., 2020) agree with published reports on declines of POPs in other Arctic biota since ~2000 (AMAP, 2016b; Riget et al., 2010, 2019; Braune et al., 2005; Braune, 2007). Recent declines of approximately 1%–8% per year are apparent in bowhead blubber, depending on the POP class (Bolton et al., 2020). Between the 1990s and mid-2000s, concentrations of all POPs have fallen, with pronounced decreases of PCBs, CHLs, and DDTs. PCBs have declined by a factor of 4 since the early 1990s, while other POPs have declined by a factor of 2 or more. HCB is decreasing but more slowly, declining 30%–40% since the 1990s. HCH concentrations have fallen by a factor of ~2, but HCHs are now the predominant POP in BCB bowheads, as opposed to PCBs in previous measurement periods. This shift in abundance is attributable to declines in PCBs since peak levels in the environment were reached in the mid-1980s and continuing inputs of β-HCH. Since 2000 there has been a change in global transport of HCHs to the Arctic from the relatively labile α- and γ-HCH forms (via atmospheric transport), to the more recalcitrant β-HCH (via transport on ocean currents), which has approximately a decadal delay from emissions. HCH concentrations in bowhead tissues are reflective of this change, with β-HCH now being the predominant HCH present, an indication of more recent inputs. AMAP has also noted more rapid declines of α- and γ-HCH with β-HCH increasing in proportion to total HCHs in Arctic biota.

POPs patterns and stable carbon isotopes ($\delta^{13}C$) change seasonally in bowhead muscle and provide evidence that bowheads are feeding in both the Bering and Beaufort Seas. POP patterns are indicative of differences in the POPs present in surface waters between feeding grounds as reflected in prey, while the $\delta^{13}C$ signals reflect seasonal effects of input from rivers, with a more depleted signal in the spring relative to fall (Hoekstra et al., 2003a). Ratios of stable isotopes of nitrogen did not change seasonally, indicating that trophic level was consistent between spring/summer and fall/winter feeding grounds. Biotransformation of POPs also occurs in whales and may contribute to observed patterns of abundance. Both hydroxylated and methylsulfone metabolites of PCBs have been measured in plasma and blubber of bowhead whales although concentrations of these metabolites were quite low (6 ng/g lipid weight) and less than 10% of the concentrations of the corresponding parent PCB congeners (Hoekstra et al., 2003a). While the concentrations were low, cytochrome P450 2b–like metabolic pathways present in other cetaceans also appear present in the bowhead whale. A study of chiral PCB enantiomers in bowhead liver and blubber (Hoekstra et al., 2002c) demonstrated enantiomer-specific metabolism was occurring, although shifts away from a racemic signature were much less marked than in some other species. A study of enantiomers of α-HCH and of chiral chlordane metabolites (Hoekstra et al., 2003c) suggested that these compounds are taken up with similar profiles to those in bowhead prey, selectively biotransformed in the liver and then nonselectively partitioned into blubber.

Because available data on contaminants in bowhead whales come from animals from the BCB stock, there is a gap regarding the other three stocks. However, some general predictions can be made considering work done by AMAP contributors on global geographical trends of POPs in the Arctic. Geographical trends for POPs in other species (e.g., beluga, seabirds, polar bear, and Arctic fox) indicate that there should be differences in which POPs are more prevalent in other stocks. For example, HCHs are now the predominant POP in the BCB animals sampled in the 2000s (Bolton et al., 2020), but this will not be true for stocks further east, for which PCBs and CHLs should still be the predominant POPs. These POPs will also likely be somewhat higher in other stocks compared with the BCB stock; however, POP concentrations should have declined in all stocks after peak concentrations were reached during the 1970s–80s (Braune et al., 2005). Due to mutual isolation, different stocks of bowhead whales may also have genetic differences that could affect metabolism of POPs and other contaminants.

There are few direct studies on the health effects of POP exposure in whales. Most health assessments rely on postmortem tissue collection or correlational analysis between populations of whales differentially exposed to contaminants. Within these limitations, past researchers have concluded that bowhead whales receive less exposure and generally show no evidence of adverse health effects compared to beluga and killer whales found in the Arctic (Letcher et al., 2010). Advisories for human consumers of bowhead blubber or oil have been indicated, based on samples collected in 1992–93 and 2005–09 (O'Hara et al., 1999; Welfinger-Smith et al., 2011). Although present exposure levels are below effects thresholds for monitored POPs, continued surveillance is important because of recent studies in other Arctic species that demonstrate the potential for POPs to cause adverse effects on the neuroendocrine and immune systems (reviewed in Sonne et al., 2017; Dietz et al., 2019).

Conclusions

Current levels of contaminants in bowhead whale tissues pose low risk to whales and the humans consuming them. This confirms the perception that bowheads provide a safe food source for indigenous people. However, growing human activity in the Arctic is expected to increase direct releases of contaminants (Ma et al., 2016). Rapid climate change will also alter pathways for the introduction of contaminants into the environment. Changes in precipitation and melting of sea ice containing POPs deposited over decades of atmospheric transport plus increased biotransport of contaminants from migratory species are just some of possible processes affecting levels of POPs in Arctic biota (Macdonald, 2005). Thus, continued surveillance of established contaminant classes is important but additional emerging contaminants should be added. With more than 150,000 chemical compounds registered for use in the United States and Europe (AMAP, 2016a,b), it remains important to collect baseline information for bowhead whales to determine the potential risk these compounds may pose to the whales and their consumers from Native communities.

Acknowledgement

We appreciate the support and assistance of the whaling captains, the community of Utqiaġvik, and the Alaska Eskimo Whaling Commission in sampling landed bowhead whales. We are grateful for tissue and data collection provided by the NSB DWM staff and chemical analyses conducted by personnel from the NWFSC.

References

Aguilar, A., Borrell, A., Pastor, T., 1999. Biological factors affecting variability of persistent pollutant levels in cetaceans. J. Cetacean Res. Manage. (Spec. Issue) 1, 83–116.

AMAP, 2016a. Chemicals of Emerging Arctic Concern. Oslo, pp. xvi + 353.

AMAP, 2016b. AMAP Assessment 2015: Temporal Trends in Persistent Organic Pollutants in the Arctic. Arctic Monitoring and Assessment Programme (AMAP), Oslo, pp. vi + 71pp.

Balmer, J.E., Hung, H., Yong, Y., Letcher, R.J., Muir, D.C.G., 2019. Sources and environmental fate of pyrogenic polycyclic aromatic hydrocarbons (PAHs) in the Arctic. Emerg. Contamin 5, 128–142.

Beyer, J., Jonsson, G., Porte, C., Krahn, M.M., Ariese, F., 2010. Analytical methods for determining metabolites of polycyclic aromatic hydrocarbon (PAH) pollutants in fish bile: a review. Environ. Toxicol. Pharm. 30 (3), 224–244.

Binnington, M.J., Wania, F., 2014. Clarifying relationships between persistent organic pollutant concentrations and age in wildlife biomonitoring: individuals, cross-sections, and the roles of lifespan and sex. Environ. Toxicol. Chem. 33, 1415–1426.

Binnington, M.J., Lei, Y.D., Pokiak, L., Pokiak, J., Ostertag, S.K., Loseto, L.L., et al., 2017. Effects of preparation on nutrient and environmental contaminant levels in Arctic beluga whale (*Delphinapterus leucas*) traditional foods. Environ. Sci.: Process. Impacts 19 (8), 1000–1015.

Bolton, J.L., Ylitalo, G.M., Chittaro, P.M., George, J., Suydam, R.S., Person, B.T., et al., 2020. Multi-year assessment (2006–2015) of persistent organic pollutant concentrations in blubber and muscle from Western Arctic bowhead whales (*Balaena mysticetus*), North Slope. Alaska. Mar. Pollut. Bull. 151, 110857.

Bratton, G.R., Spainhour, C.B., Flory, W., Reed, M., Jayko, K., 1993. Chapter 17: Presence and potential effects of contaminants. In: Burns, J.J., Montague, J.J., Cowles, C.J. (Eds.), The Bowhead Whale. The Society for Marine Mammalogy, Lawrence, KS, pp. 701–744. Special Publication No. 2.

Braune, B.A., 2007. Temporal trends of organochlorines and mercury in seabird eggs from the Canadian Arctic, 1975-2003. Environ. Pollut. 148, 599–613.

Braune, B.M., Outridge, P.M., Fisk, A.T., Muir, D.C.G., Helm, P.A., Hobbs, K., et al., 2005. Persistent organic pollutants and mercury in marine biota of the Canadian Arctic: an overview of spatial and temporal trends. Sci. Total Environ. 351–352, 4–56.

Byrne, C.R., Balasubramanian, R., Overton, E.B., Albert, T.F., 1985. Concentrations of trace metals in the bowhead whale. Mar. Pollut. Bull. 16, 497–498.

Cockcroft, V.G., De Kock, A.C., Lord, D.A., Ross, G.J.B., 1989. Organochlorines in bottlenose dolphins *Tursiops truncatus* from the east coast of South Africa. S. Afr. J. Mar. Sci. 8, 207–217.

Colegrove, K.M., Venn-Watson, S., Litz, J., Kinsel, M.J., Terio, K.A., Fougeres, E., et al., 2016. Fetal distress and in utero pneumonia in perinatal dolphins during the Northern Gulf of Mexico unusual mortality event. Dis. Aquat. Org. 119 (1), 1–16.

Dalsøren, S.B., Endresen, Ø., Isaksen, I.S., Gravir, G., Sørgård, E., 2007. Environmental impacts of the expected increase in sea transportation, with a particular focus on oil and gas scenarios for Norway and northwest Russia. J. Geophys. Res. 112, D02310.

De Guise, S., Levin, M., Gebhard, E., Jasperse, L., Hart, L.B., Smith, C.R., et al., 2017. Changes in immune functions in bottlenose dolphins in the northern Gulf of Mexico associated with the *Deepwater Horizon* oil spill. Endang. Species Res. 33, 291–303.

Dietz, R., Letcher, R.J., Desforges, J.P., Eulaers, I., Sonne, C., Wilson, S., et al., 2019. Current state of knowledge on biological effects from contaminants on Arctic wildlife and fish. Sci. Total Environ. 133792.

Fair, P.A., Adams, J., Mitchum, G., Hulsey, T.C., Reif, J.S., Houde, M., et al., 2010. Contaminant blubber burdens in Atlantic bottlenose dolphins (*Tursiops truncatus*) from two southeastern US estuarine areas: concentrations and patterns of PCBs, pesticides, PBDEs, PFCs, and PAHs. Sci. Total Environ. 408 (7), 1577–1597.

Fossi, M.C., Panti, C., Marsili, L., Maltese, S., Coppola, D., Jimenez, B., et al., 2014. Could feeding habit and migratory behavior be the causes of different toxicological hazard to cetaceans of Gulf of California (Mexico)? Environ. Sci. Pollut. Res. 21 (23), 13353–13366.

García-Alvarez, N., Martín, V., Fernández, A., Almunia, J., Xuriach, A., Arbelo, M., et al., 2014. Levels and profiles of POPs (organochlorine pesticides, PCBs, and PAHs) in free-ranging common bottlenose dolphins of the Canary Islands, Spain. Sci. Total Environ. 493, 22–31.

George, J.C., Bada, J., Zeh, J., Scott, L., Brown, S.E., O'Hara, S.E.T.M., et al., 1999. Age and growth estimates of bowhead whales (*Balaena mysticetus*) via aspartic acid racemization. Can. J. Zool. 77, 571–580.

Geraci, J., St. Aubin, D., 1980. Offshore petroleum resource development and marine mammals: a review and research recommendations. Mar. Fish. Rev. 42, 1–12.

Godard-Codding, C.A.J., Collier, T.K., 2018. The effects of oil exposure on cetaceans. In: Fossi, M.C., Panti, C. (Eds.), Marine Mammal Ecotoxicology: Impacts of Multiple Stressors on Population Health. Academic Press, London, pp. 75–93.

Helm, R.C., Costa, D.P., DeBruyn, T.D., O'Shea, T.J., Wells, R.W., Williams, T.M., 2015. Overview of effects of oil spills on marine mammals. Handbook of Oil Spill Science and Technology. Wiley-Blackwell, pp. 455–475.

Hoekstra, P.F., O'Hara, T.M., Pallant, S.J., Solomon, K.R., Muir, D.C.G., 2002a. Bioaccumulation of organochlorine contaminants in bowhead whales (*Balaena mysticetus*) from Barrow, Alaska. Arch. Environ. Contam. Toxicol. 42, 497–507.

Hoekstra, P.F., O'Hara, T.M., Teixeira, C., Backus, S., Fisk, A.T., Muir, D.C.G., 2002b. Spatial trends and bioaccumulation of organochlorine pollutants in marine zooplankton from the Alaskan and Canadian Arctic. Environ. Toxicol. Chem. 21, 575–583.

Hoekstra, P.F., Wong, C.S., O'Hara, T.M., Solomon, K.R., Mabury, S.A., Muir, D.C.G., 2002c. Enantiomer-specific accumulation of PCB atropisomers in the bowhead whale (*Balaena mysticetus*). Environ. Sci. Technol. 36, 1419–1425.

Hoekstra, P.F., Letcher, R.J., O'Hara, T.M., Backus, S., Solomon, K.R., Muir, D.C.G., 2003a. Hydroxylated and methylsulfone-containing metabolites of polychlorinated biphenyls in the plasma and blubber of bowhead whales (*Balaena mysticetus*). Environ. Toxicol. Chem. 22, 2650–2658.

Hoekstra, P.F., O'Hara, T.M., Fisk, A.T., Borgå, K., Solomon, K.R., Muir, D.C.G., 2003b. Trophic transfer of persistent organochlorine contaminants (OCs) within an Arctic marine food web from the southern Beaufort–Chukchi Seas. Environ. Pollut. 124, 509–522.

Hoekstra, P.F., O'Hara, T.M., Karlsson, H., Solomon, K.R., Muir, D.C.G., 2003c. Enantiomer-specific biomagnification of α-hexachlorocyclohexane and selected chiral chlordane-related compounds within and Arctic marine food web. Environ. Toxicol. Chem. 10, 2482–2491.

Hoekstra, P.F., O'Hara, T.M., Backus, S.M., Hanns, C., Muir, D.C.G., 2005. Concentrations of persistent organochlorine contaminants in bowhead whale tissues and other biota from northern Alaska: implications for human exposure from a subsistence diet. Environ. Res. 98, 329–340.

Kellar, N.M., Speakman, T.R., Smith, C.R., Lane, S.M., Balmer, B.C., Trego, M.L., et al., 2017. Low reproductive success rates of common bottlenose dolphins *Tursiops truncatus* in the northern Gulf of Mexico following the Deepwater Horizon disaster (2010-2015). Endang. Species Res. 33, 143–158.

Krahn, M.M., Stein, J.E., 1998. Peer reviewed: assessing exposure of marine biota and habitats to petroleum compounds. Anal. Chem. 70 (5), 186A–192A.

Krahn, M.M., Burrows, D.G., Ylitalo, G.M., Brown, D.W., Wigren, C.A., Collier, T.K., et al., 1992. Mass spectrometric analysis for aromatic compounds in bile of fish sampled after the Exxon Valdez oil spill. Environ. Sci. Technol. 26 (1), 116–126.

Krone, C.A., Robisch, P.A., Tilbury, K.L., Stein, J.E., Mackey, E.A., Becker, P.R., et al., 1999. Elements in liver tissues of bowhead whales (*Balaena mysticetus*). Mar. Mammal Sci. 15, 123–142.

Letcher, R.J., Bustnes, J.O., Dietz, R., Jenssen, B.M., Jorgensen, E.H., Sonne, C., et al., 2010. Exposure and effects assessment of persistent organohalogen contaminants in Arctic wildlife and fish. Sci. Total Environ. 408, 2995–3043.

Leung, C.C., Jefferson, T.A., Hung, S.K., Zheng, G.J., Yeung, L.W., Richardson, B.J., et al., 2005. Petroleum hydrocarbons, polycyclic aromatic hydrocarbons, organochlorine pesticides and polychlorinated biphenyls in tissues of Indo-Pacific humpback dolphins from south China waters. Mar. Pollut. Bull. 50 (12), 713–1719.

Lubetkin, S.C., Zeh, J.E., George, J.C., 2012. Statistical modeling of baleen and body length at age in bowhead whales (*Balaena mysticetus*). Can. J. Zool. 90, 915–931.

Ma, J., Hung, H., Macdonald, R.W., 2016. The influence of global climate change on the environmental fate of persistent organic pollutants: a review with emphasis on the Northern Hemisphere and the Arctic as a receptor. Glob. Planet. Change 146, 89–108.

Macdonald, R.W., 2005. Climate change, risks and contaminants: a perspective from studying the Arctic. Hum. Ecol. Risk Assess. 11, 1099–1104.

Matkin, C.O., Saulitis, E.L., Ellis, G.M., Olesiuk, P., Rice, S.D., 2008. Ongoing population-level impacts on killer whales *Orcinus orca* following the 'Exxon Valdez' oil spill in Prince William Sound, Alaska. Mar. Ecol. Prog. Ser. 356, 269–281.

Meador, J.P., Stein, J.E., Reichert, W.L., Varanasi, U., 1995. Bio-accumulation of polycyclic aromatic hydrocarbons by marine organisms. Reviews of Environmental Contamination and Toxicology. Springer, New York, pp. 79–165.

Migliaccio, V., Lionetti, L., Putti, R., Sica, R., Scudiero, R., 2019. Combined effects of DDE and hyperlipidic diet on metallothionein expression and synthesis in rat tissues. Environ. Toxicol. 34, 283–293.

Miller, M.W.C., Lovvorn, J.R., Matz, A.C., Taylor, R.J., Latty, C.J., Safine, D.E., 2016. Trace elements in sea ducks of the Alaskan Arctic coast: patterns of variation among species, sexes, and ages. Arch. Environ. Contam. Toxicol. 71, 297–312.

Moon, H.B., An, Y.R., Choi, S.G., Choi, M., Choi, H.G., 2012. Accumulation of PAHs and synthetic musk compound in minke whales (*Balanoptera acutorostrata*) and long-beaked common dolphins (*Delphinus capensis*) from Korean coastal waters. Environ. Toxicol. Chem. 31 (3), 477–485.

O'Hara, T.M., O'Shea, T.J., 2001. Toxicology. In: Dierauf, L.A., Gulland, F.M.D. (Eds.), CRC Handbook of Marine Mammal Medicine, second ed. CRC Press, Boca Raton, FL, pp. 471–520.

O'Hara, T.M., Krahn, M.M., Boyd, D., Becker, P.R., Philo, L.M., 1999. Organochlorine contaminant levels in Eskimo harvested bowhead whales of Arctic Alaska. J. Wildl. Dis. 35, 741–752.

O'Hara, T.M., Hoekstra, P., Rosa, C., Dehn, L., Muir, D., Woshner, V., et al., 2004. Bowhead whale contaminants: a review of current state of knowledge and possible future research directions. In: Presented to the 56th International Whaling Commission. SC/56/E1.

O'Hara, T.M., Hanns, C., Bratton, G., Taylor, R., Woshner, V., 2006. Essential and non-essential elements in eight tissue types from subsistence-hunted bowhead whale: nutritional and toxicological assessment. Int. J. Circumpolar Health 65 (3), 228–242.

Riget, F., Bignert, A., Braune, B., Stow, J., Wilson, S., 2010. Temporal trends of legacy POPs in Arctic biota, an update. Sci. Total Environ. 408, 2874–2884.

Riget, F., Bignert, A., Braune, B., Dam, M., Dietz, R., Evans, M., et al., 2019. Temporal trends of persistent organic pollutants in Arctic marine and freshwater biota. Sci. Total Environ. 649, 99–110.

Roesijadi, G., 1996. Metallothionein and its role in toxic metal regulation. Comp. Biochem. Physiol. 113C, 117–123.

Rosa, C., Blake, J.E., Bratton, G.R., Dehn, L.A., Gray, M.J., O'Hara, T.M., 2008. Heavy metal and mineral concentrations and their relationship to histopathological findings in the bowhead whale (*Balaena mysticetus*). Sci. Total Environ. 399, 165–178.

Rosa, C., Zeh, J., George, J.C., Botta, O., Zauscher, M., Bada, J., et al., 2013. Age estimates based on aspartic acid racemization for bowhead whales (*Balaena mysticetus*) harvested in 1998-2000 and the relationship between racemization rate and body temperature. Mar. Mammal Sci. 3, 424–445.

Rowe, C.L., 2008. "The calamity of so long life": life histories, contaminants, and potential emerging threats to long-lived vertebrates. Bioscience 58 (7), 623–631.

Satarug, S., Garrett, S.H., Sens, M.A., Sens, D.A., 2010. Cadmium, environmental exposure, and health outcomes. Environ. Health Persp 118 (2), 182–190.

Schwacke, L.H., Smith, C.R., Townsend, F.I., Wells, R.S., Hart, L.B., Balmer, B.C., et al., 2013. Health of common bottlenose dolphins (*Tursiops truncatus*) in Barataria Bay, Louisiana, following the *Deepwater Horizon* oil spill. Environ. Sci. Technol. 48 (1), 93–103.

Smith, C.R., Rowles, T.K., Hart, L.B., Townsend, F.I., Wells, R.S., Zolman, E.S., et al., 2017. Slow recovery of Barataria Bay dolphin health following the Deepwater Horizon oil spill (2013-2014), with evidence of persistent lung disease and impaired stress response. Endang. Species Res. 33, 127–142.

Sonne, C., Letcher, R.J., Jenssen, B.M., Desforges, J.P., Eulaers, I., Andersen-Ranberg, E., et al., 2017. A veterinary perspective on One Health in the Arctic. Acta Vet. Scand. 84, 1–11.

Spiller, H.A., 2018. Rethinking mercury: the role of selenium in the pathophysiology of mercury toxicity. Clin. Toxicol. 56 (5), 313–326.

Stimmelmayr, R., Ylitalo, G.M., Sheffield, G., Beckmen, K.B., Burek-Huntington, K.A., Metcalf, V., et al., 2018. Oil fouling in three subsistence-harvested ringed (*Phoca hispida*) and spotted seals (*Phoca largha*) from the Bering Strait region, Alaska: polycyclic aromatic hydrocarbon bile and tissue levels and pathological findings. Mar. Pollut. Bull. 130, 311–323.

Varanasi, U., Brown, D.W., Hom, T., Burrows, D.G., Sloan, C.A., 1993. Supplemental information concerning a survey of Alaskan subsistence fish, marine mammal, and invertebrate samples collected 1989-91 for exposure to oil spilled from the Exxon Valdez. In: Technical Memo (No. PB-94-134012/XAB; NOAA-TM-NMFS-NWFSC--13). National Marine Fisheries Service, Northwest Fisheries Science Center, Seattle, WA.

Venn-Watson, S., Colegrove, K.M., Litz, J., Kinsel, M., Terio, K., Saliki, J., et al., 2015. Adrenal gland and lung lesions in Gulf of Mexico common bottlenose dolphins (*Tursiops truncatus*) found dead following the *Deepwater Horizon* oil spill. PLoS One 10, e0126538.

Welfinger-Smith, G., Minholz, J.L., Byrne, S., Waghiyi, V., Gologergen, J., Kava, J., et al., 2011. Organochlorine and metal contaminants in traditional foods from St. Lawrence Island, Alaska. J. Toxicol. Environ. Health, A 74, 1195–1214.

Wetzel, D.L., 2007. Analysis of polycyclic aromatic hydrocarbons in Beaufort Sea beluga whales and ringed seals. In: Final Report to Fisheries and Oceans Canada, . Mote Marine Laboratory, Sarasota, FL, 8 p.

Wetzel, D.L., Reynolds, J.E., 2003. Hydrocarbons in tissues and other matrices of bowhead whales pilot project. In: Final Report to the North Slope Borough, Department of Wildlife Management. Mote Marine Laboratory, Sarasota, FL, 9 p.

Ylitalo, G.M., Anulacion, B.F., da Silva, D.A.M., Gates, J.B., Bolton, J.L., Rowles, T., et al. In prep. Arctic polycyclic aromatic hydrocarbon baseline study: exposure of subsistence-harvested marine mammals to petroleum-related compounds.

Zhan, F., Yu, X., Zhang, X., Chen, L., Sun, X., Yu, R.Q., et al., 2019. Tissue distribution of organic contaminants in stranded pregnant sperm whale (*Physeter macrocephalus*) from the Huizhou coast of the South China Sea. Mar. Pollut. Bull. 144, 181–188.

CHAPTER

38

Conservation and management

Robert Suydam[1], Jessica Lefevre[2], Geof H. Givens[3], J.C. George[1], Dennis Litovka[4] and H.K. Brower, Jr.[5]

[1]Department of Wildlife Management, North Slope Borough, Utqiaġvik, AK, United States [2]Alaska Eskimo Whaling Commission, Utqiaġvik, AK, United States [3]Givens Statistical Solutions, Fort Collins, CO, United States [4]ChukotTINRO, Anadyr, Chukotka, Russia [5]North Slope Borough, Mayors Office, Utqiaġvik, AK, United States

Introduction

Aboriginal subsistence hunting of bowhead whales was sustainable for millennia in part because the human population was small, whaling was logistically difficult, and whale populations were large. Nonetheless, the Inuit societies developed sophisticated and complex equipment, rules and rituals associated with bowhead whale hunting, and these assured it was conducted by well-trained hunters in a careful and reverent manner (Stoker and Krupnik, 1993; Murdoch, 1892). These traditional "regulations" likely reduced losses and increased the efficiency of the hunt (Fig. 38.1). Commercial whaling changed the situation because so many whales were killed (Chapter 33). The result was that the total harvest (i.e., commercial and subsistence) was no longer sustainable. Driven almost entirely by economic incentives and lack of regulations, commercial whalers hunted bowheads until it was no longer profitable. Commercial whaling on bowheads had all but ceased by the early 1900s. Cessation of commercial whaling occurred for several reasons including the greatly reduced population size of bowheads, the discovery of petroleum, which replaced whale oil, and the invention of several substitutes that replaced baleen such as spring steel (Bockstoce, 1986). The damage was done; however, as all stocks of bowheads were severely depleted, as were populations of other species. Bockstoce and Burns (1993), Ross (1993), and Chapter 33, provide summaries of the history of commercial whaling on bowhead whales.

Even though commercial whaling decimated bowhead stocks, subsistence hunting of those whales continued at some villages, especially along the north and west coasts of Alaska and the coast of Chukotka. Bowheads helped Inuit peoples meet nutritional and

The Bowhead Whale
DOI: https://doi.org/10.1016/B978-0-12-818969-6.00038-8

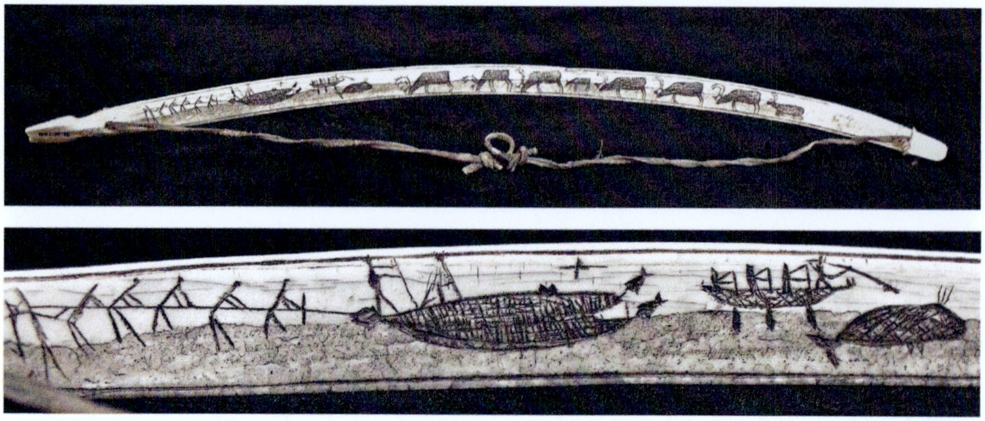

FIGURE 38.1 Upper panel—Inupiaq bow drill handle engraved with caribou and bowhead whale hunts. Lower panel—detail of bowhead whale hunt, a harpoon is thrown from an umiaq toward a whale on the right. On the left, people pull the dead whale onto the ice (or beach). Note the presence of two sealskin floats near the snout of the whale and that the whale has the characteristic elevation on the head where the blowhole is located. *Source: Specimen SJ-II-K-106 from the Sheldon Jackson Museum, Sitka, Alaska. Photo by curator Jacqueline Fernandez-Hamberg.*

cultural needs, as well as providing bones for construction material. For decades subsistence hunting occurred at a relatively low level but by the mid-1970s, the number of whales struck in northern Alaska increased dramatically; 111 whales were struck in 1977 alone (Marquette, 1979). That increase, in combination with the perception that the bowhead population was small contributed to an extremely contentious situation, referred to as the "bowhead whale problem" (Mitchell and Reeves, 1980; Tillman, 1980). The International Whaling Commission (IWC) expressed serious concern and in 1977 called for the hunt to be stopped. In 1978, a small quota was implemented to help meet the needs of the Inuit people of Alaska. In response, Inuit in northern and western Alaska created the Alaska Eskimo Whaling Commission (AEWC). One of their goals was to counter the "outside" interference with their subsistence hunting and the increased regulation that limited whaling captains from providing needed resources for the communities.

About the same time that the IWC tried to stop the bowhead hunt in Alaska, the discovery of oil and gas reserves in northern Alaska escalated exploration and development. Because some of that exploration was in marine waters within the range of bowhead whales, there was grave concern among hunters that the bowhead population would be harmed. Hunters know that bowheads are sensitive to anthropogenic sounds and that industrial operations in the offshore produce a great deal of noise. The potential effects of industrial sounds were evident to whalers and they became concerned about possible impacts to whales and that the whales might be deflected away from normal migration routes and traditional hunting areas.

Those two substantial issues in the late 1970s and early 1980s resulted in dramatic new management efforts and regulations to ensure the hunt was sustainable and that industrial impacts on bowhead whales were mitigated. More recently, rapid changes in the Arctic environment have again raised concerns about the future of bowheads. In this chapter, we discuss the conservation and management of the four stocks of bowheads. We focus on the conservation and management of Bering-Chukchi-Beaufort (BCB) bowhead whales, because efforts there have been substantial in the past 40 years because of the larger subsistence harvests (relative to other bowhead stocks) and large population size of the BCB stock.

Regulations

Many international and national laws relate to management of bowheads. Montague (1993) provides a summary of many of the international laws and those in the USA. Other domestic laws that protect bowhead whales exist in the other range states as well. Key among the international laws is (a) the 1931 League of Nations Convention that prohibited the commercial harvest of bowheads and (b) the 1946 International Convention for the Regulation of Whaling, which created the IWC (Montague, 1993). As mentioned above, the IWC initiated efforts to manage the hunt of bowheads in Alaska in the late 1970s. That effort created a new regime for the management of the subsistence harvest but it also spiraled into the creation of other regulations that protected not only bowheads but also the availability of the whales to communities to help meet cultural and nutritional needs.

East Greenland-Svalbard-Barents Sea and Okhotsk Sea stocks

The East Greenland-Svalbard-Barents Sea (EGSB) stock is showing signs of recovery (Chapter 6) and there are no directed takes; thus there is no need for harvest management. Limited monitoring is on-going (see Chapters 5 and 6) that will be useful for identifying future problems should they arise. For example, as the Arctic becomes increasingly ice-free, there may be a need to implement management actions on commercial activities, such as shipping and oil and gas exploration and development (Reeves et al., 2014).

Bowheads of the Okhotsk Sea (OKH) stock are not hunted. There are oil and gas activities and commercial fishing within the range of bowheads from this stock (Chapter 5). Climate change is likely to impact bowhead habitat. Regular monitoring of the population size and the health of whales from this population is warranted.

East Canada-West Greenland stock

The population of bowhead whales in this stock likely numbers about 6500 animals (Chapter 6). Thus, the stock is large enough to sustain the modest hunt in West Greenland and East Canada.

Greenland

In Greenland, the IWC issues quotas for multiple species (i.e., fin, minke, humpback and bowheads). Having quotas for multiple species provides flexibility for the hunters and this is useful in highly variable environmental conditions and the dramatic changes now occurring in the Arctic.

The bowhead quota for West Greenland was established by the IWC beginning in 2007 (https://iwc.int/greenland). That quota is small, allowing only two whales to be struck in a year, whether the whale is killed or not. One previously unused strike can be added to the annual quota (IWC, 2018). In 2015, the IWC Scientific Committee (SC) finished a similar process for East Canada-West Greenland (ECWG) bowhead hunting by Greenland (IWC, 2016a) as had been done for the BCB stock of bowheads (see IWC, below). A strike limit algorithm (SLA; see IWC, below for additional details) was developed and tested under assumed Canadian hunting levels but the recommended quotas from the SLA only apply to Greenland (IWC, 2016b).

Hunting is managed domestically by the Ministry of Fisheries, Hunting and Agriculture and supervised locally by the Fisheries License and Control Authority, but with the involvement of the Organization of Fishermen and Hunters, the municipalities, the Greenland Institute of Natural Resources, and the Ministry of Nature and Environment (https://iwc.int/greenland). Chapter 32 provides information about the number of bowheads landed in West Greenland in the past 40–50 years.

Canada

Bowhead harvest management in Canada occurs collaboratively between the Department of Fisheries and Oceans (DFO), Canada, and relevant Wildlife Management Boards, Game Councils and hunters. Information on population size and trend (Chapter 6), biology, natural history, and number of animals harvested are shared among the IWC, Greenland, and Canada to ensure that the subsistence harvest continues to be managed sustainably while also meeting the needs of Inuit communities. Canada and Greenland have two different and mostly independent management systems; however, sharing information on harvests and the status of the bowhead stock helps to ensure that hunts are sustainable.

Bering-Chukchi-Beaufort bowheads

Traditional management

The Inupiat of northern Alaska have a deep spiritual belief regarding whales and whale hunting (Chapter 31). Captains and their crews believe that "the whales they give themselves" to deserving crews (Brower and Brewster, 2004). Over time, this traditional management approach has helped to ensure that the harvests are sustainable in that crews take only what is needed and use what is given.

Inuit hunters in Alaska have many traditional rules surrounding whaling (Brower, 1942; Brower and Brewster, 2004; Chapters 27 and 31). The traditional rules differ slightly among locations but the rules help ensure that whales are shown respect, are available for harvest, and are shared within and among communities. Sharing is a particularly important aspect of subsistence in Inupiat communities. Most of the whale is shared within and among communities, even with some distant communities (BurnSilver et al., 2016; Kofinas et al., 2016). Whaling Captains and their families may expend $50,000 (USD) per year to field a crew for the chance to harvest a whale. Even though the Captain and their crew invest large amounts of time and resources, nearly the entire whale is shared with the community.

Alaska

In the early to mid-1970s, there was the perception that the BCB population was quite small but only index or minimum estimates were available. Some of the first efforts to count bowheads suggested there were perhaps fewer than 1000 whales (Krogman, 1980). Also, the number of whales being struck increased in part because of access to financial resources with discovery of oil in northern Alaska that allowed the formation of more whaling crews. The IWC stopped the harvest of bowheads in 1977 because of the combination of these factors. In an emergency meeting in late 1977, the IWC implemented a small quota for 1978 acknowledging that bowheads were critical to the nutritional and cultural needs of Inuit communities in Alaska.

Stopping the hunt and then implementing a small quota contributed to significant social problems in northern and western Alaska. The restrictive bowhead quota occurred concurrently with the decline of the largest caribou herd in northern Alaska and stringent harvest restrictions (Davis and Valkenburg, 1978). The loss or potential loss of two important subsistence species was devastating. This assault on the communities resulted in stress, confusion, anger, and frustration. While the impacts have not been fully described in the literature, suicides and murders were known to have resulted from the tremendous stress. This topic warrants future research.

The situation was complicated further because the whaling communities did not believe the early estimates of the bowhead population by federal scientists were accurate. The captains and elders did not think the scientists were listening to them or taking into account their perspectives on the biology of whales (Chapter 34). By the early 1980s, communities in northern and western Alaska created the AEWC. The North Slope Borough (NSB; a local municipality in northern Alaska funded primarily through property taxes on oil companies operating in Prudhoe Bay) established what would become the NSB Department of Wildlife Management (DWM). The DWM hired its own scientists who would live and work in the community to better document the abundance and biology of bowheads while working closely with elders and whaling captains.As the bowhead population grew the quota also increased (Chapter 32). Currently, the quota for Alaska is implemented by the National Oceanic and Atmospheric Administration and the AEWC manages the hunt.

Chukotka

The IWC sets the harvest quota for BCB bowheads. That quota is then shared between the United States and the Russian Federation. As in Alaska, Inuit hunters in Russia are part of the domestic management team. The Russian Ministry of Natural Resources (MNR) regulates the whale hunts and the State Border Guard Service (SBS) supervises the hunt (https://iwc.int/russian-federation). In turn, the MNR and SBS work closely with the Union of Marine Mammal Hunters of Chukotka, native communities, municipalities, the Federal Fisheries Agencies, and the Chukotka Department of Agricultural and Industrial Development. Based on applications from hunters, the Chukotka government distributes the quota among the communities each year (https://iwc.int/russian-federation).

IWC management

The IWC SC has used a variety of methods over the past 40 years to provide advice to the IWC Commissioners to inform their decisions on the setting of quotas for safe hunting levels. In the 1980s and early 1990s, the SC made some quota recommendations relying on estimation of replacement yield (i.e., the number of whales that could be removed from the population and leave the population at the same level it started the year) or other approaches. During the 1990s, the best methods to use were hotly debated (IWC, 1995a; Raftery et al., 1995; Punt and Butterworth, 1997, 1999).

In 1994, the IWC called for the development of the aboriginal whaling management procedure (AWMP; IWC, 1995b). The AWMP is comprised of stock-specific SLAs, which calculate sustainable quotas, and an overarching set of associated policies called the aboriginal whaling scheme (AWS). One of the first key SC decisions was that—unlike for commercial whaling—the calculation of strike limits should use stock-specific algorithms, primarily because the available data about different stocks was very different in terms of quality and quantity.

BCB bowheads were the first stock to be addressed by the SC. A comprehensive summary of this effort is given by IWC (2003a, b); the following is a brief overview.

An age-specific population dynamics model was developed to simulate the historical and future stock trajectory under various assumptions about biological parameters, past and future catch, past and future abundance estimates, future subsistence need, and a wide variety of other factors. A candidate SLA would be added to this framework to simulate future catches, based on periodic future abundance estimates which were generated randomly based on the simulated future abundance at the time. Thus the dynamics model and the SLA interact, with the former updating abundance and the latter imposing removals.

Hundreds of scenarios were tested for each candidate SLA to see how well the SLA satisfied aboriginal subsistence need while ensuring stock conservation and recovery. Each candidate SLA was also tested in numerous more extreme scenarios to understand better how the SLA responds to the unexpected and to attempt to avoid unintended consequences to the whale population or subsistence needs. The SLAs were evaluated by examining conservation risk, need satisfaction, and recovery. Risk statistics mostly related to reduction in the size of the stock. Need satisfaction referred to how much of the requested quota (i.e., need) was permitted. Recovery pertained to the rate of growth of the stock, how quickly it reached maximum sustainable yield level (MSYL), and how reliably it stayed above MSYL.

Five candidate SLAs for BCB bowheads were evaluated by the SC resulting in two that performed well while providing slightly different results in different scenarios.

The D-M algorithm used a population dynamics model and extended Kalman filters (Dereksdóttir and Magnússon, 2003) and the G-G algorithm used the output and internal estimates of the other competing candidate models as its input, and estimated a function of these inputs to achieve certain performance goals (Givens, 1999, 2003). The SC agreed to adopt a Bowhead SLA that took the arithmetic average of the strike limits given by D-M and G-G, which it termed the "Grand Unified Procedure (GUP)." The IWC accepted GUP as the Bowhead SLA in 2002 and began using it to set the quota for BCB bowheads (IWC, 2003c). The Bowhead SLA and relevant data are reviewed regularly by the SC. Such reviews, and potentially additional SLA testing, can also be initiated by the SC if it appears warranted.

The AWS included provisions related to survey frequency, data requirements and standards, carryover (i.e., reserving unused strikes in one year for future use in a subsequent year), interim quotas when an abundance estimate is delayed, and periodic reviews of SLA management. The latter are termed Implementation Reviews, and the primary purpose is to determine whether any new information suggests that the current state of affairs (e.g., survey precision, biological parameters, need) was not adequately covered in the previous simulation testing. A key requirement of the AWS is that a new abundance estimate for any stock subject to subsistence whaling should be presented to the IWC at least every 10 years. The AWS was adopted by the IWC in 2018.

The development of the entire AWMP spanned 25 years and represents a landmark in wildlife management science. It could not have been accomplished without the leadership of Greg Donovan, IWC Head of Science, who chaired the relevant SC working group for the entire period, and the contributions of SC members from around the globe.

Over the years, the BCB bowhead quota has been used as a political football. The quota has at times not been approved by the IWC due to efforts by some members to influence other decisions being considered by the IWC. This was devastating to the AEWC and their member villages as the bowhead quota was much more than a political tool to leverage a political outcome. The ability to harvest bowheads, process them (*aġviuq*, MacLean, 2014), and share them with many crews (*autaak*) and different communities, is the essence of Inuit culture in western and northern Alaska and provided a great deal of nutritious food resources needed to keep communities healthy (Burnsilver et al., 2016). In 2018, the AEWC and the US government, in partnership with other aboriginal whaling countries, succeeded in getting an "automatic renewal" for aboriginal quotas at the IWC. This new approach allowed for the quotas of bowheads and other species in Alaska, Chukotka, Greenland, and St. Vincent and the Grenadines to be renewed every 6 or 7 years if the quota request remains the same as the previous requests, the biological data on the relevant whale stocks supports the continued sustainable harvest, and if timely reporting demonstrates a "status quo" continuation of the hunt (IWC, 2018). Fig. 38.2 is a photo of a portion of the Alaskan/US and Chukotkan/Russian Federation delegations that worked hard to obtain the automatic quota.

Oil and gas

In the 1970s, oil and gas activities increased dramatically in northern Alaska. In 1979, two years before the AEWC's formal incorporation in 1981, the US held its first offshore oil

FIGURE 38.2 A portion of the Alaska/US and Chukotkan/Russian Federation delegations at the 2018 meeting of the IWC in Florianopolis, Brazil, shortly after the automatic quota renewal was approved. Acting US Commissioner Ryan Wulff and Russian Federation Commissioner Irina Fominykh, and many of their staff are not pictured. *Source: Photo from Jenny Evans.*

and gas lease sale in the Beaufort Sea (https/www.boem.gov/sites/default/files/documents/environment/2020_0225_HistoricalLeaseSales).

By 1985, oil and gas exploration activity in the Beaufort Sea was threatening both the hunting success and personal safety of whaling crews, especially for those hunters from Kaktovik and Nuiqsut (Fig. 32.2). The skippers of the large ocean-going drill ships and geophysical exploration vessels were plying the same waters as subsistence whaling crews in small skiffs, unaware that the small vessels were on the water. The AEWC had to act quickly to protect the lives of their hunters from this threat.

Fighting to preserve their bowhead harvest on the international front at the IWC, the young AEWC now faced serious threats on the domestic front. Shockingly there were no provisions in law that could be brought to bear to protect the hunters' lives, much less their subsistence harvest opportunities, from industrial operations. Without legal protection, the AEWC turned to private action, with a direct appeal to the oil and gas companies. To address the risk to small vessels, the AEWC and representatives of the offshore operators reached agreement to install, at the operators' expense, a radio tower at Deadhorse, at the site of the Prudhoe Bay oil fields. The stakeholders then drew up a communications protocol to be followed by both oil and gas and subsistence whalers.

In an effort to ensure adherence to this protocol by all offshore operators in northern Alaska, in 1986 the AEWC turned to a political solution. Drawing on support from the Alaska US Congressional Delegation, the AEWC succeeded in amending the section of the

Marine Mammal Protection Act providing for 5-year Letters of Authorization to allow small takes of marine mammals incidental to certain offshore activities [Marine Mammal Protection Act, Section 101(a)(5)(A), 16 USC 1371(a)(5)(A)]. The language added to the statute requires that authorized activities "will not have an unmitigable adverse impact ... on the availability of such species or stock for subsistence uses" [Public Law 99-659, Nov. 14, 1986, 100 STAT. 3706, Section 411].

While somewhat inelegant, this language was deliberately crafted to maintain the opportunity for development activities while requiring that potentially adverse impacts of those activities be mitigated. The whaling captains of the AEWC saw development as an opportunity to bring jobs and other economic benefits to their communities, but were clear that the whales, their habitat, and the subsistence harvest could not be adversely affected. Since the AEWC and offshore development operators had recently launched a collaborative process, the intent and hope were that future offshore operators, as well as government regulators, would look to this nascent collaborative process as the means of developing mutually acceptable mitigation measures.

The collaborative agreement entered between whaling captains and offshore operators in 1985 became the basis for the Open Water Season Conflict Avoidance Agreement (CAA). Renewed and revised annually, the CAA enables whaling captains to take a direct role in managing activities affecting the habitat and migration of the BCB bowhead whale stock. Stakeholders in this process have collaboratively developed logistical and environmental measures, including time-area closures and discharge restrictions that facilitate mutually productive access to the ocean and both its renewable and nonrenewable resources. The CAA has been so successful in meeting its objectives that federal agencies look to the Agreement in evaluating compliance with statutory safeguards required for authorizing oil and gas operations in marine mammal habitat where subsistence harvests exist (Lefevre, 2013).

Climate change

Major and rapid changes are occurring within the range of bowhead whales. The most evident and striking is the reduction of sea ice, but closely related and of importance are increased human activities in the Arctic. Bowheads are adapted to living in cold, ice-covered oceans. The diminishing extent and thickness of sea ice is one of the most obvious impacts from a warming climate, but there are many others, some of which we are aware of and others that remain unknown. The question of the effects of climate warming on bowheads is the subject of several publications and is a particular concern of indigenous people who depend on the bowhead whales (Laidre et al., 2008; Moore, 2018; Huntington et al., 2020; Chapter 27).

Moore et al. (Chapter 27) evaluated basic life history metrics (e.g., population size, range, behavior, health) for the four bowhead stocks relative to the environmental changes occurring in each respective area. The BCB and ECWG stocks were of least concern because of recent information about the size of the populations (Chapter 6) and evidence of resilience to large ecosystem changes. Much less is known about the life history of the EGSB stock and the estimated abundance is very low; however, there are some encouraging indicators of population resilience. Their habitat has relatively few anthropogenic threats; currently there is little or no commercial fishing in the range of that stock and little other commercial activity. The Okhotsk Sea stock is the least studied stock and is considered the most at risk from climate and other anthropogenic effects. There are fewer than 300 animals (Chapter 6) and shipping, fishing, and predation from killer whales pose substantial threats to this stock.

Of the four extant bowhead stocks, the best known, based on extensive Indigenous Knowledge and science, is the BCB stock. Over the past four decades, the stock appears resilient to recent sea ice retreat and the associated changes in the marine ecosystem (Moore, 2016; Huntington et al., 2020). The most recent estimate of the population rate of increase (Givens et al., 2016) and abundance level of the BCB stock suggest it is robust to significant ice retreat over the period 1978–2011. Also George et al. (2015) provide body condition analyses suggesting BCB bowheads have possibly benefitted from sea ice reduction. Reduced heavy sea ice cover in summer allows more sunlight to penetrate the water. That in combination with increased mixing of sea water from waves, and coastal upwelling within the summer feeding range, likely results in increased productivity and thus more bowhead prey. George et al. (2015) point out that it is not possible to extrapolate their findings regarding future trends due to the massive ecological perturbations underway.

Competition from other subarctic baleen whale species (e.g., fin and humpback whales) moving into the bowhead's feeding areas, coupled with increased predation (Chapter 29) and increasing human activities in the range of bowheads may push the whales to or past a "tolerance threshold." If this occurs, increased vulnerability measured as a decline in health indices will likely occur. The recent detection of kidney worms (i.e., *Crassicauda* spp.) in harvested whales (Chapter 30), continued detection of scarring on harvested whales associated with fishing-gear entanglement (including harvested whales actively entangled in crab pot line, Chapter 36), and a major perturbation to the fall 2019 migration near Point Barrow (Huntington et al., 2020; Chapter 27) are affecting bowhead health to varying degrees. Nonetheless, Stimmelmayr et al. (2018) summarized 10 health indices including key indicators such as high calf production over the past decade, a low disease rate in harvested whales, and relatively few carcass or beach-cast whale detections, and concluded the indices were consistent with a healthy stock.

In addition to environmental changes, human activities in the Arctic are increasing. Commercial shipping, oil and gas activities, mining, commercial fishing, tourism, and increased scientific research all have the potential to impact bowhead whales or their availability to subsistence communities. Potential impacts may be from anthropogenic sounds as bowheads are very sensitive to such sounds (McDonald et al., 2012; Chapters 23 and 35). Ship strikes, and possible oil or other product spills may impact bowheads directly or indirectly (through food). Currently, Arctic governments are considering different routes and measures for managing Arctic shipping through the International Maritime Organization.

Sustainability of the hunt

The bowhead hunts are sustainable and the removal rate is less than 0.5% of the population (Chapter 32) while helping to meet the needs of subsistence communities. The hunt of whales in the BCB stock is managed internationally by the IWC and nationally by the

Federal governments of the USA and Russian Federation, and locally by hunter groups. The hunt of whales in the ECWG stock is managed internationally by the IWC for west Greenland and by the Canadian government in eastern Canada. Hunts are managed locally by respective hunter groups.

Monitoring

It is likely that these whale stocks and hunts will continue to be sustainable if the appropriate monitoring occurs and information about the status of the stocks and harvested whales is shared. The hunts will continue to meet the nutritional and cultural needs of Inuit communities if bowheads are resilient to climate change and commercial and industrial activities are appropriately managed and mitigated.

Summary and conclusions

The largest impact on bowhead whales historically was commercial whaling. Commercial harvest levels and historical population size have been estimated through evaluation of commercial whaling logs (Bockstoce, 1986). That information in conjunction with recent population surveys has shown that it has taken approximately 100 years for two stocks (BCB and ECWG) to be nearing recovery. In addition to the cessation of commercial whaling, recovery of bowheads is ongoing due to efforts to manage the subsistence harvests of the whales, enabling populations to grow while still meeting the needs of communities and cultures who have used bowheads for millennia.

The IWC and DFO Canada are managing bowhead hunts sustainably in collaboration with other government partners and subsistence hunters and whalers. Quota levels are based on science and indigenous knowledge and are established so that the whale populations can continue to grow.

Rapid change in the Arctic and subarctic is imposing additional stresses on bowheads and bowhead hunters. Those changes include rapidly diminishing sea ice, which currently does not seem to be impacting bowheads. Other possible impacts to whales include oil and gas activities, other commercial activities (e.g., fishing, shipping, etc.), tourism, and scientific research. In order to understand and respond to the impacts, it is necessary to continue to examine harvested whales and conduct population surveys to assess the impacts on individual animals and to assess the efficacy of management efforts (i.e., mitigation measures) on the population.

Bowhead populations are increasing, although possibly not for the Okhotsk Sea stock. Flexible management systems are in place to ensure bowheads are hunted sustainably into the future. Some management systems are in place (i.e., the CAA) to mitigate impacts from industrial activities on whales and hunters but across the ranges of bowheads, there is likely a need for additional monitoring and mitigation. Results of the management of bowhead whales have mostly been a positive story, but vigilance is needed to keep this story positive. Hunters, scientists, and managers working together should help ensure continued conservation of bowheads and their availability to Inuit communities. Communication, really listening to one another respectfully is vitally important. Elders and hunters need to continue to be a meaningful part of the science, monitoring, and management. Science has made great strides to "catch up" to Indigenous Knowledge about bowhead whales (Chapter 34) but elder hunters and whaling captains continue to know more about many aspects of bowhead behavior and biology, than scientists.

Acknowledgments

Hunters and their communities play a crucial role in the conservation and management of bowheads. Many scientists, managers and their organizations also contribute substantially. The key to success is embedded in the collaboration among these individuals. We thank the many people who have contributed in numerous different ways to the conservation and management of bowheads. Please also, see the book acknowledgements.

References

Bockstoce, J.R., 1986. Whales, Ice, and Men: The History of Whaling in the Western Arctic. University of Washington Press, Seattle.

Bockstoce, J.R., Burns, J.J., 1993. Commercial whaling in the North Pacific sector. In: Burns, J.J., Montague, J.J., Cowles, C.J. (Eds.), The Bowhead Whale. Allen Press, Lawrence, KS.

Brower, C.D., 1942. Fifty years below zero. A lifetime of adventure in the far north. Dodd Mead and Company, New York.

Brower Sr., H., Brewster (Eds.), 2004. The Whales They Give Themselves: Conversations With Harry Brower Sr. University of Alaska Press, Fairbanks, AK, 232 pp.

Burnsilver, S.J., Magdanz, R., Stotts, M., Berman, Kofinas, G., 2016. Are mixed economies persistent or transitional? Evidence using social networks from Arctic Alaska. Am. Anthropol. 118 (1), 121–129. Available from: https://doi.org/10.1111/aman.12447.

Davis, J.L., Valkenburg, P., 1978. Western Arctic Caribou Studies. Alaska Department of Fish and Game. Federal Aid in Restoration, Project W-17-8 and W-17-9. Juneau, AK, 27 pp.

Dereksdóttir, E.H., Magnússon, K.G., 2003. A strike limit algorithm based on adaptive Kalman filtering with an application to aboriginal whaling of bowhead whales. J. Cetacean Res. Manage. 5, 29–37.

George, J.C., Druckenmiller, M.L., Lairdre, K.L., Suydam, R., Person, B., 2015. Bowhead whale body condition and links to summer sea ice and upwelling in the Beaufort Sea. Prog. Oceanogr. 136, 250–262. Available from: https://doi.org/10.1016/j.pocean.2015.05.001.

Givens, G.H., 1999. Multicriterion decision merging: competitive development of an aboriginal whaling management procedure. J. Am. Stat. Assoc. 94, 1003–1014.

Givens, G.H., 2003. Empirical estimation of safe aboriginal whaling limits for bowhead whales. J. Cetacean Res. Manage. 5, 39–44.

Givens, G.H., Edmondson, S.L., George, J.C., Tudor, B., DeLong, R.A., Suydam, R., 2016. Horvitz-Thompson whale abundance estimation adjusting for uncertain recapture, temporal availability variation, and intermittent effort. Envirometrics (wileyonlinelibrary.com; DOI: 10.1002/env.2379).

Huntington, H.P., Danielson, S.L., Wiese, F.K., Baker, M., Boveng, P., Citta, J.J., et al., 2020. Evidence suggests potential transformation of the Pacific Arctic ecosystem is underway. Nat. Clim. **Change** 10, 342–348. Available from: https://doi.org/10.1038/s41558-020-0695-2.

IWC, 1995a. Report of the scientific committee. Rep. Int. Whal. Commn. 45, 53–95.

IWC, 1995b. Chairman's report of the 46th annual meeting. Rep. Int. Whal. Commn. 45, 42–43.

IWC, 2003a. Report of the scientific committee. J. Cetacean Res. Manage. 5 (Suppl), 1–92.

IWC, 2003b. Report of the scientific committee, Annex E: report of the standing working group on the development of an aboriginal subsistence whaling management procedure. J. Cetacean Res. Manage. 5 (Suppl), 154–225.

IWC, 2003c. Annual Report of the International Whaling Commission 2002. <www.iwc.int>, 159 pp.

IWC, 2016a. Report of the scientific committee. J. Cetacean Res. Manage. 17 (Suppl), 1–91.

IWC, 2016b. Report of the AWMP workshop on developing strike limit algorithms (SLAs) for the Greenlandic Hunts (SC/66a/Rep03). J. Cetacean Res. Manage. 17 (Suppl), 473–483.

IWC, 2018. Chairman's Report of the 67th Meeting. <www.iwc.int>, 46 pp.

Kofinas, G., BurnSilver, S.B., Magdanz, J., Stotts, R., Okada, M., 2016. Subsistence Sharing Networks and Cooperation: Kaktovik, Wainwright, and Venetie, Alaska. BOEM Report 2015-023. AFES Report MP 2015-02. School of Natural Resources and Extension, University of Alaska Fairbanks.

Krogman, B.D., 1980. Sampling strategy for enumerating the western arctic population of the bowhead whale. Mar. Fish. Rev. 42 (9–10), 30–36.

Laidre, K.L., Stirling, I., Lowry, L.F., Wiig, Ø., Heide-Jørgensen, M.P., Ferguson, S.H., 2008. Quantifying the sensitivity of arctic marine mammals to climate-induced habitat change. Ecol. Appl. 18, S97–S125.

Lefevre, J.S., 2013. A Pioneering Effort in the Design of Process and Law Supporting Integrated Ocean Management, Environmental Law Reporter, 43 ELR 10-2013, pp. 10893–10908.

MacLean, E., 2014. Inupiaq to English Dictionary. University of Alaska Press, Fairbanks, AK, p. 1018.

Marquette, W.M., 1979. The 1977 catch of bowhead whales (*Balaena mysticetus*) by Alaskan Eskimos (SC/30/Doc 35). Rep. Int. Whal. Commn. 29, 281–289.

McDonald, T.L., Richardson, W.J., Greene Jr., C.R., Blackwell, S.B., Nations, C.S., Nielson, R.M., et al., 2012. Detecting changes in the distribution of calling bowhead whales exposed to fluctuating anthropogenic sounds. J. Cetacean Res. Manage. 12 (1), 91–106.

Mitchell, E.D., Reeves, R.R., 1980. The Alaska bowhead problem: a commentary. Arctic 33 (4), 686–723.

Montague, J.J., 1993. Introduction. In: Burns, J.J., Montague, J.J., Cowles, C.J. (Eds.), The Bowhead Whale. Allen Press, Lawrence, KS.

Moore, S.E., 2016. Is it 'boom times' for baleen whales in the Pacific Arctic region? Biol. Lett. 12. Available from: https://doi.org/10.1098/rsbl.2016.0251.

Moore, S.E., 2018. Climate change. In: Würsig, B., Thewissen, J.G.M., Kovacs, K. (Eds.), Encyclopedia of Marine Mammals, third ed. Elsevier, pp. 194–197.

Murdoch, J., 1892. Ethnological Results of the Point Barrow Expedition. Smithsonian Institution Press, Washington, DC, p. 441.

Punt, A.E., Butterworth, D.S., 1997. Assessments of the Bering-Chukchi-Beaufort Seas stock of bowhead whales (*Balaena mysticetus*) using maximum likelihood and Bayesian methods. Rep. Int. Whal. Commn. 47, 603–618.

Punt, A.E., Butterworth, D.S., 1999. On assessment of the Bering-Chukchi-Beaufort Seas stock of bowhead whales (*Balaena mysticetus*) using a Bayesian approach. J. Cetacean Res. Manage. 1, 53–71.

Raftery, A.E., Givens, G.H., Zeh, J.E., 1995. Inference from a deterministic population dynamics model for bowhead whales. J. Am. Stat. Assoc. 90 (430), 402–430.

Reeves, R.R., Ewins, P.J., Agbayani, S., Heide-Jørgensen, M.P., Kovacs, K.M., Lydersen, C., et al., 2014. Distribution of endemic cetaceans in relation to hydrocarbon development and commercial shipping in a warming Arctic. Mar. Policy 44, 375–389.

Ross, W.G., 1993. Commercial whaling in the North Atlantic sector. In: Burns, J.J., Montague, J.J., Cowles, C.J. (Eds.), The Bowhead Whale. Allen Press, Lawrence, KS.

Stimmelmayr, R., George, J.C., Willoughby, A., Brower, A., Clarke, J., Ferguson, M., et al., 2018. 2017 Health Report for the Bering-Chukchi-Beaufort Seas Bowhead Whales. Paper SC/67b/AWMP08 Presented to the Scientific Committee of the International Whaling Commission.

Stoker, S.W., Krupnik, I.I., 1993. Subsistence whaling. In: Burns, J.J., Montague, J.J., Cowles, C.J. (Eds.), The Bowhead Whale. Allen Press, Lawrence, KS.

Tillman, M.F., 1980. Introduction: a scientific perspective of the bowhead whale problem. Mar. Fish. Rev. 42 (9–10), 2–5.

CHAPTER

39

Past, present, and future

J.G.M. Thewissen[1] and J.C. George[2]

[1]Department of Anatomy and Neurobiology, Northeast Ohio Medical University, Rootstown, OH, United States [2]Department of Wildlife Management, North Slope Borough, Utqiaġvik, AK, United States

Bowhead whales and humans

Genetic evidence indicates that the split between the lineages of modern bowhead (*Balaena mysticetus*) and right whales (*Eubalaena australis*, *Eubalaena glacialis*, and *Eubalaena japonica*) occurred at least 10 million years ago (Chapters 1, and 2). Fossil evidence relevant to the origin of bowheads dates from around 6 to 7 million years ago, when the whale family that includes right whales and bowheads (Balaenidae) radiated and the earliest members of the genus *Balaena* can be recognized. At that time, *Balaena* ranged as far south as Virginia (United States) and Tuscany (Italy), and this range overlapped widely with that of *Eubalaena* (Chapter 2; Fig. 39.1). At that time, the world was several degrees warmer than it is today, and there was no sea ice in the Arctic (Zachos et al., 2001; Marx and Uhen, 2010). The early members of the genus *Balaena* were large, and they shared the ocean with a radiation of smaller bodied balaenids. Less than 3 million years ago, the planet cooled and the ice ages began: the Arctic Ocean developed ice cover. The smaller balaenids went extinct, leaving the two modern genera.

As the polar ice cap grew and shrank in glacial and interglacial periods, the range of *Balaena* and *Eubalaena* adapted, leading the genera to occupy ranges that mostly did not overlap. The continents and the polar ice cap separated groups of bowheads, giving rise to the four modern stocks of the species (Chapter 3). Of these, the Okhotsk Sea (OKS) stock is the most distinct genetically and the least diverse. The Bering–Chukchi–Beaufort Seas (BCB) stock and East Canada–West Greenland (ECWG) stocks are genetically most similar, and while separated by a sea ice plug, the data suggest a recent history of gene exchange along the northern edge of North America, the Northwest Passage. In fact, individuals from these two stocks now occasionally overlap in the Canadian Arctic Archipelago (Heide-Jørgensen et al., 2012). Furthermore, due to recent genetic and satellite telemetry work, two putative former stocks, the Davis Strait and Hudson Bay stocks

The Bowhead Whale
DOI: https://doi.org/10.1016/B978-0-12-818969-6.00039-X

FIGURE 39.1 A bowhead whaling makes a fluke-up sounding dive along the lead edge off Utqiaġvik. Such dives accounts for only about 5-10% of dives during spring migration. Most often just the peduncle (Inupiat: *itiġruq)* flexes prior to a deep dive. Sounding dive times vary but average about 10 minutes. The longest dives may approach one hour in length. *Source: Photo by Kate Stafford (North Slope Borough).*

(Moore and Reeves, 1993), are now reclassified into the single ECWG stock. Genetic bottlenecks have been documented in the BCB stock around 15,000 years ago, just after the last ice age when the planet warmed (Lamb, 1971; Brook, 2013; Phillips et al., 2013) and suggest a period of extinction that could be related to the warming ocean.

In general, baleen whale species migrate north and south over the course of a year between low-latitude breeding grounds and high-latitude feeding grounds; three of the four stocks of bowhead whales live their entire life in Arctic and Subarctic waters, migrating north through leads in spring and south as the ocean freezes (Chapters 4, and 5). Breeding and birthing, for the BCB population, take place at the start and during the northward migration through ice-choked seas to their feeding grounds (Chapters 7, and 13). Feeding grounds are abandoned in fall, as the sea ice closes in and prey descends to overwinter in deep water. The whales spend the winter within the ice at the more southern parts of their range, where they continue to feed. A similar pattern of migration and feeding occurs for the ECWG stock and possibly the East Greenland–Svalbard–Barents Sea (EGSB) stock as well. Wintering areas for Okhotsk Sea bowheads remain to be identified. However, scattered records of wintering bowheads have been recorded in a number of areas, possibly because Okhotsk Sea ice cover rarely exceeds 80% even in the most severe winters (Minervin et al., 2015). The degree to which they feed in winter is not known but could be substantial, as in other stocks (see Chapter 7).

Adaptations of bowhead whales have usually been explained as modifications to living in cold ice-covered seas with relatively low prey abundance. Consistent with this, flukes and flippers have countercurrent heat exchangers: arteries surrounded by veins that prevent heat loss (Chapter 15). Also, the blowhole is located on an elevated area of the head that is reinforced by tough connective tissue and is used to break through ice to create breathing holes (Chapter 7). While those are clear adaptations to Arctic Seas, some other morphological features are not. For instance, the enormous baleen rack presumably serves to feed efficiently in areas of low prey densities.

Bowhead blubber is thicker than that of other mysticetes, and this is sometimes proposed as primarily an adaptation for cold water living. However, the blubber of other arctic cetaceans, such as the beluga, is much thinner, even though they inhabit the same seas as bowheads. Also, bowhead blubber is only slightly thicker than that of right whales that live in much warmer water. Instead, bowhead blubber can be thought of as a massive energy storage organ. Even during summer feeding, the four stocks of bowhead whales inhabit waters of the High Arctic that have episodes when prey is lacking and feeding is minimal. The Beaufort and Chukchi Seas, where BCB bowheads do most of their feeding, are not as productive as seas where other mystietes feed, such as balaenopterids in Antarctica. Primary production in the Arctic is highly variable, and blubber likely serves to sustain the whale during periods of food scarcity.

Other features are also inconsistent with the view that thick blubber mainly serves thermoregulation. Bowheads and right whales have a vascularized organ on their palate that appears designed to dissipate heat (Chapter 14), suggesting that during exercise the animal could potentially overheat even in subfreezing water. Body temperatures are lower than any other large whale, and quite similar to those of hibernating bears. The lower temperatures may be linked to the thick blubber. Indeed, thermal and energetic data suggest that filter-feeding or fleeing bowheads may need to shed heat (Chapter 16), and while they appear overinsulated, the whale is quite capable of regulating body temperature, and there is no direct evidence of overheating based on temperature measurements made with the assistance of native hunters.

Of course, insulation provided by blubber is important for bowheads, and heat loss prevention is part of a strategy for conservative energy use. Bowheads can endure sub-0°C temperatures through the winter with little energetic stress or consequence. Other features consistent with energetic efficiency include the body temperature, the delay of sexual maturation until the third decade of life (Chapter 7), the slow swimming speeds, and the presence of relatively small brains (brain tissue is energetically expensive).

Possibly to mitigate the effects of poorly productive waters, bowheads do exploit scarce food sources efficiently. The baleen rack is used to filter prey from the water, and bowheads have baleen longer than any other species, with lengths of 3.5 m being common yet can filter the tiniest copepods (Chapter 14). The head makes up more than one-third of the body length in adults, and the baleen rack can makes up more than a quarter of the length of the body (Chapter 9). Baleen grows slowly, less than 20 cm/year in adults, and it wears with use. When young are weaned, their baleen rack is small and they ability to feed effectively is limited. However, the blubber layers of these whales, termed *ingutuqs* by Inuit hunters, are thick, thanks to a period of nursing that may have lasted 9 months (Chapters 7 and 13). For the next few years, whales live mostly on their reserves, the *qairaliq* stage of bowhead development. Their head and baleen grow, but the remainder of the body grows little or not at all, and this changes their body proportions (Chapter 7). Fat reserves decrease and so does the amount of bone in their ribs. Those mineral resources may be used to make baleen, and their bone loss can be used to regulate buoyancy, counteracting the shrinking fat reserves. Eventually, when the baleen reaches a critical length (perhaps 1.5 m), feeding is able to sustain the whale and growth resumes.

Just like many other cetaceans, bowheads grow until long after sexual maturity. Males attain sexual maturity at a smaller size than females. Possibly related to the enormous energetic demands on females from the developing fetus and nursing young, females grow much larger than males. The slow pace of life and the low metabolic rate may be related to extremely low cancer rates of the species (Chapter 20) and to their great age. Bowheads may live 200 years, the oldest mammal on the planet (Chapter 21). These features are consistent with other Arctic vertebrates such as narwhal and Greenland sharks, which live in environments where food is limited and which are also extremely long lived (Nielsen et al., 2016).

Bowhead whales have had a long and deep relation with humans. Archeologists estimate that between 8000 and 4000 years ago, the ancestors of Inuit and Yupik people crossed into Alaska from Asia, bringing their hunting technologies with them. While these peoples were likely to have used parts from beach-cast bowheads, planned systematic whale hunting can be documented approximately 1000 years ago (Tremayne et al., 2018). The Inuit developed sophisticated hunting tools, developing the toggle-head harpoon, line, and float apparatus, and by around CE 900, bowhead whaling became an integral part of the indigenous people along the Alaskan coast. Whaling in Chukotka probably predated it.

Around CE 1200, the Inuit spread to Eastern Canada and Greenland where they hunted a different stock of bowheads. With no metal weapons and tools, bowhead hunting was difficult and dangerous, and harvest levels by indigenous bowhead hunters did not affect population size. Then as now, the bowhead hunt was an anchor to several Inuit and Yupik cultures (Chapter 31), and indigenous hunters came to understand the behavior of these whales better than anyone else (Chapter 35).

Beginning in the 16th century, Europeans hunted Atlantic bowheads nearly to extinction (Chapter 33). The slaughter started in the North Atlantic (ECWG and EGSB stocks) and later moved to the North Pacific (BCB and OKS stocks). The BCB and ECWG stocks are in the process of recovery, with currently around 17,000 and 6500 individuals, respectively, but the EGSB and OKS stock never have recovered, consisting of a few hundred individuals each (Chapter 6). By the first decade of the 20th century, commercial whaling of bowhead whales ceased, as too few animals remained and replacements for whale oil and baleen developed.

Bowhead whaling was prohibited by the League of Nations Convention in 1931 and in 1946 by the International Whaling Commission (IWC, Chapter 32, and this chapter). Prohibitions on commercial hunting were enforced, but an "aboriginal whale" exemption was recognized by IWC. In the United States the Marine Mammal Protection Act of 1972 allowed indigenous people to take bowheads. In 1977 concerns about population size and overhunting led to a moratorium on the indigenous hunt of the BCB stock in the United States. To determine sustainable harvest quotas, accurate estimates of stock size were critical and research programs were launched to study the whales, which is discussed in Chapter 6. The BCB stock flourished, and its management was celebrated as an effective model, respecting needs of local people to harvest bowheads while conserving the species via sustainable quotas. The quota from the IWC Aboriginal Whaling Scheme (AWS) allows 56 BCB whales per year to be harvested from 2019 to 2025 from an annual average strike quota of 67 (Chapter 32). The annual quota of the ECWG stock is two whales (https://iwc.int/greenland); however, whales are not harvested every year (Chapter 34). Of the BCB stock, an average of 44 whales has been harvested each year over the past 10 years under the AWS. This accounts for less than 0.5% of the population, which is currently increasing at about 3.7%. The recovery of the BCB and ECWG stocks is among the great conservation success stories.

It is now 2020, and the planet is undergoing significant climate change, especially in the Arctic. Bowheads are facing new threats in part due to climate warming and sea ice retreat; in the fall of 2019, BCB bowheads migrated 1600 km North of their normal migration paths (Chapter 27). The location of food sources is changing, prey species are being replaced by others, diseases previously unknown are appearing, and predators are moving in (Chapter 29, and Chapter 30). Human use of the Arctic is increasing with regard to shipping, fishing, mining, and oil exploration and development (Chapters 36–38). The Okhotsk Sea stock is the southernmost of the extant stocks of bowheads and lives in an increasing warming sea, in areas of higher human population densities, and faces a number of threats as a consequence. The three remaining stocks live in more northern areas and face similar threats, less severe at present. But that is expected to change.

Epilog

In 1993 Burns et al. published "The Bowhead Whale" that summarized much of what was known about bowhead whales then. In his epilog, John Burns outlined some of the problems that faced bowheads, the people who depend on them, and the research needed to understand and mitigate those problems. He pointed out the need for a better understanding of the whales' digestion, distribution, and a need for integrative, ecological approaches to the study. Now, nearly 30 years later, we can report significant strides in all of these areas. We have a better understanding of their migration, the species they interact with and their physical environment, as well as their anatomy and physiology. A variety of techniques that did not exist of were rudimentary when Burns et al. wrote have now been applied to bowheads, from age estimation to metabolomics to telemetry to DNA sequencing. And the status of the species has changed too, with two stocks growing and at low conservation risk.

Undoubtedly, in the decades to come, much will change again. Climate change and arctic industrial development (shipping, oil and gas, fishing) are probably the most significant forces of change. It will affect prey, predators, competitors, and disease patterns of bowhead whales. Migration routes and behavior patterns may change. Interactions with humans will change too, as fisheries and shipping increase in the arctic. Indigenous people may find that bowheads are not dependable sources of food in the future, and it will be to the detriment of their culture. The bowhead has proven to be a robust species surviving severe changes in climate of the past 4 million years; however, we do not know what will happen with the species and its relation to people in the future.

Bowhead research can be used to document what is happening, and its results should inspire management decisions that will mitigate threats and impacts to the species, and the people who depend on them. In addition, bowhead research may contribute to solving problems that go far beyond the Arctic. The bowhead genome is being explored in depth and may yield clues to the great age and cancer-resistance of the species that may be useful for human health. Research on bowhead communication and interpretation of their songs remains elusive (Chapter 22). Many other questions regarding bowheads will serve as fascinating research topics for generations of scientists. This book is a testament to the collaboration between indigenous people, researchers, and administrators. The coastal Inupiat view *aġviq* as their spirit animal, providing food and uniting their people. We hope this book contributes to that encompassing view of life.

References

Brook, E.J., 2013. Leads and lags at the end of the last ice age. Science 339, 1042–1043.

Heide-Jørgensen, M.P., Laidre, K.L., Quakenbush, L.T., Citta, J.J., 2012. The Northwest Passage opens for bowhead whales. Biol. Lett. 8 (2), 270–273.

Lamb, H.H., 1971. Climates and circulation regimes developed over the northern hemisphere during and since the last ice age. Palaeogeogr. Palaeoclimatol. Palaeoecol. 10, 125–162.

Marx, F.G., Uhen, M.D., 2010. Climate, critters, and cetaceans: Cenozoic drivers of the evolution of modern whales. Science 327, 993–996.

Minervin, I.G., Romanyuk, V.A., Pischchal'nik, V.M., Truskov, P.A., Pokrashenko, S.A., 2015. Zoning and ice cover of the Sea of Okhotsk and the Sea of Japan. Herald of the Russ. Acad. Sci. 85, 132–139.

Moore, S.E., Reeves, R.R., 1993. Chapter 9: Distribution and movement. In: Burns, J.J., Montague, J.J., Cowles, C.J. (Eds.), Bowhead Whale. Special Publication No. 2. The Society for Marine Mammalogy, 787pp.

Nielsen, J., Hedeholm, R.B., Heinemeier, J., Bushnell, P.G., Christiansen, J.S., Olsen, J., et al., 2016. Eye lens radiocarbon reveals centuries of longevity in the Greenland shark (*Somniosus microcephalus*). Science 353 (6300), 702–704.

Phillips, C.D., Hoffman, J.I., George, J.C., Suydam, R.S., Huebinger, R.M., Patton, J.C., et al., 2013. Molecular insights into the historic demography of bowhead whales: understanding the evolutionary basis of contemporary management practices. Ecol. Evol. 3, 18–37.

Tremayne, A.H., Darwent, C.M., Darwent, J., Eldridge, K.A., Rasic, J.T., 2018. Iyatayet revisited: a report on renewed investigations of a stratified middle-to-late Holocene coastal campsite in Norton Sound, Alaska. Arctic Anthrop. 55, 1–23.

Zachos, J., Pagani, M., Sloan, L., Thomas, E., Billups, K., 2001. Trends, rhythms, and aberrations in global climate 65Ma to present. Science 292, 686–693.

Index

Note: Page numbers followed by "*f*" and "*t*" refer to figures and tables, respectively.

E

CPI Antony Rowe
Eastbourne, UK
October 05, 2020